STO

Essentials of Communication Electronics

THIRD EDITION

Essentials of Communication Electronics

Morris Slurzberg, B.S. in E.E., M.A.

William Osterheld, B.S. in E.E., M.A.

Authors of
Essentials of
Electricity–Electronics
Third Edition

McGraw–Hill Book Company

New York
St. Louis
San Francisco
Düsseldorf
Johannesburg
Kuala Lumpur
London
Mexico
Montreal
New Delhi
Panama
Rio de Janeiro
Singapore
Sydney
Toronto

Library of Congress Cataloging in Publication Data

Slurzberg, Morris.
Essentials of communication electronics.

First ed. published in 1948 under title: Essentials of Radio.
1. Electronics. 2. Telecommunication.
I. Osterheld, William, joint author. II. Title.
TK7815.553 1973 621.381 72-13839
ISBN 0-07-058309-9

ESSENTIALS OF COMMUNICATION ELECTRONICS

1 2 3 4 5 6 7 8 9 0 KPKP 7 9 8 7 6 5 4 3

The editors for this book were Alan W. Lowe and Cynthia Newby, the designer was Marsha Cohen, and its production was supervised by James E. Lee. It was set in Modern No. 8 by The Maple Press Company.
It was printed and bound by Kingsport Press, Inc.

Contents

Preface

The purpose of this text is to present at an intermediate level, in a similar manner to its previous editions, a comprehensive study of the principles of operation of solid-state devices and electron tubes, their basic circuits, and the application of these circuits to electronic communication systems.

This book is intended for (1) students studying communication electronics, and/or basic electronics in technical institutes, community colleges, trade and vocational schools, industrial training programs, and military training programs; (2) persons not attending any regular school but who wish to study the subject at home on an intermediate level; and (3) providing the background necessary for further study of electronics in fields such as high-frequency, very high frequency, and ultrahigh-frequency circuit applications which presuppose a knowledge of the basic active devices and their circuit applications as presented in this text.

During the past decade the field of electronics has expanded at a phenomenal pace, and predictions are that it will continue to grow at a rapid rate for some time to come. Applications of electronic principles cover an expansive field including (1) communications, (2) scientific research, (3) therapeutics, (4) business, (5) industry, (6) automation, (7) safety control, (8) computers, (9) data processing, (10) guided missiles, (11) telemetry, (12) aeronautics, (13) astronautics, and (14) cybernetics.

Although many of the circuits used in the applications of electronics are quite diversified and complex, they generally have two points in common: (1) They employ one or more of the basic circuit elements such as resistors, inductors, capacitors, solid-state devices, and electron tubes. (2) They employ one or more of the basic circuit applications of these basic circuit elements. To understand the many complex circuits used in electronics, it is essential to have a thorough knowledge of the basic circuit elements and their basic circuit applications. With this background it is then only necessary to study the new combination of circuit elements and their circuit applications in order to understand the operation of the complex circuits that are used in modern applications of electronics.

The following features, not generally found in any one book, have been incorporated in this text:

1. A minimum knowledge of mathematics is required. Most of the mathematics involves only the use of addition, subtraction, multiplication, division, and square root. Examples involving algebra, trigonometry, and logarithms are explained in detail using a step-by-step procedure.

2. Examples are used throughout the book to illustrate the applications of major equations and principles presented in the text. The values used in these examples, for both simple and complex circuits, represent values

of circuit elements that are used in practical applications of these types of circuits.

3. For ease in understanding the solution of complex examples the solutions are explained in a detailed step-by-step manner.

4. As an aid to the instructor and a challenge to the more interested student, there are numerous challenging questions and/or problems at the ends of the chapters. The values used in the problems have been carefully selected and represent practical values.

5. Answers for approximately one-half of the problems are provided in the appendix as a guide to conscientious students and to those persons using the book without the benefit of a classroom instructor.

6. The principle of operation of the various circuit elements and the analysis of the operation of electric, electron-tube, solid-state, and integrated circuits are explained according to the electron theory.

7. A glossary of solid-state terms, transistor parameter terms, and electron-tube parameter terms is presented in the chapters pertaining to these subjects.

8. In recognition of the value of visual instruction, drawings are used to illustrate each principle as it is presented.

9. As electronic communication systems involve radio, sound, and light waves the first chapter presents the units of measurement and the relation among these three types of waves as applied to communication systems.

10. A large portion of communication and other types of electronic equipment uses solid-state devices. Hence, in the analysis of the various types of electronic circuits the circuits using solid-state devices are presented first, after which a shorter explanation of circuits using electron tubes is presented.

11. Separate chapters are used to describe the basic construction, principle of operation, and circuit parameters of (1) solid-state diodes, (2) vacuum-tube diodes, (3) transistors (bipolar), (4) electron tubes (triode, pentode, and thyratron), and (5) other solid-state devices (tunnel-, back-, varactor-, and compensating-diode; JFET, MOSFET, and dual-gate FET; phototransistor, thyristor, SCR, triac, diac, and unijunction).

12. An entire chapter is devoted to coupling and filter circuits since all practical electronic circuits are made by coupling one circuit to another with proper filtering to prevent undesired interference.

13. The performance of an active device is dependent on its quiescent operating point, which in turn is dependent on its bias. Before presenting the analysis of circuits using active devices a complete chapter is used to explain the importance of proper bias, and the methods used to obtain and stabilize the required bias.

14. Practically every electronic circuit uses some form of amplifier circuit; therefore, five chapters are used to describe the various types of amplifiers: (1) basic amplifier circuits, (2) basic voltage-amplifier circuits, (3) frequency response of amplifiers, (4) other amplifier circuits (feedback, tuned, emitter- and cathode-follower, limiting, differential, operational, Darlington pair), (5)

power-amplifier circuits (single-ended, push-pull, complementary symmetry). Transistor amplifier circuits are analyzed by using both conventional and h-parameters.

15. Three chapters are used to explain the circuit operation of the remaining basic active-device circuits, namely, (1) tuning, (2) oscillator, (3) modulation and demodulation.

16. All electronic equipment requires some form of power supply. One chapter is used to describe the basic power sources (primary cells, secondary cells, power line), and another chapter to describe specialized power sources (solar cell, fuel cell, light-sensitive devices, voltage multipliers, vibrator, regulated, modulated).

17. The correct performance of an electronic circuit is dependent on the shape of the waveform at various points in its circuit. The types of waveforms most commonly used and methods employed to obtain them are described in one chapter.

18. The physical size of electronic equipment has become very important. Miniaturization is obtained by use of integrated circuits as described in a complete chapter.

19. Three appendixes provide reference data and serve as useful tools when working with electronic-circuit problems.

MORRIS SLURZBERG
WILLIAM OSTERHELD

Essentials of
Communication
Electronics

Chapter 1

Radio-, Sound-, and Light-wave Measurements

Electricity and electronics perform a very important part in our modern communications systems. A wide variety of both visual and auditory means of electronic communication systems has been developed such as (1) telegraph, (2) wireless telegraph, (3) telephone, (4) radio, (5) public-address systems, (6) disk recorders, (7) tape recorders, (8) facsimile, (9) television, (10) motion pictures, (11) radar, (12) sonar, (13) loran, (14) shoran, and (15) telemetry. These are broad classifications and there are many subdivisions for each of these categories. Since the medium used to transmit energy in all forms of electronic communications systems may be radio waves, sound waves, or light waves, it is essential to know the basic characteristics and measurements used in these types of wave propagation.

1-1 Wave Propagation

Need for Concept of Wave Motion. In the study of electricity and electronics, frequent reference is made to the principles of wave motion as is indicated by such terms as (1) sine-wave alternating currents, (2) sound waves, (3) radio waves, (4) light waves, (5) carrier waves, (6) ultrasonic waves, etc. A simple analogy of wave motion in electricity and electronics is the wave motion produced by dropping a pebble in a small body of water, namely, that the energy moves away from the source in ripples or waves. An alternating current in a low-frequency electric power circuit and the high-frequency radio signals sent through space both follow the principles of wave motion.

Alternating Current in an Electric Circuit. An alternating current (abbreviated a-c) in an electric circuit reverses its direction at fixed intervals, hence the name *alternating current.* During each interval the current will rise from zero to its maximum value, then diminish to zero. Figure 1-1 shows that an a-c wave goes through two similar sets of changing values, one in a positive direction and one in a negative direction. The interval required for one set of values, in either direction, is called an *alternation* and corresponds to 180 electrical degrees. After two successive alternations, or 360 electrical degrees, the a-c wave has completed one *cycle.* This a-c wave is followed by a continuing succession of such waves. The number of cycles occurring in one second is called the *frequency* of the power source. In Fig. 1-1, I is the symbol denoting current and *max* is an abbreviation for maximum; I_{max} indicates the maximum

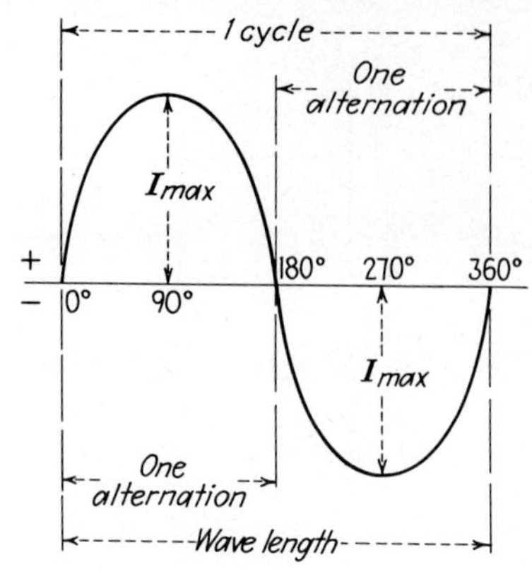

Fig. 1-1 An a-c wave.

amount of current which occurs at each 90-, 270-, 450-, etc., degree instant of an a-c sine wave.

Electric and Magnetic Waves in Space. Light, heat, X-ray, radio, and television waves are forms of radiant energy considered to be electromagnetic oscillatory disturbances in space. The frequency of oscillation determines whether a particular signal is evident as light, heat, X-ray, radio, or television waves. These forms of energy are referred to as *electromagnetic waves* because they emanate from a transmitter with a progressive action similar to waves in a body of water. The overall range of frequencies of these waves is referred to as the *electromagnetic spectrum.* The division of the electromagnetic spectrum is shown in Fig. 1-2. The division of the radio-frequency (abbreviated r-f) spectrum and the descriptive names applied to each portion of the spectrum are shown in Table 1-1. The division of the communication spectrum for radio and television broadcasting is shown in Table 1-2.

Electric and Magnetic Fields. The medium in which electromagnetic waves move is free space, commonly called the *ether*. The waves move out from their

Table 1-1 Division of the Radio-frequency Spectrum

FREQUENCY, MHz	WAVELENGTH, METERS	DESCRIPTION	ABBREVIATION
0.01–0.03	30,000–10,000	Very low frequency	vlf
0.03–0.3	10,000–1,000	Low frequency	l-f
0.3–3	1,000–100	Medium frequency	m-f
3–30	100–10	High frequency	h-f
30–300	10–1	Very high frequency	vhf
300–3,000	1–0.1	Ultrahigh frequency	uhf
3,000–30,000	0.1–0.01	Superhigh frequency	shf

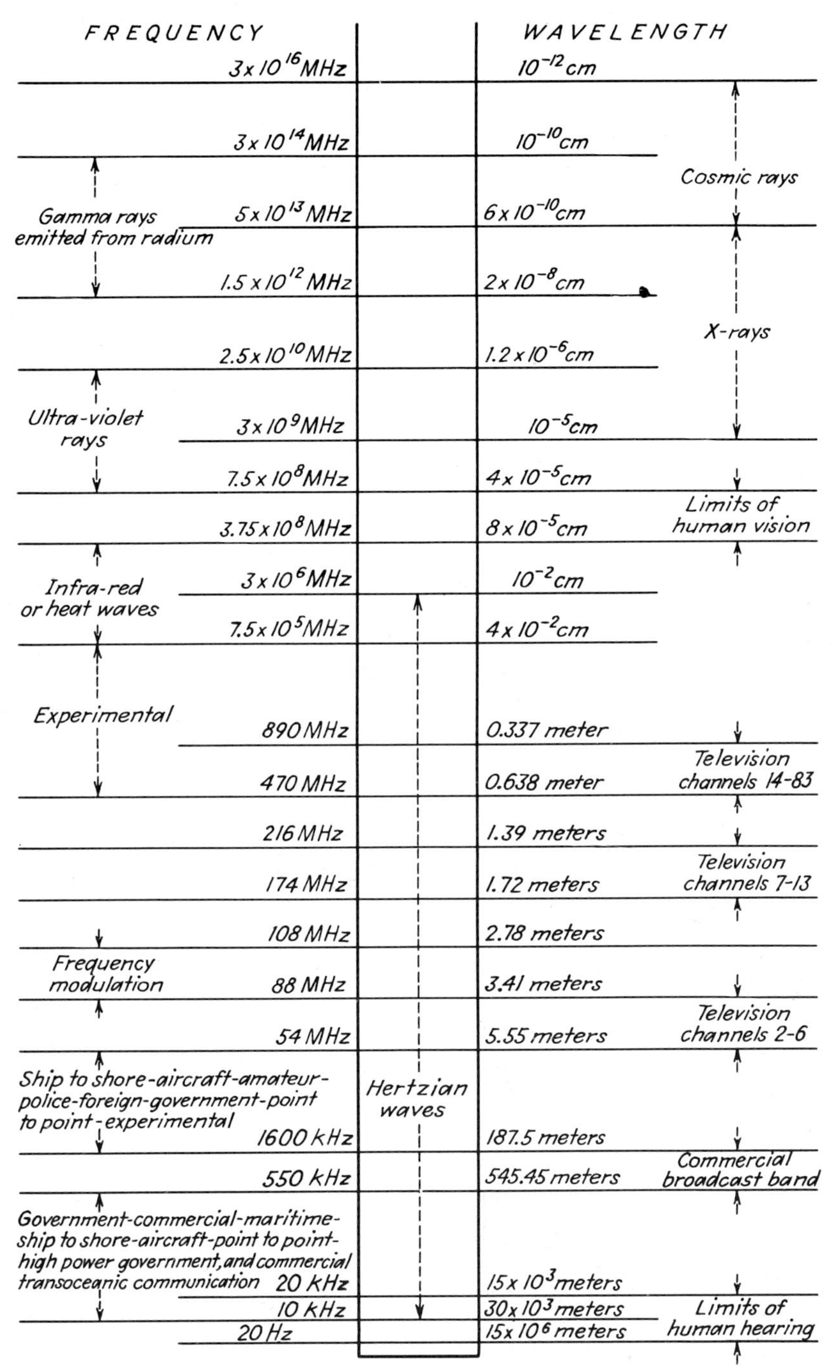

Fig. 1-2 The electromagnetic spectrum.

Table 1-2 Division of the Communication Spectrum

TYPE	NUMBER OF CHANNELS	FREQUENCY RANGE, MHz
A-m broadcast	107	0.535–1.605
F-m broadcast	100	88–108
TV vhf (low band) channels 2-6	5	54–88
vhf (high band) channels 7-13	7	174–216
uhf channels 14-83	70	470–890
Amateur	16	1.8–22,000
International broadcast	7	5.95–29.1
Citizens	3	26.9600–27.2300
		462.5375–462.7375
		467.5375–467.7375
Land mobile (industrial)	12	25.01–49.60
	7	151.49–173.40
	6	451.00–470.00

source in all directions, vertically as well as horizontally. The presence of physical bodies such as earth or wood alters but does not entirely prevent the propagation of electromagnetic waves. Although the electric and magnetic fields are not visible to the eye, they may be represented as shown in Fig. 1-3. For this case, it is assumed that the direction of propagation of the wave is straight into the paper. There is no physical motion of particles in electromagnetic wave motion, but rather there is an oscillation of the electric and magnetic fields which are oriented in a plane perpendicular to the direction of motion of the wave. The electric lines are marked E, and the magnetic lines are marked H. At one instant they would appear as in Fig. 1-3*a* and a half-cycle later, as in Fig. 1-3*b*. Both fields have reversed in direction, but the magnitude is unchanged and they are still perpendicular to each other. The magnitude of the electric and magnetic fields is so related that half the power transmitted

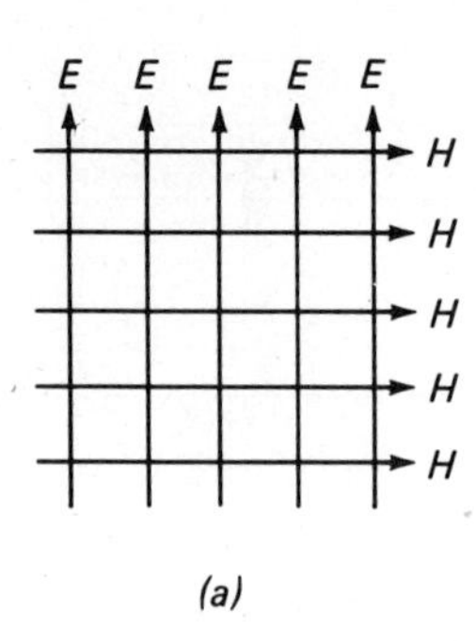

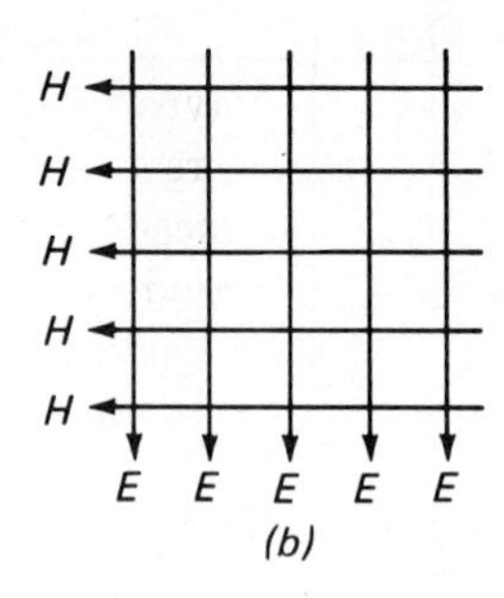

Fig. 1-3 Electric and magnetic fields. (a) At one instant. (b) One half-cycle later.

by the wave motion is contained in the electric field and half in the magnetic field.

1-2 Radio Waves

Types. Radio and television transmitting stations convert sound waves and light waves to electrical impulses. The electrical impulses representing the original sound and light waves are sent out by the use of high-frequency alternating currents. These currents produce magnetic and electric fields that radiate in all directions over long distances. The magnetic and electric fields produced by high-frequency currents are called *radio waves.* The strength and frequency of a radio wave are dependent on the high-frequency alternating current producing it and will vary in the same manner as the alternating current.

In addition to their application for the transmission of sound and light signals, radio waves are also used by other types of transmitters employing frequencies in the r-f spectrum to send out information. For example, a radar transmitter sends out high-frequency pulses at regular intervals and the echo pulses, reflected from a target, are picked up by a receiving antenna.

Speed of Radio Waves. Radio waves travel at the same speed as light waves, or 186,000 miles per sec. In some calculations the metric system is used, and the speed of the radio waves is then expressed in meters per second.

Example 1-1 Radio waves travel at the rate of 186,000 miles per sec. What is the rate in (*a*) feet per second, (*b*) meters per second? (One mile = 5,280 ft; also, 1 meter = 3.28 ft.)

GIVEN: Miles per sec = 186,000 Ft per mile = 5,280 Ft per m = 3.28

FIND: (*a*) Ft per sec (*b*) M per sec

SOLUTION:

(*a*)
$$\text{Ft per sec} = 186{,}000 \times 5{,}280 \cong 982{,}000{,}000$$

(*b*)
$$\text{M per sec} = \frac{982{,}000{,}000}{3.28} \cong 300{,}000{,}000$$

Note: $\cong$ means *is approximately equal to.*

Wavelength and Frequency Definitions

Wavelength. The distance that the radio wave travels in the time of one cycle is called its *wavelength*; it is expressed in meters and is represented by the symbol λ, a letter of the Greek alphabet pronounced "lambda."

Frequency. The number of cycles per second of a radio wave is called its *frequency* and is represented by the letter f. The basic unit of frequency is the *hertz*, abbreviated Hz; it represents a frequency of one cycle per second of an alternating-current wave. The frequency was formerly referred to as the number of cycles, which was understood to mean the number of cycles per second.

Wavelength and Frequency Calculations

Wavelength. If the frequency of a wave is known, the distance it will travel in the time required for one cycle can be calculated by

$$\lambda = \frac{300{,}000{,}000}{f} \tag{1-1}$$

where λ = wavelength, meters (m)
f = frequency, Hz

Kilohertz and Megahertz. The frequencies of the common radio waves are of high values, that is, in the hundreds of thousands or millions of cycles per second. For convenience these frequencies are generally expressed in kilohertz or megahertz and abbreviated as kHz and MHz respectively. *Kilo-* is a prefix meaning one thousand; hence a kilohertz is equal to 1,000 hertz. The prefix *mega-* means one million; hence a megahertz is equal to 1,000,000 hertz.

When radio frequencies are expressed in kilohertz or megahertz, Eq. (1-1) becomes

$$\lambda = \frac{300{,}000}{f \text{ (in kHz)}} \tag{1-1a}$$

$$\lambda = \frac{300}{f \text{ (in MHz)}} \tag{1-1b}$$

Example 1-2 What is the wavelength of a radio station which operates on an assigned frequency of 570 kHz?

GIVEN: f = 570 kHz

FIND: λ

SOLUTION:

$$\lambda = \frac{300{,}000}{f} = \frac{300{,}000}{570} = 526.3 \text{ m}$$

Example 1-3 What is the wavelength of a television video carrier wave whose frequency is 77.25 MHz?

GIVEN: f = 77.25 MHz

FIND: λ

SOLUTION:

$$\lambda = \frac{300}{f} = \frac{300}{77.25} = 3.88 \text{ m}$$

Frequency. Equations (1-1), (1-1*a*), and (1-1*b*) can be transposed to solve for frequency instead of wavelength and become

$$f \text{ (in Hz)} = \frac{300{,}000{,}000}{\lambda} \tag{1-2}$$

$$f \text{ (in kHz)} = \frac{300{,}000}{\lambda} \tag{1-2a}$$

$$f \text{ (in MHz)} = \frac{300}{\lambda} \tag{1-2b}$$

Example 1-4 If by definition a short radio wave is one whose wavelength does not exceed 200 m, what is the lowest frequency at which a short-wave radio receiver may operate?

GIVEN: $\lambda = 200$ m

FIND: f

SOLUTION:

$$f = \frac{300{,}000}{\lambda} = \frac{300{,}000}{200} = 1{,}500 \text{ kHz}$$

If the wavelength in meters is changed to our more commonly used units of feet and inches, it should provide a better understanding of the length of the radio waves.

Example 1-5 What is the length in feet of one radio wave of the broadcast station referred to in Example 1-2?

GIVEN: $\lambda = 526.3$ m

FIND: Length, ft

SOLUTION:

$$\text{Length} = \text{meters} \times 3.28 = 526.3 \times 3.28 = 1{,}726 \text{ ft}$$

Example 1-6 A video carrier wave used in uhf transmission has a frequency of 867.25 MHz. What is its length in inches? (One meter = 39.37 inches.)

GIVEN: $f = 867.25$ MHz

FIND: Length, inches

SOLUTION:

$$\lambda = \frac{300}{f} = \frac{300}{867.25} = 0.346 \text{ m}$$

$$\text{Length} = 0.346 \times 39.37 = 13.62 \text{ inches}$$

The solutions of Examples 1-5 and 1-6 indicate that each wave transmitted by a 570-kHz radio station is 1,726 ft long, or approximately one-third of a mile, and each wave transmitted by a 867.25-MHz video carrier has a length of 13.62 inches, or approximately 1 ft.

Knowing that radio waves travel 186,000 miles per sec, the time required for a radio wave to get from one place to another can be readily calculated.

Example 1-7 How long does it take a radio wave to travel from New York to San Francisco, a distance of approximately 2,600 miles?

GIVEN: Distance = 2,600 miles Rate = 186,000 miles per sec

FIND: Time, t

SOLUTION:

$$t = \frac{\text{miles}}{186{,}000} = \frac{2{,}600}{186{,}000} \doteq 0.0139 \text{ sec}$$

Example 1-7 shows that it takes only about $\frac{1}{70}$ sec for a radio program broadcast from New York to travel to San Francisco.

1-3 Sound

Characteristics of Sound. *Sound* is the sensation produced in the brain by sound waves. It makes use of one of our five fundamental senses, namely, that of hearing. The air in a room in which no sound is present is in a static condition; in other words, it is motionless. If a sound is made by a person, by a musical instrument, or by any other means, the air about it is set into vibration. These vibrations are transmitted to adjacent layers of air and so on until all the original energy is expended. Such air vibrations are called *sound waves.* When these vibrations strike the eardrum of a person, the eardrum too will vibrate in a similar manner. The auditory nerves will be stimulated and will communicate the sensation of sound to the brain (Fig. 1-4).

Sound waves are produced by the mechanical vibration of any material

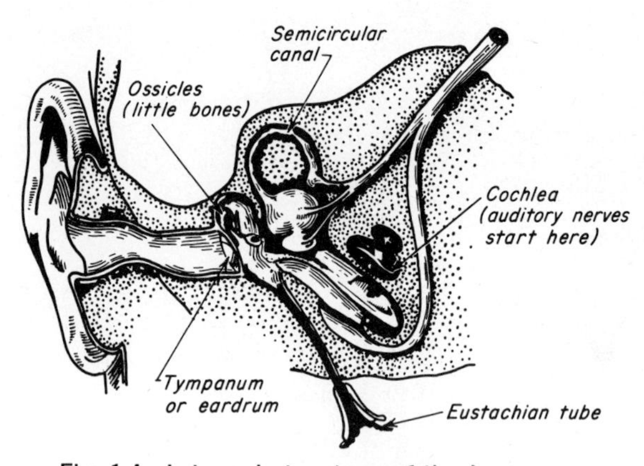

Fig. 1-4 Internal structure of the human ear.

in elastic media such as gases, liquids, and some solids, but they will not travel in a vacuum. Sound waves are longitudinal waves and travel outward in all directions from the source. A longitudinal wave may be defined as one in which the vibrating molecules or particles of the transmitting medium move back and forth in the same direction in which the sound wave is traveling.

Intensity. The *intensity*, or *loudness*, of a sound depends upon the energy of motion imparted to the vibrating molecules of the medium transmitting the sound. A greater amount of energy causes more violent movement of the molecules, which in turn exerts a greater pressure upon the eardrum, thus causing the auditory nerves to send the sensation of a louder sound to the brain. Loudness is affected by the distance between the listener and the source of the sound, and its intensity varies inversely with the square of this distance. For example, if the distance between the listener and the source of the sound is doubled, the intensity is reduced to one-quarter. Also, if the distance between the listener and the source of the sound is decreased to one-half the original amount, the intensity of the sound will be four times as great.

Speed of Sound. The speed at which sound waves travel varies with the kind of material through which it is traveling. For air, the most common medium used for transmitting sound waves, the speed is 1,130 ft per sec at the normal room temperature of 68°F.

Frequency, Pitch, and Wavelength. The vibration of the reeds in a harmonica, of the skin on a drum, of the strings on a violin, or of the cone of a radio loudspeaker will all send out various sound waves. These waves will produce different sounds, depending on the number of vibrations that the wave makes per second. The number of complete waves or vibrations created per second is known as the *frequency* of the sound and is generally expressed as the number of hertz. For example, a sound wave that is making 2,000 vibrations per second is the same as a sound whose frequency is 2,000 Hz; this is also commonly referred to as a 2,000-hertz sound or a 2,000-hertz note.

If the sound is loud enough to be heard by the human ear, it is said to be *audible*. Its *pitch* will vary with the frequency. High frequencies produce sounds having a high pitch, and low frequencies produce sounds of low pitch.

Sound waves may also be referred to in terms of the length of a wave. Knowing that sound waves travel 1,130 ft per sec in air, the length of one wave can be calculated by dividing the number 1,130 by the frequency of the sound. Figure 1-5 illustrates a tuning fork producing sound waves whose frequency is 256 Hz and whose wavelength is 4.41 ft.

Example 1-8 The frequency range of a piano is from 25 to 8,000 hertz. (*a*) What is the range of wavelengths in feet? (*b*) In meters? (*c*) If the sound waves are converted to electrical waves by a microphone, what is the frequency range of the electric currents?

GIVEN: Sound waves = 25–8,000 hertz

FIND: (*a*) Wavelengths, ft (*b*) Wavelengths, m (*c*) Frequency range of electric currents, hertz

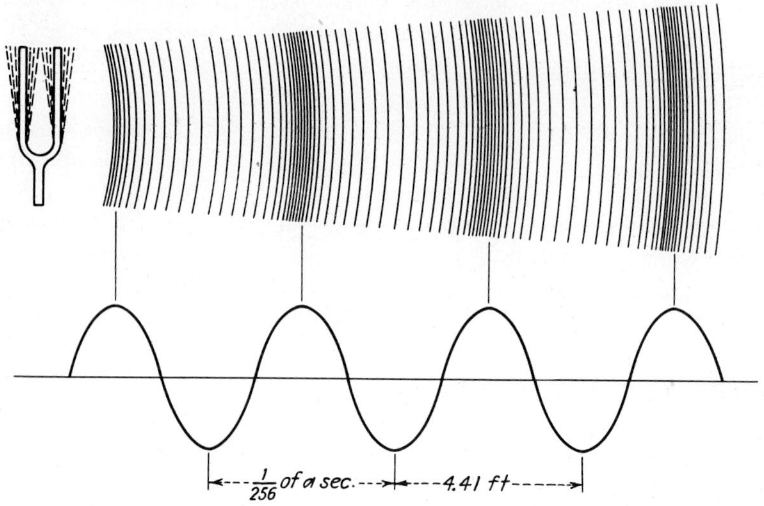

Fig. 1-5 Propagation of a sound wave of 256 hertz in air.

SOLUTION:

(*a*) $$\text{Wavelength, 25 hertz} = \frac{\text{ft per sec}}{\text{Hz}} = \frac{1{,}130}{25} = 45.2 \text{ ft}$$

$$\text{Wavelength, 8,000 hertz} = \frac{\text{ft per sec}}{\text{Hz}} = \frac{1{,}130}{8{,}000} = 0.14125 \text{ ft}$$

(*b*) $$\text{Wavelength, 25 hertz} = \frac{\text{wavelength, ft}}{3.28} = \frac{45.2}{3.28} = 13.7 \text{ m}$$

$$\text{Wavelength, 8,000 hertz} = \frac{\text{wavelength, ft}}{3.28} = \frac{0.14125}{3.28} = 0.043 \text{ m}$$

(*c*) 25 to 8,000 hertz (same frequencies as the sound waves)

Frequency Ranges of Sound Waves. The range of frequencies that the human ear is capable of hearing will vary with the individual, the lower limit being approximately 20 hertz and the upper limit 20,000 hertz. Some persons are able to hear the low-pitch sounds but cannot hear those of high pitch, while others can hear the high-pitch sounds but cannot hear those of low pitch. However, there are people able to hear sounds covering a wide range of frequencies. The frequency range of various audible sound waves is illustrated in Fig. 1-6.

1-4 Characteristics of Sound

Musical Sounds and Noise. When sound waves are produced repeatedly at regular intervals, the result is a *musical sound* at some definite pitch which is more or less pleasant to the ear. The orderly repetition produces rhythm which is also a requirement to obtain a musical note. When sound waves of constant or varying frequencies are produced at haphazard irregular intervals, the result is an unpleasant sound called *noise*.

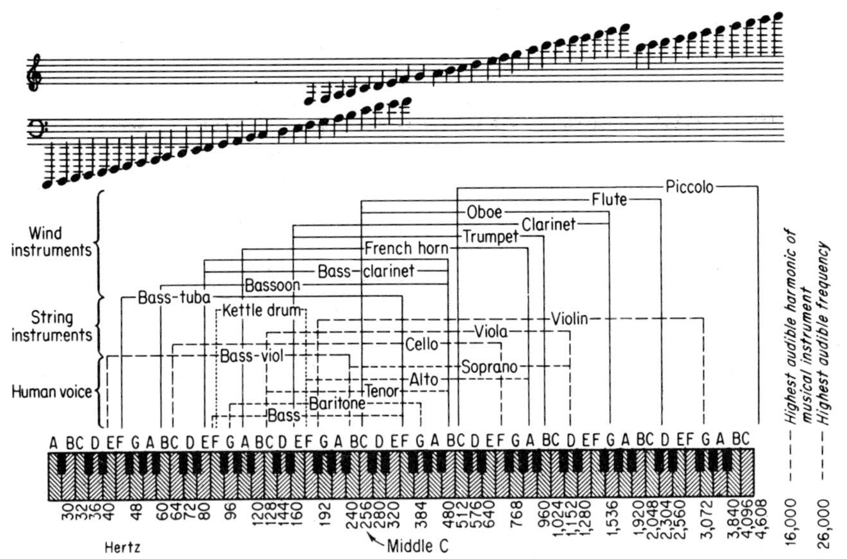

Fig. 1-6 Frequency ranges of human voices and some musical instruments.

Sound Levels. The relative level of any sound may be expressed as (1) the pressure (dynes per sq cm), (2) the power (watts per sq cm), or (3) the change in volume (decibels), see Art. 1-6, at a person's ear. Table 1-3 lists these levels for some of the most common sounds, ranging from the threshold of hearing to beyond the threshold of pain.

Quality, Fundamentals, and Overtones. The middle C of a piano has a frequency of 256 Hz. A corresponding note of 256 hertz can also be produced on other musical instruments such as a violin, clarinet, or harmonica. The notes from the various instruments differ in *quality*, which depends upon the number and relative amplitude of the overtones blended with the fundamental.

The *fundamental* note is the lowest tone produced, which in the above example would be the 256-hertz note. The overtones, which are higher-pitched notes, blend with the fundamental and give each instrument (and each human voice) its individuality. The *overtones* are vibrations whose frequencies are multiples of the fundamental. The frequencies of the overtones of a 256-hertz note would be 512, 768, 1,024, 1,280, etc., hertz. The overtones are also often referred to as *harmonics*.

Reflection of Sound. When sound waves strike a solid object such as a wall of a building, the side of a cliff, or the wall of a room, the sound will be reflected and may cause an echo. An *echo* is the effect produced when a reflected sound returns to the ear a fraction of a second after the original source of the sound has ceased. If the interval between the original sound and the reflected sound is $\frac{1}{10}$ sec or greater, an echo is likely to result. Echoes do not appear in small rooms because the reflected sounds return to the ear too soon after the original sound to be distinguished. Large rooms or auditoriums, where the reflecting

Table 1-3 Relative Approximate Values of Sound Levels

SOUND	PRESSURE, DYNES/SQ CM	POWER, WATTS/SQ CM	DECIBELS
Threshold of hearing	0.0002	10^{-16}	0
Soundproof room	0.0005	8×10^{-16}	8
Leaves rustling	0.0006	10^{-15}	10
Quiet sound studio	0.001	6×10^{-15}	16
Quiet whisper (at 5 ft)	0.0015	8×10^{-15}	18
Very soft music	0.008	2×10^{-13}	32
Average residence	0.035	5×10^{-12}	45
Quiet residential street	0.1	5×10^{-11}	55
Department store	0.3	4×10^{-10}	64
Conversation (at 3 ft)	0.35	5×10^{-10}	65
Average automobile	0.55	9×10^{-10}	69
Loud music	0.65	10^{-9}	70
Very loud music	2	10^{-8}	80
Heavy street traffic	3.5	5×10^{-8}	85
Riveting machine (at 35 ft)	20	10^{-6}	100
Thunder	50	8×10^{-6}	108
Airplane (at 18 ft)	250	2×10^{-4}	122
Pneumatic chipper (at 5 ft)	350	5×10^{-4}	125
Threshold of pain	650	10^{-3}	130
50-hp siren (at 100 ft)	1,600	8×10^{-3}	138

surfaces are more than 50 ft away from the source of the sound, often produce echoes. In such cases the walls and ceilings may be decorated or padded with tapestries or soft materials to eliminate or reduce the production of echoes.

Sympathetic Vibration. The sound waves set up by one sound-producing object can cause a nearby object to start vibrating and thereby also produce sound waves if both objects have the same natural frequency. The vibrations of the second object are said to be the result of *sympathetic vibration.* An example of this phenomenon may sometimes be observed when the music from a radio receiver causes a metal vase, picture frame, or other object to start vibrating and give off sound waves. Sympathetic vibration may also cause a loose part in sound-producing equipment to start vibrating.

Sympathetic vibration may be explained in the following manner: The original sound wave strikes all nearby objects and sets them in motion, even though the amount of movement is very slight. If the second wave strikes at the precise instant which causes it to add its motion in perfect unison to that of the preceding wave, the movement will be increased. If all the succeeding waves strike at the corresponding precise instants, the movement will be cumulative and the object will vibrate at the frequency of the source of the sound. If, however, successive waves do not strike the object at precisely the proper instant, some of the waves will neutralize the motion imparted to the object by some preceding waves and no sympathetic vibration will result.

Forced Vibrations. An object in the area near a source of sound may be set into vibration by the sound waves of frequencies other than the natural frequency of the object if the intensity of the sound is great enough. This phenomenon is called *forced vibration.* Examples illustrating forced vibrations are the sounding board of a piano, sound reflectors of musical instruments, and the loudspeaker of a radio or television receiver.

Resonance. When a sound wave sent out by an object is reflected in such a manner that it returns to the object at the proper instant, it will produce sympathetic vibration within that object and thereby increase the intensity of the sound. This phenomenon is called *resonance.* An example of resonance can be observed by causing air waves to flow through pipes as in a pipe organ. The resonance of a pipe (for a fixed value of frequency) depends upon the length of the pipe and whether it is open or closed. A pipe closed at one end will produce resonance when the length of the pipe is one-quarter of the wavelength of the sound. A pipe open at both ends will produce resonance when the length of the pipe is one-half of the wavelength of the sound.

Beats. In the preceding paragraphs on Sympathetic Vibration and Resonance it was shown that two sounds of the same frequency would reinforce each other. If two sounds of different frequency are considered, it can be shown that the two sounds when started at the same instant of time will at first reinforce each other and then, after a number of cycles have been completed, will be out of step and will neutralize each other. Over a relatively long span of time there will be periods of reinforcements producing a strong sound and periods of neutralizing effect during which no sound will be produced. The sounds produced in this manner are called *beats.* If the number of beats appears frequently and at regular intervals, a new sound or beat note will be produced. The frequency of the beat note will be equal to the difference between the frequencies of the two sounds; for example, sounds of 256 and 200 hertz will produce a beat note of 56 hertz.

1-5 Use of Logarithms in Sound Measurements

Relation of Sound Energy to Ear Response. The operating characteristics of an audio amplifier are generally expressed in terms of its gain or loss in volume. The unit used to express this change is based on the ability of the human ear to respond to these changes.

In the rendition of a musical program, a symphonic orchestra will produce varying amounts of sound energy. The amount of energy used in producing the loudest note may be many thousand times as great as that used to produce the lowest note. However, the ear does not respond to these sounds in proportion to the energy used; the loudest note is not heard many thousands of times as loud as the lowest note. Research has shown that the response will vary logarithmically. In radio terminology, the ratio of any two levels of power is expressed in a unit called the *decibel,* commonly abbreviated db. The number of decibels is equal to 10 times the logarithm of the ratio of the two levels or values of power. A knowledge of logarithms is essential in order to understand problems involving sound as related to amplifiers, speakers, microphones, etc.

Logarithms. Logarithms are commonly used in engineering mathematics to facilitate mathematical computations. Although numerous systems (or bases) of logarithms can be used, the common logarithm, that is, the logarithm to the base 10, is used most frequently and is the system used with the decibel. By definition, *a common logarithm of a number is the exponent* (or power) *to which* 10 (called the base) *must be raised to produce the number.* The logarithm of 100 will therefore be equal to 2, as 10^2 equals 100. This may be expressed as

$$\log_{10} 100 = 2$$

As only the common system of logarithms is used in sound-level calculations, the reference to the base 10 can be omitted. The expression may then be written as

$$\log 100 = 2$$

Since the logarithm of 10 is equal to 1 and the logarithm of 100 is equal to 2, it is evident that the logarithm of any number between 10 and 100 must be greater than 1 and less than 2. Consequently the logarithm of a number will consist of two parts: (1) a whole number called the *characteristic,* and (2) a decimal called the *mantissa.* For example, the logarithm of 50, which is equal to 1.699, is made up of the characteristic whose value is 1 and the mantissa whose value is 0.699.

The characteristic of any number greater than 1 is always positive; numerically it is equal to 1 less than the number of figures to the left of the decimal point. The characteristic of any number less than 1 is always negative; numerically it is equal to 1 more than the number of zeros between the decimal point and the first significant figure.

Example 1-9 What is the characteristic of the following numbers: (*a*) 18.3, (*b*) 183, (*c*) 18,300, (*d*) 1.83, (*e*) 0.183, (*f*) 0.00183?

GIVEN: 18.3 183 18,300 1.83 0.183 0.00183

FIND: Characteristic

SOLUTION:

	NUMBER	CHARACTERISTIC
(*a*)	18.3	1
(*b*)	183	2
(*c*)	18,300	4
(*d*)	1.83	0
(*e*)	0.183	−1
(*f*)	0.00183	−3

In solutions (*e*) and (*f*) of Example 1-9 the characteristic is shown as a negative quantity. The negative sign must be associated only with the characteristic, as the mantissas are always positive. One method of indicating that the negative sign applies only to the characteristic is to place the minus sign over the number instead of in front of it, as $\bar{1}$, $\bar{3}$, etc, Another more commonly used method is to add 10 to the characteristic and then write −10 after the logarithm (see Example 1-10).

The mantissa or decimal part of a logarithm is found by reference to a table of logarithms, which is a tabulation of mantissas. The mantissa is always a positive number and is determined from the significant digits in the number. A table of common logarithms of numbers is provided in Appendix 3. It is to be understood that a decimal point is assumed in front of each of the values of mantissas in the table of Appendix 3.

Example 1-10 What is the logarithm of: (*a*) 18,300? (*b*) 18.3? (*c*) 0.00183? (*d*) 650,000? (*e*) 1.25?

GIVEN: 18,300 18.3 0.00183 650,000 1.25

FIND: Logarithm

SOLUTION: See Table 1-4

Table 1-4

NUMBER	CHARACTERISTIC	MANTISSA FROM APPENDIX 3	LOGARITHM
(*a*) 18,300	4	.2625	4.2625
(*b*) 18.3	1	.2625	1.2625
(*c*) 0.00183	−3	.2625	$\bar{3}$.2625 (or 7.2625 − 10)
(*d*) 650,000	5	.8129	5.8129
(*e*) 1.25	0	.0969	0.0969

It is sometimes necessary to find the number corresponding to a logarithm. The number corresponding to a logarithm is called an *antilogarithm* or *antilog* and may be found by working in the reverse order of finding the logarithm of a number. For example, the common logarithm of 100 is 2 and hence it is said that the antilog of 2 is 100.

Example 1-11 What is the antilog of the common logarithm: (*a*) 3.8751? (*b*) 0.0645? (*c*) 7.3711 − 10?

GIVEN: (*a*) Log = 3.8751 (*b*) Log = 0.0645 (*c*) Log = 7.3711 − 10

FIND: Antilog

SOLUTION:

(*a*) Antilog of 3.8751

The logarithm should be divided into its characteristic (3) and its mantissa (0.8751). From Appendix 3, the mantissa corresponds to the number 750. The characteristic 3 indicates that the number will have 3 + 1 or 4 figures to the left of the decimal point. Thus, the antilog of 3.8751 is found to be 7,500.

(*b*) Antilog of 0.0645

Characteristic = 0; mantissa = 0.0645. Number corresponding to the mantissa = 116 (from Appendix 3). Number of places to the left of the decimal point = 0 + 1 = 1. Thus, the antilog of 0.0645 is 1.16.

(*c*) Antilog of 7.3711 − 10

Characteristic = 7 − 10 = −3; mantissa = 0.3711. Number corresponding to the mantissa = 235 (from Appendix 3). With a characteristic of $\bar{3}$, there must be two zeros between the decimal point and the first significant number. Thus, the antilog of 7.3711 − 10 is 0.00235.

1-6 Sound Measurements

The Decibel. The operating characteristics of an audio amplifier are generally expressed in terms of its gain or loss in volume. The unit used to express this change in volume is based on the ability of the human ear to respond to these changes. The unit most frequently used is the *bel*, named in honor of Alexander Graham Bell. The bel is defined as the common logarithm of the ratio between two quantities and is a relative unit of measurement; it does not specify any definite amount of sound, power, voltage, or current. The bel is too large a unit for general use and the *decibel*, which is one-tenth of a bel, is commonly used. The decibel may be used to express the ratio between two values of either sound, power, voltage, or current. The change in volume in any circuit, expressed in decibels, abbreviated db, can be found by the equation

$$\text{db} = 10 \log \frac{P_1}{P_2} \tag{1-3}$$

where P_1 and P_2 express the ratio of two values of power.

The decibel is a logarithmic unit and therefore represents a logarithmic change. As the response of the human ear to sounds of varying intensity is logarithmic, regardless of the power level, the decibel provides a good means of expressing variations in sound measurements. The smallest change in sound intensity that can be detected by the human ear is approximately 1 db, although the average person does not ordinarily detect changes under 2 or 3 db.

Example 1-12 A certain radio receiver delivers a power output of 3.2 watts. What decibel gain will be obtained by increasing the power output to: (*a*) 4.8 watts? (*b*) 6.4 watts?

GIVEN: $P = 3.2$ watts $\quad P_a = 4.8$ watts $\quad P_b = 6.4$ watts

FIND: (*a*) Decibel gain (*b*) Decibel gain

SOLUTION:

(*a*)
$$\text{Gain} = 10 \log \frac{P_a}{P} = 10 \log \frac{4.8}{3.2} = 10 \log 1.5 = 10 \times 0.1761 = 1.761 \text{ db}$$

$$\log 1.5 = 0.1761 \text{ (from Appendix 3)}$$

(*b*)
$$\text{Gain} = 10 \log \frac{P_b}{P} = 10 \log \frac{6.4}{3.2} = 10 \log 2 = 10 \times 0.301 = 3.01 \text{ db}$$

$$\log 2 = 0.301 \text{ (from Appendix 3)}$$

Example 1-12 shows that doubling the power increases the volume by only 3 db. As this change is barely perceptible to the average listener, it would not be practical to increase the power output as suggested in Example 1-12 in order to obtain a gain in volume.

Example 1-13 What power output is required in order to produce a gain of 10 db over the 3.2 watts of output used in Example 1-12?

GIVEN: $P_1 = 3.2$ watts $\quad$ Gain $= 10$ db

FIND: P_2

SOLUTION:

$$\text{db} = 10 \log \frac{P_2}{P_1}$$

$$\log \frac{P_2}{P_1} = \frac{\text{db}}{10} = \frac{10}{10} = 1$$

Antilog of 1 = 10 (from Appendix 3).

$$\frac{P_2}{P_1} = 10$$

$$P_2 = P_1 \times 10 = 3.2 \times 10 = 32 \text{ watts}$$

Voltage and Current Ratios. The decibel is fundamentally a measure of power ratio; however, as voltage and current are functions of power, Eq. (1-3) can be transformed to express the change in volume in decibels for two different values of voltage or current output.

Substituting $\frac{E^2}{R}$ for P in Eq. (1-3), then

$$\text{db} = 10 \log \frac{\dfrac{E_1^2}{R_1}}{\dfrac{E_2^2}{R_2}} = 10 \log \left(\frac{E_1}{E_2}\right)^2 \frac{R_2}{R_1} \tag{1-4}$$

or
$$\text{db} = 10 \log \left(\frac{E_1}{E_2}\right)^2 + 10 \log \frac{R_2}{R_1} \tag{1-4a}$$

or
$$\text{db} = 20 \log \frac{E_1}{E_2} + 10 \log \frac{R_2}{R_1} \tag{1-4b}$$

or
$$\text{db} = 20 \log \frac{E_1 \sqrt{R_2}}{E_2 \sqrt{R_1}} \tag{1-4c}$$

Substituting I^2R for P in Eq. (1-3), then

$$\text{db} = 10 \log \frac{I_1{}^2 R_1}{I_2{}^2 R_2} \tag{1-5}$$

or
$$\text{db} = 10 \log \left(\frac{I_1}{I_2}\right)^2 + 10 \log \frac{R_1}{R_2} \tag{1-5a}$$

or
$$\text{db} = 20 \log \frac{I_1}{I_2} + 10 \log \frac{R_1}{R_2} \tag{1-5b}$$

or
$$\text{db} = 20 \log \frac{I_1 \sqrt{R_1}}{I_2 \sqrt{R_2}} \tag{1-5c}$$

For conditions where the impedances are equal, Eqs. (1-4*c*) and (1-5*c*) can be simplified as

$$\text{db} = 20 \log \frac{E_1}{E_2} \tag{1-6}$$

$$\text{db} = 20 \log \frac{I_1}{I_2} \tag{1-7}$$

Example 1-14 The characteristics of a certain audio amplifier are such that a voltage amplification of 5 is obtained at 50 hertz, 15 at 1,500 hertz, and 30 at 5,000 hertz. Assuming the voltage amplification at 1,500 hertz as the reference level, what is the loss or gain in decibels at the other frequencies?

GIVEN: $f_1 = 50$ hertz, $A_v = 5$ $f_2 = 1{,}500$ hertz, $A_v = 15$ $f_3 = 5{,}000$ hertz, $A_v = 30$

FIND: Change at 50 hertz, db Change at 5,000 hertz, db

SOLUTION:

$$\text{Change at 50 hertz} = 20 \log \frac{A_v \text{ at } f_1}{A_v \text{ at } f_2} = 20 \log \frac{5}{15} = 20 \log 0.333$$
$$= 20(9.5224 - 10) = -9.552 \text{ db}$$
$$\text{Change at 5,000 hertz} = 20 \log \frac{A_v \text{ at } f_3}{A_v \text{ at } f_2} = 20 \log \frac{30}{15} = 20 \log 2$$
$$= 20 \times 0.301 = 6.02 \text{ db}$$

Example 1-14 indicates that this amplifier has poor fidelity. There is a loss of almost 10 db at the low frequencies and a gain of 6 db at the high frequencies. Such differences in volume are easily detected by the average listener.

1-7 Light

Characteristics of Light. *Light* is the sensation produced in the brain by light rays. It makes use of one of our five fundamental senses, namely, that of sight. In the study of physics light is considered as a form of energy that may be derived from mechanical, electrical, chemical, heat, or light energy. Objects are *visible* because light rays from them reach the eyes, which in turn stimulate the optic nerves and send the sensation of sight to the brain. When the visible object is the source of light energy, it is *luminous*. The electric light, the sun, and the picture tube of a television receiver are examples of luminous objects. Most objects are not luminous, and the light rays that reach the eyes from such objects are actually the reflected light from some luminous body. Objects visible because of the light they reflect are called *illuminated bodies*.

Light is transmitted by *transverse waves*, that is, waves in which the motion is transmitted in a direction at right angles to the vibrations. Light waves can be transmitted only by a transparent medium, including vacuum. In terms of the ability to transmit light, matter may be divided into three classifications: transparent, translucent, and opaque materials. Materials that transmit light so well that objects can be seen clearly through the material are called *transparent*. Materials that transmit light so poorly that objects cannot be seen clearly through the material are called *translucent*. Materials through which light cannot pass are called *opaque*.

Brightness and Intensity of Illumination. The first consideration in regard to the measure of light is usually its brightness. The *brightness* of a source of light is expressed in candlepower, and a light source of one candlepower produces the same amount of light as that emitted by a candle of standard dimensions. A more useful measure of light is the *intensity of illumination*, which expresses the rate of flow of light energy upon a unit of surface. The unit of the intensity of illumination is the foot-candle (abbreviated ft-c), which is the amount of

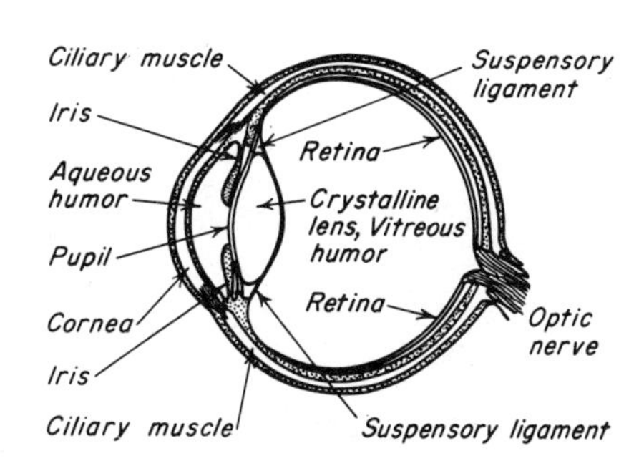

Fig. 1-7 Internal structure of the human eye.

illumination produced on a surface that is one foot away from a standard candle of one candlepower (abbreviated 1cp).

The intensity of illumination varies inversely as the square of the distance between the light source and the surface to be illuminated. For example, a 100-cp lamp will provide an intensity of illumination of 100 ft-c at a distance of 1 ft, 25 ft-c at 2 ft, 6.25 ft-c at 4 ft, and 1 ft-c at 10 ft. In practice, the intensity of illumination is generally determined by use of a photometer.

Speed of Light. For general purposes, the speed of light through air is taken as 186,000 miles per sec. The speed of light varies with the medium through which it travels. Its speed through various media depends upon the density of the medium to light rays, as is indicated by the fact that light will travel faster through vacuum than through air, water, glass, etc.

Frequency, Wavelength, and Color. If light is considered as wave motion, it can be expressed in terms of frequency and wavelength the same as sound waves. The color of light varies with its frequency just as the pitch of a sound varies with the frequency of the sound waves. Also as in the case of sound, when the velocity and frequency of the light waves are known, it is possible to calculate the wavelength of the light waves. The values listed in Table 1-5 indicate the average frequency and wavelengths of the various colors of light that make white light when combined.

From this table it can be observed that the frequency of light waves is very high compared with sound waves and that the wavelength of light waves is much shorter than sound waves. Furthermore, examination of the frequency spectrum chart of Fig. 1-2 reveals that the frequency of light waves is also much higher than radio waves. It may be observed that white is not included in Table 1-5; it is omitted because white light contains all seven colors listed in

Table 1-5 Average Frequency and Wavelength of Various Colors of Light

COLOR	FREQUENCY, MHz	WAVELENGTH	
		MICRONS	MICRO-INCHES
Red	423,000,000	0.71	28
Orange	483,000,000	0.62	24
Yellow	525,000,000	0.57	22
Green	576,000,000	0.52	20
Blue	639,000,000	0.47	18
Indigo	682,000,000	0.44	17
Violet	732,000,000	0.41	16

Note: One meter = 1,000,000 microns. One inch = 1,000,000 microinches.

the table. The spectrum of visible light consists of a band of colors changing gradually from violet at one end to deep red at the other, just as in a rainbow.

The phenomenon of color is explained by the fact that different materials may transmit (or reflect) lights of different colors. For example, a red piece of glass appears red because it transmits only the red and absorbs all the other colors contained in white light. If a material transmits two or more colors, a new color results. When a material transmits all seven colors, it appears colorless, as illustrated by ordinary window glass.

Frequencies just below those of visible red light are classed as *infrared*, and frequencies just above those of violet are classed as *ultraviolet*. These classifications are being used in a rapidly increasing number of applications that include electronic equipment.

Propagation of Light. As in radio- and sound-wave motion, *propagation* of light refers to the transmitting or spreading out of wave motion. The outstanding factor concerning the propagation of light is that light rays travel in straight lines when the medium transmitting the light is homogeneous (which means that the transmitting medium must be uniform). That light travels in a straight line may be observed when rays of sunlight enter a darkened room through a small opening. The straight path of the light rays becomes visible owing to the illumination of the dust particles in the air.

Reflection of Light. When light energy from a source strikes the surface of an object, some of the light energy is reflected. The amount and the color of the light reflected will depend on the condition of the surface and the color of the reflecting body. Smooth bodies reflect light better than irregular ones; also light bodies reflect more light than dark ones. This further explains the theory of color as illustrated by the fact that an opaque red body appears red because it reflects only the red light and absorbs all others. In the case of black, an object appears black because it absorbs all colors and hence reflects none.

When the reflecting surface is smooth and flat, the reflected rays will be reproductions of the original and the reflection is *regular*. When the reflecting surface is irregular, the reflected rays will not have the same relation to one another as the original and the reflection is *diffused*.

Refraction of Light. Refraction is the name used to describe the effect that causes rays of light to bend when they pass from one medium to another of different optical density. For example, air and water have different optical densities, which is another way of saying that light travels through air and water at different rates of speed. (In air the speed of light is 186,000 miles per sec; in water it is approximately 140,000 miles per sec.) Thus if an object such as a pencil is placed in a glass of water, the refraction of the light causes the pencil or object to appear bent or broken.

Lenses. A *lens* is a piece of transparent substance denser than the surrounding medium and with at least one of the two surfaces ground to conform to a definite curvature. Lenses are classified in terms of the curvature being convex or concave. *Convex lenses* are thicker at the center than at the edges, and *con-*

cave lenses are thinner at the center than at the edges. Although lenses are usually made of glass, they can also be made of quartz crystals, water or other clear liquids, and a variety of other materials.

Fundamentally, the function of a lens is to change the direction of rays of light. Convex lenses cause the light rays to converge, that is, to come to a common point from different directions; the image may be enlarged or decreased in size depending upon the distances between the lens, the object, and the image. Concave lenses cause the light rays to diverge, that is, to extend from a common point in different directions; the size of the image is always decreased.

There are many applications of lenses in everyday life such as eyeglasses, magnifying glasses, cameras, motion-picture projectors, microscopes, and telescopes.

Persistence of Vision. An important property of the human eye that makes television and motion pictures possible is persistence of vision. The eye cannot observe or follow any sequence of changes that occur at a rate of 10 or more times per second. Anything in excess of this rate produces the effect of a continuous picture. This phenomenon is called *persistence of vision.*

In a television receiver the image appearing on the picture tube is not a steady picture but is actually a sequence of 60 individual pictures per second, each separated from the preceding one by a short interval of time during which the screen of the picture tube is dark. The viewing area of the picture tube is coated with a fluorescent substance, usually a phosphor, that will glow for only a very short period of time after the electron beam strikes the screen; hence the tube is said to have low persistency. Because the persistence of vision of the human eye is much greater than the persistence of the picture tube, the image at the television receiver appears as a continuous picture that also reproduces the movements of the objects being viewed.

QUESTIONS

1. Name 15 classifications of electronic communication systems.

2. Name three media that may be used to transmit energy in electronic communication systems.

3. Describe a simple analogy of electrical and electronic wave motion.

4. Define (*a*) alternating current, (*b*) electrical degree, (*c*) alternation, (*d*) cycle, (*e*) frequency, and (*f*) I_{max}.

5. Name five forms of radiant energy that are electromagnetic oscillatory disturbances in space.

6. Using the chart of Fig. 1-2, determine the frequency band for the following rays: (*a*) heat, (*b*) light, (*c*) ultraviolet, (*d*) *X*, (*e*) gamma, and (*f*) cosmic.

7. What are the names and frequency limits of the seven divisions of the radio-frequency spectrum?

8. What are the names and frequency limits of the nine divisions of the communication spectrum?

9. Describe the movement of the electric and magnetic fields in the propagation of radio waves.
10. What is the speed of radio waves in (*a*) miles per second? (*b*) Feet per second? (*c*) Meters per second?
11. (*a*) How are radio waves produced? (*b*) What two factors make up a radio wave? (*c*) Name three applications of radio waves.
12. Define (*a*) wavelength, (*b*) hertz, (*c*) kilohertz, (*d*) megahertz.
13. What is sound?
14. Explain what occurs when sound waves strike the human ear and produce the sensation of sound.
15. How are sound waves produced?
16. What factors affect the intensity at which a sound is heard?
17. Define the following terms as used with sound: (*a*) frequency, (*b*) pitch, (*c*) wavelength, (*d*) audible.
18. How does the speed of radio waves compare with the speed of sound?
19. Name three methods used to express relative sound levels.
20. What are the relative approximate values of sound levels in pressure, power, and decibels for (*a*) threshold of hearing, (*b*) soft music, (*c*) loud music, (*d*) very loud music, (*e*) threshold of pain?
21. What are the relative approximate values of sound levels in pressure, power, and decibels for (*a*) leaves rustling, (*b*) average residence, (*c*) conversation at 3 ft, (*d*) heavy street traffic, (*e*) thunder?
22. Define the following terms as used with sound: (*a*) quality, (*b*) fundamental, (*c*) overtone, (*d*) harmonic.
23. Explain what is meant by (*a*) reflection of sound waves, (*b*) echoes, (*c*) sympathetic vibrations, (*d*) forced vibrations.
24. What is meant by resonance?
25. How are beat notes produced?
26. (*a*) In what manner does the human ear respond to sounds of various levels of energy? (*b*) What unit is used to express such variations?
27. (*a*) Define a common logarithm. (*b*) How is the characteristic of a logarithm determined? (*c*) How is its mantissa determined? (*d*) What is an antilog?
28. (*a*) Define the unit called the *bel*. (*b*) Define the decibel. (*c*) Explain the advantages of using the decibel.
29. Explain how the decibel may be used to express ratios of voltage and current as well as power.
30. What is light?
31. What is the speed of light waves in (*a*) miles per second? (*b*) Feet per second? (*c*) Meters per second?
32. Explain how objects are made visible.
33. Define the following terms: (*a*) transparent, (*b*) translucent, (*c*) opaque.
34. (*a*) What is meant by the intensity of illumination? (*b*) What is its unit of measurement? (*c*) What is its relation to distance from the source of light?
35. (*a*) What is the relation between the color of light and its frequency? (*b*) Why is white omitted from the frequency-color-spectrum chart?
36. Explain the phenomenon of color.
37. What is meant by (*a*) propagation of light? (*b*) Reflection of light? (*c*) Refraction of light?

38. (*a*) What is a lens? (*b*) How are lenses usually classified? (*c*) Describe each of the classifications named in (*b*).

39. (*a*) What materials are used in making lenses? (*b*) Name six applications of lenses.

40. (*a*) What is meant by persistence of vision? (*b*) Explain two commercial applications of this phenomenon.

PROBLEMS

1. What is the wavelength of a carrier wave of a transmitter whose frequency is 1,200 kHz?

2. What is the wavelength of a carrier wave of a transmitter whose frequency is 660 kHz?

3. What is the frequency of the radio waves from a transmitter operating on a 10-m wavelength?

4. What is the frequency of the radio waves of a transmitter if its wavelength is 75 m?

5. The frequency of the video carrier wave for channel 42 is 639.25 MHz. Find its wavelength in (*a*) meters, (*b*) feet, (*c*) inches.

6. The frequency of the audio carrier wave for channel 77 is 853.75 MHz. Find its wavelength in (*a*) meters, (*b*) feet, (*c*) inches.

7. What is the frequency of the video carrier wave whose wavelength is 21.044 inches?

8. What is the frequency of the audio carrier wave whose wavelength is 15.464 inches?

9. A certain f-m radio station operates on an assigned frequency of 95.5 MHz. Find its wavelength in (*a*) meters, (*b*) feet, (*c*) inches.

10. A certain f-m radio station operates on an assigned frequency of 101.1 MHz. Find its wavelength in (*a*) meters, (*b*) feet, (*c*) inches.

11. An experimental radio wave has a frequency of 60,000 MHz. Find its wavelength in inches.

12. An experimental radio wave has a frequency of 150,000 MHz. Find its wavelength in inches.

13. A certain radio station located in New York operates on a carrier frequency of 30 MHz. How long does it take for an audio signal being transmitted to reach (*a*) Honolulu, Hawaii (approximately 5,000 miles)? (*b*) Melbourne, Australia (approximately 10,000 miles)?

14. A certain radio station located in Chicago operates on a carrier frequency of 1,210 kHz. How long does it take for an audio signal being transmitted to reach (*a*) New York (approximately 800 miles)? (*b*) San Francisco (approximately 2,200 miles)?

15. How long does it take the radio carrier waves from a television transmitter to travel a distance of 100 miles?

16. The time between the sending out of a radar pulse and the receiving of its echo pulse is 0.001 sec. How far away is the target?

17. If the frequency of middle C on a piano is 256 hertz, what is its wavelength in (*a*) meters? (*b*) Feet?

18. If the frequency of high C on a piano is 4,096 hertz, what is its wavelength in (*a*) meters? (*b*) Feet?

19. If an a-f wave of 256 hertz is superimposed on a carrier wave of 1,200 kHz, how many cycles does the carrier wave make during the time it takes the a-f wave to complete 1 cycle?

20. If an a-f wave of 4,096 hertz is superimposed on a carrier wave of 600 kHz, how many cycles does the carrier wave make during the time it takes the a-f wave to complete 1 cycle?

21. The creaking of a door makes a sound of approximately 15,000 hertz. What is its wavelength?

22. What is the wavelength of the sound waves being produced by an insect if the frequency of the sound is 12,000 hertz?

23. How many cycles will a carrier wave of 1,200 kHz make during the time required for (*a*) 1 cycle of a 256-hertz a-f wave? (*b*) 1 cycle of a 15,000-hertz sound wave?

24. How many cycles will a carrier wave of 100 MHz make during the time required for (*a*) 1 cycle of a 256-hertz a-f wave? (*b*) 1 cycle of a 12,000-hertz sound wave?

25. Radio programs are often presented to studio audiences as well as to the radio audience. (*a*) How long does it take the sound waves to reach a listener in the studio audience seated 100 ft away? (*b*) How long does it take the program to reach a listener at the loudspeaker of a radio receiver 200 miles away? (*c*) Which listener hears the program first?

26. A certain public event is being broadcast from a large arena. (*a*) How long does it take the sound waves to reach a listener who is seated in the arena 250 ft from the microphone? (*b*) How long does it take for the program to reach a listener at the loudspeaker of a radio receiver 2,500 miles away? (*c*) Which listener hears the program first?

27. How far would a radio wave travel in the time it takes for a sound wave to travel 10 ft?

28. How far would a sound wave travel in the time it takes for a radio wave to travel around the earth (approximately 25,000 miles)?

29. A pipe of an organ is measured and found to be 0.565 ft long. At what frequency is it resonant if the pipe is (*a*) closed at one end? (*b*) Open at both ends?

30. What length of organ pipe, open at both ends, is required to produce resonance for a frequency of (*a*) 5,000 hertz? (*b*) 500 hertz? (*c*) 50 hertz?

31. What is the frequency of the beat note produced when a sound having a frequency of 400 hertz is combined with one of (*a*) 300 hertz? (*b*) 350 hertz? (*c*) 450 hertz? (*d*) 500 hertz?

32. If it is desired to produce a beat note of 250 hertz, what frequency sound wave must be added if the original sound has a frequency of (*a*) 100 hertz? (*b*) 200 hertz? (*c*) 500 hertz?

33. Find the logarithms of the following numbers: (*a*) 180, (*b*) 2,750, (*c*) 8.75, (*d*) 12.5, (*e*) 5, (*f*) 98.5, (*g*) 35,000, (*h*) 0.00307, (*i*) 18.3, (*j*) 0.000986.

34. Find the logarithms of the following numbers: (*a*) 4.82, (*b*) 675, (*c*) 0.0000548, (*d*) 750,000, (*e*) 0.0377.

35. Find the antilog of the following common logarithms: (*a*) 2.4771, (*b*) 5.8779, (*c*) 0.7782, (*d*) 8.0294 − 10, (*e*) 6.3385 − 10.

36. Find the antilog of the following common logarithms: (*a*) 3.7007, (*b*) 0.9832, (*c*) 9.6085 − 10, (*d*) 7.3181 − 10, (*e*) 1.3444.

37. What decibel gain is obtained by increasing the power output of an amplifier circuit from 2.2 to 4.3 watts?

38. A tube operated with certain parameters has a power output of 0.035 watt. What is the decibel gain when the operating parameters are changed to produce an output of 0.2 watt?

39. What power output is required to produce a power gain of 4 db over the 2.2 watts of output power obtained with the amplifier circuit used in Prob. 37?

40. The power output of a certain audio amplifier is 25 watts. If a gain of 3 db is required, what must be the power output of the audio amplifier used?

41. Determine the decibel gain of a vacuum tube whose voltage amplification is (*a*) 20, (*b*) 60.

42. Determine the decibel gain of a transistor whose current gain is (*a*) 0.98, (*b*) 49.

43. The characteristics of a certain audio amplifier are such that a voltage amplification of 24 is obtained at 50 hertz and 48 at 1,500 hertz. Determine the loss in decibels at 50 hertz compared to the gain obtained at 1,500 hertz.

44. The characteristics of a certain audio amplifier are such that a voltage amplification of 48 is obtained at 1,500 hertz and 60 at 5,000 hertz. Determine the gain in decibels at 5,000 hertz compared to the gain obtained at 1,500 hertz.

45. A transistor has an r-f power output of 23 watts at 225 MHz and 15 watts at 400 MHz. Determine the loss in gain at 400 MHz as compared to the gain obtained at 225 MHz.

46. Determine the decibel gain of a field-effect power transistor if its input is 0.04 watt and its output is 10 watts.

47. What is the intensity of illumination from a 200-cp lamp at a distance of (*a*) 2 ft? (*b*) 5 ft? (*c*) 20 ft?

48. What size lamp must be used to produce an intensity of illumination of 10 ft-c: (*a*) 10 ft from the lamp? (*b*) 5 ft from the lamp? (*c*) 2 ft from the lamp?

49. A source of light having a color between red and orange has a frequency of 450,000,000 MHz. What is its wavelength in (*a*) centimeters? (*b*) Inches? (*c*) Microinches?

50. A source of blue light has a wavelength of 17.5 microinches (abbreviated μin.). What is its frequency in (*a*) megahertz? (*b*) Hertz?

Chapter 2

Solid-state Diodes

A diode is a device that may be used to control a voltage, protect a circuit element, act as a switch, or rectify an alternating current. There are numerous applications of diodes, using one or more of these functions, in all types of electronic circuits. Because of the excellent operating, mechanical, physical, and electrical characteristics of solid-state diodes they are used extensively in electronic circuit applications.

2-1 Semiconductors

A solid-state diode consists of a two-layer crystal. However, the layers are not separate pieces but areas within the crystal which have slightly different electrical characteristics. Solid-state diodes are made from a *semiconductor* whose electrical properties are midway between that of a conductor such as silver and an insulator such as porcelain. A semiconductor under one condition may act as a conductor allowing an easy flow of current, while under another condition it may act as an insulator and virtually block the flow of current.

In a conductor, an electric current is thought of only as the movement of free electrons along the conductor. In a semiconductor, current consists of the movement of free electrons and holes. When an electron leaves an atom, the resulting charge on the atom is positive. The absence of an electron from the atomic structure in a crystal is referred to as a *hole*. The application of an electric field will cause electrons and holes to drift in opposite directions. The holes being positively charged will drift toward the negative terminal, while the electrons being negatively charged will drift toward the positive terminal.

2-2 Physical Concepts of Solids

Conductors and Insulators. Solids having a low electrical resistivity at room temperature are called *conductors*. The solids in this group include most of the common metals such as copper, aluminum, and silver. The resistivity of copper is of the order of 1.6×10^{-6} ohm per centimeter cube. Solids having a high electrical resistivity at room temperature are called *insulators*. Their resistivity is of the order of 10^9 to 10^{18} ohms per centimeter cube. Examples of solids in this group are porcelain, quartz, glass, and mica.

Semiconductors. Solids having a value of resistivity midway between that of conductors and insulators are called *semiconductors*. Their resistivity varies

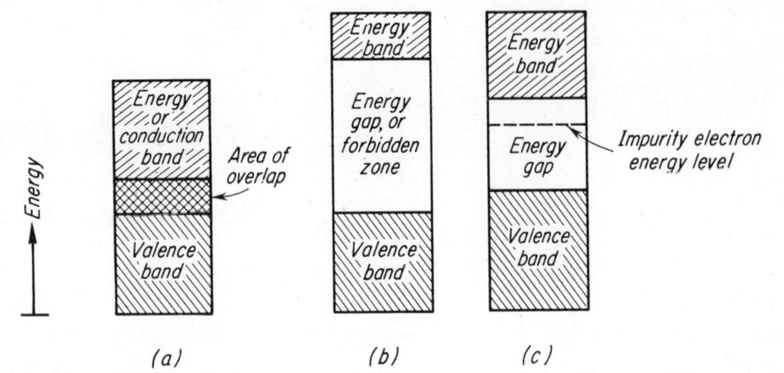

Fig. 2-1 Energy and valence bands. (a) Conductor. (b) Insulator. (c) Semiconductor.

nonlinearly with temperature changes, and these solids may possess either a positive or negative temperature coefficient. The resistance characteristics depend largely upon the amount of impurities in the material. Examples of solids in this group are elements such as germanium and silicon and compounds such as zinc oxide and copper oxide.

When the particle concentration in a given semiconductor consists of both electrons and holes in approximately equal numbers, the material is called an *intrinsic semiconductor*. When the holes predominate, the material is called a *positive* or *P-type semiconductor*. When the electrons predominate, the material is called a *negative* or *N-type semiconductor*. An intrinsic semiconductor may be given P- or N-type characteristics by adding various substances called *impurities* to the base material.

Energy Levels. In order for conduction to take place, there must be a movement of electrons. In atomic theory, the electrical properties of the elements are explained by the concept of energy bands. The electrons in the outer orbit of an atom can be moved with the least amount of energy and are called *valence electrons*. Valence electrons have definite energy levels or bands (Fig. 2-1), and the conductivity of an element is determined by the energy required to move its valence electrons from their normal energy level, or *valence band*, to their highest energy level, the *conduction* or *energy band*. The distance a valence electron moves in its travel from the valence band to the energy band varies with each type of atom. The energy gap separating the valence and conduction bands in an insulator is extremely large, and it is very difficult for a valence electron to reach the energy band. In a conductor the valence and energy bands overlap, and the valence electrons are available for conduction. In a semiconductor the energy gap is very small, and the thermal energy of the valence electrons at room temperature is sufficient to permit an appreciable amount of conduction. Because an electron cannot remain in the space between the valence and energy bands, this region is sometimes referred to as the *forbidden zone*.

2-3 Conduction in Crystals

Conduction. In a semiconductor the atoms of the material form a lattice structure called a *crystal*. A crystal may be (1) intrinsic, that is, having approximately equal numbers of positive and negative charges, (2) N-type, that is, having an excess of negative charges or electrons, or (3) P-type, that is, having an excess of positive charges or holes (Fig. 2-2). Conduction may take place in a crystal by the movement of either electrons or holes. In negative conduction the electrons flow toward the positive terminal of the power source, while in positive conduction the holes flow toward the negative terminal of the power source.

Carbon as a Crystal. Carbon in its common (noncrystalline) form is a conductor; its atomic structure is shown in Fig. 2-3. Carbon is also found in crystallized form as the *diamond*. Although carbon and the diamond have the same atomic structure, carbon is a conductor and the diamond is an insulator. The difference in their electrical conductivity is explained by the actions of

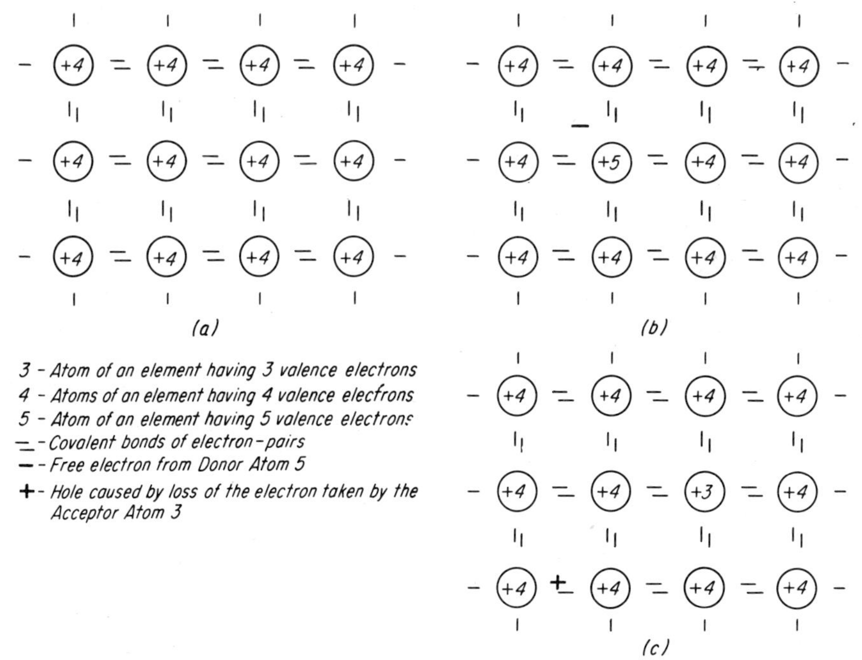

Fig. 2-2 Lattice structure of a crystal. (a) Pure crystal. (b) Crystal containing an impurity having five valence electrons. (c) Crystal containing an impurity having three valence electrons.

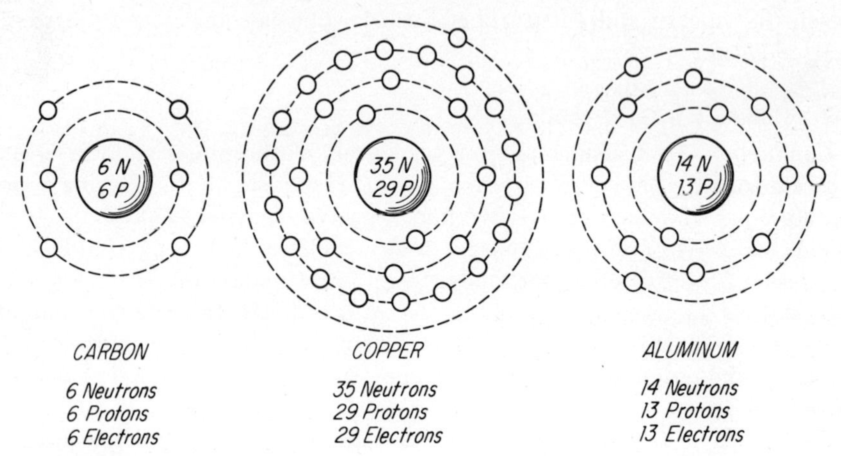

Fig. 2-3 Structure of carbon, copper, and aluminum atoms.

their valence electrons. In ordinary carbon the atom has four valence electrons that can move quite freely from one atom to another to constitute a flow of current. In a diamond the arrangement of the atoms forms a lattice-type structure and the valence electrons of neighboring atoms form covalent bonds (Figs. 2-2*a* and 2-4). Because of these characteristics, the valence electrons have a strong bond with their nuclei and cannot easily be dislodged from their orbits. Therefore, the diamond is a good insulator.

Covalent Bonds. In a crystalline structure each valence electron of an atom, orbiting around its nucleus, coordinates its motion with that of a valence electron from an adjacent atom (Figs. 2-2 to 2-6). The association of these electrons with each other is called a *covalent bond.* Figure 2-4 shows that valence electron *a* from atom *A* and valence electron *b* from atom *B* form a covalent bond between atoms *A* and *B*. In a similar manner covalent bonds are formed between atoms *A* and *C*, *A* and *D*, *A* and *E*, *F* and *C*, *F* and *G*, etc.

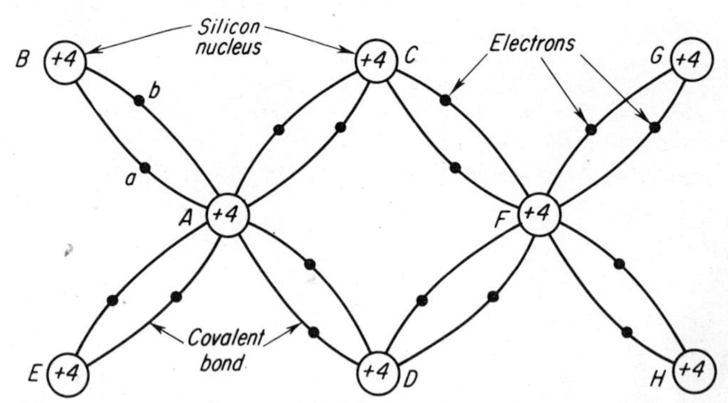

Fig. 2-4 Lattice structure of a pure silicon crystal illustrating the covalent bonds between adjacent atoms.

Lattice Structures. The atoms of a pure crystalline material form a lattice-type pattern as represented by Fig. 2-2*a*. Figure 2-4 is one of several methods of presenting an enlargement of a section of the lattice pattern of Fig. 2-2*a*. Although carbon is used here to illustrate the principles of the lattice-type structure and covalent bonding, there are certain other elements having atomic structures that will act in the same manner. Two elements frequently used in the construction of transistors, namely silicon and germanium, also have four valence electrons and can be produced in crystalline form. The preceding explanations also apply to these materials. The following discussion, although referring only to silicon, will also apply to germanium.

Conductivity of a Pure Crystalline Material. In a pure crystalline material all atoms have four valence electrons and form complete sets of valence-bond pairs (Figs. 2-2*a* and 2-4). Under this condition each atom behaves as though its outer ring contained eight electrons and all electrons had a strong bond with their nuclei. The resulting lattice structure is in equilibrium as there are no excess positive or negative charges and the material acts as an insulator.

Conductivity of an Impure Crystalline Material. If a suitable second material, hereafter called an *impurity*, is added to a pure crystalline material, it will produce a semiconductor material that will perform satisfactorily in a transistor. Adding an impurity to a pure material is also referred to as *doping* the material.

Silicon crystals are produced with an impurity ratio of less than one part per 100 million. The amount of impurity that is added is very small and also very critical. For example, if added in the ratio of one atom of impurity to every 100 million atoms of silicon, the resistivity drops from 60 to 3.8 ohms per centimeter cube. This value is satisfactory for transistor use. However, if the ratio is increased to 10 impurity atoms for every 100 million atoms of silicon, the resistivity drops to 0.38 ohm per centimeter cube. This value is too low for transistor use.

Donor Impurities. If an impurity material having five valence electrons (for example, arsenic or antimony) is added to pure silicon, four valence electrons of each impurity atom will form four covalent bonds with one valence electron from each of four nearby silicon atoms in order to keep the lattice structure intact. The fifth valence electron of the impurity atom is free to wander about the crystal in a manner much like the movement of free electrons in an ordinary metallic conductor (Figs. 2-2*b* and 2-5). As even a very small crystal will contain many billion atoms, an impurity ratio as low as one part per 100 million produces a sufficient number of free electrons to carry the amount of current required for transistor circuit applications. If the impurity ratio is increased, more electrons will be released to wander about the crystal; consequently, the conductivity of the silicon increases, and its resistivity decreases. Silicon having an excess of electrons is called *N-type silicon*. When the impurity donates electrons to the crystal conductivity, it is called a *donor impurity*. After contributing a free electron to the crystal structure, a donor impurity atom has a positive charge.

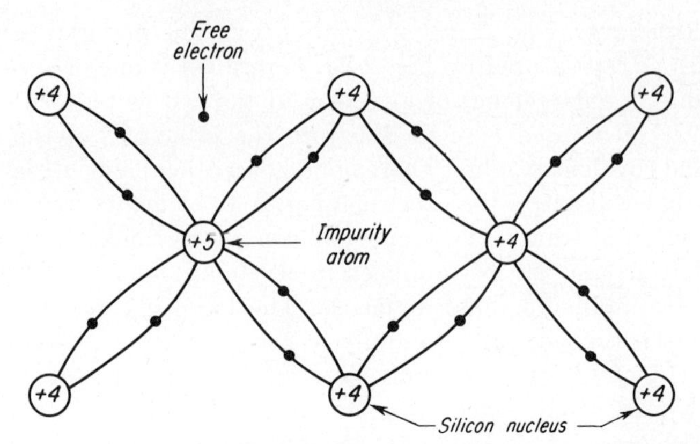

Fig. 2-5 Effect of adding a donor impurity to pure silicon.

Acceptor Impurities. If an impurity material having three valence electrons (for example, gallium or indium) is added to pure silicon, the three valence electrons of each impurity atom will form three complete covalent bonds with one valence electron from each of three adjacent silicon atoms. If the crystal lattice structure is to remain intact, the impurity atom will have to borrow (or *accept*) a valence electron from a nearby silicon atom in order to form a fourth covalent bond with its fourth adjacent silicon atom (Figs. 2-2*c* and 2-6). The nearby silicon atom that provides the electron for the fourth covalent bond pair of the impurity atom now suffers a deficiency of one electron in one of its covalent bond pairs. This deficiency of an electron in a covalent bond group is called a *hole*. Holes have a positive charge, can move about from atom to atom, and can conduct a current just as negative electron charges do. If the impurity ratio is increased, more holes are formed and the conductivity of the crystal increases. Silicon having an excess of holes is called a *P-type silicon*.

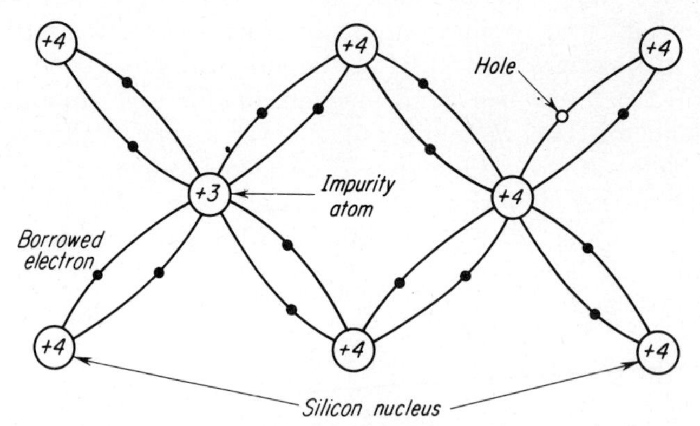

Fig. 2-6 Effect of adding an acceptor impurity to pure silicon.

When the impurity accepts electrons in order to keep the lattice structure of the crystal intact, it is called an *acceptor impurity*. An acceptor impurity has a negative charge.

Negative and Positive Conduction. An N-type crystal has a large number of free electrons that can become current carriers. Current resulting from electron movement is called *negative conduction*. A P-type crystal has a large number of hole charges that also can become current carriers. Current resulting from hole movement is called *positive conduction*.

Majority and Minority Carriers. An N-type crystal will contain some stray hole charges and a P-type crystal will contain some stray electrons. When a crystal is connected in an electric circuit there will be some electron movement and some hole movement in both the N-type and P-type crystals. In the N-type crystal the electrons will constitute the *majority carriers* and the holes the *minority carriers*. In the P-type crystal the holes will constitute the *majority carriers* and the electrons the *minority carriers*.

Hole Injection. In the operation of certain transistors, a forward bias is applied to one portion of its crystal and a reverse bias to another portion. When an N-type material is involved, the forward bias causes free electrons to move from one portion of the transistor to another. The movement of an electron out of an N-type portion of the transistor temporarily creates a hole in that part. The creation of holes in this manner, called *hole injection*, plays an important part in the performance of the transistor. The holes will drift along the material, some as positive conduction current, others losing their identity owing to their recombining with available free electrons.

Solid-state Conduction. Units of N-type and P-type crystals have an excess of electrons and holes respectively, and hence are capable of transmitting electrons or holes through a solid material. The principles of conduction through solids are referred to as *solid-state physics*. The application of solid-state devices is referred to as *solid-state circuitry*.

2-4 The PN Junction or Diode

Characteristics. A PN junction is formed when sections of P-type and N-type silicon are joined together to produce a *silicon diode* (Fig. 2-7). The circles with negative symbols in the P-type silicon represent *acceptor atoms*. The holes, created by missing electrons in some covalent bond pairs, are positive charges and are represented by the positive symbols. The circles with positive symbols in the N-type silicon represent *donor atoms*. The free electrons, being negative charges, are represented by negative symbols.

Potential Barrier. Free electrons in the N section will be attracted by the holes in the P section, and vice versa. However, as soon as any electrons from the N section cross the junction they will be repelled by the negative charge of the acceptor atoms in the P section. Likewise, hole charges from the P section that cross the junction will be repelled by the positive charge of the donor atoms in the N section. As a result, the potential barrier soon established across the

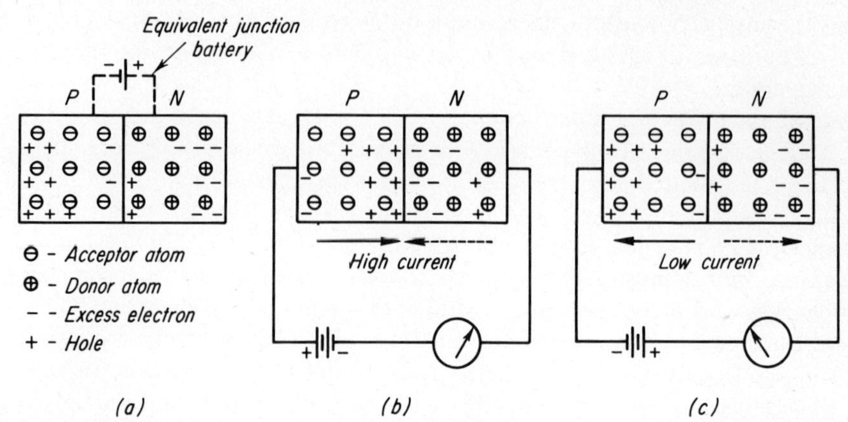

Fig. 2-7 Effect of junction bias voltage. (a) Neutral or zero bias. (b) Forward bias. (c) Reverse bias. The solid-line arrows indicate the direction of hole current, and the broken-line arrows indicate the direction of electron current.

two surfaces of the junction prevents any large movement of either electrons or holes across the junction. The potential established at the junction depends on (1) the type of crystal such as germanium or silicon and (2) the amount of impurity in the crystal. This potential is in the order of 0.2 volt for germanium and 0.7 volt for silicon. This voltage is called a *potential barrier* or a *potential hill*, and its effect is the same as if a battery was connected across the junction as shown in Fig. 2-7*a*.

Forward Bias. In order to transmit a current through the junction, the potential barrier across the junction must be neutralized. This may be done by connecting a power source across the two sections of silicon as shown in Fig. 2-7*b*. The voltage applied in this manner is called a *forward potential* or *forward bias*. Free electrons in the N section will be repelled by the negative force set up by the power source and will move toward the junction. At the same time the holes in the P section will be repelled by the positive force set up by the power source and will move toward the junction. The voltage of the power source imparts sufficient energy to these carriers to overcome the potential barrier at the junction and enables them to cross through the junction. Once the junction is crossed, free electrons from the N section will combine with holes in the P section, and holes from the P section will combine with free electrons in the N section. This action decreases the potential barrier at the junction. For each hole in the P section that combines with an electron from the N section, an electron from an electron-pair bond leaves the crystal and enters the positive terminal of the power source. This action creates a new hole that is forced to move toward the junction because of the electric field produced by the power source. For each electron in the N section that combines with a hole from the P section, an electron enters the crystal from the negative terminal of the power source. This constant movement of electrons toward the positive terminal and holes toward the negative terminal produces a high forward current I_f (Fig. 2-9*a*). It should be noted that the relatively high current in the external circuit is the sum of the positive and negative conduction currents.

Reverse Bias. If the polarity of the power source connected across the two sections of silicon is reversed, as shown in Fig. 2-7*c*, a *reverse potential* or *reverse bias* is obtained. The excess holes in the P section will be attracted by the negative force of the power source and will move away from the junction. At the same time the free electrons in the N section will be attracted by the positive force of the power source and will also move away from the junction. The net effect will be the same as increasing the voltage of the potential barrier and, as a result, theoretically the current through the junction will be zero. However, in actual practice a small amount of reverse current I_r will flow as is indicated by Fig. 2-9*a*.

2-5 Types of Junction

Methods of Construction. The characteristics of a semiconductor diode will depend on the material and the type of PN junction used. The materials most often used are germanium and silicon. Germanium has higher electrical conduction than silicon and is used in low- and medium-power diodes (and numerous transistors). Silicon can be operated at higher temperatures than germanium and is therefore more suitable for high-power applications than germanium. A material called *gallium arsenide* combines the desirable features of both germanium and silicon and is now being used in many new applications. The PN junction may be made in various ways, among which are (1) point contact, (2) grown, (3) diffused, (4) recrystallized, (5) alloyed, (6) surface barrier, (7) drift field, and (8) epitaxial (Fig. 2-8). There are many variations of these eight methods, and in the commerical manufacture of diodes (and transistors) more than one method is generally used.

Point-contact Junction. Basically the point-contact type of junction (Fig. 2-8*a*) consists of a fine pointed wire that makes pressure contact with the face of a N-type germanium wafer. The PN junction is formed by applying a relatively high current momentarily across the pointed wire and the crystal face. The heat generated during this interval drives some of the electrons from the atoms in a small region around the point of contact, leaving holes. This region in the N-type germanium is thus converted to P-type germanium.

Grown Junction. There are several methods of growing crystals; one of the methods generally used is the Czochralski technique. In this method a single crystal is slowly pulled out of a vat of molten silicon. The melted silicon crystallizes in exactly the right order on the single crystal, or seed. If a donor impurity is added to the molten silicon the grown crystal will be N type, and if an acceptor impurity is added the grown crystal will be P type. Using this idea, a grown junction is made by starting with an N-type material. At a certain point, a small pellet of an acceptor impurity is dropped into the melt, and a P-type silicon crystal starts to grow. After a definite period, the crystal is

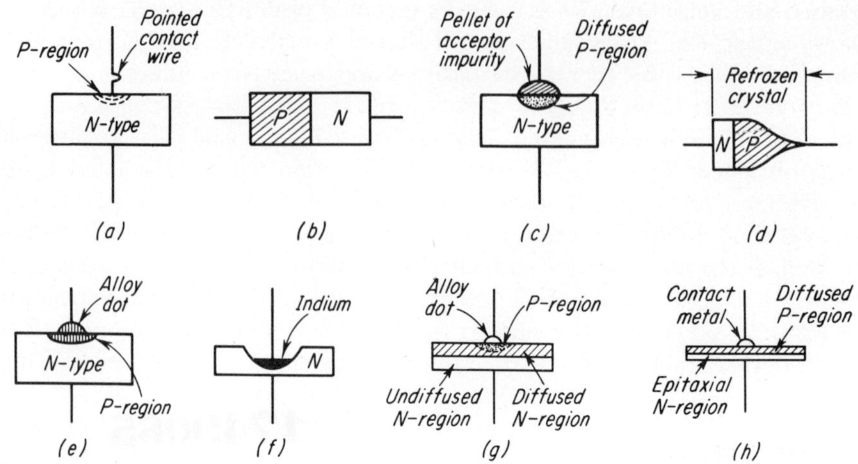

Fig. 2-8 PN junctions. (a) Point contact. (b) Grown. (c) Diffused. (d) Recrystallized. (e) Alloy technique. (f) Surface barrier. (g) Drift field. (h) Epitaxial.

removed and the junction located so that the crystal can be cut into thin wafers. Each wafer possesses the properties of a PN junction (Fig. 2-8*b*).

Diffused Junction. The diffused junction (Fig. 2-8*c*) is made by placing a small pellet of an acceptor impurity on one face of an N-type wafer. The combination is then heated in order to melt the impurity. A portion of the impurity will diffuse a short distance into the wafer, creating a P-type region that is in close contact with an N-type region.

Recrystallized Junction. The recrystallized junction (Fig. 2-8*d*) is made by placing a piece of N-type silicon between hot and cold temperature areas. The silicon will melt back partway so that it is part molten and part solid. The temperature is then reduced, and the silicon refreezes into a single crystal having a P region and a N region in close contact.

Alloyed-technique Junction. The alloyed junction is the one most commonly used. It is known by various names, such as *fused junction*, *fusion alloy*, and *diffused alloyed*. In its simplest form, a small dot of an indium alloy is placed on one face of an N-type silicon wafer. The combination is heated until the alloy melts and dissolves some of the silicon. The temperature is then lowered, and the silicon refreezes to form a single crystal having a PN junction (Fig. 2-8*e*).

Surface-barrier Junction. The structure of the surface-barrier junction is similar to the alloyed junction. Instead of alloying a dot deep into the base material to obtain a narrow base width, a well is electrochemically etched into the base material until the thickness of the base is only a few tenths of a mil. Indium is then plated to the well, and a PN junction is formed (Fig. 2-8*f*). Microetching techniques as used with surface-barrier junctions are also used to produce bonded-barrier and microalloy junctions.

Drift-field Junction. The drift-field junction is a modification of the alloyed junction. In this type of structure the N-type crystal wafer is diffused, or graded, as shown in Fig. 2-8*g*. Two advantages are derived from this type of construction: (1) The resultant built-in voltage (drift field) speeds up the current flow. (2) The capacitive charging times are minimized by using a low impurity concentration on one side of the N-type wafer, and a high impurity concentration on the other side of the wafer.

Epitaxial Junction. The epitaxial junction (Fig. 2-8*h*) differs from the diffused junction only in the manner in which the junction is fabricated. Photolithographic and masking techniques and a single impurity diffusion are used. Epitaxial junctions are grown on top of an N-type crystal wafer in a high-temperature reaction chamber. The growth proceeds atom by atom, and is a perfect extension of the crystal lattice of the wafer on which it is grown. This type of construction offers the advantages of low junction resistance, and an easy means of controlling impurity spacings.

2-6 Diode Characteristics

Static Volt-Ampere Characteristics. The PN junction possesses the property of a rectifier since the current resulting from a voltage applied in one direction across the junction is different from the current resulting from the same amount of voltage applied in the opposite direction across the junction. The static volt-ampere characteristic curve for a typical solid-state diode is shown in Fig. 2-9*a*. The forward- and reverse-current curves are drawn to different voltage and current scales, and both are nonlinear over a considerable portion of their ranges. The forward-current curve has a sharp upward swing at a relatively low value of voltage, less than 0.5 volt for the diode of Fig. 2-9*a*. The point on the curve at which the sharp upswing occurs is called the *knee*, and the potential at this point is called the *knee voltage* V_K. The barrier potential and the knee voltage are approximately equal in value. The reverse-current curve indicates that a relatively high voltage (more than 10 volts) is required to produce even a very low value of current.

Static Volt-Ohm Characteristics. The static voltage-resistance curve for a typical solid-state diode is shown in Fig. 2-9*b*. This curve was obtained by dividing the voltage at a number of points on the curve of Fig. 2-9*a* by the current at these voltages. The forward- and reverse-resistance curves are drawn to different voltage and resistance scales. Increasing the forward voltage decreases the resistance to a low value, usually less than 100 ohms. Decreasing the forward voltage increases the resistance, and at zero voltage the resistance is in the order of hundreds of thousands of ohms. As the reverse voltage is increased the resistance decreases from a peak value to almost zero.

Breakdown. If the applied reverse voltage to a solid-state diode is gradually increased, a point is reached where the valence bonds start to break up, thus releasing a large number of holes and electrons. This point is indicated on the curve (Fig. 2-9*a*) by a sharp increase in current, and the voltage causing this

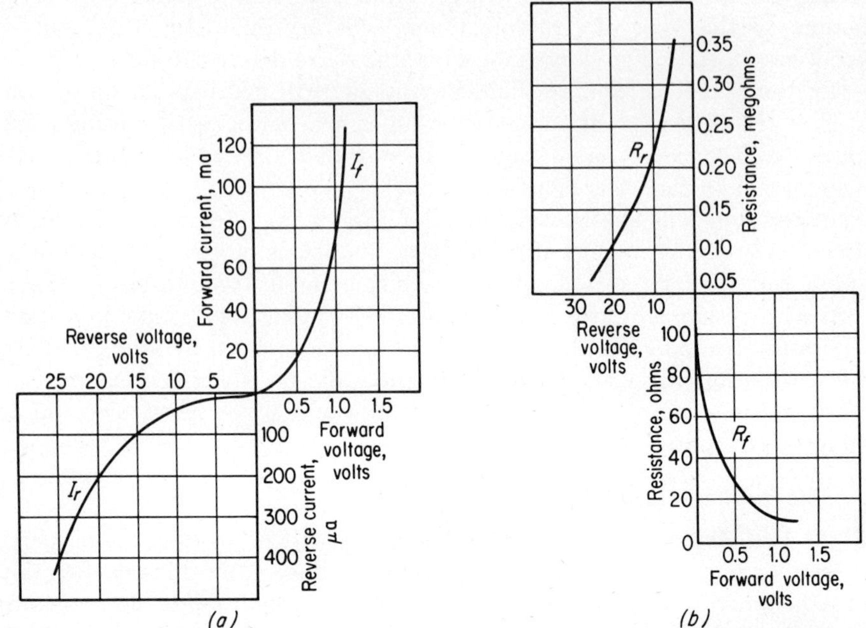

Fig. 2-9 Static characteristics of a PN junction diode. (a) Volt-ampere. (b) Volt-ohm.

breakdown of the crystalline structure is called the *Zener voltage*. The Zener voltage will therefore indicate the maximum reverse voltage that can be applied to a semiconductor without producing an excessive reverse current. The Zener voltage of the diode junction used for obtaining the characteristic curve of Fig. 2-9*a* is 15 volts. From this curve it can be seen that the initial increase of 15 volts reverse voltage (0 to 15) produces an increase of 100 μa in the reverse current, while an additional increase of only 5 volts produces the same amount of increase, and the next 5-volt increase produces an increase of 200 μa.

A-C Resistance of a Diode. The a-c resistance of a solid-state diode is due to two factors: (1) the bulk resistance and (2) the junction resistance. The resistance of the P- and N-type materials used is referred to as the *bulk resistance* and is represented by the equation

$$r_B = \frac{dV}{dI} \tag{2-1}$$

When the current-voltage curve of a diode is practically straight, as is often true at voltages above the knee of the curve, the diode acts as a linear device and the bulk resistance is practically a constant value.

Example 2-1 What is the bulk resistance of the diode represented by Fig. 2-9*a* with (*a*) a forward voltage? (*b*) A reverse voltage?

GIVEN: Fig. 2-9a

FIND: (a) r_{B-F} (b) r_{B-R}

SOLUTION:

(a) From the curve in Fig. 2-9a

$$V_1 = 0.6 \text{ volt} \qquad I_1 = 20 \text{ ma} \qquad V_2 = 1 \text{ volt} \qquad I_2 = 75 \text{ ma}$$

$$r_{B-F} = \frac{dV}{dI} = \frac{1 - 0.6}{(75 - 20)10^{-3}} \cong 7.3 \text{ ohms}$$

(b) From the curve in Fig. 2-9a

$$V_1 = 15 \text{ volts} \qquad I_1 = 100 \ \mu\text{a} \qquad V_2 = 20 \text{ volts} \qquad I_2 = 200 \ \mu\text{a}$$

$$r_{B-R} = \frac{dV}{dI} = \frac{20 - 15}{(200 - 100)10^{-6}} = 50{,}000 \text{ ohms}$$

The manufacturers' data sheet for a diode sometimes lists the forward current at 1 volt. Assuming that the diode is a linear device, the bulk resistance may be found by

$$r_{B-F} = \frac{1 - V_K}{I} \tag{2-2}$$

Example 2-2 What is the bulk resistance of a silicon diode whose forward current is 10 ma at 1 volt?

GIVEN: $V = 1$ volt $I = 10$ ma ($V_K = 0.7$ volt for silicon)

FIND: r_{B-F}

SOLUTION:

$$r_{B-F} = \frac{1 - V_K}{I} = \frac{1 - 0.7}{10 \times 10^{-3}} = 30 \text{ ohms}$$

The junction resistance of a perfect junction at an average room temperature of 25°C is theoretically equal to

$$r_j = \frac{25 \text{ mv}}{I \text{ ma}} \tag{2-3}$$

Because there is no perfect junction, the junction resistance will be higher than indicated by Eq. (2-3). For practical purposes, Eq. (2-3) can be stated as

$$r_j = \frac{25 \text{ mv}}{I \text{ ma}} \quad \text{to} \quad \frac{50 \text{ mv}}{I \text{ ma}} \tag{2-3a}$$

where r_j = a-c resistance of diode junction, ohms

Example 2-3 If the diode of Example 2-1 is operated at an average current of 10 ma, what is its probable range of junction resistance?

GIVEN: $I = 10$ ma

FIND: Range of r_j

SOLUTION:

$$r_{j\cdot\min} = \frac{25}{I} = \frac{25}{10} = 2.5 \text{ ohms}$$

$$r_{j\cdot\max} = \frac{50}{I} = \frac{50}{10} = 5 \text{ ohms}$$

Range of r_j = 2.5 to 5 ohms

The a-c resistance of a solid-state diode is the combined effect of the bulk and junction resistances, or

$$r_{a\text{-}c} = r_B + r_j \tag{2-4}$$

Example 2-4 What is the probable range of a-c resistance for the diode of Examples 2-1 and 2-3?

GIVEN: $r_B = 7.3$ ohms $\quad r_{j\cdot\min} = 2.5$ ohms $\quad r_{j\cdot\max} = 5$ ohms

FIND: Range of $r_{a\text{-}c}$

$$r_{a\text{-}c\cdot\min} = r_B + r_{j\cdot\min} = 7.3 + 2.5 = 9.8 \text{ ohms}$$
$$r_{a\text{-}c\cdot\max} = r_B + r_{j\cdot\max} = 7.3 + 5 = 12.3 \text{ ohms}$$

Range of $r_{a\text{-}c}$ = 9.8 to 12.3 ohms

2-7 Load-line Analysis

The diode forward current may be limited to a predetermined value by using a *series-limiting* or *load resistor* (Fig. 2-10a). The value of the load resistance required to establish the operating point Q_1 to limit the forward current to a desired value may be found by

$$R_L = \frac{E_b - E_D}{I_D} \tag{2-5}$$

where R_L = resistance of limiting resistor, ohms
E_b = voltage of power source, volts
E_D = voltage drop across the diode, volts
I_D = diode forward current, amperes

Example 2-5 A 3-volt d-c power source, a series-limiting resistor, and a diode are connected as shown in Fig. 2-10a. If the circuit characteristics of the diode are as illustrated for diode A in Fig. 2-10c, determine the value of resistance required to limit the forward current to 2 milliamperes.

GIVEN: $E_b = 3$ volts $\quad I_D = 2$ ma $\quad$ Circuit characteristics of diode A (Fig. 2-10c)

FIND: R_L

SOLUTION: Draw a load line from the 3-volt point on the abscissa reference line through the point where the forward-characteristic curve for diode A intersects the 2.0-ma

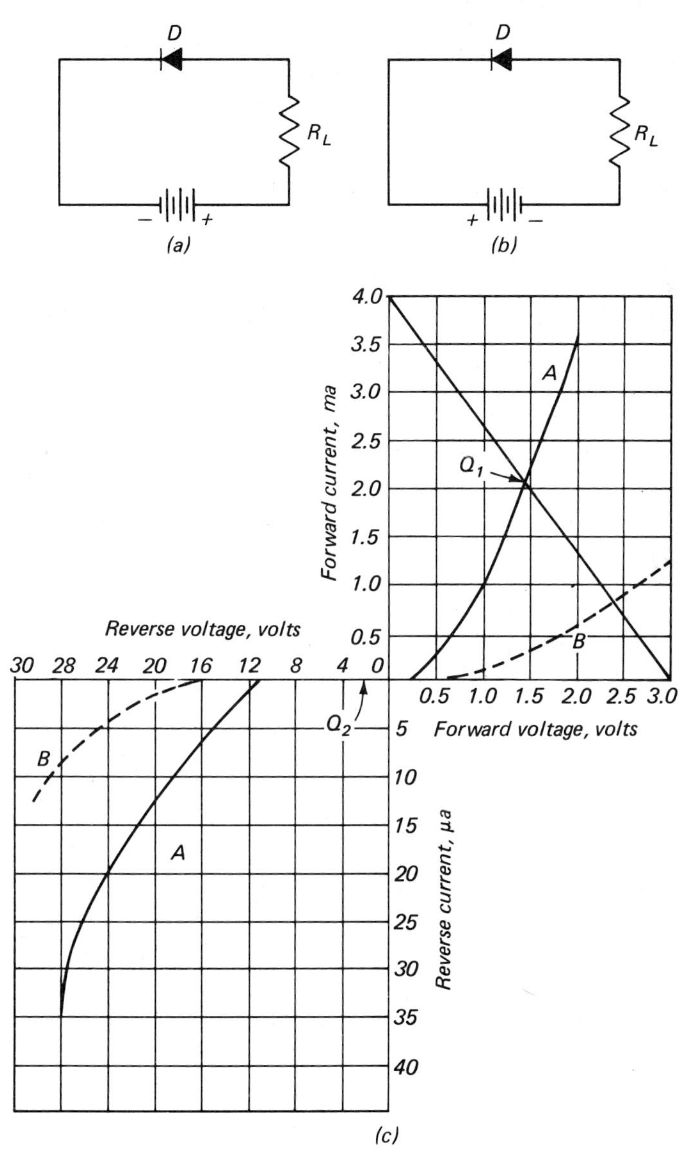

Fig. 2-10 Diode biasing. (a) Forward bias. (b) Reverse bias. (c) Static forward and reverse characteristics of two diodes.

ordinate (Fig. 2-10*c*). It will be noted that at this operating point Q_1 the voltage across the diode is 1.45 volts.

$$R_L = \frac{E_b - E_D}{I_D} = \frac{3.0 - 1.45}{0.002} = \frac{1.55}{0.002} = 775 \text{ ohms}$$

Reversing the power source, as in Fig. 2-10*b*, causes the operating point Q_2 of the diode to be shifted to the reverse quadrant. The characteristic curve for diode *A* shows that the value of diode current is practically zero for reverse biases up to 11 volts. For practical purposes (1) the diode current is zero with a 3-volt bias, (2) the voltage drop across the limiting resistor is also zero, and (3) the voltage across the diode is equal to the full voltage of the power source, or 3 volts.

The unidirectional property of diode *A* of Fig. 2-10*c* is characteristic of all solid-state diodes. Thus, a solid-state diode may be used as a switch permitting a high forward current with a low forward voltage, and a low reverse current with a high reverse voltage.

2-8 Diode Specifications

Manufacturers usually specify the various operating parameters and characteristics of solid-state diodes according to the following definitions. The symbols listed are used to identify these parameters.

Forward Voltage Drop V_{FM}*:* The maximum value of the forward voltage drop permissible during the conduction of load current.

Supply Voltage V_{RMS}*:* The effective value of the power-source voltage.

Peak Reverse Voltage V_{RM}*:* The highest amount of reverse voltage which can be applied before the breakdown point is reached. Two values for this rating are usually listed: (1) for nonrepetitive conditions such as a transient or surge, and (2) for repetitive conditions such as those which occur when a diode is used in counter or computor circuits. This rating is also referred to as the *peak inverse voltage.*

Breakdown Voltage BV: The reverse voltage at which the leakage current starts to increase by many orders of magnitude.

Blocking Voltage V_M*:* The maximum amount of d-c blocking voltage that can be applied during the reversal period.

Forward Current I_{FAV}*:* The average current in the forward direction, also equal to the maximum value of forward direct current.

Reverse Current I_{RM}*:* Also called *leakage current,* the maximum value of reverse current for stated conditions of temperature and voltage.

Peak Current i_{PM}*:* There are two listings, (1) for the peak repetitive forward current, and (2) for the peak surge forward current.

In addition to these operating parameters for general-purpose diodes, two other specifications are usually listed for logic-purpose diodes:

Reverse Recovery Time t_{rr}*:* The maximum time taken for the forward-bias diode (ON state) to recover its reverse bias (OFF state).

Shunt Capacitance C_s: The value of capacitance at the maximum frequency of operation.

Note: The name *hertz*, abbreviated Hz, is now used to denote the frequency of voltages and currents instead of cycles as had been used formerly. Thus, a power frequency is expressed as 60 hertz, 60 Hz, rather than 60 cycles; a commercial a-m broadcast frequency as 800 kilohertz, 800 kHz, rather than 800 kilocycles; and a uhf broadcast frequency as 1,500 megahertz, 1,500 MHz, rather than 1,500 megacycles.

2-9 Rectifier Circuits

Purpose of the Rectifier. The purpose of the rectifier is to change an alternating current to a unidirectional current. The action of a rectifier that permits current to flow through it more easily in one direction than the other makes possible its use for changing alternating current to direct current.

The Diode Rectifier. The principle of operation of a rectifier may be mechanical, thermal, chemical, or electronic. Of these, the electronic principle, as performed by a diode, is the one most generally used in the rectifier circuit of d-c power supplies associated with electronic equipment.

Semiconductor Diodes. A semiconductor offers a very low resistance to the flow of current in one direction (low forward resistance) and a very high resistance to the flow of current in the opposite direction (high reverse resistance). A semiconductor diode may be made of a PN junction of silicon, germanium, or selenium. Advantages of semiconductor-diode rectifiers are (1) high ratio of reverse resistance to forward resistance (in the order of 100,000,000 to 1), (2) high efficiency, (3) low forward-voltage drop, (4) high peak-reverse-voltage ratings, (5) wide range of current ratings, (6) low heat radiation, (7) small size, and (8) long life.

2-10 Half-wave Rectifier Circuit

Circuit Operation. This type of rectifier circuit can be operated directly from the power source, or from the secondary of a step-up transformer when the input voltage required is higher than the voltage of the a-c power source. In a half-wave rectifier circuit a single diode is used (Fig. 2-11*a*). During the half-cycle when terminal T_1 is positive, the diode has a forward bias, the diode acts as a conductor, and current flows in the output circuit. During this half-cycle electrons flow through the path T_2-T_4-R-T_3-D-T_1. During the opposite half-cycle, T_1 will be negative, the diode will have a reverse bias, the diode will act as an insulator, and no current will flow in the output circuit.

The relation between the input voltage and the output voltage or current for the single-diode rectifier circuit is shown in Fig. 2-11*b* and *c*. Because output current flows during only one-half of the input cycle, the single-diode rectifier is commonly referred to as a *half-wave rectifier*.

Peak Inverse Voltage. The peak inverse voltage is equal to the maximum voltage at the transformer secondary. For a sine-wave input this peak value is

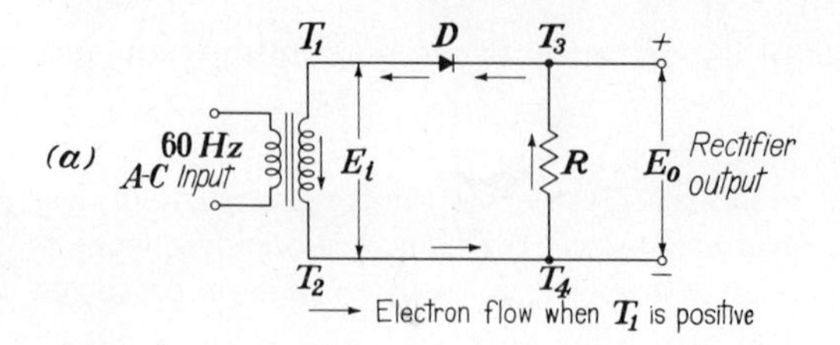

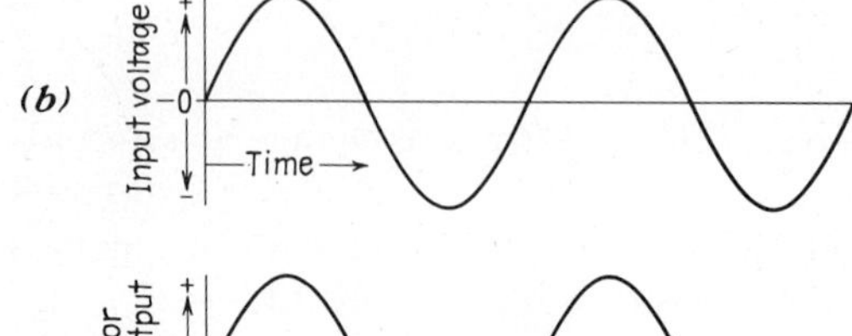

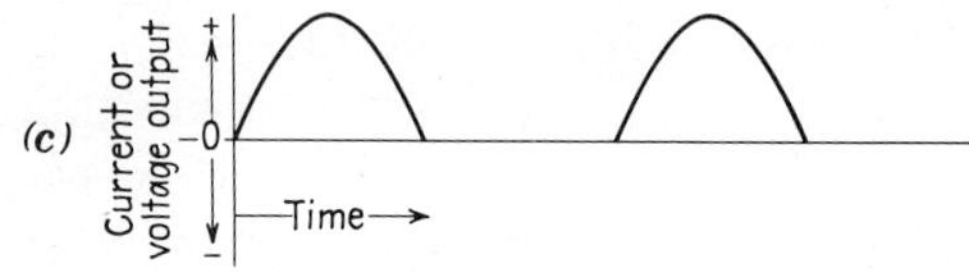

Fig. 2-11 A half-wave diode rectifier. (a) The circuit. (b) Waveform of the input voltage. (c) Waveform of the output current or voltage.

equal to 1.41 times the rms value of the secondary output voltage, or 1.41 times E_i.

Output Voltage. The output voltage of a half-wave rectifier is a unidirectional voltage as represented by Fig. 2-11*c*. The fundamental frequency of the ripple voltage of the half-wave rectifier is equal to the input frequency. The output voltage, as recorded with a d-c voltmeter, is the average value of the pulsations. For sine-wave input voltages and resistance loads, the output voltage for half-wave rectification is

$$E_o = \frac{0.637}{2 \times 0.707} E_i = 0.45E_i \tag{2-6}$$

Example 2-6 Determine the value of the circuit output voltage from a half-wave rectifier when the rms value of the input voltage is (*a*) 120 volts, (*b*) 90 volts.

GIVEN: (*a*) $E_i = 120$ volts (*b*) $E_i = 90$ volts

FIND: E_o

SOLUTION:

(*a*) $E_o = 0.45E_i = 0.45 \times 120 = 54$ volts

(*b*) $E_o = 0.45E_i = 0.45 \times 90 = 40.5$ volts

2-11 Full-wave Rectifier Using a Center-tapped Transformer

Circuit Operation. The operating characteristics of a rectifier will be improved if current is made to flow in the output circuit for the entire period of the input cycle of the input circuit. This is accomplished by using two single semiconductor diodes, or a duplex diode (two single diodes in one unit) and a

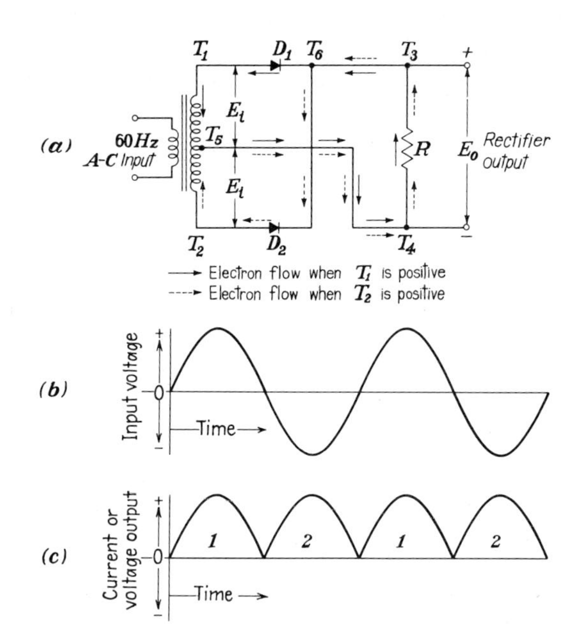

Fig. 2-12 A full-wave diode rectifier using a center-tapped transformer. (a) The circuit. (b) Waveform of the input voltage. (c) Waveform of the output current or voltage.

center-tapped power transformer (Fig. 2-12*a*). Because the *P* sections of the diodes D_1 and D_2 are connected to opposite ends of the secondary winding of the transformer, their polarities are always opposite to one another. Thus, during the first half of the input cycle one diode is conducting and the other diode is not conducting. During the second half-cycle the second diode is conducting and the first diode is not conducting. Under this condition, current flows in the output circuit during both halves of the input cycle. Figure 2-12*c* shows that both halves of the input cycle (Fig. 2-12*b*) have been rectified, hence the name *full-wave rectifier*. Since each diode supplies energy for one-half of the cycle, the diodes of a full-wave rectifier carry only one-half of the diode current required by a half-wave rectifier for equal amounts of load current.

The action of this type of full-wave rectifier is explained in the following manner: During one-half of the input cycle, T_1 is positive and T_2 negative. During this half-cycle, diode D_1 has a forward bias and diode D_2 a reverse bias. While D_1 is conducting, electrons will flow through the path T_5-T_4-R-T_3-T_6-D_1-T_1. During the next half-cycle, T_2 is positive and T_1 negative. During this half-cycle, diode D_2 has a forward bias and diode D_1 a reverse bias. While D_2 is conducting, electrons will flow through the path T_5-T_4-R-T_3-T_6-D_2-T_2.

Peak Inverse Voltage. While one section of the full-wave center-tapped transformer rectifier circuit is conducting, the peak inverse voltage across the other section is 1.41 times the rms value of the full secondary voltage, or 1.41 E_i.

Output Voltage. The output voltage of a full-wave rectifier is a unidirectional voltage as represented by Fig. 2-12*c*. The fundamental frequency of the ripple voltage of the full-wave rectifier is equal to twice the input frequency. The output voltage, as recorded with a d-c voltmeter, is the average value of the pulsations. For sine-wave input voltages and resistance loads, the output voltage for full-wave rectification is

$$E_o = \frac{0.637}{0.707} E_i = 0.9E_i \tag{2-7}$$

Comparison of Eqs. (2-6) and (2-7) indicates that with the same value of input voltage applied to its conducting circuit, the output voltage of a half-wave rectifier is one-half that of a full-wave rectifier. The reason for the double value of output voltage is that the half-wave rectifier supplies current and voltage for only one-half the amount of time that a full-wave rectifier supplies current and voltage to the load.

2-12 Full-wave Rectifier Using a Bridge Circuit

Circuit Operation. The bridge-type full-wave rectifier circuit (Fig. 2-13) is another method of obtaining a continuous flow of rectified current in the output circuit of a rectifier. During one-half of the input cycle, T_1 is positive and T_2 negative. During this half-cycle diodes D_2 and D_3 have a forward bias and diodes D_1 and D_4 a reverse bias. When diodes D_2 and D_3 are conducting, electrons flow through the path T_2-T_8-D_3-T_5-T_4-R-T_3-T_7-D_2-T_6-T_1. During the

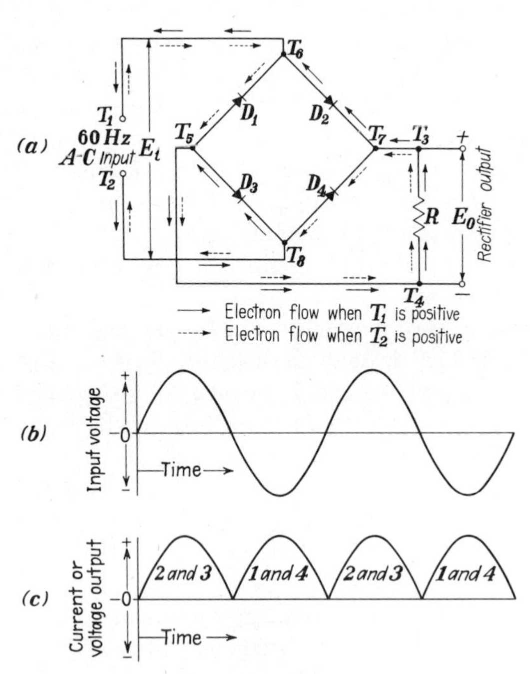

Fig. 2-13 A full-wave rectifier using four solid-state diodes. (a) Simple bridge-type rectifier circuit. (b) Waveform of the input voltage. (c) Waveform of the output voltage or current.

next half-cycle T_2 is positive and T_1 negative. During this half-cycle diodes D_1 and D_4 have a forward bias and diodes D_2 and D_3 a reverse bias. When diodes D_1 and D_4 are conducting, electrons flow through the path T_1-T_6-D_1-T_5-T_4-R-T_3-T_7-D_4-T_8-T_2.

Circuit Characteristics. The bridge-type rectifier does not require a transformer with a center-tapped secondary. It can be operated directly from an a-c power source, or from the secondary of a step-up transformer when the input voltage required is higher than the a-c power source. This circuit utilizes the total input voltage for the entire period of each cycle. For sine-wave input voltages and resistance loads, the output voltage is

$$E_o = \frac{0.637}{0.707} E_i = 0.9E_i \tag{2-8}$$

The output voltage waveform, the peak reverse voltage, and the fundamental frequency of the ripple voltage are the same as for the full-wave center-tapped transformer rectifier circuit.

2-13 Gating Circuits

Uses. A computer circuit may use thousands of solid-state diodes to perform logical functions (such as addition and subtraction) in a relatively short period of time. In this type of operation, the diode changes from a high-current conduction state to a low-current conduction state in a few microseconds. The diode is thus operating as a switch, and it can therefore be used in the design of the functional circuits of a computer for logical operations. In the operation of this type of circuit, an input signal *may* or *may not* pass through it depending on conditions existing at the time the signal is applied. A control signal will determine whether an input signal will appear as an output current or voltage. Because of this gating action, this type of circuit is called a *gate*.

Gating Terminology. The following terms are those which are commonly used to describe the operating functions of gating circuits.

Gating Circuit: A circuit that can either (1) *pass* or (2) *block* a signal is called a *gating circuit*.

Logic Circuit: A control circuit that can automatically select a specific one of a number of different outputs for each possible combination of inputs is called a *logic circuit*. The *decision* of a logic circuit must always be correct.

Sensing Signal: A voltage or current keyed at a station or for a condition, and applied to the input of a logic circuit is called a *sensing signal*. The sensing signal is usually bistable; that is, it has two definite signal values; for example, 0 volt and +10 volts. In logic-circuit analysis, the two conditions are designated as 0 and 1.

AND *Gate:* A logic circuit, or portion thereof, that requires each of its input-sensing signals to be present in order to produce a change in the output signal is called an AND *gate*.

OR *Gate:* A logic circuit, or portion thereof, that will produce a change in the output signal when any one of the input sensing signals is present is called an OR *gate*.

Bistable Device: A device that must be in either one of two possible conditions is called a *bistable device,* for example, a door that is either closed or open. Logical gates are bistable devices.

Crystal-diode Gate: A crystal diode is a bistable circuit element in that it offers a very low resistance to the flow of current with one condition of polarity of an applied voltage, and a very high resistance to the flow of current with the opposite polarity of applied voltage. More briefly, a diode is in the condition of either (1) conducting or (2) not conducting.

2-14 Simple Diode Gate

Figure 2-14 illustrates the basic principle of the gate circuit. The input is a sensing signal having values of either zero or +10 volts, as might be produced by the closing and opening of a switch. When the sensing signal is zero, the diode D will have a forward bias, and if the value of R is high compared to the resistance of the diode the voltage drop across R will be approximately 10 volts, with X_1 negative and X_2 positive. The output voltage E_o will then be zero. When the sensing signal is +10 volts, the voltage across the diode D will be zero volts, and the diode will have the same effect as with reverse bias; namely, it will have a high resistance and thus will be nonconducting. The output voltage will then be +10 volts.

2-15 AND Gate Circuit

Circuit Operation. Figure 2-15 illustrates a three-circuit AND gate. This circuit differs from the simple circuit of Fig. 2-14 only in that there are three sensing signals that the circuit must handle rather than one. Either signal condition may be present or absent at any one circuit regardless of the signal conditions for the other two circuits. The preceding article has shown that with zero signal voltage, the diode is conducting and E_o is zero. In the three-circuit gate, as long as any one sensing signal is zero, the output signal will remain at +10

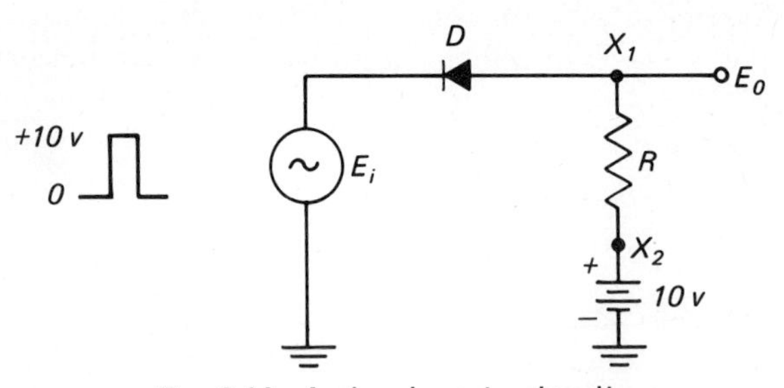

Fig. 2-14 A simple gate circuit.

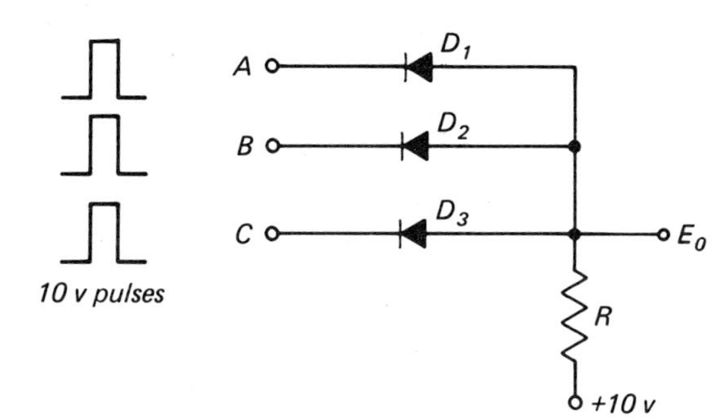

Fig. 2-15 A simple three-circuit AND gate.

volts. Thus, in an AND circuit all input circuits must have a signal applied to each one of them in order to reduce the output voltage to zero.

Truth Table. In logic-circuit analysis, the circuit condition for all combinations of sensing-signal conditions is usually set up in tabular form, and it is referred to as a *truth table*. For example, for the circuit of Fig. 2-15, (1) a sensing signal at zero volts is labeled 0, (2) a sensing signal of +10 volts is labeled 1, (3) a +10-volt output signal is labeled 0, and (4) a zero-output signal is labeled 1; then the action of the circuit can be represented by Table 2-1.

2-16 OR Gate Circuit

Circuit Operation. Figure 2-16 illustrates a three-circuit OR gate. As with the AND gate of Art. 2-15, there are three independent sensing signals. The output circuit differs in that the load resistor R is now connected to ground. When all three sensing signals are zero, all three diodes are nonconducting, have a very high resistance, and E_o is zero. When sensing signal A is +10 volts, D_1 will conduct, its resistance will be very low, and the output signal will be +10 volts. When either sensing signal B or C is +10 volts, the output will again be +10 volts. Thus, in an OR gate circuit when any one or a combination of two or more sensing signals are present, the output voltage will be increased to +10 volts.

Table 2-1 Truth Table for AND Gate Circuit of Fig. 2-15

A	B	C	E_o
0	0	0	0
0	0	1	0
0	1	1	0
0	1	0	0
1	0	0	0
1	0	1	0
1	1	0	0
1	1	1	1

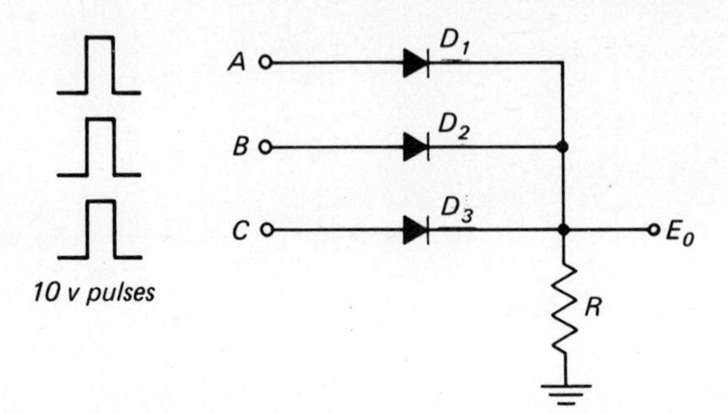

Fig. 2-16 A simple three-circuit OR gate.

Truth Table. The truth table for the OR gate circuit of Fig. 2-16 is shown in Table 2-2. This table is based on (1) zero-volt sensing signal as 0, (2) +10-volt sensing signal as 1, (3) zero-volt output signal as 0, and (4) +10-volt output signal as 1.

QUESTIONS

1. Name four uses of a diode.
2. Name four desirable characteristics of solid-state diodes.
3. Explain the movement of electrons and holes in a semiconductor.
4. Define the following terms: (*a*) conductor, (*b*) insulator, (*c*) semiconductor.
5. What is meant by (*a*) intrinsic semiconductor? (*b*) P-type semiconductor? (*c*) N-type semiconductor?
6. Define the following terms: (*a*) valence electron, (*b*) valence band, (*c*) conduction band, (*d*) energy band, (*e*) forbidden zone.
7. Explain the lattice structure of a silicon or germanium crystal.
8. (*a*) What is meant by impurity conduction? (*b*) How does the relative amount of impurity addition affect the resistance of a semiconductor?
9. Explain the effect on a semiconductor of adding (*a*) a donor impurity, (*b*) an acceptor impurity.
10. Describe the characteristics of a PN junction.

Table 2-2 Truth Table for OR Gate Circuit of Fig. 2-16

A	B	C	E_o
0	0	0	0
0	0	1	1
0	1	1	1
0	1	0	1
1	0	0	1
1	0	1	1
1	1	0	1
1	1	1	1

11. Describe how a high forward current is obtained from a PN junction.
12. What is meant by (*a*) reverse bias? (*b*) Reverse current? (*c*) Potential hill?
13. Name and describe the construction of four basic types of PN junction.
14. Name and describe the construction of four complex types of PN junction.
15. Compare the forward- and reverse-current static volt-ampere characteristics of a semiconductor diode.
16. What is meant by the Zener voltage?
17. Compare the forward- and reverse-resistance static volt-ohm characteristics of a semiconductor diode.
18. (*a*) What is meant by the knee voltage of a semiconductor diode? (*b*) What factors determine the value of this voltage?
19. Define the following semiconductor diode terms: (*a*) bulk resistance, (*b*) junction resistance, (*c*) a-c resistance.
20. What is the purpose of using a series-limiting resistor in the circuit of a diode?
21. What effect is produced on the output circuit when the power source across a diode and the load resistor is reversed?
22. Describe the following terms: (*a*) forward voltage drop, (*b*) supply voltage, (*c*) peak reverse voltage, (*d*) breakdown voltage, (*e*) blocking voltage.
23. Describe the following terms: (*a*) forward current, (*b*) reverse current, (*c*) peak current, (*d*) reverse recovery time, (*e*) shunt capacitance.
24. Describe the following terms: (*a*) hertz, (*b*) kilohertz, (*c*) megahertz.
25. Describe the purpose of a rectifier.
26. Name four principles of operation of a rectifier.
27. Explain why a semiconductor diode can be used as a rectifier.
28. Name three materials that are generally used to make solid-state diodes for use in power supplies.
29. Name eight advantages of solid-state diodes.
30. Explain the operation of a half-wave rectifier circuit using a single diode.
31. Describe the following terms in regard to a half-wave rectifier using a single diode: (*a*) peak reverse voltage, (*b*) output voltage, (*c*) ripple frequency.
32. Explain the operation of a full-wave rectifier circuit using a center-tapped transformer and two diodes.
33. Describe the following terms in regard to a full-wave rectifier using a center-tapped transformer and two diodes: (*a*) peak inverse voltage, (*b*) output voltage, (*c*) ripple frequency.
34. Explain the operation of a full-wave bridge-type rectifier circuit using four diodes.
35. Describe the following terms in regard to a full-wave bridge-type rectifier circuit using four diodes: (*a*) peak inverse voltage, (*b*) output voltage, (*c*) ripple frequency.
36. What is meant by logical functions?
37. Explain how a diode is used as a switch in logic circuits.
38. Explain why a diode switching circuit as used in a logic circuit is called a gate.
39. Describe the following terms: (*a*) gating circuit, (*b*) logic circuit, (*c*) sensing signal.
40. Describe the following terms: (*a*) AND gate, (*b*) OR gate, (*c*) bistable device.
41. Explain why a crystal diode is a bistable circuit element.
42. Explain the operation of a simple diode gate circuit controlled by a sensing signal that varies from zero to +8 volts.
43. Describe the operation of an AND gate circuit that is controlled by three sensing input signals.

44. What is the basic relationship between the input-sensing signals and the output signal in an AND gate circuit?
45. What is meant by a truth table?
46. What values are used in an AND gate-circuit truth table for (*a*) a zero-sensing signal? (*b*) A positive-voltage sensing signal? (*c*) A positive-voltage output signal? (*d*) A zero output signal?
47. Describe the operation of an OR gate circuit that is controlled by three sensing input signals.
48. What is the basic relationship between the input-sensing signals and the output signal in an OR gate circuit?
49. What values are used in an OR gate-circuit truth table for (*a*) a zero-sensing signal? (*b*) A positive-voltage sensing signal? (*c*) A positive-voltage output signal? (*d*) A zero output signal?

PROBLEMS

1. A silicon diode has a forward current of 80 ma at 1 volt, a reverse current of 0.1 μa at 40 volts, and its knee voltage is 0.6 volt. Find (*a*) the bulk resistance, (*b*) the reverse resistance.
2. A silicon diode has a forward current of 90 ma at 1.0 volt, a reverse current of 0.15 μa at 45 volts, and its knee voltage is 0.7 volt. Find (*a*) the bulk resistance, (*b*) the reverse resistance.
3. A germanium diode has a forward current of 80 ma at 1.0 volt, a reverse current of 200 μa at 20 volts, and its knee voltage is 0.2 volt. Find (*a*) the bulk resistance, (*b*) the reverse resistance.
4. Assuming a perfect junction for the diode of Prob. 1 find the junction resistance for the following values of forward current: (*a*) 80 ma, (*b*) 0.8 ma.
5. Assuming a perfect junction for the diode of Prob. 2 find the junction resistance for the following values of forward current: (*a*) 90 ma, (*b*) 45 ma, (*c*) 4.5 ma.
6. Assuming a perfect junction for the diode of Prob. 3 find the junction resistance for the following values of forward current: (*a*) 80 ma, (*b*) 40 ma, (*c*) 4.0 ma.
7. The diode of Prob. 1 does not have a perfect junction. Find the junction resistance when the forward current is 80 ma and the voltage at the junction is (*a*) 30 mv, (*b*) 40 mv, (*c*) 50 mv.
8. Assuming the diode junction of Prob. 2 is not perfect find the junction resistance when a value of 36 mv is used in the junction-resistance equation, and the forward current is (*a*) 90 ma, (*b*) 45 ma, (*c*) 4.5 ma.
9. Assuming the diode junction of Prob. 3 is not perfect find the junction resistance when a value of 40 mv is used in the junction-resistance equation, and the forward current is (*a*) 80 ma, (*b*) 40 ma, (*c*) 4.0 ma.
10. Determine the a-c resistance of the diode of Prob. 1 for the junction conditions of Prob. 4.
11. Determine the a-c resistance of the diode of Prob. 2 for the junction conditions of Prob. 5.
12. Determine the a-c resistance of the diode of Prob. 3 for the junction conditions of Prob. 6.
13. Determine the a-c resistance of the diode of Prob. 1 for the junction conditions of Prob. 7.

14. Determine the a-c resistance of the diode of Prob. 2 for the junction conditions of Prob. 8.

15. Determine the a-c resistance of the diode of Prob. 3 for the junction conditions of Prob. 9.

16. What is the probable range of a-c resistance for the diode of Prob. 1 for the conditions (*a*) Prob. 10? (*b*) Prob. 13?

17. What is the probable range of a-c resistance for the diode of Prob. 2 for the conditions (*a*) Prob. 11? (*b*) Prob. 14?

18. What is the probable range of a-c resistance for the diode of Prob. 3 for the conditions (*a*) Prob. 12? (*b*) Prob. 15?

19. Using the static volt-ampere characteristic curve for the diode shown in Fig. 2-9*a*, determine the forward current with a forward bias of (*a*) 0.5 volt, (*b*) 1.0 volt.

20. Using the static volt-ampere characteristic curve for the diode shown in Fig. 2-9*a*, determine the reverse current with a reverse bias of (*a*) 5 volts, (*b*) 10 volts, (*c*) 20 volts.

21. Using the static volt-ohm characteristic curve for the diode shown in Fig. 2-9*b*, determine the forward resistance with a forward bias of (*a*) 0.5 volt, (*b*) 1.0 volt.

22. Using the static volt-ohm characteristic curve for the diode shown in Fig. 2-9*b*, determine the reverse resistance with a reverse bias of (*a*) 5 volts, (*b*) 10 volts, (*c*) 20 volts.

23. A 2.8-volt d-c power source, a series-limiting resistor, and a diode are connected as shown in Fig. 2-10*a*. If the circuit characteristics of the diode are as illustrated for diode *A* in Fig. 2-10*c*, determine the values of resistance required to limit the forward current to 1.6 ma.

24. A 3.0-volt d-c power source, a series-limiting resistor, and a diode are connected as shown in Fig. 2-10*a*. If the circuit characteristics of the diode are as illustrated for diode *B* in Fig. 2-10*c*, determine the value of resistance required to limit the forward current to 0.8 ma.

25. Determine the value of the direct output voltage of a rectifier when the rms value of the input voltage applied to its conducting circuit is 117 volts and the rectifier is (*a*) a half-wave rectifier, (*b*) a full-wave rectifier using a center-tapped transformer, (*c*) a bridge-type rectifier.

26. Determine the value of the direct output voltage of a rectifier when the rms value of the input voltage applied to its conducting circuit is 208 volts, and the rectifier is (*a*) a half-wave rectifier, (*b*) a full-wave rectifier using a center-tapped transformer, (*c*) a bridge-type rectifier.

27. (*a*) Draw a diagram for a two-circuit AND gate. (*b*) Prepare a truth table for the circuit.

28. (*a*) Draw a diagram for a four-circuit AND gate. (*b*) Prepare a truth table for the circuit.

29. (*a*) Draw a diagram for a two-circuit OR gate. (*b*) Prepare a truth table for the circuit.

30. (*a*) Draw a diagram for a four-circuit OR gate. (*b*) Prepare a truth table for the circuit.

Chapter 3

Electron-tube Diodes

Electron-tube diodes are used in various types of electronic equipment. These tubes may be either of the vacuum type as used in radio and television receivers, or of the gas-type as used in some applications of industrial equipment. The vacuum tube is generally used for high-voltage and low-current outputs. The gas tube is generally used for high-current outputs. Electron-tube diodes may be considered as a circuit element in the same manner as a solid-state diode.

3-1 The Cathode

Purpose of the Cathode. When any substance is heated, the speed of the electrons revolving about their nucleus is increased and some electrons acquire sufficient speed to break away from the surface of the material and go off into space. This action, which is accelerated when the substance is heated in a vacuum, is utilized in vacuum tubes to produce the necessary electron supply. When used for this purpose, the substance is called the *cathode*. All vacuum tubes contain a cathode and one or more electrodes mounted in an evacuated envelope, which may be a glass bulb or a compact metal shell.

Purpose of the Heater. The cathode is an essential part of a vacuum tube because it supplies the electrons necessary for operation of the tube. The electrons are generally released by heating the cathode. The purpose of a heater in a vacuum tube is to radiate heat when an electric current flows through it. The amount of heat that is radiated is dependent on the material of which the conductor is made and the amount of current flowing in the conductor. The source of power used to supply current for heating the cathode is called the *heater power supply*.

Directly Heated Cathodes. A directly heated cathode, called the *filament type*, is one in which the heater is also the cathode (Fig. 3-1*a*). Materials that are good conductors are found to be poor electron emitters; therefore directly heated cathodes must be operated at high temperatures in order to emit a sufficient number of electrons. Directly heated cathodes require a comparatively small amount of heating power and are used in almost all the tubes designed for battery operation. Alternating-current-operated tubes seldom use the filament-type cathode.

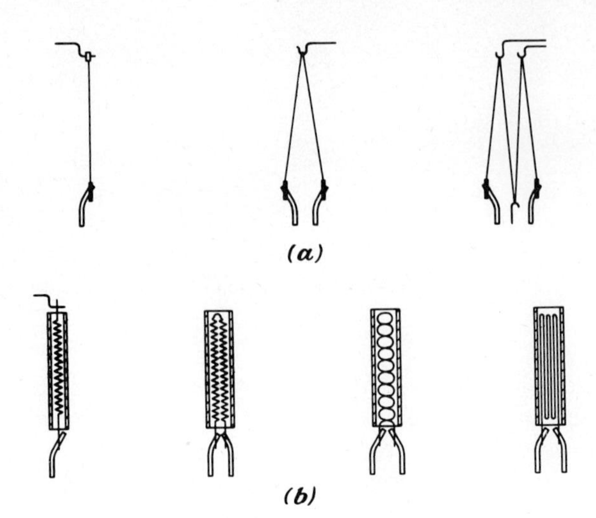

Fig. 3-1 Types of cathodes. (a) Directly heated cathodes. (b) Indirectly heated cathodes.

Indirectly Heated Cathodes. An *indirectly heated cathode* is one in which the emitting material is coated on a thin sleeve which is heated by radiation from a heater placed inside the sleeve and insulated from it (Fig. 3-1*b*). It is thus possible to use a material for the heater that will radiate the maximum amount of heat with the minimum amount of current and to use a material for the cathode that is a good electron emitter.

3-2 Diodes

The Plate. A vacuum tube having a cathode and one other electrode is called a *diode*. The second electrode is called the *plate* or *anode*. If a positive voltage is applied to the plate, the electrons emitted from the cathode, being negative, will be attracted to the plate. These electrons will flow through the external plate battery circuit as indicated in Fig. 3-2. This flow of electrons is called the *plate current*. If the polarity of the plate is made negative, the electrons will be forced back to the cathode and no current will flow in the plate circuit.

When an alternating current is applied to the plate, its polarity will be positive during every other half-cycle. As electrons will flow to the plate only

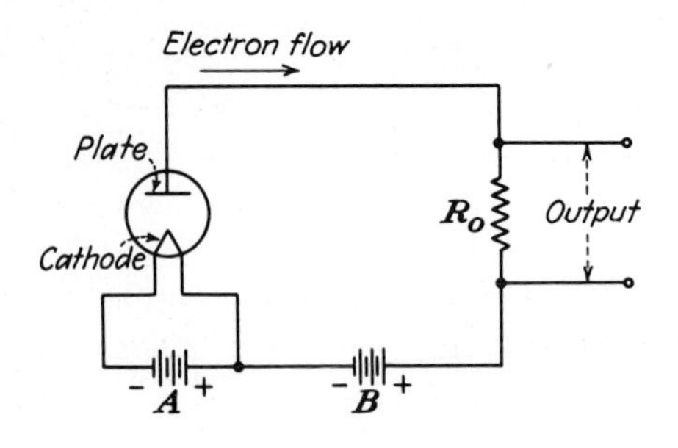

Fig. 3-2 Circuit diagram for a diode showing the electron flow, and the proper connections for the filament and B power supplies.

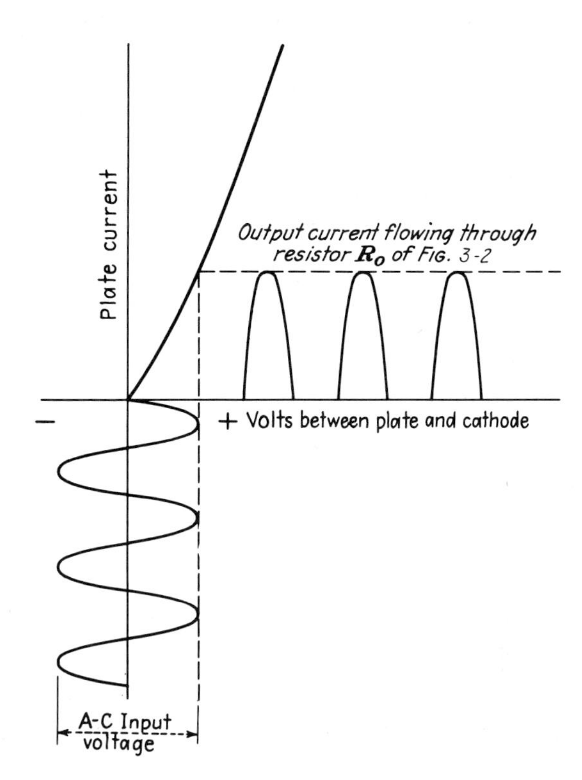

Fig. 3-3 The diode as a rectifier.

when it is positive, the current in the plate circuit will flow in only one direction, and the current is called a *rectified current* (Fig. 3-3).

Diodes are used in radio and television receivers as detectors and in a-c operated electronic equipment to convert alternating current to direct current. Rectifier tubes having one plate and one cathode are called *half-wave rectifiers* because the rectified current flows only during one-half of the cycle.

Duodiodes. Rectifier tubes having two plates and one or two cathodes have their plates connected into the external circuit so that each plate will be positive for opposite halves of the cycle. As one or the other of the plates will always be positive, current will flow in the external circuit during each half-cycle. These tubes are called *full-wave rectifiers* because current flows during both half-cycles.

3-3 Diode Characteristics

Space Charge. The number of electrons drawn to the plate depends on the number given off by the cathode and also on the plate voltage. If the plate voltage is not high enough to draw off all the electrons emitted by the cathode, those not drawn off will remain in space. These electrons form a repelling force to the other electrons being given off by the cathode, thus impeding their flow to the plate. This repelling action is called the *space charge.*

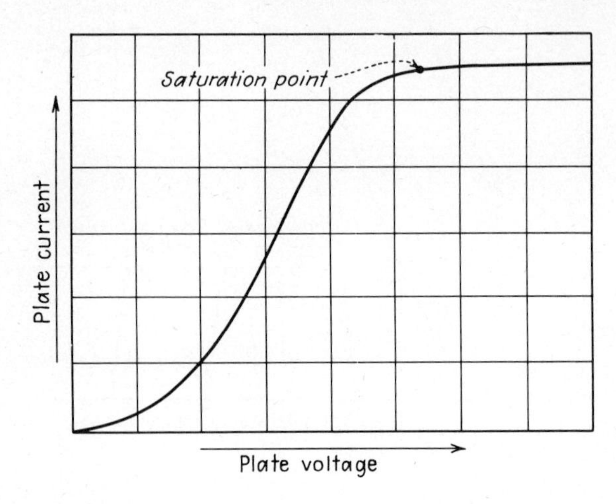

Fig. 3-4 Variation of plate current with changes in plate voltage.

Characteristic Curve. The characteristic curve for a typical vacuum-tube diode is shown in Fig. 3-4. Increasing the plate voltage will increase the plate current until all the electrons given off by the cathode are drawn to the plate. Further increase in plate voltage will not increase the plate current, as no more electrons can be drawn off to it. The point on the curve at which the current has reached its highest value is called the *saturation point,* and the plate current for this condition is called the *saturation current* or *emission current.*

Ratings of Rectifier Tubes. Rectifier tubes are generally rated according to (1) the alternating voltage per plate, (2) the peak inverse voltage, (3) the peak plate current, (4) the load current. These ratings are not fixed values but vary with the rectifier tube and filter circuits with which the tube is associated.

The *alternating voltage per plate* is the highest rms value of voltage that can safely be applied between the plate and cathode of the tube. In a half-wave rectifier circuit the open circuit voltage across the secondary winding of the power transformer should not exceed this value. In a full-wave rectifier circuit the open circuit voltage between either end of the secondary winding of the power transformer and its center tap should not exceed this value.

The *peak inverse voltage* rating is the maximum value of voltage that a rectifier tube can safely withstand between its plate and cathode when the tube is not conducting. During the portion of the a-c input cycle when the plate is negative with respect to the cathode, no voltage drops will exist in the rectifier circuit, and hence the full secondary voltage will be impressed between the plate and cathode. For normal operating conditions this voltage is equal to the peak value of the transformer secondary voltage. For transient conditions this peak value of voltage may be greatly exceeded.

The *peak plate current* represents the maximum amount of electron emission that the cathode can supply. It is the maximum instantaneous value of current that can safely flow through the rectifier tube.

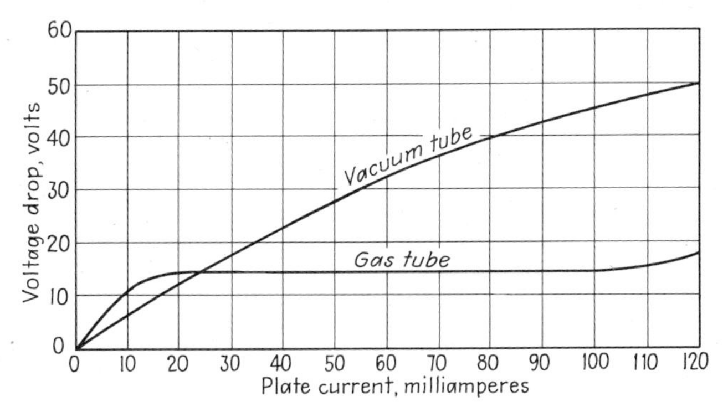

Fig. 3-5 Internal voltage drop between the cathode and plate for comparable gas and vacuum tubes.

The *load current* is the maximum safe value of direct current that the tube can deliver. Since current flows through a plate circuit of a rectifier tube during only half of the input cycle, the average value of its d-c output will be less than one-half of its peak plate current.

Voltage Regulation. During the portion of the cycle in which the tube is conducting, voltage drops will be produced at the tube and the secondary winding of the transformer. Since these voltage drops reduce the voltage available across the output load they should be kept low.

The internal voltage drop of a high-vacuum-type rectifier tube will vary in almost direct proportion to its load current (see Fig. 3-5). A varying load current will cause the voltage drop across the tube to vary, thus also causing a variation in the voltage across the output load. The voltage regulation of high-vacuum-type rectifier tubes is therefore very poor. The internal voltage drop of directly heated (filament-type) high-vacuum rectifier tubes is high, in some cases as much as 50 to 60 volts.

3-4 Gas Tubes

Types of Cathodes. In the manufacture of high-vacuum tubes, as much of the air as possible is removed from inside the envelope. In the manufacture of gas-filled tubes, a high-vacuum tube is first produced and then gas is added under low pressure. The gases most frequently used are argon, neon, xeon, and mercury vapor.

Two basic types of gas-filled tubes are used: (1) the *hot-cathode type* and (2) the *cold-cathode type*. In the hot-cathode type, the cathode is heated in the same manner as thermionic vacuum tubes; in the cold-cathode type, the cathode is not purposely heated but may become hot as it is bombarded by moving particles. The symbolic representation of several types of gas-filled tubes is shown in Fig. 3-6. The small dot within the circle indicates that the tube is gas-filled.

Electrical Conduction in Gas-filled Tubes. In a gas-filled tube, the electron stream from the cathode will encounter gas molecules on its way to the plate. When an electron collides with a gas molecule, the energy transmitted by the collision may cause the molecule to release an electron. This free electron may then join the original stream of electrons moving toward the plate and thus is capable of releasing other electrons through collision with other gas molecules. This process, which is cumulative, is called *ionization.* The molecule that has lost an electron is called a *positive ion.* The tube in its ionized state contains gas molecules, positive ions, and free electrons within its envelope. The heavier positive ions are attracted toward the negative cathode, and while moving toward it they attract electrons from the space charge. The positive gas ions in the vicinity of the cathode also neutralize a portion of the space charge. Electrons will therefore flow from cathode to plate in a gas tube with less opposition than in a high-vacuum tube.

A voltage source between the plate and cathode supplies the energy required to dislodge electrons from their atomic orbits to produce ionization. There is a definite voltage value for each type of gas-filled tube at which ionization starts. At the instant ionization occurs, large currents will flow at relatively low values of plate voltage. The voltage at which ionization commences is called *ionization potential, striking potential,* or *firing point.* After ionization has started, the action is maintained even at a voltage considerably lower than the firing point. However, a minimum amount of voltage is required to maintain ionization. Should the voltage across the tube drop below this minimum value, the gas will deionize and conduction will stop. The voltage at which current ceases to flow is called the *deionizing potential* or the *extinction potential.*

A gas-filled tube has almost infinite resistance before ionization and a very low resistance after ionization. This characteristic makes possible its use as an electronic switch that closes at a certain value of voltage and permits current to flow and opens at some lower value of voltage and blocks the flow of current.

Cold-cathode Gas Diodes. In this type of tube, the structure of the electrodes may be such that (1) current is permitted to flow in only one direction, as in

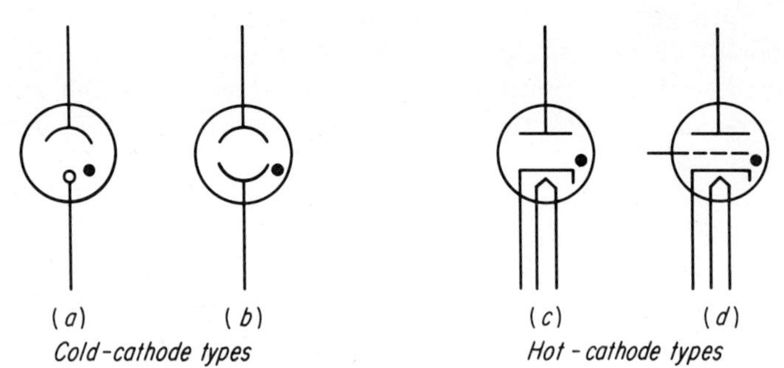

Fig. 3-6 Symbolic representation of several types of gas-filled tubes. (a) Voltage-regulator. (b) Neon-glow. (c) Gas-diode. (d) Thyratron.

the voltage-regulator tube of Fig. 3-6*a*, or (2) the cathode and plate may have the same size and shape to permit the tube to conduct in either direction, as in the neon-glow lamp of Fig. 3-6*b*. In this type of tube, the cathode is not heated; hence no electrons are emitted to aid in the ionization process. However, some ions and free electrons are always present in gases in their normal state. When a positive potential is applied to the plate, the free electrons and negative ions will drift toward the plate while the positive ions will move toward the cathode. During the movement of these charged particles, numerous collisions will take place, resulting in the release of additional electrons and the production of additional ions. Since this process is cumulative, ionization eventually takes place. However, the firing potential for this type of tube will be much higher than for a hot-cathode gas tube.

Hot-cathode Diodes. Most high-vacuum diodes draw currents of less than 1 amp, while some gas-filled diodes are capable of handling currents of several hundred amperes. The gas-filled diode is therefore used whenever large quantities of rectified current are required. The gas used in this type of rectifier may be any one of the gases previously mentioned; however, mercury vapor is most commonly used. A small amount of liquid mercury is vaporized by the hot cathode, causing mercury vapor to form inside the tube. These tubes are not capable of supplying their rated output until all the mercury has been completely vaporized. Before the tube starts conducting, the relatively high voltage between the plate and cathode causes a large increase in the electron velocity. These high-velocity electrons cause the gas ions to acquire a higher positive charge, thus causing them to bombard the cathode with greater impact. If this action is permitted to continue for even a short period of time, the force of this bombardment is high enough to disintegrate the surface of the emitter. Therefore, the plate voltage should not be applied until the tube has had sufficient time for the mercury to become completely vaporized.

Voltage Regulation. The characteristics of the cold-cathode gas diode may be used to obtain voltage regulation (see Fig. 3-5). The voltage-regulator tube of Fig. 3-7 will maintain the output voltage relatively constant regardless of

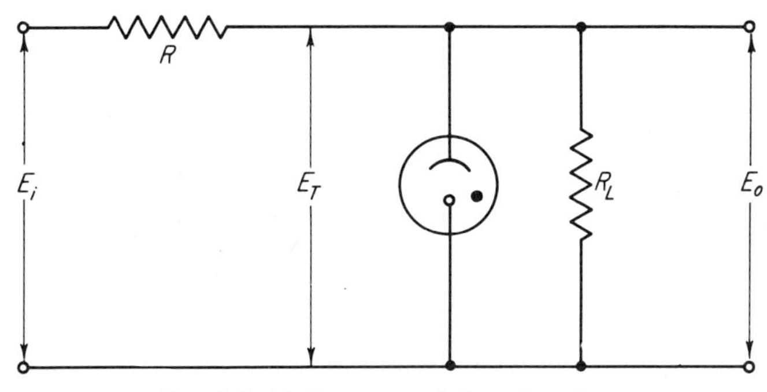

Fig. 3-7 Voltage-regulator circuit.

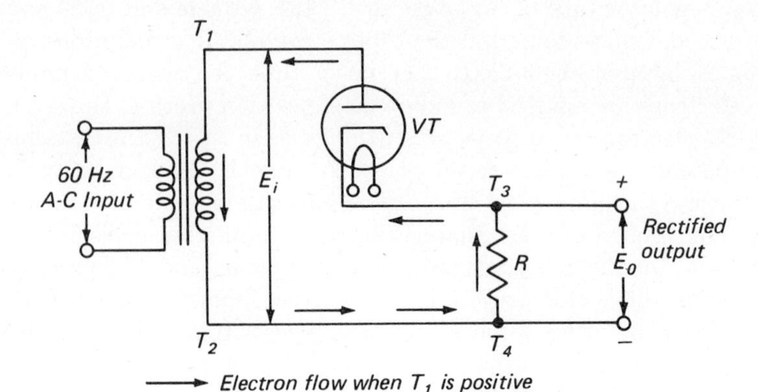

Fig. 3-8 A half-wave vacuum-tube diode rectifier circuit.

variations in load current or input voltage. When the load current increases, the voltage drop across the limiting resistor R also increases and reduces the voltage across the diode. A reduction of the diode voltage will cause a decrease in the tube current, which in turn, decreases the voltage drop across R. The voltage across the diode will therefore remain almost constant. When the input voltage varies, the degree of ionization and the tube resistance will change, which in turn maintains a constant voltage across the diode. Since the diode is across the output resistor R_L, the voltage of E_T will be equal to E_o.

3-5 Half-wave Rectifier

In a half-wave rectifier circuit a single-diode vacuum or gas tube is used as shown in Fig. 3-8. This circuit operates in the same manner as the half-wave solid-state rectifier circuit (Art. 2-10). The value of the output voltage, its waveform, and its relation to the input voltage are also the same as for the solid-state rectifier circuit. Depending on the input voltage and output requirements, this circuit may be operated directly from an a-c power source, or from the secondary of a power transformer.

During the half-cycle when T_1 is positive, the plate of the tube is positive with respect to the cathode, the tube acts as a conductor, and current flows in the output circuit. During this half-cycle electrons flow through the path T_2-T_4-R-T_3-VT-T_1. During the opposite half-cycle, T_1 will be negative with respect to the cathode, the tube acts as an insulator, and no current flows in the output circuit.

3-6 Full-wave Rectifier Circuit Using a Center-tapped Transformer

In this type of full-wave rectifier circuit two single-diode tubes, or a duplex diode (two single diodes in one envelope) and a power transformer with its secondary center-tapped are used (Fig. 3-9). This circuit operates in the same manner as the full-wave solid-state rectifier circuit using a center-tapped

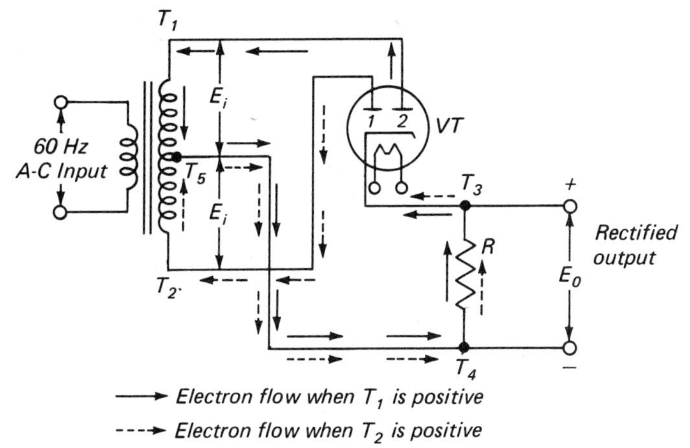

Fig. 3-9 A full-wave vacuum-tube diode rectifier circuit using a power transformer with a center-tapped secondary.

transformer (Art. 2-11). The value of the output voltage, its waveform, and its relation to the input voltage are also the same as for the solid-state rectifier circuit. Because the two plates of the rectifier tube are connected to opposite ends of the secondary winding, their polarity with respect to the cathode will always be opposite to each other. Thus, during one-half of the input cycle one plate will be conducting and the other plate will not be conducting; and during the second half-cycle the second plate will be conducting and the first plate will not be conducting. Under this condition, current will flow in the output circuit during both halves of the input cycle.

During the half-cycle that T_1 is positive, plate 2 will be positive with respect to the cathode, this diode acts as a conductor, and current flows in the output circuit. During this half-cycle electrons flow through the path T_5-T_4-R-T_3-$VT_{sec\cdot 2}$-T_1. During the opposite half-cycle T_2 is positive, plate 1 will be positive with respect to the cathode, this diode acts as a conductor, and current flows in the output circuit. During this half-cycle electrons flow through the path T_5-T_4-R-T_3-$VT_{sec\cdot 1}$-T_2.

3-7 Full-wave Rectifier Using a Bridge Circuit

In this type of full-wave rectifier circuit, two duplex diodes having separate cathodes are used (Fig. 3-10). This circuit operates in the same manner as the full-wave solid-state bridge circuit (Art. 2-12). The value of the output voltage, its waveform, and its relation to the input voltage are also the same as for the solid-state rectifier circuit.

During the half-cycle when T_1 is positive, electrons will flow through the path T_2-T_8-VT_3-T_6-T_4-R-T_3-VT_2-T_5-T_1. During the opposite half-cycle when T_2 is positive, electrons will flow through the path T_1-T_5-VT_1-T_6-T_4-R-T_3-VT_4-

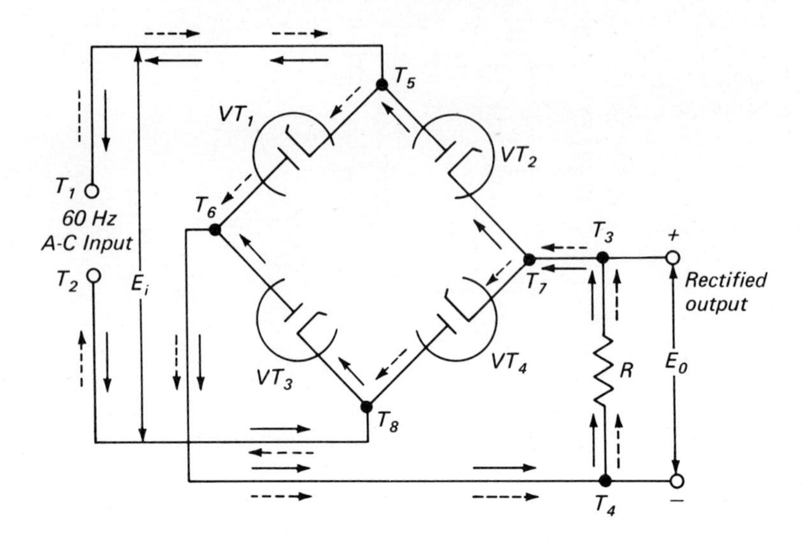

Fig. 3-10 A full-wave bridge-type rectifier circuit using four vacuum-tube diodes.

T_8-T_2. Under this condition, current will flow in the output circuit during both halves of the input cycle.

QUESTIONS

1. For what type of circuit output are the following diodes generally used: (*a*) Vacuum tube? (*b*) Gas tube?
2. Define the following terms: (*a*) cathode, (*b*) filament, (*c*) heater, (*d*) directly heated cathode, (*e*) indirectly heated cathode.
3. Explain the purpose and theory of operation of (*a*) the cathode, (*b*) the heater.
4. (*a*) What are the advantages of directly heated cathodes? (*b*) Where are they generally used?
5. What are the advantages of indirectly heated cathodes?
6. Explain the following terms: (*a*) plate, (*b*) diode, (*c*) duodiode.
7. Explain the rectifier action of a diode.
8. Explain the following terms: (*a*) space charge, (*b*) emission current, (*c*) saturation current.
9. (*a*) What four factors are generally used in rating rectifier tubes? (*b*) Are these ratings fixed values? (*c*) Explain your answer to part (*b*).
10. Define (*a*) alternating voltage per plate, (*b*) peak inverse voltage, (*c*) peak plate current, (*d*) load current.
11. Explain the variation of the internal voltage drop with load variations, for (*a*) a vacuum-tube diode, (*b*) a gas-tube diode.
12. Describe what is meant by (*a*) a hot-cathode gas tube, (*b*) a cold-cathode gas tube.

13. Explain the following terms: (*a*) ionization, (*b*) positive ion, (*c*) firing point, (*d*) extinction potential.
14. Describe the electric conduction in a cold-cathode gas diode.
15. Explain how a cold-cathode gas diode may be used as a voltage regulator.
16. Describe the electric conduction in a hot-cathode gas diode.
17. Explain the principle of operation of a half-wave rectifier circuit using a single vacuum-tube diode.
18. What are the output current and inverse peak voltage characteristics of the rectifier circuit of question 17?
19. Explain the principle of operation of a full-wave rectifier circuit using two diodes and a power transformer having a center-tapped secondary.
20. What are the output current and inverse peak voltage characteristics of the rectifier circuit of question 19?
21. Compare the output voltage of a half-wave and a full-wave rectifier for a sine-wave input. Assume that the output is applied directly to a resistive load and that the voltage drops at the transformer secondaries and the rectifier tube are negligible.
22. Explain the principle of operation of a full-wave bridge rectifier circuit using four vacuum-tube diodes.
23. What are the output current and inverse peak voltage characteristics of the rectifier circuit of question 22?

Chapter 4

Transistors

A transistor is basically a resistor that amplifies electrical impulses as they are transferred through it from its input to its output terminals. The name *transistor* is derived from the words *transfer* and *resistor*. The transistor has many advantages, among which are its (1) low current requirement, (2) small size, (3) light weight, (4) long operating and shelf life, (5) elimination of warm-up time, (6) mechanical strength, and (7) photosensitivity.

4-1 Transistors

The transistor is a solid-state electronic device whose operation depends on the flow of electric charges carried within the solid. Two fundamental actions are involved: (1) generation of carriers within the solid and (2) control of these carriers within the solid.

A triode junction transistor consists of a single crystal, the leads, and an envelope. The crystal is essentially a three-layer unit. However, the layers are not separate pieces but are areas within the crystals which have slightly different electrical characteristics. Transistors are made from a semiconductor whose electrical properties are described in Art. 2-1.

4-2 Junction Transistors

Types of Junction. The characteristics of a *semiconductor* diode, triode, or tetrode will depend on the material and the type of PN junction used. Germanium is usually better than silicon at high frequencies, while at high power levels silicon is generally better than germanium. The PN junction can be made in various ways as described in Art. 2-5.

The Triode. When a layer of N-type germanium is joined to a PN junction, an NPN junction is formed (Fig. 4-2*a*), and when a layer of P-type germanium is joined to a PN junction, a PNP junction is formed (Fig. 4-3*a*). Either of these two junctions can be used as a triode junction transistor, and they are generally referred to as NPN *junction transistors* or PNP *junction transistors*. The center section is called the *base*, one of the outside sections the *emitter*, and the other outside section the *collector*.

Transistor Biasing. In order to use the NPN transistor as an amplifier, the emitter-base junction is biased with a forward voltage and the base-collector junction with a reverse voltage (Fig. 4-2*b*). A relatively high current will flow

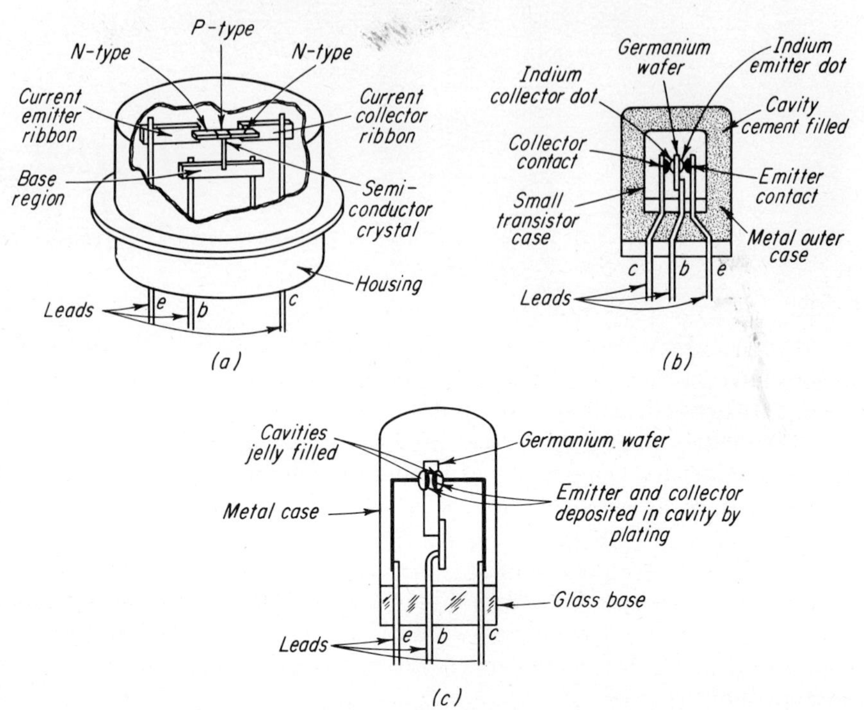

Fig. 4-1 Cutaway view of commercial transistors. (a) Grown junction. (b) Diffused junction. (c) Surface barrier.

through the emitter-base junction and a relatively low current through the base terminal. The base current is very low, as only a small number of emitter electrons (usually less than 5 per cent) will combine with the base holes. Since the remaining electrons flow toward the collector, the emitter-collector current is high.

The operation of a PNP transistor is similar to the operation of an NPN transistor except that the bias-voltage polarities are reversed and the current

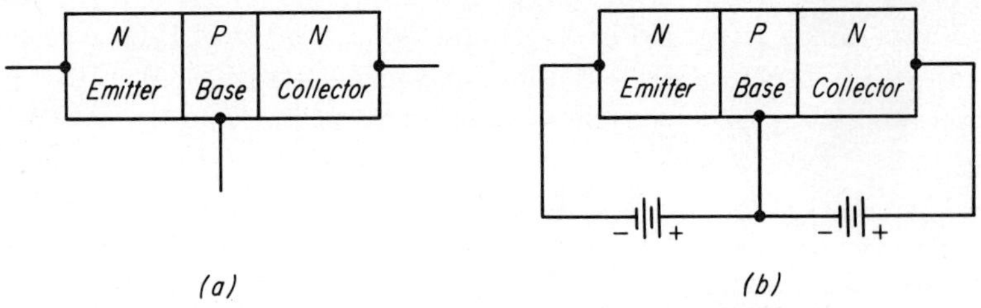

Fig. 4-2 NPN junction transistor. (a) The junction and the name of each section. (b) Biasing potentials.

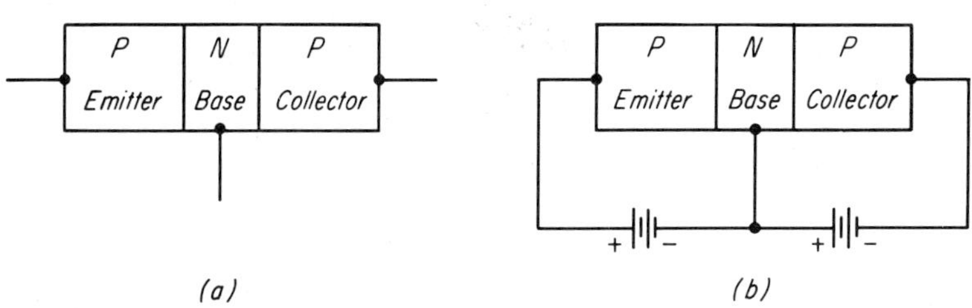

Fig. 4-3 PNP junction transistor. (a) The junction and the name of each section. (b) Biasing potentials.

carriers are holes instead of electrons. The emitter-base junction is biased with a forward voltage and the base-collector junction with a reverse voltage (Fig. 4-3*b*). The holes from the emitter will diffuse through the base to the base-collector junction, where the applied negative voltage will draw them to the collector terminal.

For normal transistor operation, the first two letters of the NPN and PNP designation indicate the manner of connecting the power source to the emitter and collector terminals of the transistor. With an NPN transistor, the emitter is connected to the negative terminal of its power source and the collector is connected to the positive terminal of its power source. With a PNP transistor, the emitter is connected to the positive terminal of its power source and the collector is connected to the negative terminal of its power source. This method of connection provides forward bias to the emitter-base circuit and reverse bias to the collector-base circuit, and thus the transistor can produce amplification.

Symbol and Lead Identification. The symbols for triode transistors are shown in Fig. 4-4. Although there are several variations of symbols in use, those shown here are frequently used and have been adopted by the IEEE. The horizontal line represents the base, and the two angular lines represent the emitter and collector. The arrowhead drawn on the emitter indicates the direction of current flow (not electron flow). In the PNP type, the arrow is drawn pointing toward the base (Fig. 4-4*a*), and in the NPN type, the arrow is drawn pointing away from the base (Fig. 4-4*b*).

The lead identification of a triode transistor varies with the manufacturer. Three systems in general use are shown in Fig. 4-5. When the leads of a transistor are in the same plane but unevenly spaced (Fig. 4-5*a*), they are identified

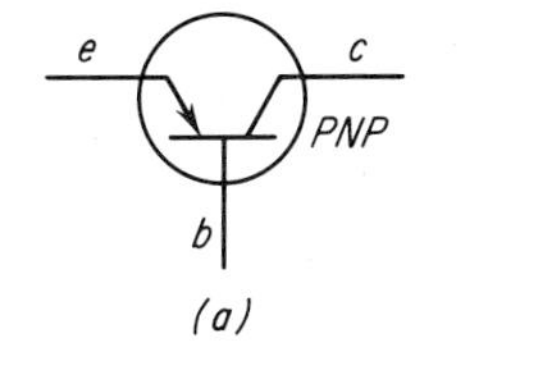

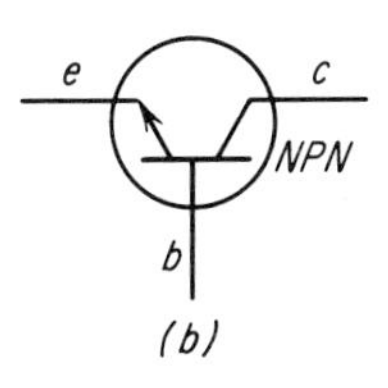

Fig. 4-4 Symbol identification for triode transistors. (a) PNP. (b) NPN.

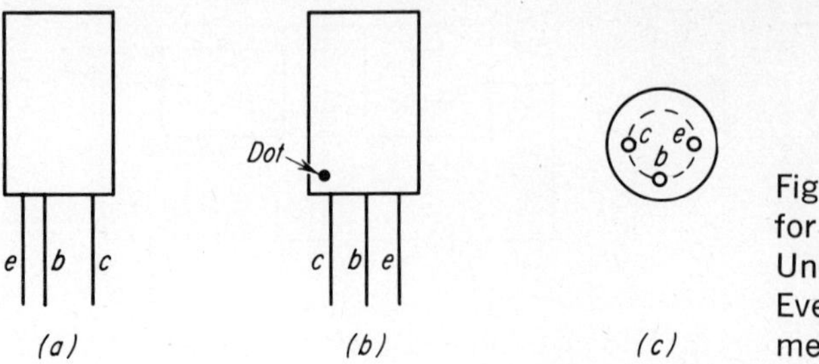

Fig. 4-5 Lead identification for triode transistors. (a) Unevenly spaced leads. (b) Evenly spaced leads. (c) Symmetrically spaced leads.

by the position and spacing of the leads. The center lead is the base lead, and the emitter and collector leads are on either side of this lead. The collector lead is identified by the larger spacing existing between it and the base lead. When the leads on a transistor are in the same plane and evenly spaced (Fig. 4-5*b*), the center lead is the base, the lead identified by the dot the collector, and the remaining lead the emitter. When the leads are spaced around the circumference of a circle as shown in Fig. 4-5*c*, the center lead is the base lead. The collector and emitter leads are identified by their positions relative to the base lead, when viewed looking into the base of the transistor. Sometimes a fourth lead is added and it is electrically connected to the mounting base, or case, of the transistor. Another variation is to connect the common electrode to the base, or case, and use only two leads.

4-3 Methods of Operation

Types of Connections. Basically there are three types of connection (also called *configuration*) for operating the triode transistor: (1) common base, (2) common emitter, and (3) common collector (Fig. 4-6). The term *common* is used to denote the electrode that is common to both the input and output circuits. Because the common electrode is often grounded, these modes (methods) of operation are frequently referred to as *grounded base, grounded emitter,* and *grounded collector.*

Circuit Characteristics. The three modes of operation provide distinct individual circuit characteristics (Table 4-1). The common-base connection

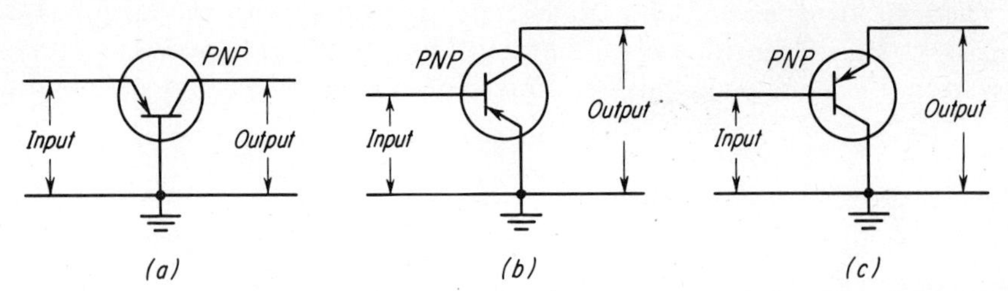

Fig. 4-6 Transistor configurations. (a) Common base. (b) Common emitter. (c) Common collector.

Table 4-1 Comparison of Transistor Characteristics

COMMON ELECTRODE	INPUT RESISTANCE	OUTPUT RESISTANCE	CURRENT GAIN	VOLTAGE GAIN	POWER GAIN
Base	Lowest	Highest	Lowest	High	High
Emitter	Low	High	High	High	Highest
Collector	Highest	Low	High	Lowest	Low

has a low input impedance and a high output impedance. The common-collector connection has a high input impedance and a low output impedance. The common-emitter connection has a low-to-medium input impedance and a medium-to-high output impedance. The common-emitter and common-base circuits are the types most frequently used.

4-4 Action of the Transistor

Figure 4-7 illustrates a basic common-emitter transistor circuit using an NPN transistor Q_1. The power supply V_{EE} provides a forward bias to the emitter-base input circuit, and the signal voltage v_i is also applied to the input circuit. The power supply V_{CC} provides a reverse bias to the collector-emitter output circuit, and the load resistor R_L provides a means of obtaining a varying output voltage v_o. The common-emitter mode of operation, with forward bias at the input circuit and reverse bias at the output circuit, produces a low-impedance input circuit and a high-impedance output circuit. With this set of circuit conditions, amplification of the input signal will result.

When the input signal is zero, the currents in the input and output circuits, and the output voltage v_o, will be determined by the characteristics and values of the circuit components. When the input voltage rises in a positive direction, it increases the forward bias on the transistor, thereby increasing the current in both the emitter and collector circuits. When the input-signal voltage rises in the negative direction, the forward bias on the transistor decreases, thereby decreasing the current in both the emitter and collector circuits. Thus, (1) any change in the input-signal voltage will be reflected in the output signal, and (2) because the emitter and collector currents are of practically equal values and the output resistance is many times greater than the input resistance, the

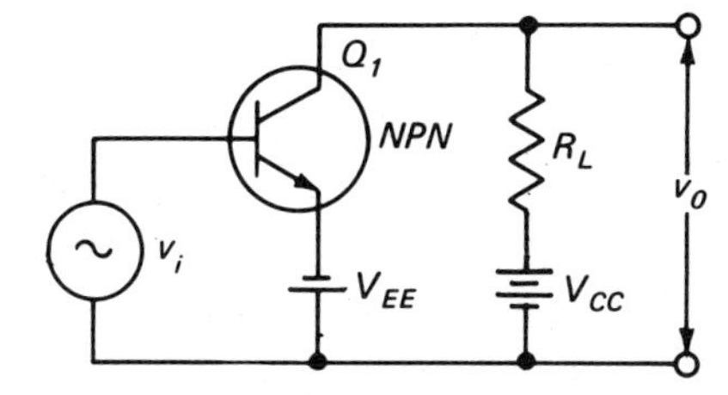

Fig. 4-7 Basic common-emitter transistor circuit.

changes in the output voltage will be many times greater than any changes in the input-signal voltage.

4-5 Phase Relation between the Input-signal Voltage and the Output-signal Voltage

Figure 4-8 illustrates the phase relation among the various voltages and currents in the basic transistor circuit of Fig. 4-7. All variations are initiated by the input-signal voltage v_i (Fig. 4-8*a*). The emitter-base fixed bias component V_{EB} (Fig. 4-8*b*) and the varying signal voltage v_i result in a varying emitter-base bias v_{eb} (Fig. 4-8*c*) that is 180° out of phase with the input voltage v_i. The variations in v_{eb} will cause variations in the emitter and collector currents i_e

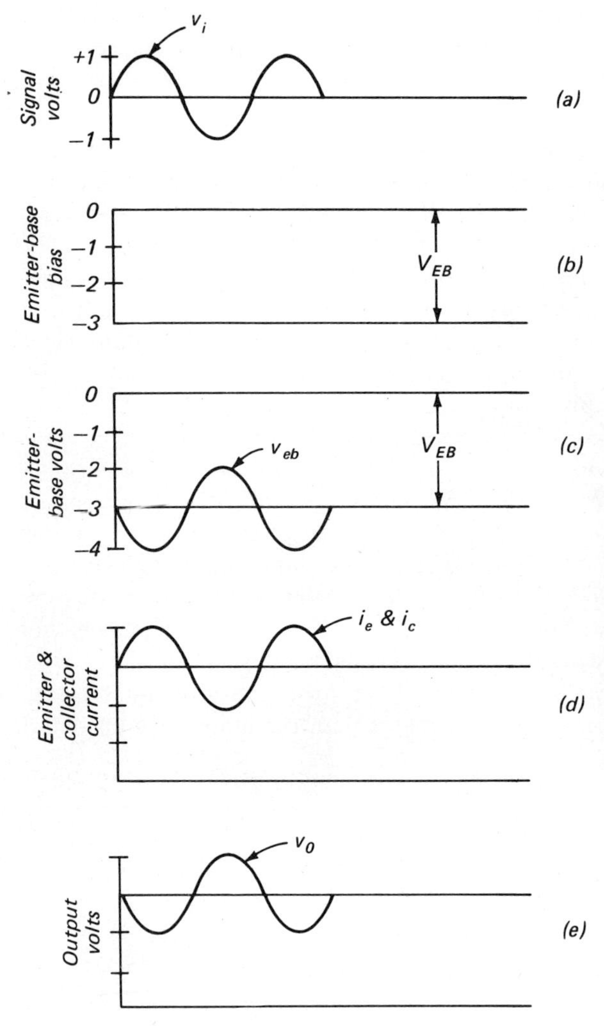

Fig. 4-8 Effect of signal voltage on the voltages and currents in the common-emitter transistor circuit of Fig. 4-7. (a) Input signal voltage v_i. (b) Emitter-base bias provided by V_{EE}. (c) Combined effect v_{eb} of the input-signal voltage v_i and the emitter-base bias V_{EE}. (d) Variation in the emitter and collector currents i_e and i_c, caused by the variations in the emitter-base voltage v_{eb}. (e) Variation in the output voltage v_o caused by the variation in the input-signal voltage v_i.

and i_c (Fig. 4-8*d*) that will be in phase with the input-signal voltage v_i. The output voltage v_o (Fig. 4-8*e*) is equal to the collector supply voltage V_{CC} less the amount of the voltage drop at the load resistor R_L. During the periods when the collector current is increasing, the voltage drop at the load resistor will also be increasing, and as the fixed voltage V_{CC} is thus being reduced by a greater amount, the output voltage v_o will be decreasing. By applying similar reasoning, it becomes evident that during the periods when the collector current is decreasing, the output voltage v_o will be increasing. Thus, the output voltage v_o will be 180° out of phase with the collector current, and also 180° out of phase with the input signal voltage v_i.

The development of the phase relation between the input and output voltages presented above is for the common-emitter circuit arrangement. By similar reasoning, it can be shown that for a common-base circuit the output voltage will be in phase with the input voltage. Also, for a common-collector circuit the output and input voltages will be in phase.

4-6 Letter Symbols

For the purpose of brevity, many letter symbols and abbreviations have been introduced, some frequently used and others only rarely used. The symbols presented here are commonly used.

Maximum, average (d-c), and effective (root-mean-square) values are represented by the uppercase letter of the proper symbol, as

I = current V = voltage R = resistance P = power

Instantaneous values that vary with time are represented by the lowercase letter of the proper symbol:

i = current v = voltage r = resistance p = power

Abbreviations frequently used as subscripts are

E, e = emitter electrode i = input
B, b = base electrode o = output
C, c = collector electrode f = forward
X, x = circuit node r = reverse

Where two subscripts are used, the first designates the electrode at which the current is measured, or where the electrode potential is measured with respect to the reference electrode or circuit node, which is designated by the second subscript. When the reference electrode or circuit node is understood, the second subscript may be omitted if its use is not required to preserve the meaning of the symbol.

Direct-current values and instantaneous total values are indicated by uppercase subscripts, as

$$I_C,\ V_{EB},\ P_C,\ i_C,\ v_{EB},\ p_C$$

Varying component values are indicated by lowercase subscripts, as

$$I_c,\ V_{eb},\ P_e,\ i_c,\ v_{eb},\ p_c,\ v_i,\ v_o$$

To distinguish among maximum, average, and effective values the maximum value is represented by the subscript *m* or *M*, and the average value by the subscript av or AV:

$$i_{c\cdot m},\ I_{c\cdot m},\ I_{C\cdot M},\ I_{C\cdot \mathrm{AV}},\ i_{C\cdot \mathrm{AV}}$$

Supply voltages may be indicated by repeating the electrode subscript. The reference electrode may then be designated by the third subscript:

$$V_{EE},\ V_{CC},\ V_{BB},\ V_{EEB},\ V_{CCB},\ V_{BBC}$$

In devices having more than one electrode of the same type, the electrode subscripts are modified by adding a number following the subscript, and on the same line:

$$V_{B1},\ V_{b2},\ I_{B2}$$

In multiple-unit devices, the electrode subscripts are modified by a number preceding the electrode subscript, as

$$1b,\ 2B,\ 1c,\ 2C$$

4-7 Transistor Terms

Alpha (α)*:* The forward-current transfer ratio for the common-base configuration.

Beta (β)*:* The forward-current transfer ratio for the common-emitter configuration.

Breakdown Voltage: The amount of voltage that (1) causes a change in the crystal structure, and/or (2) causes a rapid increase in current. There are two breakdown points: (1) the initial breakdown voltage, and (2) the sustaining breakdown voltage.

Collector Cutoff Current: In a reversed-biased collector-to-base circuit, the amount of direct current flowing in the collector circuit when the emitter-to-base circuit is open.

Cutoff Currents: The small d-c reverse currents (saturation and leakage currents) present when a transistor has a reverse bias; the symbol is I_{CO}.

Emitter Cutoff Current: In a reversed-biased emitter-to-base circuit, the amount of direct current flowing in the emitter circuit when the collector-to-base circuit is open.

Forward-current Transfer Ratio: The ratio of the output current to the input current of a transistor.

Frequency Cutoff: The frequency at which the current gain has decreased to 0.707 of its low-frequency reference value, usually 1,000 hertz. For the common-base configuration, the cutoff frequency is that value of frequency where alpha is 0.707 times its value at 1 kHz. For the common-emitter configuration, the cutoff frequency is that value of frequency where beta is 0.707 times its value at 1 kHz.

Gain-Bandwidth Product: This term is used for the common-emitter configuration to represent the frequency at which its value of beta is 1. The gain-bandwidth product f_T is a measure of the useful frequency range of a circuit and is a factor in determining the choice of configuration to be used. It is sometimes called the *figure of merit.*

Leakage Current: The d-c reverse current present when a transistor has a reverse bias. It is caused by surface characteristics of the transistor materials, and is normally only a few microamperes.

Saturation Current: The d-c reverse current present when a transistor has a reverse bias. It is caused by concentrations of impurities in the transistor materials, and is normally only a few microamperes.

Saturation Voltage: The minimum amount of collector-to-emitter voltage that will maintain full conduction with a given amount of base input current is called the *collector-to-emitter saturation voltage.* Increasing the voltage beyond this point produces relatively little increase in current.

Stored Base Charge: The amount of charge in the base area of a transistor when the forward bias is removed.

Transit Time: The time taken by the electrons, or holes, to pass from one electrode to another. For example, in a common-base circuit the transit time is the amount of time taken by the electrons or holes to pass from the emitter to the collector.

4-8 Hybrid Parameters

A *parameter,* also called a *network constant,* is a constant that enters into a functional equation and corresponds to some characteristic of a circuit such as resistance, inductance, capacitance, or any other property value in a network. In order to make comparative studies of transistors and their circuit characteristics, four types of comparative measurements are generally used, each relating to the slope of a certain type or form of characteristic curve. Because one is measured in ohms, one in mhos, and two are dimensionless, these four parameters are called *hybrid parameters* or *h-parameters.*

The four h-parameters are (1) h_i—input impedance, ohms; (2) h_o—output admittance, mhos; (3) h_f—forward-current transfer ratio, dimensionless; (4) h_r—reverse-voltage transfer ratio, dimensionless. A second subscript is added to indicate which electrode is common. For example, h_{FB} indicates the static or d-c forward-current transfer ratio of a transistor operated with a common base,

Table 4-2 Common h-parameters

PARAMETER FOR	COMMON BASE	COMMON EMITTER	COMMON COLLECTOR
Output admittance	h_{ob}	h_{oe}	h_{oc}
Input resistance	h_{ib}	h_{ie}	h_{ic}
Forward-current transfer ratio	h_{fb}	h_{fe}	h_{fc}
Reverse-voltage transfer ratio	h_{rb}	h_{re}	h_{rc}

and h_{fe} indicates the small-signal forward-current transfer ratio of a transistor operated with a common emitter. Table 4-2 illustrates the small-signal h-parameters for the three modes of operation.

Other Transistor Parameters. In order to meet the needs of various types of circuit conditions and circuit analysis, other forms of parameters are available, such as

r-parameters—for resistance measurements with an open-circuit condition and at low frequencies

z-parameters—for impedance measurements with an open-circuit condition and at high frequencies

g-parameters—conductance measurements with a short-circuit condition and at low frequencies

Y-parameters—admittance measurements with a short-circuit condition and at high frequencies

Equations for converting from r-parameters to h-parameters, and other combinations, are available in some engineering-level texts.

4-9 Transistor Specifications

Although many properties of a transistor could be specified, manufacturers list only some of them. The properties listed vary with (1) the manufacturer, (2) the type of transistor, and (3) the application for which the transistor is designed. The values are listed under three general headings: (1) general data, (2) maximum ratings, and (3) typical operation. Under general data are listed (1) application or class of service, (2) type, (3) outline dimensions, and (4) lead arrangement. The maximum ratings are the direct voltage and current values that must not be exceeded in the operation of the unit. Maximum ratings usually include the direct (1) collector-to-base voltage, (2) emitter-to-base voltage, (3) collector current, (4) emitter current, and (5) collector power dissipation. The typical operating values are presented as a guide, since the values vary widely, as they are dependent on the operating voltages and which electrode is used as the common circuit element. The values listed may include the (1) common electrode, (2) collector-emitter volts, (3) collector current, (4) current-transfer ratio, (5) input resistance, (6) load resistance, (7) power gain, (8) noise factor, and (9) alpha cutoff frequency. Some manufacturers also list for the common-emitter circuit the small-signal (1) hybrid-π parameters, (2) h-parameters, and (3) T-parameters.

4-10 Transistor Characteristic Curves

The major operating characteristics of a transistor are used to identify the electrical features and operating values of the transistor. These values may be listed in tabular form or plotted on graph paper to form a curve. There is a wide variety of types of curves, each showing the relation between a specific set of parameters. The two families of curves most commonly used are (1) collector current versus collector voltage for constant values of emitter current (Fig. 4-9), and (2) collector current versus collector voltage for constant values of base current (Fig. 4-10). The curves of Fig. 4-9 are used for common-base connected transistors, and those of Fig. 4-10 for common-emitter connected transistors. The collector current in a common-base connected transistor is relatively independent of collector voltage and is primarily dependent on the emitter current. When the transistor is connected for common-emitter operation, the collector voltage has very little effect on the collector current for low values of base current. However, as the base current increases, the effect of the collector voltage on the collector current also increases.

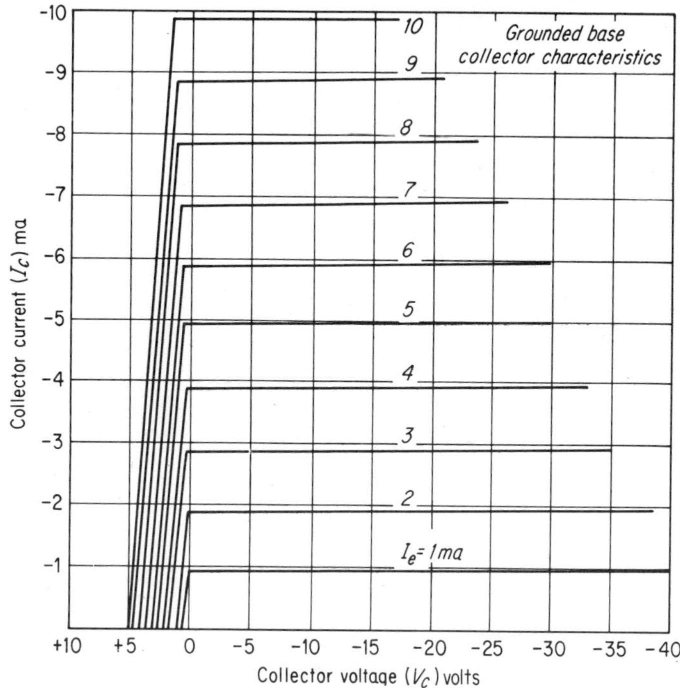

Fig. 4-9 Static common-base collector characteristics of a PNP triode junction transistor.

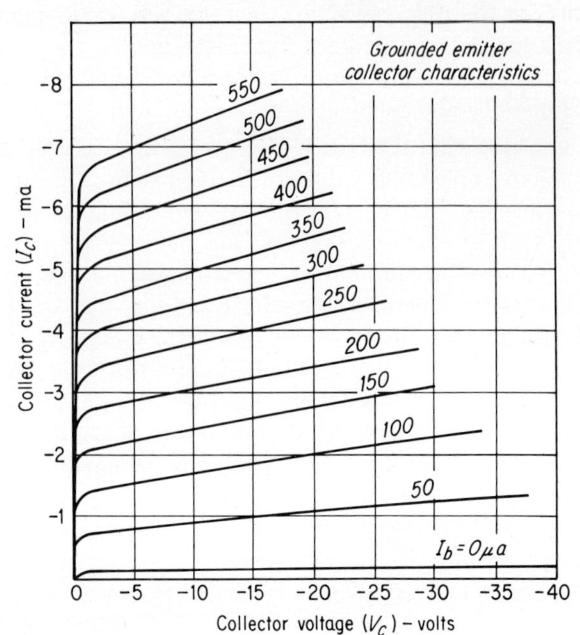

Fig. 4-10 Static common-emitter collector characteristics of a PNP triode junction transistor.

4-11 Current Gain

Forward-current Transfer Ratio. One of the most important characteristics of the transistor is its *current gain* or *forward-current transfer ratio.* The current gain of a transistor is basically the ratio of the current in its output circuit to the current in its input circuit, or

$$A_i = \frac{I_o}{I_i} \tag{4-1}$$

where A_i = current gain (or amplification), a dimensionless ratio
I_o = current in the output circuit
I_i = current in the input circuit

The current gain for a given transistor will depend on whether it is being operated in the common-base, common-emitter, or common-collector configuration. Also, the current gain may be different for the static or zero-signal condition than with a varying-signal input.

The current gain when expressed in its h-parameter is called the *forward-current transfer ratio* and is defined as the ratio of the a-c variation in the output current to the a-c variation in the input current, with a specific control voltage kept constant.

Current Distribution in a Transistor. It is important to first understand the relation among the relative amounts of current in the base, emitter, and collector of a transistor. The common-base transistor circuit shown in Fig. 4-11 is used here to illustrate the current distribution in the transistor electrodes. Because the power source V_{EE} applies a forward bias to the emitter-base junction, a current I_E flows from V_{EE} to the emitter of the transistor. The electrons flowing through the N-type emitter are majority carriers, and upon crossing the junction to the P-type base become minority carriers (Fig. 4-11*b*). In the base, a small number of electrons unite with holes in the P-type material, flow through the base lead, and return to the power source V_{EE}; this current I_B is in the order of 2 to 5 per cent of the emitter current and is called the *recombination current.* The remainder of the emitter current diffuses across the base-collector junction and becomes the major portion of the collector current I_C which flows through V_{CC} and returns to the power source V_{EE}. However, the power source V_{CC} reverse-biases the base-collector junction, which causes a small amount of current to flow around the circuit formed by V_{CC}, the base, and the collector. This current is called the *saturation current* (also *leakage current*) and is represented by the symbol I_{CO} or I_{CBO}; its value is less than 1 per cent of the collector current. The effect of the saturation current is to cause a very small increase in the collector current and to reduce the base current by a small amount.

4-12 Current Characteristics of the Common-base Configuration

Current Gain. For the common-base configuration, the forward-current transfer ratio is expressed as

$$\alpha = h_{FB} = \frac{\text{change in collector current}}{\text{change in emitter current}} = \frac{dI_C}{dI_E} \qquad (V_{CB}\text{—constant}) \qquad (4\text{-}2)$$

This parameter is called the *alpha direct-current gain* and is represented by the symbol α. The value of α for junction transistors is always less than 1, as the collector current can never be higher than the emitter current; its value ranges from 0.94 to 0.99. The value of h_{FB} can be calculated from the transistor characteristic curves, such as Fig. 4-9.

Example 4-1 Determine the current gain for the transistor represented by the curves of Fig. 4-9 when used in the common-base configuration. Use -10 volts for V_C, and 7 and 5 ma for emitter currents.

GIVEN: Fig. 4-9 $\quad V_C = -10$ volts $\quad I_{e.1} = 7$ ma $\quad I_{e.2} = 5$ ma

FIND: h_{FB}

SOLUTION: From Fig. 4-9 when $V_C = -10$ volts and $I_e = 7$ ma, then $I_C = 6.9$ ma; and when $V_C = -10$ volts and $I_e = 5$ ma, then $I_C = 4.95$ ma.

$$h_{FB} = \frac{dI_C}{dI_e} = \frac{6.9 - 4.95}{7 - 5} = 0.975$$

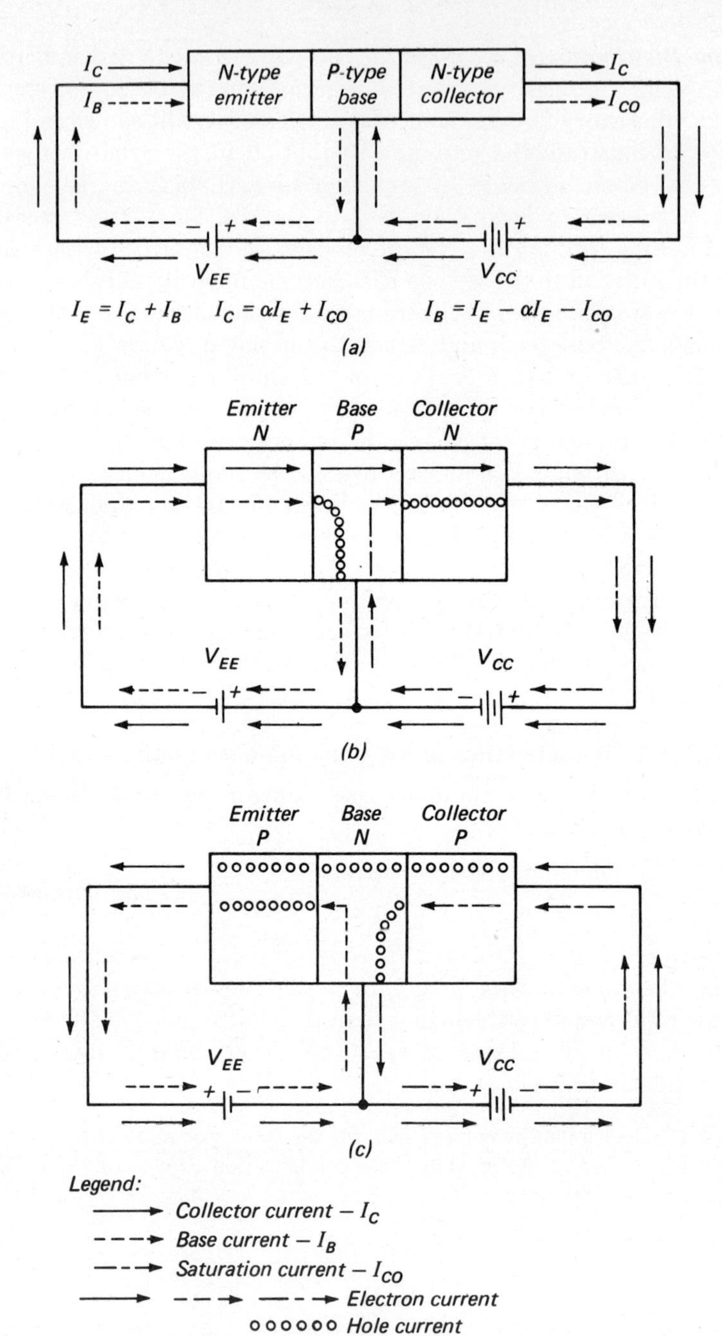

Fig. 4-11 Current distribution in a transistor with a common-base configuration. (a) External circuit currents for an NPN transistor. (b) Internal currents in an NPN transistor. (c) Internal currents in a PNP transistor.

Current Distribution. Figure 4-11 illustrates the current distribution in a transistor circuit using the common-base configuration. The relations among the currents are expressed by

$$I_C = \alpha I_E + I_{CO} \tag{4-3}$$

$$I_B = I_E - \alpha I_E - I_{CO} \tag{4-4}$$

where I_B = base current
I_C = collector current
I_E = emitter current
I_{CO} = saturation current
α = a transistor constant

Example 4-2 A transistor used in the common-base configuration has an α value of 0.975 and saturation current of 10 μa. If the emitter current is 2 ma, what are: (*a*) the apparent collector and base currents if the saturation current is ignored? (*b*) The collector and base currents when the saturation current is taken into account?

GIVEN: $\alpha = 0.975$ $I_E = 2$ ma $I_{CO} = 10$ μa

FIND: (*a*) Approximate values of I_C and I_B (*b*) I_C and I_B

SOLUTION:

(*a*) $I_C \cong \alpha I_E \cong 0.975 \times 2 \cong 1.95$ ma
$I_B \cong I_E - I_C \cong 2 - 1.95 \cong 0.05$ ma, or 50 μa

(*b*) $I_C = \alpha I_E + I_{CO} = (0.975 \times 2) + 0.01 = 1.96$ ma
$I_B = I_E - \alpha I_E - I_{CO} = 2 - (0.975 \times 2) - 0.01 = 0.04$ ma, or 40 μa

4-13 Current Characteristics of the Common-emitter Configuration

Current Gain. When a transistor is used in the common-emitter configuration, its current gain is the ratio of the change in collector current to the change in base current for a constant value of collector voltage. When the current gain is specified for static or relatively high values of collector and base currents, it is called the *beta direct-current gain*. It is represented by the symbol β or the *h*-parameter symbol h_{FE} and is expressed as

$$\beta = h_{FE} = \frac{\text{change in collector current}}{\text{change in base current}} = \frac{dI_C}{dI_B} \qquad (V_{CE}\text{—constant}) \tag{4-5}$$

The value of β is always greater than unity and ranges from about ten to several hundred. The value of β can be calculated from the data available on curves such as Fig. 4-10.

Example 4-3 Determine the current gain for the transistor represented by the curves of Fig. 4-10 when used in the common-emitter configuration. Use -7 volts for V_{CE}, and 50 and 100 μa for the base currents.

GIVEN: Fig. 4-10 $V_{CE} = -7$ volts $I_{B1} = 50$ μa $I_{B2} = 100$ μa

FIND: β

SOLUTION: From Fig. 4-10 when $V_C = -7$ volts and $I_B = 50$ μa, then $I_C = 0.8$ ma; and when $V_C = -7$ volts and $I_B = 100$ μa, then $I_C = 1.6$ ma.

$$\beta = \frac{dI_C}{dI_B} = \frac{(1.6 - 0.8) \text{ ma}}{(100 - 50)\ \mu\text{a}} = \frac{0.8 \times 10^{-3} \text{ amp}}{50 \times 10^{-6} \text{ amp}} = 16$$

For the common-emitter configuration, the realtion between β and α is

$$\beta = \frac{\alpha}{1 - \alpha} \tag{4-6}$$

also

$$\alpha = \frac{\beta}{1 + \beta} \tag{4-7}$$

Example 4-4 A junction transistor operated in the common-emitter configuration is biased so that α is 0.941. What is β for these conditions?

GIVEN: $\alpha = 0.941$

FIND: β

SOLUTION:

$$\beta = \frac{\alpha}{1 - \alpha} = \frac{0.941}{1 - 0.941} \cong 16$$

Example 4-5 What is the alpha value of a transistor if its value of beta is 50?

GIVEN: $\beta = 50$

FIND: α

SOLUTION:

$$\alpha = \frac{\beta}{1 + \beta} = \frac{50}{1 + 50} = \frac{50}{51} = 0.980$$

Current Distribution. Figure 4-12 illustrates the current distribution in a transistor circuit using the common-emitter configuration. It should be observed that the collector-to-base leakage current I_{CO} flows in the base-collector circuit and consequently the effect of I_{CO} will be increased by the beta current gain of the transistor. The relations among the currents are then expressed by

$$I_C = \alpha I_E + I_{CO} + \beta I_{CO} \tag{4-8}$$
$$I_B = I_E - \alpha I_E - I_{CO} - \beta I_{CO} \tag{4-9}$$

Example 4-6 A transistor used in the common-emitter configuration has an alpha of 0.98 and a beta of 50, and its leakage current is 10 μa. If its emitter current is 2 ma, determine (*a*) the collector current, (*b*) the base current.

GIVEN: $\alpha = 0.98$ $\quad \beta = 50$ $\quad I_{CO} = 10\ \mu$a $\quad I_E = 2$ ma

FIND: (*a*) I_C $\quad$ (*b*) I_B

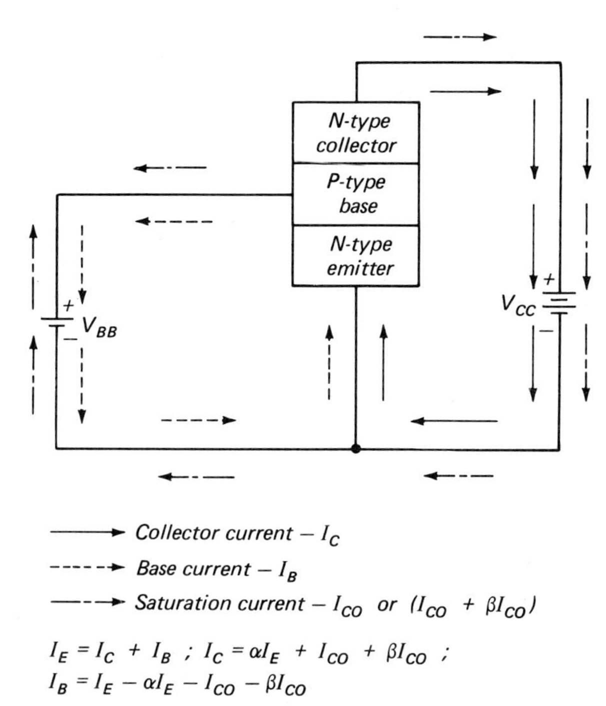

Fig. 4-12 Current distribution in an NPN transistor in the common-emitter configuration.

SOLUTION:

(a) $I_C = \alpha I_E + I_{CO} + \beta I_{CO} = (0.98 \times 2) + 0.01 + (50 \times 0.01) = 2.47$ ma

(b) $I_B = I_E - \alpha I_E - I_{CO} - \beta I_{CO}$

$= 2 - (0.98 \times 2) - 0.01 - (50 \times 0.01) = -0.47$ ma

When the value of β is 50 or more, the effect of the current βI_{CO} can cause instability in the operation of the transistor circuit; this effect is called *thermal runaway*. Control of the stability of the circuit is obtained by appropriate biasing procedures as presented in Chap. 8.

4-14 Current Characteristics of the Common-collector Configuration

Current Gain. When a transistor is used in the common-collector configuration, its current gain is expressed by

$$h_{FC} = \frac{\text{change in emitter current}}{\text{change in base current}} = \frac{dI_E}{dI_B} \qquad (V_{EC}\text{—constant}) \qquad (4\text{-}10)$$

The curves for the transistor in the common-collector configuration are practically the same as for the common-emitter configuration, as the emitter current is only slightly higher than the collector current in a transistor.

Example 4-7 Determine the forward-current transfer ratio for a transistor in the common-collector configuration by assuming that the curves of Fig. 4-10 are satisfactory for this purpose. Use −15 volts for V_{CE} and a base current range of 50 to 150 μa.

GIVEN: Fig. 4-10 $V_{CE} = -15$ volts $I_{B1} = 50\ \mu a$ $I_{B2} = 150\ \mu a$

FIND: h_{FC}

SOLUTION: From Fig. 4-10 it is found that when E_{CE} is kept at −15 volts a change in base current from 50 to 150 μa will be accompanied by a change in emitter current from 1 to 2.65 ma. Therefore,

$$h_{FC} = \frac{dI_E}{dI_B} = \frac{(2.65 - 1)\text{ ma}}{(150 - 50)\ \mu\text{a}} = \frac{1.65 \times 10^{-3}\text{ amp}}{100 \times 10^{-6}\text{ amp}} = 16.5$$

In general, the forward-current transfer ratio for the common-collector configuration will be slightly higher than for the common-emitter configuration.

Current Distribution. Figure 4-13 illustrates the current distribution in a transistor circuit using the common-collector configuration. The relations among the currents are expressed by

$$I_C = \alpha I_E + I_{CO} \tag{4-11}$$
$$I_B = I_E - \alpha I_E - I_{CO} \tag{4-12}$$

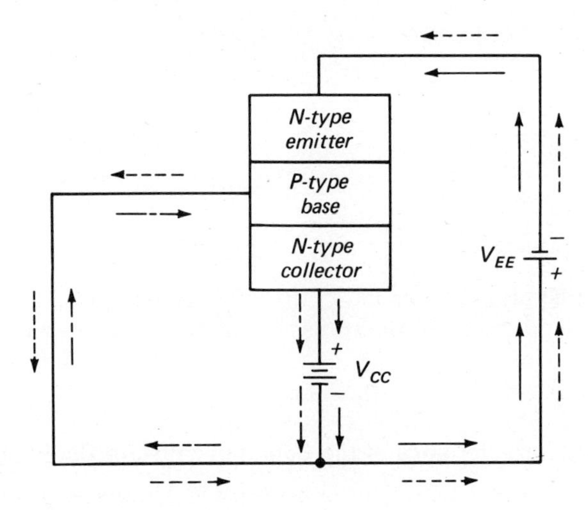

Fig. 4-13 Current distribution in an NPN transistor in the common-collector configuration.

Example 4-8 What are the collector and base currents for a transistor used in the common-collector configuration if beta is 16.5, the emitter current is 1.8 ma, and the saturation current is 12 μa?

GIVEN: $\beta = 16.5 \qquad I_E = 1.8$ ma $\qquad I_{CO} = 12\ \mu$a

FIND: $I_C \qquad I_B$

SOLUTION:

$$\alpha = \frac{\beta}{1+\beta} + \frac{16.5}{1+16.5} = \frac{16.5}{17.5} = 0.943$$

$$I_C = \alpha I_E + I_{CO} = (0.943 \times 1.8) + 0.012 \cong 1.70 + 0.012 \cong 1.71 \text{ ma}$$
$$I_B = I_E - \alpha I_E - I_{CO} = 1.8 - (0.943 \times 1.8) - 0.012 \cong 1.80 - 1.70 + 0.012$$
$$\cong 0.088 \text{ ma, or } 88\ \mu\text{a}$$

4-15 Voltage Gain

The direct-voltage gain in a transistor is equal to the product of the current gain and the resistance gain, or

$$A_v = A_i \times A_r \tag{4-13}$$

where A_v = voltage gain (or amplification), a dimensionless ratio
A_r = resistance gain (ratio of output to input resistance)

High voltage gains are obtained with the common-base and common-emitter configurations, largely because of the high gain in resistance. With the common-collector configuration, the voltage gain is approximately 1.

Example 4-9 What is the voltage gain of a transistor used in the common-base configuration if: α = 0.96, the collector-base resistance = 750,000 ohms, and the emitter-base resistance = 500 ohms?

GIVEN: $A_i = 0.96 \qquad R_o = 750{,}000$ ohms $\qquad R_i = 500$ ohms

FIND: A_v

SOLUTION:

$$A_v = A_i \times A_r = 0.96 \times \frac{750{,}000}{500} = 1{,}440$$

Example 4-10 What is the voltage gain of a transistor used in the common-emitter configuration if: β = 25, collector-emitter resistance = 150,000 ohms, emitter-base resistance = 2,500 ohms?

GIVEN: $A_i = 25 \qquad R_o = 150{,}000$ ohms $\qquad R_i = 2{,}500$ ohms

FIND: A_v

SOLUTION:

$$A_v = A_i \times A_r = 25 \times \frac{150{,}000}{2{,}500} = 1{,}500$$

Example 4-11 What is the voltage gain of a transistor used in the common-collector configuration if: $\beta = 25$, input resistance = 275,000 ohms, and output resistance = 10,000 ohms?

GIVEN: $\beta = 25$ $\quad R_o = 10{,}000$ ohms $\quad R_i = 275{,}000$ ohms

FIND: A_v

SOLUTION:

$$A_v = A_i \times A_r = 25 \times \frac{10{,}000}{275{,}000} = 0.909$$

The high voltage gains indicated in Examples 4-9 and 4-10 can be obtained only if the transistor is operated into a very high impedance circuit. In practical amplifier circuits, the voltage gain is much lower because of the lower impedances that must be used to satisfy the impedance-matching requirements.

4-16 Power Gain

The high ratio of output resistance to input resistance combined with the current gain of a transistor produces a direct power gain of

$$A_p = A_i^2 \times A_r \tag{4-14}$$

where A_p = power gain (or amplification), a dimensionless ratio

The power gain of a transistor is more frequently expressed in decibels (Art. 1-6), and accordingly

$$G_p = 10 \log A_p \tag{4-15}$$

where G_p = power gain, decibels

Example 4-12 What is the power gain in decibels of a transistor used in the common-base configuration if: $\alpha = 0.96$, collector-base resistance = 750,000 ohms, and emitter-base resistance = 500 ohms?

GIVEN: $\alpha = 0.96$ $\quad R_o = 750{,}000$ ohms $\quad R_i = 500$ ohms

FIND: G_p

SOLUTION:

$$A_p = A_i^2 A_r = 0.96^2 \times \frac{750{,}000}{500} = 1{,}382$$

$$G_p = 10 \log A_p = 10 \log 1{,}382 = 10 \times 3.14 = 31.4 \text{ db}$$

Example 4-13 What is the decibel power gain of a transistor used in the common-emitter configuration if: $\beta = 25$, collector-emitter resistance = 150,000 ohms, and emitter-base resistance = 2,500 ohms?

GIVEN: $\beta = 25$ $\quad R_o = 150{,}000$ ohms $\quad R_i = 2{,}500$ ohms

FIND: G_p

SOLUTION:

$$A_p = A_i^2 A_r = 25^2 \times \frac{150{,}000}{2{,}500} = 37{,}500$$

$$G_p = 10 \log A_p = 10 \log 37{,}500 = 10 \times 4.574 = 45.74 \text{ db}$$

Table 4-3 Power Gain of a Transistor

COMMON ELECTRODE	INPUT RESISTANCE, OHMS	LOAD RESISTANCE, OHMS	POWER GAIN, db
Emitter.	1,980	100,000	44.1
Emitter.	2,670	2,670	34.5
Collector.	500,000	10,000	17
Base.	215	500,000	32.5

Example 4-14 What is the decibel power gain of a transistor used in the common-collector configuration if: $\beta = 25$, input resistance = 275,000 ohms, and output resistance = 10,000 ohms?

GIVEN: $\beta = 25$ $R_o = 10{,}000$ ohms $R_i = 275{,}000$ ohms

FIND: G_p

SOLUTION:

$$A_p = A_i{}^2 A_r = 25^2 \times \frac{10{,}000}{275{,}000} = 22.7$$

$$G_p = 10 \log A_p = 10 \log 22.7 = 10 \times 1.356 = 13.56 \text{ db}$$

The high power gain indicated in Examples 4-12 and 4-13 can be obtained only if the transistor is operated into a very high impedance circuit. In practical amplifier circuits, the power gain is much lower because of the lower impedances that must be used to satisfy the impedance-matching requirements.

Since the power gain of a transistor is dependent on the (1) configuration used, (2) input resistance, and (3) load resistance, these values are usually listed in the characteristic chart for the power gain specified. The variation of the power gain of a transistor for the different configurations and the input and output resistances are listed in Table 4-3. This table shows that (1) the greatest power gain is obtained by using the common-emitter configuration and a high load resistance, (2) the least power gain is obtained with the common-collector configuration, and (3) a low load resistance with the common-emitter configuration produces a power gain less than that obtained with a high load resistance but greater than that obtained with either the common-base or common-collector configurations. These conditions are typical for all transistors.

4-17 Graphical Analysis of the Operation of a Common-emitter Transistor Amplifier

Advantages. The operating characteristics of a transistor amplifier can also be determined by use of the characteristic curves of the transistor. With this method, it is possible to (1) determine the current, voltage, and power gains

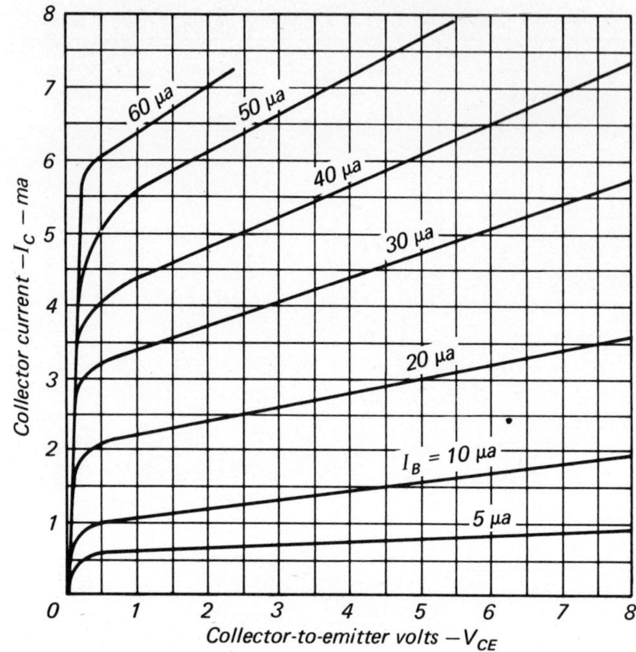

Fig. 4-14 Typical static characteristic curves for the common-emitter transistor configuration.

of the amplifier circuit, and (2) graphically show the relationship between the input and output signals.

Output Static Characteristic Curves. Figure 4-14 shows a *family* of output static characteristic curves for a transistor used in the common-emitter configuration. These curves show the variations in collector current (plotted vertically) for varying amounts of collector-emitter voltage (plotted horizontally) with constant amounts of base current as the running parameter. Data for plotting these curves are listed in Table 4-4 and were obtained by (1) adjusting the base current to a desired value and maintaining this current constant, (2) adjusting the collector-emitter voltage to a series of progressively higher amounts, and (3) reading and recording the value of the collector current for each value of collector-emitter voltage. This process was repeated for a series of values of base current ranging from 5 to 60 μa.

The static characteristic curves can be used to determine the current, voltage, and power gains of the amplifier circuit in the following manner:

1. Plot a load line on the curves to represent the specific voltage and resistance values of the circuit.

2. Locate the operating base current q_1, called the *quiescent value,* and the peak base currents q_2 and q_3 on the characteristics curves.

Table 4-4 Laboratory Data for Plotting the Output Characteristic Curves of a Transistor Connected in the Common-emitter Configuration

I_B, μa	5	10	20	30	40	50	60
E_{CE}, VOLTS	COLLECTOR CURRENT, ma						
0.5	0.55	1	2.15	3.2	4.1	5.2	6.1
1	0.55	1.1	2.25	3.4	4.4	5.6	6.4
2	0.6	1.2	2.45	3.75	4.8	6.15	7
3	0.65	1.3	2.6	4.1	5.3	6.65	7.7
4	0.7	1.4	2.8	4.4	5.7	7.15	
5	0.75	1.55	3.05	4.8	6.15	7.65	
6	0.8	1.7	3.25	5.1	6.5		
7	0.85	1.8	3.45	5.4	6.95		
8	0.9	1.9	3.65	5.8	7.4		

3. Determine the highest and the lowest values of collector current produced by the values of base currents at q_2 and q_3.
4. Calculate the current gain of the amplifier.
5. Determine the variation in the collector-emitter voltage corresponding to points q_2 and q_3.
6. Calculate the variation in the input voltage from the given input-signal current variation and the input circuit resistance.
7. Calculate the voltage gain of the amplifier.
8. Calculate the power gain of the amplifier.

Plotting the Load Line. The *load line* is a line added to the characteristic curves that defines the variations in currents and voltage for (1) a specific value of load resistance R_L, and (2) collector-to-emitter voltage. Figure 4-16 shows a load line X-Y added to the characteristic curves for the transistor represented by Fig. 4-14 for the circuit values shown on Fig. 4-15. For clearness of presentation, Fig. 4-16 includes only those curves of Fig. 4-14 that apply to the circuit of Fig. 4-15.

The load line is a straight line and therefore can be drawn if any two points are available. Two points that can easily be determined are (1) *collector-current cutoff*, namely the point where the collector current is zero, and (2) *collector-current saturation*, namely the point where the collector current is at its maximum value. Because at collector-current saturation the collector-emitter voltage will be zero, this point can also be called the *collector-emitter voltage cutoff*. The collector current will be at its maximum value when all the voltage appears across the resistor R_L in the collector circuit. Therefore, the saturation current will be equal to $V_{EE} + V_{CC}$ divided by R_L, hence 7.5 divided by 1,000, or 7.5 ma; this point is indicated on Fig. 4-16 as Y. Connecting points X and Y with a straight line completes the load line.

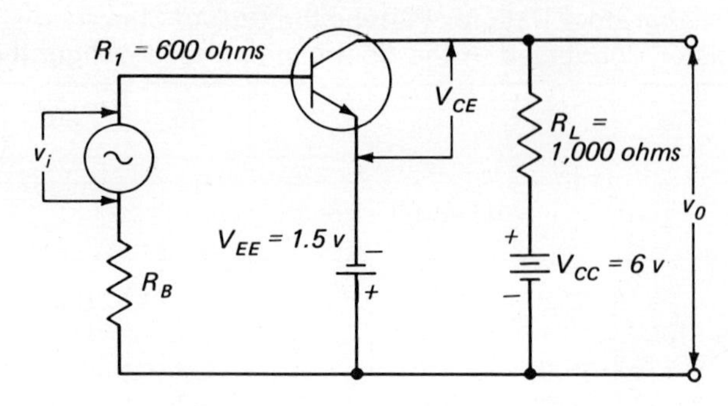

Input signal — i_B = 20 μa (peak-to-peak)
Quiescent point — I_B = 20 μa

Fig. 4-15 Typical common-emitter transistor amplifier circuit.

Locating the Operating Points. The given data for the amplifier (Fig. 4-15) indicate that the operating point about which the input-signal current (base current) varies is 20 μa. Thus the quiescent point q_1 is located at the intersection of the load line and the curve for the condition of $I_B = 20$ μa.

The given data also state that the peak-to-peak value of the input-signal current is 20 μa. Consequently, the base current will vary from 10 μa above the operating point q_1 to 10 μa below the operating point. The base current will therefore vary between (20 + 10) μa and (20 − 10) μa, or between 30 and 10 μa. On Fig. 4-16, the intersection of the load line and the 30-μa curve is labeled q_2, and the intersection of the load line and the 10-μa curve is labeled q_3.

Collector-current Variations. The collector currents corresponding to points q_1, q_2, and q_3 are obtained from the curves by projecting horizontal lines through points q_1, q_2, and q_3 and reading their values on the ordinate scale. For Fig. 4-16, the collector current for operating point q_1 is 2.95 ma, for q_2 it is 4.2 ma, and for q_3 it is 1.7 ma.

Current Gain. The current gain of the amplifier is the ratio of the output-current change to the input-current change, or

$$A_i = \frac{di_C}{di_B} \tag{4-16}$$

where A_i = current gain of the amplifier, a dimensionless ratio
di_C = peak-to-peak variation of the collector current, amp
di_B = peak-to-peak variation of the base current, amp

Note: i_C and i_B may also be in ma or μa, but both must be in the same kind of units.

Example 4-15 What is the current gain of the amplifier circuit represented by Figs. 4-15 and 4-16?

GIVEN: $di_B = (30 - 10)$ μa $\quad di_C = (4.2 - 1.7)$ ma

FIND: A_i

SOLUTION:

$$A_i = \frac{di_C}{di_B} = \frac{(4.2 - 1.7)10^{-3}}{(30 - 10)10^{-6}} = \frac{2.5}{20} \times 10^3 = 125$$

Collector-to-Emitter Voltage Variations. The collector-emitter voltages corresponding to points q_1, q_2, and q_3 are obtained from the curves by projecting vertical lines through these points and reading their values on the abscissa scale. For Fig. 4-16, the collector-emitter voltage for operating point q_1 is 4.55 volts, for q_2 it is 3.3 volts, and for q_3 it is 5.8 volts.

Input-voltage Variation. The variation in the input-signal voltage can be calculated by use of Ohm's law since the current variation and the resistance of the input circuit are known.

$$dv_i = di_B R_i \tag{4-17}$$

where dv_i = variation in the input voltage, volts
di_B = variation in input current, amp
R_i = resistance of the input circuit, ohms

Example 4-16 What is the peak-to-peak variation of the input voltage of the amplifier circuit represented by Figs. 4-15 and 4-16?

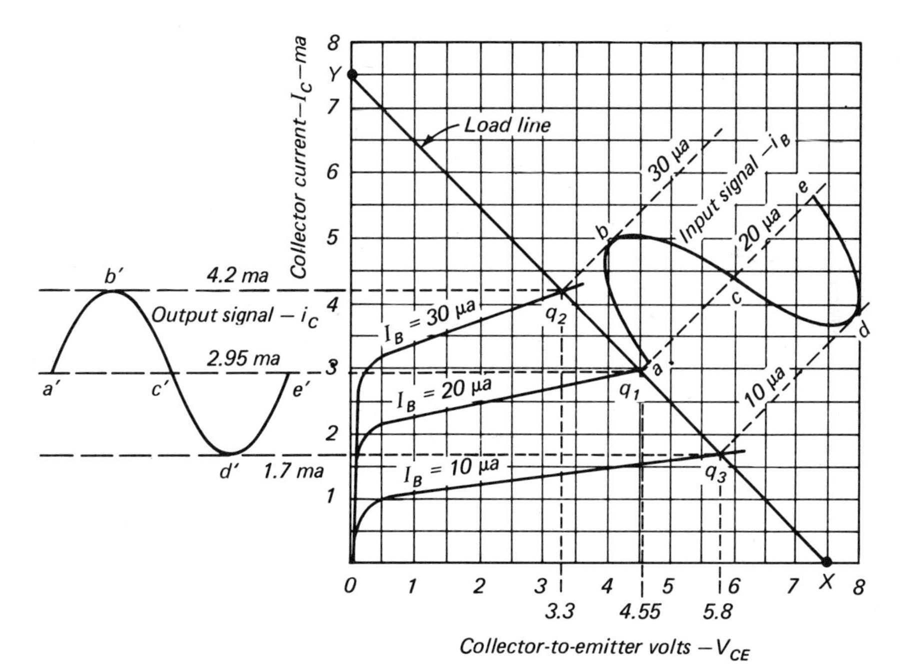

Fig. 4-16 Dynamic operating characteristics of the transistor amplifier circuit of Fig. 4-15.

GIVEN: $di_B = 20\ \mu a \qquad R_i = 600$ ohms

FIND: dv_i

SOLUTION:

$$dv_i = di_B R_i = 20 \times 10^{-6} \times 600 = 0.012 \text{ volt}$$

Voltage Gain. The voltage gain of the amplifier is the ratio of the output-voltage variation to the input-voltage variation, or

$$A_v = \frac{dv_o}{dv_i} = \frac{dv_{CE}}{dv_i} \tag{4-18}$$

where A_v = voltage gain of the amplifier, a dimensionless ratio
dv_{CE} = variation in the collector-to-emitter (or output) voltage, volts

Example 4-17 What is the voltage gain of the amplifier circuit represented by Figs. 4-15 and 4-16?

GIVEN: $dv_{CE} = (5.8 - 3.3)$ volts; from Fig. 4-16 $\qquad dv_i = 0.012$ volt; from Example 4-16

FIND: A_v

SOLUTION:

$$A_v = \frac{dv_{CE}}{dv_i} = \frac{5.8 - 3.3}{0.012} = \frac{2.5}{0.012} = 208$$

Power Gain. The power gain can be calculated from the current gain and voltage gain that have already been determined.

$$A_p = A_v A_i \tag{4-19}$$

where A_p = power gain of the amplifier, a dimensionless product

Example 4-18 What is the power gain of the amplifier circuit represented by Figs. 4-15 and 4-16 expressed (*a*) as a product, (*b*) in decibels?

GIVEN: $A_v = 208$; from Example 4-17 $\qquad A_i = 125$; from Example 4-15

FIND: (*a*) A_p $\qquad$ (*b*) G_p

SOLUTION:

(*a*) $\qquad A_p = A_v A_i = 208 \times 125 = 26{,}000$

(*b*) $\qquad G_p = 10 \log A_p = 10 \log 26{,}000 = 10 \times 4.415 = 44.15$ db

Graphical Representation of the Input and Output Signals. Figure 4-16 contains representations of the input-signal variations i_B and the corresponding output-signal variations i_C. (It is assumed that the input signal has a sine waveform.) The input signal is constructed by extending lines perpendicular to the load line from points q_1, q_2, and q_3. The input sine wave thus has a peak-to-peak value extending from 10 to 30 μa. The output signal is constructed by extending horizontal lines outward from points q_1, q_2, and q_3. The output waveform also will be a sine wave and its peak-to-peak values extend from 1.7 to 4.2 ma.

4-18 Transconductance

One of the specifications sometimes given for a transistor used in the common-emitter configuration is the *extrinsic transconductance*. This parameter can be determined from suitable characteristic curves as it expresses the ratio of a small change in the collector current to the change in emitter-base voltage producing it, with the collector-emitter voltage kept constant, or

$$g_m = \frac{di_c}{dv_{eb}} \qquad (v_{ce}\text{—constant}) \tag{4-20}$$

where g_m = small signal transconductance, mhos
di_c = change in collector current, amp
dv_{eb} = change in emitter-base voltage, volts

Example 4-19 What is the extrinsic transconductance of a transistor used in the common-emitter configuration if a 0.1-volt change in the emitter-base voltage causes a 2.5-ma change in the collector current while the collector-emitter voltage is kept constant?

GIVEN: $dv_{eb} = 0.1$ volt $\quad di_c = 2.5$ ma

FIND: g_m

SOLUTION:

$$g_m = \frac{di_c}{dv_{eb}} = \frac{2.5 \times 10^{-3}}{0.1} = 25 \times 10^{-3} \text{ mho} = 0.025 \text{ mho, or } 25 \text{ mmhos}$$

In terms of h-parameters, the transconductance of a transistor is

$$g_m = \frac{\beta}{h_{ie}} \tag{4-21}$$

where h_{ie} = resistance of the forward-biased emitter-base junction, ohms

Example 4-20 What is the transconductance of a transistor that has a β of 200 and a small-signal input impedance of 600 ohms?

GIVEN: $\beta = 200 \quad h_{ie} = 600$ ohms

FIND: g_m

SOLUTION:

$$g_m = \frac{\beta}{h_{ie}} = \frac{200}{600} = 0.333 \text{ mho}$$

4-19 Gain-bandwidth Product

One of the specifications sometimes given for a transistor used in the common-emitter configuration is the *gain-bandwidth product,* also called the *figure of merit;* its symbol is f_T and its unit is the hertz, with ratings usually in megahertz. The gain-bandwidth product is the frequency at which β for a common-emitter transistor is equal to 1. Figure 4-17 is a typical curve showing how the value of β is affected by the operating frequency. This curve shows that when β is equal to 1, the corresponding frequency is 7.5 MHz; therefore, the gain-bandwidth product of the transistor represented by this curve is 7.5 MHz. Knowing the value of the gain-bandwidth product is sometimes helpful in selecting the transistor to be used and the configuration that will most closely provide the desired circuit characteristics. The higher values of f_T indicate (1) a greater bandwidth capability of the transistor circuit, and (2) a faster rise time for the circuit.

The gain-bandwidth product is expressed mathematically by

$$f_T = \frac{g_m}{2\pi C_{be}} \tag{4-22}$$

where f_T = gain-bandwidth product, hertz
g_m = extrinsic transconductance, mho
C_{be} = input (base-emitter) capacitance, farads

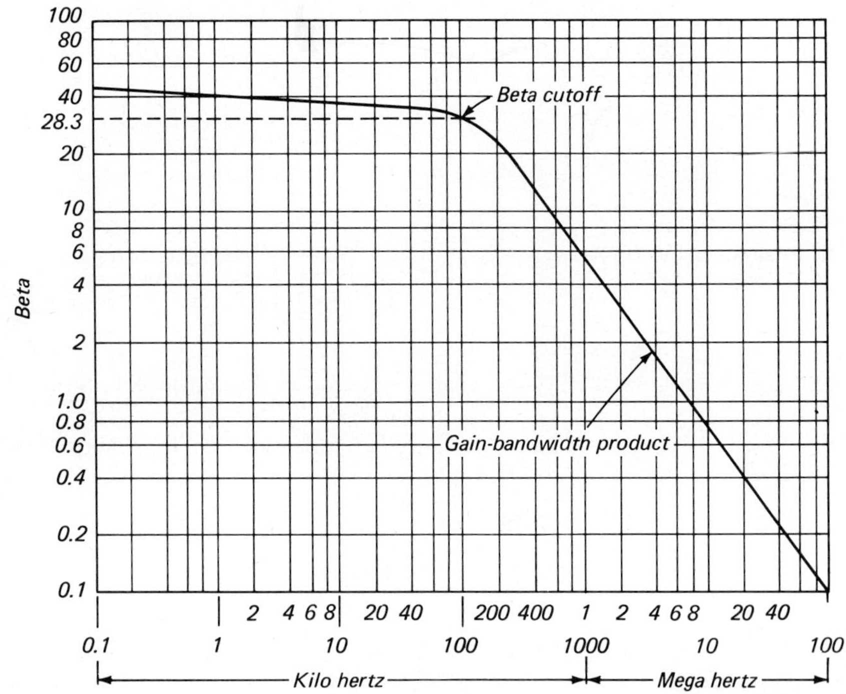

Fig. 4-17 Transistor characteristic curve showing beta as a function of frequency.

Example 4-21 What is the gain-bandwidth product of an audio-frequency transistor that has a transconductance of 8,000 μmhos and whose input capacitance is 50 pf?

GIVEN: $g_m = 8{,}000\ \mu\text{mhos} \qquad C_{be} = 50\ \text{pf}$

FIND: f_T

SOLUTION:

$$f_T = \frac{g_m}{2\pi C_{be}} = \frac{8{,}000 \times 10^{-6}}{6.28 \times 50 \times 10^{-12}} = \frac{8{,}000 \times 10^{6}}{314} = 25.5\ \text{MHz}$$

4-20 Frequency Cutoff

The upper frequency limit of a transistor is determined by its transit time, which is the time taken by the electrons or holes to pass from the emitter to the collector in a common-base configuration. Transistors are rated according to their (1) alpha cutoff frequency or (2) beta cutoff frequency depending on the circuit configuration being used. The *alpha cutoff frequency* is that value of frequency at which the current gain has decreased to 0.707 of its low-frequency reference value, usually 1,000 hertz. The *beta cutoff frequency* is that value of frequency at which the value of β is 0.707 times its low-frequency reference value, usually 1,000 hertz.

The curve of Fig. 4-17 shows that at 1 kHz the value of β is 40 and that at 0.707×40, or 28.3, the corresponding frequency value is 100 kHz; thus, the cutoff frequency for the transistor represented by this curve is 100 kHz.

Example 4-22 It is stated that the operating frequency in the various stages of a radio receiver should be approximately 20 per cent of the cutoff frequency of the transistor in its circuit. What is the minimum cutoff frequency required of the transistor in the audio section if the a-f range of the receiver is 50 to 20,000 hertz?

GIVEN: $f_{a\cdot\max} = 20\ \text{kHz} \qquad \text{Proportion} = 20$ per cent, or 0.2

FIND: f_c

SOLUTION:

$$f_c = \frac{f_{a\cdot\max}}{\text{proportion}} = \frac{20}{0.2} = 100\ \text{kHz}$$

The beta cutoff frequency for a transistor used in the common-emitter configuration can be expressed in terms of beta and the gain-bandwidth product by

$$f_\beta = \frac{f_T}{\beta} \tag{4-23}$$

where f_β = beta cutoff frequency, same units as f_T

Example 4-23 (*a*) What is the beta cutoff frequency of a transistor if its gain-bandwidth product is 25 MHz and its β value is 125? (*b*) What percentage of the cutoff frequency is a 20-kHz audio signal?

GIVEN: $f_T = 25$ MHz $\beta = 125$

FIND: (*a*) f_β (*b*) Per cent

SOLUTION:

(*a*)
$$f_\beta = \frac{f_T}{\beta} = \frac{25}{125} = 0.2 \text{ MHz, or } 200 \text{ kHz}$$

(*b*)
$$\text{Per cent signal} = \frac{\text{audio frequency}}{\text{cutoff frequency}} \times 100 = \frac{20}{200} \times 100 = 10 \text{ per cent}$$

4-21 High-frequency Effects

At high frequencies, the operation of transistors is adversely affected by (1) charge-carrier diffusion, (2) phase shift, and (3) base-collector junction capacitance. *Charge-carrier diffusion* is caused by the differences in the length of the paths taken by the emitter current carriers in traveling through the base from the emitter to the collector. The resulting variations in traveling time cause some of the current carriers that left the emitter at the same instant to arrive at the collector at different times. When the differences in the arrival time represent a large portion of the input-signal cycle (for example, 180°) some of the current carriers will cancel each other, thus causing distortion. Since there is no potential across the base to accelerate the current carriers from the emitter junction to the collector junction, they diffuse relatively slowly through the base to the collector. This action results in a time delay between the emitter and collector currents, which represents a phase shift between the two currents at the signal frequency.

The reverse voltage at the base-collector junction causes current carriers of both the base and the collector to be repelled from the junction. The charge at the junction will then be the result of the donor and acceptor atoms on each side of the junction. The density of the charge carriers increases with the distance from the junction, and decreases with an increase in voltage. Since the charge at the junction is a function of the voltage, the base-collector junction indicates the presence of capacitance. The capacitance of the junction varies with (1) voltage, (2) current, (3) area, and (4) type of junction.

4-22 Noise Figure

A specification sometimes given for a transistor is the signal-to-noise ratio, or *noise figure*, abbreviated as NF and usually rated in decibels. In most transistor applications the input signal contains some thermal noise such as is generated in resistors and some other circuit components. The amount of noise in a circuit is expressed as the ratio of signal power to noise power. The noise figure of an amplifier circuit is expressed as the ratio of the signal power to the noise power at the input, to the signal power to the noise power at the output, or

$$\text{NF} = \frac{S_i/N_i}{S_o/N_o} \tag{4-24}$$

where NF = noise factor, dimensionless ratio
S_i = input signal power, watts
N_i = input noise power, watts
S_o = output signal power, watts
N_o = output noise power, watts

If the ratio of signal-to-noise power is the same in both the input and output circuits, an amplifier has apparently not added any noise and (1) the amplifier is said to be noiseless, (2) the noise figure has a value of one, and (3) the noise figure is zero decibels. However, in practical amplifier circuits noise is added by the components of the amplifier circuit, such as transistors and resistors.

The primary source of internal noise in a transistor is the molecular agitation resulting from the movement of electrons and holes through a semiconductor. The magnitude of the noise currents in a transistor is dependent upon the (1) average current, (2) frequency, and (3) bandwidth. With zero bias current, the noise output of a transistor is equal to the thermal noise obtained from a resistor of equivalent resistance. The amount of noise produced by a transistor is called its noise figure which is generally specified in decibels above the thermal input power for a one-hertz bandwidth at one kilohertz. The output power noise per one-hertz bandwidth varies approximately inversely with frequency to about 100 kHz, where it attains a value several times that of its equivalent resistor thermal noise, and then remains relatively constant for the higher frequencies. Transistors used with low frequencies (50 kHz or less) have a noise figure in the order of 5 to 10 decibels, while transistors used with high frequencies have a noise figure in the order of 5 decibels.

Example 4-24 What is the noise figure of a transistor circuit that has an input signal-to-noise power ratio of 20 and an output signal-to-noise power ratio of 10 when (*a*) expressed as a ratio? (*b*) Expressed in decibels?

GIVEN: $S_i/N_i = 20 \qquad S_o/N_o = 10$

FIND: (*a*) NF as a ratio (*b*) NF in decibels

SOLUTION:

(*a*) $$\text{NF (ratio)} = \frac{S_i/N_i}{S_o/N_o} = \frac{20}{10} = 2$$

(*b*) $$\text{NF} = 10 \log \text{NF ratio} = 10 \log 2 = 10 \times 0.302 \cong 3 \text{ db}$$

Example 4-25 If the noise figure of a transistor circuit is 20 decibels, by what factor has the noise ratio increased?

GIVEN: NF = 20 db

FIND: Increase in noise

SOLUTION: (See Art. 1-6)

$$\text{db} = 10 \log \frac{P_1}{P_2} \qquad \text{and} \qquad \frac{P_1}{P_2} = \text{NF ratio}$$

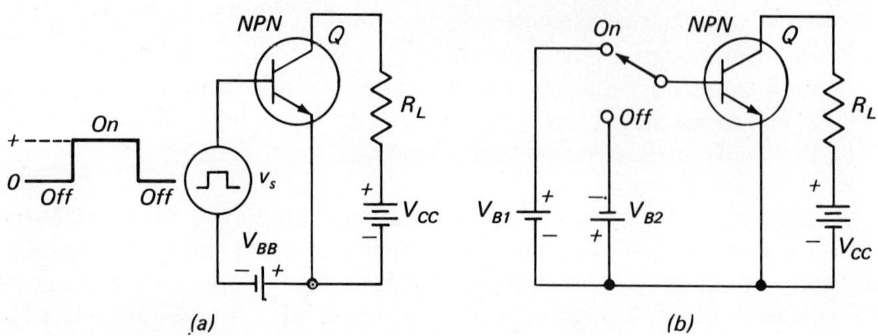

Fig. 4-18 Transistor switching circuit. (a) Action initiated by a pulse signal. (b) Simplified mechanical equivalent of pulse switching.

therefore,

$$\log \text{NF} = \frac{\text{db}}{10} = \frac{20}{10} = 2$$

$$\text{antilog } 2 = 100$$

hence NF = 100, or the noise ratio has increased 100 times.

4-23 Switching Transistors

An important application of transistors is performing switching operations in computers, data-processing, telemetry, communications, and industrial-processing equipment. Switching is generally accomplished by applying a pulse-type signal to the input circuit of the transistor, usually in the common-emitter configuration. Figure 4-18*a* illustrates a pulse-switching circuit and Fig. 4-18*b* shows its mechanical equivalent. In the circuit of Fig. 4-18*a*, when the input signal is zero the emitter-base junction of the NPN transistor is reverse-biased by V_{BB} and no current will flow in the transistor input or output circuits. When a positive signal whose voltage is greater than V_{BB} is applied, the emitter-base junction becomes forward-biased and current will flow in the transistor input and output circuits.

Although the response to changes in the transistor currents and voltages takes place very rapidly, they are not instantaneous. The study of switching circuits requires knowledge of the various components of the switching action. The following list of switching terms is illustrated in Fig. 4-19, where I_B is the *ideal* waveform of the current in the input circuit of the transistor and I_C represents the corresponding output-current variations.

1. The *delay time* t_d is the length of time between the start of the input pulse signal and the time that the output current I_C reaches 10 per cent of its maximum value.

2. The *rise time* t_r, also called the *buildup time*, is the elapsed time during which the output current I_C rises from 10 to 90 per cent of its maximum value.

3. The *storage time* t_s is the elapsed time during which the input signal

returns to zero and the output current I_C decreases to 90 per cent of its maximum value.

4. The *fall time* t_f, also called the *decay time*, is the elapsed time during which the output current falls from 90 to 10 per cent of its maximum value.

5. The *input pulse duration* T_i is the length of time that the input current I_B is at its maximum value.

6. The *output pulse duration* T_o is the length of time that the output current I_C is 90 per cent or more of its maximum value.

7. The *turn-on time* t_{on} is the sum of the delay time and the rise time.

8. The *turn-off time* t_{off} is the sum of the storage time and the fall time.

The advantages of transistors as switching devices include (1) high reliability, (2) low power requirement, (3) ease of miniaturization, (4) low rise time, and (5) low cost. Technical data specifications for transistors designed for computers, data processing, telemetry, logic circuits, etc., frequently list values

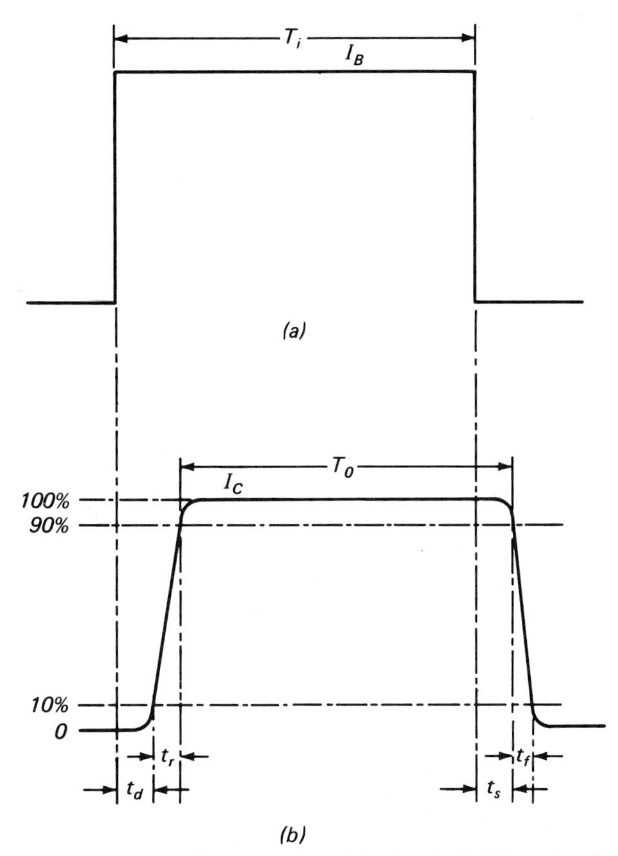

Fig. 4-19 Base current and collector current in the switching circuit of Fig. 4-18a. (a) Ideal waveform of the base current. (b) The response in the collector current.

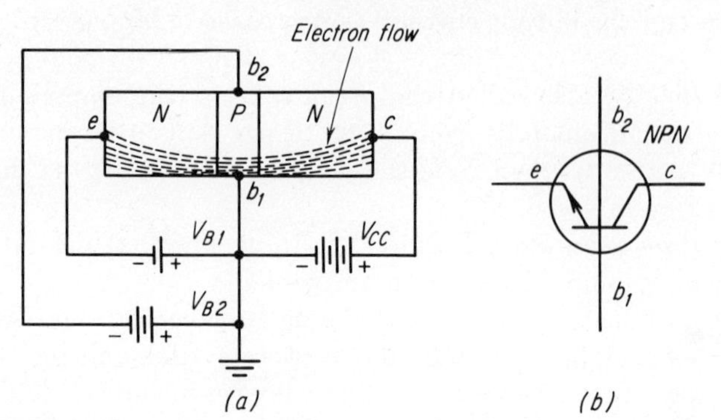

Fig. 4-20 Tetrode transistor. (a) D-c biasing voltages. (b) Schematic symbol.

for t_d, t_r, t_s, t_f, or turn-on time and turn-off time. Values of delay, rise, storage, and fall time range in the order of one microsecond to only a few nanoseconds.

4-24 Tetrodes

The high-frequency operation of a transistor can be improved with the addition of a fourth lead to an NPN transistor. Figure 4-20*a* shows the additional lead b_2 attached to the base region at a position that is on the side opposite to the original base connection b_1. This type of transistor is called a *tetrode,* and its schematic symbol is shown in Fig. 4-20*b*. As with an NPN triode transistor, the normal emitter-base junction is forward-biased by V_{B1} and the collector-base junction is reverse-biased by V_{CC}. The additional lead b_2 is forward-biased by V_{B2} with a negative potential higher than the normal emitter-base voltage. This negative potential restricts the electrons flowing through the base and causes them to flow through a relatively narrow area of the base region (Fig. 4-20*a*). The improvement in high-frequency operation is obtained because (1) the reduced effective area of each region adjacent to the base decreases the emitter and collector capacitances, (2) the shorter path taken by the base current decreases the base resistance. The base resistance of a tetrode may be decreased in the order of 10 to 1 over that of a triode.

When the electrons are forced through a narrow channel in the base region, the collector-current capabilities of the transistor are reduced, thus also reducing its power-handling capabilities. The tetrode is primarily intended for high-frequency use as an r-f amplifier, i-f amplifier, mixer, and oscillator. The second base connection may be used for automatic gain control, as this connection will cause very little detuning of the collector circuit.

QUESTIONS

1. How was the name transistor derived?
2. Name seven favorable characteristics of transistors.
3. Describe the construction of (*a*) an NPN transistor, (*b*) a PNP transistor.

4. (*a*) How is the emitter-base junction biased? (*b*) How does it affect the flow of current?
5. (*a*) How is the base-collector junction biased? (*b*) How does it affect the flow of current?
6. Describe the method used for biasing transistors used as amplifiers for (*a*) NPN transistors, (*b*) PNP transistors.
7. Compare the amount of current in the emitter of a transistor with the amount of current in the (*a*) base, (*b*) collector.
8. Describe three systems used for identifying the leads of a triode transistor.
9. Name three basic methods of transistor connection.
10. Describe the input and output circuit characteristics of a common-emitter transistor configuration.
11. Describe the input and output circuit characteristics of a common-base transistor configuration.
12. Describe the input and output circuit characteristics of a common-collector transistor configuration.
13. Describe the action that takes place when an input signal is applied to a transistor in the common-emitter configuration.
14. Explain why with the common-emitter configuration the output voltage of a transistor is 180° out of phase with the input voltage.
15. Develop the phase relation between the output voltage and the input voltage for a transistor used in the common-base configuration.
16. Develop the phase relation between the output voltage and the input voltage for a transistor used in the common-collector configuration.
17. Define (*a*) alpha, (*b*) beta, (*c*) forward-current transfer ratio.
18. Define (*a*) cutoff currents, (*b*) leakage currents, (*c*) saturation current.
19. Define (*a*) frequency cutoff, (*b*) gain-bandwidth product, (*c*) transit time.
20. Define (*a*) breakdown voltage, (*b*) saturation voltage, (*c*) collector cutoff current, (*d*) emitter cutoff current, (*e*) stored base charge.
21. (*a*) What is a parameter? (*b*) What is a hybrid parameter?
22. What are the four common *h*-parameters and what does each represent?
23. What are the maximum ratings of a transistor that are generally included in the manufacturers' specifications?
24. What operating values of a transistor are generally included in manufacturers' specifications?
25. Describe the characteristic curves for (*a*) collector current versus collector voltage for a constant value of emitter current, (*b*) collector current versus collector voltage for a constant value of base current.
26. (*a*) Define the *current gain* of a transistor. (*b*) What is its relation to the forward-current transfer ratio?
27. What are the approximate proportions of currents in the emitter, collector, and base of a transistor?
28. What is meant by recombination current?
29. (*a*) How is the current gain for the common-base configuration expressed mathematically? (*b*) What is it called? (*c*) What symbol is used to represent it?
30. (*a*) How is the current gain for the common-emitter configuration expressed mathematically? (*b*) What is it called? (*c*) What symbol is used to represent it?

31. (*a*) What is meant by thermal runaway? (*b*) What is its cause? (*c*) How can it be controlled?
32. How is the current gain for the common-collector configuration expressed mathematically?
33. Define (*a*) direct voltage gain, (*b*) direct power gain.
34. How is the power gain of a transistor affected by (*a*) its method of connection? (*b*) Its input resistance? (*c*) Its load resistance?
35. What operating characteristics of a transistor circuit can be determined by graphical analysis?
36. (*a*) What procedure is used to obtain test data for plotting characteristic curves? (*b*) What is meant by a running parameter? (*c*) What is meant by a family of curves?
37. Define (*a*) load line, (*b*) quiescent point, (*c*) minimum and maximum operating values.
38. What data about a transistor circuit must be known in order to plot a load line?
39. Describe how a load line is constructed.
40. How are the operating points of the collector current located on the characteristic curves?
41. Explain how the current gain of a transistor circuit can be determined from the characteristic curves.
42. Explain how the output-voltage variations can be determined from the characteristic curves.
43. Explain how the input-voltage variations of a transistor circuit may be determined from the input-signal data, the circuit parameters, and the characteristic curves.
44. How is the voltage gain of the transistor circuit determined from the data related to questions 42 and 43?
45. How is the power amplification of the transistor circuit determined from the data related to questions 41 and 44?
46. How is the power gain in decibels determined when the power-amplification ratio is known?
47. Explain how the output signal of a transistor circuit can be shown graphically when the input-signal characteristics and the transistor characteristic curves are available.
48. How is the extrinsic transconductance of a transistor expressed mathematically in terms of (*a*) changes in current and voltage, (*b*) beta and one of its h-parameters?
49. (*a*) Define the gain-bandwidth product for the common-emitter configuration. (*b*) By what other name is it known? (*c*) What is its symbol? (*d*) In what unit is it expressed?
50. Define (*a*) alpha cutoff frequency, (*b*) beta cutoff frequency.
51. What factors affect the operation of a transistor at high frequencies?
52. (*a*) How is the noise figure of a transistor expressed mathematically? (*b*) Name two ways in which the noise factor is expressed.
53. (*a*) What is meant if an amplifier is said to be noiseless? (*b*) Can this condition be achieved? (*c*) Why?
54. Name some important applications of switching transistors.
55. Explain the operation of a simple pulse-switching circuit.
56. Define (*a*) delay time, (*b*) rise time, (*c*) storage time, (*d*) fall time.
57. Name five advantages of transistors as switching devices.

58. (*a*) What is a tetrode transistor? (*b*) Describe how the tetrode improves the high-frequency operation of a transistor.

PROBLEMS

1. From the curves of Fig. 4-9, determine the approximate values of collector current when (*a*) $V_C = -15$ volts and $I_e = 8$ ma, (*b*) $V_C = -15$ volts and $I_e = 6$ ma.
2. From the curves of Fig. 4-9, determine the approximate values of collector current when (*a*) $V_C = -20$ volts and $I_e = 7$ ma, (*b*) $V_C = -20$ volts and $I_e = 5$ ma.
3. From the curves of Fig. 4-10, determine the approximate values of collector current when (*a*) $V_C = -15$ volts and $I_b = 250$ μa, (*b*) $V_C = -15$ volts and $I_b = 50$ μa.
4. From the curves of Fig. 4-10, determine the approximate values of collector current when (*a*) $V_C = -20$ volts and $I_b = 200$ μa, (*b*) $V_C = -20$ volts and $I_b = 100$ μa.
5. From the curves of Fig. 4-10, determine the approximate values of collector current when (*a*) $V_C = -5$ volts and $I_b = 300$ μa, (*b*) $V_C = -5$ volts and $I_b = 200$ μa.
6. From the curves of Fig. 4-10, determine the approximate values of collector current when (*a*) $V_C = -22.5$ volts and $I_b = 150$ μa, (*b*) $V_C = -22.5$ volts and $I_b = 50$ μa.
7. A transistor used in the common-base configuration has an alpha value of 0.98 and a saturation current of 10 μa. If the emitter current is 5 ma, determine (*a*) the apparent collector and base current when neglecting the saturation current, (*b*) the collector and base currents with the effect of the saturation current included.
8. A certain transistor used in an amplifier circuit in the common-base configuration has the following values: $\alpha = 0.985$, $I_{CO} = 12$ μa, and $I_E = 1.5$ ma. Determine (*a*) the apparent collector and base currents by neglecting the saturation current, (*b*) the collector and base currents with the effect of the saturation current included.
9. Determine the alpha current gain for the transistor represented by the curves of Fig. 4-9 for the common-base configuration using the values from Prob. 1.
10. Determine the alpha current gain for the transistor represented by the curves of Fig. 4-9 for the common-base configuration using the values from Prob. 2.
11. Determine the beta current gain for the transistor represented by the curves of Fig. 4-10 for the common-emitter configuration using the values from Prob. 3.
12. Determine the beta current gain for the transistor represented by the curves of Fig. 4-10 for the common-emitter configuration using the values from Prob. 4.
13. A junction transistor operated in the common-emitter configuration is biased so that α is 0.95. What is β for these conditions?
14. A junction transistor operated in the common-emitter configuration is biased so that α is 0.96. What is β for these conditions?
15. What is α for a transistor whose β is 20?
16. What is α for a transistor whose β is 99?
17. A transistor used in the common-emitter configuration has an α of 0.987, a β of 74, and a leakage current of 20 μa. If the emitter current is 10 ma, determine (*a*) the collector current, (*b*) the base current.
18. A transistor used in the common-emitter configuration has a β of 35 and a leakage current of 8 μa. If the emitter current is 1.5 ma, determine (*a*) the collector current, (*b*) the base current.
19. Determine the forward-current transfer ratio for a transistor in the common-collector configuration using the values from Prob. 5. (The curves of Fig. 4-10 are considered satisfactory for this purpose because I_E and I_C are practically equal.)

20. Determine the forward-current transfer ratio for a transistor in the common-collector configuration using the values from Prob. 6. (The curves of Fig. 4-10 are considered satisfactory for this purpose because I_E and I_C are practically equal.)

21. What are the collector and base currents for a transistor used in the common-collector configuration if β is 25, the saturation current is 8 μa, and the emitter current is 1.4 ma?

22. A transistor has an α of 0.975 and its saturation current is 15 μa. If the transistor is used in the common-collector configuration and its emitter current is 3 ma, what is (*a*) the collector current? (*b*) The base current?

23. What is the direct voltage gain of a transistor used in the common-base configuration if α is 0.95, the collector-base resistance is 1.2 megohms, and the emitter-base resistance is 600 ohms?

24. What is the direct voltage gain of a transistor used in the common-base configuration if α is 0.97, the collector-base resistance is 3.5 megohms, and the emitter-base resistance is 1,400 ohms?

25. A certain transistor has a β of 23.5, an output resistance of 100,000 ohms, and an input resistance of 2,000 ohms. What is the direct voltage gain of this transistor when used in the common-emitter configuration?

26. A certain high-frequency transistor (200 MHz) has an input resistance of 75 ohms and an output resistance of 6,000 ohms. What is its direct voltage gain when used in the common-emitter configuration and its β is 50?

27. What is the direct voltage gain of a transistor when used in the common-collector configuration if $\beta = 35$, the input resistance = 400,000 ohms, and the output resistance = 10,000 ohms?

28. What is the direct voltage gain of a transistor when used in the common-collector configuration if $\beta = 45$, the input resistance = 600,000 ohms, and the output resistance = 12,000 ohms?

29. What is the direct power gain of the transistor used in Prob. 23?

30. What is the direct power gain of the transistor used in Prob. 24?

31. What is the direct power gain of the transistor used in Prob. 25?

32. What is the direct power gain of the transistor used in Prob. 26?

33. What is the direct power gain of the transistor used in Prob. 27?

34. What is the direct power gain of the transistor used in Prob. 28?

35. (*a*) Using the data in Table 4-4, plot curves for the running parameters of I_B = 10 μa, I_B = 20 μa, and I_B = 30 μa. (*b*) With the value of V_{CC} in the circuit of Fig. 4-15 decreased to 4.5 volts, and all other parameters unchanged, draw a load line on the curves plotted in (*a*).

36. (*a*) Using the data in Table 4-4, plot curves for the running parameters of I_B = 20 μa, I_B = 30 μa, and I_B = 40 μa. (*b*) With the circuit parameters the same as in Fig. 4-15, draw a load line on the curves plotted in (*a*).

37. For the transistor circuit of Prob. 35 find (*a*) the current gain, (*b*) the voltage gain, (*c*) the db power gain.

38. For the transistor circuit of Prob. 36 and Fig. 4-15, but with the quiescent point shifted to 30 μa, find (*a*) the current gain, (*b*) the voltage gain, (*c*) the db power gain.

39. What is the extrinsic transconductance of a transistor if it is being operated so that a 10-ma change in collector current is caused by a 0.08-volt change in the emitter-base voltage while the emitter-collector voltage is kept constant?

40. What is the transconductance of a transistor if it is being operated so that a 0.2-volt change in the emitter-base voltage causes a 1.5-ma change in the collector current while the emitter-collector voltage is kept constant?

41. What is the transconductance of a transistor if its small-signal impedance is 480 ohms and β is 40?

42. What is the transconductance of a transistor if its beta value ranges from 70 to 270, and its small-signal input impedance is 1,500 ohms? (*Note:* Use the mid value of beta.)

43. What is the gain-bandwidth product of a transistor used in a video amplifier (4.5 MHz) if its transconductance is 15 mmhos (millimhos) and its input capacitance is 30 pf?

44. What is the gain-bandwidth product of a transistor used in a high-power high-frequency (200 MHz) r-f amplifier if its transconductance is 376 mmhos (millimhos) and its input capacitance is 100 pf?

45. What should be the minimum cutoff frequency of a transistor used in the r-f circuit of a radio receiver that has an upper frequency limit of 1,700 kHz? (*Note:* It is desired that the operating frequency should be approximately 20 per cent of the cutoff frequency.)

46. What should be the minimum cutoff frequency of a transistor used in the i-f circuit of an f-m radio receiver that has an operating frequency of 10.7 MHz? (*Note:* It is desired that the operating frequency should be approximately 20 per cent of the cutoff frequency.)

47. What is the cutoff frequency of a transistor used in a small-signal low-power a-f circuit if its gain-bandwidth product is 400 MHz and β is 100?

48. What is the cutoff frequency of a transistor used in a computer circuit if its gain-bandwidth product is 350 MHz and β is 25?

49. What is the noise figure of a transistor circuit that has an input signal-to-noise power ratio of 400 and an output signal-to-noise power ratio of 20 when expressed (a) as a ratio? (*b*) In decibels?

50. What is the noise figure of a transistor circuit that has an input signal-to-noise power ratio of 350 and an output signal-to-noise power ratio of 25 when expressed (*a*) as a ratio? (*b*) In decibels?

51. If the noise figure of a transistor circuit is 6 db, by what proportion has the signal-to-noise power ratio increased?

52. If the noise figure of a transistor circuit is 10 db, by what proportion has the signal-to-noise power ratio increased?

53. What is the turn-on time of a control transistor if its delay time is 150 ns (nanoseconds) and its rise time is 500 ns?

54. What is the turn-off time of the control transistor of Prob. 53 if its storage time is 800 ns and its fall time is 500 ns?

55. A transistor used for high-speed switching applications in a computer has the following constants: (1) delay time = 6 ns, (2) rise time = 7 ns, (3) fall time = 6 ns, and (4) storage time = 10 ns. What is its (*a*) turn-on time? (*b*) Turn-off time?

Chapter 5

Electron Tubes

The importance of the electron tube lies in its ability to operate efficiently over a wide range of frequencies and to control almost instantaneously the flow of millions of electrons. Before the development of transistors, the electron tube was the only device available to perform many of the circuit operations used in the broad field of electronics. Although transistors have largely replaced vacuum tubes in the radio and television fields, there are still many uses for electron tubes. They are used in various types of electronic equipment such as communication, industrial, scientific, therapeutic, computer, data processing, etc. These tubes may be either of the vacuum type as used in communications, the gas type as used in some types of industrial electronic equipment, or the photo type as used in the reproduction of soúnd on motion-picture film and light-operated relay applications. Electron tubes may be considered as a circuit element in the same manner as are resistors, inductors, and capacitors.

5-1 The Triode

The Control Grid. When a third electrode is used, the tube is called a *triode.* The third electrode is called the *control grid* and consists of a spiral winding or a fine-mesh screen extending the length of the cathode and placed between the cathode and the plate. The circuit connections and the direction of electron flow for a triode are shown in Fig. 5-1. If the grid is made more negative than the cathode, some of the electrons going toward the plate will be repelled by the grid, thus reducing the plate current. By making the grid still more negative, it is possible to reduce the plate current to zero. This condition is called *cutoff*.

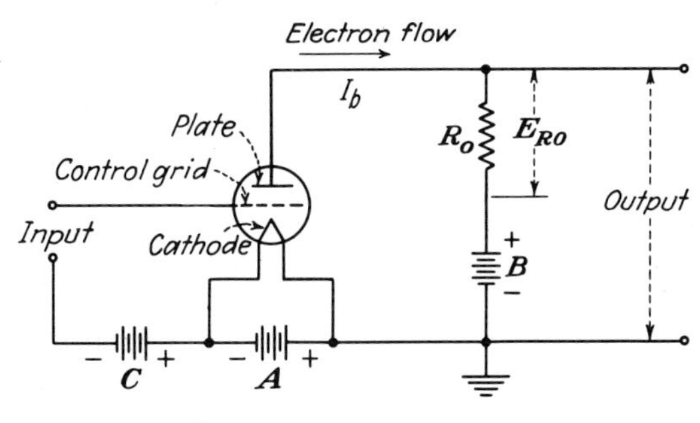

Fig. 5-1 Circuit diagram for a triode showing the electron flow and the connections for the cathode, grid, and plate power supplies.

Grid Bias. The third electrode is called the *control grid* because it controls the number of electrons allowed to flow from the cathode to the plate. The amount of voltage used to keep the grid negative with respect to the cathode is called the *grid bias.*

Action of the Triode with Normal Grid Bias. If a varying signal voltage is applied to the grid, the number of electrons flowing in the plate circuit will vary in the same manner as the signal voltage. The triode in the circuit of Fig. 5-1 operates with a grid bias provided by battery C, and in the absence of a signal voltage a steady plate current I_b will flow. If an alternating signal voltage, whose maximum value does not exceed the value of the grid bias, is impressed across the input terminals, the voltage at the control grid will vary in the same manner as the signal voltage between the values of (1) E_C minus the signal voltage when the signal voltage is positive, and (2) E_C plus the signal voltage when the signal voltage is negative. As the voltage at the grid becomes less negative during the positive portion of the input signal, the plate current will increase to values higher than I_b. During the negative portion of the signal voltage the grid becomes more negative and the plate current decreases to values below I_b. It should be observed that the instantaneous variations in the plate current will be in phase with the variations in the signal voltage. The output voltage is equal to the voltage of battery B minus the voltage drop $E_{R\cdot O}$; therefore, as the plate current increases the output voltage decreases, and vice versa. Thus, the output-voltage variations will be 180° out of phase with the signal-voltage variations.

Action of the Triode with Positive Grid Voltage. If the value of the signal voltage is such that it will make the grid positive, the grid will act in the same manner as a plate and will draw electrons to it, thus causing current to flow in the grid circuit. It is generally desirable to avoid this condition because (1) current flowing in the grid circuit can distort the shape of the output voltage with respect to the shape of the input voltage, and (2) positive grid voltage can cause excessive plate current and result in damage to the tube.

5-2 Characteristic Curves of a Triode

The major operating characteristics of an electron tube are used to identify the electrical features and operating values of the tube. These values may be listed in tabular form or plotted as *characteristic curves.* These curves are used to determine the performance of a tube under any operating condition.

Characteristic curves are classified as to the condition of the circuit under which the values for the curves are obtained. *Static characteristics* are obtained by varying the direct voltages applied to the electrodes, and with no load applied to the plate. *Dynamic characteristics* approximate the performance of a tube under actual working conditions. They are obtained by applying an alternating voltage to the grid circuit and inserting a load resistance in the plate circuit, the direct voltages on all electrodes being adjusted to the desired values.

Curves showing the variation in plate current for changes in plate voltage

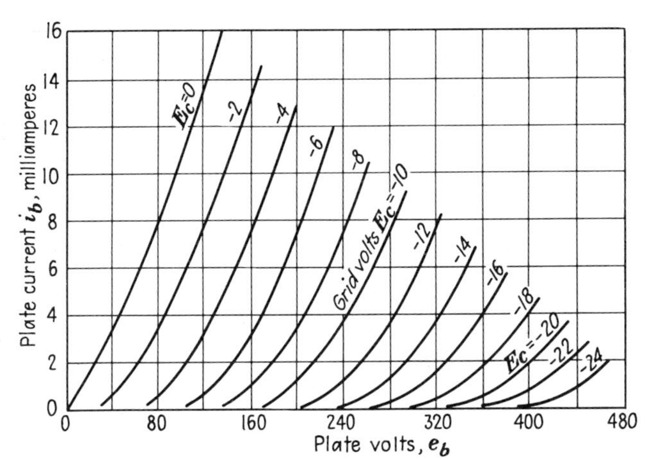

Fig. 5-2 Plate characteristic curves of a triode showing the variation in plate current for changes in plate voltage.

with constant steps of grid voltage are called *plate characteristic curves* (Fig. 5-2). Curves showing the variation in plate current for changes in grid voltage with constant steps of plate voltage are called *grid-plate-transfer characteristic curves* (Fig. 5-3).

5-3 Vacuum-tube Constants

Certain vacuum-tube characteristics are often referred to as the constants of the tube. The constants most commonly used are (1) *amplification factor* μ (mu),

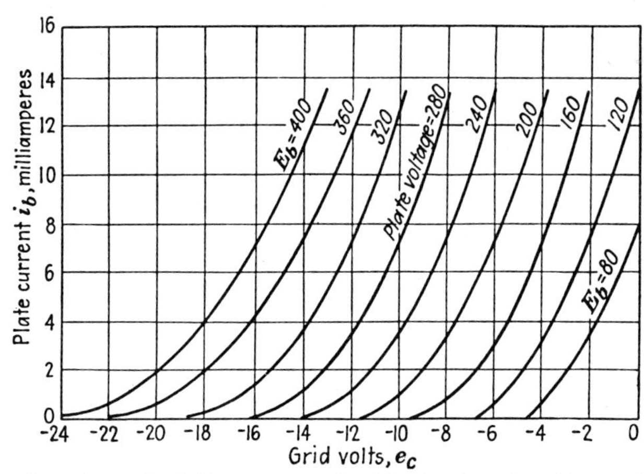

Fig. 5-3 Transfer characteristic curves of a triode showing the variation in plate current for changes in grid voltage. These curves have been cross-plotted from Fig. 5-2.

(2) *plate resistance* r_p, and (3) *control-grid-to-plate transconductance* g_m. Control-grid-to-plate transconductance is usually referred to as just *transconductance*. Transconductance is also known as *mutual conductance*.

5-4 Amplification Factor

The amplification factor of a tube is a measure of the relative ability of the grid and the plate to produce an equal change in the plate current. Mathematically it is equal to the ratio of the change in plate voltage to a change in control-grid voltage for a constant value of plate current, with the voltages applied to all other electrodes maintained constant; or

$$\mu = \frac{de_b}{de_c} \qquad (i_b\text{—constant}) \tag{5-1}$$

where μ = amplification factor
d = change or variation in value
de_b = change in instantaneous total plate voltage, volts
de_c = change in instantaneous total control-grid voltage necessary to produce the same effect upon the plate current as would be produced by de_b, volts
i_b = instantaneous total plate current, amp

The amplification factor may be calculated from values obtained from plate characteristic curves such as those shown in Fig. 5-2.

Example 5-1 What is the amplification factor of a tube whose characteristic curves are shown in Fig. 5-2 when operated with a plate voltage of 160 volts and a grid bias of 4 volts?

GIVEN: $e_b = 160$ volts $\quad E_c = -4$ volts

FIND: μ

SOLUTION: From Fig. 5-2, $i_b = 7.5$ ma when $e_b = 160$ volts and $E_c = -4$ volts. Figure 5-2 shows also that if the plate voltage is increased from 160 to 200 volts, the grid bias will have to be increased from -4 to -6 volts in order to maintain the plate current at 7.5 ma. Thus the amplification factor is

$$\mu = \frac{de_b}{de_c} = \frac{200 - 160}{6 - 4} = \frac{40}{2} = 20$$

5-5 Plate Resistance

The plate resistance of a tube is the resistance to the flow of alternating current offered by the path between the cathode and the plate. The value of the plate resistance will depend upon the values of the grid and plate voltages being applied to the tube. Mathematically it is equal to the ratio of a change in plate voltage to the corresponding change produced in the plate current with the grid voltage maintained constant, or

$$r_p = \frac{de_b}{di_b} \qquad (e_c\text{—constant}) \tag{5-2}$$

where r_p = dynamic plate resistance, ohms
di_b = change in instantaneous total plate current, amp
e_c = instantaneous total grid voltage, volts

The plate resistance may be calculated from values obtained from characteristic curves such as those shown in Figs. 5-2 and 5-3.

Example 5-2 What is the plate resistance of a tube whose characteristic curves are shown in Figs. 5-2 and 5-3 when the tube is operated with a grid bias of 4 volts and the plate-voltage change is from 160 to 200 volts?

GIVEN: $e_c = -4$ volts $\quad e_b = 160$ to 200 volts

FIND: r_p

SOLUTION: From Fig. 5-3, $i_b = 7.4$ ma when $e_b = 160$ volts, and $i_b = 12.8$ ma when $e_b = 200$ volts. Therefore,

$$r_p = \frac{de_b}{di_b} = \frac{200 - 160}{(12.8 - 7.4) \times 10^{-3}} = \frac{40 \times 10^3}{5.4} = 7{,}400 \text{ ohms}$$

The solution of Example 5-2 gives the dynamic or a-c plate resistance of the tube. It should be noted that this is not the same value as would be obtained by dividing the operating plate voltage by its corresponding plate current. Such a value is known as the *static resistance* or the *d-c resistance* of the tube, but it is seldom used in the study of tubes and their circuits. The plate resistance is sometimes referred to as the *impedance* of the tube, the *internal resistance* of the tube, the *dynamic plate resistance*, or the tube's *a-c resistance*.

5-6 Transconductance

The transconductance of a tube is the ratio of the change in plate current to a change in the control-grid voltage when all other tube-element voltages are kept constant. Expressed mathematically,

$$g_m = \frac{di_b}{de_c} \qquad (e_b\text{—constant}) \tag{5-3}$$

where g_m = transconductance, mhos
e_b = instantaneous total plate voltage, volts

The transconductance may be calculated from values obtained from characteristic curves such as those shown in Figs. 5-2 and 5-3.

Example 5-3 What is the transconductance of a tube whose characteristic curves are shown in Figs. 5-2 and 5-3 when the tube is operated with 160 volts at its plate and the grid-bias variation is from -4 to -6 volts?

GIVEN: $e_b = 160$ volts $e_c = -4$ to -6 volts

FIND: g_m

SOLUTION: From Fig. 5-2, $i_b = 7.6$ ma when $e_c = -4$ volts, and $i_b = 3.2$ ma when $e_c = -6$ volts. Therefore,

$$g_m = \frac{di_b}{de_c} = \frac{(7.6 - 3.2) \times 10^{-3}}{6 - 4} = \frac{4.4 \times 10^{-3}}{2} = 0.0022 \text{ mho}$$

The unit of conductance is the mho, which is ohm spelled backward. For convenience of numbers, the transconductance is often expressed in micromhos. Thus, the answer to Example 5-3 may be expressed as 2,200 micromhos (abbreviated as 2,200 μmhos).

Equation (5-3) shows that a tube that produces a relatively large change in plate current for a small change in grid voltage will have a relatively high value of transconductance, and since such conditions are generally desired, tubes having a high value of transconductance are preferred.

5-7 Relationship among Amplification Factor, Plate Resistance, and Transconductance

A definite relation exists among the three tube constants; this relation is developed mathematically in the following steps:

Step 1:

$$\mu = \frac{de_b}{de_c} \tag{5-1}$$

Step 2:

$$\mu = \frac{de_b}{de_c} \times \frac{di_b}{di_b} \tag{5-4}$$

$$\mu = \frac{di_b}{de_c} \times \frac{de_b}{di_b} \tag{5-4a}$$

Step 3: Substituting Eqs. (5-3) and (5-2) in (5-4*a*)

$$\mu = g_m r_p \tag{5-5}$$

and

$$g_m = \frac{\mu}{r_p} \tag{5-6}$$

and

$$r_p = \frac{\mu}{g_m} \tag{5-7}$$

Example 5-4 What is the transconductance of the triode section of a 6AV6 twin diode-triode tube if its amplification factor is 100 and its place resistance is 80,000 ohms?

GIVEN: Tube = 6AV6 $\mu = 100$ $r_p = 80{,}000$ ohms

FIND: g_m

SOLUTION:

$$g_m = \frac{\mu}{r_p} = \frac{100}{80{,}000} = 0.00125 \text{ mho, or } 1{,}250\ \mu\text{mhos}$$

Example 5-5 What is the plate resistance of a 6AF4 high-frequency triode when operated at such values that its amplification factor is 13.5 and its transconductance is 6,500 μmhos?

GIVEN: Tube = 6AF4 $\mu = 13.5$ $g_m = 6{,}500$ μmhos

FIND: r_p

SOLUTION:

$$r_p = \frac{\mu}{g_m} = \frac{13.5}{6{,}500 \times 10^{-6}} = 2{,}077 \text{ ohms}$$

Example 5-6 What is the amplification factor of a 12AT7 high-mu twin triode if its plate resistance is 15,000 ohms and its transconductance is 4,000 μmhos?

GIVEN: Tube = 12AT7 $r_p = 15{,}000$ ohms $g_m = 4{,}000$ μmhos

FIND: μ

SOLUTION:

$$\mu = g_m r_p = 4{,}000 \times 10^{-6} \times 15{,}000 = 60$$

5-8 Relationship between the Transconductance and the Operating Performance of a Tube

Vacuum-tube amplifiers are ordinarily operated so that small variations in voltage of the grid or input circuit will produce large current variations in the plate or output circuit. Since it is usually desired to keep the plate resistance low, the transconductance must be high if a high value of amplification factor

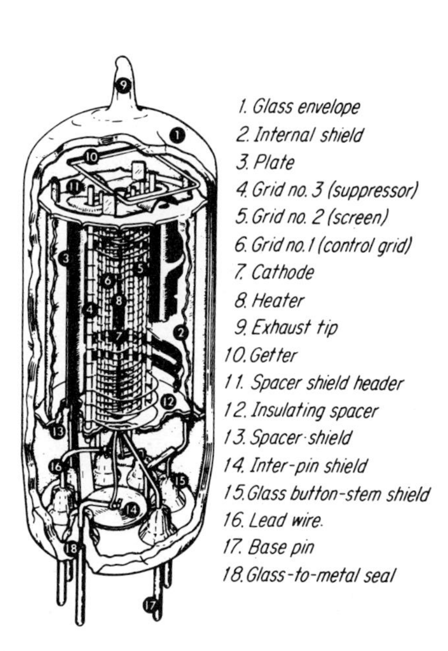

Fig. 5-4 Miniature-type tube with a button base. (*RCA*)

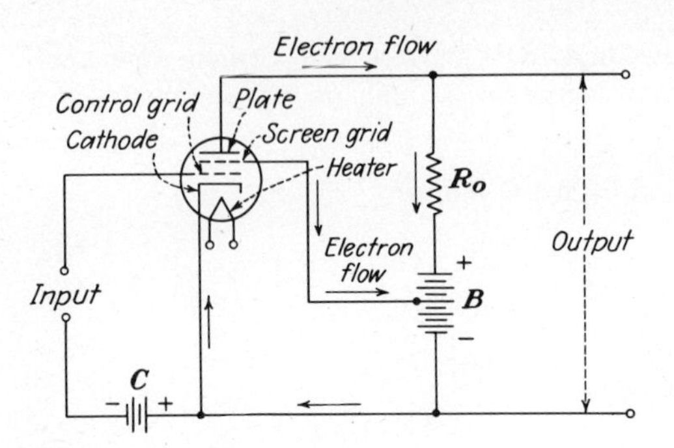

Fig. 5-5 Circuit diagram for a tetrode showing the electron flow and the proper connection for the screen grid.

is desired. The transconductance is very useful when comparing the relative merits and performance capabilities of tubes designed for the same service. A comparison of the transconductance of a power-output tube with that of a tube used as a converter would, however, have no practical value. Generally, the value of transconductance is accepted as the best single figure of merit for vacuum-tube performance.

5-9 The Tetrode

The Screen Grid. When a fourth electrode is used, the tube is called a *tetrode;* it is also referred to as a *screen-grid tube* or a *four-electrode tube.* The fourth electrode is known as the *screen grid* and consists of a spiral-wound wire or a screen, slightly coarser than that used for the control grid, placed between the plate and the control grid of the tube. The circuit connections and the direction of electron flow for a tetrode are shown in Fig. 5-5.

Interelectrode Capacitance and Its Effects. In any tube, the electrodes act as conductors and the space between the electrodes acts as an insulator; therefore, a capacitance will exist between each pair of electrodes. These capacitances, known as *interelectrode capacitances,* may form undesired paths through which current can flow (Fig. 5-6).

The capacitance between the grid and plate is generally the most troublesome. This capacitance causes some of the energy of the plate circuit to be applied to the grid circuit. The energy transferred from the output to the

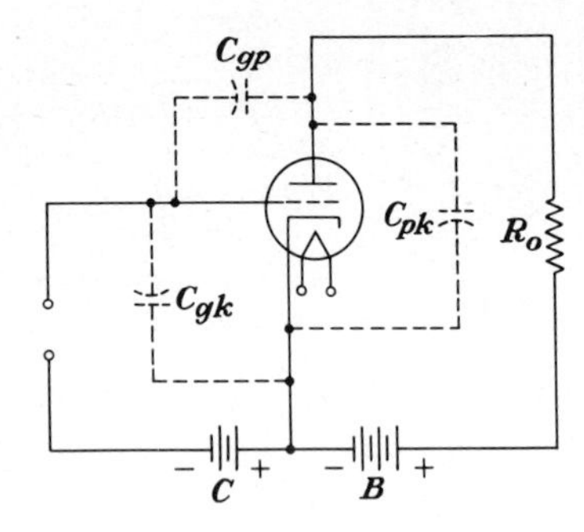

Fig. 5-6 Interelectrode capacitances of a triode.

input circuit is called *feedback*. If the output circuit contains resistance load, the feedback voltage will be 180° out of phase with the input voltage and therefore will reduce the effect of the input signal; this is called *degeneration* or *negative feedback*. When the output circuit load is inductive, the feedback voltage will be in phase with the input voltage and therefore will increase the effect of the input signal; this is called *regeneration* or *positive feedback*.

The amount of feedback caused by the interelectrode capacitance is usually very small at audio frequencies because the capacitive reactance at these frequencies is very high. At radio frequencies, however, the capacitive reactance becomes much lower and the feedback may reach an amount sufficient to cause trouble in the operation of the circuit.

Positive feedback has the advantage of increasing the gain of a circuit but has the disadvantage of causing distortion. Negative feedback has the advantage of reducing distortion but has the disadvantage of causing a reduction in the voltage gain of the circuit. If either regeneration or degeneration gets out of control, the tube no longer operates successfully as an amplifier.

Elimination of Feedback by Use of a Screen Grid. The capacitance between the control grid and the plate of a triode can be reduced to a negligible amount by adding a fourth electrode or *screen grid*. The screen grid is mounted between the control grid and the plate and acts as an electrostatic shield between the two, thus reducing the control-grid-to-plate capacitance. Connecting a bypass capacitor between the screen grid and the cathode will increase the effectiveness of the shielding action.

The screen grid also reduces the space charge. Since the screen grid is situated between the control grid and the plate and has a positive potential, the electrons coming from the cathode receive added acceleration on their way to the plate and the tendency to form a space charge is reduced.

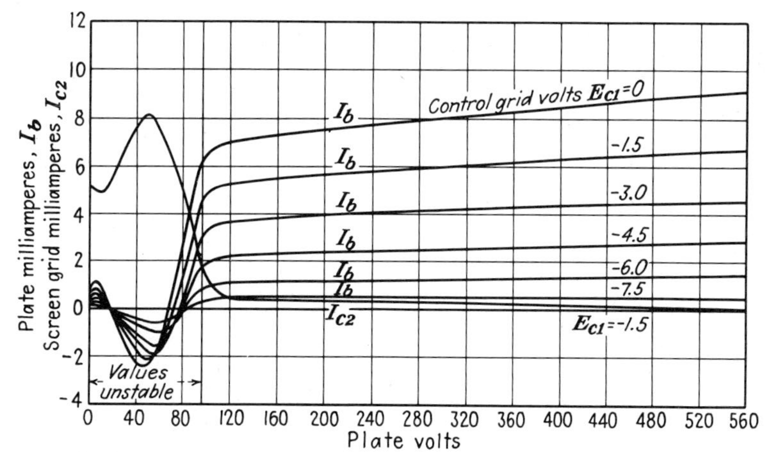

Fig. 5-7 Family of plate characteristic curves for a tetrode.

5-10 The Pentode

The Suppressor Grid. A tube with five electrodes—namely, a cathode, three grids, and a plate—is called a *pentode* or *five-electrode tube.* The fifth electrode is an extra grid called the *suppressor grid.* The electrodes of the pentode are arranged with the cathode at the center and surrounded by the control grid, the screen grid, the suppressor grid, and the plate in the order named. The circuit connections and the direction of electron flow for the pentode are shown in Fig. 5-8.

The suppressor grid consists of a spiral-wound wire or a coarse-mesh screen placed between the screen grid and the plate. When the various grids of the tube are in the form of a screen, the control grid is of a very fine mesh so that small changes in control-grid voltage will produce relatively large changes in plate current and consequently will produce a high value of transconductance for the tube. The screen grid is of a somewhat coarser mesh so that it will not appreciably affect the flow of electrons to the plate, its purpose being largely to reduce the control-grid-to-plate capacitance. The suppressor grid is of a coarser mesh so that it will not retard the flow of electrons to the plate while serving its function of preventing the secondary emission from reaching the screen grid. In some pentodes the suppressor grid is internally connected to the cathode, while in others the suppressor grid is brought out as a separate terminal.

Action of the Suppressor Grid. In pentodes, the suppressor grid is added to prevent the secondary electrons from traveling to the screen grid. In order to accomplish this the suppressor grid is connected directly to the cathode. Being at cathode potential, the suppressor grid is negative with respect to the plate, and because it is close to the plate, it will repel the secondary electrons and drive them back into the plate.

Characteristics of a Pentode. A family of curves for a typical pentode are shown in Fig. 5-9. The curves show that the plate current in a pentode is relatively independent of the plate voltage and primarily dependent on the grid voltage. The shapes of the characteristic curves for the pentode are quite similar to those for a transistor (see Figs. 4-9 and 4-10).

Uses of Pentodes. Pentode tubes can be used as either voltage or power amplifiers. In power pentodes, a higher power output is obtained with lower

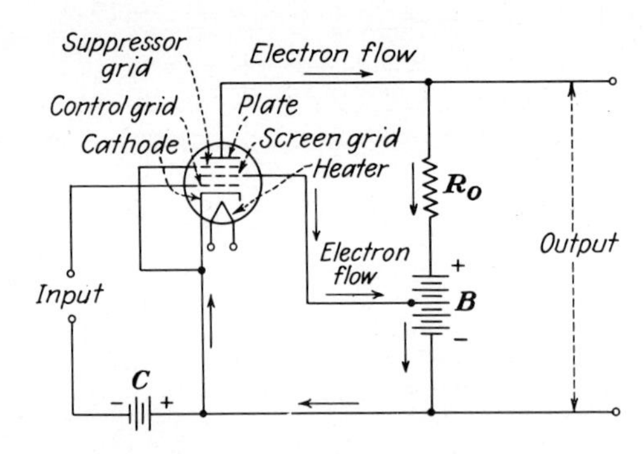

Fig. 5-8 Circuit diagram for a pentode showing the electron flow and the proper connection for the screen grid and the suppressor grid.

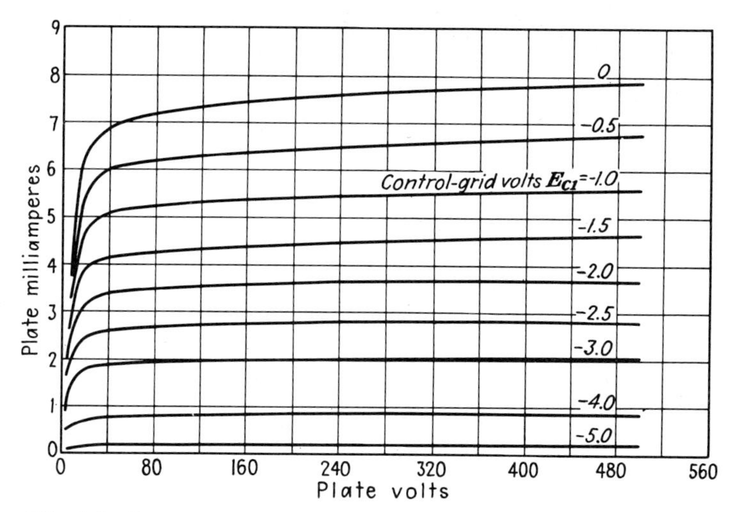

Fig. 5-9 Family of plate characteristic curves for a pentode.

grid voltages. Pentodes used as r-f amplifiers give a high voltage amplification when used with moderate values of plate voltage.

5-11 Multiunit Tubes

When a tube contains within one envelope two or more groups of electrodes associated with independent electron streams, it is called a *multiunit tube*. In general, the combinations are easily identified by the names given to the tubes, such as duplex-diode, twin-triode, duplex-diode-triode, diode-triode-pentode, and rectifier-beam power amplifier (Fig. 5-10). In most cases a single cathode

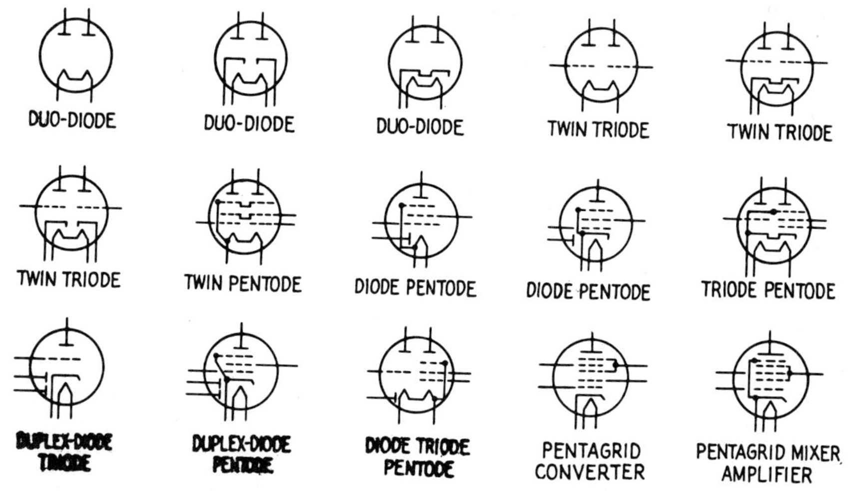

Fig. 5-10 Base diagrams for various types of multiunit tubes.

is used for all the units in a tube, although in some types separate cathodes are provided for each unit.

5-12 Tube Bases and Socket Connections

Methods of Identifying Socket Connections. The method of numbering the socket or tube-base connections for the early tube types is shown in Fig. 5-11. In this system, the filament or heater pins of the tube and the corresponding holes in the socket are larger in diameter than the others and are generally shown at the bottom of the diagram. The lower left-hand pin is designated as number 1, and the remaining pins are numbered consecutively in a clockwise direction. The order in which the tube elements are arranged varies with the tube types and may be obtained from a tube manual.

Metal and other octal-base tubes all use the same eight-pin socket. The *octal socket,* as it is commonly called, has eight equally spaced holes arranged in a circle. All the holes in the socket are of the same size, and in order to ensure correct placement of the tube in the socket, a large center hole with an extra notch is provided. The socket connection to the left of the centering notch is designated as number 1, and the remaining connections are numbered consecutively up to 8. The pins in the base of the tube are all of equal size and are arranged in a circle. A large insulated pin, provided with a centering key to fit the notch on the socket, is located in the center of the tube base. Some tubes using octal sockets have only six or seven pins, while others have eight. The six or seven pin bases merely omit one or two of the eight pins according to the number used. This, however, does not alter the numbering system for the socket connections. The order in which the tube elements are arranged varies with

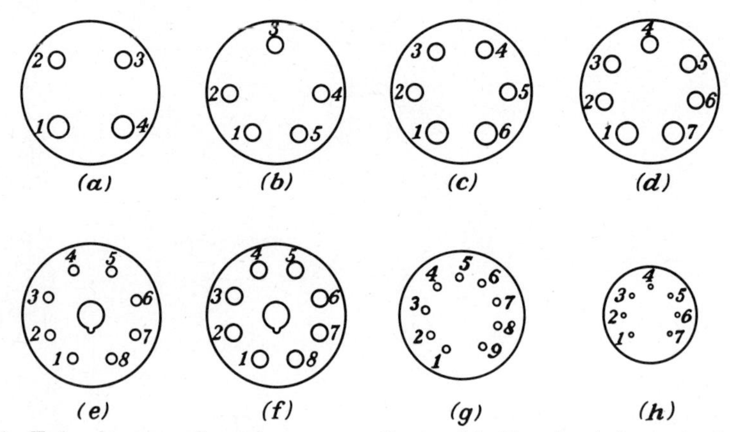

Fig. 5-11 Tube base and socket connections. (a) Standard four-pin base. (b) Standard five-pin base. (c) Standard six-pin base. (d) Standard seven-pin base. (e) Eight-pin loktal base. (f) Eight-pin octal base. (g) Nine-pin miniature base. (h) Seven-pin miniature base.

tube types and may be obtained from a tube manual. The loktal socket has an eight-pin base similar to the octal socket with the following differences: (1) the holes are smaller, and (2) provision is made for locking the tube in the socket. Miniature tubes are made with either a seven- or nine-pin base (Fig. 5-11). The miniature tube (Fig. 5-4) has a thin glass base, called a *button base,* and its socket is called a *button socket.*

5-13 Variable-mu or Supercontrol Tubes

A triode or pentode with its control grid constructed in such a manner that the amplification factor of the tube will vary with a change in grid bias is called a *variable-mu tube, supercontrol tube,* or a *remote-cutoff tube.*

Cutoff with Ordinary Grid Structure. If the grid bias on a tube is steadily increased, it will eventually reach a value that will reduce the plate current to zero, and it is then said that the tube has reached *cutoff.* In a tube with the ordinary grid construction, that is, one in which the turns of the spiral-wound control grid are equally spaced, increasing the grid bias causes the plate current to decrease very rapidly to cutoff (Fig. 5-12). This type of grid construction produces a tube with a practically constant amplification factor for all values of grid bias and is called a *sharp-cutoff* or *constant-mu tube.*

Effects of Sharp Cutoff. A tube with a sharp cutoff is limited to use in circuits with relatively small changes in grid voltage. In circuits that have a large signal voltage, sharp cutoff would produce distortion in the form of cross modulation and modulation distortion.

Cross modulation or *cross talk* is the effect produced when the signal from a second station is heard in addition to the signal of the desired station. Cross modulation is generally caused in the first stage of r-f amplification.

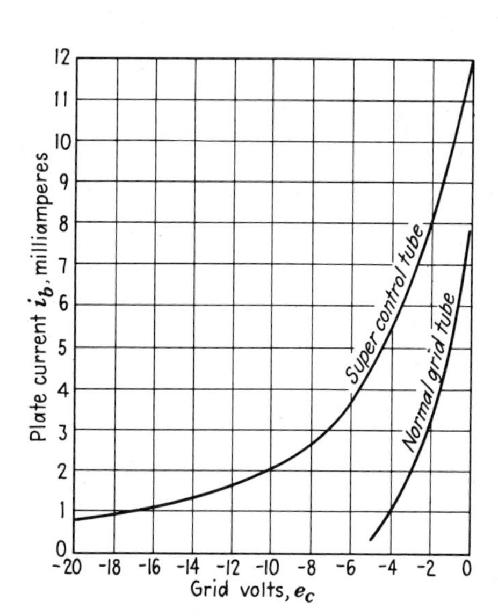

Fig. 5-12 Comparative characteristic curves of tubes with supercontrol grid and with uniformly spaced grid.

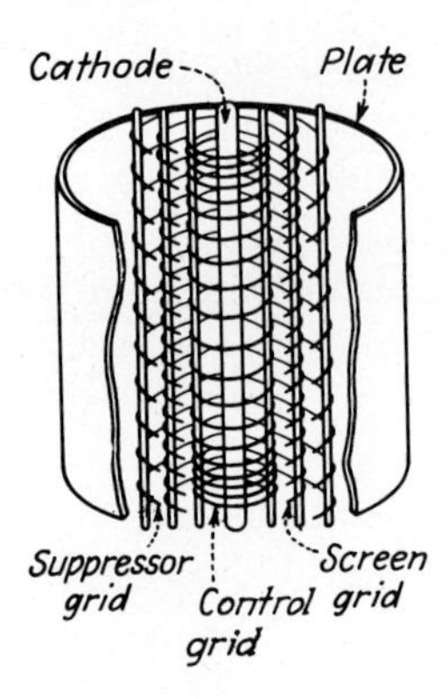

Fig. 5-13 Cross-sectional view showing the construction of the electrodes of a supercontrol amplifier tube.

Modulation distortion is the effect produced when the signal voltage drives the tube beyond cutoff and thereby distorts the desired audio signal. Modulation distortion is generally caused by the last stage of i-f amplification.

Action of the Supercontrol Tube. The characteristics of the supercontrol tube enable the tube to handle both large and small input signals with a minimum amount of distortion over a wide range of signal voltage. The control grid is spiral-wound with its turns close together at the ends but with considerable space between turns at the middle (Fig. 5-13). For weak signals and a low grid bias, the tube operates practically the same as one with a uniformly spaced grid. With larger signal inputs, more grid bias is required. The increased grid bias will produce a cutoff effect at the section of the cathode enclosed by the ends of the grid because of the close spacing of the grid wires. The plate current and other tube characteristics are now dependent on the electron flow through the center section of the grid, where the turns are spaced farther apart. The wide spacing changes the gain of the tube and enables it to handle large signals with a minimum amount of distortion.

Figure 5-12 shows the characteristic curves for a sharp cutoff and a supercontrol tube. The rate of plate-current change is approximately the same in both tubes for small values of grid-bias voltage, while for large values of grid bias the plate current decreases at a much lower rate for the supercontrol tube. This low rate of change enables the tube to handle large signals satisfactorily. The variable mu permits the tube to be used in automatic-volume-control (avc) circuits.

5-14 The Beam Power Tube

The beam power tube is constructed so that the electrons flowing to the plate are made to travel in concentrated beams. As it is capable of handling larger amounts of power it is called a *beam power tube*. This tube contains a cathode, a control grid, a screen grid, and a suppressor.

Action of the Beam Power Tube. The control grid and the screen grid are of the spiral-wound wire construction and their respective turns are placed so that each turn of the screen grid is shaded from the cathode by a turn of the control grid. This arrangement of the grid wires causes the electrons to flow in directed paths between the turns of the screen grid. Beam-forming plates (Fig. 5-14) are added to aid in producing the desired beam effect and to prevent stray electrons from the plate from flowing into the screen-grid circuit. This results in the tube having a relatively low value of screen-grid current. The beam-forming plates are operated at cathode potential by connecting them directly to the cathode. By increasing the spacing between the screen grid and the plate, a space charge is set up in this area. This space charge repels the secondary electrons emitted from the plate and forces them back into the plate.

Characteristics of a Beam Power Tube. The characteristic curves of the beam power tube (Fig. 5-15) are similar to those of the pentode. The beam power tubes provide a straighter curve at the lower plate voltages than the pentode and hence will have less chance of producing distortion. The amplification factor is high when compared with triodes and low when compared with tetrodes and pentodes. The plate resistance is high but not so high as that of pentodes. The transconductance is generally higher than for any other type of tube.

The combined effects of the directed concentrated beam of electrons, the suppressor action of the space charge, and the low value of the screen-grid current result in a tube of high power output, high efficiency, and high power sensitivity.

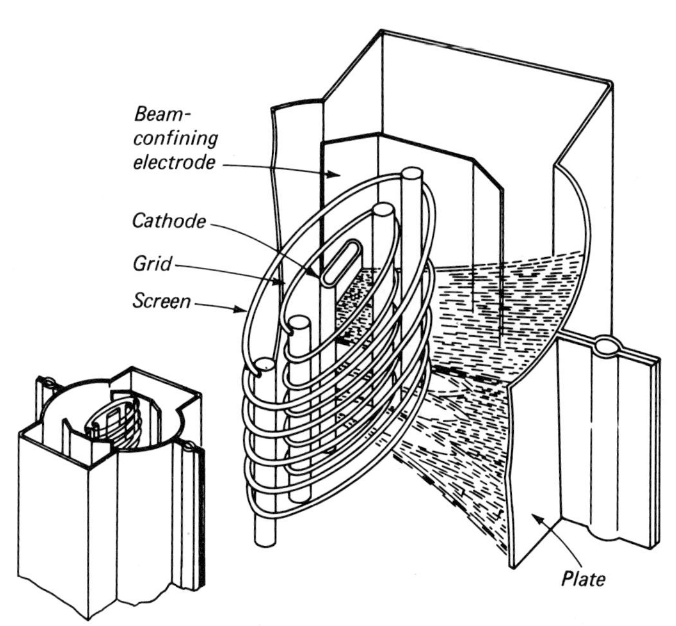

Fig. 5-14 Internal structure of a beam power tube.

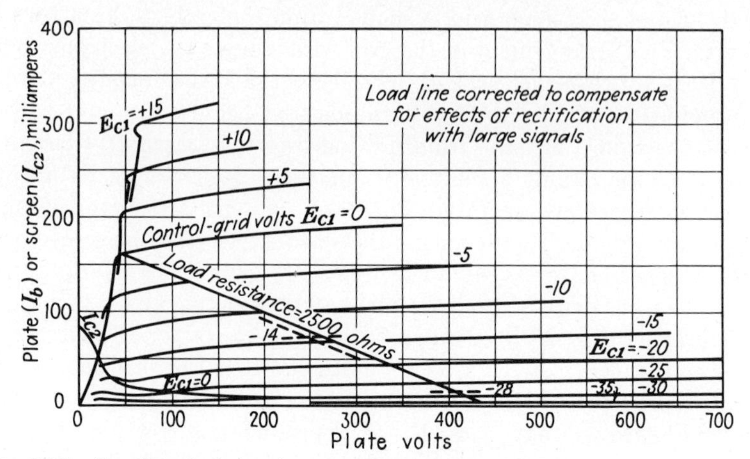

Fig. 5-15 Family of plate characteristic curves for a beam power tube.

5-15 The Thyratron

Adding a control grid to the hot-cathode gas diode (Art. 3-4) produces a gas-filled triode, called a *thyratron*. A cylindrical grid almost completely surrounds the cathode and electrons can flow from the cathode to the plate only through an aperture in the grid. The ability of the plate to attract electrons from the cathode can be canceled by a relatively low grid bias. The plate is thus electrostatically shielded from the cathode. Just as there is a definite voltage value at which ionization (also referred to as *firing*) takes place in a gas-filled diode, there is a combination of plate voltage and grid bias at which ionization takes place in the thyratron. Once ionization takes place, the thyratron continues conducting down to very low values of plate voltage. In the event that conduction ceases because of a great decrease in plate voltage, conduction cannot be restored until the original ionization voltages are again restored. As with the gas-filled diode, the thyratron has almost infinite resistance before ionization takes place and a very low resistance after ionization.

Characteristics of the thyratron are: (1) ionization (and conduction) takes place only after a fixed critical combination of plate voltage and grid bias is applied; (2) the change from nonconduction to high conduction is very sudden; (3) once conduction starts, the grid no longer has any effect on the operation of the tube; (4) conduction can be cut off only by reducing the plate voltage to nearly zero, usually by opening a switch in the plate-supply circuit; (5) effectiveness of the grid can be restored only by first stopping ionization and thereby extinguishing the arc; (6) when alternating voltage is applied to the plate, the grid regains control after each positive half-cycle is completed.

The grid-control action of a typical thyratron is shown in Fig. 5-16. This curve shows the critical plate and grid voltages that will initiate ionization or firing of the tube. From this curve it can be seen that with 800 volts applied to the plate, any grid-bias values in excess of 8 volts will keep the tube from

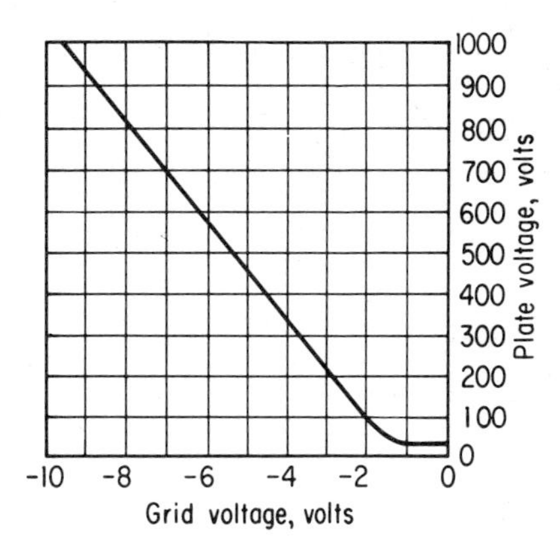

Fig. 5-16 Grid-control characteristics of a typical thyratron.

firing, but a grid voltage of approximately −8 volts (or slightly less negative) will initiate ionization. Decreasing the plate voltage to 600 volts reduces the critical bias voltage to approximately −6 volts. Any combination of plate and grid voltage to the left of the curve will keep the tube from starting conduction, and any combination to the right of the curve will initiate conduction.

Because of their grid-control action, thyratrons have many practical applications in industry. They may be used as (1) controlled rectifiers to supply variable direct current or voltage from an a-c power source; (2) inverters to supply a-c power from a d-c power source; (3) electronic switches in counting, triggering, and control circuits; (4) frequency changers; (5) sawtooth generators.

QUESTIONS

1. What characteristics of the electron tube make it an important component in electronic circuits?

2. Name three types of electron tubes and give some uses of each type.

3. Explain the following terms: (*a*) triode, (*b*) control grid, (*c*) grid bias.

4. Draw the circuit diagram for a triode showing the connections for the cathode, control grid, and plate power supplies. Indicate the direction of the electron flow on the diagram.

5. Explain the action of the triode.

6. Explain the action of a triode having grid bias but no signal input.

7. Explain the action of a triode having grid bias and a varying signal input.

8. Explain the action of a triode having a positive grid voltage.

9. In the triode circuit of Fig. 5-1, what is the phase relation between the input signal voltage and (*a*) voltage variations at the control grid, (*b*) plate-current variations, (c) output-voltage variations?

10. Why should the fixed grid bias of a tube always be greater than the maximum value of the input signal voltage?

11. What are the advantages of a family of characteristic curves over the tabular listing of tube characteristics?

12. Explain the difference between the static and dynamic characteristics of a vacuum tube.

13. How are the static characteristics of a tube obtained?
14. How are the dynamic characteristics of a tube obtained?
15. (*a*) What is meant by the tube constants? (*b*) Name the three constants.
16. (*a*) What does the amplification factor of a tube represent? (*b*) How is it obtained?
17. What factors affect the amplification factor of a tube?
18. (*a*) What does the plate resistance of a tube represent? (*b*) How is it obtained?
19. What factors affect the plate resistance of a tube?
20. (*a*) What does the transconductance of a tube represent? (*b*) How is it obtained?
21. What factors affect the transconductance of a tube?
22. What is the mathematical relation among the amplification factor, transconductance, and the plate resistance of a tube?
23. Explain how the relationship among the amplification factor, transconductance, and plate resistance is developed mathematically.
24. What is the relation between the transconductance of a tube and its operating characteristics?
25. (*a*) What is a tetrode? (*b*) By what other names is it also known?
26. Describe the construction and location of the screen grid in a tetrode.
27. (*a*) What is meant by interelectrode capacitance? (*b*) Why is it necessary to keep these capacitances at a minimum?
28. What is meant by (*a*) regeneration? (*b*) Degeneration?
29. How does the addition of a screen grid eliminate feedback?
30. What is a pentode?
31. Describe the construction and location of the suppressor grid in a pentode.
32. Describe the action of the suppressor grid in a pentode.
33. (*a*) What is meant by multiunit tubes? (*b*) What are their advantages?
34. Name six combinations used in multiunit tubes.
35. Describe two systems of tube-pin and socket-connection numbering.
36. (*a*) What is a supercontrol tube? (*b*) What are its constructional features?
37. Explain the meaning of the following terms: (*a*) sharp cutoff, (*b*) remote cutoff, (*c*) cross modulation, (*d*) modulation distortion.
38. Explain the action of the supercontrol tube.
39. What are the advantages of variable-mu tubes and where are they generally used?
40. Describe the beam power tube.
41. Describe the action of a beam power tube.
42. What are the characteristics of the beam power tube?
43. In a beam power tube, what factors are responsible for its high power output and high efficiency?
44. Explain the operation of a thyratron.
45. (*a*) Will the thyratron represented by Fig. 5-16 start conducting with 500 volts applied to its plate and −8 volts at its grid? (*b*) If the answer to (*a*) is no, what change or changes in electrode voltages will initiate conduction?
46. Name five applications of the thyratron.

PROBLEMS

1. For the triode circuit of Fig. 5-1, plot a series of graphs showing the waveform of the following quantities: (*a*) sine-wave input signal having a maximum value of 1 volt, (*b*) grid bias provided by a 3-volt d-c source, (*c*) total voltage at the grid of the tube, (*d*) typical waveform of the plate current, (*e*) typical waveform of the output voltage.

2. For the triode circuit of Fig. 5-1, plot a series of graphs showing the waveform of the following quantities: (*a*) sine-wave input signal having a maximum value of 3 volts, (*b*) grid bias provided by a 3-volt d-c source, (*c*) total voltage at the grid of the tube, (*d*) typical waveform of the plate current, (*e*) typical waveform of the output voltage.

3. For the triode circuit of Fig. 5-1, plot a series of graphs showing the waveform of the following quantities: (*a*) sine-wave input signal having a maximum value of 4 volts, (*b*) grid bias provided by a 3-volt d-c source, (*c*) total voltage at the grid of the tube.

4. From the curves of Fig. 5-2, determine the values of i_b when (*a*) $e_b = 80$ and $E_C = 0$, (*b*) $e_b = 180$ and $E_C = -4$, (*c*) $e_b = 360$ and $E_C = -16$.

5. From the curves of Fig. 5-2, determine the values of e_b when (*a*) $E_C = -2$ and $i_b = 10$, (*b*) $E_C = -18$ and $i_b = 4$, (*c*) $E_C = -7$ and $i_b = 10$.

6. From the curves of Fig. 5-2, determine the values of E_C when (*a*) $e_b = 105$ and $i_b = 6$, (*b*) $e_b = 260$ and $i_b = 10$, (*c*) $e_b = 300$ and $i_b = 4$.

7. From the curves of Fig. 5-3, determine the values of e_c when (*a*) $E_b = 240$ and $i_b = 10$, (*b*) $E_b = 160$ and $i_b = 7$, (*c*) $E_b = 400$ and $i_b = 7.2$.

8. From the curves of Fig. 5-3, determine the values of i_b when (*a*) $E_b = 320$ and $e_c = -12$, (*b*) $E_b = 240$ and $e_c = -10$, (*c*) $E_b = 160$ and $e_c = -4$.

9. From the curves of Fig. 5-3, determine the values of E_b when (*a*) $e_c = -14$ and $i_b = 11$, (*b*) $e_c = -2$ and $i_b = 3.5$, (*c*) $e_c = -12$ and $i_b = 5$.

10. The data in Table 5-1 were obtained from a test of a type 6AT6 tube operated at its rated heater voltage. Plot the plate characteristic curves from these test data.

Table 5-1 Test Data—Type 6AT6

e_b, VOLTS	i_b, ma		
	$E_c = 0$	$E_c = -1$	$E_c = -2$
0			
25	0.6		
50	1.3	0.15	
75	2.1	0.45	
100	2.9	0.9	0.1
125	3.7	1.4	0.3
150	4.6	2.05	0.65
175	5.4	2.75	1.0
200		3.5	1.5
225		4.3	2.0
250		5.1	2.65
275			3.4
300			4.15
325			5.0

11. A high-mu triode is operated at its normal heater voltage, a grid bias of 1.75 volts, and a plate potential of 220 volts. Determine the amplification factor if the characteristics of this tube are such that (1) the plate current is 2 ma when $e_b = 200$ volts and $e_c = -1.5$ volts, and (2) when the plate voltage is increased to 240 volts the grid bias has to be increased to 2 volts in order to maintain the plate current at 2 ma.

12. A sharp-cutoff beam triode is operated at its normal heater voltage, a grid bias of 4.5 volts, and a plate potential of 22,500 volts. Determine the amplification factor if the characteristics of this tube are such that (1) the plate current is 2 amp when $e_b = 20{,}000$ volts and $e_c = -3.25$ volts, and (2) when the plate voltage is increased to 25,000 volts the grid bias has to be increased to 5.75 volts in order to maintain the plate current at 2 amp.

13. What is the amplification factor of the tube represented by the curves of Fig. 5-2 if it is operated with 200 volts on its plate, and with a grid bias of 6 volts?

14. What is the amplification factor of the tube used in Prob. 13 when the grid bias is increased to 8 volts and the plate voltage is kept at 200 volts?

15. Determine the plate resistance of the electron tube used in Prob. 11 if the characteristics of this tube are such that (1) if the plate voltage is decreased to 200 volts, the plate current will be 1.6 ma; (2) if the plate voltage is increased to 240 volts, the plate current will increase to 2.4 ma; and (3) the grid bias is maintained at 1.75 volts.

16. termine the plate resistance of the electron tube used in Prob. 12 if the characteristics of this tube are such that (1) if the plate voltage is decreased to 20,000 volts, the plate current will be 0.75 amp; (2) if the plate voltage is increased to 25,000 volts, the plate current will increase to 1.5 amp; and (3) the grid bias is maintained at 7.0 volts.

17. What is the plate resistance of the tube represented by the curves of Fig. 5-2 if it is operated with 200 volts on its plate, and with a grid bias of 6 volts?

18. What is the plate resistance of the tube used in Prob. 17 when the grid bias is increased to 8 volts and the plate voltage is kept at 200 volts?

19. Determine the transconductance of the electron tube used in Prob. 11 if the characteristics of this tube are such that (1) if the grid bias is decreased to 1.5 volts, the plate current will be 2.2 ma; (2) if the grid bias is increased to 2.0 volts, the plate current will decrease to 1.5 ma; and (3) the plate voltage is maintained at 220 volts.

20. Determine the transconductance of the electron tube used in Prob. 12 if the characteristics of this tube are such that (1) if the grid bias is decreased to 3.75 volts, the plate current will be 2.12 amp; (2) if the grid bias is increased to 5.25 volts, the plate current will decrease to 1.62 amp; and (3) the plate voltage is maintained at 22,500 volts.

21. (*a*) What is the transconductance of the tube represented by the curves of Figs. 5-2 and 5-3 if it is operated with 200 volts on its plate, and with a grid bias of 6 volts? (Solve by obtaining data directly from the curves of Fig. 5-3.) (*b*) Check the answer to part (*a*) by use of Eq. (5-6) and the answers to Probs. 13 and 17.

22. (*a*) What is the transconductance of the tube used in Prob. 21 when its grid bias is increased to 8 volts and the plate voltage is kept at 200 volts? (Solve by obtaining data directly from the curves of Fig. 5-3.) (*b*) Find the value by use of Eq. (5-6) and the answers of Probs. 14 and 18.

23. The type 6AT6 tube, whose curves were plotted in Prob. 10, when operated with 200 volts at its plate and with a grid bias of 2 volts has a plate resistance of 50,000 ohms and an amplification factor of 70. What is the transconductance of the tube under these conditions?

24. What is the transconductance of each section of a twin triode when it is operated at such values that its amplification factor is 20 and its plate resistance is 6,450 ohms?

25. When the tube of Prob. 23 is operated with 100 volts at its plate and with a grid bias of 1 volt, its plate resistance is 53,000 ohms and its transconductance is 1,400 μmhos. What is the amplification factor of the tube under these conditions?

26. What is the amplification factor of each section of a twin triode when it is operated at such values of voltage that its plate resistance is 15,000 ohms and its transconductance is 4,000 μmhos?

27. When the tube of Prob. 23 is operated with 200 volts at its plate and with a grid bias of 1 volt, its transconductance is 2,500 μmhos and its amplification factor is 80. What is the plate resistance of the tube under these conditions?

28. What is the plate resistance of each section of a twin triode when it is operated at such values of voltage that its transconductance is 6,000 μmhos and its amplification factor is 36?

29. From the curves for a tetrode (Fig. 5-7) determine the approximate value of I_b when (*a*) $E_{C1} = 0$ and $E_b = 300$ volts, (*b*) $E_{C1} = -3$ volts and $E_b = 220$ volts, (*c*) $E_{C1} = -3$ volts and $E_b = 500$ volts, (*d*) $E_{C1} = -1.5$ volts and $E_b = 300$ volts.

30. From the curves of a pentode (Fig. 5-9) determine the approximate value of I_b when (*a*) $E_{C1} = -2$ volts and $E_b = 80$ volts, (*b*) $E_{C1} = -2$ volts and $E_b = 160$ volts, (*c*) $E_{C1} = -2$ volts and $E_b = 320$ volts, (*d*) $E_{C1} = -0.5$ volt and $E_b = 40$ volts, (*e*) $E_{C1} = -0.5$ volt and $E_b = 240$ volts.

31. From the curves of Fig. 5-12, what is the approximate grid-voltage change corresponding to a variation in plate current of from 1 to 3 ma for (*a*) the normal grid tube? (*b*) The supercontrol tube?

32. From the curves of Fig. 5-12, what is the approximate grid-voltage change corresponding to a variation in plate current of from 5 to 2 ma for (*a*) the normal grid tube? (*b*) The supercontrol tube?

33. From the curves of the beam power tube (Fig. 5-15) determine the approximate value of I_b when E_b is 400 volts and E_{C1} is (*a*) -5 volts, (*b*) -10 volts, (*c*) -15 volts, (*d*) -20 volts.

34. From the curves of the beam power tube (Fig. 5-15) determine the approximate value of I_b when $E_{C1} = -10$ volts and E_b is (*a*) 100 volts, (*b*) 200 volts, (*c*) 300 volts, (*d*) 400 volts, (*e*) 500 volts.

35. From Fig. 5-16, determine the approximate maximum value of grid voltage at which the thyratron will start conducting if the plate voltage is (*a*) 900 volts, (*b*) 750 volts, (*c*) 400 volts, (*d*) 200 volts.

36. From Fig. 5-16, determine the approximate minimum value of plate voltage at which the thyratron will start conducting if the grid voltage is (*a*) -3 volts, (*b*) -5 volts, (*c*) -6.5 volts, (*d*) -7.5 volts.

Chapter 6

Power Sources

All types of communications equipment and practically all electronic devices require some amount of d-c power in order to perform their functions satisfactorily. When the amount of power required is small and/or the device is portable, some type of cell or battery is generally used as the source of energy. However, if the device requires an amount of power beyond that which is economically feasible from a battery, and/or the application of the device does not require portability, the energy is generally obtained from a power line or from an auxiliary power source.

6-1 Primary Cells

Definition. A primary cell is one in which the chemical action decomposes one of the electrodes when the cell delivers current. When a relatively large portion of the electrode has been eaten away, the cell is no longer capable of delivering its rated voltage and current and has to be replaced.

Fundamental Principles. If a strip of copper and a strip of zinc (the *electrodes*) are placed in a jar of water containing a small amount of sulfuric acid (the *electrolyte*), chemical action will make electrical energy available at the open ends of the copper and zinc electrodes (Fig. 6-1). The basic characteristics of a cell are:

1. The two electrodes must be of different materials.
2. The electrodes must be conductors of electricity.
3. The electrolyte must contain an acid, alkali, or salt that will conduct the current.
4. The voltage of the cell will vary with the materials used as electrodes and electrolyte but will not exceed approximately 2 volts.
5. The voltage of the cell is not affected by the size or spacing of its electrodes.
6. The current capacity of a cell may be raised by increasing the surface area of that part of its electrodes actually making contact with the electrolyte.

Chemical Action. If a cell as shown in Fig. 6-2 is made by placing a carbon rod and a strip of zinc into a jar containing water to which a small amount of ammonium chloride (sal ammoniac) has been added, certain chemical actions will result. When the ammonium chloride is placed in the water, many of the atoms become separated from their original molecules and ammonia and chlo-

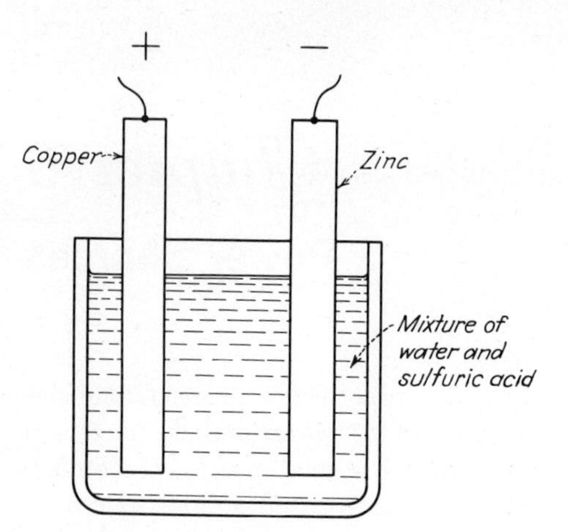

Fig. 6-1 Simple cell.

rine molecules are formed. Another action that takes place during this process is an unbalancing of the number of electrons and protons in each atom. When an atom is unbalanced, the number of electrons and protons no longer equal one another and it is then called an *ion.* If the protons are greater in number, the atom is called a *positive ion;* if the electrons are greater in number, the atom is called a *negative ion.* The process of unbalancing the atoms is called *ionization* of the electrolyte.

While chemical action is going on in the cell, zinc ions unite with chlorine ions to form zinc chloride, which remains in the solution. As some zinc is continually being given off to the solution, the zinc electrode is being eaten away, and this device is therefore a primary cell.

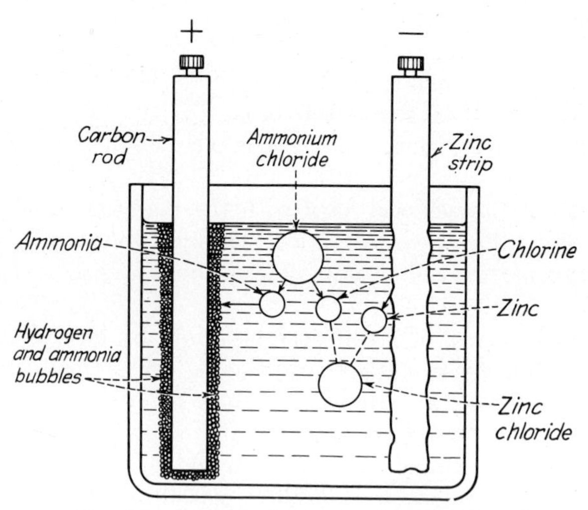

Fig. 6-2 Chemical action of a simple cell.

Types of Primary Cells. The earliest cell consisted basically of carbon, zinc, and ammonium chloride and is commonly referred to as a *carbon-zinc cell.* Many different types of cells have since been developed to meet specific needs. The following types of cells are most generally used: (1) the carbon-zinc cell, (2) the mercury cell, (3) the silver oxide cell, and (4) the alkaline cell.

Cell Life. The life of a primary cell is determined by the following factors: (1) *initial drain,* the current that the cell is expected to deliver at full voltage; (2) *operating schedule,* the daily time interval (or intervals) during which the cell is required to deliver current; (3) *cutoff voltage,* the voltage below which the cell is no longer useful; (4) *temperature*—this factor is dependent on its operating characteristics at both low and high temperatures; (5) *shelf life*—a cell that is not put into use will gradually deteriorate because of slow chemical reactions and changes in moisture content which take place in the cell.

6-2 The Carbon-zinc Cell

Construction. The cross section of a typical carbon-zinc cell is shown in Fig. 6-3*a*. The cell is built in a cylindrical zinc container which also acts as the anode or negative electrode. A cathode mix, which acts as the positive electrode, occupies most of the cell. The cathode mix is a powder, consisting of manganese dioxide, zinc chloride, and graphite; a carbon rod is placed in

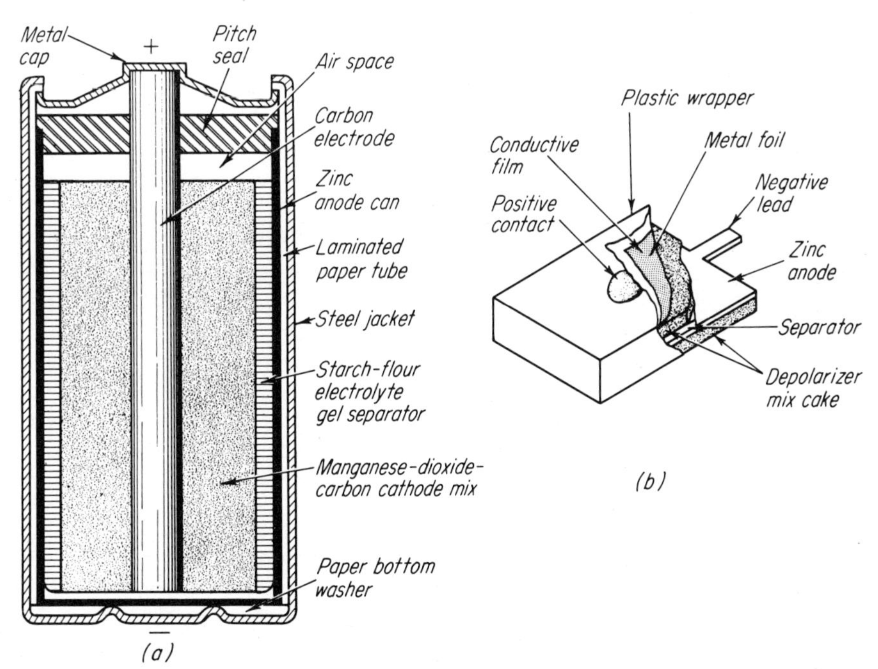

Fig. 6-3 (a) Cross-sectional view of a cylindrical-type carbon-zinc cell. (b) Cutaway view of a flat-type carbon-zinc cell. (*RCA*)

the center of the cathode mix to serve as a connection for the positive terminal. A gelatinous paste, which contains the electrolyte, separates the cathode mix from the zinc and functions as the ion-transfer medium between the positive and negative electrodes. The electrolyte is a solution of ammonium chloride, zinc chloride, and water. As the zinc container is also the negative electrode, it is protected with an insulating covering of paper or cardboard and in some types of cell also with a steel jacket.

Figure 6-3*b* illustrates a compact, flat-type, carbon-zinc cell used extensively in portable electronic equipment.

Action of the Cell. The carbon-zinc cell is fundamentally the same as the simple cell described in Art. 6-1, and its action, therefore, will be the same. The action of the water and the ammonium chloride in the paste, together with the zinc and carbon electrodes, produces the electric current. The manganese dioxide is added to reduce polarization due to hydrogen, and the zinc chloride is added to reduce polarization due to the ammonia. The gelatinous paste contains materials, such as cornstarch and flour, to permit the electrolyte to filter through it slowly.

Characteristics of the Cell. The carbon-zinc cell has a rated output of 1.5 volts and a low discharge rate; its terminal voltage may decrease appreciably with the use of the cell. For similar ratings, it is both larger in size and heavier in weight than the newer types of primary cell. This cell is generally used where size, weight, and voltage fluctuation are not of great importance. Carbon-zinc cells are available in a wide variety of sizes and shapes with a broad range of operating characteristics. Most primary cells used in communications and other electronic applications are of this type.

6-3 The Mercury Cell

Construction of the Mercury Cell. The fundamental components of the mercury cell (Fig. 6-4) are (1) a *depolarizing cathode* consisting of small pellets of mercuric oxide and a small percentage of graphite, (2) an *anode* of high-purity amalgamated zinc powder, (3) a *concentrated aqueous electrolyte* of potassium hydroxide and zinc oxide, and (4) a *sealed steel container*. It should be noted that the polarity of the electrodes of the mercury cell is reversed from that of the carbon-zinc cell.

In the cylindrical type of cell (Fig. 6-4*a*), the anode is pressed into a hollow cylindrical rod that forms the center portion of the cell. The cathode is pressed into a cylindrical sleeve that forms the outer portion of the cell. In the flat-type cell (Fig. 6-4*b*), the cathode is pressed into a flat shape and is consolidated with the bottom of the cell case. The anode is also pressed into a flat shape and forms the upper portion of the cell. In both types of cell the two electrodes are separated by an absorbent material that contains the electrolyte. A permeable barrier prevents the migration of any solid particles in the cell, thereby contributing effectively to long shelf and service life. The insulating and sealing gaskets are molded polyethylene or neoprene, depending on the

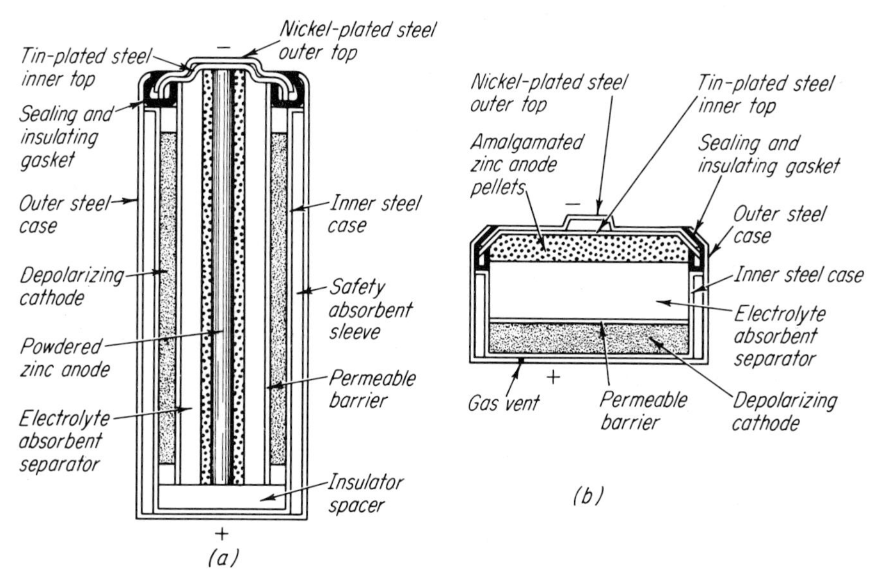

Fig. 6-4 Cross-sectional views of mercury cells. (a) Cylindrical type. (b) Flat-type. *(Union Carbide Corporation, Consumer Products Division)*

cell application. The inner-cell tops are plated with a material that provides an inner surface to which the zinc forms a zinc amalgam bond. The cell cases and outer tops are made of nickel-plated steel in order to (1) resist corrosion and (2) provide greater inactivity to internal cell materials. The outer nickel-plated steel jacket is a necessary component for self-venting construction. This vent provides a means for releasing the excessive gas in a cell that occurs during a reverse current or short circuit.

Characteristics of the Mercury Cell. The mercury cell may be obtained with a rating of either 1.35 or 1.40 volts. This voltage is relatively stable at rated current, having a voltage regulation of 0.5 per cent for long periods of time and 0.1 per cent for short intervals. A high-current drain or a momentary short circuit will not permanently damage the cell, and there is a complete recovery to its full open-circuit voltage within minutes.

In comparison with the carbon-zinc cell the mercury cell has (1) a much higher rate of discharge, (2) a more constant ampere-hour capacity that is approximately three times greater per unit volume, (3) a much better performance at high temperatures, (4) a lower and more substantially constant internal impedance, (5) a much longer shelf life, (6) less frequent replacement, and (7) a higher initial cost.

Applications of the Mercury Cell. Because of its superior characteristics, the mercury cell has replaced the carbon-zinc cell in some applications. Because of its relatively constant potential, the mercury cell may be used as a secon-

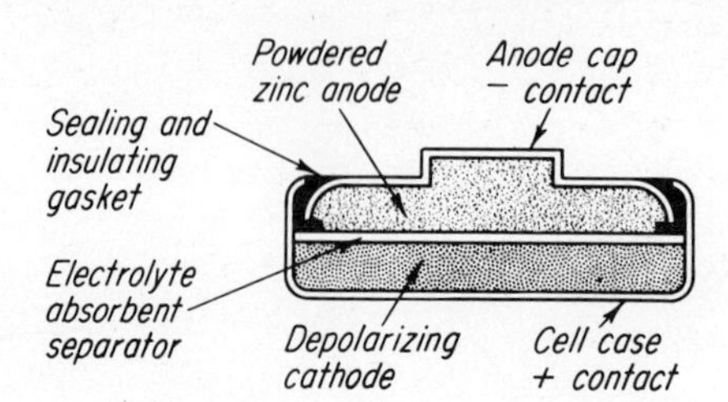

Fig. 6-5 Cross-section view of a silver oxide cell. (*Union Carbide Corporation, Consumer Products Division*)

dary standard of voltage. Some of its applications as a reference source are regulated power supplies, radiation detector meters, portable potentiometers, electronic computers, and voltage recorders. The flat-type cell is used in compact electronic equipment where physical size is an important factor.

6-4 The Silver Oxide Cell

Construction of the Silver Oxide Cell. A silver oxide cell basically consists of a depolarizing silver oxide cathode, a zinc anode, and a highly alkaline electrolyte (Fig. 6-5). It should be noted that the polarity of the electrodes of the silver oxide cell is reversed from that of the carbon-zinc cell. The depolarizing cathode is a mixture of silver oxide and manganese dioxide that is pressed into a flat shape and consolidated with the bottom of the cell case. The percentage of the manganese dioxide depolarizer that is used will depend upon whether the cell is to provide a flat-discharge current or increased service hours. Powdered zinc, which is pressed into a disk having a relatively large surface area, forms the upper portion of the cell. The two electrodes are separated by an absorbent material containing the electrolyte. This electrolyte may be either (1) potassium hydroxide for cells requiring maximum power density at rated currents, such as cells used in hearing aids, or (2) sodium hydroxide for cells requiring long-term reliability, such as cells used in electric timepieces.

Characteristics of the Silver Oxide Cell. The silver oxide cell is an excellent miniature power source. Its relatively high open-circuit and operating voltages, 1.6 and 1.5 volts, respectively, and its flat discharge characteristic enable the silver oxide cell to provide more service life or power than is available from other miniature cells of the same size. Its internal impedance is low and nearly constant. The silver oxide cell will withstand severe abuses with no hazard to personnel or equipment. It has an excellent shelf life, having approximately 90 per cent capacity after one year of storage at 70°F. Because of the relatively large surface area of the anode, this type of cell is capable of supplying a service capacity that is much greater than can be obtained from other types of cell of comparable size. The silver oxide cell is primarily used for hearing aids, electric watches, and reference voltage sources.

6-5 The Alkaline Manganese-zinc Cell

Construction of the Alkaline Cell. The alkaline-manganese cell is commonly referred to as an *alkaline cell;* its basic construction is shown in Fig. 6-6. The alkaline cell is similar to the carbon-zinc cell in that they both have zinc negative electrodes and manganese dioxide positive electrodes. They differ only in the structure of the electrodes and the material used for the electrolyte. The cathode consists of a manganese dioxide cylindrical sleeve that forms the outer portion of the cell. A steel can, which makes direct contact to the outside surface of this electrode, serves as a cathode collector. The upper portion of this can makes contact to the metal cap, or positive terminal. The zinc anode is also a cylindrical sleeve and forms the inner portion of the cell. A cylindrical anode collector makes contact to the inner surface of this electrode and also to the outer metal bottom of the cell, or negative terminal. The alkaline cell uses potassium hydroxide to produce a highly alkaline electrolyte. The construction of this cell permits the electrolyte to make contact with a very large surface area of the zinc anode. To prevent leakage the cells are hermetically sealed and encased in a steel jacket.

Characteristics of the Alkaline Cell. The alkaline cell is designed to meet a need for a power source having a high-rate source of energy, with a high-service capacity. This cell is rated at 1.5 volts and has a relatively constant ampere-hour capacity over a wide range of current drains. The primary advantage of the alkaline cell is its ability to perform with a high degree of efficiency under continuous high-current service. Under certain conditions, the alkaline cell will provide more than ten times the service of a similar size carbon-zinc cell. The alkaline cell has an extremely low internal impedance,

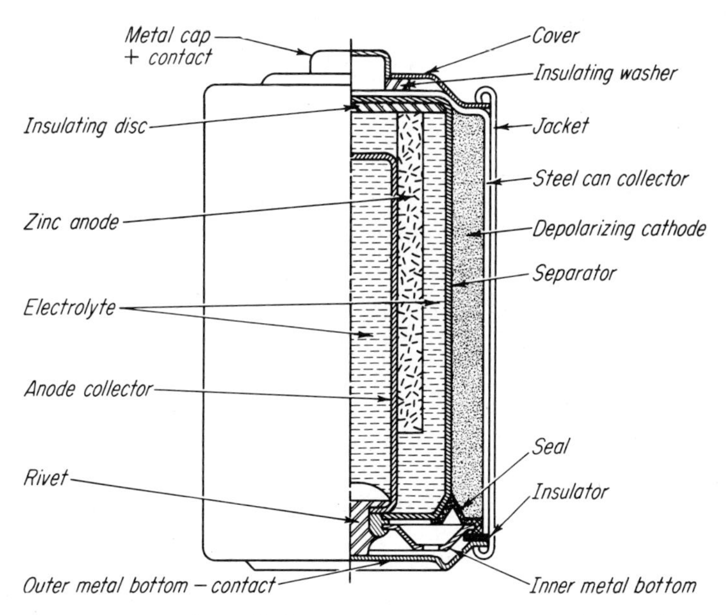

Fig. 6-6 Cutaway view of an alkaline manganese-zinc cell. (*Union Carbide Corporation, Consumer Products Division*)

and its low-temperature characteristic results in good service life even under outdoor winter conditions. The alkaline cell has a relatively long shelf life at normal temperatures and is better than the carbon-zinc cell for both high- and low-temperature applications.

Applications of the Alkaline Cell. Before the alkaline cell was developed, many devices were considered to be impractical to operate from a battery because a suitable power source was not available. These devices can now be operated from alkaline batteries which are ideal for high-drain heavy-discharge schedules such as (1) heater element applications; (2) bicycle lights and horns; (3) heavy-duty lighting; (4) motion-picture cranking; (5) glo-plug ignition; (6) toys; (7) model boats, airplanes, and automobiles; (8) electric shavers; and (9) electronic photoflash units.

6-6 Secondary Cells

Definition. A secondary cell is one in which the electrodes and electrolyte are altered by chemical action that takes place when the cell delivers current and may be restored to their original condition by sending a current through them in the opposite direction. The automobile storage battery and the nickel-cadmium rechargeable battery are examples of the secondary cell. Charging these batteries after they have become discharged represents restoring the cells to their original state.

Fundamental Principles. Secondary cells operate on the same principles as primary cells but differ in the method in which they may be renewed. Some of the materials of a primary cell are used up in the process of changing chemical energy to electrical energy, and they must be replaced to renew the cell. In the secondary cell, the materials are merely transferred from one electrode to the other, and they may be restored to their original state by sending an electric current from some other source through the cell in the opposite direction.

6-7 Lead-acid Cell

Construction. Figure 6-7 illustrates the construction of a commercial type of lead-acid cell. The negative electrodes are made of spongy lead, and the positive electrodes are made of lead peroxide. The electrolyte is a sulfuric acid solution. The lead-acid cell produces approximately 2.2 volts.

The capacity of a lead-acid cell is expressed in ampere-hours, and this rating is proportional to the amount of active surface area of the electrodes, or *plates* as they are commonly called. The total active area of the plates is therefore an important factor in the rating of a cell. In a multiple-plate cell a positive plate is placed between two negative plates, and both sides of the positive plate become active and the cell can be made smaller. To conserve space, the plates are placed close to each other, and insulators are placed between them to prevent a positive plate from making contact with either of its ad-

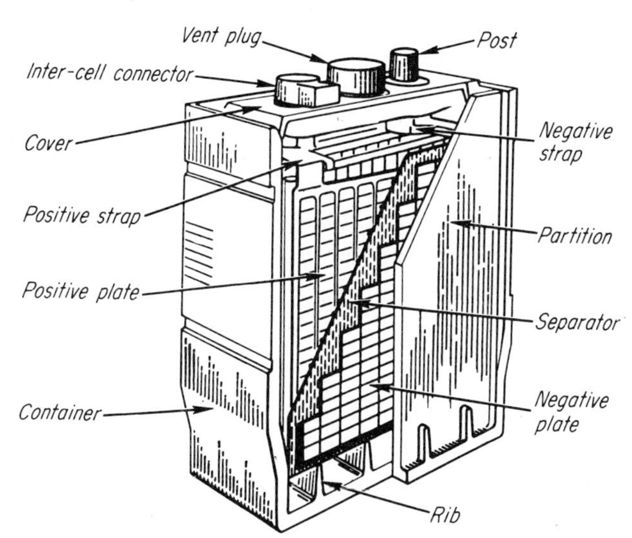

Fig. 6-7 A cutaway view of a lead-acid cell showing details of construction. (*Electric Storage Battery Co.*)

jacent negative plates. These insulators, called *separators*, are generally made of hard rubber, wood, or plastics.

Each cell is placed in a container that is usually made of hard rubber. After the plates are placed in the container, a hard-rubber cover fitted with a filler tube and vent cap is set over the cell and sealed with a compound. The electrolyte is a dilute sulfuric acid solution mixed in such proportions that with a fully charged cell its specific gravity will be approximately 1,300.

Characteristics. The advantages of the lead-acid cell are its (1) relatively high voltage per cell, (2) high current capacity, (3) relatively long life, and (4) relatively low initial cost. The disadvantages are (1) it is relatively large and heavy, (2) it cannot be hermetically sealed, (3) it has poor low-temperature characteristics, and (4) it cannot remain in the discharge state too long without damage.

6-8 Nickel-iron Cell

The nickel-iron cell is usually called the *Edison cell* after its inventor. This cell basically consists of an iron anode, a nickel oxide cathode, and a potassium hydroxide alkaline electrolyte. In a commercial cell, the positive plate consists of a number of long nickel tubes filled with nickel oxide. As nickel oxide is a poor conductor, very fine nickel flakes are mixed with it to produce the required conductivity. The negative plate consists of a number of flat perforated nickel-plated steel stampings containing finely divided metallic iron. The electrolyte is potassium hydroxide. As with the lead-acid cell, the capacity of this cell can be increased by connecting a number of negative plates and a number of positive plates in parallel.

The nickel-iron cell is relatively light in weight and extremely rugged and can withstand abuses such as overcharges and overdischarges. In addition, it can remain discharged for a long period of time or be subjected to freezing temperatures without chemical deterioration.

The Edison cell is rated on the basis of a 5-hr charging rate and has an average emf of 1.2 volts during discharge. The nickel-iron cell has a cycle service of 1,800 cycles, a float charging life of 7 to 12 years in heavy discharge service, and 14 to 25 years as a standby source of power. This type of cell is employed mainly in heavy-duty industrial and railway applications.

6-9 Nickel-cadmium Cell

Action of the Nickel-Cadmium Cell. Basically, the nickel-cadmium cell is similar to the nickel-iron cell except for the use of cadmium as the negative electrode. In the charged condition, the cathode is nickelic hydroxide, the anode cadmium, and the electrolyte potassium hydroxide. In the discharge condition, the positive electrode becomes nickel oxide, the negative electrode cadmium hydroxide, and the electrolyte does not change and remains as potassium hydroxide.

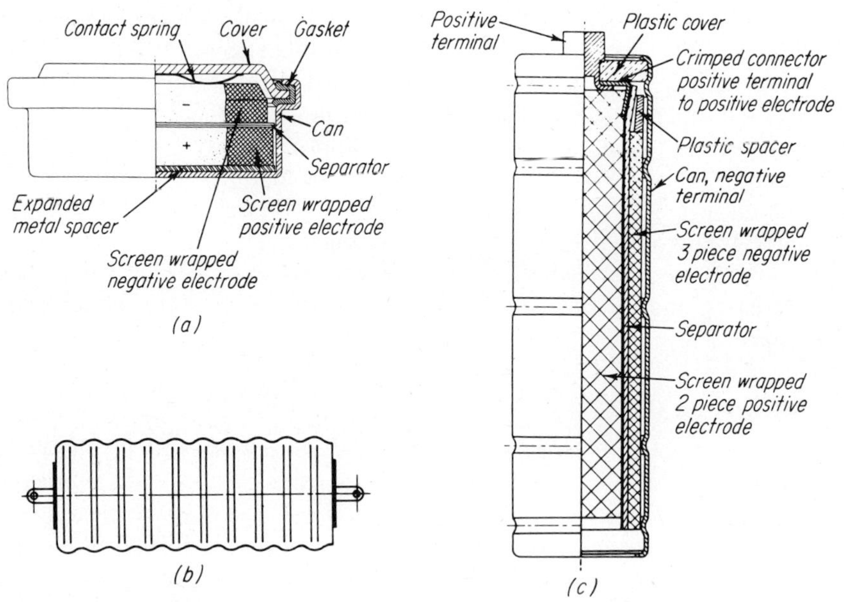

Fig. 6-8 Nickel-cadmium cells. (a) Cutaway view of a button cell. (b) Button stack assembly. (c) Cutaway view of a cylindrical cell. (*Union Carbide Corporation, Consumer Products Division*)

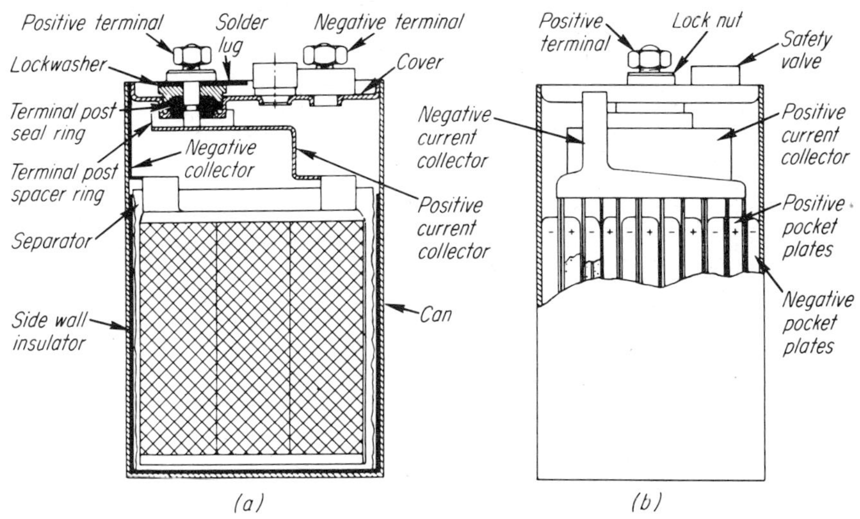

Fig. 6-9 Cutaway views of a rectangular nickel-cadmium cell. (a) Side view. (b) End view. (*Union Carbide Corporation, Consumer Products Division*)

Construction of the Nickel-Cadmium Cell. The nickel-cadmium cell is a sealed unit that is available in three basic types of construction: (1) button, (2) cylindrical, and (3) rectangular. The electrodes used in the button and cylindrical cells (Fig. 6-8) consist of molded screen-encased active materials. In the rectangular cell the plates consist of pressed active materials held in perforated steel pockets that are locked into welded frames (Fig. 6.9).

During the latter part of a recommended charge cycle and also during an overcharge, the nickel-cadmium cell generates gases. When the cell becomes fully charged, hydrogen is formed at the cadmium electrode, and after it is fully charged, oxygen is generated at the nickelic hydroxide electrode. In a conventional *vented-type cell* the hydrogen and oxygen gases, plus any entrained fumes from the electrolyte, are liberated through a valve. In a hermetically sealed cell these gases are used up inside the cell. This is accomplished in the following manner: (1) The cadmium electrode is constructed with an excess ampere-hour capacity. (2) The positive electrode will reach full charge first and will start to generate oxygen; however, the negative electrode will not reach full charge as yet, and therefore no hydrogen will be generated. (3) The cell is so designed that the oxygen reaches the surface of the cadmium electrode, where it reacts to form electrochemical equivalents of cadmium oxide. (4) During overcharge the cadmium is oxidized at a rate that offsets the input energy, thus keeping the cell in equilibrium at full charge. This process can continue for long periods.

Each of the three types of nickel-cadmium cell come in various sizes and may be used individually or connected in series to obtain a desired voltage. The button cell is made in series stacks (Fig. 6-8*b*), the cylindrical cell in assemblies of varying configurations, and the rectangular cell in a five-cell (6 volts) pack in a plastic case. The cylindrical cell incorporates a different electrode arrangement from the button cell (Fig. 6-8*c*). Its terminal connections are similar to those of small cylindrical dry cells, and some sizes may be used as a direct replacement for these primary cells. The rectangular cell is a heavier-duty cell than the other two types and covers a range of 1.5 to 23 amp-hr. The steel case is polarized positive and contains a safety vent that is activated only in case of severe misuse of the cell (Fig. 6.9).

Characteristics of the Nickel-Cadmium Cell. Like the nickel-iron cell, the nickel-cadmium cell is mechanically rugged and has a long potential life. During discharge the average voltage is approximately 1.2 volts. Under conditions of very light or casual service the expected life of a cell is several years. With normal service and discharge of rated capacity, the cycle life is in excess of 100 for the button and cylindrical cells and in excess of 400 for the rectangular cell. The nickel-cadmium cell has a high effective capacitance; however, its impedance is very low. Although the sealed nickel-cadmium cell will lose some of its charge during storage, this type of cell has a lower self-discharge rate than any other cell. More important, the cell is not harmed chemically even if not used for long periods of time.

Applications of the Nickel-Cadmium Cell. Because the nickel-cadmium cell is (1) relatively small, (2) rechargeable, (3) hermetically sealed, (4) economical, and (5) trouble-free, it is ideal for use in many types of battery-operated equipment. Some of these applications are (1) alarm systems, (2) amplifiers, (3) dictating machines, (4) electric shavers, (5) electronic photoflash, (6) emergency lighting, (7) instruments, (8) hearing aids, (9) motion-picture cameras, (10) radio receivers, (11) tape recorders, (12) telemetery, and (13) transmitters.

6-10 Cadmium–Silver Oxide Cell

The cadmium-silver oxide cell uses a cadmium anode for long operating life, a silver oxide cathode for high watthour capacity, and a potassium hydroxide electrolyte. This type of cell is capable of providing higher currents, a more level voltage, and up to three times greater watthour capacity per unit weight

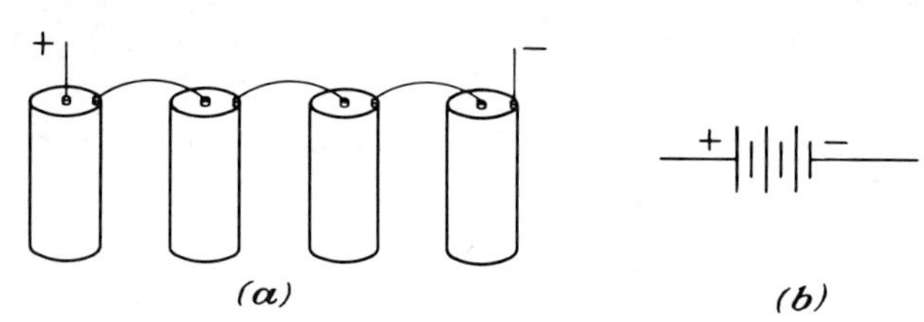

Fig. 6-10 Cells connected in series. (a) Pictorial representation. (b) Schematic-diagram representation.

and volume than any other type of secondary cell. It also has a high efficiency on extended shelf life in either a charged or discharged condition, is mechanically rugged, and operates satisfactorily over a wide temperature range. Its use has been limited to applications where cost is not a factor and space and weight are the prime considerations. Cadmium–silver oxide cells can be used (1) to power a portable television receiver, (2) as storage batteries in satellite programs, and (3) in conjunction with solar-cell systems.

6-11 Batteries

Definition. A battery consists of two or more cells that are connected to each other and are usually placed in a common container. For example, the 12-volt batteries used in automobiles consist of six 2-volt cells connected in series, and a 4½-volt battery used in a transistor radio receiver may consist of six 1½-volt cells connected in a parallel-series arrangement of two groups of cells connected in parallel with each group having three cells connected in series.

Need for Extra Cells. Many applications require a higher voltage, a higher current, or both, than a single cell can provide. To meet these needs, a number of cells are joined into a series-connected group, a parallel-connected group, or a series-parallel-connected group, depending upon the voltage and current requirements.

Cells Connected in Series. Whenever the voltage required exceeds that of a single cell, it becomes necessary to use more than one cell and the cells must be connected in series as shown in Fig. 6-10*a*. In a series circuit the battery voltage will equal the sum of the separate cell voltages.

$$E_B = E_{c1} + E_{c2} + E_{c3}, \text{ etc.} \tag{6-1}$$

If each cell has the same voltage, then

$$E_B = \text{number of cells} \times \text{volts per cell} \tag{6-2}$$

also

$$\text{Number of cells required} = \frac{\text{voltage of battery}}{\text{volts per cell}} \tag{6-3}$$

In the series arrangement of cells, the current rating of the battery will be the same as that of an individual cell; hence this arrangement can be used

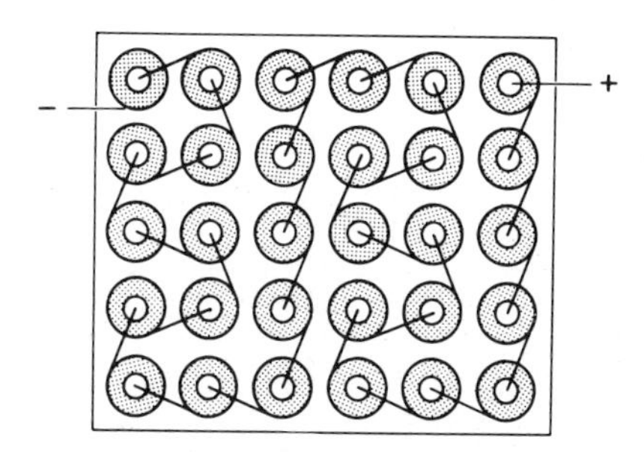

Fig. 6-11 Arrangement of cells in a 45-volt battery using cylindrical cells.

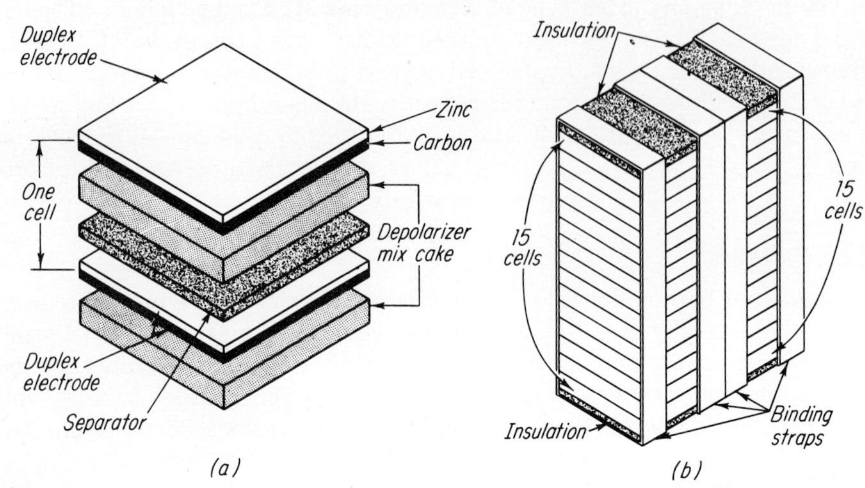

Fig. 6-12 A miniature 45-volt battery. (a) Arrangements of the parts of a cell. (b) 30 cells assembled in two stacks of 15 each. (*Union Carbide Corporation, Consumer Products Division*)

only where the continuous current requirement does not exceed the rated current of one cell.

Cells Connected in Parallel. Whenever a continuous current greater than the rated current of one cell is required, it becomes necessary to use more than one cell and the cells must be connected in parallel. The parallel arrangement of cells is shown in Fig. 6-13. In the parallel circuit, the current rating of the battery is equal to the sum of the rated currents of the separate cells.

$$I_B = I_{c1} + I_{c2} + I_{c3}, \text{ etc.} \tag{6-4}$$

If each cell has the same current rating, then

$$I_B = \text{number of cells} \times \text{current per cell} \tag{6-5}$$

also

$$\text{Number of cells required} = \frac{\text{current of battery}}{\text{current per cell}} \tag{6-6}$$

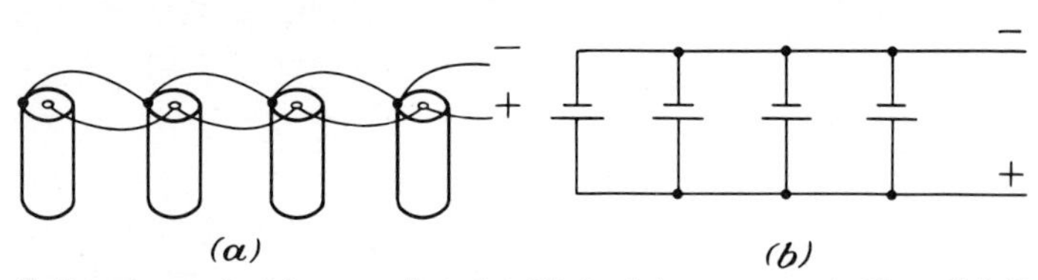

Fig. 6-13 Cells connected in parallel. (a) Pictorial representation. (b) Schematic-diagram representation.

In the parallel arrangement of cells, the voltage of the battery will be the same as that of a single cell; hence this arrangement can be used only where the voltage requirement is that of one cell.

Cells Connected in Series and Parallel Combinations. When both the voltage and current required exceed the rated voltage and current of a single cell, it becomes necessary to use four or more cells connected in a parallel-series combination. To get the higher voltage, a number of cells must be connected in series, and to get the higher current rating, a number of series-connected groups must be connected in parallel.

$$\text{Number of cells in each series-connected group} = \frac{\text{voltage of battery}}{\text{volts per cell}} \tag{6-7}$$

$$\text{Number of parallel groups} = \frac{\text{current of battery}}{\text{current per cell}} \tag{6-8}$$

$$\text{Number of cells required} = \text{number of cells in a series group} \times \text{number of parallel groups} \tag{6-9}$$

Example 6-1 A load of ½ ampere at 4½ volts is to be supplied by a battery composed of No. 6 dry cells; each cell is rated at ⅛ ampere and 1½ volts. Determine (*a*) the number of cells in each series-connected group, (*b*) the number of parallel groups required, (*c*) the number of cells required, (*d*) the circuit diagram.

GIVEN: Load = ½ amp at 4½ volts

FIND: (*a*) Cells in series (*b*) Parallel groups (*c*) Number of cells
(*d*) Circuit diagram

SOLUTION:

(*a*) $$\text{Cells in series} = \frac{\text{voltage of battery}}{\text{volts per cell}} = \frac{4.5}{1.5} = 3$$

(*b*) $$\text{Parallel groups} = \frac{\text{current of battery}}{\text{current per cell}} = \frac{0.5}{0.125} = 4$$

(*c*) Number of cells = cells in series × parallel groups = 3 × 4 = 12

(*d*) The same as Fig. 6-14.

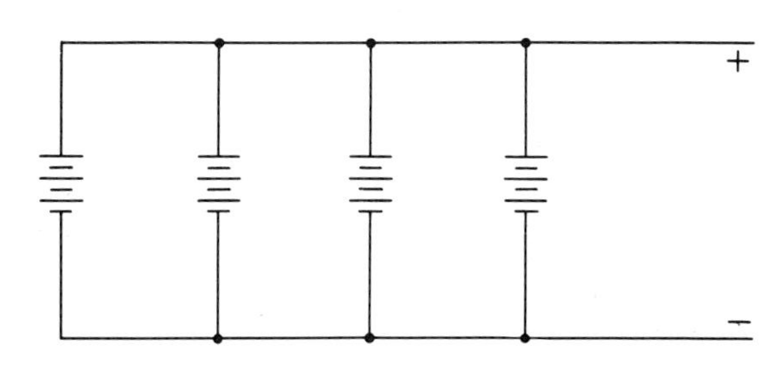

Fig. 6-14 Cells connected in parallel-series.

6-12 Power-line Energy Source

The most common source of power is that of an alternating-current power line. Such power sources generally have a frequency rating of 60 hertz and a nominal voltage rating of 117 volts, although commonly referred to as 110, 115, or 120 volts.

Electronic devices using transistors and/or vacuum tubes require a direct-current energy source that has little or no variation in its terminal voltage. In order to operate such devices from an a-c power line, it is necessary to (1) convert the alternating voltage into a unidirectional voltage, and (2) reduce the variations in terminal voltage to an acceptable amount. The circuit used to convert the a-c to d-c is called the *rectifier*, and the circuit used to reduce the variations in the terminal voltage is called the *filter*.

The Rectifier. The rectifier may be (1) a solid-state diode, as described in Chap. 2, (2) an electron tube, as described in Chap. 3, or (3) a mechanical device, as described in Chap. 7. The rectifier circuit may be either (1) half-wave or (2) full-wave.

6-13 Filters

Purpose of the Filter. Although the output of a rectifier is unidirectional, this output is not steady but is pulsating. Because of the variations in magnitude, called *ripples*, of the current of a rectifier circuit, the current cannot be used in this form for most electronic circuits.

Ripple Voltage. The unidirectional output voltage of a power-supply unit may be considered as a steady voltage having a pulsating voltage superimposed upon it. The pulsating component of the output voltage is called the *ripple voltage*. The frequency of the ripple voltage will depend upon the frequency of the input voltage and the type of rectifier. The ripple voltage does not vary in the same manner as a perfect sine-wave voltage, and it is therefore considered as consisting of a fundamental and a series of harmonics. In general, the relative effect of the harmonics is negligible as compared to the fundamental, and the harmonics can usually be ignored. The fundamental frequency of the ripple voltage is equal to the input frequency for half-wave rectifiers and twice the input frequency for full-wave rectifiers.

The effectiveness of a filter is measured by the ratio of the effective (rms) value of the fundamental component of the ripple voltage to the output voltage. This ratio is called the *ripple factor*.

$$k_r = \frac{E_r}{E_{\text{d-c}}} \tag{6-10}$$

where k_r = ripple factor

E_r = rms value of fundamental component of ripple voltage

$E_{\text{d-c}}$ = average value of output voltage

The ripple voltage is often expressed in terms of its percentage of the output voltage, as

$$\text{Per cent } E_r = \frac{E_r}{E_{\text{d-c}}} \times 100 \tag{6-11}$$

The type of service for which a power supply is to be used determines its allowable value of ripple voltage. For the plate-supply voltages of a radio receiver, a ripple voltage of 0.25 per cent or less is required in order to reduce the hum to a negligible amount. The ripple voltage for a microphone circuit should be less than 0.003 per cent. In cathode-ray oscilloscopes, a ripple voltage as high as 1 per cent is sometimes permitted.

Example 6-2 The output voltage of a power-supply unit is 300 volts and the rms value of the ripple voltage is 0.6 volt. What is the per cent of ripple voltage?

GIVEN: $E_{\text{d-c}} = 300$ volts $E_r = 0.6$ volt

FIND: Per cent E_r

SOLUTION:

$$\text{Per cent } E_r = \frac{E_r}{E_{\text{d-c}}} \times 100 = \frac{0.6}{300} \times 100 = 0.2$$

Operation of the Filter Circuit. Filter circuits associated with rectifier units use the energy-storing properties of capacitors and inductors to smooth out the ripple in the rectified output. The capacitor smooths out the voltage variations and also increases the value of the output voltage. The inductor smooths out the variations in current. The capacitor will store electrons during a portion of each cycle that the voltage increases, indicated as 1 to 2, 3 to 4, and 5 to 6 on Fig. 6-15*a*. During the portion of the cycle that the voltage decreases

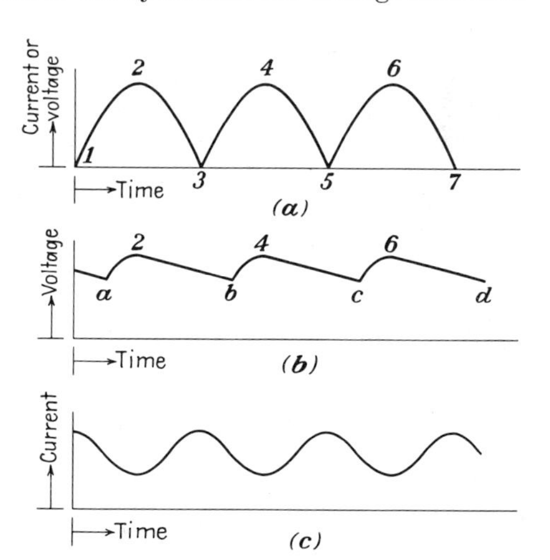

Fig. 6-15 Filter action of a capacitor and an inductor. (a) Current or voltage output of a full-wave rectifier. (b) Output from a capacitor. (c) Output from an inductor.

(2 to 3, 4 to 5, and 6 to 7 on Fig. 6-15*a*), the capacitor will slowly discharge some of its stored electrons. The voltage across the capacitor is thus made more uniform as indicated by Fig. 6-15*b*. Because electrolytic capacitors provide high voltage and high capacitance ratings in comparatively small-size units, they are generally used in power-supply filter circuits associated with electronic circuit applications. The capacitance of electrolytic capacitors used for this purpose generally ranges from 10 to 100 μf, and the d-c voltage rating may be as high as 800 volts.

A characteristic of inductors is that they oppose any change in the amount of current that flows through them. Thus, when the output current of a rectifier flows through an inductor, the variations in current (both increases and decreases) will be opposed by the action of the inductor. The output will thus be more uniform as indicated in Fig. 6-15*c*. Inductors used in power-supply filter circuits are called *filter chokes* and are wound on a soft-iron core. In order to maintain a high value of inductance for a wide variation in current flow, some chokes use an iron core with a small air gap to prevent saturation. The inductance of the filter chokes used in power-supply units ranges from 10 to 30 henrys.

Resistors may be used with a capacitor to form an *RC* filter circuit. The time constant of *RC* filters must be large compared to the time of one cycle of the lowest frequency to be attenuated. Because of this, the d-c resistance of this type of filter is comparatively high and thus the voltage drop, voltage regulation, and heat dissipation are great. The development of electrolytic capacitors having a high capacitance makes it possible to use lower values of resistance with this type of circuit. *RC* filters are used when the requirements

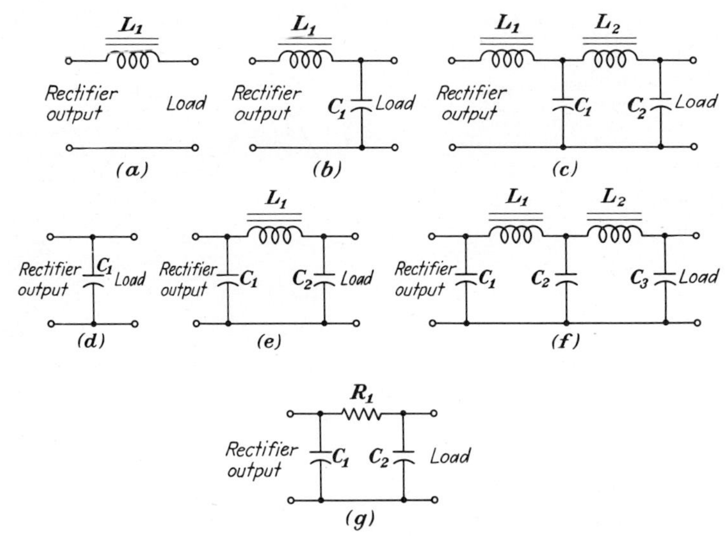

Fig. 6-16 Types of filter circuits used with rectifiers.

of low cost and compactness outweigh the desirability of a high degree of filtering.

Types of Filter Circuits. Power-supply filter circuits are of the low-pass type, using one or more series inductors and one or more shunt capacitors. These filter circuits are usually referred to as being *choke input* or *capacitor input* filters depending upon whether an inductor or a capacitor is the first element in the filter network. A number of different types of filter circuits are shown in Fig. 6-16, the choke-input filters being represented by (*a*), (*b*), and (*c*), and the capacitor-input filters by (*d*), (*e*), (*f*), and (*g*). These filter circuits may be further classified as single-section filters represented by (*a*), (*b*), and (*d*); two-section filters as in (*c*), (*e*), and (*g*); and three-section filters as in (*f*). The number of sections of filtering that is required will depend upon the rectifier, the type of filter circuit, and the allowable ripple voltage.

6-14 Capacitor-input Filter

Theory of Operation. A capacitor-input filter is a filter circuit in which the first element is a capacitor connected in parallel with the input from the rectifier. During the time that the rectifying element is conducting, energy will be stored in the capacitor, and when the rectifying element is not conducting, part of the stored energy will be discharged through the filter network to the load. The capacitor increases the average value of the output voltage. The waveform of the capacitor voltage shown in Fig. 6-15*b* may be considered as consisting of two parts: (1) the portion during the charging period represented by *a* to 2, *b* to 4, *c* to 6, etc.; (2) the portion during the discharge period represented by 2 to *b*, 4 to *c*, 6 to *d*, etc. If the capacitor were to be discharged directly through a resistor, the discharge portion of the curve would decrease exponentially. If the capacitor is discharged through an inductor, which helps further to smooth out the output current, the discharge portion of the curve will decrease linearly, and the resultant voltage wave that is applied to the inductor will be a wave with a sawtooth characteristic. The ripple component of the voltage across the input capacitor is prevented from reaching the output circuit by the combined actions of any inductors and capacitors that follow the input capacitor in the filter circuit.

Ripple Voltage. The per cent of ripple voltage developed across the input capacitor will vary inversely with the frequency of the rectified output, the effective load resistance, and the capacitance of the input capacitor. Thus, increasing any of these factors in a filter circuit will decrease the per cent of ripple voltage. The effect of a variation in any of these factors on the d-c output can be seen by observation of Fig. 6-15*b*. Increasing the frequency decreases the time that the input capacitor is permitted to discharge. The capacitor will lose less of its charge, thus maintaining the voltage across it more nearly uniform. The extent to which the voltage across the input capacitor drops off is also affected by the time constant of the RC circuit, consisting of the input capacitor C_1 and the effective load resistance R_o. Increasing either of these two values will increase the time constant of the R_oC_1 circuit, thereby decreasing the rate of discharge. The voltage across the input capacitor will thus be maintained more nearly uniform.

Although it is difficult to obtain accurate calculations of the per cent of ripple voltage, the following equations will provide reasonable results for ripple voltages of 10 per cent or less; beyond this amount the accuracy decreases. The per cent of ripple voltage at the output of a single-section capacitor-input filter as shown in Fig. 6-16d is

$$\text{Per cent } E_{r\cdot 1} \cong \frac{10^8 \sqrt{2}}{2\pi f_r R_o C_1} \cong \frac{2{,}245 \times 10^4}{f_r R_o C_1} \tag{6-12}$$

where f_r = frequency of ripple voltage, hertz
R_o = resistance of load, ohms
C_1 = capacitance of input filter capacitor, μf

This equation can also be used to calculate the per cent of ripple voltage at the output of the first section (per cent of ripple voltage at C_1) for any multi-section capacitor-input filter, examples of which are shown as (e), (f), and (g) of Fig. 6-16.

Example 6-3 A power-supply unit has a 60-hertz input to its rectifier, uses a single-section capacitor-input filter (Fig. 6-16d), and delivers 40 ma direct current at 320 volts to the load. Determine the per cent of ripple voltage for (a) half-wave rectification and a 10-μf filter capacitor, (b) full-wave rectification and a 10-μf filter capacitor, (c) half-wave rectification and a 20-μf filter capacitor, (d) full-wave rectification and a 20-μf filter capacitor.

GIVEN: $E_{d\text{-}c} = 320$ volts $I_{d\text{-}c} = 40$ ma $C_1 = 10\ \mu$f (a), (b)
$C_1 = 20\ \mu$f (c), (d) $f_r = 60$ (a), (c) $f_r = 120$ (b), (d)

FIND: (a) Per cent $E_{r\cdot 1}$, half-wave (b) Per cent $E_{r\cdot 1}$, full-wave
(c) Per cent $E_{r\cdot 1}$, half-wave (d) Per cent $E_{r\cdot 1}$, full-wave

SOLUTION:

$$R_o = \frac{E_{d\text{-}c}}{I_{d\text{-}c}} = \frac{320}{40 \times 10^{-3}} = 8{,}000 \text{ ohms}$$

(a) $$\text{Per cent } E_{r\cdot 1} \cong \frac{2{,}245 \times 10^4}{f_r R_o C_1} = \frac{2{,}245 \times 10^4}{60 \times 8{,}000 \times 10} = 4.67$$

(b) $$\text{Per cent } E_{r\cdot 1} \cong \frac{2{,}245 \times 10^4}{f_r R_o C_1} = \frac{2{,}245 \times 10^4}{120 \times 8{,}000 \times 10} = 2.34$$

(c) $$\text{Per cent } E_{r\cdot 1} \cong \frac{2{,}245 \times 10^4}{f_r R_o C_1} = \frac{2{,}245 \times 10^4}{60 \times 8{,}000 \times 20} = 2.34$$

(d) $$\text{Per cent } E_{r\cdot 1} \cong \frac{2{,}245 \times 10^4}{f_r R_o C_1} = \frac{2{,}245 \times 10^4}{120 \times 8{,}000 \times 20} = 1.17$$

The results of Example 6-3 show that the per cent of ripple voltage for a single-section filter is considerably higher than is usually acceptable for most electronic circuits. The additional filtering necessary to reduce the ripple voltage to an acceptable value can be obtained by providing additional filter sections (Fig. 6-16*e*, *f*, and *g*). The per cent of ripple voltage at the output of a two-section capacitor-input filter similar to that of Fig. 6-16*e* can be found by use of Eq. (6-13). The results obtained with this equation are not extremely accurate but provide reasonable accuracy for low percentages of ripple voltage.

$$\text{Per cent } E_{r\cdot 2} \cong \frac{\% \; E_{r\cdot 1}}{[10^{-6}(2\pi f_r)^2 L_1 C_2] - 1} \tag{6-13}$$

where $E_{r\cdot 2}$ = ripple voltage at capacitor C_2 (Fig. 6-16*e*)
$\% \; E_{r\cdot 1}$ = obtained by use of Eq. (6-12), which is the ripple at capacitor C_1 (Fig. 6-16*e*)
f_r = frequency of the ripple voltage, hertz
L_1 = inductance of L_1 (Fig. 6-16*e*), henrys
C_2 = capacitance of C_2 (Fig. 6-16*e*), μf

Example 6-4 A power-supply unit using a full-wave rectifier and a filter circuit similar to Fig. 6-16*e* has a 60-hertz input and delivers 40 ma direct current at 320 volts to the load. Capacitors C_1 and C_2 are each 20 μf, and the inductance of L_1 is 15 henrys. What is the per cent of ripple voltage at (*a*) C_1? (*b*) C_2?

GIVEN: $f_r = 120$ hertz $\quad L_1 = 15$ henrys $\quad C_1, C_2 = 20\ \mu$f

FIND: (*a*) Per cent E_r at C_1 $\quad$ (*b*) Per cent E_r at C_2

SOLUTION:

(*a*) $\quad \text{Per cent } E_{r\cdot 1} \cong 1.17 \quad$ (Same as Example 6-3*d*)

(*b*)
$$\text{Per cent } E_{r\cdot 2} \cong \frac{\% \; E_{r\cdot 1}}{[10^{-6}(2\pi f_r)^2 L_1 C_2] - 1}$$
$$= \frac{1.17}{[(2 \times 3.14 \times 120)^2 15 \times 20 \times 10^{-6}] - 1} \cong 0.007$$

An approximate value of the per cent of ripple voltage $E_{r\cdot 3}$ at the output of a three-section capacitor-input filter similar to that of Fig. 6-16*f* can be found by modifying Eq. (6-13).

$$\text{Per cent } E_{r\cdot 3} \cong \frac{\% \; E_{r\cdot 2}}{[10^{-6}(2\pi f_r)^2 L_2 C_3] - 1} \tag{6-13a}$$

In a similar manner, Eq. (6-13) can be modified to determine the per cent of ripple voltage at the output of any number of succeeding sections of a multisection capacitor-input filter of this type.

For a two-section *RC* filter similar to that of Fig. 6-16*g*, an approximate value of the per cent of ripple voltage at the output can be found as in Example 6-4 by use of Eqs. (6-12) and (6-14). Equation (6-14) can also be modified in the same manner as for Eq. (6-13) to determine the per cent of ripple voltage at the output of succeeding sections of multisection RC filters.

$$\text{Per cent } E_{r\cdot 2} \cong \frac{\%\ E_{r\cdot 1} \times 10^6}{2\pi f_r C_2 R_1} \tag{6-14}$$

where C_2 = capacitance, μf
R_1 = resistance, ohms

The results obtained with this equation are not extremely accurate but provide reasonable accuracy when the product of C_2R_1 is 10,000 or more.

Output Voltage. The output voltage of a capacitor-input filter circuit will vary with the capacitance and with changes in the effective load resistance. At no load, or with a comparatively light load, the effective load resistance is comparatively high. The time constant of R_oC_1 will also be high and the output voltage will approach the peak value (also called *crest value*) of the alternating voltage being rectified. As the load current increases, the effective load resistance will decrease, thus also decreasing the time constant of the circuit. The resulting increase in the rate of discharge of the input capacitor will lower the average value of the voltage across this capacitor. When the output current is high, the effective load resistance is low, thus causing a considerable decrease in the output voltage. For average applications of capacitor-input filters, the output voltage at full load will be approximately equal to the effective value of the alternating voltage being rectified.

The variation in output voltage with changes in current is called the *voltage regulation* of the circuit and expressed as a percentage is

$$\text{Voltage regulation} = \frac{E_{NL} - E_L}{E_L} \times 100 \tag{6-15}$$

where E_{NL} = no-load voltage
E_L = full-load voltage

Since variations in the load current of a capacitor-input filter circuit result in a wide range of output voltage, the voltage regulation of this type of circuit is very poor.

Example 6-5 It is desired to have a power-supply unit that will provide 300 volts at the output terminals when supplying its rated full-load current. What is the per cent of regulation if (*a*) the unit employs a full-wave rectifier and a capacitor-input filter, and the voltage with no load rises to 426 volts? (*b*) The unit employs a half-wave rectifier and a capacitor-input filter, and the voltage with no load rises to 480 volts?

GIVEN: E_L = 300 volts (*a*) E_{NL} = 426 volts (*b*) E_{NL} = 480 volts

FIND: (*a*) Per cent of regulation (*b*) Per cent of regulation

SOLUTION:

(*a*) $$\text{Per cent of regulation} = \frac{E_{NL} - E_L}{E_L} \times 100 = \frac{426 - 300}{300} \times 100 = 42$$

(*b*) $$\text{Per cent of regulation} = \frac{E_{NL} - E_L}{E_L} \times 100 = \frac{480 - 300}{300} \times 100 = 60$$

Characteristics of the Capacitor-input Filter. Compared to the choke-input filter, the capacitor-input filter will deliver a higher voltage at light loads, has a slightly better filtering characteristic, but has poorer voltage regulation. The current in the rectifier associated with a capacitor-input filter circuit does not flow uniformly but flows in pulses; hence the ratio of peak rectifier current to average current will be higher than in the choke-input system. The d-c voltage rating of the input capacitor should never be less than the peak input voltage, since at light loads the output voltage approaches this value. In order to provide a safety factor, it is usually desirable to use a capacitor whose working voltage rating is somewhat higher than this value. Capacitor-input filter circuits are generally used in power-supply units that are required to deliver only small amounts of power, such as radio receivers, audio amplifier systems, and testing apparatus.

6-15 Choke-input Filter

Theory of Operation. In a choke-input filter, the first element is an inductor connected in series with the input from the rectifier. The filtering action of the series input inductor of the choke-input filter circuit (Fig. 6-16) is explained in the following manner. During the portion of the rectified output cycle in which the current increases, the strength of the magnetic field about the inductor will increase, and energy will be stored in the field. The inductor also opposes the increase in current. During the portion of the rectified output cycle in which the current decreases, the magnetic field about the inductor collapses, returning part of its stored energy to the circuit, and also opposes the decrease in current. These actions of an inductor smooth out the ripple in the rectified output, as shown in Fig. 6-15*c*. This figure shows that the current through the inductor is made up of an a-c component and a d-c component. The capacitor following the input inductor will tend to short-circuit the a-c component, thus producing a practically smooth voltage at its output terminals.

If the ripple voltage from a single-section filter circuit exceeds the allowable percentage, it can be further reduced by using additional filter sections as shown in Fig. 6-16*c*. The first inductor is called the *input choke* and the second inductor is called the *smoothing choke*.

Ripple Voltage. The per cent of ripple voltage developed across the first capacitor C_1 (Fig. 6-16*b* and *c*) will vary inversely with the capacitance of C_1 and the inductance of the input choke. For most practical purposes an approximate value of the per cent of ripple voltage for a single-section filter can be obtained by the equation

$$\text{Per cent } E_{r\cdot 1} \cong \frac{144 \times 10^4}{f_r^2 L_1 C_1} \tag{6-16}$$

where f_r = ripple frequency, hertz
L_1 = inductance of input choke, henrys
C_1 = capacitance of first capacitor, μf

This equation may be further simplified for determining the approximate per cent of ripple voltage of a single-section filter whose ripple frequency is 120 hertz, as is the case when the input to the filter is obtained from a full-wave rectifier operated from a 60-hertz power source. Equation (6-16) may then be simplified to

$$\text{Per cent } E_{r\cdot 1} \cong \frac{100}{L_1 C_1} \tag{6-16a}$$

Example 6-6 Determine the approximate per cent of ripple voltage at the output of a single section choke-input filter circuit using a 15-henry choke and a 10-μf capacitor. The input to the filter circuit is obtained from a full-wave rectifier operated from a 60-hertz power source.

GIVEN: L_1 = 15 henrys C_1 = 10 μf

FIND: Per cent $E_{r.1}$

SOLUTION:

$$\text{Per cent } E_{r\cdot 1} \cong \frac{100}{L_1 C_1} = \frac{100}{15 \times 10} = 0.66$$

An approximate value of the per cent of ripple voltage at the output of a two-section choke-input filter can be obtained by the equation

$$\text{Per cent } E_{r\cdot 2} \cong \frac{1{,}350 \times 10^8}{f_r^4 L_1 L_2 (C_1 + C_2)^2} \tag{6-17}$$

where L_2 = inductance of smoothing choke, henrys
C_2 = capacitance of second capacitor, μf

This equation may be further simplified for determining the approximate per cent of ripple voltage of a two-section filter whose ripple frequency is 120 hertz, as is the case when the input to the filter is obtained from a full-wave rectifier operated from a 60-hertz power source. Equation (6-17) may then be simplified to

$$\text{Per cent } E_{r\cdot 2} \cong \frac{650}{L_1 L_2 (C_1 + C_2)^2} \tag{6-17a}$$

Example 6-7 Determine the approximate per cent of ripple voltage at the output of a two-section choke-input filter circuit using two 15-henry chokes and two 10-μf capacitors. The input to the filter circuit is obtained from a full-wave rectifier operated from a 60-hertz power source.

GIVEN: $L_1, L_2 = 15$ henrys $\quad C_1, C_2 = 10\ \mu$f

FIND: Per cent $E_{r\text{-}2}$

SOLUTION:

$$\text{Per cent } E_{r\text{-}2} \cong \frac{650}{L_1L_2(C_1 + C_2)^2} = \frac{650}{15 \times 15(10 + 10)^2} \cong 0.0072$$

Example 6-8 Determine the approximate per cent of ripple voltage at the output of the single-section choke-input filter circuit used in Example 6-6, if the input is obtained from a full-wave rectifier circuit having a 25-hertz input.

GIVEN: $L_1 = 15$ henrys $\quad C_1 = 10\ \mu$f $\quad f_{\text{in}} = 25$ hertz

FIND: Per cent $E_{r\text{-}1}$

SOLUTION:

$$f_r = 2f_{\text{in}} = 2 \times 25 = 50 \text{ hertz}$$

$$\text{Per cent } E_{r\text{-}1} \cong \frac{144 \times 10^4}{f_r^2 L_1 C_1} = \frac{144 \times 10^4}{50 \times 50 \times 15 \times 10} \cong 3.8$$

The Input Choke. The input choke of a filter circuit serves two functions: (1) to maintain a continuous flow of current from the rectifier, (2) to prevent the output voltage from increasing above the average value of the alternating voltage applied to the rectifier. The output voltage and the peak plate current of the rectifier are both dependent upon the inductance of the input choke and the d-c resistance of the load. The minimum value of inductance required to maintain the output voltage at the average value of the alternating voltage being rectified is called the *critical inductance*. For a rectified output having a 120-hertz ripple frequency, an approximate value of the critical inductance may be obtained by use of the following equation:

$$L_c = \frac{R_o}{1{,}000} \tag{6-18}$$

where L_c = critical value of inductance, henrys
R_o = output load resistance, ohms

If the inductance of the input choke is less than its critical value, its impedance to the varying component of the rectified output will be so small that the filter circuit will tend to operate as a capacitor-input filter. Increasing the inductance of the input choke to more than its critical value will further decrease the ratio of peak to average current, thus maintaining a more nearly uniform flow of current through the inductor. Increasing the value of the inductance beyond twice the critical value does not correspondingly improve the operating characteristics of the filter. The *optimum* value of inductance L_o is thus equal to twice the critical value of inductance L_c.

Example 6-9 Determine the optimum value of inductance for the input choke of a filter circuit having a d-c load resistance of 4,000 ohms. The frequency of the rectified input is 120 hertz.

GIVEN: $R_o = 4{,}000$ ohms $\quad f_r = 120$ hertz

FIND: L_o

SOLUTION:

$$L_c = \frac{R_o}{1{,}000} = \frac{4{,}000}{1{,}000} = 4 \text{ henrys}$$

$$L_o = 2L_c = 2 \times 4 = 8 \text{ henrys}$$

Swinging Choke. Equation (6-18) shows that the value of the inductance required for the input choke will vary directly with the effective load resistance and inversely with the load current. Thus, if the load current varies over a wide range, some means must be provided for preventing the ratio of peak to average current from becoming excessive. The inductance of the choke coil will vary inversely with the value of the direct current flowing through it. A choke coil having an inductance of 10 henrys with 100 ma flowing through it may have an inductance of 15 henrys when the current flow is reduced to practically zero. A choke designed to have a critical value of inductance at full load and an optimum value of inductance at no load is called a *swinging choke.*

Output Voltage. The average value of the output voltage at full load of a choke-input filter is in the order of 65 to 75 per cent of the rms volts at the rectifier. It should be observed that the output voltage with a choke-input filter is lower than that with a capacitor-input filter supplied with the same value of rms volts. The decrease in output voltage is due to the effect of the inductance being introduced into the circuit and to the voltage drop at the choke due to its d-c resistance. In order to reduce the drop in voltage at the filter chokes, their d-c resistance should be kept as low as possible.

Characteristics of the Choke-input Filter. Although the choke-input filter circuit delivers a lower output voltage than the capacitor-input filter circuit, its voltage regulation is much better. Another advantage of the choke-input filter is that the input choke prevents high instantaneous peak currents, thus protecting the rectifier element from being damaged.

The input inductor and the first capacitor of a choke-input filter form a series resonant circuit. If the values of these two circuit elements make the circuit resonant to the ripple frequency, high values of ripple voltage will be produced. It is therefore important that the values of inductance and capacitance used do not form a series resonant circuit that is tuned to the ripple frequency.

Choke-input filter circuits are generally used where the output current is large or where the voltage regulation must be fairly good. Because choke-input filters operate best when the current flow is sustained over the complete cycle, they are usually used only with full-wave rectifiers.

6-16 The Voltage Divider

Basic Principle. If a tapped resistor is connected across a power supply (Fig. 6-17), a number of loads requiring different amounts of voltage can be connected to the taps along the body of the resistor. The voltage between any two points will be equal to the product of the current flowing through that part of the resistor and the value of the resistance between the two points. A resistor used in this manner is called a *voltage divider*.

Bleeder Resistor. Removing the external load from a power-supply unit causes a high voltage to be developed across the filter capacitors. This voltage may result in a breakdown of the insulation in these capacitors. The voltage at zero load can be reduced to a safe value by connecting a fixed resistor, called a *bleeder resistor*, across the output terminals of the filter circuit. The amount of bleeder current varies with the requirements of the individual power-supply unit and generally ranges from 10 to 25 per cent of the total current drawn from the rectifier.

Since the bleeder resistor draws a fixed amount of current continuously from the power-supply unit, it reduces the value of the output voltage at no load and thereby improves the voltage regulation of the unit.

Typical Applications. Voltage dividers are used to divide the voltage of the power supply into such values of potential as are required by the various parts of the circuit. Voltage dividers also act as a safety load to protect the capacitors from excessive voltage.

A power-supply unit may be required to supply the high operating voltages for the plates and screen grids of electron tubes, the low grid-bias voltages of electron tubes, and other operating voltages required of electronic circuit com-

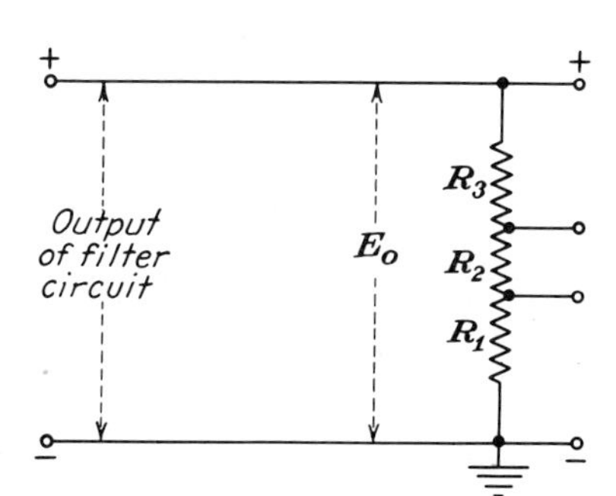

Fig. 6-17 Connection of a voltage divider used in a power-supply unit.

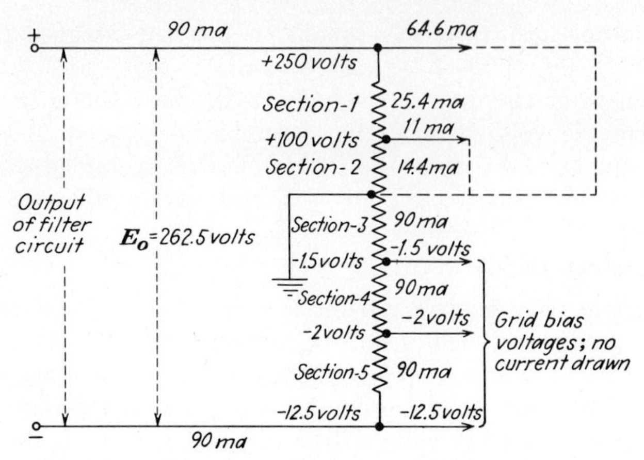

Fig. 6-18 A voltage-divider circuit used for providing plate and grid-bias voltages.

ponents. The negative voltages required for grid bias are obtained in the manner shown in Fig. 6-18.

6-17 Calculation of a Typical Voltage Divider

The resistance values and the power rating of the voltage divider may be calculated by use of Ohm's law. The following procedure should be followed:

1. Determine the voltage required at each tap and the current to be drawn from it.
2. Determine the amount of bleeder current desired. This is the difference between the total current required by the load and the current necessary to operate the power supply at approximately 90 per cent of its rated value.
3. Determine the current in each section of the divider.
4. Calculate the resistance of one section at a time.
5. Determine the power rating of the voltage divider.

Example 6-10 Determine the resistance values of a voltage divider for a small superheterodyne receiver that employs a 6BE6 oscillator-mixer tube, a 6BA6 i-f amplifier tube, a 6AV6 detector-amplifier tube, and a 6AQ5-A power-output tube. The operating

Table 6-1 Values from a Tube Manual

TUBE	6BE6	6BA6	6AV6	6AQ5-A
E_b, volts	250	250	250	250
$E_{c\text{-}2}$, volts	100	100	...	250
$E_{c\text{-}1}$, volts	−1.5	−12.5	−2	−12.5
I_b, ma	2.9	11	1.2	45
$I_{c\text{-}2}$, ma	6.8	4.2	...	4.5

voltages and currents are to be obtained from a tube manual for plate voltages of 250 volts. A power transformer rated at 100 ma is to be operated at 90 per cent of its rated value.

GIVEN: Tubes: 6BE6, 6BA6, 6AV6, 6AQ5-A $\quad I_T = 100$ ma $\quad I_S = 0.9\ I_T$

FIND: R of each section

SOLUTION: See Table 6-1.

Using the values of voltage and current listed in Table 6-1, the voltage and current at each tap of the voltage divider will be as shown in Fig. 6-18. The total voltage required from the power-supply unit will be the sum of the highest plate voltage and the highest grid-bias voltage.

$$E_o = 250 + 12.5 = 262.5 \text{ volts}$$
$$I_{250} = 2.9 + 11 + 1.2 + 45 + 4.5 = 64.6 \text{ ma}$$
$$I_{100} = 6.8 + 4.2 = 11 \text{ ma}$$
$$I_o = 64.6 + 11 = 75.6 \text{ ma}$$
$$I_{\text{bleeder}} = 0.9I_T - I_o = (0.9 \times 100) - 75.6 = 14.4 \text{ ma}$$

$$R_{\text{sec}\cdot 1} = \frac{e_1}{i_1} = \frac{250 - 100}{25.4 \times 10^{-3}} = 5{,}905 \text{ ohms}$$

$$R_{\text{sec}\cdot 2} = \frac{e_2}{i_2} = \frac{100 - 0}{14.4 \times 10^{-3}} = 6{,}944 \text{ ohms}$$

$$R_{\text{sec}\cdot 3} = \frac{e_3}{i_3} = \frac{1.5 - 0}{90 \times 10^{-3}} = 16.7 \text{ ohms}$$

$$R_{\text{sec}\cdot 4} = \frac{e_4}{i_4} = \frac{2 - 1.5}{90 \times 10^{-3}} = 5.5 \text{ ohms}$$

$$R_{\text{sec}\cdot 5} = \frac{e_5}{i_5} = \frac{12.5 - 2}{90 \times 10^{-3}} = 117 \text{ ohms}$$

Power Rating. If a single resistor of uniform wire size is to be used, the wire size must be based on the maximum current in any section of the voltage divider. However, the voltage divider could also be made with separate resistors for each section, in which case the wire size for the separate resistors is determined by the currents in the individual sections.

When the voltage divider is placed in an area having little ventilation, it is recommended that its power rating be approximately double that of the load it is to carry.

Example 6-11 Assuming that the voltage divider of Example 6-10 has poor ventilation, what power rating is recommended for the unit if it is made with (*a*) a single resistor having four taps? (*b*) Five separate resistors? (*c*) What is the total rated wattage of the five separate resistors? (*d*) How much power is actually used by the voltage divider?

GIVEN: Resistance and current values in Example 6-10

FIND: (*a*) P_R $\quad$ (*b*) P_{R1-5} $\quad$ (*c*) P_T $\quad$ (*d*) P_{actual}

SOLUTION:

(*a*) $P_R = 2I_{max}^2 R_T = 2(90 \times 10^{-3})^2 \times 12{,}988 = 210$ watts

where $R_T = 5{,}905 + 6{,}944 + 16.7 + 5.5 + 117 = 12{,}988$ ohms

(*b*) $P_1 = 2I_1^2 R_1 = 2(25.4 \times 10^{-3})^2 \times 5{,}905 = 7.62$ watts

$P_2 = 2I_2^2 R_2 = 2(14.4 \times 10^{-3})^2 \times 6{,}944 = 2.87$ watts

$P_3 = 2I_3^2 R_3 = 2(90 \times 10^{-3})^2 \times 16.7 = 0.27$ watt

$P_4 = 2I_4^2 R_4 = 2(90 \times 10^{-3})^2 \times 5.5 = 0.088$ watt

$P_5 = 2I_5^2 R_5 = 2(90 \times 10^{-3})^2 \times 117 = 1.88$ watts

(*c*) $P_T = P_1 + P_2 + P_3 + P_4 + P_5$

$= 7.62 + 2.87 + 0.27 + 0.088 + 1.88 \cong 12.7$ watts

(*d*) $$P_{actual} = \frac{P_T(c)}{2} \cong \frac{12.7}{2} \cong 6.35 \text{ watts}$$

Thus, if the voltage divider is made of uniform wire size designed to withstand the maximum current flowing in any part of the unit, the wattage rating required for the voltage divider (with restricted ventilation) is 210 watts even though the total wattage actually consumed by the individual sections is only 6.35 watts. A voltage divider designed to meet the needs of only the individual sections is less expensive and more practical.

QUESTIONS

1. Why is it important to have a knowledge of power sources?
2. What is a primary cell?
3. What are the components of a simple cell?
4. State six basic characteristics of a cell.
5. What are the requirements of (*a*) the electrodes? (*b*) The electrolyte?
6. What is the approximate voltage that a cell can deliver?
7. What determines the voltage that a cell can deliver?
8. What factor determines the current capacity of a cell?
9. Describe the chemical action in a simple carbon-zinc cell.
10. Define (*a*) ion, (*b*) positive ion, (*c*) negative ion, (*d*) ionization.
11. Name four types of primary cells.
12. Describe five factors that determine the life of a cell.
13. Describe the construction of a commercial carbon-zinc cell.
14. Describe the construction of a mercury (*a*) flat-type cell, (*b*) cylindrical cell.
15. Explain why in a mercury cell (*a*) the zinc is the anode, (*b*) the mercuric oxide is the cathode.
16. Describe seven characteristics of comparison between the carbon-zinc and mercury cells.
17. Describe the construction of a commercial silver oxide cell.
18. Describe six favorable characteristics of the silver oxide cell.
19. Describe the construction of a commercial alkaline cell.
20. What are the basic differences between the carbon-zinc cell and the alkaline cell?
21. What is a secondary cell?
22. How do secondary cells differ from primary cells?
23. Describe the construction of a commercial lead-acid cell.

24. In a lead-acid cell, what is the function of (*a*) separators? (*b*) Containers? (*c*) Vent caps? (*d*) Connectors?
25. (*a*) How is a storage battery rated? (*b*) What are the factors that determine the rating of a lead-acid battery?
26. Describe four advantages of the lead-acid cell.
27. Describe four disadvantages of the lead-acid cell.
28. What are the basic materials used in an Edison cell for (*a*) the cathode? (*b*) The anode? (*c*) The electrolyte?
29. Describe the construction in a commercial nickel-iron cell (*a*) of the positive plate, (*b*) of the negative plate.
30. Describe four advantages of the nickel-iron cell.
31. Describe the construction of a nickel-cadmium (*a*) button-type cell, (*b*) cylindrical-type cell, (*c*) rectangular-type cell.
32. Explain how the gases and fumes inside a hermetically sealed nickel-cadmium cell are consumed inside the cell.
33. Describe six advantages of the nickel-cadmium cell.
34. Describe the construction of the cadmium–silver oxide cell.
35. (*a*) What are the prime advantages of a cadmium–silver oxide cell? (*b*) What are the prime factors that determine its use?
36. (*a*) What is a battery? (*b*) Why is it necessary to form batteries?
37. Describe the construction of a 45-volt battery composed of (*a*) carbon-zinc cylindrical cells, (*b*) carbon-zinc flat cells.
38. Describe the connections that must be made in order to form a battery having four cells connected in series.
39. When a number of cells are connected in series, what are the characteristics of the battery in terms of (*a*) its voltage rating, (*b*) its current rating?
40. Describe the connections that must be made in order to form a battery having four cells connected in parallel.
41. When a number of cells are connected in parallel, what are the characteristics of the battery in terms of (*a*) its voltage rating? (*b*) Its current rating?
42. What voltage and current demands of a battery would require a parallel-series connection for the cells of the battery?
43. (*a*) What two basic operations must be performed in order to operate electronic equipment from commercial power lines? (*b*) What are the circuits called that perform these two operations?
44. Define (*a*) ripple voltage, (*b*) ripple factor, (*c*) per cent of ripple voltage.
45. What is the fundamental frequency of the ripple voltage with (*a*) half-wave rectification? (*b*) Full-wave rectification?
46. What is the function of (*a*) the filter circuit? (*b*) The filter capacitor? (*c*) The filter choke?
47. Explain the filtering action of the filter capacitor.
48. Explain the filtering action of the filter choke.
49. (*a*) What are the operating requirements and characteristics of an *RC* filter circuit? (*b*) Where is this type of filter circuit used?
50. What is meant by (*a*) a low-pass filter? (*b*) A choke-input filter? (*c*) A capacitor-input filter? (*d*) A single-section filter? (*e*) A multisection filter?
51. Explain the principle of operation of a single-section capacitor-input filter circuit.
52. Explain how the per cent of ripple voltage is affected by changes in (*a*) the frequency

of the rectified output, (*b*) the effective load resistance, (*c*) the capacitance of the input capacitor.

53. Why does a capacitor-input filter circuit have poor voltage regulation?
54. How do the operating characteristics of a capacitor-input filter circuit compare with those of a choke-input filter circuit?
55. For what type of service are capacitor-input filter circuits generally used?
56. What is the purpose of connecting a resistance directly in the rectifier circuit of a power-supply unit employing a capacitor-input filter?
57. Explain the principle of operation of a single-section choke-input filter circuit.
58. Explain how the per cent of ripple voltage of a choke-input filter circuit is affected by (*a*) the input choke, (*b*) the first capacitor, (*c*) the smoothing choke, (*d*) the second capacitor.
59. What are two functions of the input choke?
60. What is meant by (*a*) the critical value of inductance? (*b*) The optimum value of inductance?
61. How does the value of inductance of the input choke affect the operation of the filter circuit?
62. (*a*) What is meant by a swinging choke? (*b*) What is its function?
63. What are the characteristics of a choke-input filter circuit?
64. For what type of service are choke-input filter circuits generally used?
65. What is the basic principle of the voltage divider?
66. (*a*) What is the bleeder resistor? (*b*) What is its function?
67. What are some applications of the voltage divider?
68. What are the advantages of using a separate resistor for each section of a voltage divider rather than a single resistor having the necessary taps?
69. Explain how the use of a bleeder resistor improves the voltage regulation of a power-supply unit.
70. Why is it sometimes advisable to select voltage-divider resistors with double wattage ratings?

PROBLEMS

1. What voltage will be produced by a battery with its carbon-zinc cells connected in series if the number of cells is (*a*) 5? (*b*) 8? (*c*) 15? (*d*) 20? (*e*) 40?
2. What voltage will be produced by a mercury battery (1.4 volts per cell) if the number of cells connected in series is (*a*) 3? (*b*) 5? (*c*) 16? (*d*) 32? (*e*) 48?
3. What voltage will be produced by a lead-acid battery (2.2 volts per cell) if the number of cells connected in series is (*a*) 3? (*b*) 6? (*c*) 15? (*d*) 40? (*e*) 60?
4. How many carbon-zinc cells are required for a battery if the voltage desired is (*a*) 4.5? (*b*) 6? (*c*) 10.5? (*d*) 22.5? (*e*) 90?
5. How many carbon-zinc cells are required for a battery if the voltage desired is (*a*) 9? (*b*) 45? (*c*) 67.5? (*d*) 120? (*e*) 150?
6. How many nickel-cadmium cells (1.3 volts per cell) are required for a battery if the voltage desired is (*a*) 6.5? (*b*) 9? (*c*) 22? (*d*) 45?
7. If the current drain on a No. 6 carbon-zinc cell is to be 200 milliamperes, how much current can be delivered by a battery if the number of cells connected in parallel is (*a*) 2? (*b*) 3? (*c*) 5? (*d*) 8? (*e*) 12?
8. If the current drain on a type-C carbon-zinc cell is to be 20 milliamperes, how much current can be delivered by a battery if the number of cells connected in parallel is (*a*) 3? (*b*) 5? (*c*) 8? (*d*) 10? (*e*) 15?

9. If the current drain on a No. 6 carbon-zinc cell for a continuous duty cycle is not to exceed 100 milliamperes, how many cells must be connected in parallel in order to supply a current of (*a*) 300 ma? (*b*) 0.5 amp? (*c*) 0.8 amp? (*d*) 1 amp?
10. If the current drain for a type M-35 mercury cell for continuous duty is not to exceed 12.5 milliamperes, how many cells must be connected in parallel in order to supply a current of (*a*) 25 ma? (*b*) 50 ma? (*c*) 62.5 ma? (*d*) 75 ma?
11. A battery power source is needed to supply a 12-volt, 300-milliampere load on a continuous duty cycle. Number 6 carbon-zinc cells are to be used and the maximum current drain to be permitted is 60 milliamperes. (*a*) How many cells must be used in each series-connected group? (*b*) How many parallel-connected groups are needed? (*c*) How many cells are needed? (*d*) Draw a diagram of the arrangement of the cells.
12. A battery power source is needed to supply a 6-volt, 15-milliampere load on a 4-hour-per-day duty cycle. Type-C carbon-zinc cells are to be used with an allowable current drain of 5 milliamperes. (*a*) How many cells must be used in each series-connected group? (*b*) How many parallel-connected groups are needed? (*c*) How many cells are needed? (*d*) Draw a diagram of the arrangements of the cells.
13. A battery power source is needed to supply a 7-volt, 37.5-milliampere load on a continuous duty cycle. Mercury cells rated at 1.4 volts and 12.5 milliamperes are to be used. (*a*) How many cells must be used in each series-connected group? (*b*) How many parallel-connected groups are needed? (*c*) How many cells are needed? (*d*) Draw a diagram of the arrangement of the cells.
14. A battery power source is needed to supply a 5-volt, 0.25-ampere load on a continuous duty cycle. D-size alkaline cells, rated at 1.25 volts and 63 milliamperes, are to be used. (*a*) How many cells must be used in each series-connected group? (*b*) How many parallel-connected groups are needed? (*c*) How many cells are needed? (*d*) Draw a diagram of the arrangement of the cells.
15. A certain power-supply unit produces an average d-c output of 90 volts and the rms value of the ripple is 5 volts. (*a*) What is the per cent ripple voltage? (*b*) What is the ripple factor?
16. A certain low-voltage power-supply unit used to operate a transistor circuit produces a steady supply of 6 volts and has a ripple peak of 0.1 volt. (*a*) What is the ripple factor? (*b*) What is the per cent ripple voltage?
17. A certain power-supply unit is to provide a load with 180 volts and its ripple factor is not to exceed 0.001. What is the maximum permissible rms value of the ripple component?
18. A certain power-supply unit is to provide a load with 22.5 volts and its ripple voltage is not to exceed 0.5 per cent. What is the maximum permissible rms value of the ripple voltage?
19. A simple power-supply unit using a single half-wave rectifier is to be operated from a 60-hertz power line. What is the approximate per cent of ripple voltage at the first capacitor of its capacitor-input filter circuit if the effective load resistance is 5,000 ohms and the value of capacitance is 20 μf?
20. A simple power-supply unit using a full-wave rectifier is to be operated from a 60-hertz power line. (*a*) What is the approximate per cent of ripple voltage at the first capacitor of its capacitor-input filter circuit if the effective load resistance is 3,000 ohms and the value of capacitance is 30 μf? (*b*) What is the magnitude of the ripple voltage if the output is 250 volts?
21. It is desired to determine whether it would be practical to use a 35W4 rectifier

tube and a single-unit filter, consisting of a single capacitor, to provide a 60-ma load at 90 volts with a maximum ripple of 0.25 per cent. The peak plate current of the 35W4 is 600 ma. (*a*) What is the effective load resistance? (*b*) What value of capacitance is required? (*c*) What is the capacitive reactance of the required capacitor to a 60-hertz current? (*d*) If the peak value of the 60-hertz input voltage to the rectifier is 165 volts, what is the magnitude of the rectifier current during the first cycle after the power is applied? (*e*) What effect will the current during this first cycle have upon the tube? (*f*) Is this circuit design practical?

22. A power-supply unit being operated from a 60-hertz power line is supplying a load with 100 ma at 250 volts. A full-wave rectifier is used and the filter circuit is similar to that of Fig. 6-16*e*. The values of C_1 and C_2 are 10 μf each and L_1 is 20 henrys. What is the per cent of ripple voltage at (*a*) C_1? (*b*) C_2?

23. A power-supply unit being operated from a 60-hertz power line is supplying a load with 120 ma at 300 volts. A half-wave rectifier is used and the filter circuit is similar to that of Fig. 6-16*e*. The values of C_1 and C_2 are 20 μf each and L_1 is 20 henrys. What is the per cent of ripple voltage at (*a*) C_1? (*b*) C_2?

24. A power-supply unit being operated from a 60-hertz power line is supplying a load with 120 ma at 360 volts. A full-wave rectifier is used and the filter circuit is similar to that of Fig. 6-16*f*. The values of C_1, C_2, and C_3 are 10 μf each; L_1 is 20 henrys and L_2 is 15 henrys. What is the per cent of ripple voltage at (*a*) C_1? (*b*) C_2? (*c*) C_3?

25. A power-supply unit being operated from a 60-hertz power line is supplying a load with 120 ma at 360 volts. A half-wave rectifier is used and the filter circuit is similar to that of Fig. 6-16*f*. The values of C_1, C_2, and C_3 are 10 μf each; L_1 is 20 henrys and L_2 is 15 henrys. What is the per cent of ripple voltage at (*a*) C_1? (*b*) C_2? (*c*) C_3?

26. A power-supply unit being operated from a 60-hertz power line is supplying a load with 50 ma at 100 volts. A half-wave rectifier is used and the filter circuit is similar to that of Fig. 6-16*g*. The values of C_1 and C_2 are 40 μf each and R_1 is 1,000 ohms. What is the per cent of ripple voltage at (*a*) C_1? (*b*) C_2?

27. A power-supply unit being operated from a 60-hertz power line is supplying a load with 50 ma at 80 volts. A half-wave rectifier is used and the filter circuit is similar to that of Fig. 6-16*g*. The value of C_1 is 30 μf, C_2 is 50 μf, and R_1 is 800 ohms. What is the per cent of ripple voltage at (*a*) C_1? (*b*) C_2?

28. A power-supply unit being operated from a 60-hertz power line is supplying a load with 14 ma at 80 volts. A half-wave rectifier is used and the filter circuit is similar to that of Fig. 6-16*g*. The current for the power tube is taken off at capacitor C_1, and hence this current does not flow through resistor R_1. The value of C_1 is 40 μf, C_2 is 20 μf, and R_1 is 1,200 ohms. What is the per cent of ripple voltage at (*a*) C_1? (*b*) C_2?

29. What is the per cent of regulation of the power-supply unit of Prob. 22 if its output voltage is 250 volts at full load and 350 volts at no load?

30. What is the per cent of regulation of the power-supply unit of Prob. 23 if its output voltage is 300 volts at full load and 450 volts at no load?

31. What is the per cent of regulation of the power-supply unit of Prob. 27 if the output voltage is 80 volts at full load and 150 volts at no load?

32. If it is desired to limit the regulation of the power-supply unit of Prob. 27 to 20 per cent by adding a bleeder resistor to the unit, to what value of no-load voltage must the circuit be limited?

33. A power-supply unit being operated from a 60-hertz power line is supplying a load with 125 ma at 300 volts. A full-wave rectifier is used and the filter circuit is similar to that of Fig. 6-16*b*. The value of L_1 is 5 henrys and C_1 is 10 μf. What is the per cent of ripple voltage at the output?

34. A power-supply unit being operated from a 60-hertz power line is supplying a load with 150 ma at 450 volts. A full-wave rectifier is used and the filter circuit is similar to that of Fig. 6-16*b*. The value of L_1 is 5 henrys and C_1 is 20 μf. What is the per cent of ripple voltage at the output?

35. A power-supply unit being operated from a 60-hertz power line is supplying a load with 170 ma at 425 volts. A full-wave rectifier is used and the filter circuit is similar to that of Fig. 6-16*c*. The values of L_1 and L_2 are 10 henrys each, and C_1 and C_2 are 10 μf each. What is the per cent of ripple voltage at (*a*) C_1? (*b*) C_2?

36. A power-supply unit being operated from a 60-hertz power line is supplying a load with 200 ma at 450 volts. A full-wave rectifier is used and the filter circuit is similar to that of Fig. 6-16*c*. The value of L_1 is 10 henrys, L_2 is 5 henrys, C_1 is 10 μf, and C_2 is 20 μf. What is the per cent of ripple voltage at (*a*) C_1? (*b*) C_2?

37. Determine the per cent of ripple voltage at the output of the power-supply unit of Prob. 33 if it is to be operated from a 25-hertz power line.

38. Determine the per cent of ripple voltage at the output of the power-supply unit of Prob. 34 if it is to be operated from a 40-hertz power line.

39. Determine the per cent of ripple voltage at the output of the power-supply unit of Prob. 35 if it is to be operated from a 50-hertz power line.

40. Determine the optimum value of inductance for the input choke of the power-supply unit in Prob. 33.

41. Determine the optimum value of inductance for the input choke of the power-supply unit in Prob. 34.

42. It is desired to have a voltage divider (Fig. 6-19) supply the following needs from a 7.5-volt source: (1) 20 ma at 6 volts, (2) 15 ma at 3 volts, (3) 5 ma at 1.5 volts, (4) −1.5 volts at zero current. It is also desired to have a bleeder current at 10 ma. Determine the resistance of each section of the voltage divider.

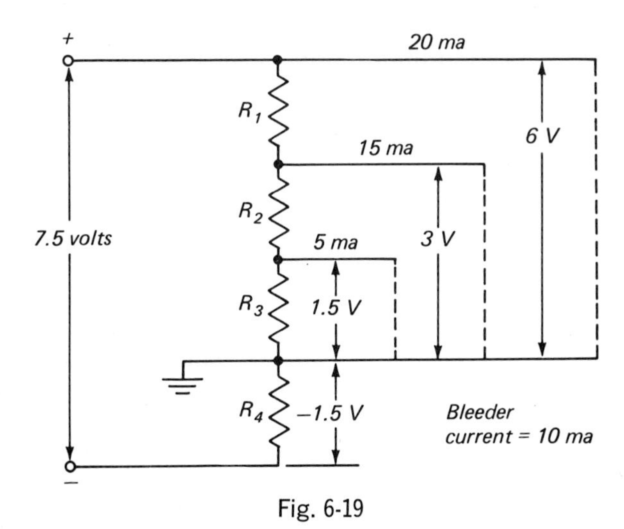

Fig. 6-19

43. Determine the power rating of the voltage divider of Prob. 42 for (*a*) a single tapped resistor, (*b*) four separate resistors. (*c*) What is the total rated wattage of the four separate resistors?

44. A power-supply unit is to provide 55 ma at 250 volts, and 5 ma at 100 volts; the bleeder current is to be 20 ma. A voltage divider is formed by two resistors connected in series across a 250-volt power source, and with 100 volts available at the junction of the two resistors. Determine the resistance of each section of the voltage divider.

45. If the voltage divider of Prob. 44 is to be located in an area having restricted ventilation, what is its power rating for (*a*) a single tapped resistor? (*b*) Two separate resistors? (*c*) What is the total rated wattage of the two separate resistors?

46. Determine the resistance values of a voltage divider for a superheterodyne receiver that employs a 6SA7 converter, a 6SG7 i-f amplifier tube, a 6SQ7 detector-amplifier tube, and a 6K6-GT/G power-output tube. The operating voltages and currents are to be obtained from a tube manual for plate voltages of 250 volts. (The screen grid of the 6SG7 is to be operated at 150 volts.) The voltage divider should have taps to supply all plate, screen-grid, and grid-bias voltages. Its power transformer, rated at 90 ma, is to be operated at 90 per cent of its rated current.

47. Determine the power rating of the voltage divider of Prob. 46 for (*a*) a single resistor having five taps, (*b*) six separate resistors.

48. Determine the resistance values of a voltage divider for a superheterodyne receiver that employs a 7B8 converter, a 7A7 i-f amplifier tube, a 7B6 detector-amplifier tube, and a 7B5 power output tube. The operating voltages and currents are to be obtained from a tube manual for plate voltages of 250 volts. All control-grid bias voltages are to be obtained by separate cathode-bias resistors, and hence these voltages are not to be provided by the voltage divider. Its power transformer, rated at 90 ma, is to be operated at 90 per cent of its rated current.

49. Determine the power rating of the voltage divider of Prob. 48 for (*a*) a single-tapped resistor, (*b*) two separate resistors.

50. (*a*) What rms value of alternating voltage must be applied at the input of the power-supply unit of Prob. 23 if the resistance of L_1 is 150 ohms and the rectifier is a solid-state diode whose voltage drop at normal load is 5 volts? (*b*) What rms input voltage is required if the solid-state diode is replaced by a vacuum-tube diode whose voltage drop at normal load is 25 volts? (Assume that the average value of the rectifier output voltage is equal to the rms value of the applied voltage when the voltage drop at the diode is neglected.)

51. What rms value of alternating voltage must be supplied to the cathode of the rectifier diode of Prob. 24 if the resistance of L_1 is 150 ohms, L_2 is 100 ohms, and the voltage drop at the diode is (*a*) 5 volts with a solid-state diode, (*b*) 30 volts with a vacuum-tube diode? (Assume that the average value of the rectifier output voltage is equal to the rms value of the applied voltage when the voltage drop at the diode is neglected.)

52. What rms value of alternating voltage must be supplied to the cathode of the rectifier diode of Prob. 28 if the internal voltage drop at the diode is (*a*) 3 volts with a solid-state diode? (*b*) 20 volts with a vacuum-tube diode? (Assume that the average value of the rectifier output voltage is equal to the rms value of the applied voltage when the voltage drop at the diode is neglected.)

Chapter 7
Specialized Power Sources

Military, medical, and space requirements have stimulated fantastic advances in the field of electronics with an accompanying need for specialized power sources. Continuous research and development have produced new chemical-electrical systems. In addition, new methods of obtaining electrical energy have been developed such as (1) the solar cell, which converts light energy from the sun to electrical energy, and (2) the fuel cell, which converts the energy from conventional fuel oxidation directly to electrical energy.

Other specialized power-supply equipment includes such devices and circuits as (1) power-driven a-c or d-c generators, (2) hand-driven generators, (3) dynamotors, (4) converters, (5) voltage multiplier circuits, (6) vibrators, and (7) voltage regulator circuits.

7-1 The Solar Cell

Basic Principle. The solar cell is based on the principle that, when a silicon crystal is exposed to light, light rays are absorbed by the crystal by liberating free-to-move negative charges, called *electrons*, and free-to-move positive charges, called *holes*. The direction of movement of the electrons and holes can be controlled to produce an electric current. The strength of this current is dependent upon the size of the crystal and the intensity of the source of light.

Basic Construction. The basic construction of a solar cell is shown in Fig. 7-1; the cell consists of an *N-type* silicon layer, or negative electrode, and a *P-type* silicon layer, or positive electrode. The two crystal layers are in direct contact with each other, and the junction formed between them exhibits the property of a strong built-in electric field. This field tends to contain the electrons in the N-silicon crystal and the holes in the P-silicon crystal, thus forming a barrier between them. The rays of light are absorbed through the thin P-type silicon crystal into this barrier region, where they liberate both electrons and holes. The built-in electric field forces the electrons to move to the N-type layer, making this crystal a negative electrode, and the holes to move to the P-type layer, making this crystal a positive electrode.

Characteristics. The efficiency of a solar cell in converting light energy to electrical energy is dependent upon the following factors: (1) the intensity of the light source, (2) the spectral content of the light waves, (3) the angle at which the light waves strike the surface of the cell (angle of incidence), (4) the

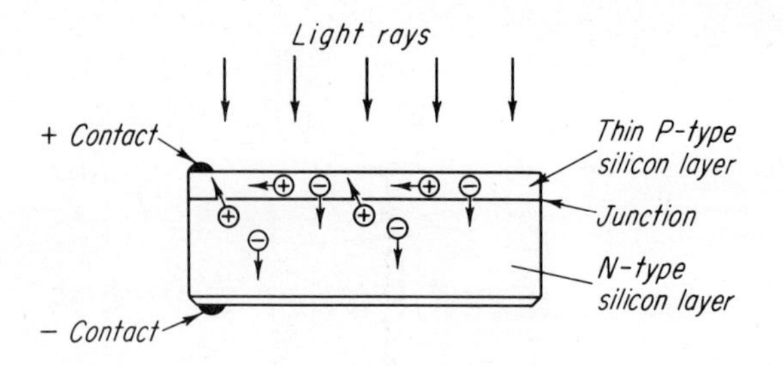

Fig. 7-1 Basic construction and operation of a silicon solar cell.

operating temperature of the cell, and (5) the characteristics of the external circuit. The solar cell has an output of approximately 0.4 volt, and cells may be connected in series to increase the voltage output and in parallel to increase the current output. The simplicity of a solar cell permits its construction to be relatively rugged. These cells have a high output level, an indefinite life, and a nonaging characteristic.

Applications. A highly publicized application of the solar cell is to provide the power requirements of satellites and space vehicles. Solar cells are used in these applications to charge storage batteries and to supply power during that portion of the orbit when the cells are in direct sunlight. The storage battery provides a continuity of power for instrumentation, transmitters, and telemetry systems during the periods of darkness when the cells are inoperative. Because of its advantages, the silicon solar cell is finding increased and varied industrial applications. This type of cell is used extensively in data-processing equipment to read punched tape, punched cards, or film. It is also used as power supplies for remote and unattended radio stations, lighthouses, beacons, and telephone repeaters in sunny areas (such as the polar regions). Other potential uses of solar cells are (1) light-operated transistor radio, (2) light-controlled toys, (3) automatic street lighting, (4) analog-to-digital encoders, (5) servomechanisms, (6) gyrocompasses, (7) light-operated burglar alarms, (8) modulated light-beam communication.

7-2 The Fuel Cell

Another source of energy is the fuel cell. This type of cell, unlike the chemical cell, does not store energy but rather converts energy from conventional fuel oxidation directly to electrical energy. The substances used are (1) of the inexpensive fossil types, such as coal and hydrocarbons, or (2) of more expensive substances, such as hydrogen, alcohol, and carbon monoxide, which may be derived from the fossil types.

The basic construction of a simple hydrocarbon fuel cell is shown in Fig. 7-2. Air and fuel enter opposite sides of the upper portion of the cell and are kept separated by the electrolyte. However, hydrogen ions, formed from the hydrocarbon fuel, flow through the electrolyte to unite with oxygen. At the same time, the electrons stripped from the hydrocarbon fuel travel through the external circuit. An electric current is thus produced *directly* without steam,

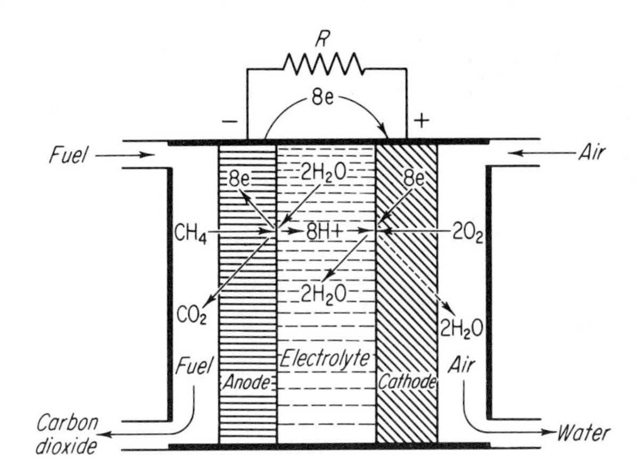

Fig. 7-2 Basic principle of operation of a hydrocarbon fuel cell. *(General Electric Co.)*

combustion, noise, vibration, or moving parts. Carbon dioxide and water are the two harmless byproducts of the chemical reactions of this type of cell.

Hydrogen fuel cells are used as a power source in the spacecraft of our outer space exploration programs. This type of cell is expensive and is not practical for commercial applications. However, experimental cells have been developed that use a variety of hydrocarbons such as propane, natural gas, octane, gasoline, and diesel oil.

7-3 Solid-state Light-sensitive Devices

Basic Principles of the Photocell. The solid-state photocell is based on the principle that, when certain materials are exposed to light, light rays are absorbed by the material by liberating free-to-move electrons and holes. The direction of movement of these negative and positive charges can be controlled to produce an electric current. The strength of the current and voltage produced is dependent upon the material used, its physical size, and the intensity of the source of light. The material used in the construction of a photocell will depend on its application. Some of the materials used are silicon, germanium, selenium, cadmium sulfide, gallium arsenide, indium antimonide, and indium arsenic.

In addition to converting light energy to electrical energy, photocells may also be used as control devices. Photocells used for control purposes are usually of a high-voltage type, such as the cadmium-sulfide cell, with ratings up to 500 volts d-c. These cells may be operated directly from a 115-volt a-c (230-volt d-c) power source with a normal power dissipation of ½ watt and a maximum power dissipation of 15 watts. The illumination from a very small neon lamp can be used to switch as much as 40 watts of power, an action representing a power gain of 80.

One of the many applications of the photocell as a control device is shown in Fig. 7-3*a*. Smooth control of the speed of the 115-volt a-c motor may be ob-

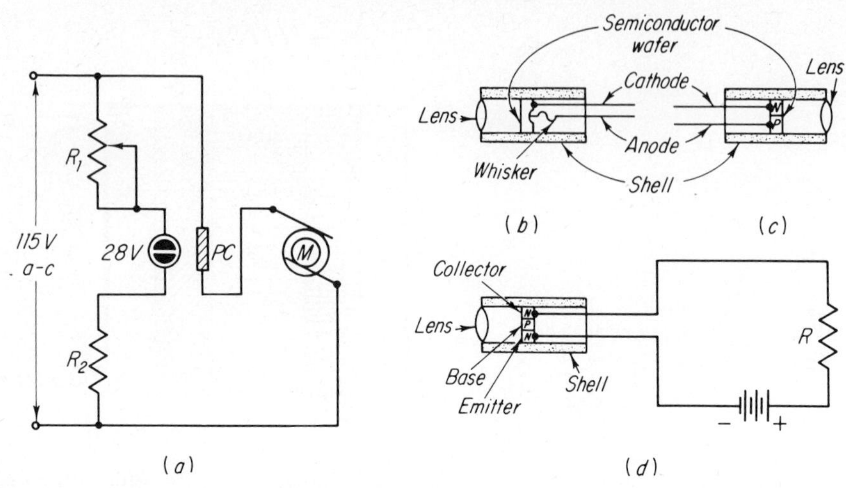

Fig. 7-3 Solid-state light-sensitive devices. (a) Photocell used to control the speed of a motor. (b) Point-contact type of photodiode. (c) Junction-type of photodiode. (d) A simple phototransistor circuit.

tained by varying the resistance of the photocell. This action is accomplished by adjusting the resistance of R_1, which in turn varies the brilliance of the neon lamp, thus also varying the resistance of the photocell. Smooth speed control may also be obtained by operating the neon lamp at maximum brilliance and controlling the illumination at the cell by rotating a variable-opacity vane between the neon lamp and the cell.

Photodiodes. The basic construction of two types of photodiodes is shown in Fig. 7-3*b* and *c*. The semiconductor wafer, made of germanium or silicon, is illuminated through a very small lens mounted at one end of the enclosed unit. Illumination of the wafer causes (1) the resistance of the diode to change from a very high value to a comparatively low value and (2) with no applied voltage, the production of a direct voltage across its output terminals. Because of its photoconductive properties the photodiode has many applications where small size and fast operation are desirable. Some of these applications are relay, counting, alarm systems, sound on film, and punched-tape and punched-card reading. The solar cell, as explained in Art. 7-1, is one of the applications of the voltaic properties of the photodiode.

Phototransistor. The construction of a phototransistor is similar to that of a junction transistor. However, only two external leads are used, one connected to the collector and the other connected to the emitter, with no connection being made to the base (Fig. 7-3*d*). The transistor is positioned so that the base is illuminated through a small lens that is mounted at one end of the enclosing shell. A direct voltage is applied between the emitter and collector in the usual manner for a common-emitter amplifier. With no illumination the wafer is darkened and the collector current is extremely low, being the normal collector cutoff current for the applied voltage. Illumination of the wafer causes current carriers to be injected into the base region. A large collector current, which is proportional to the illumination and is equal to beta times the base current, will flow through the external circuit. The phototransistor has the advantage over the photodiode in that it also provides amplification, because a relatively low light intensity will produce a comparatively high output current. Typical applications of the phototransistor are similar to those of the photodiode with the addition of optical coupling, light-beam reception, and electron-optical control service.

Miscellaneous Types of Light-sensitive Devices. The basic principles of the photocell, the photodiode, and the phototransistor are used to produce many other types of light-sensitive devices. Two of these are the light-activated switch and the Raysistor. The light-activated switch is a bistable component of the PNPN type. In a manner similar to a four-layer diode, once the unit is triggered to its ON state, it will continue to conduct current from its external power source until the power is interrupted. The Raysistor consists of a light source and a photoconductive cell that are enclosed in a lighttight housing. The light source may be either an incandescent or glow lamp. The operation of this unit is similar to that of a lamp-photocell combination. The dark (OFF) resistance is of the order of 1,800 megohms, while the light (ON) resistance may drop to values as low as 300 ohms.

7-4 Phototubes

Principle of Operation. A *photoelectric tube*, commonly called a *phototube*, is capable of converting light energy directly into electrical energy. When energy in the form of light strikes a photosensitive metal, electrons are emitted from this metal under the impact of the energy of the light rays. This method of producing electron emission is called *photoelectric emission* and is the basic principle of operation of phototubes.

Construction. The basic phototubes consists of two electrodes that are placed in an evacuated glass envelope. One of the electrodes emits electrons when light rays strike its surface, and it is called the *cathode*. The other electrode attracts these electrons to it when a positive voltage is applied to its terminal, and it is called the *plate*. The plate usually consists of a small rod that is placed in the center of the tube. The cathode is a half cylinder that is placed in the tube so that it surrounds the plate. The inside surface of the cathode is covered with multiple layers of a photosensitive metal such as cesium (Fig. 7-4).

The sensitivity of a phototube is dependent on the frequency, or color, of the light used to excite the cathode. Phototubes are manufactured to provide sensitivity characteristics for different types of application. For example, some tubes are designed to be sensitive to blue light, some to red light, while others have response characteristics that are similar to those of the human eye.

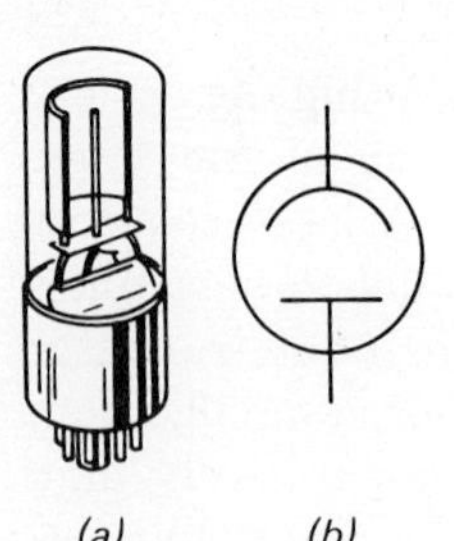

(a) (b) Fig. 7-4 (a) Phototube. (b) Symbol for phototube.

Vacuum and Gas Phototubes. Phototubes may be obtained with a high vacuum or be gas-filled. The amount of current that a phototube is capable of passing for a given amount of cathode illumination is higher for gas-filled tubes, because of ionization, than for vacuum tubes. The positive gas ions will also strike the cathode and thus produce an appreciable amount of secondary emission that will also increase the sensitivity of the tube. Gas phototubes are therefore generally used where high-sensitivity requirements are important, such as the reproduction of sound from sound-motion-picture film. Vacuum phototubes are more stable and less easily damaged, have a higher internal resistance, and are used for light-operated relay applications.

7-5 Voltage-multiplier Circuits

Purpose. Voltage-multiplier circuits make it possible to obtain d-c output voltages higher than that of the a-c input voltage without the use of a transformer; accordingly, they are sometimes called *transformerless power-supply circuits*. Multiplier circuits can raise the output voltage in multiples of the a-c input voltage and are generally restricted to relatively low current ratings. They can be designed for either half-wave or full-wave rectification. With half-wave rectification, the ripple frequency will be the same as the frequency of the input voltage and the voltage regulation of the circuit will be poor. With full-wave rectification, the ripple frequency will be twice the frequency of the input voltage and the voltage regulation will be better than with half-wave rectification. The rectifiers used may be either (1) solid-state diodes or (2) electron-tube diodes.

Disadvantages of voltage-multiplier circuits include (1) poor voltage regulation; (2) low current rating; (3) they require high values of capacitance; and (4) capacitors must have high voltage ratings.

Basic Principles. The basic principle of all voltage-multiplier circuits is that two or more capacitors are charged and discharged on alternate half-cycles of the a-c input voltage. The output voltage is the sum of two or more voltage sources (charged capacitors) connected in series.

7-6 Full-wave Voltage Doubler

The most common type of voltage-multiplier circuit is the voltage doubler, which produces an output voltage of approximately twice the a-c input voltage.

Figure 7-5 illustrates a full-wave voltage-doubling circuit which operates in the following manner: During the half-cycle of the input voltage E_i that terminal T_1 is positive, electrons flow from T_2 to C_1, and from C_1 through D_1 and R_1 to T_1; C_1 becomes charged to the peak value E_m of the input voltage and with the polarities indicated. During the next half-cycle, when T_2 is positive, electrons flow from T_1 through R_2 and D_2 to C_2, and from C_2 to T_2; C_2 becomes charged to a value E_m and with the polarities indicated. The output voltage E_o is a direct voltage equal to the sum of the voltages at C_1 and C_2, and with the polarities indicated. When the load resistance R_L is very high, the output voltage will approach a value equal to double the peak value of the alternating input voltage decreased by the voltage drops at R_1, R_2, D_1, D_2. The resistors R_1 and R_2 are of low values and are used to protect the diodes D_1 and D_2 against any possible high surges of current. When the load resistance R_L is in the range of normal load conditions, the capacitors will discharge through the load, C_1 discharging while C_2 is charging and vice versa. If the time constant of R_L, C_1, C_2 is long enough, the capacitors will lose only a small portion of their charge in the short interval required for the input voltage to reverse its polarity. During the half-cycle that a diode is conducting, the capacitor in its circuit will not start charging until the instantaneous value

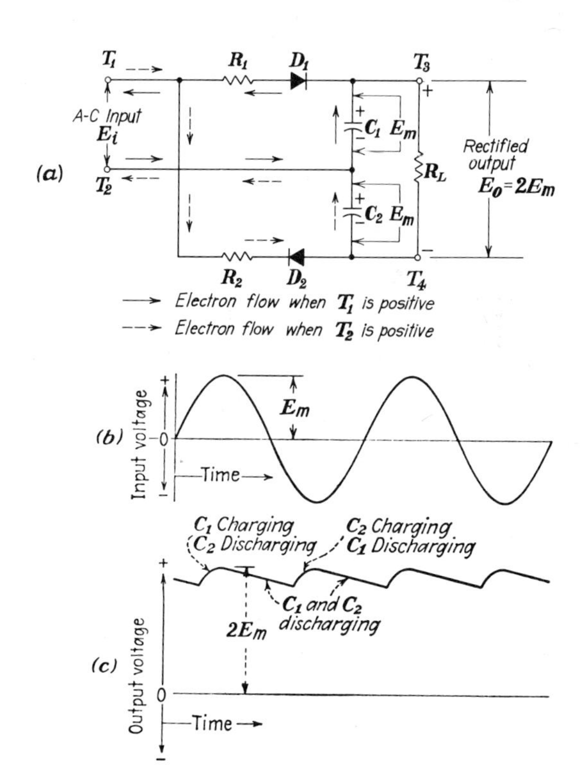

Fig. 7-5 A full-wave voltage-doubler circuit. (a) The circuit. (b) Waveform of the input voltage (c) Waveform of the output voltage.

of the input voltage exceeds the capacitor's terminal voltage. The waveform of the output voltage illustrating the charge and discharge actions of the two capacitors is shown in Fig. 7-5c. Examination of this figure will show that the frequency of the ripple voltage at the output is twice the frequency of the a-c input voltage. Also, E_o is appoximately twice the value of E_m.

The voltage ratings of the capacitors should be higher than the peak value of the input voltage. The voltage rating and the peak inverse voltage rating of the diodes should be higher than $2E_m$.

7-7 Half-wave Voltage Doubler

The half-wave voltage doubler shown in Fig. 7-6 operates in the following manner: During the half-cycle that terminal T_2 is positive, D_1 conducts and

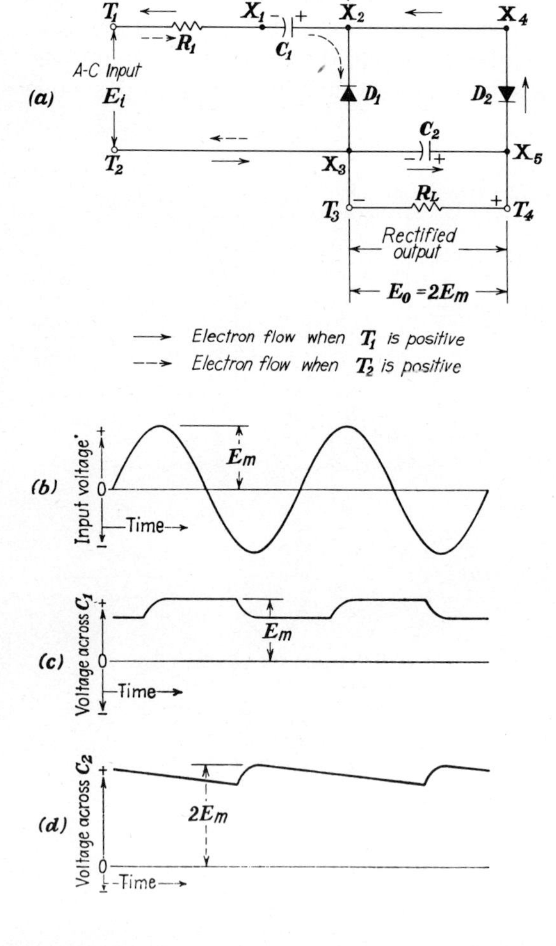

Fig. 7-6 A half-wave voltage-doubler circuit. (a) The circuit. (b) Waveform of the input voltage. (c) Waveform of the voltage across C_1. (d) Waveform of the voltage across C_2, which is also the rectified output voltage.

electrons flow from T_1 through R_1 to C_1, and from C_1 through D_1 to T_2; C_1 becomes charged to the peak value E_m of the input voltage (less any drop at D_1 and R_1) with the polarities indicated. During the next half-cycle, when T_1 is positive, D_2 conducts and electrons flow from T_2 to C_2, from C_2 through D_2 to C_1, and from C_1 through R_1 to T_1; C_2 becomes charged to a value equal to $2E_m$. This is explained by using the negative terminal (T_2 in this instance) as a zero reference level and observing that at the peak of the cycle T_1 (also X_1) is positive by the maximum amount (E_m) of the input voltage; point X_2 (also X_4 and X_5) is positive by $2E_m$ with respect to T_2 as it is the sum of the line voltage E_m and the charge E_m on capacitor C_1. Since the voltage at point X_5 is $2E_m$ above T_2, capacitor C_2 will become charged to a value of $2E_m$ and with the polarities indicated. However, the voltage at C_2 can attain this value only at no load or when the value of R_L is very high. Resistor R_1 protects the diodes D_1 and D_2 against any high surges of current.

When the circuit is supplying power to a load with normal values of R_L, the capacitors will charge and partially discharge during alternate halves of the input cycle. Capacitor C_1 will charge when T_1 is negative and discharge when T_1 is positive. Capacitor C_2 will charge when T_2 is negative and discharge when T_2 is positive. Thus when C_1 is charging, C_2 is discharging, and vice versa. Figure 7-6*c* and *d* show the charging and discharging periods of these two capacitors. Figure 7-6 also shows that the frequency of the ripple in the output voltage is the same as the frequency of the a-c input voltage.

The voltage rating of C_1 should be higher than E_m, and that of C_2 should be higher than $2E_m$. The voltage rating and the peak inverse voltage rating of the diodes should be higher than $2E_m$. Disadvantages of the half-wave voltage-doubler circuit are: (1) the voltage regulation is not as good as that of the full-wave voltage-doubler circuit, and (2) the voltage rating of C_2 must be double that of the capacitors in a full-wave circuit of the same rating.

7-8 Full-wave Voltage Tripler

Figure 7-7 illustrates a full-wave voltage-tripling circuit, which operates in the following manner: During the half-cycle that terminal T_2 is positive, diodes D_2 and D_3 will conduct and capacitors C_1 and C_3 (being operated in parallel) become charged to the peak value E_m when R_L is very high. The charging path for C_1 is T_1, C_1, D_2, R_1, T_2. The charging path for C_3 is T_1, D_3, C_3, R_1, T_2. During the half-cycle that terminal T_1 is positive, D_1 conducts and capacitor C_2 becomes charged, the path being T_2, R_1, C_2, D_1, C_1, T_1. Capacitor C_1 is already charged to a voltage E_m, and being in series with the power source voltage E_m, and with their polarities aiding, the charge on capacitor C_2 will be $2E_m$. The voltage at the output terminals T_3 and T_4 will be the sum of the voltage charges on capacitors C_2 and C_3, and hence E_o will be equal to $3E_m$ when R_L is very high.

Figure 7-7*b* illustrates the output characteristics of a typical full-wave voltage-tripler circuit. The frequency of the ripple in the output voltage will be

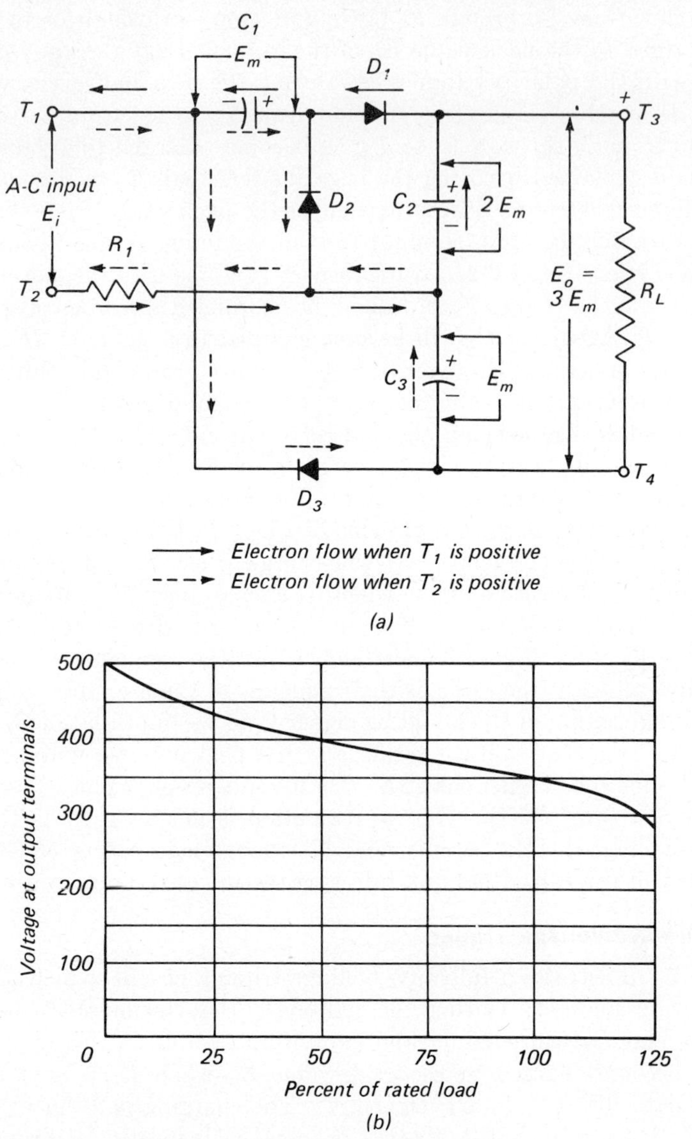

Fig. 7-7 A full-wave voltage-tripler circuit. (a) The circuit. (b) Typical voltage versus load characteristics of a full-wave voltage tripler.

double the frequency of the a-c input voltage. The voltage ratings of capacitors C_1 and C_3 should be higher than E_m, and the rating of C_2 should be higher than $2E_m$. The peak inverse voltage rating of the diodes should be higher than $2E_m$. Resistor R_1 provides protection for the diodes against any high surges of current.

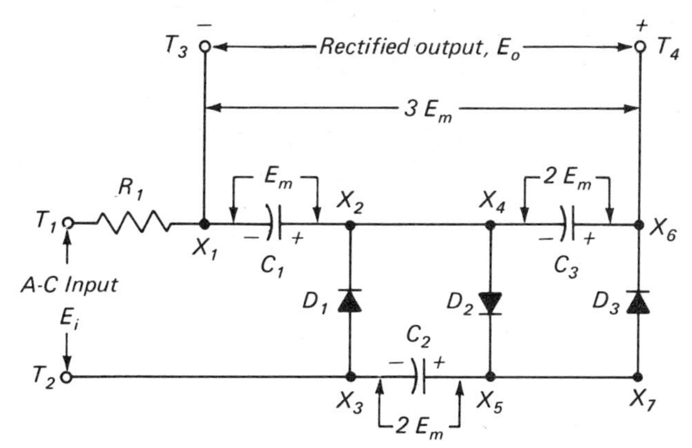

Fig. 7 8 A half-wave voltage-tripling circuit.

7-9 Half-wave Voltage Tripler

Figure 7-8 illustrates a half-wave voltage-tripling circuit. This circuit is identical to the voltage-doubling circuit of Fig. 7-6 up to the points marked X_4 and X_5. Adding the new section, consisting of the diode D_3 and capacitor C_3, converts the voltage doubler to a voltage tripler.

The operation of the circuit up to points X_4 and X_5 is the same as described in Art. 7-7. Capacitor C_3 charges during those half-cycles that terminal T_1 is negative. Point X_4 is E_m volts above the voltage at T_1 because of the charge existing on C_1. Point X_5 (also X_6 and X_7) is $3E_m$ above the voltage at T_1, as it is the sum of the power-source voltage E_m and the charge of $2E_m$ on C_2, or a total of $3E_m$. Capacitor C_3 will become charged to a voltage equal to the difference between the potentials at X_6 ($3E_m$) and at X_4 (E_m) for a net charge of $2E_m$. The output at terminals T_3 and T_4 is the sum of the voltage charges at capacitors C_1 and C_3, or $3E_m$.

The ripple frequency of the output voltage will be the same as the frequency of the a-c input voltage. The voltage rating of C_1 should be higher than E_m, and the rating of C_2 and C_3 should be higher than $2E_m$. The peak inverse voltage rating of the diodes should be higher than $2E_m$. Resistor R_1 provides protection for the diodes against any high surges of current.

7-10 Half-wave Voltage Quadrupler

Figure 7-9 illustrates a half-wave voltage-quadrupling circuit. This circuit is identical to the voltage-tripling circuit of Fig. 7-8 up to the points marked X_6 and X_7. Adding the new section, consisting of the diode D_4 and capacitor C_4, converts the voltage tripler to a voltage quadrupler.

The operation of the circuit up to points X_6 and X_7 is the same as described in Art. 7-9. Capacitor C_4 charges during those half-cycles that terminal T_2 is negative. Point X_6 (also X_8 and X_9) is $4E_m$ volts above the zero reference level

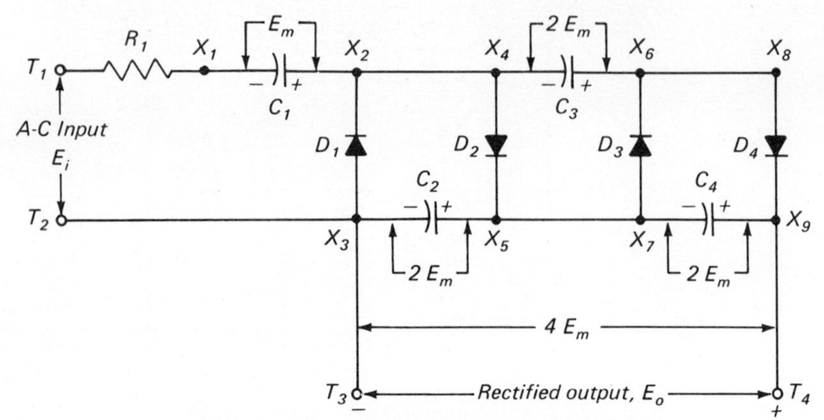

Fig. 7-9 A half-wave voltage-quadrupler circuit.

at T_2 because of the input voltage E_m, a voltage charge of E_m at C_1, and a voltage charge of $2E_m$ at C_3. Point X_7 is $2E_m$ volts above the zero reference level at T_2 because of the voltage charge of $2E_m$ at C_2. Capacitor C_4 will become charged to a voltage equal to the difference between the potentials at X_9 ($4E_m$) and X_7 ($2E_m$) for a net charge of $2E_m$. The output at terminals T_3 and T_4 is the sum of the voltage charges at C_2 and C_4, or $4E_m$.

By tracing the direction of electron flow when T_1 is negative and observing the polarities at the various capacitors, it can be seen that C_1 and C_3 will be charging, and C_2 and C_4 discharging. When T_2 is negative, C_2 and C_4 will be charging, and C_1 and C_3 discharging. The ripple frequency of the output voltage is the same as the frequency of the a-c input voltage. The voltage rating of C_1 should be higher than E_m, and the ratings of C_2, C_3, and C_4 should be higher than $2E_m$. The peak inverse voltage rating of the diodes should be higher than $2E_m$. Resistor R_1 provides protection for the diodes against any high surges of current.

7-11 Other Multiplier Circuits

A study of the voltage doubler of Fig. 7-6, the voltage tripler of Fig. 7-8, and the voltage quadrupler of Fig. 7-9 will show that circuits can readily be designed to produce an output voltage that is any multiple of the input voltage by providing the proper number of sections consisting of a diode and a capacitor. Such circuits are sometimes referred to as a *cascade voltage tripler*, a *cascade voltage quadrupler*, etc. An advantage of these types of multiplier circuits is that the voltage rating of any capacitor or diode is never greater than $2E_m$.

7-12 Vibrator Power Supply

When electronic equipment is to be operated from a low-voltage d-c power source, such as a battery, and a relatively high voltage d-c source is required by the equipment, a vibrator-type power supply can be used. The low-voltage d-c power from the battery can be changed to an alternating voltage by use of a vibrator, which is an electromagnetic device that reverses the direction of current flow during each vibration of its armature. By applying the output of the vibrator to a transformer, the voltage can be increased to the desired value. The output of the transformer is an alternating voltage and must be passed through rectifying and filtering circuits to obtain the desired high-voltage d-c output. The complete circuit consisting of the vibrator, transformer, rectifier, and filter is called a *vibrator power supply*.

7-13 Types of Vibrators

Basically, there are two types of vibrators: (1) the *synchronous* vibrator and (2) the *nonsynchronous* vibrator. The nonsynchronous vibrator interrupts the d-c circuit at a frequency that is unrelated to the other circuit constants. The resulting alternating voltage output must then be rectified by means of a semiconductor diode or a tube-type rectifying circuit. A synchronous vibrator, in addition to changing the low direct voltage to an alternating voltage, simultaneously rectifies the high a-c output at the secondary of the power transformer. Rectification is accomplished by use of an additional set of contacts, thereby eliminating the need of a semiconductor diode or tube-type rectifier.

7-14 Nonsynchronous-vibrator Power Supply

Figure 7-10 shows the circuit diagram of a nonsynchronous-vibrator power-supply unit. At the instant that the switch S_1 is closed, current flows from the battery through coil L_2, section 1-2 of the primary winding of the transformer, coil L_1, and back to the battery. Coil L_2 is wound on a soft-iron core, which becomes magnetized when current flows through the coil. The vibrating reed R is so constructed that when current flows through L_2, the reed is attracted toward the magnet. As the reed approaches the magnet it makes contact with point A, thus short-circuiting the coil L_2. Since the iron core is then no longer magnetized, the reed is released and tends to return to its normal position. However, because of the force present when it is released, the reed moves past its normal position and makes contact with point B and then returns to its

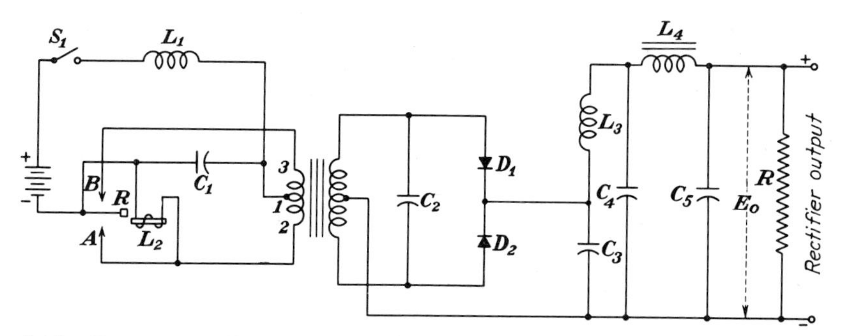

Fig. 7-10 Circuit diagram of a power-supply unit using a nonsynchronous vibrator.

normal position. At the instant that contact is being made at point B, current will flow through section 1-3 of the primary winding of the transformer. Current will flow in this circuit for only a very short period of time because contact is made at point B only instantaneously during the forward swing of the reed after it is released from point A. When the reed returns to its normal position, current again flows through L_2 and the cycle of operations is repeated. These operations occur very rapidly, and the complete cycle is repeated many times per second.

In each cycle of operations current flows alternately in opposite directions through alternate halves of the primary winding of the transformer. Since this flow of current is essentially the same as an alternating current, the voltage at the secondary terminals can be increased to any desired value by increasing the ratio of secondary to primary turns of the transformer. The output of the secondary winding is then rectified and filtered in the usual manner.

7-15 Synchronous-vibrator Power Supply

The output from the secondary of the transformer can be rectified by adding another set of points to the vibrator, thereby eliminating the need of a separate rectifier. These points are connected to opposite ends of the secondary winding as shown in Fig. 7-11. When the vibrating reed makes contact with point A, it also makes contact with point C, thus grounding terminal 2 of the primary winding and terminal 4 of the secondary winding. In a similar manner, when the reed makes contact with point B it also makes contact with point D, thus grounding terminals 3 and 5 of the transformer. Terminal 6, which is an output terminal, is always positive, and terminals 4 and 5 are alternately connected to ground during opposite halves of the cycle. Thus the output of the transformer is rectified since current flows in opposite directions in each half of the secondary winding during alternate halves of the cycle.

7-16 Filters for Vibrator Power Supplies

A surge of current occurs each time the primary winding in a vibrator power-supply unit is connected to or disconnected from the d-c power source. Connect-

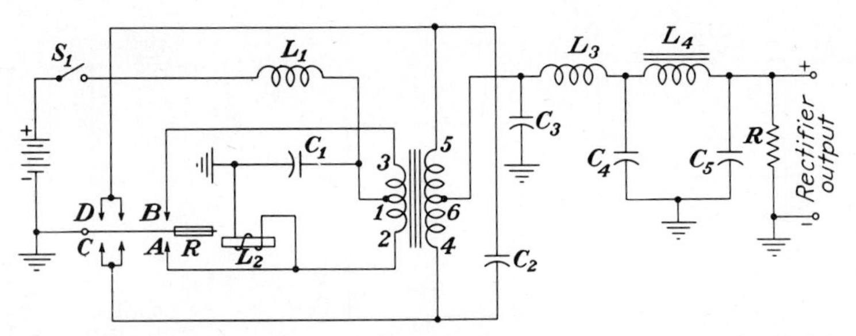

Fig. 7-11 Circuit diagram of a power-supply unit using a synchronous or self-rectifying vibrator.

ing a capacitor across the primary winding will absorb this surge of current, thus preventing the contact points or the rectifier from being damaged. Because the amount of capacitance required decreases with an increase in the applied voltage, it is more economical to connect a capacitor across the secondary winding, which has a much higher voltage than the primary. This capacitor (C_2 in Figs. 7-10 and 7-11) is called a *buffer capacitor*. As the capacitance reflected to the primary by a capacitor connected across the secondary increases as the square of the secondary to primary turns ratio, a low value of capacitance in the secondary circuit produces substantially the same result as connecting a high value of capacitance in the primary circuit.

Each time the contacts are opened, sparking will take place at the contacts. This sparking produces high-frequency transients, thus causing interference if the power supply is used with a radio receiver. This interference is commonly called *hash* and may be minimized by use of filter circuits and by shielding the vibrator unit. A hash-filter circuit is usually connected in the battery circuit and consists of a choke coil and a capacitor as represented by L_1 and C_1 in Figs. 7-10 and 7-11. Another hash filter consisting of L_3 and C_3 is shown connected in the output circuit.

7-17 Characteristics of the Vibrator Power Supply

The vibrator power supply is an inexpensive and compact means of obtaining high operating voltages from a low-voltage battery. It can be used in portable electronic equipment that is to be operated from low-voltage batteries. How-

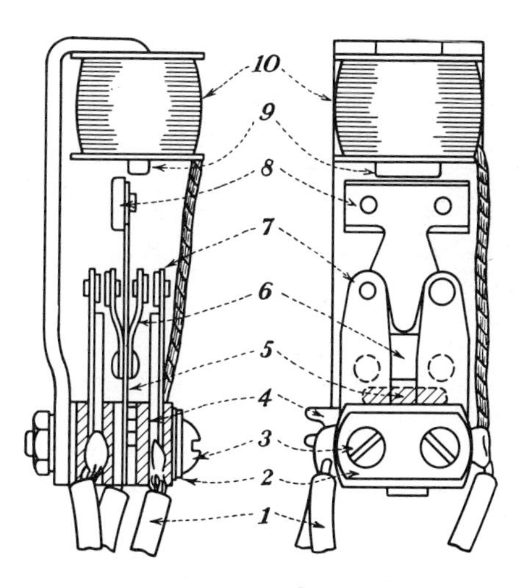

Fig. 7-12 Construction of a vibrator unit employing four sets of contactors. (Courtesy of P. R. Mallory and Co., Inc.)

ever, the vibrator has a limited life, and its associated circuits require a complex filtering and shielding system.

The type of vibrator to be used will depend upon the power-supply requirements. In addition to the choice of either the synchronous or nonsynchronous type of vibrator, the current output and the type of interrupter and rectifier circuits must also be considered. Figure 7-12 shows the construction of an eight-contact vibrator that can be connected to operate simply as an interrupter or as a self-rectifier.

7-18 Regulated Power Sources

Purpose. Because of varying load conditions, the output voltage of a power-supply unit will generally vary inversely with changes in the amount of load. Variations in the a-c input voltage will cause the output voltage to also vary. There are many power-supply applications where the voltage applied to the load must be kept practically constant regardless of any variations in the input voltage. There are also some applications where the current flowing through the load must be kept constant regardless of variations in the input voltage to the power-supply unit.

A power-supply system designed to maintain a constant output voltage regardless of changes in the output load, input line voltage, temperature conditions, etc., is called a *constant-voltage power source.* A power-supply system designed to maintain a constant output current regardless of changes in the load conditions, input line voltage, temperature conditions, etc., is called a *constant-current power source.*

Voltage Regulation. The amount of change in the output voltage of a power system may be expressed (1) directly as a voltage variation between two load conditions, usually no-load and full-load, or (2) as a percentage change in the output voltage between two load conditions as expressed by

$$\text{Per cent of regulation} = \frac{E_{N-L} - E_{F-L}}{E_{F-L}} \times 100 \qquad (7\text{-}1)$$

where E_{N-L} = voltage at no load, volts
E_{F-L} = voltage at full load, volts

Example 7-1 A power-supply unit represented by the curve of Fig. 7-7*b* supplies a load at 350 volts, and when the load is removed the voltage rises to 500 volts. What is the voltage regulation expressed as (*a*) actual voltage deviation? (*b*) Per cent?

GIVEN: $E_{N-L} = 500$ volts $\quad E_{F-L} = 350$ volts

FIND: (*a*) Voltage change (*b*) Per cent regulation

SOLUTION:

(*a*) Voltage change $= E_{N-L} - E_{F-L} = 500 - 350 = 150$ volts

(*b*) $$\text{Regulation} = \frac{E_{N-L} - E_{F-L}}{E_{F-L}} \times 100 = \frac{500 - 350}{350} \times 100 = 42.8 \text{ per cent}$$

Kinds of Regulating Circuits. A regulated power-supply unit is formed by adding a suitable auxiliary circuit to any of the power-supply circuits described earlier in this chapter. Two basic classifications of regulating circuits are (1) open loop and (2) closed loop. They are also classified as (1) series- or shunt-type regulators and (2) semiconductor or tube-type regulators.

7-19 The Zener Diode

A silicon junction diode produces the conduction characteristics shown in Fig. 7-13. In Fig. 7-13*a*, the steep slope from O to A indicates that the diode has a relatively low forward resistance. The practically horizontal section of the curve from O to B represents a very slight slope which indicates a very high (and nearly constant) resistance for this range of reverse voltage. Beyond point B a relatively small increase in the reverse voltage, as in the range of B to C, causes a very great increase, or *avalanche*, in the current. Thus within a few tenths of a volt change the back resistance decreases very sharply from several megohms to a few ohms. For this same voltage change, the current increases from a few microamperes to many milliamperes. The high conduction in the area of B to C could easily destroy the diode, and provision must

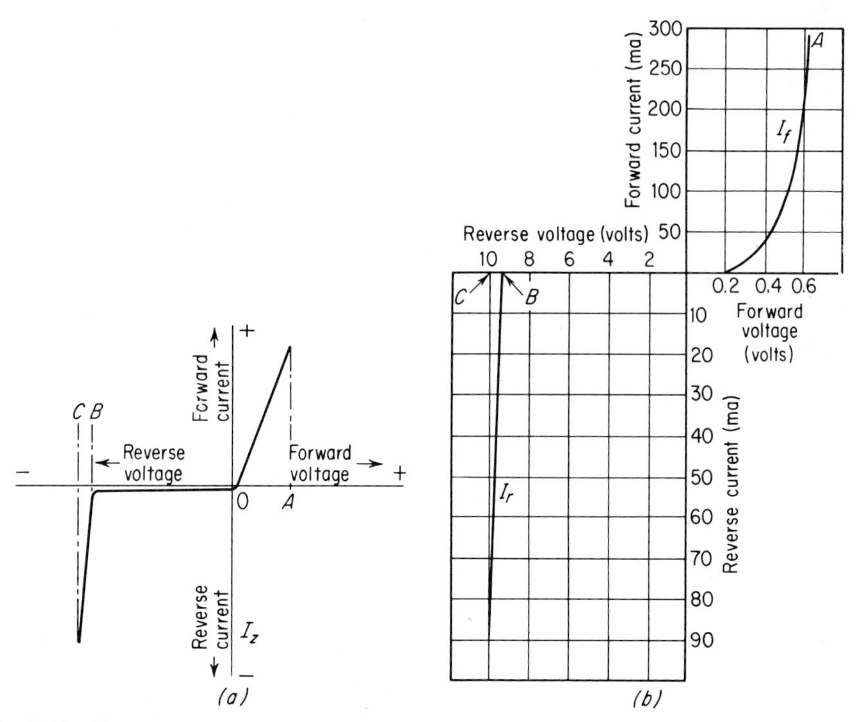

Fig. 7-13 Zener-diode static conduction characteristics. (a) Basic characteristics. (b) Characteristics of a typical diode.

be made in its circuit to prevent power dissipation beyond the rated safe limit. The voltage at point B is called the *Zener voltage,* and a diode designed to place the zener point at a definite voltage level is called a *Zener diode.*

Zener diodes are available with rated voltages from as low as several volts to as high as several hundred volts. The current ratings range from a few milliamperes to several hundred milliamperes. Standard tolerance ratings are 20, 10, and 5 per cent; however, they can be obtained with tolerances as low as 1 per cent. Zener diodes are highly adaptable to voltage-regulation applications because of their (1) sharp breakdown characteristics, (2) very high ratio of front to back resistance, (3) low resistance in the voltage range indicated by points B to C on Fig. 7-13a, (4) small size, and (5) mechanical ruggedness.

7-20 Open-loop Regulator Circuits

Basic Circuit. The simplest voltage regulator, shown in Fig. 7-14, is formed by a resistor R_1 and a suitably rated Zener diode D_1. The value of the resistor is determined by the voltage to be absorbed by the resistor and the rated maximum current of the diode (with zero output load), or

$$R_1 = \frac{E_i - E_o}{I_z} \tag{7-2}$$

where R_1 = value of the current limiting resistor, ohms
E_i = value of the unregulated supply voltage, volts
E_o = value of the regulated output voltage, volts
I_z = maximum Zener-diode current (at no load), amperes

Example 7-2 What value of series resistance must be used with a 30-volt Zener diode connected to a 45-volt d-c power source if the maximum current rating of the diode is 125 ma?

GIVEN: E_i = 45 volts E_o = 30 volts I_z = 125 ma

FIND: R_1

SOLUTION:

$$R_1 = \frac{E_i - E_o}{I_z} = \frac{40 - 35}{125 \times 10^{-3}} = \frac{15 \times 10^3}{125} = 120 \text{ ohms}$$

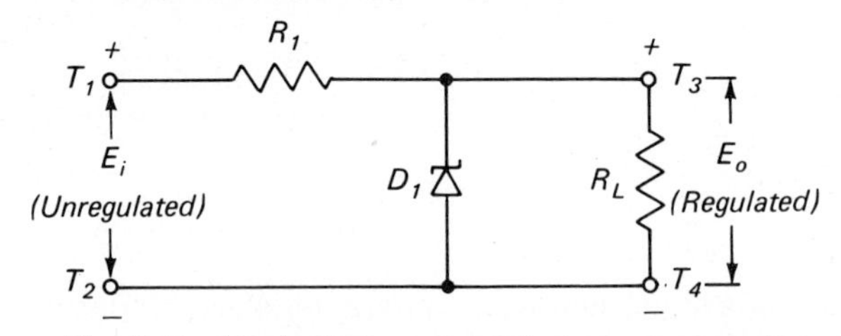

Fig. 7-14 Basic Zener-regulated power supply.

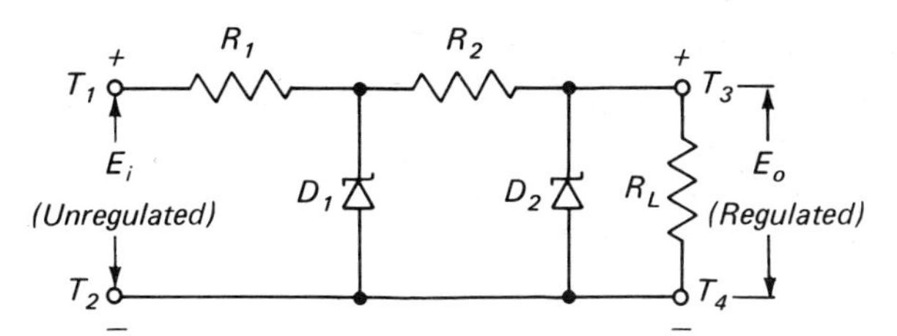

Fig. 7-15 Zener-regulated power supply with two cascaded stages.

This circuit operates in the following manner: The diode D_1 is connected for reverse bias and the value of R_1 is chosen to produce the maximum rated diode current in the Zener voltage range, point C of Fig. 7-13a. When the output load is zero, only the diode current (at its maximum value) flows through R_1 and the output voltage is equal to the Zener voltage. When a load that might vary in magnitude is applied, any increase in the load (or an increase in the input voltage) will cause an increase in the current in R_1 thereby increasing the voltage drop at R_1 and decreasing the voltage at the load and the diode. A small decrease in the voltage at D_1 results in a substantial decrease in the diode current (Fig. 7-13a) which then reduces the current through R_1 and counteracts the original cause of the increase in current through R_1. A decrease in the load current, or a decrease in the input voltage, produces the opposite chain of reactions and hence automatic regulation of the output voltage is achieved. Good regulation can be obtained as long as the diode voltage and current variations are held to the area of the curve between points B and C; regulations in the order of 5 to 10 per cent can be achieved.

The breakdown characteristics illustrated in Fig. 7-13a show that in the operating range B to C a relatively large variation in the diode current will result in only a small variation in the Zener voltage and hence the variation in the output voltage E_o will also be very small. Consequently, comparatively large changes in the input voltage, which would tend to cause large changes in the current in R_1, will now have very little effect on the current in R_1 and on the voltage at D_1 and R_L.

Cascading Stages. Figure 7-15 illustrates a two-stage voltage regulator that will provide better regulation than the single-stage circuit of Fig. 7-14. However, because the circuit also acts in the manner of a voltage divider it will be necessary to provide a higher input voltage in order to maintain the same output voltage.

Diodes in Series. When the desired regulated output voltage is higher than the rated voltage of a Zener diode, two or more diodes can be connected in series (Fig. 7-16). In such circuits care must be exercised to select diodes that have the same current ratings. In determining the value of R_1, it is necessary to use the sum of the Zener voltages.

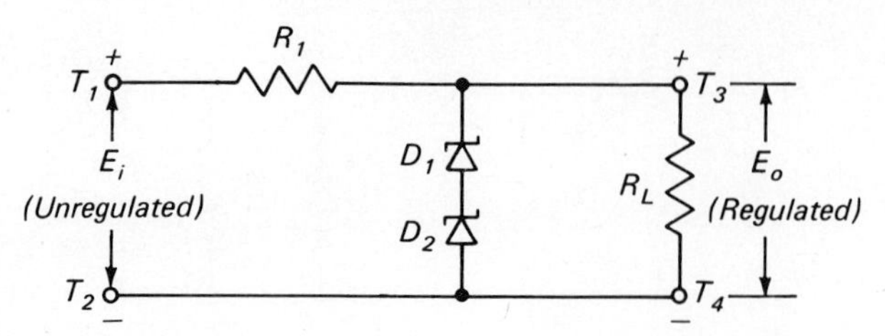

Fig. 7-16 Regulated power supply with two Zener diodes connected in series.

Example 7-3 The circuit of Fig. 7-16 uses two Zener diodes, each rated at 15 volts and 225 ma. If the circuit is connected to a 45-volt unregulated d-c power source, determine (*a*) the regulated output voltage, (*b*) the value of R_1.

GIVEN: $E_{D1}, E_{D2} = 15$ volts $\quad E_i = 45$ volts $\quad I_{z1}, I_{z2} = 225$ ma

FIND: (*a*) E_o $\quad$ (*b*) R_1

SOLUTION:

(*a*)
$$E_o = E_{D1} + E_{D2} = 15 + 15 = 30 \text{ volts}$$

(*b*)
$$R_1 = \frac{E_i - E_o}{I_z} = \frac{45 - 30}{225 \times 10^{-3}} = \frac{15 \times 10^3}{225} = 66.6 \text{ ohms}$$

7-21 Closed-loop Regulators

Basic Series-type Regulator. In a closed-loop regulator, a feedback path (or loop) is used to obtain regulation of the output voltage. Figure 7-17 shows a simple closed-loop voltage-regulator circuit. In this circuit the collector-to-emitter path of the transistor Q_1 is connected in series with the load and acts as a variable-resistance element; the transistor is therefore called a *series regulator* or a *pass element*.

The Zener diode maintains the base of the transistor at a constant voltage. The value of R_1 is selected so that with the lightest load condition (R_L at its highest anticipated value) and E_o at its highest value, the Zener-diode current will be just below its maximum allowable value. The Zener voltage will then remain practically constant for any decrease in Zener current down to the bend in its operating curve. Because of the manner in which the Zener diode is connected in the regulator circuit, the base of the transistor will remain at a fixed voltage. The base-to-emitter voltage, or base bias, is the algebraic sum of the reference voltage E_z and the output voltage E_o. Any increase in the load current as a result of a decrease in R_L will cause an increase in the voltage drop at the transistor and thereby decrease the output voltage. The voltage at the emitter of the transistor then becomes less negative (or more positive), thereby increasing the forward bias and reducing the collector-to-emitter resistance. The decrease in the resistance of this pass element decreases the voltage drop at the pass element and brings the output voltage E_o back to its former value. The

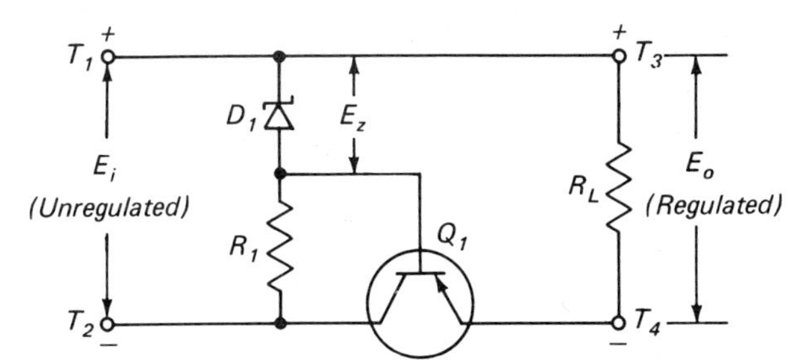

Fig. 7-17 Basic series-type voltage regulator using one transistor and one Zener diode.

amount of change in the bias voltage caused by a change in the load current is called the *error voltage* or the *error signal,* and it is this error voltage that produces the change in the transistor's collector-to-emitter resistance.

Any decrease in the load current causes an opposite chain of reactions that nullifies the change and maintains the output voltage at a constant value. Any change in the input voltage will likewise be nullified.

Good regulation can be obtained as long as the Zener-diode current variations are held to the portion of the curve (Fig. 7-13) between points B and C. Regulation in the order of from 2 to 5 per cent can be achieved.

Basic Shunt-type Regulator. The circuit of Fig. 7-18 shows a basic shunt-type regulator. In this circuit the collector-to-emitter path of the transistor is connected across the output terminals of the regulator. The Zener diode D_1 establishes a fixed constant reference voltage at the base of the transistor in the same manner as for the series-type regulator of Fig. 7-17. Any increase in the load current I_3 will cause an increase in the voltage drop I_2R_2 and thereby decreases the output voltage E_o. With the voltage at the base of Q_1 fixed by D_1, a decrease in the output voltage E_o results in a decrease in the forward bias of the transistor, thereby increasing its collector-to-emitter resistance and decreasing the transistor current I_5. The decrease in I_5 decreases the current I_2 that

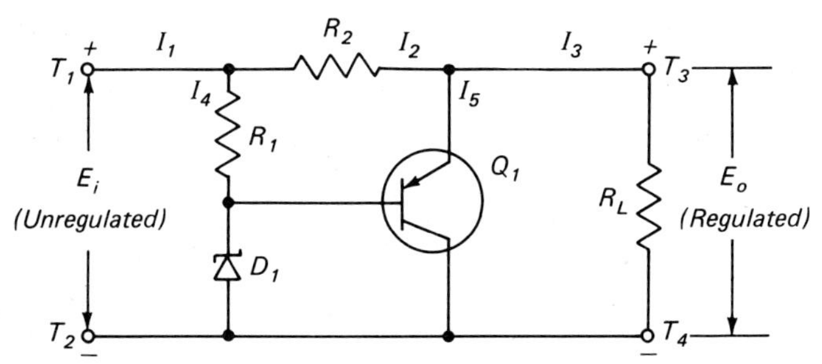

Fig. 7-18 Basic shunt-type voltage regulator using one transistor and one Zener diode.

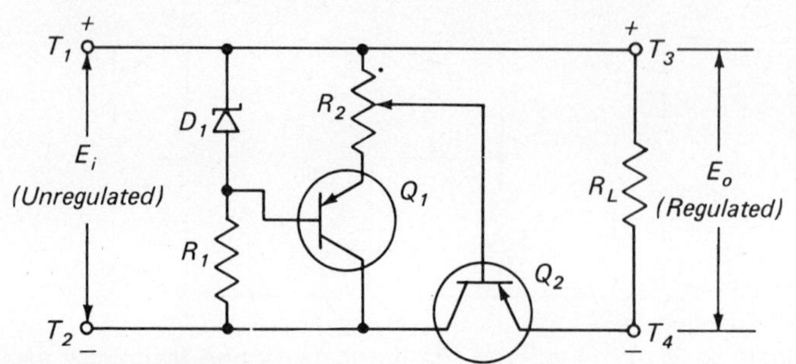

Fig. 7-19 Series-type voltage regulator incorporating an error-signal amplifier and adjustable output voltage.

flows through R_2; this causes the voltage drop I_2R_2 to decrease, thereby increasing the output voltage and thus bringing E_o back to its former value. As with the series-type regulator, the amount and the direction of the error signal establishes the chain of reactions that maintains good regulation.

Series versus Shunt Regulators. In each type of regulator, energy is lost in the regulating element. The advantage of one type of regulator over the other depends largely upon the load conditions. With the series type, the regulating element has the least amount of loss at no load, while with the shunt type the regulating element has the greatest amount of loss at no load. An advantage of the shunt-type regulator is that it affords protection for the transistor against any overload or short-circuit conditions since these conditions will only decrease or remove the operating voltages at the transistor. With a series-type regulator, an overload condition will result in excessive current through the transistor and probably ruin it.

Advanced Series-type Regulator. The modified series-type regulator of Fig. 7-19 provides improved regulation of the output voltage. Comparison of Fig. 7-19 with Figs. 7-17 and 7-18 will show similarities among the three circuits. In the advanced circuit, transistor Q_1 is operated as an emitter-follower and transistor Q_2 as a series control element. Any voltage variation at the output terminals produces an error signal at Q_1 which it amplifies and then this amplified signal is applied to Q_2. The overall result is better regulation of the output

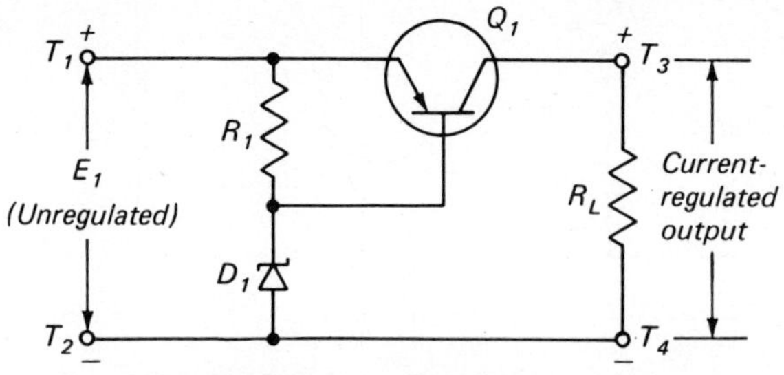

Fig. 7-20 Basic current-regulator circuit.

voltage, usually less than 1 per cent. The potentiometer R_2 permits using a portion of the stabilized voltage by varying the reference voltage for transistor Q_2 in order to establish a choice of regulated output voltage.

Basic Current Regulator. Figure 7-20 illustrates a basic current-regulated circuit. As with the voltage regulator, the Zener diode provides a constant reference voltage to the base of the transistor. In this circuit the transistor bias is made of such a value that the transistor collector current will be practically constant for a relatively wide range of bias voltage. Any change in the load current must result in a similar change in the collector current. This is accompanied by a change in the base-to-emitter bias as the error signal which restores the collector current to its former value, thus automatically maintaining the load current at a constant value.

Other Semiconductor Regulators. There are numerous variations and extensions of the basic regulator circuits that are designed to meet the needs of a wide variety of requirements. Some of these variations are (1) additional stages of error control, (2) parallel operation of transistors in the final output stage to increase the current rating of the regulator, (3) use of silicon-controlled rectifiers (SCRs), (4) combined voltage and current regulators.

7-22 Gaseous Tubes for Regulators

For low-current applications, a cold-cathode gaseous diode (Art. 3-4) having a practically constant internal voltage drop can be used as a voltage regulator.

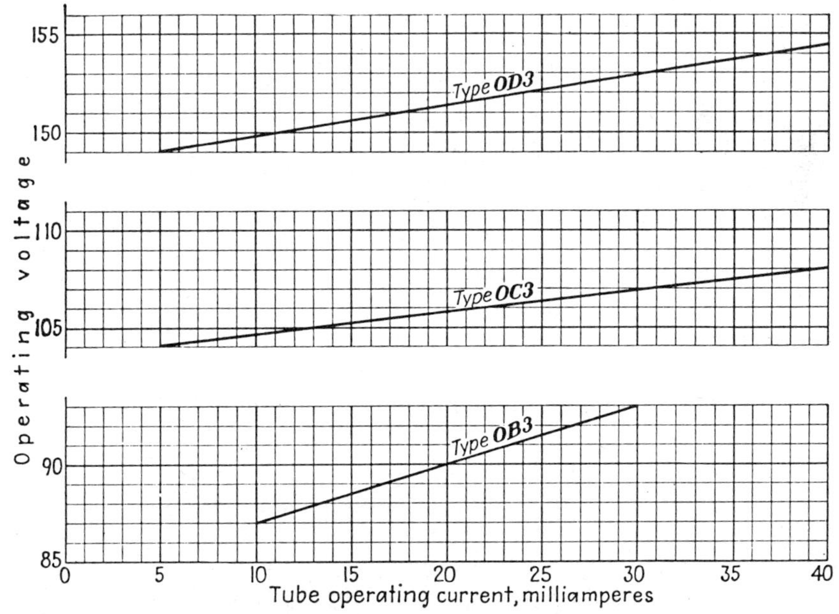

Fig. 7-21 Operating characteristics of typical voltage-regulating tubes.

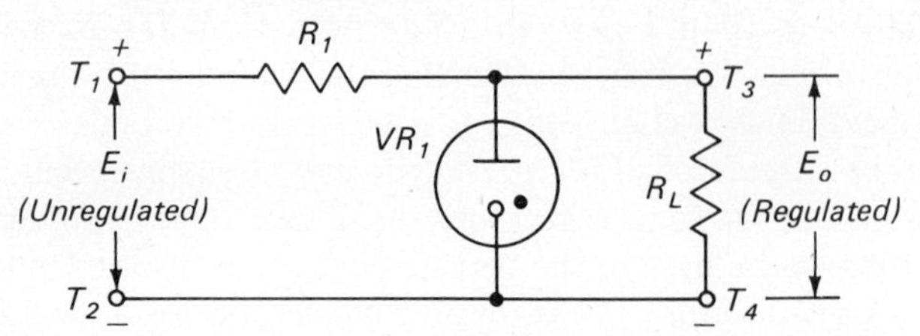

Fig. 7-22 Basic gaseous-tube voltage-regulating circuit.

The anode of this tube consists of a single wire and it is surrounded by a cylindrical cathode; the tube is filled with either an argon or a neon gas. Typical voltage-regulator tubes are the OB3/VR90, OC3/VR105, and OD3/VR150. The voltage-regulating characteristics of these tubes are illustrated by the curves of Fig. 7-21. A voltage slightly higher than the operating voltage must be used to ionize the gas inside the tube in order to have it start operating. Once the tube is started, it will continue to operate at some value of voltage within its operating range.

7-23 Gaseous-tube Regulators

The basic gaseous-tube voltage-regulating circuit (Fig. 7-22) consists of only a current-limiting resistor R_1 and a gas-filled diode tube VR_1. The function of R_1 is to protect the tube against any current in excess of its maximum safe limit; its value is determined by

$$R_1 = \frac{E_i - E_o}{I_{VR}} \tag{7-3}$$

where R_1 = series resistor, ohms
E_i = input voltage at no load, volts
E_o = output voltage at no load, volts
I_{VR} = no-load operating current, amperes

This circuit operates in the following manner: When the input voltage is applied, the voltage at the regulator tube rises to a value at which the tube suddenly ionizes, or fires. The tube then starts conducting and the voltage across the tube decreases (slightly) to its operating value, which it maintains over its operating current range. For any load variation within the rated current of the tube, the voltage variation is in the order of 4 to 6 volts and the regulation of the circuit is in the order of 4 to 8 per cent.

Tubes Operated in Series. When it is required to provide a regulated voltage that is higher than that obtainable from one tube, two or more voltage-regulating tubes can be operated in series. For example, if a 75-volt, a 105-volt, and a 150-volt tube are connected in series (Fig. 7-23), a variety of regulated voltages from 75 to 330 volts can be obtained. In calculating the value of R_1, the value of E_o is the sum of the individual tube ratings.

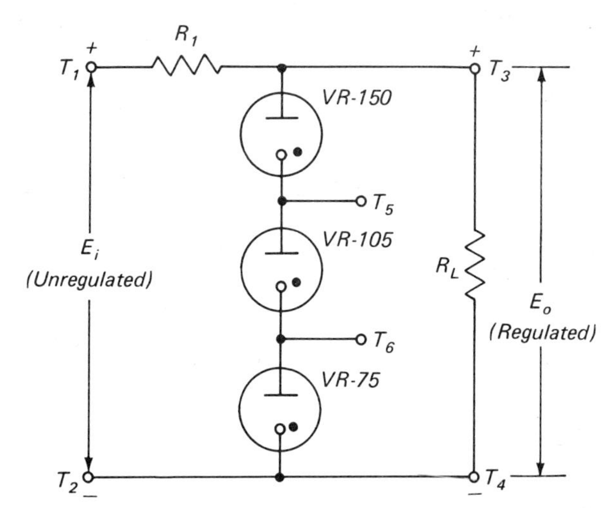

Fig. 7-23 Voltage-regulator tubes connected in series.

7-24 Electron-tube Voltage Regulator

Figure 7-24 illustrates a basic voltage-regulating circuit using two vacuum tubes and a gaseous voltage-regulating tube. The regulator tube VR_1 maintains the cathode of VT_2 at a fixed voltage, and the potentiometer R_3 of the voltage divider R_2-R_3-R_4 permits adjusting the grid bias of the feedback amplifier VT_2 so that it can control the feedback to the series-regulator tube VT_1 and thus automatically control the regulated output voltage E_o. Any increase in the load current will decrease the value of E_o and also cause a decrease in the grid voltage at VT_2. The plate current of VT_2 will decrease and therefore

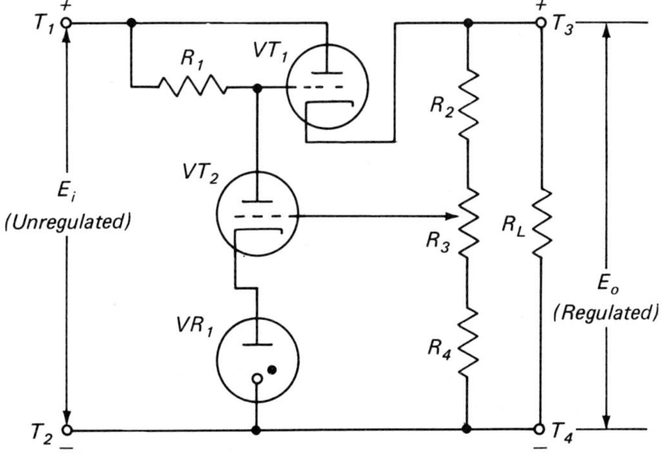

Fig. 7-24 Electron-tube voltage-regulating circuit.

the voltage drop at R_1 will also decrease. The decrease in the voltage drop at R_1 will make the grid of VT_1 more positive and thereby decrease the plate-to-cathode resistance of VT_1. The decrease in the resistance of the pass element (VT_1) will increase the output voltage E_o and thereby cancel the original drop in the output voltage.

The operation and the voltage regulation of this closed-loop circuit are better than those for the circuit described in Art. 7-23. Numerous modifications can be made to this basic circuit to further improve its regulation and to increase its current rating.

QUESTIONS

1. Describe the basic principle of a solar cell.
2. Describe the construction of a simple silicon solar cell.
3. Explain how a silicon solar cell transforms light energy to electrical energy.
4. Describe five factors that determine the efficiency of a solar cell.
5. Name some advantages of the solar cell.
6. Describe how solar cells are used in satellites and space vehicles.
7. Name six industrial applications of the solar cell.
8. (*a*) What is the basic principle of the fuel cell? (*b*) What substances are used for fuel in this type of cell?
9. Describe the construction of a simple fuel cell.
10. Explain how a fuel cell converts fuel directly to electrical energy.
11. What is the basic principle of the solid-state photocell?
12. What factors affect the amount of voltage and current produced by a solid-state photocell?
13. Describe two basic applications of a photocell.
14. (*a*) What is meant by a photodiode? (*b*) Describe its photoconductive properties.
15. Describe (*a*) the construction of a phototransistor, (*b*) its circuit operation.
16. Describe the circuit operation of (*a*) a light-activated switch, (*b*) the Raysistor.
17. Explain what is meant by (*a*) a phototube, (*b*) photoelectric emission.
18. Describe the construction of a basic phototube.
19. What are the characteristics of (*a*) vacuum phototubes? (*b*) Gas-filled phototubes?
20. Explain (*a*) the purpose of voltage-multiplier circuits, (*b*) the basic principle of voltage-multiplier circuits.
21. Explain the operation of the full-wave voltage doubler.
22. Explain the operation of the half-wave voltage doubler.
23. Explain the operation of the full-wave voltage tripler.
24. Explain the operation of the half-wave voltage tripler.
25. Explain the operation of the half-wave voltage quadrupler.
26. How can the basic half-wave voltage-doubler circuit be expanded to provide other multiples of voltage?
27. What is the relation between the frequency of the ripple in the output voltage to the frequency of the a-c input voltage for any (*a*) half-wave multiplier circuit? (*b*) Full-wave multiplier circuit?
28. (*a*) What is a vibrator power supply? (*b*) What is its purpose?
29. What is meant by (*a*) a synchronous vibrator? (*b*) A nonsynchronous vibrator?

30. Describe the principle of operation of a nonsynchronous-vibrator power-supply circuit.
31. Describe the principle of operation of a synchronous-vibrator power-supply circuit.
32. Why are filters required as part of a vibrator power supply?
33. (*a*) What is the purpose of the buffer capacitor? (*b*) How is the buffer capacitor connected in the circuit?
34. (*a*) What is meant by hash? (*b*) How is it eliminated?
35. Define (*a*) constant-voltage power source, (*b*) constant-current power source.
36. How is voltage regulation expressed?
37. Name four classifications of regulator circuits.
38. Define (*a*) Zener voltage, (*b*) Zener diode, (*c*) avalanche current.
39. What important characteristic of the Zener diode makes it so useful in regulator circuits?
40. Describe the operation of the basic open-loop regulator circuit.
41. In regulator circuits, what is the purpose of (*a*) using cascaded stages? (*b*) Using two or more Zener diodes connected in series?
42. Define (*a*) open-loop circuit, (*b*) closed-loop circuit, (*c*) pass element, (*d*) error voltage.
43. Explain the operation of the basic closed-loop series-type regulator.
44. Explain the operation of the basic closed-loop shunt-type regulator.
45. Explain how the type of load condition affects the choice between series-type and shunt-type regulators.
46. Explain how the advanced series-type regulator of Fig. 7-19 differs from the basic series-type regulator.
47. Explain the operation of the basic current regulator.
48. Describe the operation of the basic gaseous-tube regulator.
49. What is the purpose of operating two or more gaseous tubes connected in series?
50. Explain the operation of the electron-tube voltage-regulating circuit of Fig. 7-24.

PROBLEMS

1. A full-wave doubler circuit similar to Fig. 7-5 is connected to a 60-Hz, 120-volt power source. Disregarding the voltage drops at the diodes and their protecting resistors, determine (*a*) the approximate maximum output voltage, (*b*) the frequency of the ripple in the output voltage, (*c*) the minimum voltage ratings of the capacitors, (*d*) the peak inverse voltage ratings of the diodes.
2. Repeat Prob. 1 for a half-wave voltage doubler similar to Fig. 7-6.
3. Repeat Prob. 1 for a full-wave voltage tripler similar to Fig. 7-7.
4. Repeat Prob. 1 for a half-wave voltage tripler similar to Fig. 7-8.
5. Repeat Prob. 1 for a half-wave voltage quadrupler similar to Fig. 7-9.
6. (*a*) Draw a circuit diagram for a multiplier circuit that will provide a voltage of $5E_m$. (*b*) What is the value of the output voltage if the a-c input is 60 Hz and 120 volts (rms)? (*c*) What is the highest voltage rating required of any capacitor? (*d*) What is the highest peak inverse voltage rating of any diode?
7. (*a*) How many stages are required in a voltage-multiplier circuit if it is desired to obtain a 1,000-volt output from a 118-volt, 60-Hz input? (*b*) Draw a circuit diagram of the circuit. (*c*) What is the highest voltage rating required of any capacitor? (*d*) What is the highest peak inverse voltage rating of any diode?

8. What is the voltage regulation of the power supply represented by Fig. 7-7*b* if the load varies between 50 and 100 per cent?
9. What is the voltage regulation of the power supply represented by Fig. 7-7*b* if the load varies between 25 and 100 per cent?
10. What is the voltage regulation of the power supply represented by Fig. 7-7*b* if the load varies between 25 and 75 per cent?
11. What value of series resistance must be used with a 1-watt, 10-volt Zener diode connected to a 12-volt power source if the maximum current rating of the diode is 100 ma?
12. What value of series resistance must be used with a 10-watt, 22-volt Zener diode connected to a 28-volt power source if the maximum current rating of the diode is 450 ma?
13. The circuit of Fig. 7-15 is used to obtain an 18-volt regulated power supply from a 28-volt power source. Find the values of R_1 and R_2 when D_1 is rated at 1 watt, 22 volts, 45 ma and D_2 is rated at 1 watt, 18 volts, 55 ma.
14. In the circuit of Fig. 7-16, what value is required of R_1 when two 3.5-watt, 27-volt, 125-ma Zener diodes are connected in series to obtain a 54-volt regulated output from a 90-volt power source?
15. What value of series resistance is required when three 10-watt, 10-volt, 1,000-ma Zener diodes are connected in series to obtain a 30-volt regulated output from a 45-volt power source?

Chapter 8

Biasing Circuits

An important consideration in understanding the operation of amplifiers is the effect that the voltages applied to the various elements of a transistor or tube have on the operating characteristics of the amplifier. This is often referred to as the effects of the bias of a transistor or tube on the operating characteristics of the amplifier. This chapter will be concerned chiefly with (1) the modes of operation which are determined by the magnitude of the biasing voltages applied to the transistor or tube in the amplifier circuit, (2) the characteristics of the various modes of operation, and (3) the methods of providing the proper amount of bias required to achieve the desired mode of operation.

8-1 Classification of Amplifiers According to Biasing Considerations

Letter Classification. An important classification of amplifiers is based on the operating characteristics that are influenced chiefly by the biasing of the transistor or vacuum tube used in the amplifier; these classifications are also referred to as *modes of operation*. On this basis, amplifiers are identified by a code letter (or letters) as (1) Class A, (2) Class B, (3) Class AB, (4) Class AB_1, (5) Class AB_2, or (6) Class C.

Definitions of Classifications of Transistor Amplifiers. A *Class A transistor amplifier* is one in which the base bias and the alternating signal input voltages are of such values that collector current flows in the output circuit at all times, even when no signal is present. A *Class B transistor amplifier* is one in which the base is biased to produce cutoff of the collector current. In this case, the collector current will be approximately zero for zero-signal input, and collector current will flow for approximately one-half of each input cycle. A *Class AB transistor amplifier* is one in which the base bias and the signal input voltages are of such values that collector current will flow for appreciably more than one-half but less than the complete time of the input cycle. A *Class C transistor amplifier* is one in which the base bias is greater than the amount required to produce collector-current cutoff so that collector current is zero for zero-signal input, and collector current will flow for appreciably less than one-half the time of the input cycle.

8-2 Graphical Analysis of the Operating Characteristics of a Transistor Amplifier

Need for Graphical Analysis. The operating characteristics of the various modes of operation of transistor amplifiers can best be presented by use of the characteristic curves of the transistor. With this method, it is possible to show clearly (1) the distinction between the various modes of operation, and (2) how each mode of operation can be achieved.

Output Static Characteristic Curves. Figure 8-1 shows a *family* of output static characteristic curves for a transistor used in the common-emitter configuration. These curves show the variations in collector current for varying amounts of collector-to-emitter voltage and with constant amounts of base current as the running parameter.

After a load line is constructed for a specific set of circuit parameters on the static characteristic curves, the input and output signals can be plotted on the static characteristic curves (see Art. 4-17). Although the relationship between the input signal and the output signal can be shown on the static curves, they can be shown to better advantage on the dynamic characteristic curves of the transistor.

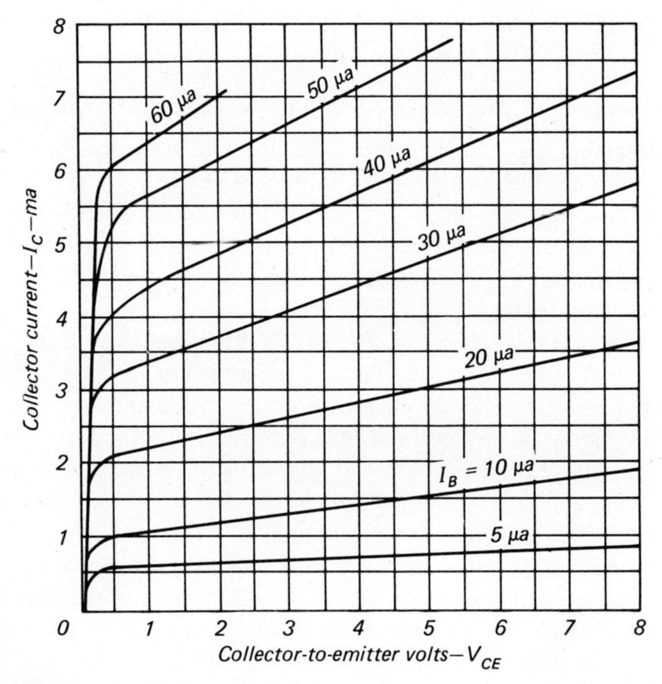

Fig. 8-1 Typical static characteristic curves for the common-emitter transistor configuration.

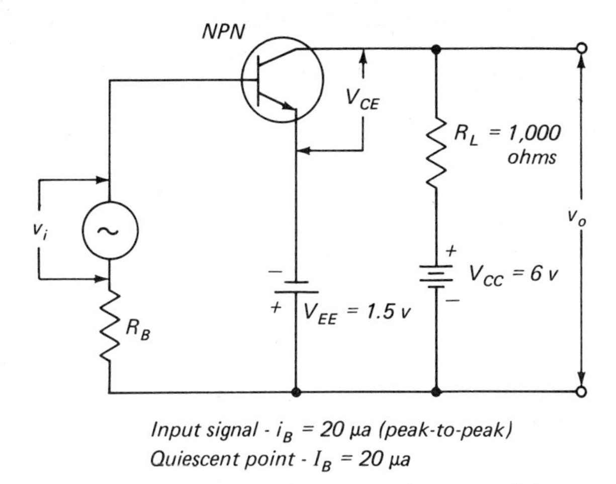

Fig. 8-2 Typical common-emitter transistor amplifier circuit.

8-3 Dynamic Transfer Characteristic Curves

Construction. The dynamic transfer characteristic curve of Fig. 8-4 shows how the output current I_C varies with the input current I_B for a specific set of circuit parameters. The transfer curve is constructed with data obtained from the static characteristic curves to which a load line has been added (Fig. 8-3). Data for plotting Fig. 8-4 were obtained by reading and recording the values of collector current at each point where the load line intersects a base-current curve on Fig. 8-3; these data are given in Table 8-1.

Characteristics. Examination of the dynamic transfer characteristic curve of Fig. 8-4 shows that the curve may be divided into three sectors: (1) section O-A, which has a slight curvature, (2) section A-B, which is practically straight, and (3) section B-C, which has a curvature that is more pronounced than that of section O-A. Whether the curve is straight or bends is important in analyzing the degree to which an amplifier faithfully reproduces the output signal with respect to the input signal.

8-4 Class A Operation of a Transistor Amplifier

Ideal. Ideally, an amplifier used in sound-reproduction applications should produce at its output terminals a signal whose waveform is an amplified exact

Table 8-1 Data Obtained from Fig. 8-3 for Use in Plotting the Dynamic Transfer Characteristic Curve of Fig. 8-4

I_B, μa.	5	10	20	30	40	50	60
I_C, ma.	0.9	1.7	2.95	4.2	5.05	5.9	6.45

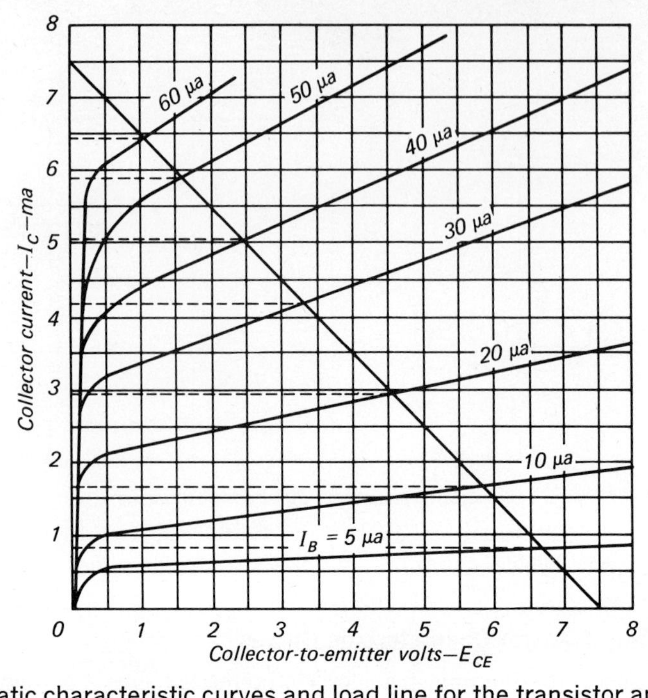

Fig. 8-3 Static characteristic curves and load line for the transistor and circuit of Figs. 8-1 and 8-2.

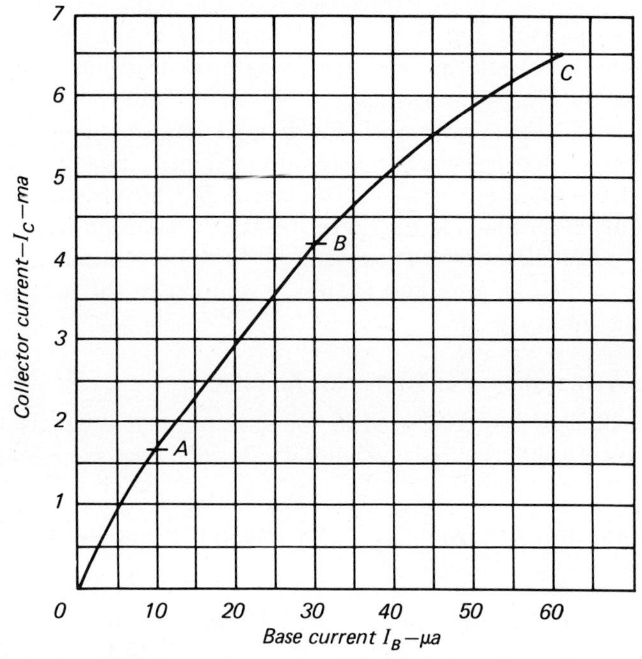

Fig. 8-4 Dynamic transfer characteristic curve for the transistor and circuit of Figs. 8-1 and 8-2.

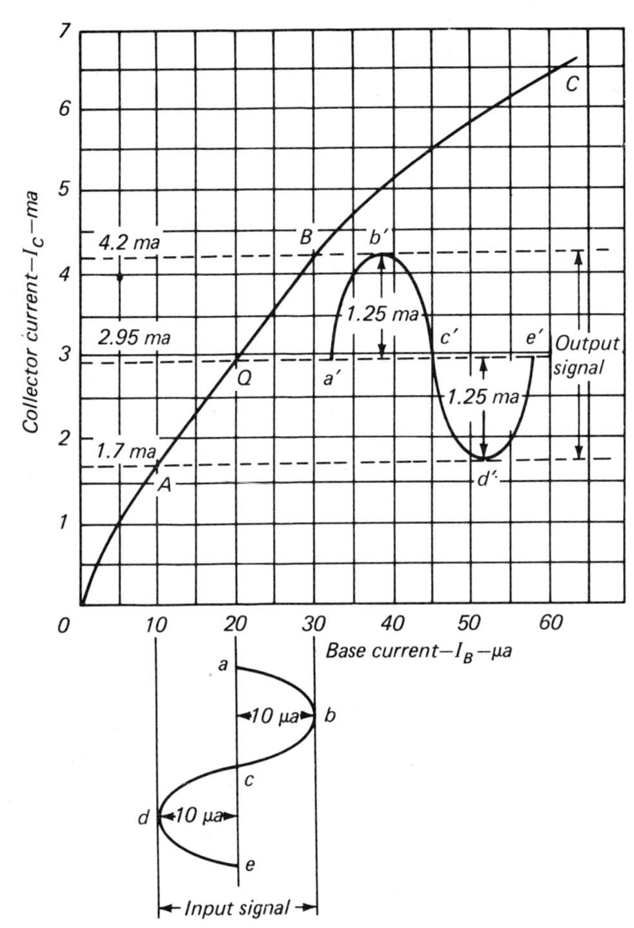

Fig. 8-5 Input and output characteristics plotted on dynamic transfer characterstic curve showing operation without distortion.

replica of the waveform of the signal input to the amplifier. Plotting the input and output signals of an amplifier on its dynamic transfer characteristic curve will show how well an amplifier faithfully performs its function. Figure 8-5 shows the input and output signals for the amplifier of Fig. 8-2 using a transistor whose output characteristics are represented by Figs. 8-1 and 8-4.

Plotting the Operating Characteristics. The input to the amplifier (Fig. 8-2) is a 20-μa peak-to-peak sine-wave signal varying around a quiescent value of 20 μa; this is shown on Fig. 8-5 as the signal *a*, *b*, *c*, *d*, *e* plotted along the base-current values of the abscissa. By projecting vertically the abscissa values for *a*, *b*, and *d* to the points of intersection along the transfer characteristic curve, the points *Q* (2.95 ma), *B* (4.2 ma), and *A* (1.7 ma) are located. Projecting point *Q* in the horizontal direction establishes the quiescent axis of the output-signal waveform. Projecting points *A* and *B* horizontally establishes the minimum and maximum peaks of the output signal. By projecting vertically a large number of points of the input waveform to their intersections with the transfer characteristic curve and then projecting these points horizontally, the waveform of the output signal can be constructed as shown by a', b', c', d', and e' on Fig. 8-5.

Characteristics. A Class A amplifier should be operated with such circuit parameters that all variations in the input signal will be reproduced at the output of the amplifier but with increased amplitude. The characteristics of the amplifier represented by Fig. 8-5 show that (1) output-signal current is flowing during 360 degrees of the input-signal time, and (2) the output-signal waveform a', b', c', d', e' is an exact replica of the input-signal waveform a, b, c, d, e. Thus, the amplifier is functioning as an ideal Class A amplifier.

Distortion. In the amplifier of Fig. 8-5 the output waveform is an exact replica of the input waveform because: (1) the operating point Q is located

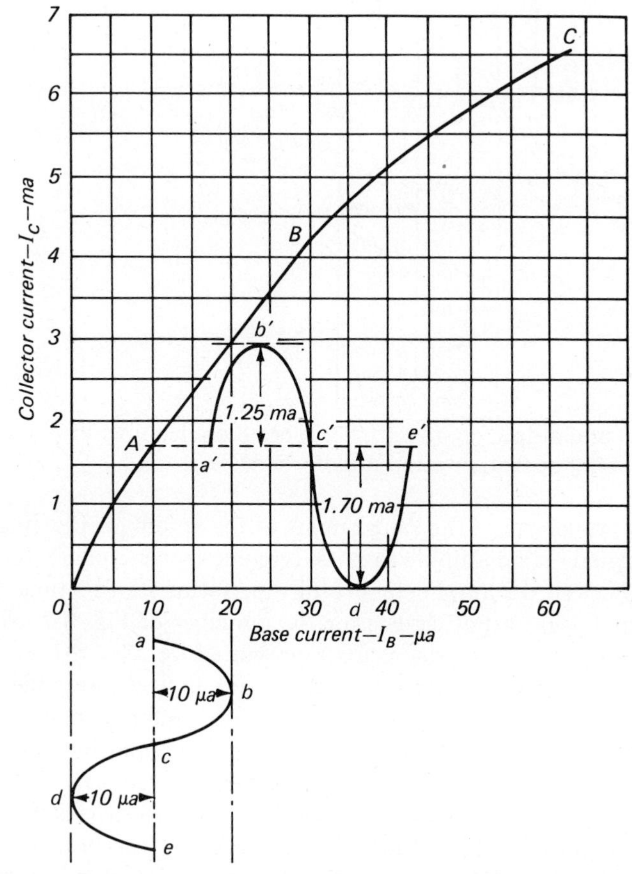

Fig. 8-6 Input and output characteristics plotted on dynamic transfer characteristic curve showing distortion of the output signal due to operating with too little bias.

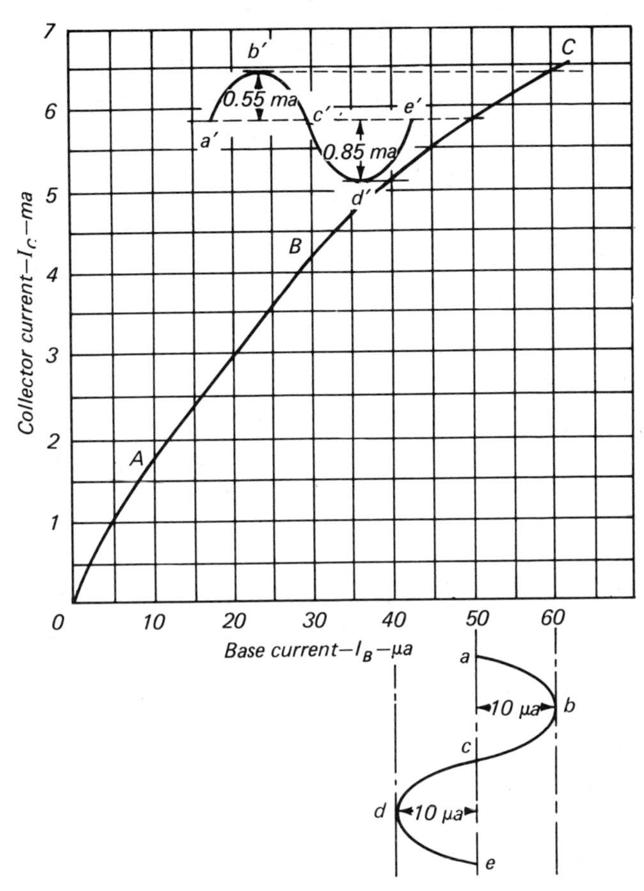

Fig. 8-7 Input and output characteristics plotted on dynamic transfer characteristic curve showing distortion of the output signal due to operating with too much bias.

at the midpoint of the straight portion *A-B* of the transfer characteristic curve, and (2) the peak-to-peak value of the input signal does not drive the output signal beyond the limits *A-B* of the straight portion of the curve. Failure to observe these two criteria may result in a distorted waveform at the output of the amplifier.

Figures 8-6 and 8-7 show the effect of using an incorrect location of the quiescent point along the curve. In Fig. 8-6, where a 20-μa peak-to-peak signal is applied at the 10-μa base-current line, the positive peak value is 1.25 ma and the negative peak value is 1.70 ma. The output signal has been distorted because the negative half of the input signal is operating on the curved portion *O-A* of the transfer characteristic curve.

In Fig. 8-7, where a 20-μa peak-to-peak signal is applied at the 50-μa base-current line, the output waveform is also distorted (as indicated by a positive peak of 0.55 ma and a negative peak of 0.85 ma) because the entire cycle of

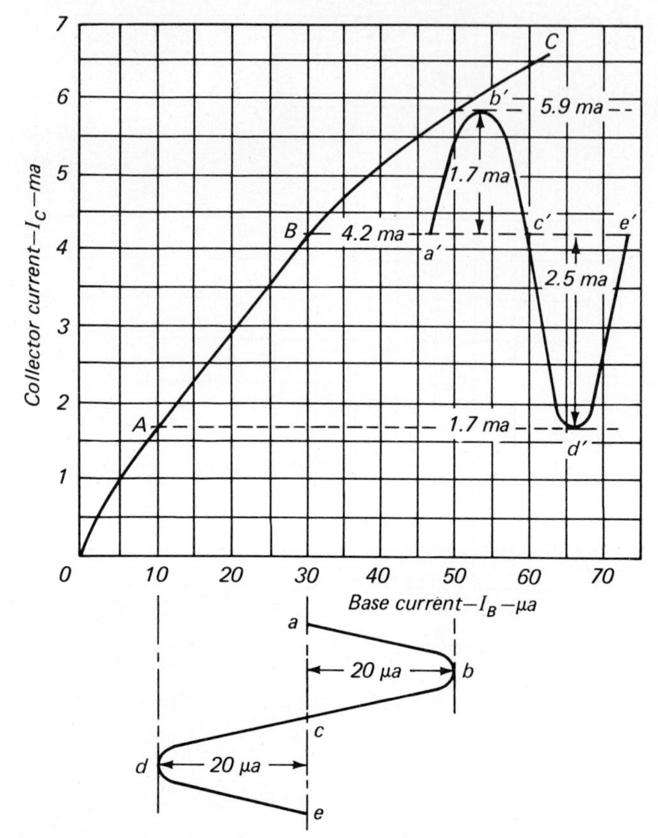

Fig. 8-8 Input and output characteristics plotted on dynamic transfer characteristic curve showing distortion of the output signal caused by too high an input signal.

the input signal is being operated on the curved portion *B-C* of the transfer characteristic curve.

In Fig. 8-8, (1) the amplifier is overdriven by a 40-μa peak-to-peak input signal, and (2) the signal is applied at the upper limit *B* of the straight portion of the transfer characteristic curve. The result is excessive distortion of the input signal as is indicated by the output signal having an unsymmetrical waveform with its positive peak value of 1.7 ma and its negative peak value of 2.5 ma. Further analysis of distortion in amplifiers is presented in Chap. 16.

8-5 Class AB Operation of a Transistor Amplifier

In Class AB operation, signal current should flow in the output circuit for more than 180 but less than 360 degrees of the time of the input cycle. This is accomplished by biasing the transistor in the area near, but not at, collector-current cutoff. Figure 8-9 shows the characteristics of an amplifier operating in the Class AB mode, in which (1) the quiescent point was chosen at 10 μa, and (2) the input is a 40-μa peak-to-peak signal. Examination of the figure will show that (1) output current will flow during all of the positive half of the input cycle, (2) output current will flow during only part of the negative half of the input cycle, (3) the output waveform during the positive half of the input cycle is undistorted, (4) the output waveform during the negative half of the input cycle is distorted, and (5) too large an input signal would result in distortion of the positive half of the input signal as well as of the negative half.

8-6 Class B Operation of a Transistor Amplifier

In Class B operation, signal current should flow in the output circuit for approximately one-half of the time of the a-c input signal. This is generally accomplished by biasing the transistor to produce collector-current cutoff, although Class B operation can also be achieved by biasing the transistor to produce collector-voltage cutoff. Figure 8-10 shows the transfer characteristic curve of a Class B amplifier in which (1) the circuit is biased for zero collector-current cutoff, and (2) a 60-μa peak-to-peak input signal is applied to the circuit. Examination of this figure will show that output current flows only

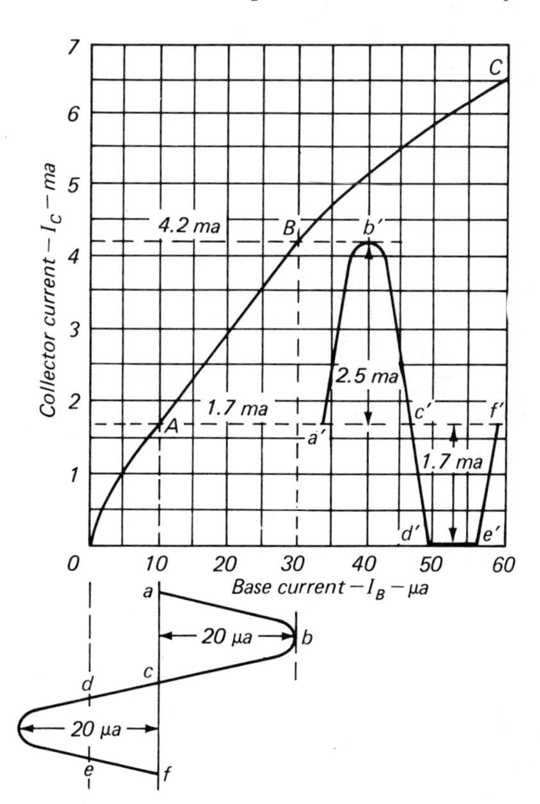

Fig. 8-9 Input and output characteristics plotted on dynamic transfer characteristic curve showing the operation of a Class AB transistor amplifier.

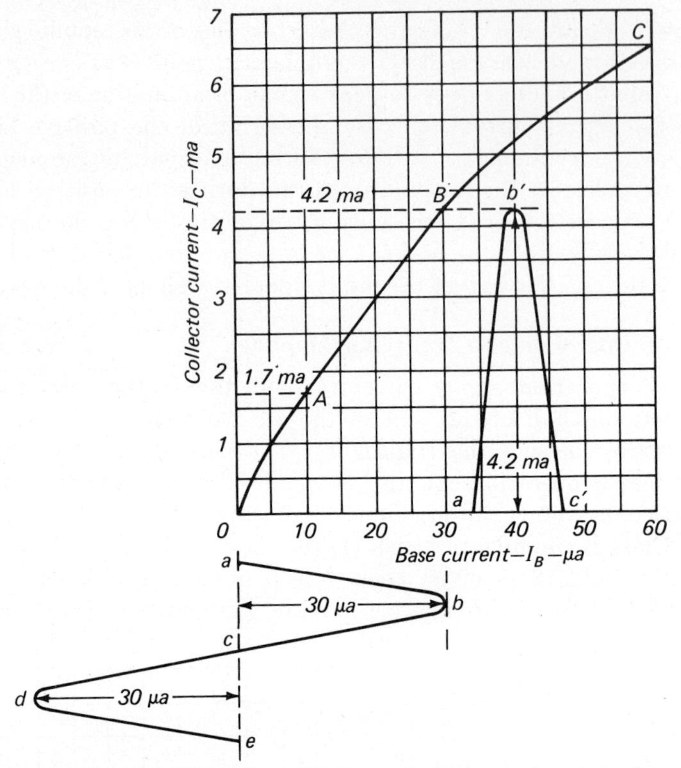

Fig. 8-10 Input and output characteristics plotted on dynamic transfer characteristic curve showing the operation of a Class B transistor amplifier.

during the positive half of the input-signal time and that the transistor is at cutoff during all of the negative half of the input-signal time. Also, the waveform of the output signal is a relatively good (though not perfect) replica of the positive half of the input-signal waveform. As was shown for the Class A amplifier, any overdrive caused by using too large a signal input will cause distortion of the output waveform.

8-7 Class C Operation of a Transistor Amplifier

In Class C operation, signal current should flow in the output circuit for less than 180 degrees of the time of the input cycle. This is accomplished by biasing the transistor beyond the point of collector-current cutoff, shown as point Q on Fig. 8-11. This figure shows the characteristics of an amplifier operating in the Class C mode, in which (1) the zero axis of the input signal falls to the left of the collector-current cutoff position on the characteristic curve, and (2) the input is a 80-μa peak-to-peak a-c signal. Examination of the figure will show that (1) output current will flow only during part of the time of the positive half of the input signal, (2) there will be no output current during any part of the time of the negative half of the input signal, (3) the output signal is not an exact replica of the input signal, and (4) too large an input signal (in this case larger than 80 μa peak-to-peak) will result in additional distortion of the output signal. Class C amplifiers cannot be used in sound-reproduction systems but they are used in r-f power amplifiers.

8-8 Transistor Biasing

Need for Biasing. Normal transistor amplifier applications require that the transistors have (1) a forward bias across the emitter-base junction, and (2) a reverse bias across the collector-base junction. In addition, the amount of bias is important in establishing the quiescent point of operation dictated by the mode of operation desired. Incorrect biasing of a transistor will cause (1) inefficient operation of the amplifier, and/or (2) distortion of the output signal. The preceding discussion of Class A, AB, B, and C amplifier operation illustrates the importance of having the correct amount of bias.

In the simple transistor amplifier circuit of Fig. 8-2, the power supplies for

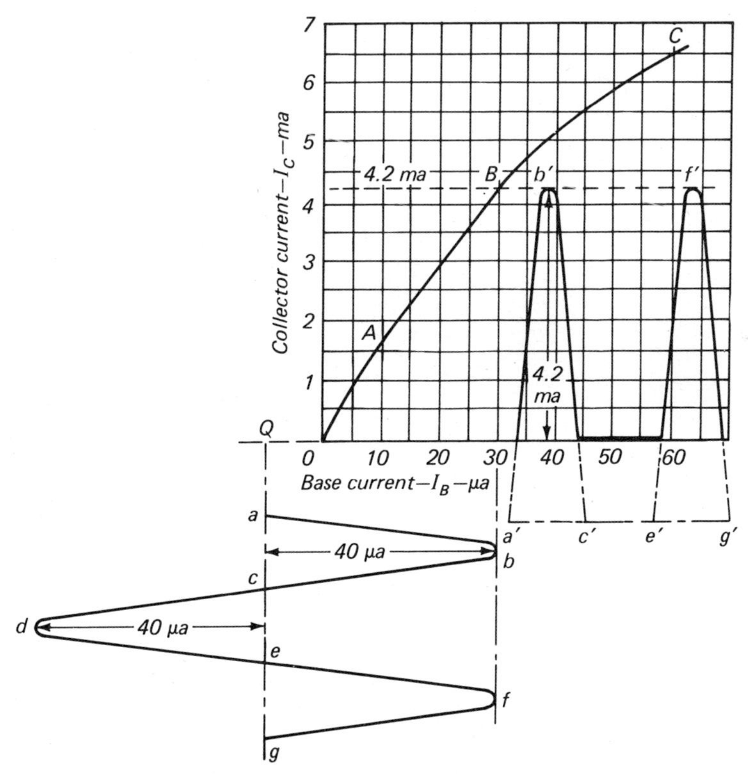

Fig. 8-11 Input and output characteristics plotted on dynamic transfer characteristic curve showing the operation of a Class C transistor amplifier.

forward and reverse bias are shown as individual battery power sources. Transistors generally obtain the power for the various circuits from (1) a single battery and a biasing network, or (2) an a-c power line which then requires the use of a rectifier, a filtering network, and a biasing network.

Classifications. The methods used to provide bias for a transistor are classified as (1) fixed bias, (2) self-bias, (3) base bias, (4) emitter bias, (5) collector bias, and (6) stabilized bias. Combinations of these methods are frequently used.

Sensitivity and Stability. Selecting the proper type of biasing arrangement and values for the circuit components has a great effect on the operation of the circuit. Two undesirable effects encountered in transistor amplifier circuits are (1) beta sensitivity (Art. 8-11), and (2) temperature versus current instability (Art. 8-12).

Relation between Bias and Operating Point. The manner in which the operating point is related to the transistor characteristics and the circuit parameters can be illustrated by applying the biasing principles to the transistor represented by Fig. 8-1 and the circuit parameters given in Fig. 8-2.

Example 8-1 If the transistor in the circuit of Fig. 8-2 has a beta value of 150, find (*a*) the collector-emitter voltage, (*b*) the voltage drop across the load resistor, (*c*) collector current, (*d*) base current, and (*e*) base resistance R_B.

GIVEN: $\beta = 150$ Figs. 8-1, 8-2, 8-3

FIND: (*a*) V_{CE} (*b*) $I_C R_L$ (*c*) I_C (*d*) I_B (*e*) R_B

SOLUTION:

(*a*) From Fig. 8-3; for the quiescent value of $I_B = 20\mu a$, $V_{CE} = 4.5$ volts

(*b*) $$I_C R_L = V_{CC} + V_{EE} - V_{CE} = 6 + 1.5 - 4.5 = 3 \text{ volts}$$

(*c*) $$I_C = \frac{I_C R_L}{R_L} = \frac{3}{1{,}000} = 3 \text{ ma}$$

(*d*) $$I_B = \frac{I_C}{\beta} = \frac{3 \times 10^{-3}}{150} = 20\ \mu a$$

(*e*) $$R_B = \frac{V_{EE}}{I_B} = \frac{1.5}{20 \times 10^{-6}} = 75{,}000 \text{ ohms}$$

8-9 Using Two Power Sources to Obtain Bias

Figure 8-12 shows the three basic circuit configurations with their bias voltages supplied by independent power sources. The paths of the base, emitter, and collector currents are indicated by the arrows on this figure. The relationship among the voltage, current, and resistance values for the common-emitter circuit of Fig. 8-12*a* is expressed by

$$I_B = \frac{V_{BB}}{R_B} \tag{8-1}$$

$$I_C = \beta I_B \tag{8-2}$$
$$I_E = I_C + I_B \tag{8-3}$$

also

$$I_E \cong I_C \qquad (\text{when } I_B \ll I_C) \tag{8-3a}$$
$$V_{CE} = V_{CC} - I_C R_L \tag{8-4}$$

Example 8-2 If the circuit parameters of Fig. 8-12*a* are: $V_{CC} = 9$ volts, $V_{BB} = 3$ volts, $\beta = 25$, $R_L = 400$ ohms, $R_B = 10{,}000$ ohms, find (*a*) I_B, (*b*) I_C, (*c*) I_E, and (*d*) V_{CE}.

GIVEN: $V_{CC} = 9$ volts $\quad V_{BB} = 3$ volts $\quad \beta = 25 \quad R_L = 400$ ohms
$R_B = 10{,}000$ ohms

FIND: (*a*) I_B (*b*) I_C (*c*) I_E (*d*) V_{CE}

SOLUTION:

(*a*) $$I_B = \frac{V_{BB}}{R_B} = \frac{3}{10{,}000} = 0.3 \text{ ma}$$

(*b*) $$I_C = \beta I_B = 25 \times 0.3 = 7.5 \text{ ma}$$

(*c*) $$I_E = I_C + I_B = 7.5 + 0.3 = 7.8 \text{ ma}$$

(*d*) $$V_{CE} = V_{CC} - I_C R_L = 9 - (7.5 \times 10^{-3} \times 400) = 6 \text{ volts}$$

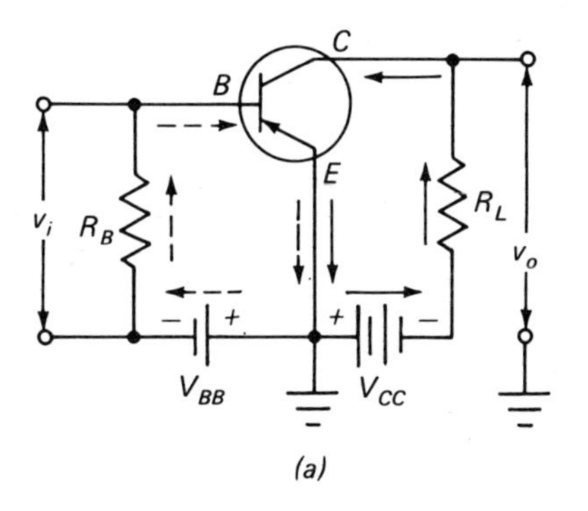

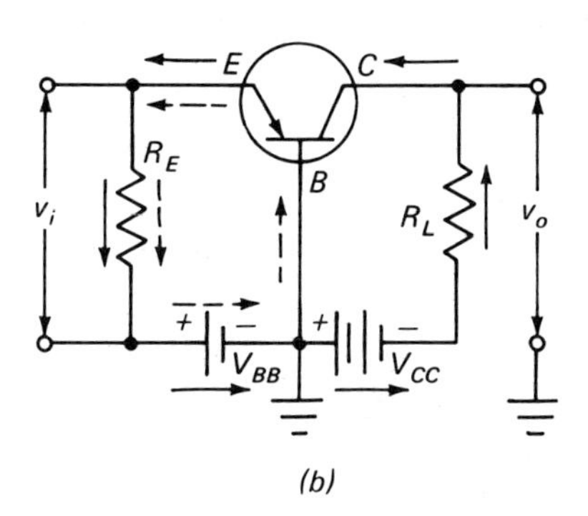

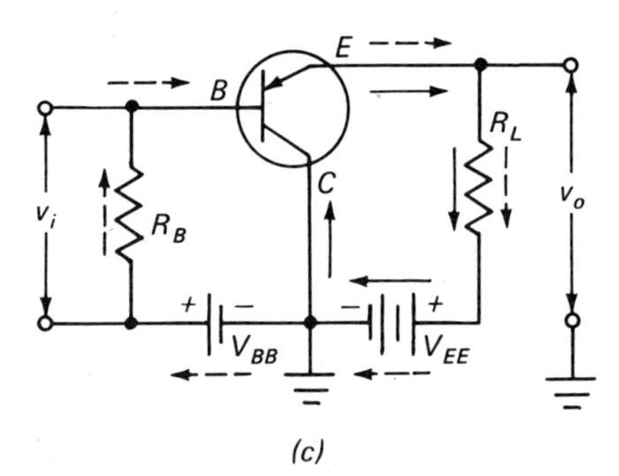

Legend:
⟶ I_C
- - → I_B

Note: $I_E = I_C + I_B$

Fig. 8-12 Biasing transistors with two power sources. (a) Common-emitter. (b) Common-base. (c) Common-collector.

The relationship among the voltage, current, and resistance values for the common-base circuit of Fig. 8-12*b* is expressed by

$$I_E = \frac{V_{BB}}{R_E} \tag{8-5}$$

$$I_C = \alpha I_E \tag{8-6}$$

also

$$I_C \cong I_E \qquad (\text{when } I_B \ll I_C) \tag{8-6a}$$

$$I_B = I_E - I_C \tag{8-7}$$

$$V_{CE} = V_{BB} + V_{CC} - I_E R_E - I_C R_L \tag{8-8}$$

Example 8-3 If the circuit parameters of Fig. 8-12*b* are: $V_{BB} = 6$ volts, $V_{CC} = 9$ volts, $\alpha = 0.97$, $R_L = 2{,}500$ ohms, and $R_E = 5{,}000$ ohms, find (*a*) I_E, (*b*) I_C, (*c*) I_B, (*d*) V_{CE}.

GIVEN: $V_{BB} = 6$ volts $V_{CC} = 9$ volts $\alpha = 0.97$ $R_L = 2{,}500$ ohms
$R_E = 5{,}000$ ohms

FIND: (*a*) I_E (*b*) I_C (*c*) I_B (*d*) V_{CE}

SOLUTION:

(*a*) $$I_E = \frac{V_{BB}}{R_E} = \frac{6}{5{,}000} = 1.2 \text{ ma}$$

(*b*) $$I_C = \alpha I_E = 0.97 \times 1.2 = 1.164 \text{ ma}$$

(*c*) $$I_B = I_E - I_C = 1.2 - 1.164 = 0.036 \text{ ma}$$

(*d*) $$\begin{aligned} V_{CE} &= V_{BB} + V_{CC} - I_E R_E - I_C R_L \\ &= 6 + 9 - (1.2 \times 10^{-3} \times 5{,}000) - (1.164 \times 10^{-3} \times 2{,}500) \\ &\cong 6.1 \text{ volts} \end{aligned}$$

The relationship among the voltage, current, and resistance values for the common-collector circuit of Fig. 8-12*c* is expressed by

$$I_E = \frac{V_{EE} - V_{BB}}{R_L + \dfrac{R_B}{\beta}} \tag{8-9}$$

$$I_C \cong I_E \tag{8-10}$$

$$I_B = \frac{I_C}{\beta} \tag{8-11}$$

$$V_{CE} = V_{EE} - I_E R_L \tag{8-12}$$

Example 8-4 If the circuit parameters of Fig. 8-12*c* are: $V_{BB} = 21$volts, $V_{EE} = 30$ volts, $\beta = 50$, $R_L = 3{,}000$ ohms, and $R_B = 57{,}000$ ohms, find (*a*) I_E, (*b*) I_C, (*c*) I_B, and (*d*) V_{CE}.

GIVEN: $V_{BB} = 21$ volts $V_{EE} = 30$ volts $\beta = 50$ $R_L = 3{,}000$ ohms
$R_B = 57{,}000$ ohms

FIND: (*a*) I_E (*b*) I_C (*c*) I_B (*d*) V_{CE}

SOLUTION:

(*a*) $$I_E = \frac{V_{EE} - V_{BB}}{R_L + \frac{R_B}{\beta}} = \frac{30 - 21}{3{,}000 + \frac{57{,}000}{50}} = 2.17 \text{ ma}$$

(*b*) $$I_C \cong I_E = 2.17 \text{ ma}$$

(*c*) $$I_B = \frac{I_C}{\beta} = \frac{2.17 \times 10^{-3}}{50} = 43.4\ \mu\text{a}$$

(*d*) $$V_{CE} = V_{EE} - I_E R_L = 30 - (2.17 \times 10^{-3} \times 3{,}000) \cong 23.5 \text{ volts}$$

8-10 Using a Single Power Source to Obtain Bias

Figure 8-13 shows the three basic circuit configurations with their voltage and current requirements supplied by a single power source. With the common-emitter circuit of Fig. 8-13*a*, the emitter is at the highest positive potential since it is connected directly to the positive terminal of the power source. The base terminal is at a lower positive potential by an amount corresponding to the voltage drop at the base resistor R_B. Therefore, the base is negative with respect to the emitter, as is required for forward-biasing the emitter-base junction of the PNP transistor. Also, the collector terminal is negative with respect to the base, which is a requirement for the reverse bias at the collector-base junction of the PNP transistor. The relationship among the voltage, current, and resistance values for the common-emitter circuit of Fig. 8-13*a* is expressed by

$$I_B = \frac{V_{EE}}{R_B} \tag{8-13}$$

$$I_C = \beta I_B \tag{8-2}$$

$$I_E = I_C + I_B \tag{8-3}$$

also $$I_E \cong I_C \quad (\text{when } I_B \ll I_C) \tag{8-3a}$$

$$V_{CE} = V_{EE} - I_C R_L \tag{8-14}$$

Example 8-5 If the circuit of Example 8-2 is converted to use a single 12-volt power source V_{EE} as in Fig. 8-13*a*, and R_B is changed to 20,000 ohms, find (*a*) I_B, (*b*) I_C, (*c*) I_E, (*d*) V_{CE}.

GIVEN: $V_{EE} = 12$ volts $\quad \beta = 25 \quad R_L = 400$ ohms $\quad R_B = 20{,}000$ ohms

FIND: (*a*) I_B (*b*) I_C (*c*) I_E (*d*) V_{CE}

SOLUTION:

(*a*) $$I_B = \frac{V_{EE}}{R_B} = \frac{12}{20{,}000} = 0.6 \text{ ma}$$

(*b*) $$I_C = \beta I_B = 25 \times 0.6 = 15 \text{ ma}$$

(*c*) $$I_E = I_C + I_B = 15 + 0.6 = 15.6 \text{ ma}$$

(*d*) $$V_{CE} = V_{EE} - I_C R_L = 12 - (15 \times 10^{-3} \times 400) = 6 \text{ volts}$$

The common-base circuit of Fig. 8-13*b* uses the voltage divider $R_B R_S$ to obtain the forward bias for the base-emitter junction. The direction of current flow through the voltage divider makes the base (point *B*) negative with respect to the emitter and the relative values of R_B and R_S determine the magnitude of the forward bias for the transistor. The relationship among the voltage, current, and resistance values for the common-base circuit of Fig. 8-13*b* is expressed by

$$V_B = V_{CC} \frac{R_B}{R_B + R_S} \tag{8-15}$$

$$I_E = \frac{V_B}{R_E} \tag{8-16}$$

$$I_C = \alpha I_E \tag{8-6}$$

$$I_C \cong I_E \qquad (\text{when } \alpha \cong 1) \tag{8-6b}$$

$$I_B = I_E - I_C \tag{8-7}$$

$$V_{CE} = V_{CC} - I_C R_L - I_E R_E \tag{8-17}$$

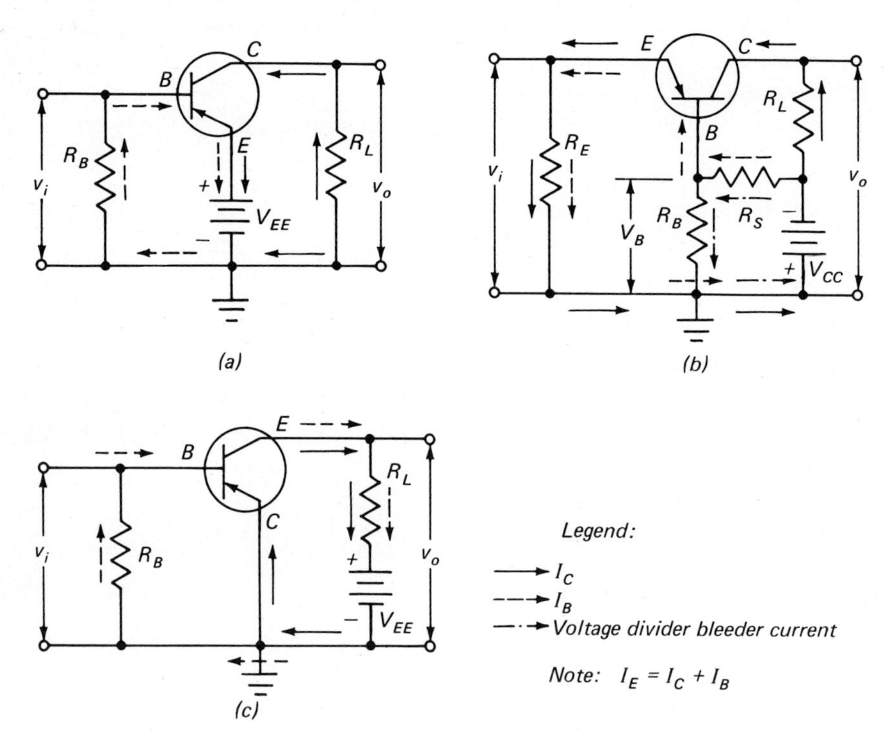

Fig. 8-13 Biasing transistors with one power source. (a) Common-emitter. (b) Common-base. (c) Common-collector.

Example 8-6 If the circuit of Example 8-3 is converted to use a single 12-volt power source V_{CC}, as in Fig. 8-13*b*, with $R_S = 400$ ohms and $R_B = 200$ ohms, find (*a*) V_B, (*b*) I_E, (*c*) I_C and (*d*) V_{CE}. (*Note:* Assume $\alpha \cong 1$.)

GIVEN: $V_{CC} = 12$ volts $R_S = 400$ ohms $R_B = 200$ ohms $R_E = 5{,}000$ ohms $R_L = 2{,}500$ ohms

FIND: (*a*) V_B (*b*) I_E (*c*) I_C (*d*) V_{CE}

SOLUTION:

(*a*) $$V_B = V_{CC}\frac{R_B}{R_B + R_S} = 12 \times \frac{200}{200 + 400} = 4 \text{ volts}$$

(*b*) $$I_E = \frac{V_B}{R_E} = \frac{4}{5{,}000} = 0.8 \text{ ma}$$

(*c*) $$I_C \cong I_E = 0.8 \text{ ma}$$

(*d*) $$V_{CE} = V_{CC} - I_C R_L - I_E R_E$$
$$= 12 - (0.8 \times 10^{-3} \times 2{,}500) - (0.8 \times 10^{-3} \times 5{,}000) = 6 \text{ volts}$$

With the common-collector circuit (Fig. 8-13*c*) the emitter current is determined by the values of (1) the base resistor R_B, (2) the emitter (or load) resistor R_L, (3) the beta of the transistor, and (4) the voltage of the power source V_{EE}. Applying Kirchhoff's law for voltages to the path of the base current will show that the base is negative with respect to the emitter and positive with respect to the collector, thereby establishing the proper forward- and reverse-bias conditions for the base-emitter junction and the base-collector junction respectively. The relationship among the voltage, current, and resistance values for the common-collector circuit of Fig. 8-13*c* is expressed by

$$I_E = \frac{V_{EE}}{R_L + \dfrac{R_B}{\beta}} \tag{8-18}$$

$$I_C \cong I_E \tag{8-10}$$

$$I_B = \frac{I_C}{\beta} \tag{8-11}$$

$$V_{CE} = V_{EE} - I_E R_L \tag{8-19}$$

Example 8-7 If the circuit of Example 8-4 is converted to use a single 27-volt power source V_{EE}, as in Fig. 8-13*c*, and all other parameters remain unchanged, find (*a*) I_E, (*b*) I_C, (*c*) I_B, and (*d*) V_{CE}.

GIVEN: $V_{EE} = 27$ volts $\beta = 50$ $R_L = 3{,}000$ ohms $R_B = 57{,}000$ ohms

FIND: (*a*) I_E (*b*) I_C (*c*) I_B (*d*) V_{CE}

SOLUTION:

(a) $$I_E = \frac{V_{EE}}{R_L + \dfrac{R_B}{\beta}} = \frac{27}{3{,}000 + \dfrac{57{,}000}{50}} = 6.5 \text{ ma}$$

(b) $$I_C \cong I_E = 6.5 \text{ ma}$$

(c) $$I_B = \frac{I_C}{\beta} = \frac{6.5}{50} = 0.13 \text{ ma}$$

(d) $$V_{CE} = V_{EE} - I_E R_L = 27 - (6.5 \times 10^{-3} \times 3{,}000) = 7.5 \text{ volts}$$

8-11 Beta Sensitivity

The *beta sensitivity* of a transistor amplifier circuit expresses the degree to which a circuit maintains uniform operating conditions with variations in the beta value of the transistor. Variations in the β value may be caused by (1) variations in the circuit operating conditions, and (2) substituting one transistor for another (even of the same type). The simple biasing circuits of Figs. 8-12*a* and 8-13*a* are not practical because of (1) poor beta sensitivity, and (2) poor temperature versus collector-current stability.

Example 8-8 If the value of beta in Example 8-2 increases to 50, find (*a*) I_B, (*b*) I_C, (*c*) I_E, (*d*) V_{CE}.

GIVEN: $V_{CC} = 9$ volts $V_{BB} = 3$ volts $\beta = 50$ $R_L = 400$ ohms
$R_B = 10{,}000$ ohms

FIND: (*a*) I_B (*b*) I_C (*c*) I_E (*d*) V_{CE}

SOLUTION:

(a) $$I_B = \frac{V_{BB}}{R_B} = \frac{3}{10{,}000} = 0.3 \text{ ma}$$

(b) $$I_C = \beta I_B = 50 \times 0.3 = 15 \text{ ma}$$

(c) $$I_E = I_C + I_B = 15 + 0.3 = 15.3 \text{ ma}$$

(d) $$V_{CE} = V_{CC} - I_C R_L = 9 - (15 \times 10^{-3} \times 400) = 3 \text{ volts}$$

Poor beta sensitivity is illustrated by comparing the values obtained in Examples 8-2 and 8-8. These values indicate that an increase in beta from 25 to 50 caused a decrease in the value of V_{CE} from 6 to 3 volts. The effect of this change is to shift the operating point of the circuit, the results of which can be observed on the curves of Fig. 8-3. Such great sensitivity to changes in beta makes this circuit impractical for most transistor applications.

As beta sensitivity is dependent on the values of β and the resistances in the base and emitter circuits, it can be represented mathematically by

$$K_\beta = \frac{1}{1 + \dfrac{\beta R_E}{R_B}} \tag{8-20}$$

where K_β = beta sensitivity, a dimensionless ratio

The value of K_β ranges from low values of almost zero to high values approaching unity. The lower values of K_β indicate less sensitivity to changes in beta. For example, with the base-bias arrangement in the circuit of Fig. 8-16, where $R_E = 0$, K_β approaches unity, which indicates very poor beta sensitivity.

8-12 Temperature versus Current Stability

Effect of Temperature on the Currents in a Transistor. An important consideration in selecting a biasing circuit for a transistor application is the effects of the ambient and operating temperatures on the magnitude of the currents in the transistor. A change in temperature can drastically affect the amount of current flowing in a transistor; this effect is called the *temperature versus current stability* and is often shortened to *temperature stability*, or simply *stability*. Special circuits are often used to improve the temperature versus current stability of a transistor circuit.

Need for Stabilization. The current flowing through the collector-base junction of a transistor consists of two components: (1) the collector current I_C that is due to the normal transistor action and is proportional to the emitter current I_E and the transistor current-amplification factor alpha, and (2) the reverse-bias collector current (sometimes referred to as cutoff current, reverse current, leakage current, or saturation current), designated as I_{CO} or I_{CBO}, which is the current that flows through the collector-base junction when the emitter current is zero. This cutoff current also consists of two parts: (1) the reverse leakage current which is dependent on the collector voltage, and (2) a current that varies exponentially with temperature and is practically independent of the collector-base voltage.

In a common-base junction transistor circuit, the total collector current I_C is determined by the emitter current I_E, the current amplification factor α, and the collector cutoff current I_{CO}, or

$$I_C = \alpha I_E + I_{CO} \tag{8-21}$$

Since the collector current is usually in milliamperes and the collector cutoff current in microamperes, any normal increase in temperature will not increase the collector cutoff current to a value that would seriously affect the total collector current. Generally the problem of stabilization of the collector current because of temperature change is not an important problem in grounded-base junction-transistor circuits.

In a common-emitter junction-transistor circuit the total collector current is determined by the base current I_B, the collector cutoff current I_{CO}, and the current amplification factor β, or

$$I_C = \beta I_B + (1 + \beta)\, I_{CO} \tag{8-22}$$

This equation shows that any increase in the collector cutoff current will be amplified $(1 + \beta)$ times. Since beta is of the order of 30 to 200, and both the

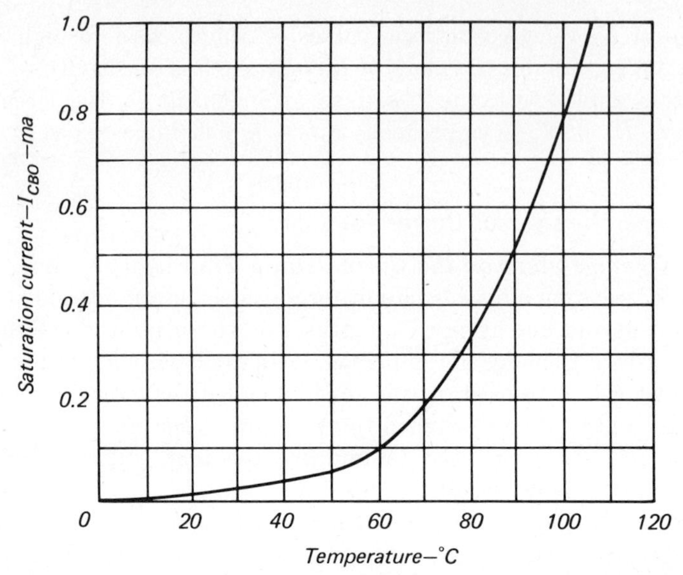

Fig. 8-14 Typical variation of collector saturation current with the temperature at the junction of a transistor.

base and collector cutoff currents are in microamperes, a slight increase in the cutoff current will have a significant effect on the total collector current. An increase in collector current increases the collector power dissipation, thus also increasing the operating temperature, which will cause a further increase in collector current. If this succession of increases is permitted to continue, the total collector current can increase beyond the safe operating current of the transistor, causing it to become damaged; this condition is called *thermal runaway*. Figure 8-14 illustrates how rapidly the collector cutoff current rises with an increase in the operating temperature beyond approximately 50°C. Analysis of Fig. 8-14 and Eq. (8-22) indicates that some form of stabilizing circuit is necessary in order to prevent thermal runaway.

The increase in collector current with an increase in temperature is due not only to the increase in collector cutoff current, but also to a decrease in the resistance of the emitter-base junction as the temperature of the transistor increases. Figure 8-15 shows that the collector current rises very rapidly with any appreciable increase in temperature.

Stability Factor. The degree to which the collector current, or emitter current, of a transistor varies due to changes in temperature is expressed by its current stability factor. The *current stability factor* is the ratio of a change in emitter current to a change in saturation current, or

$$SF = \frac{dI_E}{dI_{CBO}} \tag{8-23}$$

where SF = current stability factor, a dimensionless ratio
dI_E = change in emitter current, amp
dI_{CBO} = change in collector cutoff current, amp

The current stability factor can be expressed in terms of the circuit parameters of a stabilized circuit, such as Fig. 8-19, as

$$SF = \frac{\dfrac{1}{R_E}}{\dfrac{1}{R_B} + \dfrac{1}{R_S} + \dfrac{1-\alpha}{R_E}} \tag{8-24}$$

The terms of Eq. (8-24) can be rearranged by algebra to produce

$$SF = \frac{1}{\dfrac{R_E}{R_B} + \dfrac{R_E}{R_S} + (1-\alpha)} \tag{8-25}$$

Equation (8-25) can be applied to other biasing circuits by observing the following rules: (1) if R_S is not used in a circuit, its value is taken as infinity (∞); (2) the value of R_L is used for R_E when R_L appears in the emitter circuit; and (3) if R_B or R_E is omitted in a circuit, its value is taken as zero. Applications of this equation are given in the following articles.

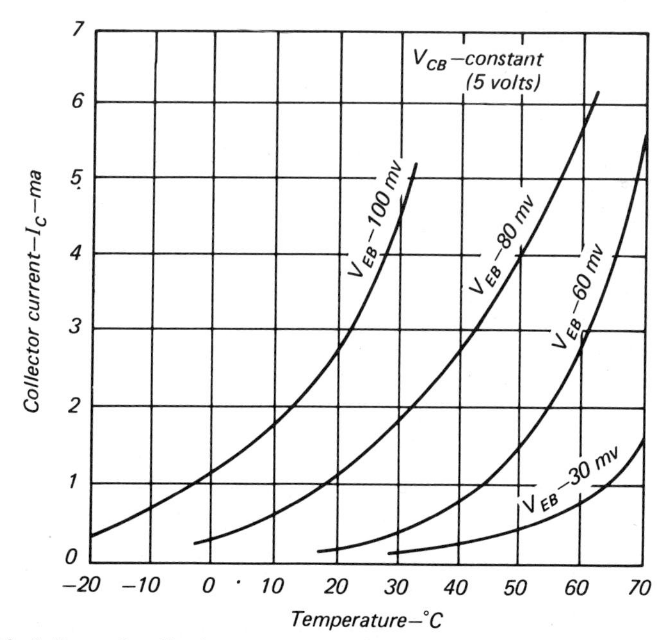

Fig. 8-15 Variation of collector current with temperature for a transistor at different values of emitter-base voltage.

8-13 Fixed Bias for the Common-emitter Transistor Amplifier

Simple Bias Arrangements. The circuit of Fig. 8-12*a* shows the simplest bias arrangement for the common-emitter transistor amplifier using two power sources. The circuit provides a *fixed bias* by means of the power source V_{BB} and the base resistor R_B. The circuit of Fig. 8-13*a* shows the simplest bias arrangement for the common-emitter transistor amplifier using a single power source. These simple biasing arrangements are not practicable because of (1) poor beta sensitivity, and (2) poor temperature versus current stability. The poor beta sensitivity is illustrated by Example 8-8. The poor temperature-current stability is caused by (1) a high external base circuit resistance (R_B), and (2) a low external emitter circuit resistance ($R_E = 0$). The stability of this circuit is inherently poor, as is indicated by analysis of Eq. (8-25) for the condition when $R_E = 0$; then

$$SF = \frac{1}{\dfrac{0}{R_B} + \dfrac{0}{R_S} + (1 - \alpha)} = \frac{1}{1 - \alpha} \tag{8-26}$$

Because alpha is generally in the order of 0.95 to 0.99, *SF* will have a high value, thereby indicating poor temperature-current stability.

Fixed Bias with a Single Power Supply. Another simple fixed-bias arrangement, sometimes called *base bias,* is shown in Fig. 8-16. This circuit differs from that of Fig. 8-13*a* in that the input and output signals are shown with respect to ground potential. Forward bias and reverse bias are obtained in the same manner as explained for the circuit of Fig. 8-13*a*. Analysis of this circuit in terms of Eq. (8-25) shows that this circuit (with $R_E = 0$) has poor temperature-current stability.

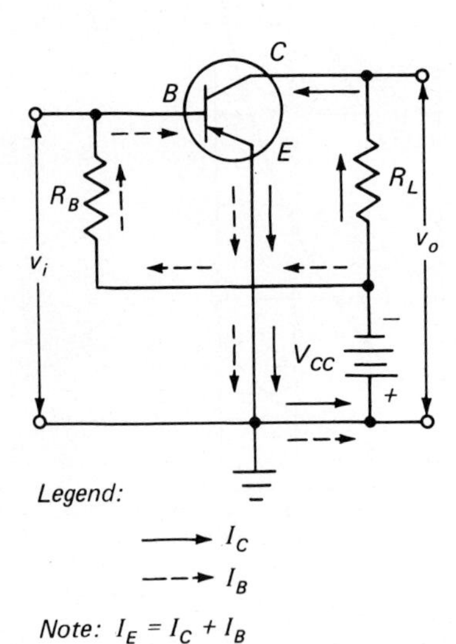

Fig. 8-16 Common-emitter transistor amplifier circuit with fixed bias and a single power source.

Example 8-9 If the parameters of the circuit of Fig. 8-16 are $V_{CC} = 28$ volts, $\beta = 80$, $R_L = 8{,}000$ ohms, and $R_B = 700{,}000$ ohms, find (*a*) I_B, (*b*) I_C, (*c*) V_{CE}.

GIVEN: $V_{CC} = 28$ volts $\beta = 80$ $R_L = 8{,}000$ ohms $R_B = 700{,}000$ ohms

FIND: (*a*) I_B (*b*) I_C (*c*) V_{CE}

SOLUTION:

(*a*) $$I_B = \frac{V_{CC}}{R_B} = \frac{28}{700{,}000} = 40\ \mu\text{a}$$

(*b*) $$I_C = \beta I_B = 80 \times 40 = 3.2\ \text{ma}$$

(*c*) $$V_{CE} = V_{CC} - I_C R_L = 28 - (3.2 \times 10^{-3} \times 8{,}000) = 2.4\ \text{volts}$$

Examination of Example 8-9 indicates that the operating point is affected by the value of beta in the same manner as for the circuit of Examples 8-2 and 8-8. Therefore, the circuit of Fig. 8-16 also exhibits the disadvantage of poor beta sensitivity and poor temperature-current stability.

8-14 Fixed Bias for the Common-base Transistor Amplifier

In the common-base circuit of Fig. 8-12*b*, the base is at ground potential for both of the d-c power sources and the a-c signal voltage. The current stability factor of this circuit can be determined with Eq. (8-25) by substituting zero and infinity for R_B and R_S, respectively; then

$$SF = \frac{1}{\dfrac{R_E}{R_B} + \dfrac{R_E}{R_S} + (1 - \alpha)} = \frac{1}{\dfrac{R_E}{0} + \dfrac{R_E}{\infty} + (1 - \alpha)}$$

$$= \frac{1}{\infty + 0 + (1 - \alpha)} = \frac{1}{\infty} = 0 \qquad (8\text{-}27)$$

The emitter current in a circuit having a stability factor of zero will not be affected by temperature variations at the transistor and hence is ideal in this respect.

In the common-base circuit of Fig. 8-13*b*, the emitter-base forward bias is equal to the difference of potential between the emitter and base terminals. The emitter is at the highest positive potential of the three terminals, as it is connected to the positive terminal of the power source through the resistor R_E. The collector is least positive, hence negative, as it is nearest the negative terminal of the power source. The base or bias potential is *fixed* by the voltage divider $R_B R_S$ connected across the power source V_{CC}. The values of R_B and R_S must be of such relative values that the voltage at their junction will provide the proper forward bias at the emitter-base junction of the transistor. When the values of the circuit parameters are known, the current stability factor can be determined by use of Eq. (8-25).

8-15 Fixed Bias for the Common-collector Transistor Amplifier

In the circuit of Fig. 8-12*c*, the collector is at ground potential for both of the d-c power sources and the a-c signal voltage. The stability factor of this circuit can be determined with Eq. (8-25) by substituting (1) infinity for R_S, and (2) zero for $(1 - \alpha)$ because the value of α is very nearly unity; then

$$SF = \frac{1}{\dfrac{R_E}{R_B} + \dfrac{R_E}{R_S} + (1 - \alpha)} = \frac{1}{\dfrac{R_E}{R_B} + \dfrac{R_E}{\infty} + 0} \cong \frac{1}{\dfrac{R_E}{R_B}} \cong \frac{R_B}{R_E} \tag{8-28}$$

This equation indicates that the stability factor of the circuit is determined by the ratio of the external base resistance to the external emitter resistance.

In the circuit of Fig. 8-13*c*, the power for the circuit is supplied by a single power source V_{EE}. The manner in which the bias is established in the common-collector configuration is the same as for the common-emitter circuit described in Art. 8-10. Reverse bias for the collector-base junction is indicated by the collector being connected directly to the negative terminal of the power source.

8-16 Self-bias

Principle of Self-biasing. Because circuits using only fixed bias generally have poor beta sensitivity and poor temperature-current stability, the principle of self-biasing is often used. *Self-bias*, sometimes called *automatic bias*, uses the effects of changes in current in the collector and/or emitter circuits to counteract automatically any changes in collector and emitter current. The basic principle of self-bias is to introduce some form of control to the base bias that will automatically counteract the original change in the collector or emitter current. Many variations of the self-bias principle have been introduced into transistor circuitry, a number of which are presented in this text.

Emitter Bias. The circuits of Fig. 8-17 provide both improved beta sensitivity and temperature-current stability because of (1) the introduction of resistance into the emitter circuit, and (2) the use of relatively low values of resistance in the base circuit. This method of biasing is generally referred to as *emitter bias* or *self-bias*. With emitter bias, an increase in the emitter current is accompanied by an increase in the voltage drop across the resistor (R_E) in the emitter circuit which in turn (1) decreases the emitter voltage, (2) decreases the forward bias at the emitter-base junction, and (3) decreases the emitter current. This circuit action, called *degeneration*, counteracts the original increase in the emitter current and results in improved beta sensitivity and temperature-current stability for the circuit. A disadvantage of emitter bias is that the resultant degeneration causes a reduction in the gain achieved by the amplifier circuit.

The sensitivity of the emitter-biased circuit to changes in beta is greatly improved over that of the fixed-bias circuit because the effective resistance of

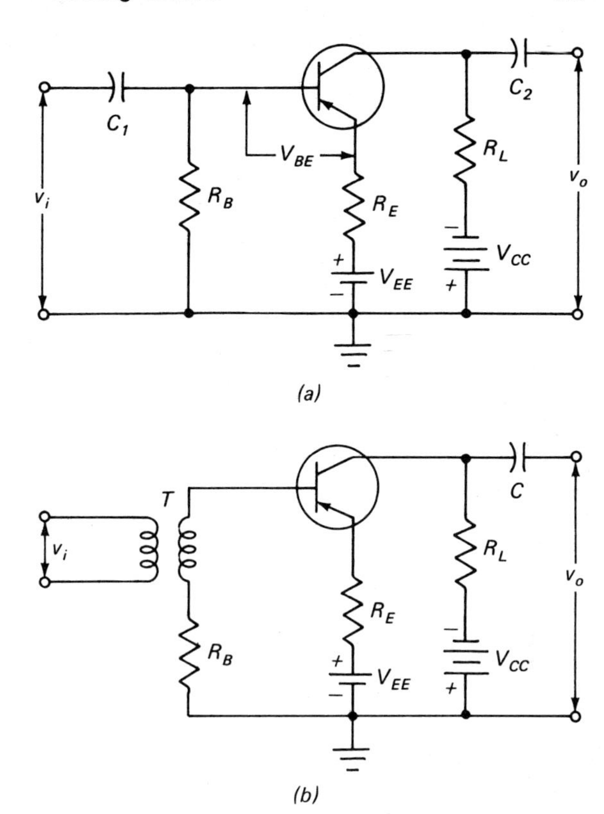

Fig. 8-17 Common-emitter transistor amplifier circuit with emitter bias. (a) Basic circuit. (b) Transformer input to permit use of low values of base resistance.

the base circuit is reduced by a factor equal to beta, as is indicated in the following equation for the emitter current for the circuit of Fig. 8-17*a*:

$$I_E = \frac{V_{EE} - V_{BE}}{R_E + \dfrac{R_B}{\beta}} \tag{8-29}$$

When V_{EE} is much greater than V_{BE} and R_B/β is much less than R_E, Eq. (8-29) can be simplified to

$$I_E \cong \frac{V_{EE}}{R_E} \tag{8-30}$$

Example 8-10 If the parameters of the circuit of Fig. 8-17*a* are $V_{EE} = 20$ volts, $V_{CC} = 28$ volts, $V_{BE} = 0.3$ volt, $R_B = 5{,}000$ ohms, $R_E = 10{,}000$ ohms, and $R_L = 5{,}000$ ohms, what is the emitter current when beta is (*a*) 50? (*b*) 100?

GIVEN: $V_{EE} = 20$ volts $\quad V_{CC} = 28$ volts $\quad V_{BE} = 0.3$ volt $\quad R_B = 5{,}000$ ohms
$R_E = 10{,}000$ ohms $\quad R_L = 5{,}000$ ohms

FIND: (*a*) I_E $\quad$ (*b*) I_E

SOLUTION:

(*a*)
$$I_E = \frac{V_{EE} - V_{BE}}{R_E + \dfrac{R_B}{\beta}} = \frac{20 - 0.3}{10{,}000 + \dfrac{5{,}000}{50}} \cong 1.95 \text{ ma}$$

(*b*)
$$I_E = \frac{V_{EE} - V_{BE}}{R_E + \dfrac{R_B}{\beta}} = \frac{20 - 0.3}{10{,}000 + \dfrac{5{,}000}{100}} \cong 1.96 \text{ ma}$$

Example 8-11 What is the approximate value of the emitter current for the circuit of Example 8-10 when found by use of Eq. (8-30)?

GIVEN: $V_{EE} = 20$ volts $\quad R_E = 10{,}000$ ohms

FIND: I_E

SOLUTION:

$$I_E \cong \frac{V_{EE}}{R_E} = \frac{20}{10{,}000} = 2 \text{ ma}$$

The results of Example 8-10 indicate that with the emitter-biased circuit changes in the value of beta have practically no effect on the emitter current. The results of Example 8-11 indicate that the emitter current can be determined by use of the simplified equation (8-30) with very little error in the indicated value of emitter current.

Emitter bias also provides reasonably good temperature-current stability and beta sensitivity.

Example 8-12 (*a*) What is the current stability factor of the circuit of Fig. 8-17*a* if the circuit parameters are $\alpha = 0.95$, $R_B = 5{,}000$ ohms, $R_E = 10{,}000$ ohms, and $R_L = 5{,}000$ ohms? (*Note:* For value of R_S use infinity.) (*b*) What is the beta sensitivity of the circuit? (*c*) What do the numerical values found in (*a*) and (*b*) indicate about this circuit?

GIVEN: $\alpha = 0.95 \quad R_B = 5{,}000$ ohms $\quad R_E = 10{,}000$ ohms $\quad R_L = 5{,}000$ ohms

FIND: (*a*) SF $\quad$ (*b*) K_β

SOLUTION:

(*a*)
$$SF = \frac{1}{\dfrac{R_E}{R_B} + \dfrac{R_E}{R_S} + (1 - \alpha)} = \frac{1}{\dfrac{10{,}000}{5{,}000} + \dfrac{10{,}000}{\infty} + (1 - 0.95)} = 0.488$$

(*b*)
$$K_\beta = \frac{1}{1 + \dfrac{\beta R_E}{R_B}} = \frac{1}{1 + \dfrac{19 \times 10{,}000}{5{,}000}} = 0.0256$$

where
$$\beta = \frac{\alpha}{1 - \alpha} = \frac{0.95}{1 - 0.95} = 19$$

(*c*) The circuit has good stability and sensitivity characteristics.

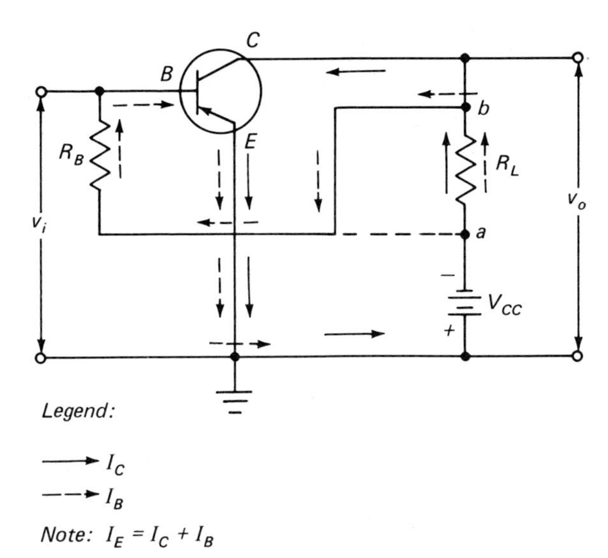

Fig. 8-18 Common-emitter transistor amplifier circuit with collector self-bias.

Collector Bias. Figure 8-18 illustrates a simple self-bias arrangement sometimes called *collector bias.* This circuit differs from the simple fixed-bias arrangement of Fig. 8-16 only in that the bias resistor R_B is connected to the collector side of R_L (point b) instead of directly to the negative terminal of the power source (point a). With this circuit, an increase in the collector current is accompanied by (1) an increase in the voltage drop across R_L, (2) a decrease in the voltage at point b, (3) a decrease in the forward bias on the transistor, and (4) a decrease in the emitter and collector currents. Thus a change in the collector current is automatically neutralized by the reactions of the circuit to the original change in collector current. Because the change in voltage at point b is reflected (or fed back) to the base terminal for the purpose of causing a change in the collector current in the opposite manner to the original collector-current change, the circuit is referred to as having *negative feedback.* For this circuit

$$I_E \cong I_C = \frac{V_{CC} - V_{BE}}{R_L + \dfrac{R_B}{\beta}} \tag{8-31}$$

$$V_{CE} = V_{CC} - I_C R_L \tag{8-32}$$

$$K_\beta = \frac{1}{1 + \dfrac{\beta R_L}{R_B}} \tag{8-33}$$

$$R_B = \frac{V_{CE} - V_{BE}}{I_B} \tag{8-34}$$

Example 8-13 If the circuit parameters of Fig. 8-18 are $V_{CC} = 12$ volts, $V_{BE} = 0.3$ volt, $R_L = 1{,}000$ ohms, $R_B = 100{,}000$ ohms, and $\beta = 100$, find (*a*) I_C, (*b*) V_{CE}, (*c*) I_B, (*d*) K_β, (*e*) *SF*.

GIVEN: $V_{CC} = 12$ volts $\quad V_{BE} = 0.3$ volt $\quad R_L = 1{,}000$ ohms
$R_B = 100{,}000$ ohms $\quad \beta = 100$

FIND: (*a*) I_C (*b*) V_{CE} (*c*) I_B (*d*) K_β (*e*) *SF*

SOLUTION:

(*a*)
$$I_C = \frac{V_{CC} - V_{BE}}{R_L + \dfrac{R_B}{\beta}} = \frac{12 - 0.3}{1{,}000 + \dfrac{100{,}000}{100}} = 5.85 \text{ ma}$$

(*b*)
$$V_{CE} = V_{CC} - I_C R_L = 12 - (5.85 \times 10^{-3} \times 1{,}000) = 6.15 \text{ volts}$$

(*c*)
$$I_B = \frac{I_C}{\beta} = \frac{5.85 \times 10^{-3}}{100} = 58.5\ \mu\text{a}$$

(*d*)
$$K_\beta = \frac{1}{1 + \dfrac{\beta R_L}{R_B}} = \frac{1}{1 + \dfrac{100 \times 1{,}000}{100{,}000}} = 0.5$$

(*e*)
$$SF = \frac{1}{1 - \alpha} = \frac{1}{1 - 0.99} = 100$$

where
$$\alpha = \frac{\beta}{1 + \beta} = \frac{100}{1 + 100} = 0.99$$

8-17 Other Fixed- and Self-biased Common-emitter Circuits

Fixed Bias and Emitter Bias. The circuit of Fig. 8-19 has a combination of (1) fixed bias provided by the voltage divider $R_B R_S$ and the power source V_{CC}, and (2) emitter bias provided by the resistor R_E. For this circuit

$$I_C \cong I_E = \frac{V_B}{R_E} \tag{8-16}$$

where
$$V_B = V_{CC} \frac{R_B}{R_B + R_S} \tag{8-15}$$

$$V_{CE} = V_{CC} - I_C R_L - I_E R_E \tag{8-17}$$

$$K_\beta = \frac{1}{1 + \dfrac{\beta R_E}{R_T}} \tag{8-35}$$

where
$$R_T = \frac{R_B R_S}{R_B + R_S} \tag{8-36}$$

The temperature-current stability factor of this circuit is expressed by Eq.

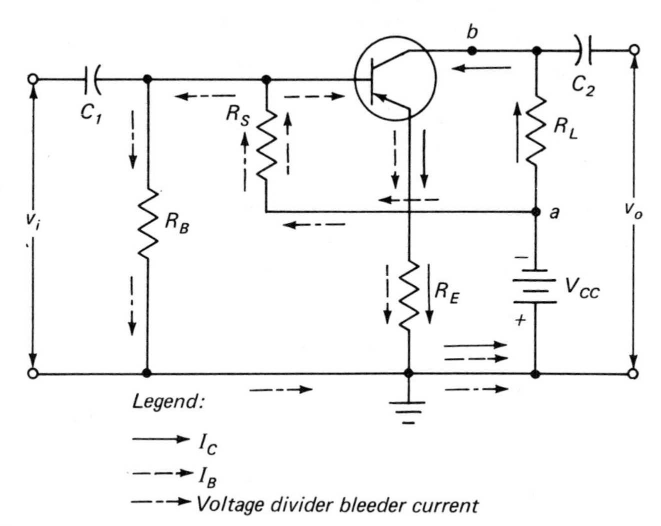

Fig. 8-19 Common-emitter transistor amplifier circuit with emitter bias and fixed-base bias.

(8-25). As the term $(1 - \alpha)$ is normally a very low value, it can be omitted from the equation without causing any appreciable error and thus

$$SF = \frac{1}{\dfrac{R_E}{R_B} + \dfrac{R_E}{R_S}} \tag{8-37}$$

or

$$SF = \frac{R_B R_S}{R_E(R_B + R_S)} \tag{8-38}$$

Equation (8-38) shows that the stability of the circuit improves (the value of SF decreases) as (1) the value of R_B decreases, and (2) the value of R_E increases.

Example 8-14 If the circuit parameters of Fig. 8-19 are $V_{CC} = 15$ volts, $\beta = 50$, $R_B = 5{,}000$ ohms, $R_S = 10{,}000$ ohms, $R_E = 5{,}000$ ohms, and $R_L = 2{,}000$ ohms, find (*a*) I_C, (*b*) V_{CE}, (*c*) K_β, (*d*) SF.

GIVEN: $V_{CC} = 15$ volts $\quad \beta = 50 \quad R_B = 5{,}000$ ohms $\quad R_S = 10{,}000$ ohms
$R_E = 5{,}000$ ohms $\quad R_L = 2{,}000$ ohms

FIND: (*a*) I_C (*b*) V_{CE} (*c*) K_β (*d*) SF

SOLUTION:

(*a*) $$I_C \cong I_E = \frac{V_B}{R_E} = \frac{5}{5{,}000} = 1 \text{ ma}$$

where $$V_B = V_{CC}\frac{R_B}{R_B + R_S} = 15 \times \frac{5{,}000}{5{,}000 + 10{,}000} = 5 \text{ volts}$$

(b) $V_{CE} = V_{CC} - I_C R_L - I_E R_E = 15 - (1 \times 10^{-3} \times 2{,}000) - (1 \times 10^{-3} \times 5{,}000)$
$= 8$ volts

(c) $$K_\beta = \frac{1}{1 + \dfrac{\beta R_E}{R_T}} = \frac{1}{1 + \dfrac{50 \times 5{,}000}{3{,}333}} \cong 0.013$$

where $$R_T = \frac{R_B R_S}{R_B + R_S} = \frac{5{,}000 \times 10{,}000}{5{,}000 + 10{,}000} = 3{,}333 \text{ ohms}$$

(d) $$SF = \frac{R_B R_S}{R_E(R_B + R_S)} = \frac{5{,}000 \times 10{,}000}{5{,}000(5{,}000 + 10{,}000)} = 0.667$$

The results of Example 8-14 indicate that the circuit of Fig. 8-19 has good beta sensitivity and good temperature-current characteristics.

The effect of different combinations of resistance values in the external base and emitter circuits on the collector-current stability is shown graphically in Fig. 8-20. The greatest instability occurs when both R_B and R_E are zero, as is indicated by curve *D*. The best circuit operation is indicated by curve *B*, which represents the condition when R_B is zero and R_E is high. Curve *C* represents the condition when R_E is zero and R_B has a high value. The horizontal reference line *A-A* represents the ideal condition.

Fixed Bias and Collector Bias. The circuit of Fig. 8-21 has a combination of (1) fixed bias provided by the voltage divider $R_B R_S$ and the voltage source

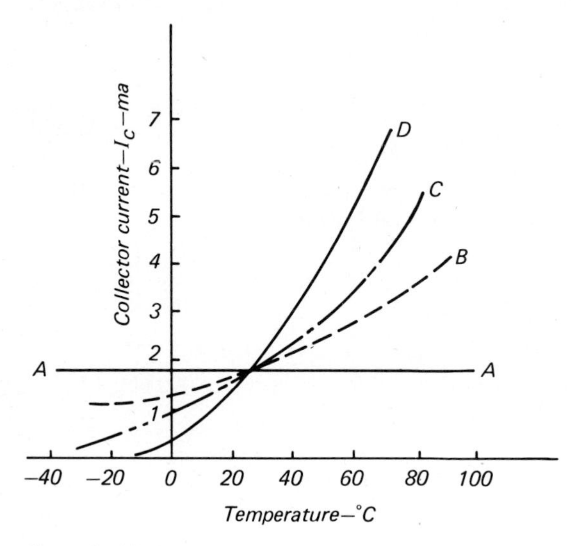

Fig. 8-20 Curves showing effect of external base and emitter circuit resistance on collector-current stability.

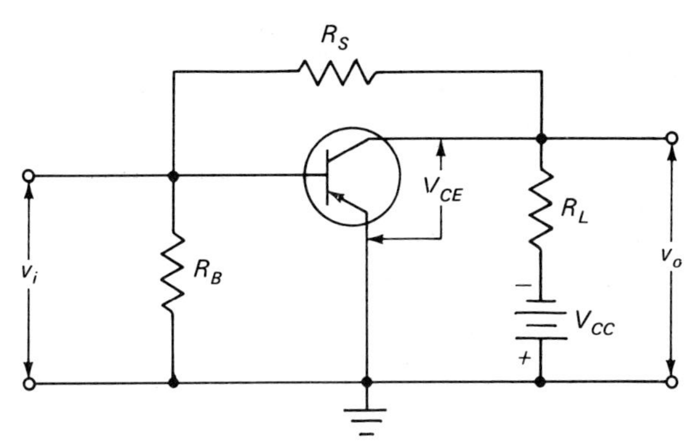

Fig. 8-21 Common-emitter transistor amplifier circuit using a voltage divider to obtain both fixed bias and self-bias.

V_{CE} and (2) collector bias provided by connecting R_S to the junction of the collector and R_L. Collector self-bias is provided by the negative feedback voltage made available at the base terminal through the feedback resistor R_S. The stabilizing action of the collector self-bias is the same as is explained in Art. 8-16.

The circuit of Fig. 8-22, which is a variation of the circuit of Fig. 8-21, illustrates another method of obtaining good current stability with a minimum sacrifice in gain. In this circuit, the d-c return resistor R_S is divided into two parts R_1 and R_2, and all a-c variations are bypassed by capacitor C. The value of R_2 is usually five to ten times that of R_1. The voltage gain and circuit stability are determining factors in selecting the circuit to be used.

Emitter and Collector Bias. The circuit of Fig. 8-23 illustrates the use of base bias with both emitter and collector feedback control. The purpose of this

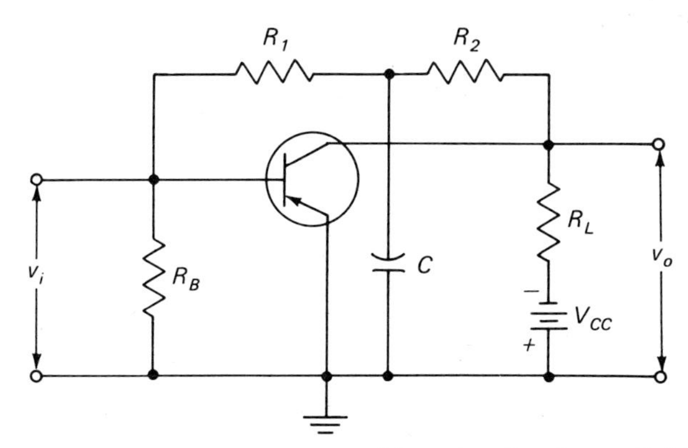

Fig. 8-22 Common-emitter transistor amplifier circuit using a divided d-c return path and a capacitor for obtaining both fixed bias and self-bias.

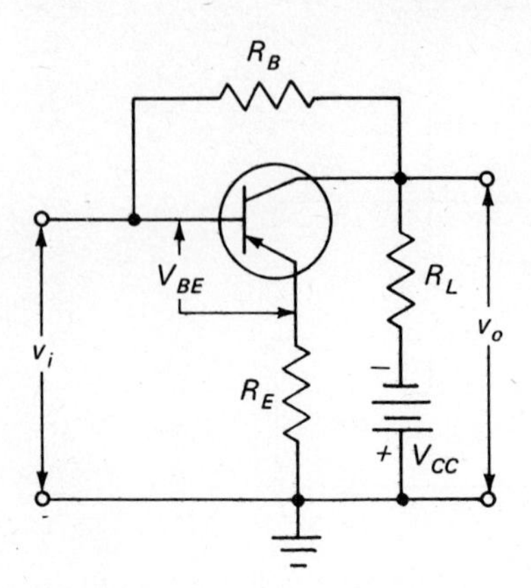

Fig. 8-23 Common-emitter transistor amplifier circuit using both emitter and collector feedback bias control.

bias arrangement is to obtain improvement of the beta-sensitivity characteristic. For this circuit

$$I_E \cong I_C = \frac{V_{CC} - V_{BE}}{R_E + R_L + \dfrac{R_B}{\beta}} \tag{8-39}$$

$$V_{CE} = V_{CC} - I_C R_L - I_E R_E \tag{8-17}$$

$$K_\beta = \frac{1}{1 + \dfrac{\beta(R_E + R_L)}{R_B}} \tag{8-40}$$

$$SF \cong \frac{R_B}{R_E} \tag{8-41}$$

Example 8-15 If the circuit parameters of Fig. 8-23 are $V_{CC} = 28$ volts, $\beta = 100$, $R_B = 500{,}000$ ohms, $R_E = 10{,}000$ ohms, and $R_L = 20{,}000$ ohms, find (*a*) I_C, (*b*) V_{CE}, (*c*) K_β, (*d*) SF. (*Note:* Assume V_{BE} is negligible.)

GIVEN: $V_{CC} = 28$ volts $\quad \beta = 100 \quad R_B = 500{,}000$ ohms $\quad R_E = 10{,}000$ ohms
$R_L = 20{,}000$ ohms

FIND: (*a*) I_C (*b*) V_{CE} (*c*) K_β (*d*) SF

SOLUTION:

(*a*) $$I_C \cong \frac{V_{CC}}{R_E + R_L + \dfrac{R_B}{\beta}} = \frac{28}{10{,}000 + 20{,}000 + \dfrac{500{,}000}{100}} = 0.8 \text{ ma}$$

(*b*) $$V_{CE} = V_{CC} - I_C R_L - I_E R_E = 28 - (0.8 \times 10^{-3} \times 20{,}000) - (0.8 \times 10^{-3} \times 10{,}000)$$

$$= 4 \text{ volts}$$

(*c*) $$K_\beta = \frac{1}{1 + \dfrac{\beta(R_E + R_L)}{R_B}} = \frac{1}{1 + \dfrac{100(10{,}000 + 20{,}000)}{500{,}000}} = 0.143$$

(*d*) $$SF \cong \frac{R_B}{R_E} = \frac{500{,}000}{10{,}000} = 50$$

8-18 Thermistor Stabilizing Circuits

The preceding articles described a number of methods and circuits used to improve the stability of transistor circuits. Since instability is due primarily to the temperature sensitivity of the collector saturation current, stability may be restored by using a thermistor, or *temperature-sensitive resistor*, in the bias control circuit.

Figure 8-24 uses two voltage dividers to provide the proper bias for the transistor. The voltage divider $R_B R_S$ is the same as that used in the circuit of Fig. 8-19, and its function and operation are the same as previously described. The second voltage divider consists of the emitter circuit resistor R_E and the thermistor RT. With an increase in temperature: (1) the resistance of RT decreases, thereby increasing the negative potential at point a, (2) the emitter becomes more negative, thereby decreasing the forward bias of the transistor, and (3) the circuit counteracts the tendency for the base and collector currents to increase.

There are numerous variations of stabilizing circuits using the negative-temperature-coefficient characteristic of the thermistor. One of the difficulties encountered in using thermistors in stabilizing circuits is that of obtaining a thermistor whose characteristics match the temperature characteristics of the transistor with which it is to be used. The ability of the thermistor to react in the desired manner to current changes in the transistor over a wide range of current values is called *tracking*. The thermistor control circuit is used chiefly to compensate for changes in current due to changes in ambient temperature. The curves of Fig. 8-25 show how well the thermistor-stabilized circuit can control the current variations. The undulations in curve B indicate that the thermistor and the transistor temperature characteristics do not track perfectly.

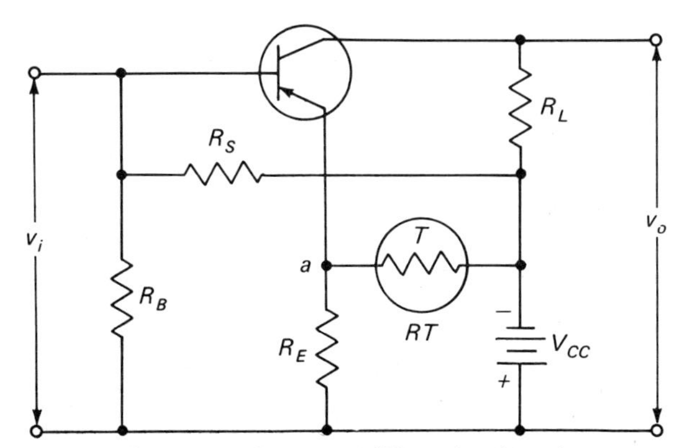

Fig. 8-24 Common-emitter transistor amplifier circuit using two voltage dividers to supply the base and emitter potentials. The voltage divider in the emitter circuit uses a thermistor.

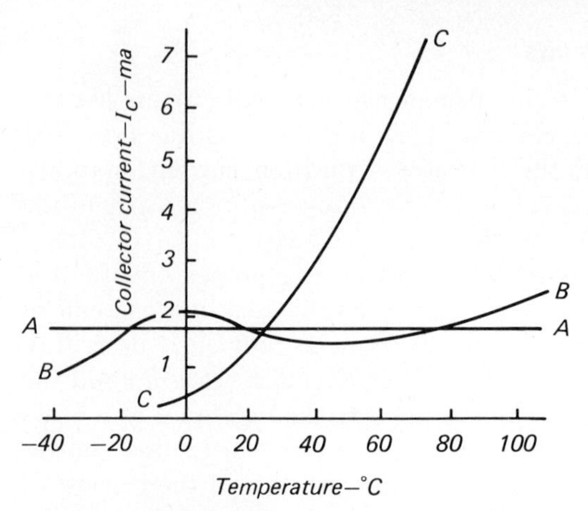

Fig. 8-25 Curves showing the stability achieved with a thermistor-stabilized transistor circuit.

8-19 Diode Stabilizing Circuits

A characteristic of the junction diode is that it has a negative temperature coefficient of resistance regardless of whether the diode is forward- or reverse-biased. Therefore, the junction diode can be used in a biasing circuit in place of either a resistor or a thermistor. If the junction diode is made of the same material as the transistor requiring stabilization, the stabilizing circuit will track more accurately than it would using either a thermistor or only resistors.

Figure 8-26 illustrates a circuit using a diode in its biasing arrangement. This circuit will make corrections for variations in the base-emitter junction resistance due to changes in temperature at the junction. In this circuit the diode D and the resistor R_S form a voltage divider that provides the negative base voltage required to forward-bias the transistor. The polarities applied to the diode terminals provide a forward bias at the diode, causing a current I_S to flow in the diode path. The current flowing through R_S consists of two components (1) the transistor base-emitter current I_B, and (2) the stabilizing circuit current I_S. When the temperature increases, the resistance at both the base-emitter junction of the transistor and the junction of the diode decreases and causes an increase in the base current I_B, collector current I_C, and stabilizing circuit current I_S. The increase in current through R_S causes an increase in the voltage drop at R_S, thereby reducing the base-biasing voltage at the transistor. The decrease in the base-biasing voltage causes (1) a reduction in the base-emitter current and (2) an increase in the base-emitter junction resistance. Thus, the action of the stabilizing circuit counteracts a change in the base-emitter junction resistance and tends to maintain it at a constant value. Curve B of Fig. 8-28 shows that this type of circuit can stabilize the collector current for temperatures up to about 60°C.

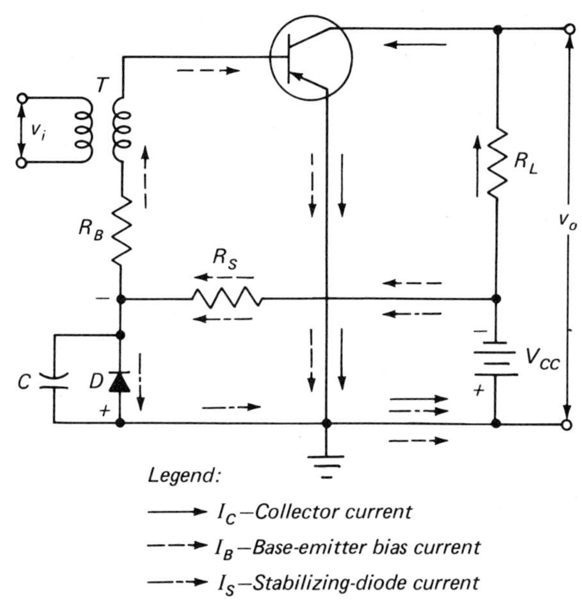

Fig. 8-26 Common-emitter transistor amplifier circuit using a diode in its biasing circuit to improve the temperature-current stability.

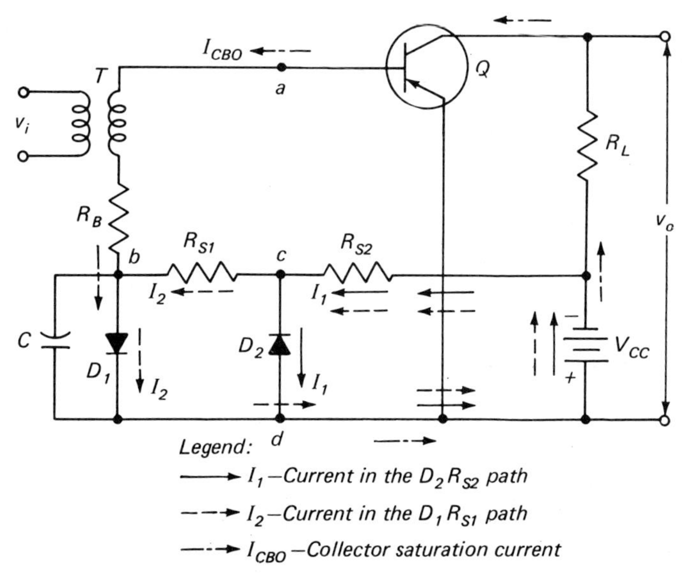

Fig. 8-27 Common-emitter transistor amplifier circuit using two stabilizing diodes: D_2 compensating for variations in the base-emitter junction resistance, and D_1 compensating for variations in the collector saturation current.

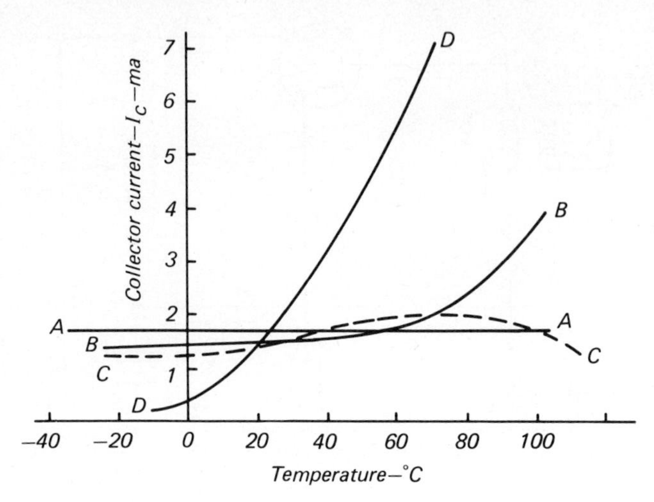

Curve A—Ideal condition
Curve B—With one stabilizing diode
Curve C—With two stabilizing diodes
Curve D—Without stabilization

Fig. 8-28 Curves showing the variation of collector current with changes in temperature for circuits using stabilizing diodes to improve stability.

Figure 8-27 illustrates a circuit using two diodes in its biasing arrangement. In this circuit, D_2 is forward-biased and D_1 reverse-biased. The voltage divider formed by D_2 and R_{S2} acts in the same manner as the voltage divider in the circuit of Fig. 8-26, and its function is to compensate for any changes in the base-emitter junction resistance of the transistor. The voltage divider formed by D_1 and R_{S1} is included in this circuit to compensate for changes in the transistor collector saturation current I_{CBO} that occur as the temperature exceeds approximately 60°C. This portion of the circuit operates in the following manner: With temperatures below 60°C, no current will flow through diode D_1 as it is reverse-biased. At increased temperatures, D_1 will conduct in its reverse direction as is indicated by the arrows for the current I_2 on Fig. 8-27. The direction of current flow through R_{S1} will make point c negative and point b positive. Point c is negative with respect to d by the action of the voltage divider D_2R_{S2}. The potential at point a with respect to d is the sum of the voltages $E_{d\text{-}c}$, $E_{c\text{-}b}$, and $E_{b\text{-}a}$. Because of the relatively low resistance of the secondary winding of the transformer T, the voltage drop $E_{b\text{-}a}$ is of very little significance. It should now be apparent that an increase in temperature will (1) cause an increase in base and collector currents due to a rapid increase in the saturation current I_{CBO}, (2) cause an increase in the reverse current in diode D_1 which will increase the voltage drop across R_{S1}, (3) reduce the voltage at point a, (4) reduce the base current, and (5) restore the collector current to its original value. Thus, this circuit will compensate for a decrease in the base-emitter junction resistance and an increase in the collector saturation current I_{CBO} due to increases in temperature. Curve C of Fig. 8-28 indicates that practically ideal stabilization characteristics can be obtained by use of this type of circuit.

8-20 Comparison of Biasing Methods

The choice of which biasing method to use depends chiefly upon the degree of stabilization required. Comparison is best indicated by the curves showing the variation of collector current with temperature for the various methods. Figures 8-20, 8-25, and 8-28 illustrate typical curves for the three basic methods using (1) resistors, (2) thermistors, and (3) diodes.

Biasing circuits used with Class A amplifiers are usually of the feedback type using resistors. Class B amplifiers generally employ stabilizing circuits using thermistors or diodes. There are numerous other stabilizing circuits, some employing (1) transistors as stabilizing elements, (2) Zener diodes, and (3) combinations of transistors and diodes.

8-21 Definitions of Classifications of Vacuum-tube Amplifiers

A *Class A vacuum-tube amplifier* is one in which the grid bias and alternating signal input voltages are of such values that plate current flows in the output circuit at all times (Fig. 8-29*a*). A *Class AB vacuum-tube amplifier* is one in which the grid bias and the signal-input voltages are of such values that plate current will flow for appreciably more than half but less than the complete time of the input cycle (Fig. 8-29*b*). A *Class B vacuum-tube amplifier* is one in which the grid bias is made approximately equal to the cutoff value (Fig. 8-29*c*). In this case the plate current will be approximately zero for zero-signal input, and plate current will flow for approximately one-half the time of the input cycle. A *Class C vacuum-tube amplifier* is one in which the grid bias is considerably greater than the cutoff value so that the plate current is zero for zero-signal input, and plate current will flow for appreciably less than one-half the time of the input cycle (Fig. 8-29*d*).

A further designation is made by adding the subscript 1 or 2 to a Class AB amplifier, in which 1 indicates that grid current does not flow during any part of the input cycle, while 2 indicates that grid current does flow during a part of the input cycle. Thus, an AB_1 amplifier does not draw grid current at any time while an AB_2 amplifier does take grid current during some part of the input cycle.

8-22 Class A Operation of a Vacuum-tube Amplifier

Use. A vacuum-tube amplifier is generally operated as Class A so that its output signal will be an amplified replica of the input signal. In the basic voltage-amplifier circuit of Fig. 8-30 the variations in the output voltage e_o

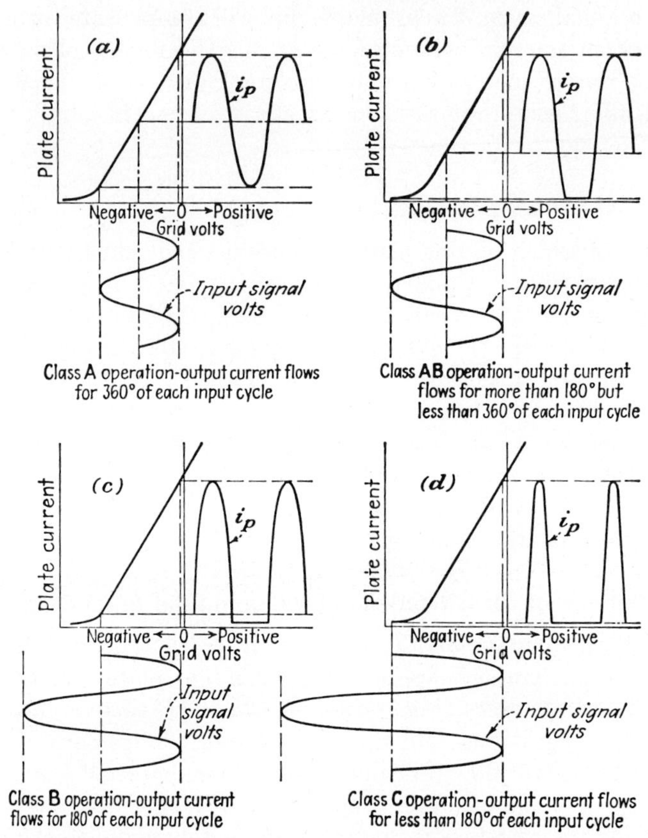

Fig. 8-29 Curves showing the relation between the current in the output circuit and the input signal for various classifications of vacuum-tube amplifier operation.

across the load impedance Z_L will be a reproduction of the variations in the signal input voltage e_i impressed across the grid, but of increased amplitude.

Operating Grid Bias for a Class A Vacuum-tube Amplifier. The grid bias required for Class A operation will depend upon the operating characteristics of the tube used and the voltage of the input signal. The operating character-

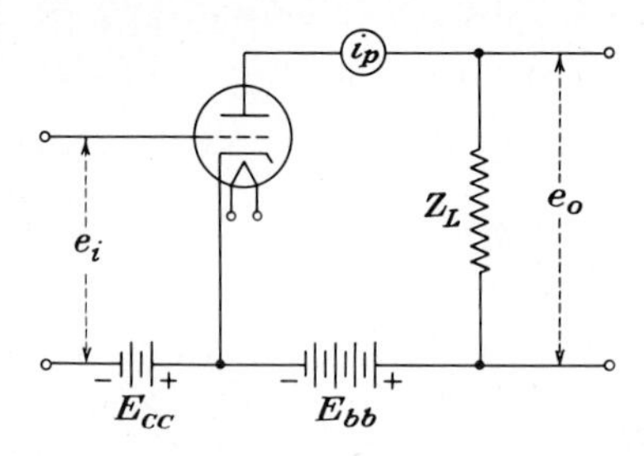

Fig. 8-30 Basic vacuum-tube voltage-amplifier circuit.

istics of the Class A amplifier shown in Fig. 8-31 indicate that the values of grid bias and input signal used will cause the tube to operate on only the straight portion A-B of the grid-plate transfer characteristic curve. The current in the plate circuit will then be an exact amplified reproduction of the input-signal variations that were applied to the grid. The maximum input signal that should be applied to the amplifier is dependent upon (1) establishing the quiescent operating point at the middle (point C) of the straight portion (A-B) of the characteristic curve and (2) using an amount of input-signal voltage that will keep the tube operating on only the straight portion A-B of the characteristic curve.

For the amplifier whose operating characteristics are shown in Fig. 8-31, the limits of grid voltage are represented by the points D and O. For maximum signal input, the tube should be operated with its quiescent value at point E. Increasing the grid bias to M or decreasing it to N will shift the quiescent point to M' or N'. In either case the limits of grid-voltage swing will be reduced, thus decreasing the amount of input-signal voltage that can be applied to the grid of the tube without causing distortion. The maximum input-signal voltage that should be applied to a tube will be equal to either (1) the amount of grid bias used (distance O-E on Fig. 8-31), or (2) the differ-

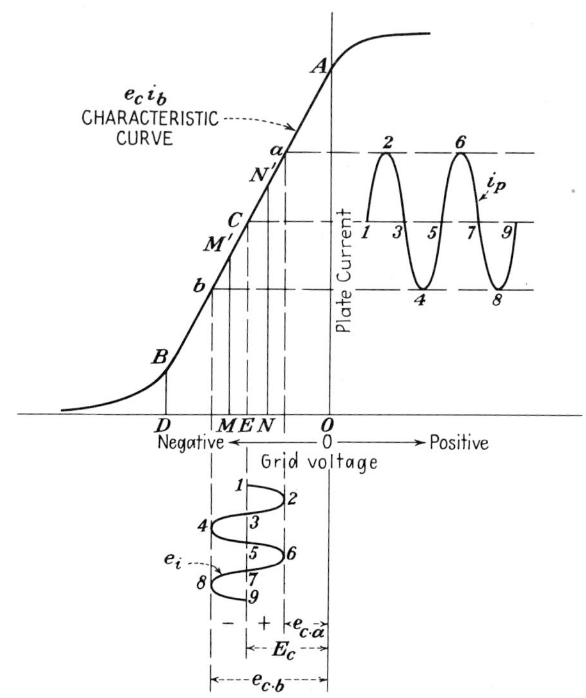

Fig. 8-31 Operating characteristics of a Class A vacuum-tube voltage amplifier.

ence between the maximum amount of negative grid voltage that will cause the tube to operate on the straight portion of its curve and the grid-bias voltage at which the tube is operated (distance *D-E*). The smaller of these two values will be the maximum input-signal voltage that can be applied without causing distortion.

Example 8-16 A triode whose characteristic curves are shown in Fig. 5-3 is being operated as a Class A amplifier with 240 volts applied to its plate. What is the maximum input-signal voltage that can be applied without producing distortion when the tube is being operated with a grid bias of (*a*) 8 volts? (*b*) 6 volts? (*c*) 4 volts?

Note: Assume that the straight portion of the curve extends beyond zero volts grid bias. This assumption is true for most Class A triode amplifier tubes.

GIVEN: $E_b = 240$ volts (*a*) $E_c = -8$ volts (*b*) $E_c = -6$ volts
(*c*) $E_c = -4$ volts

FIND: (*a*) $e_{i\cdot m}$ (*b*) $e_{i\cdot m}$ (*c*) $e_{i\cdot m}$

SOLUTION: The curve for $E_b = 240$ volts (Fig. 5-3) shows that the straight portion of the curve ends when the grid bias is approximately 10 volts. Thus

(*a*) $e_{i\cdot n} = 10 - 8 = 2$ volts
(*b*) $e_{i\cdot m} = 10 - 6 = 4$ volts
(*c*) $e_{i\cdot m} = 4 - 0 = 4$ volts

8-23 Distortion in Class A Vacuum-tube Amplifiers

Operation on the Curved Portion of the Characteristic Curve. When the tube whose operating characteristics are shown in Fig. 8-31 is operated with a grid bias whose value is E_c, the variation in the input-signal voltage e_i will produce a grid-voltage swing whose limits are $e_{c\cdot a}$ and $e_{c\cdot b}$. With this variation in grid voltage, the tube will operate on the straight portion of the curve between *a* and *b*. The varying plate current i_p will therefore change in the same manner as the input-signal voltage e_i.

When the tube is operated with a grid bias whose value is too near the negative bend of the curve, the output signal will become distorted (Fig. 8-32). The grid bias $E_{c\cdot 1}$ causes the tube to operate about point c_1 on the curve. The input signal voltage e_i will produce a grid-voltage swing whose limits are $e_{c\cdot a1}$ and $e_{c\cdot b1}$. With this variation in grid voltage, the tube will operate on the portion of the curve between a_1 and b_1. The negative halves of each cycle of the input-signal voltage will operate on the curved portion of the curve between b_1 and c_1. The plate current resulting from each of the negative half-cycles will therefore be distorted as is illustrated by the shaded area on Fig. 8-32.

When the tube is operated with a grid bias whose value is too near the positive bend of the curve, it will also cause distortion as is illustrated on Fig. 8-33. The grid bias $E_{c\cdot 2}$ causes the tube to operate about point c_2 on the curve. The input signal e_i will produce a grid-voltage swing whose limits are $e_{c\cdot a2}$ and $e_{c\cdot b2}$. This variation in grid voltage causes the tube to operate on the

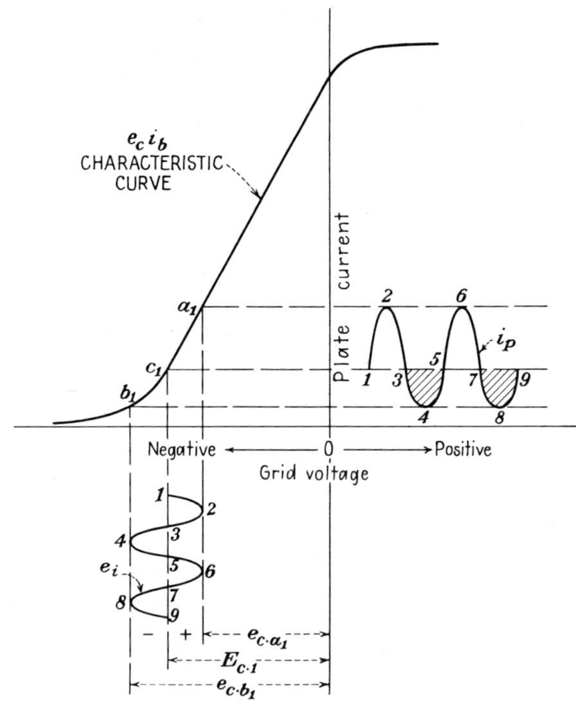

Fig. 8-32 Distortion in a Class A vacuum-tube voltage amplifier caused by too much bias.

portion of the curve between a_2 and b_2. The positive halves of each cycle of the input-signal voltage will operate on the curved portion of the curve between c_2 and a_2. The plate current resulting from each of the positive half-cycles will therefore be distorted as is illustrated by the shaded area of Fig. 8-33.

Distortion Caused by Driving the Grid Positive. When the grid is made positive with respect to the cathode, it will act in the same manner as the plate. Some of the electrons emitted by the cathode will be attracted to the grid, causing a current i_g to flow in the external grid circuit (Fig. 8-34). This current must flow through the resistance R_g or any other circuit element connected in this path. The current i_g flowing through R_g produces a voltage drop v_g in this circuit each instant that the grid is positive. The effective grid voltage at these instants is equal to the applied voltage e_i minus the voltage drop v_g. The two operating conditions shown in Figs. 8-35 and 8-36 demonstrate how the input signal is distorted when the voltage on the grid is made positive. This distortion of the input-signal voltage will therefore produce distortion in the plate current as is illustrated by these diagrams.

This type of distortion is produced by operating the tube with an incorrect amount of grid bias (Fig. 8-35) or by applying too great a signal voltage to the

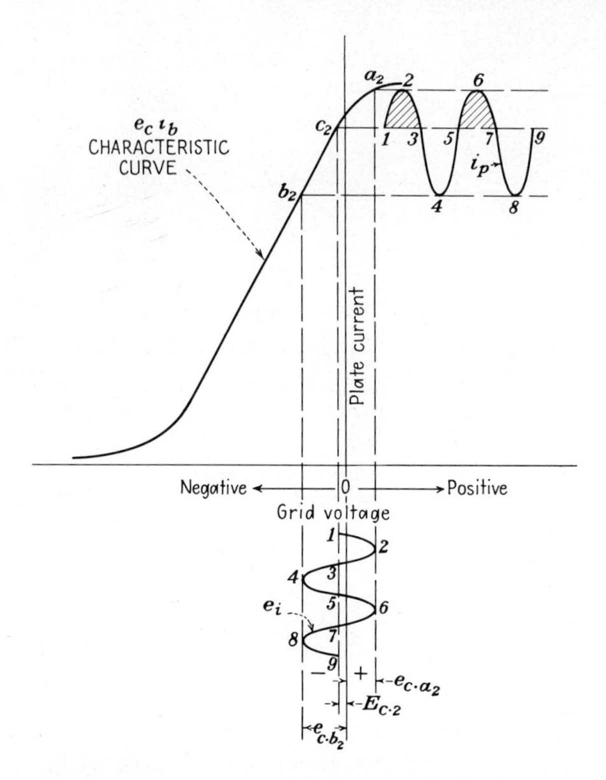

Fig. 8-33 Distortion in a Class A vacuum-tube voltage amplifier caused by too little bias.

input circuit (Fig. 8-36). For purposes of comparison the two characteristic curves used in Figs. 8-35 and 8-36 are the same as the one used in Fig. 8-31.

Distortion Caused by Operating with an Incorrect Bias. When the tube is operated with a grid bias whose value is less than the maximum value of the input-signal voltage, the grid voltage will be driven positive during the portion of each cycle in which the positive value of the input signal is greater than the grid bias (Fig. 8-35). With a grid-bias voltage as indicated at $E_{c \cdot 3}$, the input signal e_i causes the tube to operate on the straight portion of the curve between a_3 and b_3. However, under this condition the grid is made positive at some instants. The voltage drop due to the current in the grid circuit during these intervals reduces the effective grid voltage during these instants,

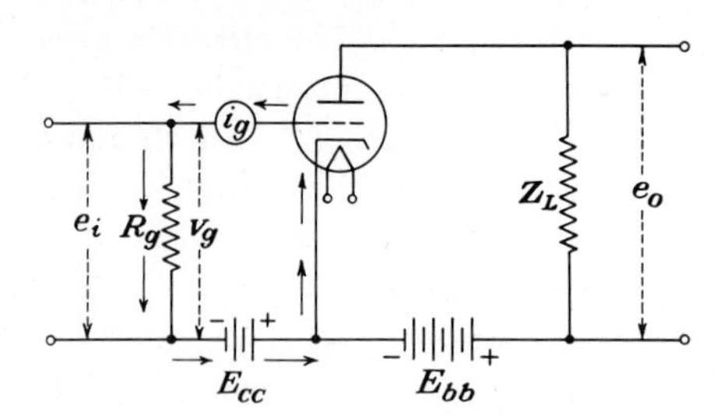

Fig. 8-34 Flow of current in the grid circuit of a vacuum-tube amplifier when the grid is positive with respect to its cathode.

thus reducing the maximum positive grid voltage from $e_{c \cdot a3}$ to $e_{c \cdot a}$. This distortion of the input-signal voltage causes the tube to operate between b_3 and a on the curve. The output current will therefore be distorted in a similar manner, as is illustrated by the distorted output wave shown in Fig. 8-35.

Distortion Caused by Too Large an Input Signal. When the tube is operated with its correct value of grid bias but the applied signal is too large, either half or both halves of the output signal may be distorted (Fig. 8-36). The variation in input-signal voltage $3e_i$ produces a grid-voltage swing whose limits are $e_{c \cdot a4}$ and $e_{c \cdot b4}$. With this variation in grid voltage, the tube will operate between points a_4 and b_4. During a part of each negative half-cycle the tube will operate on the negative bend of the curve, thus causing the plate current to be distorted during these intervals. During a part of each positive half-cycle of the input signal, the grid is driven positive. The voltage drop due to the current flowing in the grid circuit reduces the effective grid voltage, thus reducing the maximum positive grid-voltage swing from $e_{c \cdot a4}$ to $e_{c \cdot a}$. This

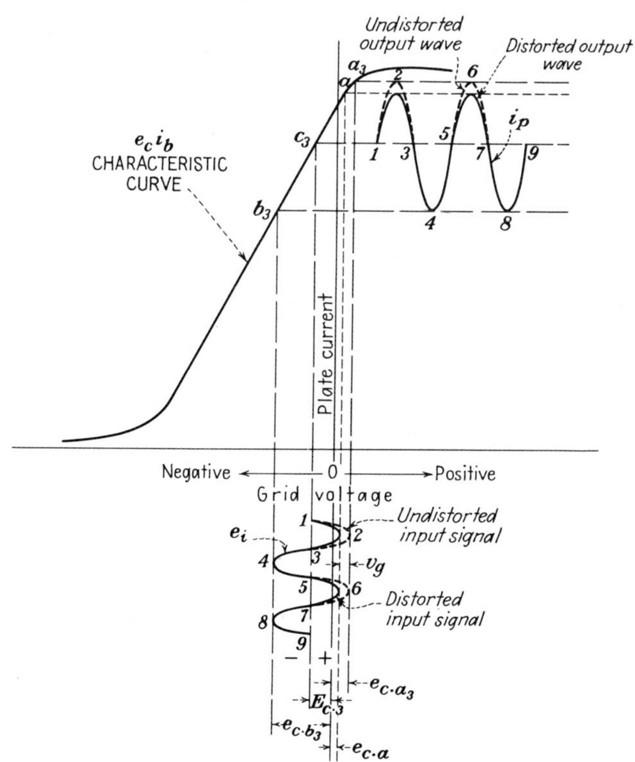

Fig. 8-35 Distortion in a Class A vacuum-tube voltage amplifier when operated with incorrect grid bias so that the grid is driven positive during a part of each cycle.

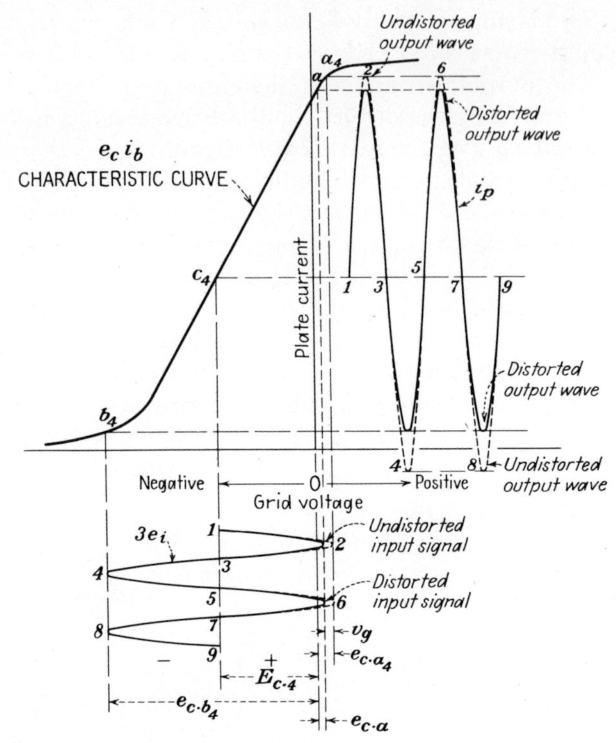

Fig. 8-36 Distortion in a Class A vacuum-tube voltage amplifier caused by applying too high an input signal.

distortion of the input signal causes the tube to operate between c_4 and a on the curve during a part of each positive half-cycle, thereby distorting the plate current. Thus, for this operating condition distortion occurs during both the negative and positive half-cycles.

8-24 Class AB Operation of a Vacuum-tube Amplifier

Class AB amplifiers are generally used in push-pull power-amplifier circuits. This type of operation is accomplished by increasing the grid bias of the tube to an amount greater than is used for Class A operation but less than that required to produce plate-current cutoff with zero signal input (Fig. 8-29*b*). Increasing the grid bias of a tube decreases the plate current at zero signal input. This decrease in the quiescent value of the plate current permits the use of higher screen-grid and plate voltages, and also increases the plate efficiency of the tube. Because of these factors, a greater power output can be obtained by operating two tubes as Class AB push-pull than by using the same tubes operated as Class A push-pull.

Class AB amplifiers may be operated with or without grid current flowing.

As Class AB_1, the grid bias is always greater than the peak value of the input-signal voltage. There will be no grid current as the potential on the grid will not be positive during any part of the input cycle. In Class AB_2, the grid bias is always less than the peak value of the input-signal voltage. There will therefore be some grid current during the portion of the input cycle in which the grid is positive.

The grid current in a Class AB_2 amplifier represents a loss of power. This loss plus the power loss in the input transformer represents the total amount of driving power required by the grid circuit. In order to minimize the amount of distortion set up in the grid circuit, the power of the driving stage is generally made considerably higher than the minimum required amount. The distortion caused by operating a tube on the lower bend of its characteristic curve, which is the region of high distortion, is eliminated by using a push-pull circuit (Chap. 16). It is therefore possible to operate two tubes as Class AB amplifiers without producing distortion by connecting them in a push-pull circuit.

8-25 Class B Operation of a Vacuum-tube Amplifier

Class B operation is generally accomplished by biasing the tube so that with zero signal input the plate current will be at cutoff. Figure 8-29*c* shows the tube being operated with slightly less grid bias than is required for plate-current cutoff in order to avoid operating on the sharp bend at the lower end of the characteristic curve.

In a Class B amplifier the plate-current variations will be greater than with Class A and Class AB amplifiers. The tube will require higher plate voltage and better regulation of its power source. Because the plate current with zero input-signal voltage will be practically zero, Class B amplifiers have a high plate-circuit efficiency and moderate power amplification. The grid is usually driven positive and the power output will be high in proportion to the size of the tube. However, because the grids are driven positive and draw a considerable amount of power when operated as Class B, a greater amount of input-signal power must be supplied from the driver stage to compensate for this loss of power.

To avoid the use of large fixed sources of biasing voltage, a number of tubes have been designed especially for Class B operation. These tubes have a high amplification factor and a very low plate current when the grid voltage is zero; they do not require a bias supply as they can be operated as Class B at zero volts bias.

8-26 Class C Operation of a Vacuum-tube Amplifier

Class C operation is generally accomplished by biasing the tube with a bias voltage greater than that required to produce plate-current cutoff with zero input-signal voltage (Fig. 8-29*d*). Class C amplifiers operate in the same manner as Class B amplifiers except that plate current flows for a shorter period of the input cycle. Consequently, the distortion of the output signal will be greater in the Class C amplifier. The subscripts 1 and 2 may be added, in which case Class C_1 indicates that grid current does not flow during any part of the time of the input cycle, and Class C_2 indicates that grid current does flow during some part of the time of the input cycle.

Class C amplifiers have (1) high power output, (2) plate efficiency in the order of 75 per cent, (3) a low ratio of power amplification, and (4) great distortion. Class C amplifiers are not used with sound-amplifier (low-frequency) applications. They are used in some high-frequency applications where sharply tuned resonant circuits can aid in suppressing harmonics and distortion. Tubes operated in Class C are used in transmitters as (1) r-f amplifiers, (2) oscillators, and (3) modulators.

8-27 Vacuum-tube Biasing Methods

Need for Biasing. It has been established in the preceding articles that the best location of the operating point along the characteristic curve of a tube differs with the mode of operation being used, that is, Class A, Class B, etc. Also, the mode of operation is dependent largely on the grid bias at which the tube is being operated. Thus the method of providing the proper grid bias for the tube is of great importance.

Methods of Biasing a Tube. Establishing the proper grid bias for a tube can be accomplished in several ways, commonly called (1) *fixed bias*, (2) *self-bias*, and (3) *grid-leak bias*.

Fixed Bias. Fixed bias can be provided by (1) a battery or (2) a tap on a d-c voltage-divider network. The simplest method of providing a fixed bias to one or more tubes is by means of one or more voltage sources obtained from a battery; such a battery is often referred to as a *C battery*. The battery E_{cc} in Fig. 8-30 illustrates an application of fixed bias. With a complex circuit, such as a radio receiver, several values of direct voltage (including one or more values of grid bias) can be supplied from a single d-c power source with the aid of a voltage-divider network. The voltage-divider circuit of Fig. 6-18 provides three values of grid bias, namely, -1.5, -2, and -12.5 volts, to meet the bias requirements of three different types of tubes used in the circuit to which the voltage divider supplies power.

Self-bias. The grid bias for a tube can also be obtained by causing the cathode current to flow through a resistor placed in the cathode circuit (Fig. 8-37). Grid bias obtained in this manner is called *self-bias*. It is sometimes also called *cathode bias*, and *automatic grid bias*.

In the circuit of Fig. 8-37, the grid bias for the tube is obtained by placing the resistor R_k between the cathode and ground. Grid bias is obtained by making the d-c component of the cathode current flow through R_k. (In a triode the cathode current is equal to the plate current, while in a pentode the cathode current is equal to the sum of the plate and screen-grid currents.) The voltage drop across R_k is determined by the value of its resistance and the

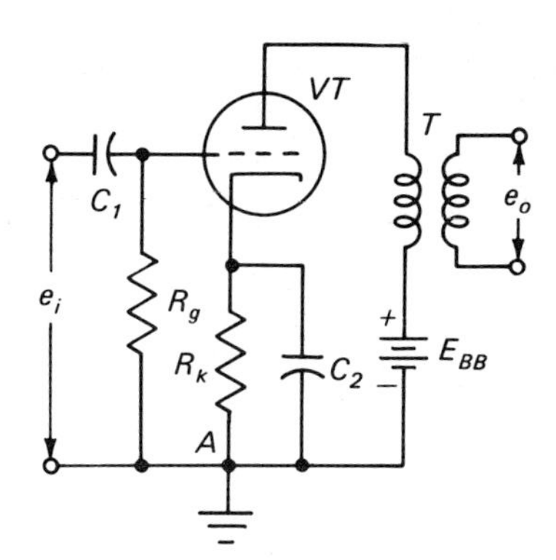

Fig. 8-37 Method of obtaining grid bias by use of a cathode-bias resistor.

amount of current flowing through it. The value of resistance required for R_k can be obtained by

$$R_k = \frac{E_c}{I_k} \tag{8-42}$$

where R_k = value of the cathode bias resistor, ohms
E_c = grid bias, volts
I_k = quiescent value of the cathode current, amp

From Fig. 8-37, it can be seen that point A will be negative with respect to the cathode. The grid, being connected to point A through the grid resistor R_g, will be negative with respect to the cathode by the amount of the voltage drop across R_k.

Example 8-17 The grid bias for the tube of Fig. 8-37 is to be obtained by use of the cathode-bias resistor R_k. The grid bias required is 12.5 volts and the cathode current is 50 ma. (*a*) What value of resistance is required? (*b*) How much power is consumed by the resistor? (*c*) What power rating should the resistor have for a restricted ventilation application (see Art. 6-17).

GIVEN: $E_c = 12.5$ volts $\quad I_k = 50$ ma

FIND: (*a*) R_k $\quad$ (*b*) $P_{R\cdot k}$ $\quad$ (*c*) Power rating

SOLUTION:

(*a*)
$$R_k = \frac{E_c}{I_k} = \frac{12.5}{50 \times 10^{-3}} = 250 \text{ ohms}$$

(*b*)
$$P_{R\cdot k} = I_k^2 R_k = (50 \times 10^{-3})^2 \times 250 = 0.625 \text{ watt}$$

(*c*)
$$\text{Rating} = 2 \times P_{R\cdot k} = 2 \times 0.625 = 1.25 \text{ watts}$$

A-C Bypass Capacitor. The cathode-bias resistor must be shunted by a capacitor in order to bypass the a-c component of the cathode current, which would otherwise cause the voltage across the grid circuit to vary continually. This capacitor is indicated as C_2 in Fig. 8-37.

Example 8-18 (*a*) What amount of bypass capacitance should be used with the cathode-bias resistor of Example 8-17 if the resistor is to offer at least 20 times more impedance to a 500-hertz a-f current than the capacitor? (*b*) What standard rating and type of capacitor is recommended for this application?

GIVEN: $R_k = 250$ ohms $\quad f = 500$ Hz $\quad \dfrac{R_k}{Z_c} = 20$

FIND: (*a*) C $\quad$ (*b*) Rating, type

SOLUTION:

(*a*)
$$X_C \cong Z_c = \frac{R_k}{20} = \frac{250}{20} = 12.5 \text{ ohms}$$

$$C = \frac{159{,}000}{fX_C} = \frac{159{,}000}{500 \times 12.5} = 25.4\ \mu\text{f}$$

(*b*) A 25-μf, 15-volt electrolytic capacitor

Grid-leak Bias. When an amplifier is operated in such a manner that grid current will flow during some part of the input cycle, such as Class A_2, AB_2, B_2, or C_2, grid-bias voltage can be obtained from a resistor connected into the grid circuit as shown in Fig. 8-38. This method of obtaining grid bias is generally used only with high-frequency power amplifiers. When the input signal is positive, grid current will flow from the cathode to the grid, through R_g, and back to the cathode. With the current flowing through the resistor from A toward B, point A will be negative with respect to ground. As a result of the grid-current flow, capacitor C_1 becomes charged because of the voltage developed across resistor R_g. When the input-signal voltage becomes zero, or negative, capacitor C_1 discharges through the path formed by C_1, R_g, and the resistance of the input-signal source, and again point A is negative. Using a sufficiently high value of capacitance at C_1 will result in some charge remaining at the capacitor during the entire time of the input cycle, provided grid current

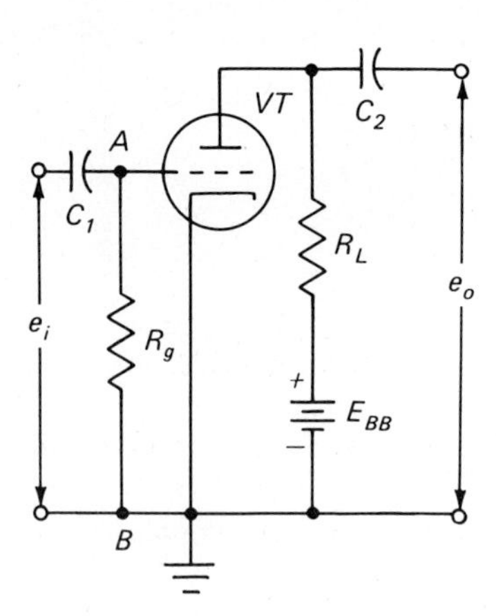

Fig. 8-38 Method of obtaining grid bias by use of a grid circuit resistor in an amplifier that has grid current flowing during some part of the time of the input signal.

flows during some part of each cycle of the input signal. Thus, the average grid potential will be negative.

QUESTIONS

1. Define (*a*) Class A transistor amplifier, (*b*) Class AB transistor amplifier.
2. Define (*a*) Class B transistor amplifier, (*b*) Class C transistor amplifier.
3. Define (*a*) static characteristic curves of a transistor, (*b*) dynamic transfer characteristic curve of a transistor.
4. Describe the ideal operation of a Class A transistor amplifier.
5. Explain how characteristic curves can be used to determine whether an amplifier will produce an accurate replica of the original input signal.
6. (*a*) Define distortion in relation to the operation of an amplifier. (*b*) How is distortion indicated in the graphical analysis of the operation of an amplifier?
7. What two factors largely determine the mode of operation of a transistor amplifier?
8. (*a*) What are the basic operating characterisics of a Class A transistor amplifier? (*b*) How is this mode or operation achieved?
9. (*a*) What are the basic operating characteristics of a Class AB transistor amplifier? (*b*) How is this mode of operation achieved?
10. (*a*) What are the basic operating characteristics of a Class B transistor amplifier? (*b*) How is this mode of operation achieved?
11. (*a*) What are the basic operating characteristics of a Class C transistor amplifier? (*b*) How is this mode of operation achieved?
12. (*a*) What are the two important bias requirements for transistor operation? (*b*) What two types of power sources are commonly used for transistor circuits?
13. Name six methods of providing bias currents for transistors.
14. How does the amount of forward bias on a transistor affect the operation of a transistor amplifier circuit?
15. Explain how forward bias and reverse bias are obtained for the common-emitter transistor circuit of Fig. 8-12*a*.
16. Explain how forward bias and reverse bias are obtained for the common-base transistor circuit of Fig. 8-12*b*.
17. Explain how forward bias and reverse bias are obtained for the common-collector transistor circuit of Fig. 8-12*c*.
18. Explain how forward bias and reverse bias are obtained for the common-emitter transistor circuit of Fig. 8-13*a*.
19. Explain how forward bias and reverse bias are obtained for the common-base transistor circuit of Fig. 8-13*b*.
20. Explain how forward bias and reverse bias are obtained for the common collector transistor circuit of Fig. 8-13*c*.
21. (*a*) What is meant by the beta sensitivity of a transistor circuit? (*b*) Describe the possible causes of variation in the beta value in a transistor circuit. (*c*) What is the undesirable effect in a circuit having poor beta sensitivity?
22. Define (*a*) temperature-current instability, (*b*) thermal runaway, (*c*) stability factor.
23. What is the cause of temperature-current instability?
24. (*a*) Is the stability rated good or poor for the common-emitter transistor circuit of Fig. 8-13*a*? (*b*) What is the reason for this?

25. (*a*) In what manner does the common-emitter transistor circuit of Fig. 8-16 differ from that of Fig. 8-13*a*? (*b*) Compare the bias stability of these two circuits. (*c*) Compare the collector-current stability for these two circuits.
26. (*a*) Is the collector-current stability of the common-base transistor circuit rated good or poor? (*b*) Why?
27. (*a*) What circuit parameters of the common-collector transistor circuit (Fig. 8-13*c*) largely determine its stability factor? (*b*) What mathematical expression represents the stability factor of this circuit?
28. (*a*) Describe what is meant by self-bias. What is its relation to (*b*) negative feedback? (*c*) Degeneration?
29. (*a*) Describe the principle of emitter bias. (*b*) What is a disadvantage of emitter-bias?
30. (*a*) How does the collector-current stability of the common-emitter transistor circuit of Fig. 8-17 compare with that of Fig. 8-13*a* and 8-16? (*b*) What is the reason for this?
31. (*a*) How does the beta sensitivity of the common-emitter circuit of Fig. 8-17 compare with that of Fig. 8-13*a* and 8-16? (*b*) What is the reason for this?
32. (*a*) Describe the principle of collector bias. (*b*) What is a disadvantage of collector bias?
33. (*a*) In what manner does the common-emitter transistor circuit of Fig. 8-18 differ from that of Fig. 8-16? (*b*) How does this change affect the forward bias of the transistor?
34. Describe how the bias currents are obtained for the common-emitter transistor circuit of Fig. 8-19.
35. (*a*) Is the beta sensitivity for the common-emitter transistor circuit of Fig. 8-19 rated good or poor? (*b*) What factors are responsible for this result?
36. (*a*) Is the collector-current stability for the common-emitter transistor circuit of Fig. 8-19 rated good or poor? (*b*) What effects do the values of R_B and R_E have on the stability factor and the collector-current stability of this circuit?
37. (*a*) What is the advantage of using both fixed-bias and self-bias in the common-emitter transistor circuit? (*b*) Explain how the circuit of Fig. 8-21 provides both fixed bias and self-bias.
38. Explain the operation of the common-emitter transistor circuit of Fig. 8-22.
39. Explain the operation of the common-emitter transistor circuit of Fig. 8-23.
40. What is the advantage of using a thermistor in the biasing circuit of a transistor?
41. Explain the operation of the circuit of Fig. 8-24.
42. What is the advantage of using a diode in the biasing circuit of a transistor?
43. Explain the operation of the circuit of Fig. 8-26.
44. Explain the operation of the circuit of Fig. 8-27.
45. What is meant by *tracking* in terms of transistor biasing circuits?
46. Compare the tracking characteristics of the circuits of Figs. 8-19, 8-24, 8-26, and 8-27.
47. Name and define four classifications of vacuum-tube amplifiers based upon their operating characteristics.
48. Compare the meaning of Class A, AB, B, and C vacuum-tube amplifiers with similar classifications of transistor amplifiers.
49. What is the fundamental difference between a Class AB_1 and a Class AB_2 vacuum-tube amplifier?

50. (*a*) Which class of amplifier operation is most commonly used in sound-amplifier applications? (*b*) Why?

51. At what value of grid bias should a tube be operated when being used in a Class A amplifier?

52. What is the maximum value of input-signal voltage that should be applied to a tube in terms of the grid bias and the shape of the characteristic curve?

53. What is meant by distortion in a vacuum-tube amplifier-circuit?

54. What are the causes of distortion of the positive portion of the input signal in a vacuum-tube amplifier?

55. What are the causes of distortion of the negative portion of the input signal in a vacuum-tube amplifier?

56. What is the effect on the fidelity of the output signal of a vacuum-tube amplifier if the input signal drives the grid voltage to a positive value?

57. Explain how an incorrect value of zero-signal bias can cause distortion in a vacuum-tube amplifier.

58. Explain how too large a signal input can cause distortion in a vacuum-tube amplifier.

59. What are the operating characteristics of a Class AB vacuum-tube amplifier?

60. How is Class AB operation of a vacuum-tube amplifier accomplished?

61. What are the operating characteristics of a Class B vacuum-tube amplifier?

62. How is Class B operation of a vacuum-tube amplifier accomplished?

63. What are the operating characteristics of a Class C vacuum-tube amplifier?

64. How is Class C operation of a vacuum-tube amplifier accomplished?

65. (*a*) What is meant by biasing a vacuum tube? (*b*) What is the purpose of biasing a vacuum tube?

66. What is meant by fixed bias for a vacuum tube?

67. Describe two methods of obtaining fixed bias for a vacuum tube.

68. What is meant by self-bias for a vacuum tube?

69. How is self-bias generally obtained for a vacuum tube?

70. Explain the functions of R_k and C_2 of Fig. 8-37 in establishing self-bias.

71. What is meant by grid-leak bias for a vacuum tube?

72. Explain how grid bias is obtained in the vacuum-tube circuit of Fig. 8-38.

PROBLEMS

1. Plot a dynamic transfer characteristic curve for the transistor represented by the curves of Fig. 8-1 and the circuit of Fig. 8-2 except that V_{CC} is 4.5 volts.

2. Plot a dynamic transfer characteristic curve for the transistor represented by the curves of Fig. 8-1 and the circuit of Fig. 8-2 except that R_L is 1,500 ohms.

3. If the transistor circuit of Prob. 1 has a 20-μa peak-to-peak input signal with a quiescent value of 20 μa, what are the values of collector current when the input signal is (*a*) zero? (*b*) +10 μa? (*c*) −10 μa? (*d*) What is the maximum positive output-signal current? (*e*) What is the maximum negative output-signal current? (*f*) Is the output signal distorted? (*g*) What is the mode of operation for this circuit?

4. If the transistor circuit of Prob. 2 has a 20-μa peak-to-peak input signal with a quiescent value of 30 μa, what are the values of collector current when the input signal is (*a*) zero? (*b*) +10 μa? (*c*) −10 μa? (*d*) What is the maximum positive output-signal current? (*e*) What is the maximum negative output-signal current? (*f*) Is the output signal distorted? (*g*) What is the mode of operation for this circuit?

5. If the transistor circuit of Prob. 1 has a 40-μa peak-to-peak input signal with a quiescent value of 10 μa, what are the values of collector current when the input signal is (*a*) zero? (*b*) +20 μa? (*c*) −20 μa? (*d*) What is the maximum positive output-signal current? (*e*) What is the maximum negative output signal current? (*f*) Is the output signal distorted? (*g*) What is the mode of operation for this circuit?
6. If the transistor circuit of Prob. 2 has a 60-μa peak-to-peak input signal with a quiescent value of 20 μa, what are the values of collector current when the input signal is (*a*) zero, (*b*) +30 μa, (*c*) −30 μa? (*d*) What is the maximum positive output-signal current? (*e*) What is the maximum negative output-signal current? (*f*) Is the output signal distorted? (*g*) What is the mode of operation for this circuit?
7. If the transistor circuit of Prob. 1 has a 50-μa peak-to-peak input signal with a quiescent value of zero, what are the values of collector current when the input signal is (*a*) zero? (*b*) +25 μa? (*c*) −25 μa? (*d*) What is the maximum positive output-signal current? (*e*) What is the maximum negative output-signal current? (*f*) Is the output signal distorted? (*g*) What is the mode of operation for this amplifier?
8. If the transistor circuit of Prob. 2 has a 60-μa peak-to-peak input signal with a quiescent value of −10 μa, what are the values of collector current when the input signal is (*a*) zero? (*b*) +30 μa? (*c*) −30 μa? (*d*) What is the maximum positive output-signal current? (*e*) What is the maximum negative output-signal current? (*f*) Is the output signal distorted? (*g*) What is the mode of operation for this amplifier?
9. What value of base resistance is required in the circuit of Fig. 8-2 if it is desired to have a 20-μa peak-to-peak signal operate around a quiescent base current of 30 μa? The value of beta is 150.
10. If the base resistance in the circuit of Fig. 8-2 is 60,000 ohms and beta is 100, find (*a*) I_B, (*b*) I_C, and (*c*) V_{CE}.
11. If the parameters in the common-emitter circuit of Fig. 8-12*a* are $V_{CC} = 24$ volts, $V_{BB} = 12$ volts, $\beta = 100$, $R_L = 10{,}000$ ohms, and $R_B = 1$ megohm, find (*a*) I_B, (*b*) I_C, (*c*) I_E, (*d*) V_{CE}.
12. What value of load resistance would be required in the circuit of Prob. 11 in order that $V_{CE} = 9$ volts?
13. What value of base resistance is required for the circuit of Prob. 11 in order to produce a collector current of 3 ma?
14. If it is desired that the circuit of Prob. 11 should have a collector current of 2.5 ma and a collector-emitter voltage of 15 volts, what value of resistance is required for (*a*) the base resistor? (*b*) The load resistor?
15. If the parameters in the common-base circuit of Fig. 8-12*b* are $V_{CC} = 30$ volts, $V_{BB} = 15$ volts, $\alpha = 0.96$, $R_E = 7{,}500$ ohms, and $R_L = 10{,}000$ ohms, find (*a*) I_E. (*b*) I_C, (*c*) I_B, (*d*) V_{CE}.
16. If the parameters in the common-base circuit of Fig. 8-12*b* are $V_{CC} = 24$ volts, $V_{BB} = 18$ volts, $R_E = 4{,}500$ ohms, and $R_L = 4{,}000$ ohms, find (*a*) I_E, (*b*) I_C, (*c*) V_{CE}. (*Note:* Assume $\alpha \cong 1$.)
17. What value of load resistance is required for the circuit of Prob. 15 in order to produce a collector-emitter voltage of 15 volts?

18. What values of R_E and R_L are required for the circuit of Prob. 16 in order to produce a collector-emitter voltage of 18 volts if the collector current is to be 3 ma?

19. If the parameters in the common-collector circuit of Fig. 8-12*c* are $V_{EE} = 24$ volts, $V_{BB} = 9$ volts, $\beta = 80$, $R_L = 20{,}000$ ohms, and $R_B = 100{,}000$ ohms, find (*a*) I_E, (*b*) I_C, (*c*) I_B, (*d*) V_{CE}.

20. If the parameters in the common-collector circuit of Fig. 8-12*c* are $V_{EE} = 30$ volts, $V_{BB} = 20$ volts, $\beta = 100$, $R_L = 2{,}000$ ohms, and $R_B = 50{,}000$ ohms, find (*a*) I_E, (*b*) I_C, (*c*) I_B, (*d*) V_{CE}.

21. If the circuit of Prob. 11 is converted to use a single 24-volt power source as in Fig. 8-13*a* and all other circuit parameters remain unchanged, find (*a*) I_B, (*b*) I_C, (*c*) V_{CE}.

22. If the circuit of Prob. 11 is converted to use a single 24-volt power source as in Fig. 8-13*a*, what new value of R_L is required in order that V_{CE} will be 12 volts?

23. If the circuit of Prob. 11 is converted to use a single 24-volt power source as in Fig. 8-13*a*, what new value of R_B is required in order that V_{CE} will be 12 volts?

24. Determine the effect that a change in the voltage of the power source in the circuit of Prob. 21 will have on the collector-emitter voltage by substituting a power supply having a voltage of (*a*) 12 volts, (*b*) 48 volts.

25. If the parameters in the common-base circuit of Fig. 8-13*b* are $V_{CC} = 20$ volts, $R_E = 10{,}000$ ohms, $R_L = 5{,}000$ ohms, $R_B = 10{,}000$ ohms, and $R_S = 10{,}000$ ohms, find (*a*) V_B, (*b*) I_E, (*c*) I_C, (*d*) V_{CE}.

26. If the parameters in the common-base circuit of Fig. 8-13*b* are $V_{CC} = 9$ volts, $R_E = 2{,}000$ ohms, $R_L = 1{,}000$ ohms, $R_B = 2{,}500$ ohms, and $R_S = 5{,}000$ ohms, find (*a*) V_B, (*b*) I_E, (*c*) I_C, (*d*) V_{CE}.

27. For the circuit of Prob. 25, what value of R_L will produce 7 volts across the collector-emitter terminals?

28. What will be the voltage across the collector-emitter terminals in the circuit of Prob. 26 if R_E is changed to 1,000 ohms, R_L is changed to 2,000 ohms, and all other values remain unchanged?

29. If the parameters in the common-collector circuit of Fig. 8-13*c* are $V_{EE} = 20$ volts, $\beta = 80$, $R_L = 10{,}000$ ohms, and $R_B = 480{,}000$ ohms, find (*a*) I_E, (*b*) I_C, (*c*) I_B, (*d*) V_{CE}.

30. If the parameters in the common-collector circuit of Fig. 8-13*c* are $V_{EE} = 30$ volts, $\beta = 50$, $R_L = 5{,}000$ ohms, and $R_B = 50{,}000$ ohms, find (*a*) I_E, (*b*) I_C, (*c*) I_B, (*d*) V_{CE}.

31. If the transistor used in Prob. 29 has a beta of 40, what is the value of V_{CE} in that circuit?

32. If the load resistor in the circuit of Prob. 30 is changed to 2,000 ohms and all other values remain unchanged, what is the new value of V_{CE}?

33. What is the beta sensitivity of the common-emitter circuit of Prob. 21?

34. What is the beta sensitivity of the common-collector circuit of Prob. 29?

35. What is the total collector current in a common-base transistor circuit if $I_E = 15$ ma, $I_{CBO} = 50$ μa, and $\alpha = 0.96$?

36. What is the total collector current in a common-base transistor circuit if $I_E = 100$ ma, $I_{CO} = 100$ μa, and $\alpha = 0.95$?

37. What is the total collector current in a common-emitter transistor circuit if $I_B = 4$ ma, $I_{CO} = 15$ μa, and $\beta = 80$?

38. What is the total collector current in a common-emitter transistor circuit if $I_B = 1$ ma, $I_{CBO} = 25\ \mu$a, and $\beta = 100$?

39. From the curve of Fig. 8-14, determine the per cent increase in the saturation current when the temperature rises from 60 to 90°C.

40. From the curve of Fig. 8-14, determine the per cent increase in the saturation current when the temperature rises from 50 to 100°C.

41. From the curves of Fig. 8-15, determine the approximate per cent of increase in the collector current for the condition when $V_{EB} = 80$ mv and the temperature increase from 18 to 50°C?

42. From the curves of Fig. 8-15, for the 50°C temperature, what is the approximate per cent increase in collector current when the emitter-base voltage increases from 60 to 80 mv?

43. What is the current stability factor of a transistor if the collector cutoff current increases by 0.8 ma while the emitter current increases by 1 ma?

44. What is the current stability factor of a transistor if the collector cutoff current increases by 500 μa while the emitter current increases by 4 ma?

45. What is the stability factor of the circuit of Prob. 25? (*Note:* Assume $\alpha = 1$.)

46. What is the stability factor of the circuit of Prob. 29? (*Note:* Assume $\alpha = 1$.)

47. What is the stability factor of the circuit of Prob. 30?

48. What is the stability factor of the circuit of Prob. 15?

49. What is the stability factor of the common-emitter circuit of Fig. 8-13*a* if $\alpha = 0.98$?

50. What is the stability factor of the common-emitter circuit of Fig. 8-13*a* if $\alpha = 0.95$?

51. If the parameters of the common-emitter circuit of Fig. 8-16 are $V_{CC} = 9$ volts, $\beta = 60$, $R_B = 300{,}000$ ohms, and $R_L = 2{,}500$ ohms, find (*a*) I_B, (*b*) I_C, (*c*) V_{CE}.

52. If the parameters of the common-emitter circuit of Fig. 8-16 are $V_{CC} = 28$ volts, $\beta = 75$, $R_B = 1{,}000{,}000$ ohms, and $R_L = 7{,}500$ ohms, find (*a*) I_B, (*b*) I_C, (*c*) V_{CE}.

53. If it is desired that the circuit of Prob. 51 should have a collector current of 2.4 ma and a collector-emitter voltage of 4.2 volts, what values will be required for R_B and R_L?

54. If a replacement transistor for the circuit of Prob. 52 has a beta of 100, what effect will it have on the collector-emitter voltage?

55. If the parameters of the common-emitter circuit of Fig. 8-17*a* are $V_{EE} = 10$ volts, $V_{CC} = 20$ volts, $V_{BE} = 0.3$ volt, $\beta = 40$, $R_B = 600$ ohms, $R_E = 470$ ohms, and $R_L = 600$ ohms, find (*a*) I_E, (*b*) I_B, (*c*) V_{CE}, (*d*) *SF*, (*e*) K_β.

56. If the parameters of the common-emitter circuit of Fig. 8-17*a* are $V_{EE} = 20$ volts, $V_{CC} = 20$ volts, $V_{BE} = 0.7$ volt, $\beta = 80$, $R_B = 20{,}000$ ohms, $R_E = 30{,}000$ ohms, and $R_L = 10{,}000$ ohms, find (*a*) I_E, (*b*) I_B, (*c*) V_{CE}, (*d*) *SF*, (*e*) K_β.

57. Repeat Prob. 55 disregarding V_{BE} and R_B/β when finding the emitter current.

58. Repeat Prob. 56 disregarding V_{BE} and R_B/β when finding the emitter current.

59. If the parameters of the common-emitter circuit of Fig. 8-17*b* are $R_B = 5$ ohms, $R_E = 500$ ohms, and $\alpha = 0.95$, find (*a*) the stability factor, (*b*) the beta sensitivity.

60. What are (*a*) the stability factor and (*b*) the beta sensitivity of the circuit of Prob. 59 if the external base resistance is increased to 50 ohms?

61. If the parameters in the common-emitter circuit of Fig. 8-18 are $V_{CC} = 24$ volts, $V_{BE} = 0.7$ volt, $\beta = 50$, $R_L = 2{,}000$ ohms, and $R_B = 100{,}000$ ohms, find (*a*) I_C, (*b*) I_B, (*c*) V_{CE}, (*d*) K_β, (*e*) *SF*.

62. What are (*a*) the collector current and (*b*) the collector-emitter voltage in the circuit of Prob. 61 if a new transistor having a beta of 75 is substituted for the original transistor?

63. If the parameters of the common-emitter circuit of Fig. 8-19 are $V_{CC} = 24$ volts, $\beta = 66.7$, $R_B = 5{,}000$ ohms, $R_S = 10{,}000$ ohms, $R_E = 5{,}000$ ohms, and $R_L = 5{,}000$ ohms, find (*a*) I_C, (*b*) V_{CE}, (*c*) K_β, (*d*) *SF*.

64. If the parameters of the common-emitter circuit of Fig. 8-19 are $V_{CC} = 10$ volts, $\beta = 40$, $R_B = 2{,}000$ ohms, $R_S = 2{,}000$ ohms, $R_E = 2{,}000$ ohms, and $R_L = 1{,}000$ ohms, find (*a*) I_C, (*b*) V_{CE}, (*c*) K_β, (*d*) *SF*.

65. The curves of Fig. 8-20 are shown intersecting at approximately 25°C and with a common value of collector current of 1.8 ma. At a temperature of 60°C, what is the collector current for the condition represented by curve: (*a*) *B*? (*b*) *C*? (*c*) *D*?

66. For the conditions of Prob. 65, what is the per cent increase in collector current when the temperature changes from 25 to 60°C for the conditions represented by curve: (*a*) *B*? (*b*) *C*? (*c*) *D*?

67. If the parameters in the circuit of Fig. 8-21 are $V_{CC} = 20$ volts, $R_B = 5{,}000$ ohms, $R_S = 10{,}000$ ohms, and $R_L = 1{,}000$ ohms, what is the base-emitter voltage when the collector current is (*a*) 5 ma? (*b*) 8 ma? (*c*) 2 ma?

68. If the parameters in the circuit of Fig. 8-22 are $V_{CC} = 24$ volts, $R_B = 10{,}000$ ohms, $R_1 = 2{,}000$ ohms, $R_2 = 18{,}000$ ohms, and $R_L = 2{,}000$ ohms, what is the base-emitter voltage when the collector current is (*a*) 4.5 ma? (*b*) 3 ma? (*c*) 1.5 ma?

69. If the circuit parameters of Fig. 8-23 are $V_{CC} = 12$ volts, $\beta = 50$, $R_B = 100{,}000$ ohms, $R_E = 2{,}000$ ohms, and $R_L = 2{,}000$ ohms, find (*a*) I_C, (*b*) V_{CE}, (*c*) K_β, (*d*) *SF*.

70. If the circuit parameters of Fig. 8-23 are $V_{CC} = 18$ volts, $\beta = 75$, $R_B = 150{,}000$ ohms, $R_E = 10{,}000$ ohms, and $R_L = 6{,}000$ ohms, find (*a*) I_C, (*b*) V_{CE}, (*c*) K_β, (*d*) *SF*.

71. What is the quiescent value of the output voltage in the circuit of Fig. 8-30 if $E_{BB} = 90$ volts, $I_p = 10$ ma, and $Z_L = 5{,}000$ ohms?

72. If the quiescent plate current in the circuit of Fig. 8-30 is 8 ma and E_{BB} is 180 volts, what value of resistance is required to produce a quiescent value for e_o of 100 volts?

73. A triode whose characteristic curves are shown in Fig. 5-3 is being operated as a Class A amplifier with 200 volts applied to its plate. What is the maximum amount of signal voltage that can be applied without producing distortion when the tube is operated with a grid bias of (*a*) 7 volts? (*b*) 6 volts? (*c*) 5 volts? (*d*) 4 volts? (*Note:* Assume that the straight portion of the curve ends at $E_c = -9$ volts.)

74. What is the maximum signal that can be applied to the tube of Prob. 73 without causing distortion if the tube is operated with 120 volts applied to its plate, and the grid bias is (*a*) 4 volts? (*b*) 3 volts? (*c*) 2 volts? (*d*) 1 volt? (*Note:* Assume that the straight portion of the curve ends at $E_c = -5$ volts.)

75. If in the vacuum-tube amplifier circuit of Fig. 8-34 R_g is 1 megohm and poor choice of operating conditions causes a peak grid current of 0.5 μa, what is (*a*) the amount of voltage introduced into the grid circuit by this grid current? (*b*) The percent distortion to the positive half of a 3-volt peak-to-peak input signal?

76. If in the vacuum-tube amplifier circuit of Fig. 8-34 R_g is 1 megohm and poor choice of operating conditions causes a peak grid current of 1 μa, what is (*a*) the amount of voltage introduced into the grid circuit by this grid current? (*b*) The per cent distortion to the positive half of a 3-volt peak-to-peak input signal? (*c*) The per cent distortion to the negative half of the input signal?

77. (*a*) What is the grid bias on the tube in the circuit of Fig. 8-37 when the plate current is 1.25 ma and the cathode bias resistor is 1,600 ohms? (*b*) How much power is consumed by the resistor?

78. In order to operate a certain pentode vacuum tube as a Class AB_1 amplifier a grid bias of 15 volts is required. (*a*) What value of cathode resistor is needed if the tube operates with a plate current of 70 ma and a screen grid current of 5 ma? (*b*) How much power is consumed by the resistor?

79. (*a*) What amount of bypass capacitance should be used with the cathode-bias resistor of Prob. 77 if the resistor is to offer at least 20 times more impedance to a 500-hertz a-f current than the capacitor? (*b*) What standard rating and type of capacitor are recommended for this application?

80. (*a*) What amount of bypass capacitance should be used with the cathode-bias resistor of Prob. 78 if the resistor is to offer at least 20 times more impedance to a 500-hertz a-f current than the capacitor? (*b*) What standard rating and type of capacitor are recommended for this application?

Chapter 9

Basic Amplifier Circuits

The values of voltage, current, and power available for operating electronic equipment are frequently very low and consequently amplification of these quantities is a very important function in most electronic equipment. Amplification is the process whereby small amounts of voltage, current, and/or power at the input side of a circuit or system are increased so that larger amounts of voltage, current, and/or power are available at the output side of the circuit or system. The most common types of amplifier circuits accomplish their function by use of an active circuit element such as a transistor or a vacuum tube.

Amplifiers are widely used for such a great variety of applications that for the purpose of an orderly presentation it becomes advantageous to classify them into several categories. Amplifiers may be classified according to (1) use, that is, voltage, current, and power amplification; (2) circuit configuration; (3) biasing voltages which determine the operating characteristics; (4) operating frequency; (5) frequency response; and (6) specialized applications. This chapter will be concerned chiefly with (1) use of amplifiers to achieve gains in voltage, current, and/or power; and (2) the basic circuit configurations and their individual circuit characteristics.

9-1 Transistor versus Vacuum-tube Amplifiers

A transistor is a current-operated device in which the current in the input circuit controls the current in the output circuit. Because the transistor does not provide isolation between its input and output circuits, the output-circuit parameters affect the input-circuit parameters, and vice versa. In a transistor amplifier, a varying signal voltage applied to the input of the amplifier circuit causes the current in the input circuit of the transistor to also vary, which in turn controls the current in the output circuit of the transistor. Transistor amplifiers generally have a low-to-medium input impedance and a moderate-to-high output impedance. The current flowing through these impedances determines the voltage and power gains of a transistor amplifier circuit. The gain of a transistor amplifier is usually specified in decibels of power.

A vacuum tube is a voltage-operated device in which the input voltage controls the output voltage, current, and/or power. In a vacuum-tube amplifier, a varying signal voltage applied to the grid of the tube controls the current in the plate circuit which in turn controls the voltage (and power) in the output circuit of the amplifier. Vacuum-tube amplifiers generally have high input and high output impedances. The gain of a vacuum-tube amplifier is usually specified as a multiplication of voltage.

There are three types of transistor amplifier circuits: (1) common or grounded emitter, (2) common or grounded base, and (3) common or grounded collector. Also, there are three types of vacuum-tube amplifier circuits: (1) common or grounded cathode, (2) common or grounded grid, and (3) common or grounded plate. The terms common and grounded are both in general use, although a common element does not necessarily have to be connected to ground.

9-2 Transistor Circuit Parameters

In the analysis of transistor circuits, the principal transistor parameters and symbols are:

α = common-base small-signal a-c current gain (also called h_{fb})
$\alpha_{\text{d-c}}$ = common-base static or d-c current gain (also called h_{FB})
β = common-emitter small-signal a-c current gain (also called h_{fe})
$\beta_{\text{d-c}}$ = common-emitter static or d-c current gain (also called h_{FE})
C_{be} = base-emitter interelement capacitance
C_{bc} = base-collector interelement capacitance
C_{ce} = collector-emitter interelement capacitance
G_{pe} = common-emitter small-signal a-c power gain
h_f = small-signal forward-current transfer ratio
h_{FE} = common-emitter static or d-c forward-current transfer ratio
h_{fe} = common-emitter small-signal or a-c forward-current transfer ratio
h_{ie} = common-emitter small-signal or a-c input resistance
h_{oe} = common-emitter small-signal or a-c output conductance
h_{re} = common-emitter small-signal reverse-voltage transfer ratio
I_B = steady or d-c component of the base current
I_b or i_b = varying or a-c component of the base current
I_C = steady or d-c component of the collector current
I_c or i_c = varying or a-c component of the collector current
I_{CBO} = reverse-bias current flow between collector and base, also called *saturation* or *leakage current*
I_{CO} = same as I_{CBO}
I_E = steady or d-c component of the emitter current
I_e or i_e = varying or a-c component of the emitter current
P_b = average power dissipation of the base
P_c = average power dissipation of the collector
P_e = average power dissipation of the emitter
P_g = small-signal power gain (also A_p, G, or G_p)
r_e = emitter a-c resistance; also emitter-base junction resistance
V_{BE} = steady or d-c component of the base-emitter voltage

v_{be} = varying or a-c component of the base-emitter voltage
V_{CE} = steady or d-c component of the collector-emitter voltage
v_{ce} = varying or a-c component of the collector-emitter voltage

The external circuit into which the transistor operates may contain the following parameters and symbols:

A_i = current gain, expressed as a dimensionless ratio
A_p = power gain, expressed as a dimensionless ratio
A_v = voltage gain, expressed as a dimensionless ratio
G_p = power gain, expressed in decibels
R_B = base resistor
R_C = collector resistor
R_E = emitter resistor
R_G = generator or signal source resistance
r_i = a-c input resistance (seen by the signal source)
r_i' = a-c input resistance (looking into the base)
R_L = load resistor
R_o = output-circuit resistor
r_o = a-c output resistance; also, a-c load resistance
r_o' = a-c output resistance looking into the collector
SF = temperature-current stability factor
V = d-c voltages
v = a-c voltages
V_{BB} = base supply voltage
V_{CC} = collector supply voltage
V_{EE} = emitter supply voltage
V_G = signal or source generator voltage
v_i = input-signal voltage (also called v_g, v_{in}, v_s)
v_o = output-signal voltage
G_s = signal or source generator
k_β = beta sensitivity

9-3 Vacuum-tube Circuit Parameters

In the analysis of vacuum-tube circuits, the principal vacuum-tube parameters and symbols are:

C_{gk} = grid-to-cathode capacitance
C_{gp} = grid-to-plate capacitance
C_{pk} = plate-to-cathode capacitance
E_C = cathode voltage
E_G = steady or d-c component of the grid voltage
E_g or e_g = varying or a-c component of the grid voltage
E_B = steady or d-c component of the plate voltage

E_p or e_p = varying or a-c component of the plate voltage
g_m = transconductance of the tube
I_P = steady or d-c component of the plate current
I_p or i_p = varying or a-c component of the plate current
r_i = input resistance of the tube in the common-grid connection
r_p = plate resistance of the tube; also, a-c plate resistance
μ = amplification factor of the tube

The external circuit into which the vacuum tube operates may contain the following parameters and symbols:

A_i = current amplification, a dimensionless ratio
A_p = power amplification, a dimensionless ratio
A_v = voltage amplification, a dimensionless ratio
C_i = total effective input capacitance
E_{BB} = plate supply voltage
E_{CC} = grid supply voltage
e_i = input-signal voltage (also called e_g, e_{in}, e_s)
e_o = output-signal voltage
G_p = power gain, decibels
P_o = power output
R_g = grid-circuit resistor
R_G = generator or signal source resistance
R_k = cathode-circuit resistor
R_L = load resistor
R_o = output-circuit resistor
VA = same as A_v
Z_L = load impedance

9-4 Amplifier Configurations

Transistor Amplifier-circuit Configurations. The basic transistor amplifier circuits are (1) the common-emitter circuit (Fig. 9-1*a*), (2) the common-base circuit (Fig. 9-2*a*), and (3) the common-collector circuit (also called an *emitter follower*) (Fig. 9-3*a*).

Vacuum-tube Amplifier-circuit Configurations. The basic vacuum-tube amplifier circuits are (1) the common-cathode circuit (Fig. 9-1*b*), (2) the common-grid circuit (Fig. 9-2*b*), and (3) the common-plate circuit (also called a *cathode follower*) (Fig. 9-3*b*).

Comparison of Transistor and Vacuum-tube Amplifier Circuits. Both transistors and vacuum tubes are used as amplifiers of voltage, current, and power. The operation of transistor amplifier circuits and vacuum-tube amplifier circuits is similar in some respects and dissimilar in others. A transistor cannot be substituted directly into a vacuum-tube amplifier circuit and vice versa.

The transistor may be considered to be similar to a vacuum tube in the

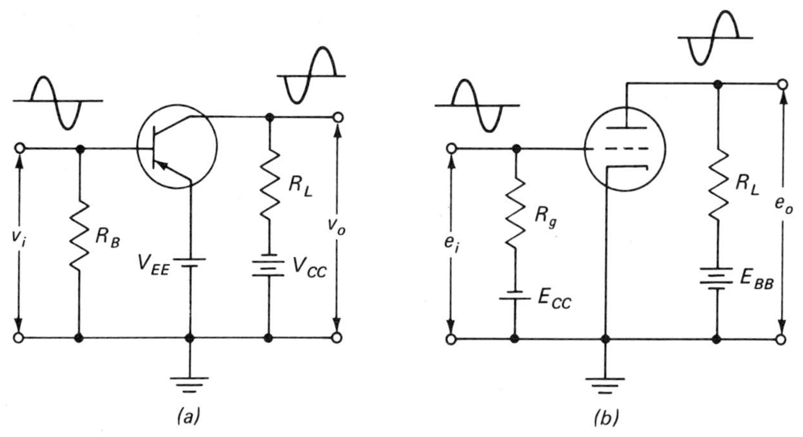

Fig. 9-1 Basic amplifier circuits. (a) Common-emitter transistor amplifier using a PNP transistor. (b) Common-cathode vacuum-tube amplifier using a triode.

following respects: (1) the emitter and the cathode each serve as the source of electron flow; (2) the base and the control grid each serve to control the electron flow through the unit; (3) the collector and the plate each are normally a part of the output circuit. These similarities can be observed in Figs. 9-1, 9-2, and 9-3 which show also a similarity in certain pairs of configurations, namely (1) the common-emitter transistor amplifier and the common-cathode vacuum-tube amplifier (Fig. 9-1), (2) the common-base transistor amplifier and the common-grid vacuum-tube amplifier (Fig. 9-2), and (3) the common-collector transistor amplifier and the common-plate vacuum-tube amplifier (Fig. 9-3).

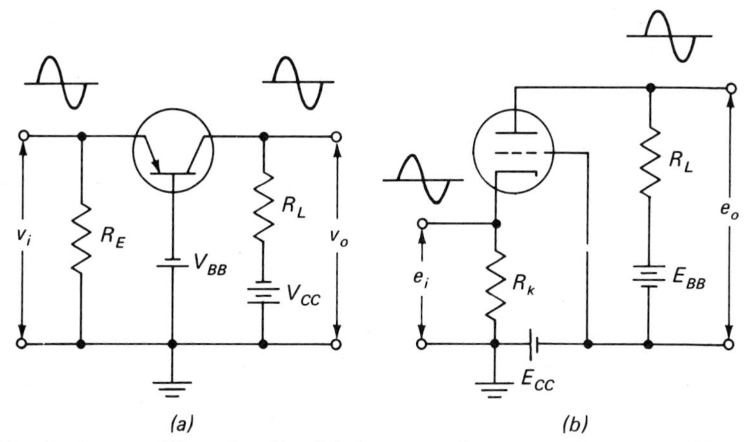

Fig. 9-2 Basic amplifier circuits. (a) Common-base transistor amplifier using a PNP transistor. (b) Common-grid vacuum-tube amplifier using a triode.

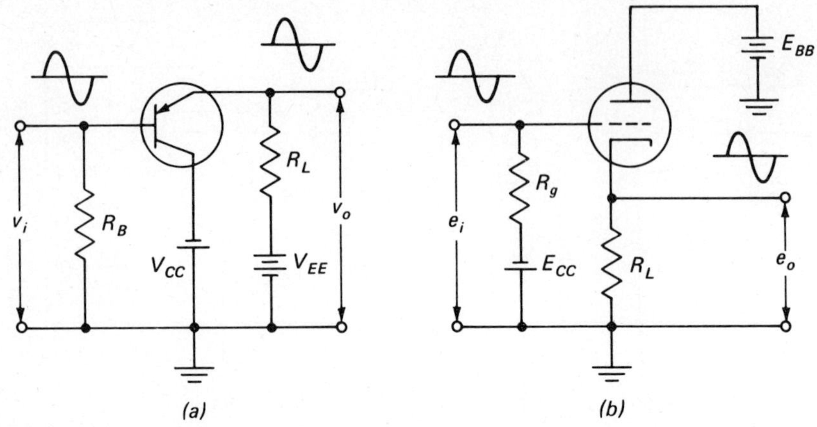

Fig. 9-3 Basic amplifier circuits. (a) Common-collector transistor amplifier using a PNP transistor. (b) Common-plate vacuum-tube amplifier using a triode.

9-5 The Common-emitter Transistor Amplifier

Advantages. Amplifiers employing transistors most frequently use the common-emitter configuration because (1) the common-emitter amplifier produces large gains in current, voltage, and power; (2) the current gain of a common-base amplifier is always less than 1; (3) the voltage gain of the common-collector amplifier is always less than 1; and (4) the input and output impedance characteristics are suitable for many applications.

Circuit Operation. In the common-emitter amplifier (Fig. 9-4*a*) the base is the driven element and the input signal is injected into the base-to-emitter circuit; the output signal is taken from the collector-to-emitter circuit; the

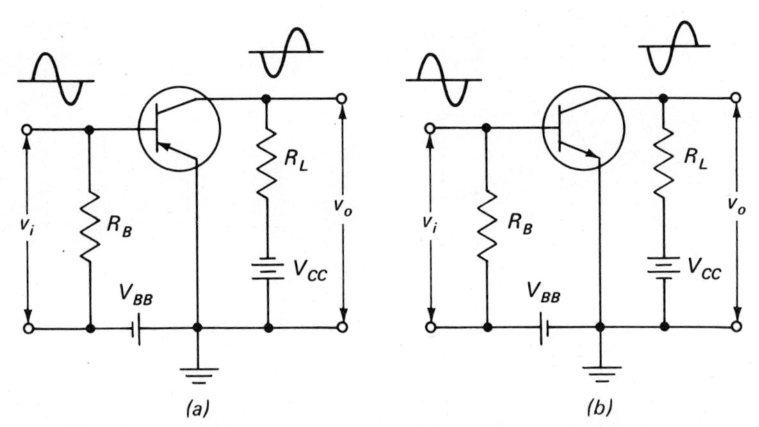

Fig. 9-4 Common-emitter transistor amplifier. (a) Basic circuit using a PNP transistor. (b) Basic circuit using an NPN transistor.

emitter is common to both the input and output circuits. The power source V_{BB} provides a forward bias to the base-emitter circuit and the power source V_{CC} provides a reverse bias to the collector-emitter circuit; for practical purposes, the resistance of the power sources may be considered to be zero.

With zero input signal, power sources V_{BB} and V_{CC}, together with resistors R_B and R_L, establish the quiescent values of the base current I_B and the collector current I_C. When the input signal is positive-going, that is, increasing in the positive direction, its polarity is in opposition to that of V_{BB}, thereby causing (1) a decrease in the forward bias at the base-emitter junction, (2) a smal decrease in the base current, (3) a larger decrease in the emitter and collector currents, (4) a decrease in the voltage drop at R_L, and (5) an increase in the output voltage v_o. Because of the polarity of V_{CC}, an increase in the output voltage establishes a negative-going output signal v_o, and consequently a phase reversal takes place between the input and output signals as is indicated on Fig. 9-4*a*. A similar step-by-step analysis will show that a negative-going input signal will result in a positive-going output signal.

Characteristics. The input resistance of the common-emitter transistor amplifier is relatively low, usually in the order of 1,000 to 2,000 ohms. The output resistance of the amplifier is relatively high, usually in the order of 50,000 ohms. The input and output signals are 180 degrees out of phase with each other.

The characteristic curves of Figs. 4-10 and 8-1 show that a small change in base current produces a relatively large change in collector current; β values of typical transistors range in the order of 20 to 100. This large change in collector current, plus the high output resistance, produces a high voltage gain, in the order of 1,500. As this circuit produces substantial gains in both current and voltage, it also produces high power gains; typical gains may be in the order of 10,000 times, or 40 db.

A common-emitter amplifier circuit using an NPN transistor is shown in Fig. 9-4*b*. The operation and the characteristics of this circuit are the same as for the circuit using the PNP transistor. It should be noted, however, that the polarities of the biasing voltages are opposite to those with the PNP transistor.

9-6 The Common-base Transistor Amplifier

Circuit Operation. In the common-base amplifier (Fig. 9-5) the input signal is injected into the emitter-base circuit and the output signal is taken from the collector-base circuit, making the base the common element to both the input and output circuits.

In the PNP common-base amplifier circuit of Fig. 9-5a, the emitter-base junction is forward-biased by the power source V_{BB} and the collector-base junction is reverse-biased by the power source V_{CC}. With zero input signal, power sources V_{BB} and V_{CC}, together with resistors R_E and R_L, establish the quiescent values of the base current I_B and the collector current I_C. When the input signal is positive-going: (1) the forward bias increases, (2) the emitter and collector currents increase, (3) the voltage drop at R_L increases, and (4) the

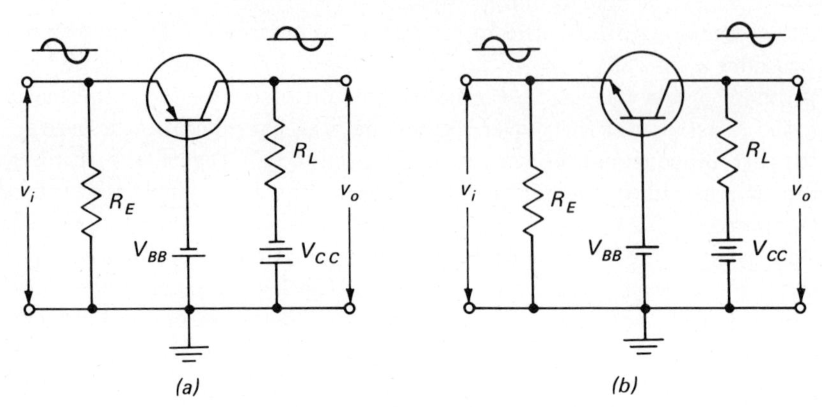

Fig. 9-5 Common-base transistor amplifier. (a) Basic circuit using a PNP transistor. (b) Basic circuit using an NPN transistor.

output voltage decreases. However, because the output signal is negative with respect to ground, the decrease in the output voltage caused by the positive-going input signal makes the output less negative, and in effect produces a positive-going output signal. Consequently, there is no phase reversal between the input and output signals. A similar step-by-step analysis will show that a negative-going input signal will result in a negative-going output signal.

Characteristics. The common-base transistor amplifier circuit has (1) a low input resistance, (2) a high output resistance, (3) a current gain of less than 1, (4) a maximum voltage gain in the order of 1,500, and (5) a power gain in the order of 20 to 30 db. The input and output signals are in phase, as no phase reversal takes place in this type of amplifier circuit. The stability of collector current with temperature change for the common-base amplifier is very good. An important use of the common-base amplifier is in matching a low-impedance circuit to a high-impedance circuit.

A common-base amplifier circuit using an NPN transistor is shown in Fig. 9-5*b*. The operation and the characteristics of this circuit are the same as for the circuit using a PNP transistor. It should be noted, however, that the polarities of the biasing voltages are opposite to those with the PNP transistor.

9-7 The Common-collector Transistor Amplifier

Circuit Operation. In the common-collector amplifier (Fig. 9-6) the input signal is injected into the base-collector circuit and the output signal is taken from the emitter-collector circuit, making the collector the common element to both the input and output circuits.

In the PNP common-collector amplifier circuit of Fig. 9-6*a*, the emitter-base junction is forward-biased by the power source V_{EE} and the collector-base junction is reverse-biased by the power source V_{CC}. With zero input signal, power sources V_{EE} and V_{CC}, together with resistors R_B and R_L, establish the quiescent values of the base current I_B and the emitter current I_E. When the input signal is positive-going: (1) the forward bias decreases, (2) the emitter current decreases, (3) the voltage drop at R_L decreases, and (4) the output voltage increases. Thus, a positive-going input signal results in a positive-going output signal, and consequently the input and output signals are in phase. A similar step-by-step analysis will show that a negative-going input signal will result in a negative-going output signal.

Characteristics. The common-collector transistor amplifier circuit has (1) a high input resistance, (2) a low output resistance, (3) a current gain in the order of 30 to 250, (4) a voltage gain of less than 1, and (5) a power gain in the order of 10 to 20 db. The input and output signals are in phase, as no phase reversal takes place in this type of amplifier circuit. The stability of the collector current with temperature change for the common-collector amplifier is not as good as that of the common-base circuit; also, the stability of the circuit is dependent upon the ratio of the base resistance R_B to the emitter resistance R_L. Two important uses of the common-collector amplifier are (1) impedance matching, and (2) circuit isolation. Also, because the common-collector circuit can pass a signal in either direction, it can be used for (1) a two-way amplifier, and (2) switching circuits.

A common-collector amplifier circuit using an NPN transistor is shown in Fig. 9-6*b*. The operation and the characteristics of this circuit are the same as for the circuit using a PNP transistor. It should be noted, however, that the polarities of the biasing voltages are opposite to those with the PNP transistor.

Emitter Follower. The common-collector amplifier is frequently referred to as an *emitter follower*. This name originates from the fact that (1) the output signal is taken from across the resistor R_L in the emitter circuit (Fig.

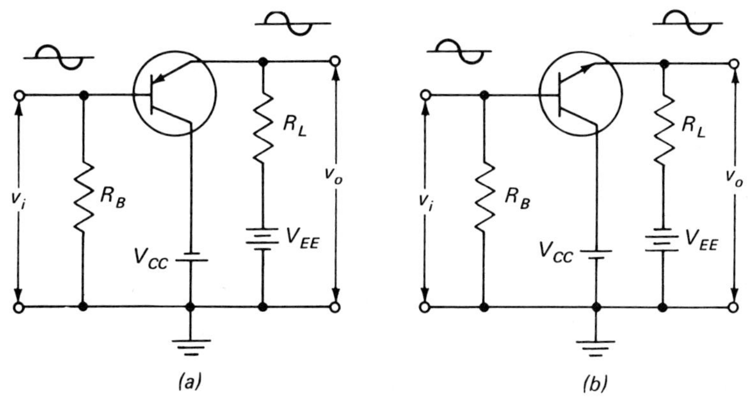

Fig. 9-6 Common-collector transistor amplifier. (a) Basic circuit using a PNP transistor. (b) Basic circuit using an NPN transistor.

9-6*a*), and (2) the output signal follows the signal changes at the base of the transistor.

9-8 Mathematical Analysis of the Basic Common-emitter Transistor Amplifier Circuit

Methods of Analysis. The purpose of the following analysis of the amplifier circuit is to develop an understanding of the characteristics of the transistor amplifier. Two methods of determining the operating characteristics are by use of (1) the four *h*-parameters of the transistor and the values of the circuit components, and (2) the beta of the transistor and the values of the circuit components. Although the method using the *h*-parameters can produce more accurate results, the following difficulties may be encountered: (1) the values of the *h*-parameters are not always readily available or easily obtained, (2) the values can vary considerably with individual transistors—even of the same type number, and (3) the values can vary considerably with a given transistor when operating with different voltages and currents. The second method is presented because (1) the values required are always readily available, (2) the procedures followed are more easily understood, and (3) the results obtained are sufficiently accurate for a practical study of the amplifier-circuit characteristics.

Typical Amplifier Circuit and Its Principal Operating Characteristics. A simple amplifier circuit that produces an increase in voltage from its input terminals to its output terminals in one increment or step is called a *single-stage voltage amplifier*. Figure 9-7*a* illustrates a typical single-stage common-emitter transistor amplifier. The common-emitter configuration is most frequently used and may employ either a PNP- or an NPN-type transistor. The principal operating characteristics of this circuit are (1) the input resistance, (2) the output resistance, (3) the current gain, (4) the voltage gain, and (5) the power gain. These characteristics are for conditions with an a-c input signal, sometimes referred to as *small-signal operation*, in which the a-c signal voltages and currents are in the order of 10 per cent or less of the quiescent voltage and current values.

A-C Equivalent Circuit. Figure 9-7*b* shows a simplified equivalent circuit for the a-c signal in the amplifier of Fig. 9-7*a*. For this simplified circuit, (1) the d-c power source is omitted because its impedance to the a-c signal is negligible, and (2) capacitors C_1 and C_2 are omitted because their impedances to the a-c signal are so low that the capacitors can be ignored. (The function of these capacitors is to prevent the d-c of the power source from reaching the signal source G_s and the output resistor R_o.)

Figure 9-7*c* illustrates the a-c equivalent circuit with the active element (transistor) of Fig. 9-7*b* replaced with equivalent passive elements expressed in their *h*-parameters.

Figure 9-7*d* illustrates the a-c equivalent circuit with the active element replaced with equivalent passive elements expressed in terms of beta and resistance values.

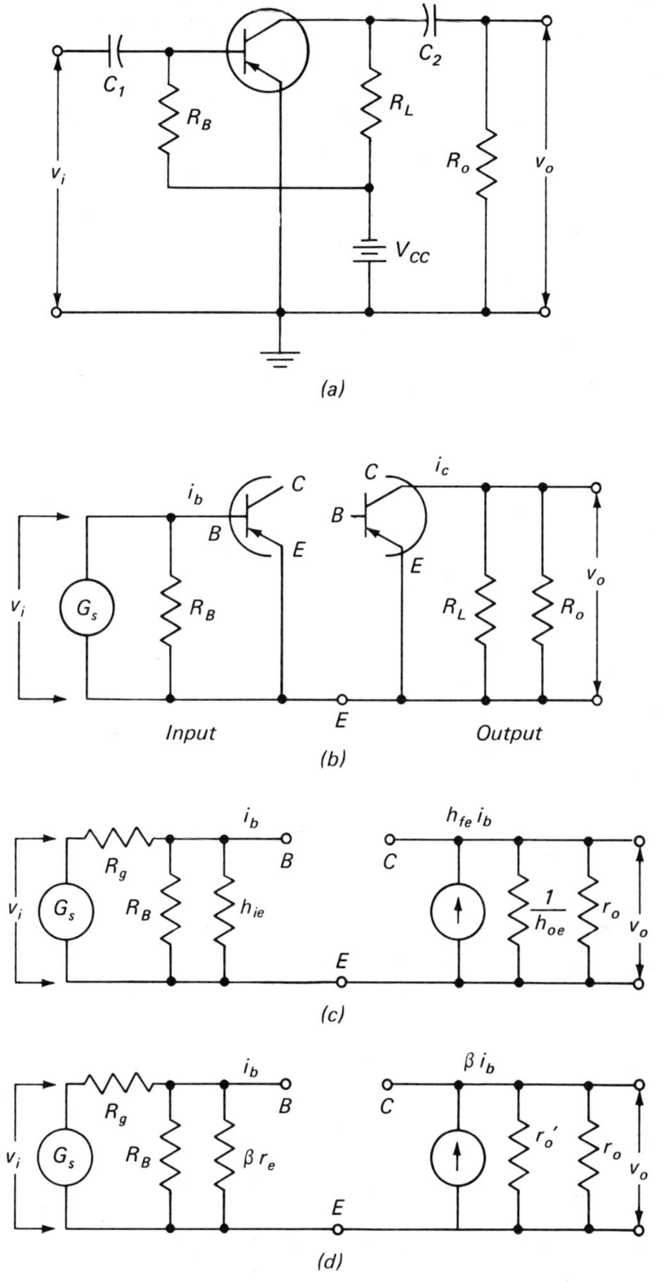

Fig. 9-7 Equivalent circuits of the common-emitter transistor amplifier. (a) The basic circuit. (b) A-c equivalent circuit divided into input and output sections. (c) A-c equivalent circuit in terms of h-parameters. (d) A-c equivalent circuit in terms of beta and resistance values.

9-9 Circuit Analysis of the Common-emitter Amplifier with h-Parameters

The h-Parameters. The four h-parameters (defined in Art. 4-8) are used to determine the transistor amplifier characteristics in the following examples. The h-parameters are found from the characteristic curves of the transistor, as is indicated by Eqs. (9-1) through (9-4). These parameters are sometimes given in transistor manuals.

Input resistance, ohms

$$h_{ie} = \frac{dv_{be}}{di_b} \qquad (V_{CE}\text{—constant}) \tag{9-1}$$

Output conductance, mhos

$$h_{oe} = \frac{di_c}{dv_{ce}} \qquad (I_B\text{—constant}) \tag{9-2}$$

Forward-current transfer ratio

$$h_{fe} = \frac{di_c}{di_b} \qquad (V_{CE}\text{—constant}) \tag{9-3}$$

Reverse-voltage transfer ratio

$$h_{re} = \frac{dv_{be}}{dv_{ce}} \qquad (I_B\text{—constant}) \tag{9-4}$$

Circuit Characteristics. When the four h-parameters are available, the circuit characteristics of the common-emitter amplifier can be determined with Eqs. (9-5) through (9-9).

Input Resistance. The a-c resistance seen by the signal-source generator is

$$r_i = \frac{h_{ie} + (h_{oe}h_{ie} - h_{fe}h_{re})r_o}{1 + h_{oe}r_o} \tag{9-5}$$

A close approximation can be obtained by

$$r_i \cong h_{ie} - \frac{h_{fe}h_{re}r_o}{1 + h_{oe}r_o} \tag{9-5a}$$

where r_o is the a-c load resistance.

Output Resistance. The a-c resistance looking into the collector is

$$r_o' = \frac{h_{ie} + R_g}{h_{oe}h_{ie} - h_{fe}h_{re} + h_{oe}R_g} \tag{9-6}$$

Current Gain. The current gain of the amplifier circuit is

$$A_i = \frac{-h_{fe}}{h_{oe}r_o + 1} \tag{9-7}$$

Voltage Gain. The voltage gain of the amplifier circuit is

$$A_v = \frac{-h_{fe}r_o}{(h_{ie}h_{oe} - h_{fe}h_{re})r_o + h_{ie}} \tag{9-8}$$

A close approximation can be obtained with

$$A_v \cong \frac{-h_{fe}r_o}{r_i(h_{oe}r_o + 1)} \tag{9-8a}$$

Power Gain. The power gain of the amplifier circuit is

$$A_p = A_iA_v \tag{9-9}$$

and

$$G_p = 10 \log A_p \tag{9-10}$$

Example 9-1 Determine the input resistance of the amplifier circuit of Fig. 9-8 by use of (*a*) Eq. (9-5), (*b*) (9-5*a*).

GIVEN: $h_{ie} = 1{,}140$ ohms $\quad h_{oe} = 2 \times 10^{-5}$ mho $\quad h_{fe} = 80 \quad h_{re} = 16 \times 10^{-5}$

FIND: (*a*) r_i $\quad$ (*b*) r_i

SOLUTION: Examination of Eqs. (9-5) and (9-5*a*) shows that the value of the term r_o is required. From Fig. 9-7*b* and *d* it can be seen that r_o is the equivalent of the parallel combination of R_L and R_o; thus

$$r_o = \frac{R_LR_o}{R_L + R_o} = \frac{8{,}000 \times 24{,}000}{8{,}000 + 24{,}000} = 6{,}000 \text{ ohms}$$

Note 1: Because $1/h_{oe}$ is much greater than r_o, its effect is being ignored in this solution. *Note* 2: Because R_B is much greater than h_{ie}, the effect of R_B is being ignored in this solution.

(*a*)

$$r_i = \frac{h_{ie} + (h_{oe}h_{ie} - h_{fe}h_{re})r_o}{1 + h_{oe}r_o}$$

$$= \frac{1{,}140 + (2 \times 10^{-5} \times 1{,}140 - 80 \times 16 \times 10^{-5})6{,}000}{1 + 2 \times 10^{-5} \times 6{,}000}$$

$$= 1{,}071.4 \text{ ohms}$$

(*b*)

$$r_i \cong h_{ie} - \frac{h_{fe}h_{re}r_o}{1 + h_{oe}r_o}$$

$$\cong 1{,}140 - \frac{80 \times 16 \times 10^{-5} \times 6{,}000}{1 + 2 \times 10^{-5} \times 6{,}000} = 1{,}071.5 \text{ ohms}$$

Example 9-2 Determine the output resistance r_o' of the transistor in the amplifier circuit of Fig. 9-8.

GIVEN: $h_{ie} = 1{,}140$ ohms $\quad h_{oe} = 2 \times 10^{-5}$ mho $\quad h_{fe} = 80 \quad h_{re} = 16 \times 10^{-5}$
$R_g = 0$

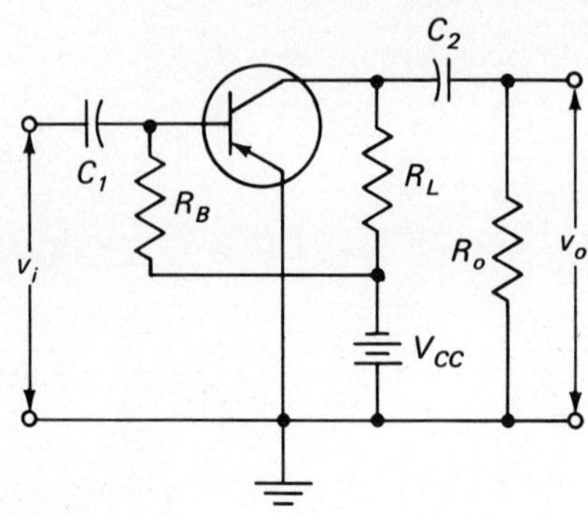

V_{CC} = 28 volts
β = 80
R_B = 1.27 megohms
R_L = 8,000 ohms
R_o = 24,000 ohms

h_{ie} = 1,140 ohms
h_{fe} = 80
$h_{oe} = 2 \times 10^{-5}$ mho
$h_{re} = 16 \times 10^{-5}$

(The resistance of the signal source is low enough to be ignored.)

Fig. 9-8 Typical common-emitter transistor amplifier circuit with practical values of *h*-parameters and circuit components.

FIND: r_o'

SOLUTION:

$$r_o' = \frac{h_{ie} + R_g}{h_{oe}h_{ie} - h_{fe}h_{re} + h_{oe}R_g}$$

$$= \frac{1{,}140 + 0}{2 \times 10^{-5} \times 1{,}140 - 80 \times 16 \times 10^{-5} + 2 \times 10^{-5} \times 0} = 114{,}000 \text{ ohms}$$

Example 9-3 Determine the current gain of the amplifier circuit of Fig. 9-8.

GIVEN: $h_{fe} = 80$ $h_{oe} = 2 \times 10^{-5}$ mho $r_o = 6{,}000$ ohms

FIND: A_i

SOLUTION:

$$A_i = \frac{-h_{fe}}{h_{oe}r_o + 1} = \frac{-80}{2 \times 10^{-5} \times 6{,}000 + 1} = -71.4$$

Note: The negative sign indicates phase reversal.

Example 9-4 Determine the voltage gain of the amplifier circuit of Fig. 9-8 by use of (*a*) Eq. (9-8), (*b*) Eq. (9-8*a*).

GIVEN: $h_{ie} = 1{,}140$ ohms $h_{oe} = 2 \times 10^{-5}$ mho $h_{fe} = 80$ $h_{re} = 16 \times 10^{-5}$
$r_o = 6{,}000$ ohms $r_i \cong 1{,}070$ ohms

FIND: (*a*) A_v (*b*) A_v

SOLUTION:

(a)
$$A_v = \frac{-h_{fe}r_o}{(h_{ie}h_{oe} - h_{fe}h_{re})r_o + h_{ie}}$$
$$= \frac{-80 \times 6{,}000}{(1{,}140 \times 2 \times 10^{-5} - 80 \times 16 \times 10^{-5})6{,}000 + 1{,}140}$$
$$= -400$$

(b)
$$A_v \cong \frac{-h_{fe}r_o}{r_i(h_{oe}r_o + 1)} = \frac{-80 \times 6{,}000}{1{,}070(2 \times 10^{-5} \times 6{,}000 + 1)} = -400$$

Example 9-5 Determine the power gain for the amplifier circuit of Fig. 9-8 expressed in (*a*) power amplification, (*b*) decibels.

GIVEN: $A_i = 71.4$ (Example 9-3) $\quad A_v = 400$ (Example 9-4)

FIND: (*a*) A_p $\quad$ (*b*) G_p

SOLUTION:

(a) $$A_p = A_iA_v = 71.4 \times 400 = 28{,}560$$
(b) $$G_p = 10 \log A_p = 10 \log 28{,}560 = 10 \times 4.456 \cong 44.6 \text{ db}$$

9-10 Circuit Analysis of the Common-emitter Amplifier with Beta and Circuit Resistance Values

Basis. In this presentation a number of assumptions are made that permit (1) elimination of the need for *h*-parameters, (2) simplification of calculations, and (3) ease of understanding the procedures involved. The assumptions made are

1. The emitter and collector currents are equal.
2. The current gain is equal to beta.
3. The transistor input circuit acts as a diode.
4. The transistor output circuit acts as a current source having a value equal to the emitter current.
5. The resistance of the signal-source generator is negligible.
6. The resistance $R_B \gg \beta r_e$ (Fig. 9-7*d*), hence can be ignored.
7. The resistance $r_o' \gg r_o$ (Fig. 9-7*d*), hence can be ignored.
8. The resistance $r_b \ll \beta r_e$, hence can be ignored.

Input Resistance. The a-c input resistance of the common-emitter amplifier circuit is

$$r_i = \beta r_e + r_b \tag{9-11}$$

Because $\beta r_e \gg r_b$, an acceptable approximate value for the a-c input resistance is

$$r_i \cong \beta r_e \tag{9-11a}$$

where
$$r_e = \frac{25}{I_E} \quad \text{from Eq. (2-3)} \tag{2-3b}$$

r_i = a-c input resistance, ohms
r_e = a-c junction resistance of the input diode of the transistor, ohms (see Art. 2-6)
r_b = *base-spreading resistance* of the transistor, ohms (this is similar to the bulk resistance, Art. 2-6)

Substituting Eq. (2-3*b*) in Eq. (9-11*a*),

$$r_i = \beta \frac{25}{I_E} \tag{9-11b}$$

where I_E = emitter current, ma

Example 9-6 What is the a-c input resistance of the amplifier circuit of Fig. 9-8?

GIVEN: $\beta = 80$ $V_{CC} = 28$ volts $R_B = 1.27$ megohms

FIND: r_o

SOLUTION:

$$r_i = \beta \frac{25}{I_E} = 80 \times \frac{25}{1.76} = 1{,}140 \text{ ohms}$$

where $$I_B = \frac{V_{CC}}{R_B} = \frac{28}{1.27 \times 10^{-6}} \cong 22\ \mu\text{a}$$

$$I_E = \beta I_B = 80 \times 22 \times 10^{-6} = 1.76 \text{ ma}$$

The a-c input resistance as seen by the signal-source generator G_s consists of the external base resistance R_B and the a-c input resistance of the transistor (βr_e) connected in parallel as shown in Fig. 9-7*d*. As the value of R_B is normally much greater than βr_e, the value of r_i is approximately equal to βr_e and thus justifies the use of Eq. (9-11*a*).

Output Resistance. From Fig. 9-7*b*, it is apparent that the path of the output current i_c (as seen by the collector) consists of R_L and R_o connected in parallel. The a-c output resistance is therefore

$$r_o = \frac{R_L R_o}{R_L + R_o} \tag{9-12}$$

Example 9-7 What is the a-c output resistance of the amplifier circuit of Fig. 9-8?

GIVEN: $R_L = 8{,}000$ ohms $R_o = 24{,}000$ ohms

FIND: r_o

SOLUTION:

$$r_o = \frac{R_L R_o}{R_L + R_o} = \frac{8{,}000 \times 24{,}000}{8{,}000 + 24{,}000} = 6{,}000 \text{ ohms}$$

As r_o' (Fig. 9-7d) is normally much higher than r_o, the a-c output circuit may be considered as simply the a-c output resistance r_o and a current generator supplying a current βi_b.

Current Gain. The current gain for the typical single-stage common-emitter amplifier circuit (Fig. 9-8) is approximately equal to the beta value of the transistor given in the manufacturers' specifications for the type of transistor being used. This applies under the condition that the output resistance of the transistor (r_o') is much higher than the output resistance of the amplifier circuit (r_o), which is true for the typical amplifier. Thus,

$$A_i \cong \beta_{\text{a-c}} \cong \beta \tag{9-13}$$

Example 9-8 What is the current gain of the amplifier circuit of Fig. 9-8?

GIVEN: $\beta = 80$

FIND: A_i

SOLUTION:

$$A_i \cong \beta = 80$$

Voltage Gain. The voltage gain of the common-emitter amplifier circuit of Fig. 9-7a is the ratio of the a-c output-signal voltage to the a-c input-signal voltage, or

$$A_v = \frac{v_o}{v_i} \tag{9-14}$$

The a-c input-signal voltage expressed mathematically is

$$v_i = i_b r_i \tag{9-15}$$

and the output-signal voltage expressed mathematically is

$$v_o = \beta i_b r_o \tag{9-16}$$

Substituting Eqs. (9-15) and (9-16) in Eq. (9-14), then

$$A_v = \frac{v_o}{v_i} = \frac{\beta i_b r_o}{i_b r_i} = \beta \frac{r_o}{r_i} \tag{9-17}$$

Substituting Eq. (9-11a) in Eq. (9-17),

$$A_v = \frac{r_o}{r_e} \tag{9-17a}$$

Example 9-9 (a) What is the voltage gain of the amplifier circuit of Fig. 9-8? (b) What is the output voltage when the input signal is 5 mv (rms)?

GIVEN: $\beta = 80$ $r_i = 1{,}140$ ohms (Example 9-6) $r_o = 6{,}000$ ohms (Example 9-7)
$v_i = 5$ mv

FIND: (a) A_v (b) v_o

SOLUTION:

(*a*) $$A_v = \beta \frac{r_o}{r_i} = 80 \times \frac{6{,}000}{1{,}140} \cong 420$$

(*b*) $$v_o = A_v v_i = 420 \times 5 \times 10^{-3} = 2.1 \text{ volts}$$

Power Gain. The power amplification is simply the product of the current gain and the voltage gain, or

$$A_p = A_i A_v \tag{9-18}$$

The power gain in decibels is

$$G_p = 10 \log A_p \tag{9-19}$$

Example 9-10 What is the power gain of the amplifier circuit of Fig. 9-8 when expressed (*a*) as a power-amplification ratio? (*b*) In decibels?

GIVEN: $A_i = 80$ (Example 9-8) $A_v = 420$ (Example 9-9)

FIND: (*a*) A_p (*b*) G_p

SOLUTION:

(*a*) $A_p = A_i A_v = 80 \times 420 = 33{,}600$

(*b*) $G_p = 10 \log A_p = 10 \log 33{,}600 = 10 \times 4.526 \cong 45.2$ db

9-11 Effects of R_B and R_L on the Current Gain

Effect of R_B on Current Gain. Example 9-8 indicates that under ideal conditions the current gain of the amplifier circuit is equal to the current gain β of the transistor. However, when R_B is not 10 or more times greater than βr_e, the current gain of the circuit will be decreased by an appreciable amount, as is indicated by

$$A_i' = \beta \frac{R_B}{R_B + \beta r_e} \tag{9-20}$$

where A_i' = current gain at the output of the transistor, and including the effect of R_B

Example 9-11 What is the current gain of the amplifier circuit of Fig. 9-8 when the effect of R_B is taken into consideration?

GIVEN: $\beta = 80$ $R_B = 1.27$ megohms $\beta r_e = 1{,}140$ ohms (Example 9-6)

FIND: A_i'

SOLUTION:

$$A_i' = \beta \frac{R_B}{R_B + \beta r_e} = 80 \times \frac{1{,}270{,}000}{1{,}270{,}000 + 1{,}140} \cong 80$$

Effect of R_L on Current Gain. Equations (9-13) and (9-20) are based on the output current of the transistor, namely, the collector current. If the current gain of the amplifier circuit is to be based on the input current at the signal source and the output current in the output resistor (R_o of Fig. 9-8), the values of R_L and r_o' must be considered. As r_o' is usually much greater than r_o, the effect of r_o' will be disregarded in this discussion. However, when R_L is not 10 or more times greater than R_o, the overall current gain of the circuit will be decreased appreciably as is indicated by

$$A_i'' = \beta \frac{R_L}{R_L + R_o} \tag{9-21}$$

where A_i'' = current gain at R_o, including the effect of R_L.

Example 9-12 What is the current gain with reference to the output resistor R_o for the amplifier circuit of Example 9-8? (Assume R_B is high enough to make its effect on the current gain negligible.)

GIVEN: $\beta = 80$ $\quad R_L = 8{,}000$ ohms $\quad R_o = 24{,}000$ ohms

FIND: A_i''

SOLUTION:

$$A_i'' = \beta \frac{R_L}{R_L + R_o} = 80 \times \frac{8{,}000}{8{,}000 + 24{,}000} = 20$$

9-12 Effect of R_g on the Voltage Gain

The voltage gain for the amplifier circuit of Fig. 9-8, as determined in Example 9-9, is based on the assumption that the resistance R_g of the signal source is zero. The voltage gain of the amplifier will decrease as the resistance of the signal source increases. The effect of any appreciable resistance at the signal source can be seen by solving for the voltage gain for the circuit of Fig. 9-9, which is the same as Fig. 9-8 except that the signal source has a resistance of 500 ohms and the input signal is specified as 25 mv.

Example 9-13 (*a*) What is the voltage at the base of the transistor for the circuit of Fig. 9-9? (*b*) What is the voltage gain of this amplifier circuit? (*Note:* Use value of A_v from Example 9-9.)

GIVEN: $v_i = 25$ mv $\quad R_g = 500$ ohms $\quad \beta r_e = 1{,}140$ ohms (Example 9-6)
$A_v = 420$ (Example 9-9)

FIND: (*a*) v_{BE} $\quad$ (*b*) A_v'

SOLUTION:
(*a*) Because R_B is many times greater than βr_e, the a-c equivalent circuit can be simplified as in Fig. 9-9*c*. Then

$$i_i = \frac{v_i}{R_g + \beta r_e} = \frac{25 \times 10^{-3}}{500 + 1{,}140} \cong 15.2\ \mu\text{a}$$

$$v_{BE} = i_i \beta r_e = 15.2 \times 10^{-6} \times 1{,}140 \cong 17.3 \text{ mv}$$

(*b*)

$$v_o = v_{BE} A_v = 17.3 \times 10^{-3} \times 420 \cong 7.26 \text{ volts}$$

$$A_v' = \frac{v_o}{v_i} = \frac{7.26}{25 \times 10^{-3}} \cong 290$$

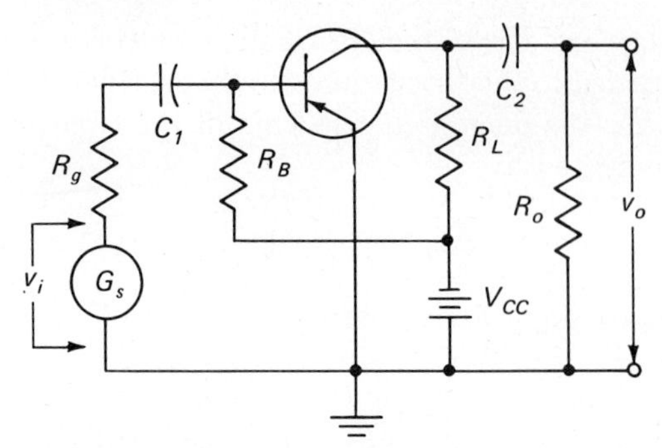

V_{CC} = 28 volts, v_i = 25 mv, β = 80, R_B = 1.27 megohms,
R_g = 500 ohms, R_L = 8,000 ohms, R_o = 24,000 ohms

(a)

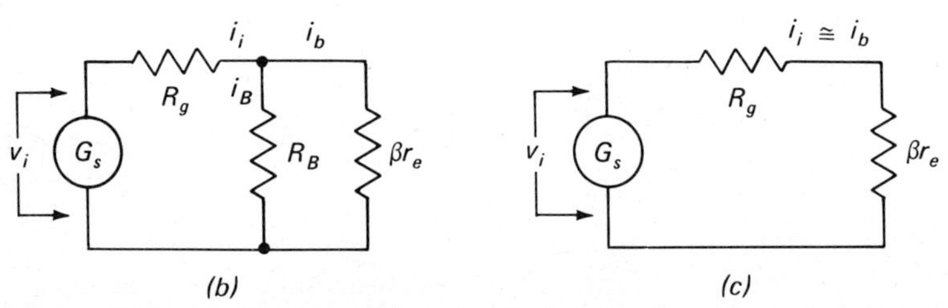

(b) (c)

Fig. 9-9 Common-emitter amplifier circuit showing the effect of resistance at the signal source. (a) Basic circuit. (b) A-c equivalent circuit for the input side of the amplifier. (c) Simplified a-c equivalent circuit for the input side of the amplifier.

9-13 Load Lines for the Common-emitter Amplifier

Purpose of Load Lines. Load lines are presented here to illustrate the difference in the load lines for the d-c or static condition and the a-c or small-signal condition. To fulfill this purpose it is not necessary that the load lines be superimposed on the family of characteristic curves as in Arts 4-17 and 8-3.

D-C Operating Characteristics. The d-c power supply in a transistor amplifier circuit, whether provided by one or two power sources, has the functions of (1) establishing the forward bias, (2) providing the reverse bias, and (3) providing the proper bias to ensure the operation of the amplifier circuit in the class desired, namely, Class A, AB, B, or C. The explanation of the d-c operating characteristics has been described in detail in Chap. 8.

A-C Operating Characteristics. Important a-c operating characteristics are (1) the input and output resistances, and (2) the gains in current, voltage, and power. These characteristics have been presented in Arts. 9-9 and 9-10. Plotting the a-c load line for an amplifier circuit provides a graphical representation of additional characteristics of the amplifier.

D-C Load Line. In order to plot the a-c load line it is desirable to first plot the d-c load line in the manner described in Art. 4-17. Chapter 8 describes how the location of the quiescent point along the d-c load line determines the class of operation.

Example 9-14 (*a*) Plot the d-c load line for the amplifier circuit of Fig. 9-8. (*b*) Locate the quiescent point on the load line.

GIVEN: Fig. 9-8

FIND: (*a*) Load line (*b*) Point q

SOLUTION:
(*a*) See Art. 4-17

Collector-current cutoff occurs at $V_{CE} = 28$ volts
Collector-current saturation occurs at $V_{CE} = 0$ volts

hence, $$I_{C\cdot sat} = \frac{V_{CC}}{R_L} = \frac{28}{8{,}000} = 3.5 \text{ ma}$$

See Fig. 9-10 for plot of the d-c load line.

(*b*) $$I_B = \frac{V_{CC}}{R_B} = \frac{28}{1.27 \times 10^6} = 22\ \mu\text{a}$$
$$I_C = \beta I_B = 80 \times 22 \times 10^{-6} = 1.76 \text{ ma}$$
$$V_{CE\cdot q} = V_{CC} - I_C R_L = 28 - (1.76 \times 10^{-3} \times 8{,}000) \cong 14 \text{ volts}$$

Quiescent point is shown at q on Fig. 9-10.

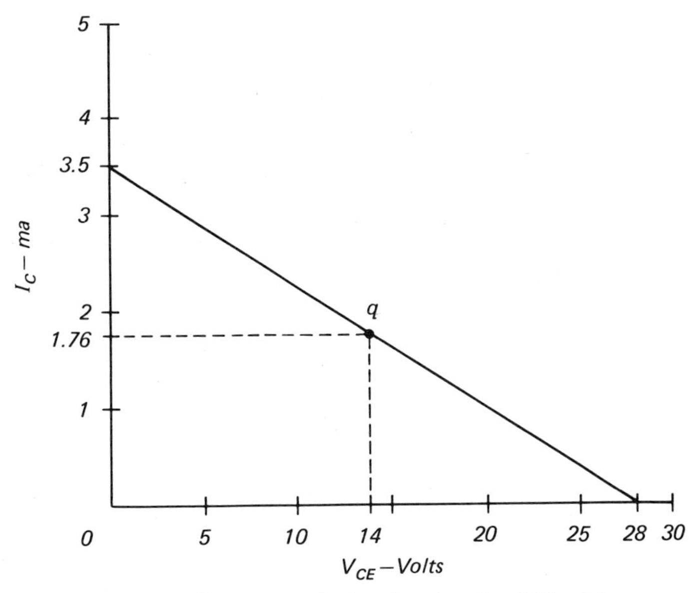

Fig. 9-10 D-c load line for the circuit of Fig. 9-8.

A-C Load Line. A review of Fig. 9-8 and Example 9-7 shows that the d-c load resistance R_L is 8,000 ohms and the a-c load resistance r_o is 6,000 ohms. Because of the difference in the d-c and a-c load resistances, the a-c signal variations will not follow along the d-c load line, but rather will follow along a line related to the a-c load resistance.

The a-c load line has the same quiescent point as the d-c load line but has different values of collector-current cutoff and collector-current saturation. The collector-current cutoff voltage is

$$V_{CE\cdot\text{a-c}} = V_{CE\cdot q} + I_{C\cdot q} r_o \tag{9-22}$$

The collector-current saturation point for the a-c load line is

$$I_{C\cdot\text{a-c}} = I_{C\cdot q} + \frac{V_{CE\cdot q}}{r_o} \tag{9-23}$$

Example 9-15 (*a*) Plot the a-c load line for the amplifier of Fig. 9-8 and Examples 9-7 and 9-14. (*b*) What is the maximum a-c output-signal voltage that can be accommodated without going beyond cutoff of the output-signal voltage?

GIVEN: Fig. 9-8 Examples 9-7 and 9-14 $V_{CE\cdot q} = 14$ volts $I_{C\cdot q} = 1.76$ ma $r_o = 6{,}000$ ohms

FIND: (*a*) a-c load line (*b*) $v_{o\cdot\max}$

SOLUTION:

(*a*) $$V_{CE\cdot\text{a-c}} = V_{CE\cdot q} + I_{C\cdot q} r_o = 14 + (1.76 \times 10^{-3} \times 6{,}000) \cong 24.6 \text{ volts}$$

$$I_{C\cdot\text{a-c}} = I_{C\cdot q} + \frac{V_{CE\cdot q}}{r_o} = 1.76 + \frac{14}{6{,}000} \cong 4.1 \text{ ma}$$

See Fig. 9-11 for plot of the a-c load line.

(*b*) $$v_{o\cdot\max} = V_{CE\cdot\text{a-c}} - V_{CE\cdot q} = 24.6 - 14 = 10.6 \text{ volts (peak)}$$

In order that the maximum input-signal voltage can be applied to the amplifier without causing cutoff at either the positive or negative input peaks, the quiescent point must be located at the midpoint of the a-c load line. To satisfy this condition,

$$V_{CE\cdot q} = I_{C\cdot q} r_o \tag{9-24}$$

and

$$I_{C\cdot q} = \frac{V_{CC}}{r_o + R_L} \tag{9-25}$$

Example 9-16 (*a*) What value of base resistance should be substituted in the circuit of Fig. 9-8 in order that the amplifier of Example 9-15 will be capable of handling the maximum possible output signal? (*b*) What is the maximum output signal that can be used without causing any cutoff of the output signal?

GIVEN: $V_{CC} = 28$ volts $r_o = 6{,}000$ ohms (Example 9-7) $R_L = 8{,}000$ ohms

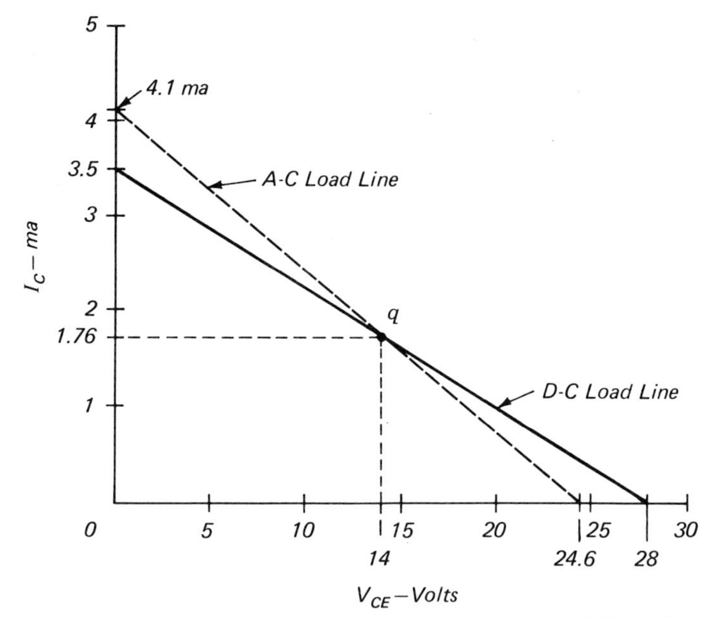

Fig. 9-11 A-c and d-c load lines for the circuit of Fig. 9-8.

FIND: (*a*) R_B (*b*) $v_{o\cdot\max}$

SOLUTION:

(*a*)

$$I_{C\cdot q} = \frac{V_{CC}}{r_o + R_L} = \frac{28}{6{,}000 + 8{,}000} = 2 \text{ ma}$$

$$I_{B\cdot q} = \frac{I_{C\cdot q}}{\beta} = \frac{2 \times 10^{-3}}{80} = 25\ \mu\text{a}$$

$$R_B = \frac{V_{CC}}{I_{B\cdot q}} = \frac{28}{25 \times 10^{-6}} = 1.12 \text{ megohms}$$

(*b*)

$$v_{o\cdot\max} = I_{C\cdot q} r_o = 2 \times 10^{-3} \times 6{,}000 = 12 \text{ volts (peak)}$$

Example 9-17 For the amplifier circuit of Example 9-16: (*a*) plot the a-c load line, (*b*) determine the maximum a-c signal input voltage.

GIVEN: $V_{CC} = 28$ volts $\beta = 80$ $R_B = 1.12$ megohms (Example 9-16)
$R_L = 8{,}000$ ohms $r_o = 6{,}000$ ohms (Example 9-7)

FIND: (*a*) a-c load line (*b*) $v_{i\cdot\max}$

SOLUTION:

(*a*) For d-c load line

$$V_{CE\cdot\max} = 28 \text{ volts} \qquad I_{C\cdot\max} = 3.5 \text{ ma} \qquad I_{C\cdot q} = 2 \text{ ma}$$

$$V_{CE\cdot q} = V_{CC} - (I_{C\cdot q} R_L) = 28 - (2 \times 10^{-3} \times 8{,}000) = 12 \text{ volts}$$

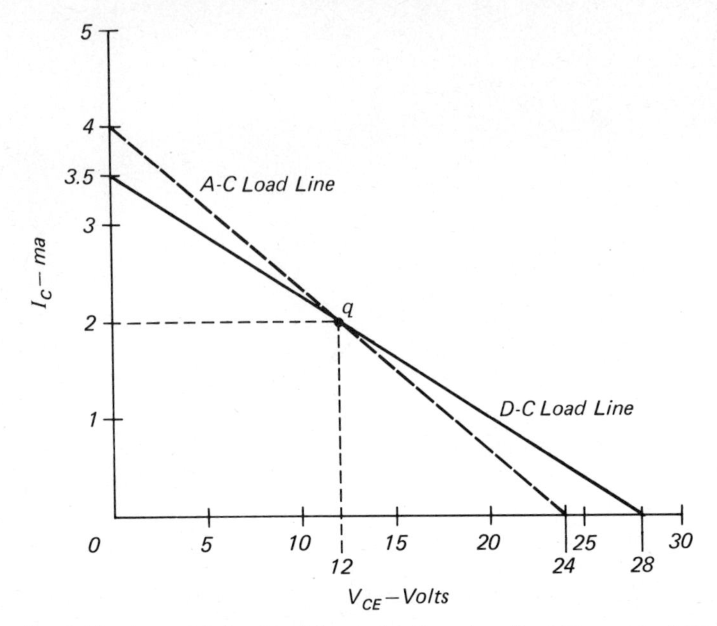

Fig. 9-12 A-c and d-c load lines for the circuit of Example 9-17.

For a-c load line

$$V_{CE\cdot\text{a-c}} = V_{CE\cdot q} + I_{C\cdot q}r_o = 12 + (2 \times 10^{-3} \times 6{,}000) = 24 \text{ volts}$$

$$I_{C\cdot\text{a-c}} = I_{C\cdot q} + \frac{V_{CE\cdot q}}{r_o} = 2 + \frac{12}{6{,}000} = 4 \text{ ma}$$

See Fig. 9-12 for plot of the a-c load line

(b)

$$r_i = \frac{\beta \times 25}{I_{C\cdot q}} = \frac{80 \times 25}{2} = 1{,}000 \text{ ohms}$$

$$A_v = \beta \frac{r_o}{r_i} = 80 \times \frac{6{,}000}{1{,}000} = 480$$

$$v_{i\cdot\max} = \frac{v_{o\cdot\max}}{A_v} = \frac{12}{480} = 25 \text{ mv (peak)}$$

9-14 Common-emitter Transistor Amplifier Circuit with Base and Emitter Bias

The circuit of Fig. 9-13 illustrates the use of a resistor R_E in the emitter circuit to provide emitter bias (Art. 8-16). Making the resistance of R_E much greater than r_e improves the beta sensitivity and the temperature-current stability. Under this condition, a large increase in r_e (for example, 50 per cent) will produce only a relatively small per cent of change in the total emitter-circuit resistance. Because of this effect, R_E is sometimes called a *swamping resistor*. The capacitor C_3 bypasses the a-c signal-current variations around the emitter resistor, thereby eliminating the degenerative effect that would be

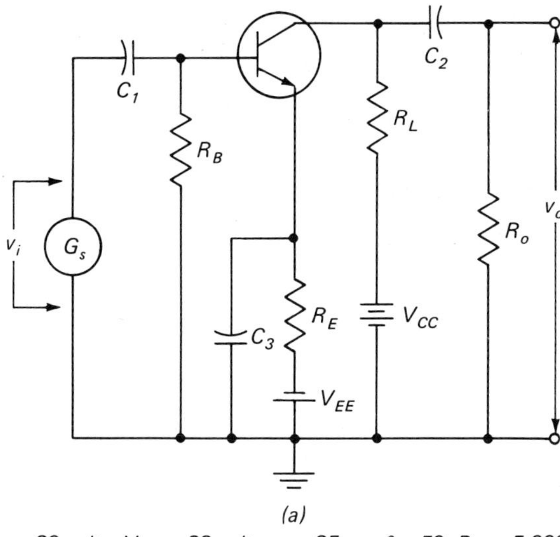

V_{EE} = 20 volts, V_{CC} = 28 volts, v_i = 25 mv, β = 50, R_B = 5,000 ohms, R_E = 10,000 ohms, R_L = 10,000 ohms, R_o = 5,000 ohms

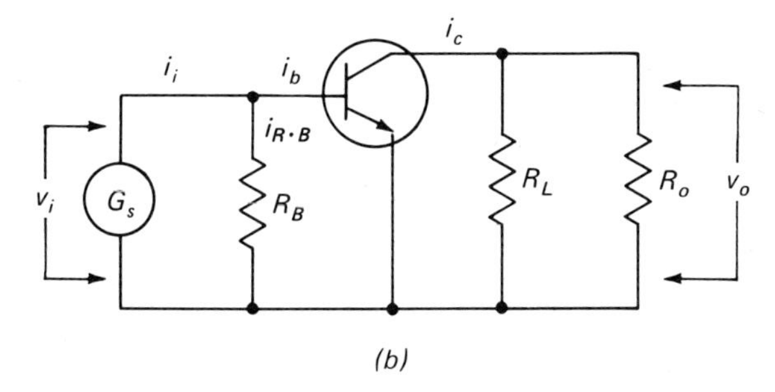

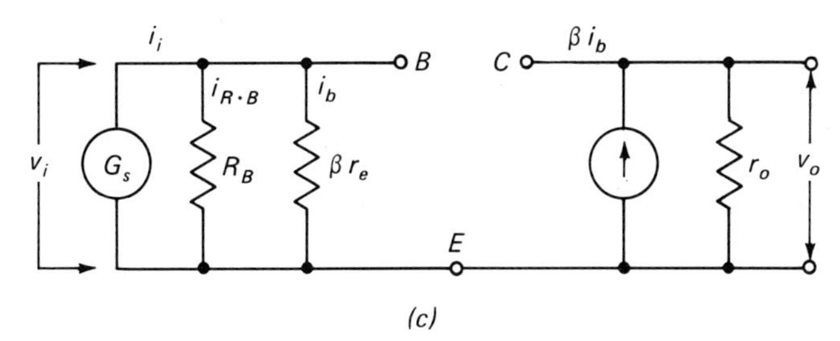

Fig. 9-13 Common-emitter amplifier circuit with base and emitter bias. (a) Basic circuit. (b) A-c equivalent circuit. (c) Simplified a-c equivalent circuit.

present if the signal currents were to pass through R_E. Because the impedance of C_3 is very low at the signal frequencies to be accommodated, it short-circuits R_E and V_{EE} to the a-c signals. Accordingly, R_E does not appear on the a-c equivalent circuits, and it should be observed that the equivalent circuits of Figs. 9-8 and 9-13 are similar.

Example 9-18 For the circuit and the parameters given on Fig. 9-13, find (*a*) input resistance, (*b*) output resistance, (*c*) voltage gain, (*d*) output voltage, (*e*) current gain of the circuit, (*f*) power gain. (*Note:* Consider R_g to be negligible.)

GIVEN: $V_{EE} = 20$ volts $\quad V_{CC} = 28$ volts $\quad v_i = 25$ mv $\quad \beta = 50$
$R_B = 5{,}000$ ohms $\quad R_E = 10{,}000$ ohms $\quad R_L = 10{,}000$ ohms
$R_o = 5{,}000$ ohms

FIND: (*a*) r_i (*b*) r_o (*c*) A_v (*d*) v_o (*e*) A_i'' (*f*) G_p

SOLUTION:

(*a*)
$$I_E = \frac{V_{EE}}{R_E + \dfrac{R_B}{\beta}} = \frac{20}{10{,}000 + \dfrac{5{,}000}{50}} \cong 2 \text{ ma} \qquad \text{(see Example 8-11)}$$

$$r_e = \frac{25}{I_E\text{ (ma)}} = \frac{25}{2} = 12.5 \text{ ohms}$$

$$r_i = \beta r_e = 50 \times 12.5 = 625 \text{ ohms}$$

(*b*)
$$r_o = \frac{R_L R_o}{R_L + R_o} = \frac{10{,}000 \times 5{,}000}{10{,}000 + 5{,}000} \cong 3{,}330 \text{ ohms}$$

(*c*)
$$A_v = \frac{r_o}{r_e} = \frac{3{,}330}{12.5} \cong 266$$

(*d*)
$$v_o = A_v v_i = 266 \times 25 \times 10^{-3} = 6.65 \text{ volts}$$

(*e*)
$$A_i' = \beta \frac{R_B}{R_B + \beta r_e} = 50 \times \frac{5{,}000}{5{,}000 + 625} \cong 44.5$$

$$A_i'' = A_i' \frac{R_L}{R_L + R_o} = 44.5 \times \frac{10{,}000}{10{,}000 + 5{,}000} \cong 29.6$$

(*f*)
$$A_p = A_i'' A_v = 29.6 \times 266 \cong 7{,}870$$
$$G_p = 10 \log A_p = 10 \log 7{,}870 = 10 \times 3.896 \cong 39 \text{ db}$$

9-15 Common-emitter Transistor Amplifier Circuit with Voltage-divider-controlled Base Bias and Emitter Bias

The circuit of Fig. 9-14 is similar to that of Fig. 8-19. The a-c equivalent circuit differs from those of Figs. 9-7 and 9-13 only by the addition of R_S to the input side. The characteristics of the circuit can be calculated in the same manner as in Examples 8-14 and 9-18.

Example 9-19 For the circuit and the parameters given on Fig. 9-14, find (*a*) emitter current, (*b*) input resistance, (*c*) output resistance, (*d*) voltage gain, (*e*) output voltage. (*Note:* This circuit is similar to Fig. 8-19 and Example 8-14.)

GIVEN: $V_{CC} = 15$ volts $\quad v_i = 17.5$ mv $\quad \beta = 50 \quad R_B = 5{,}000$ ohms
$R_S = 10{,}000$ ohms $\quad R_E = 5{,}000$ ohms $\quad R_L = 2{,}000$ ohms
$R_o = 5{,}000$ ohms

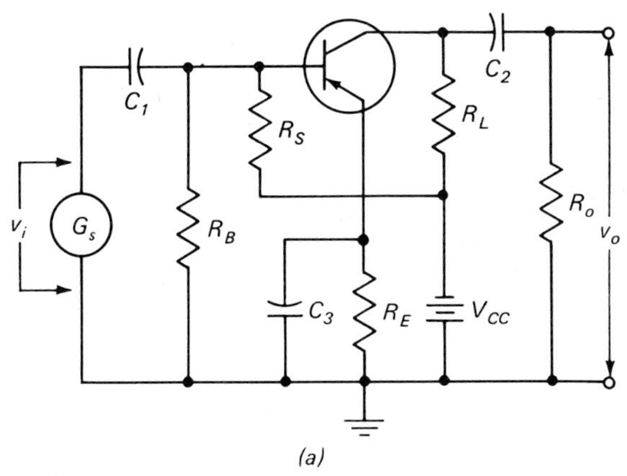

(a)

V_{CC} = 15 volts, v_i = 17.5 mv, β = 50, R_B = 5,000 ohms, R_S = 10,000 ohms, R_E = 5,000 ohms, R_L = 2,000 ohms, R_o = 5,000 ohms.

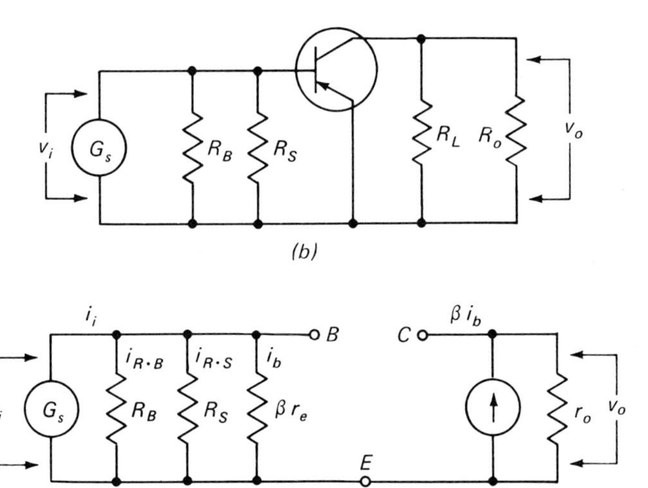

(c)

Fig. 9-14 Common-emitter amplifier circuit with base and emitter bias. (a) Basic circuit. (b) A-c equivalent circuit. (c) Simplified a-c equivalent circuit.

FIND: (a) I_E (b) r_i (c) r_o (d) A_v (e) v_o

SOLUTION:

(a)

$$I_E = \frac{V_B}{R_E} = \frac{5}{5{,}000} = 1 \text{ ma}$$

where

$$V_B = V_{CC}\frac{R_B}{R_B + R_S} = 15 \times \frac{5{,}000}{5{,}000 + 10{,}000} = 5 \text{ volts}$$

(b) $$r_e = \frac{25}{I_E} = \frac{25}{1} = 25 \text{ ohms}$$

$$\beta r_e = \beta \times r_e = 50 \times 25 = 1{,}250 \text{ ohms}$$

$$r_i = \frac{1}{\frac{1}{R_B} + \frac{1}{R_S} + \frac{1}{\beta r_e}} = \frac{1}{\frac{1}{5{,}000} + \frac{1}{10{,}000} + \frac{1}{1{,}250}} \cong 910 \text{ ohms}$$

(c) $$r_o = \frac{R_L R_o}{R_L + R_o} = \frac{2{,}000 \times 5{,}000}{2{,}000 + 5{,}000} \cong 1{,}430 \text{ ohms}$$

(d) $$A_v = \frac{r_o}{r_e} = \frac{1{,}430}{25} \cong 57$$

(e) $$v_o = A_v v_i = 57 \times 17.5 \times 10^{-3} \cong 1 \text{ volt}$$

9-16 Common-emitter Transistor Amplifier Circuit with Collector Bias

The circuit of Fig. 9-15 is similar to that of Fig. 8-18. The effect of feedback is indicated in the a-c equivalent circuit (Fig. 9-15*b*) by showing R_B connected to both the input and output circuits. When R_B is much greater than r_e and r_o, which is generally true, the a-c equivalent circuit can be simplified as in Fig. 9-15*c*. In Fig. 9-15*c*

$$R_B' = \frac{R_B}{A_v} = \frac{R_B r_e}{r_o} \tag{9-26}$$

Example 9-20 For the circuit and the parameters given on Fig. 9-15, find (*a*) emitter current, (*b*) output resistance, (*c*) input resistance, (*d*) voltage gain. (*Note:* This circuit is similar to Fig. 8-18 and Example 8-13.)

GIVEN: $V_{CC} = 12$ volts $\quad V_{BE} = 0.3$ volt $\quad \beta = 100 \quad R_B = 100{,}000$ ohms
$R_L = 1{,}000$ ohms $\quad R_o = 4{,}000$ ohms

FIND: (*a*) I_E (*b*) r_o (*c*) r_i (*d*) A_v

SOLUTION:

(a) $$I_E \cong I_C = \frac{V_{CC} - V_{BE}}{R_L + \frac{R_B}{\beta}} = \frac{12 - 0.3}{1{,}000 + \frac{100{,}000}{100}} = 5.85 \text{ ma}$$

(b) $$r_o = \frac{R_L R_o}{R_L + R_o} = \frac{1{,}000 \times 4{,}000}{1{,}000 + 4{,}000} = 800 \text{ ohms}$$

(c) $$r_e = \frac{25}{I_E} = \frac{25}{5.85} \cong 4.27 \text{ ohms}$$

$$\beta r_e = \beta \times r_e = 100 \times 4.27 = 427 \text{ ohms}$$

$$R_B' = \frac{R_B r_e}{r_o} = \frac{100{,}000 \times 4.27}{800} \cong 535 \text{ ohms}$$

$$r_i = \frac{R_B' \beta r_e}{R_B' + \beta r_e} = \frac{535 \times 427}{535 + 427} \cong 238 \text{ ohms}$$

(d) $$A_v = \frac{r_o}{r_e} = \frac{800}{4.27} \cong 187$$

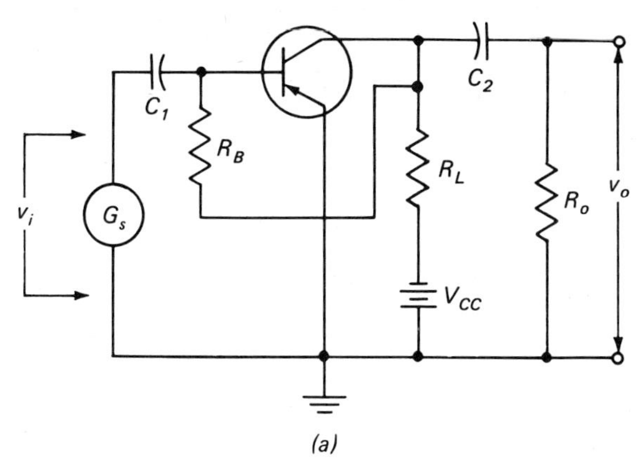

V_{CC} = 12 volts, V_{BE} = 0.3 volts, β = 100, R_B = 100,000 ohms
R_L = 1,000 ohms, R_o = 4,000 ohms

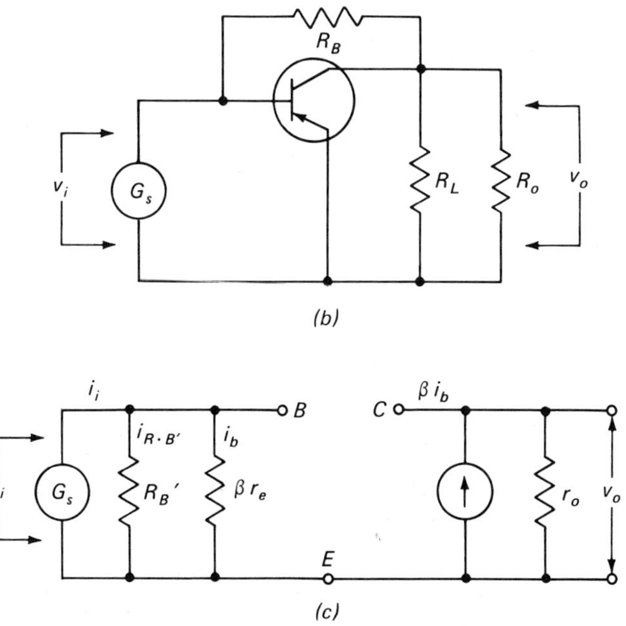

Fig. 9-15 Common-emitter amplifier circuit with collector feedback bias. (a) Basic circuit. (b) A-c equivalent circuit. (c) Simplified a-c equivalent circuit.

9-17 Circuit Analysis of the Common-base Transistor Amplifier

Typical Circuit. Figure 9-16 illustrates a typical single-stage common-base transistor amplifier. Resistors R_B and R_S do not appear on the a-c equivalent circuits because capacitor C_3 short-circuits these resistors to any varying signal currents.

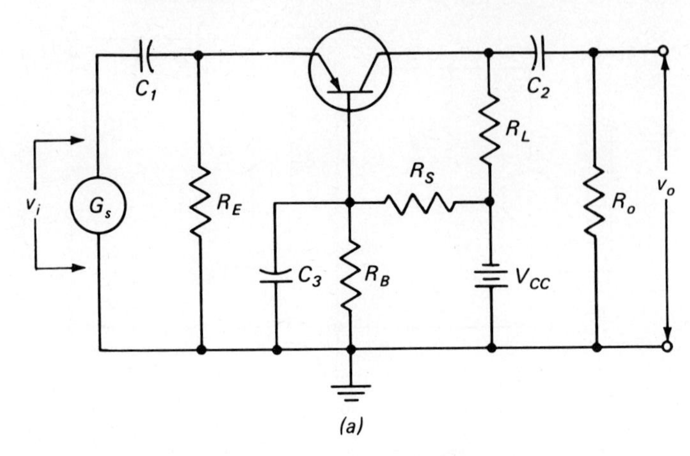

V_{CC} = 27 volts, v_i = 20 mv, $\alpha \cong 1$, R_E = 5,000 ohms, R_B = 2,000 ohms, R_S = 4,000 ohms, R_L = 2,500 ohms, R_o = 2,500 ohms

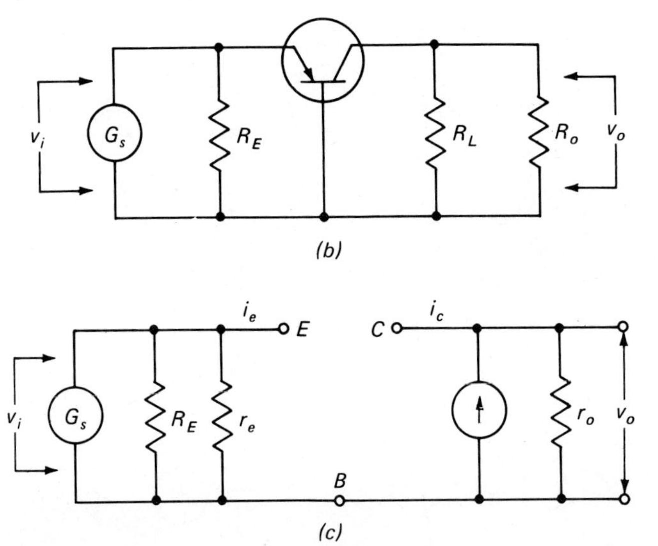

Fig. 9-16 Common-base transistor amplifier circuit. (a) Basic circuit. (b) A-c equivalent circuit. (c) Simplified a-c equivalent circuit.

Input Resistance. The a-c input resistance of the transistor in the common-base configuration is theoretically the same as the junction resistance of the base-to-emitter diode portion of the transistor; thus

$$r_e = \frac{25}{I_E\ (\text{ma})} \tag{2-3b}$$

The input resistance of the amplifier circuit (Fig. 9-16*c*) is

$$r_i = \frac{R_E r_e}{R_E + r_e} \tag{9-27}$$

Output Resistance. The output resistance of the amplifier circuit as indicated in Fig. 9-16*b* and *c* is

$$r_o = \frac{R_L R_o}{R_L + R_o} \tag{9-28}$$

Current Gain. The current gain for the ideal common-base transistor amplifier is

$$A_i \cong \alpha \tag{9-29}$$

Voltage Gain. The voltage gain for the common-base transistor amplifier is

$$A_v = \frac{r_o}{r_e} \tag{9-30}$$

Power Gain. The power gain for the common-base transistor amplifier is

$$A_p = A_i A_v \tag{9-9}$$
$$G_p = 10 \log A_p \tag{9-10}$$

Example 9-21 For the circuit and the parameters given in Fig. 9-16*a*, find (*a*) emitter current, (*b*) input resistance, (*c*) output resistance, (*d*) current gain, (*e*) voltage gain, (*f*) output voltage, (*g*) power gain. Assume $\alpha = 1$.

GIVEN: $V_{CC} = 27$ volts $\quad v_i = 20$ mv $\quad \alpha = 1 \quad R_B = 2{,}000$ ohms
$R_S = 4{,}000$ ohms $\quad R_E = 5{,}000$ ohms $\quad R_L = 2{,}500$ ohms
$R_o = 2{,}500$ ohms

FIND: (*a*) I_E (*b*) r_i (*c*) r_o (*d*) A_i (*e*) A_v (*f*) v_o (*g*) G_p

SOLUTION:

(*a*)
$$I_E = \frac{V_B}{R_E} = \frac{9}{5{,}000} = 1.8 \text{ ma}$$

where
$$V_B = V_{CC}\frac{R_B}{R_B + R_S} = 27 \times \frac{2{,}000}{2{,}000 + 4{,}000} = 9 \text{ volts}$$

(*b*)
$$r_e = \frac{25}{I_E} = \frac{25}{1.8} = 13.9 \text{ ohms}$$

$$r_i = \frac{R_E r_e}{R_E + r_e} = \frac{5{,}000 \times 13.9}{5{,}000 + 13.9} \cong 13.9 \text{ ohms}$$

(*c*)
$$r_o = \frac{R_L R_o}{R_L + R_o} = \frac{2{,}500 \times 2{,}500}{2{,}500 + 2{,}500} = 1{,}250 \text{ ohms}$$

(*d*)
$$A_i \cong \alpha \cong 1$$

(*e*)
$$A_v = \frac{r_o}{r_e} = \frac{1{,}250}{13.9} \cong 90$$

(*f*)
$$v_o = A_v v_i = 90 \times 20 \times 10^{-3} = 1.8 \text{ volts}$$

(*g*)
$$A_p = A_i A_v = 1 \times 90 = 90$$
$$G_p = 10 \log A_p = 10 \log 90 = 10 \times 1.954 \cong 19.5 \text{ db}$$

9-18 h-Parameter Analysis of the Common-base Transistor Amplifier

Principle. Equations (9-5) to (9-8a) can be used to determine the characteristics for the common-base amplifier circuit if the h-parameters are available for the common-base configuration, namely, h_{ib}, h_{ob}, h_{fb}, and h_{rb}.

Conversion of h-Parameters. Table 9-1 lists the sets of equations used to convert the h-parameters for one configuration to the h-parameters for another configuration.

Example 9-22 Find the common-base h-parameters for the transistor used in Example 9-4.

GIVEN: $h_{ie} = 1{,}140$ ohms $\quad h_{oe} = 2 \times 10^{-5}$ mho $\quad h_{fe} = 80 \quad h_{re} = 16 \times 10^{-5}$

FIND: (a) h_{ib} (b) h_{ob} (c) h_{fb} (d) h_{rb}

SOLUTION:

(a) $$h_{ib} = \frac{h_{ie}}{1 + h_{fe}} = \frac{1{,}140}{1 + 80} \cong 14 \text{ ohms}$$

(b) $$h_{ob} = \frac{h_{oe}}{1 + h_{fe}} = \frac{2 \times 10^{-5}}{1 + 80} \cong 0.247 \times 10^{-6} = 0.247\ \mu\text{mho}$$

(c) $$h_{fb} = \frac{-h_{fe}}{1 + h_{fe}} = \frac{-80}{1 + 80} \cong -0.988$$

(d) $$h_{rb} = \frac{h_{ie}h_{oe}}{1 + h_{fe}} - h_{re} = \frac{1{,}140 \times 2 \times 10^{-5}}{1 + 80} - 16 \times 10^{-5} \cong 12.2 \times 10^{-5}$$

Example 9-23 Using the transistor having the h-parameters found in Example 9-22 and the circuit parameters given on Fig. 9-16, find (a) input resistance, (b) current gain, (c) voltage gain, (d) power gain.

Table 9-1 Conversion of h-Parameters

FROM CE TO CB	FROM CE TO CC	FROM CB TO CE
$h_{ib} = \dfrac{h_{ie}}{1 + h_{fe}}$	$h_{ic} = h_{ie}$	$h_{ie} = \dfrac{h_{ib}}{1 + h_{fb}}$
$h_{ob} = \dfrac{h_{oe}}{1 + h_{fe}}$	$h_{oc} = h_{oe}$	$h_{oe} = \dfrac{h_{ob}}{1 + h_{fb}}$
$h_{fb} = \dfrac{-h_{fe}}{1 + h_{fe}}$	$h_{fc} = -(1 + h_{fe})$	$h_{fe} = \dfrac{-h_{fb}}{1 + h_{fb}}$
$h_{rb} = \dfrac{h_{ie}h_{oe}}{1 + h_{fe}} - h_{re}$	$h_{rc} = 1 - h_{re} \cong 1$	$h_{re} = \dfrac{h_{ib}h_{ob}}{1 + h_{fb}} - h_{rb}$

GIVEN: $h_{ib} = 14$ ohms $h_{ob} = 0.247\ \mu$mho $h_{fb} = -0.988$ $h_{rb} = 12.2 \times 10^{-5}$
$R_L = 2{,}500$ ohms $R_o = 2{,}500$ ohms

FIND: (a) r_i (b) A_i (c) A_v (d) G_p

SOLUTION:

(a)
$$r_o = \frac{R_L R_o}{R_L + R_o} = \frac{2{,}500 \times 2{,}500}{2{,}500 + 2{,}500} = 1{,}250 \text{ ohms}$$

$$r_i = \frac{h_{ib} + (h_{ob}h_{ib} - h_{fb}h_{rb})r_o}{1 + h_{ob}r_o}$$

$$= \frac{14 + (0.247 \times 10^{-6} \times 14 - 0.988 \times 12.2 \times 10^{-5})1{,}250}{1 + (0.247 \times 10^{-6} \times 1{,}250)}$$

$$\cong 14 \text{ ohms}$$

(b)
$$A_i = \frac{-h_{fb}}{h_{ob}r_o + 1} = \frac{-0.988}{(0.247 \times 10^{-6} \times 1{,}250) + 1} \cong -0.988$$

(c)
$$A_v \cong \frac{-h_{fb}r_o}{r_i(h_{ob}r_o + 1)} = \frac{-0.988 \times 1{,}250}{14(0.247 \times 10^{-6} \times 1{,}250 + 1)} \cong 88$$

(d)
$$A_p = A_iA_v = 0.988 \times 88 \cong 87$$
$$G_p = 10 \log A_p = 10 \log 87 = 10 \times 1.940 = 19.4 \text{ db}$$

9-19 Circuit Analysis of the Basic Common-collector Transistor Amplifier

Typical Circuit. Figure 9-17*a* illustrates a basic single-stage common-collector transistor amplifier. The a-c equivalent circuits are shown in Fig. 9-17*b* and *c*.

Input Resistance. The a-c input resistance of the transistor, that is, the a-c resistance looking into the base, is

$$r_i' = \beta(r_e + r_o) \tag{9-31}$$

The a-c input resistance of the amplifier circuit, that is, the a-c resistance seen by the signal source, is

$$r_i = \frac{R_B\beta(r_e + r_o)}{R_B + \beta(r_e + r_o)} \tag{9-32}$$

Output Resistance. The a-c output resistance of the common-collector amplifier circuit seen by the emitter is

$$r_o = \frac{R_E R_o}{R_E + R_o} \tag{9-33}$$

Current Gain. The current gain for the ideal common-collector amplifier circuit is

$$A_i \cong \beta \tag{9-34}$$

Voltage Gain. The voltage gain for the common-collector amplifier circuit is

$$A_v = \frac{r_o}{r_o + r_e} \tag{9-35}$$

and, when $r_o \gg r_e$, as is generally true

$$A_v \cong 1 \tag{9-36}$$

Power Gain. The power gain of the common-collector amplifier can be determined by use of Eqs. (9-9) and (9-10).

Example 9-24 For the circuit and the parameters given in Fig. 9-17*a*, find (*a*) emitter current, (*b*) output resistance, (*c*) base-junction resistance, (*d*) input resistance at the base of the transistor, (*e*) input resistance of the amplifier circuit, (*f*) current gain, (*g*) voltage gain.

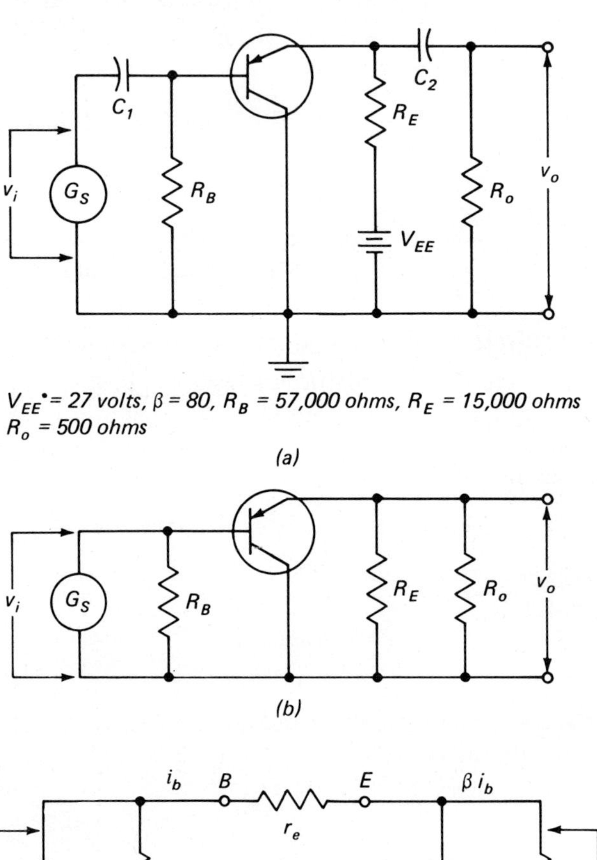

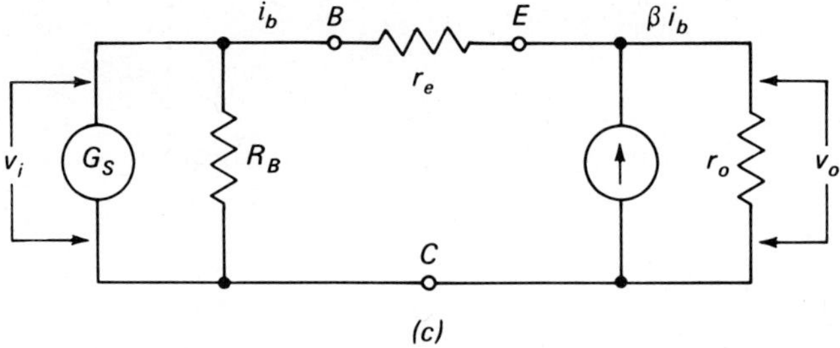

Fig. 9-17 Common-collector transistor amplifier circuit. (a) Basic circuit. (b) A-c equivalent circuit. (c) Simplified a-c equivalent circuit.

GIVEN: $V_{EE} = 27$ volts $\beta = 80$ $R_B = 57{,}000$ ohms $R_E = 15{,}000$ ohms $R_o = 500$ ohms

FIND: (a) I_E (b) r_o (c) r_e (d) r_i' (e) r_i (f) A_i (g) A_v

SOLUTION:

(a) $$I_E = \frac{V_{EE}}{R_E + \dfrac{R_B}{\beta}} = \frac{27}{15{,}000 + \dfrac{57{,}000}{80}} \cong 1.72 \text{ ma}$$

(b) $$r_o = \frac{R_E R_o}{R_E + R_o} = \frac{15{,}000 \times 500}{15{,}000 + 500} \cong 485 \text{ ohms}$$

(c) $$r_e = \frac{25}{I_E} = \frac{25}{1.72} \cong 14.5 \text{ ohms}$$

(d) $$r_i' = \beta(r_e + r_o) = 80(14.5 + 485) \cong 40{,}000 \text{ ohms}$$

(e) $$r_i = \frac{R_B\beta(r_e + r_o)}{R_B + \beta(r_e + r_o)} = \frac{57{,}000 \times 40{,}000}{57{,}000 + 40{,}000} \cong 23{,}500 \text{ ohms}$$

(f) $$A_i \cong \beta \cong 80$$

(g) $$A_v = \frac{r_o}{r_o + r_e} = \frac{485}{485 + 14.5} \cong 0.97$$

9-20 *h*-Parameter Analysis of the Common-collector Transistor Amplifier

Equations (9-5) to (9-8*a*) can be used to determine the characteristics of the common-collector amplifier circuit if the *h*-parameters are available for the common-collector configuration. Table 9-1 gives the equations for converting to the *h*-parameters for the common-collector configuration when the *h*-parameters are known for the common-emitter configuration.

Example 9-25 Find the common-collector *h*-parameters for the transistor used in Example 9-4.

GIVEN: $h_{ie} = 1{,}140$ ohms $h_{oe} = 2 \times 10^{-5}$ mho $h_{fe} = 80$ $h_{re} = 16 \times 10^{-5}$

FIND: (a) h_{ic} (b) h_{oc} (c) h_{fc} (d) h_{rc}

SOLUTION:

(a) $$h_{ic} = h_{ie} = 1{,}140 \text{ ohms}$$
(b) $$h_{oc} = h_{oe} = 2 \times 10^{-5} = 20\ \mu\text{mhos}$$
(c) $$h_{fc} = -(1 + h_{fe}) = -(1 + 80) = -81$$
(d) $$h_{rc} = 1 - h_{re} = 1 - (16 \times 10^{-5}) \cong 1$$

Example 9-26 Using the transistor having the *h*-parameters found in Example 9-25 and the circuit parameters given on Fig. 9-17*a*, find (*a*) input resistance, (*b*) current gain, (*c*) voltage gain, (*d*) power gain.

GIVEN: $h_{ic} = 1{,}140$ ohms $h_{oc} = 20\ \mu$mhos $h_{fc} = -81$ $h_{rc} = 1$

FIND: (a) r_i (b) A_i (c) A_v (d) G_p

SOLUTION:

(*a*)
$$r_i = \frac{h_{ic} + (h_{oc}h_{ic} - h_{fc}h_{rc})r_o}{1 + h_{oc}r_o}$$
$$= \frac{1{,}140 + [20 \times 10^{-6} \times 1{,}140 - (-81) \times 1]\,485}{1 + (20 \times 10^{-6} \times 485)}$$
$$\cong 40{,}000 \text{ ohms}$$

where
$$r_o = \frac{R_E R_o}{R_E + R_o} = \frac{15{,}000 \times 500}{15{,}000 + 500} \cong 485 \text{ ohms}$$

(*b*)
$$A_i = \frac{-h_{fc}}{h_{oc}r_o + 1} = \frac{-(-81)}{(20 \times 10^{-6} \times 485) + 1} \cong 80$$

(*c*)
$$A_v = \frac{-h_{fc}r_o}{r_i(h_{oc}r_o + 1)} = \frac{-(-81) \times 485}{40{,}000(20 \times 10^{-6} \times 485 + 1)} \cong 0.97$$

(*d*)
$$A_p = A_i A_v = 80 \times 0.97 = 77.5$$
$$G_p = 10 \log A_p = 10 \log 77.5 = 10 \times 1.889 \cong 19 \text{ db}$$

9-21 The Common-cathode Vacuum-tube Amplifier

Circuit Operation. Figure 9-18 shows the basic circuit and a simplified equivalent circuit for a common-cathode vacuum-tube amplifier using a triode. In the common-cathode amplifier, the input signal is injected into the grid-to-cathode circuit and the output signal is taken from the plate-to-cathode circuit, making the cathode the common element to both the input and output circuits. The configuration of the common-cathode vacuum-tube amplifier is similar to the basic common-emitter transistor amplifier (see Fig. 9-1).

In the common-cathode vacuum-tube amplifier (Fig. 9-18) the power source E_{CC} makes the grid of the tube negative with respect to the cathode, and the

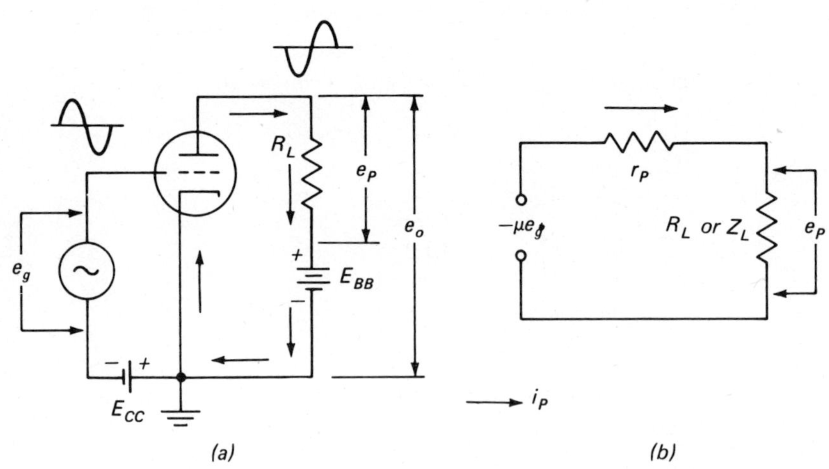

Fig. 9-18 Common-cathode vacuum-tube triode amplifier circuit. (a) Basic circuit. (b) Equivalent plate circuit in the constant-voltage-generator form.

power source E_{BB} makes the plate of the tube positive with respect to the cathode. With zero input signal, power sources E_{BB} and E_{CC}, together with resistor R_L, establish the quiescent values of grid voltage E_C and plate current I_P. When the input signal is positive-going: (1) the grid bias decreases, (2) the plate current increases, (3) the voltage drop at R_L increases, and (4) the output voltage decreases. Thus, a positive-going input signal produces a negative-going output signal. As the output signal is 180 degrees out of phase with the input signal, phase reversal takes place. A similar step-by-step analysis will show that a negative-going input signal will result in a positive-going output signal. Because a small change in the input signal will produce a relatively large change in the output-signal voltage, amplification of the signal takes place (see Art. 5-4).

Characteristics. The tube and circuit parameters for the common-cathode amplifier vary over a very wide range depending on the application intended for the circuit. For example, the output resistance of (1) a triode power amplifier may be in the order of 1,000 ohms, (2) a high-mu triode voltage amplifier in the order of 100,000 ohms, and (3) a pentode voltage amplifier in the order of 1,000,000 ohms. The input resistance is high, generally in the order of 100,000 to 1,000,000 ohms. The voltage gain varies over a wide range; for example, in the order of 5 for a triode power amplifier to several hundred for a pentode voltage amplifier. The power output ranges from several milliwatts in voltage amplifiers to many watts in power amplifiers. The power gain of a power amplifier may be in the order of 20 to 30 db. Applications of the common-cathode-type vacuum-tube amplifier circuit cover the entire field of electronics such as communications, industrial, computer, and medical.

9-22 Equivalent Circuit Analysis for a Common-cathode Vacuum-tube Amplifier Using a Triode

The Equivalent Circuit. In Chap. 5 it was shown that the grid of a tube is μ times as effective in controlling the plate current as is the plate. When an alternating signal voltage e_g is applied to the grid of a tube, the plate circuit may be considered as containing a generator of $-\mu e_g$ volts in series with the plate resistance r_p and the load impedance Z_L. This principle is a basis for calculating the operating characteristics of an amplifier circuit. The output voltage $-\mu e_g$ is negative because of the phase reversal described in Art. 9-21.

The vacuum tube used in an amplifier circuit can therefore be considered as a generator whose output voltage is equal to $-\mu e_g$. An equivalent electric circuit can be drawn for an amplifier circuit by substituting a generator for the vacuum tube. The equivalent electric circuit for the plate circuit of the basic amplifier of Fig. 9-18*a* is shown in Fig. 9-18*b*. This type of equivalent circuit is referred to as the *constant-voltage-generator* form and is very convenient for studying the operating characteristics of amplifier circuits using triodes.

Voltage Amplification. The equations for determining the voltage amplification for the circuit of Fig. 9-18*a* are derived in the following manner: The a-c component of the plate current flowing in this circuit is

$$i_p = \frac{\mu e_g}{Z_L + r_p} \tag{9-37}$$

The voltage developed across the load impedance by this current will then be

$$e_p = i_p Z_L \tag{9-38}$$

Substituting Eq. (9-37) in (9-38),

$$e_p = \frac{\mu e_g Z_L}{Z_L + r_p} \tag{9-39}$$

The voltage amplification of the circuit then becomes

$$A_v = \frac{e_p}{e_g} = \frac{\dfrac{\mu e_g Z_L}{Z_L + r_p}}{e_g} \tag{9-40}$$

and

$$A_v = \frac{\mu Z_L}{Z_L + r_p} \tag{9-41}$$

When the load impedance consists only of resistance, resistance-capacitance-coupled (*RC*) amplifier circuits, or a tuned amplifier circuit in which the reactive effects cancel one another so that the resultant impedance is only resistance, the load impedance Z_L will be equal to R_L. Under these conditions, Eqs. (9-37), (9-39), and (9-41) become

$$i_p = \frac{\mu e_g}{R_L + r_p} \tag{9-42}$$

$$e_p = \frac{\mu e_g R_L}{R_L + r_p} \tag{9-43}$$

$$A_v = \frac{\mu R_L}{R_L + r_p} \tag{9-44}$$

When the terms in Eq. (9-44) are rearranged, the load resistance required to produce a specific amount of voltage amplification may be found by

$$R_L = \frac{A_v r_p}{\mu - A_v} \tag{9-45}$$

Example 9-27 A certain triode, used as a voltage amplifier, has a plate resistance of 85,000 ohms and an amplification factor of 100 when operated with 100 volts on its plate and a grid bias of 1 volt. What is the voltage amplification of this circuit when the load resistance is 65,000 ohms?

GIVEN: $r_p = 85{,}000$ ohms $\quad R_L = 65{,}000$ ohms $\quad \mu = 100$

FIND: A_v

SOLUTION:

$$A_v = \frac{\mu R_L}{R_L + r_p} = \frac{100 \times 65{,}000}{65{,}000 + 85{,}000} = 43.3$$

Example 9-28 A 1-volt a-c signal (peak value) is applied to the input side of the amplifier circuit of Example 9-27. (*a*) What is the a-c component of the plate current? (*b*) What amount of voltage will be developed across the load resistor?

GIVEN: $e_g = 1$ volt (max) $\mu = 100$ $r_p = 85{,}000$ ohms $R_L = 65{,}000$ ohms
$A_v = 43.3$ (from Example 9-27)

FIND: (*a*) i_p (*b*) e_p

SOLUTION:

(*a*)
$$i_p = \frac{\mu e_g}{R_L + r_p} = \frac{100 \times 1}{65{,}000 + 85{,}000} = 0.666 \text{ ma (max)}$$

(*b*) $e_p = A_v \times e_g = 43.3 \times 1 = 43.3$ volts (max)

or $e_p = i_p \times R_L = 0.666 \times 10^{-3} \times 65{,}000 = 43.3$ volts (max)

Example 9-29 What is the voltage amplification of the amplifier of Example 9-27 if the value of the load resistance is increased to 130,000 ohms?

GIVEN: $r_p = 85{,}000$ ohms $R_L = 130{,}000$ ohms $\mu = 100$

FIND: A_v

SOLUTION:

$$A_v = \frac{\mu R_L}{R_L + r_p} = \frac{100 \times 130{,}000}{130{,}000 + 85{,}000} = 60.4$$

Example 9-30 What value of load resistance is required in order to obtain a voltage amplification of 50 for the amplifier of Example 9-27?

GIVEN: $A_v = 50$ $r_p = 85{,}000$ ohms $\mu = 100$

FIND: R_L

SOLUTION:

$$R_L = \frac{A_v r_p}{\mu - A_v} = \frac{50 \times 85{,}000}{100 - 50} = 85{,}000 \text{ ohms}$$

Relation among r_p, Z_L, ***and*** A_v. The maximum theoretical value of voltage amplification per amplifier stage is indicated by the amplification factor of the tube. However, this theoretical maximum value cannot be obtained in practical amplifier circuits, as the voltage amplification is limited by the plate resistance of the tube and the impedance of the load.

In order to obtain the maximum voltage amplification per stage with tubes having approximately the same amplification factor, a tube having the lowest plate resistance should be used. This is shown by Eqs. (9-41) and (9-44), which indicate that for a definite plate load impedance the voltage amplification will increase as the plate resistance is decreased. This should not be confused with the maximum power output, which occurs when Z_L or R_L is equal to r_p. These two equations also show that for given values of plate resistance and amplification factor of a tube, the voltage amplification of a circuit will increase if the value of the load impedance is increased.

9-23 Equivalent Circuit Analysis for a Common-cathode Vacuum-tube Amplifier Using a Pentode

The Equivalent Circuit. The basic circuit for a common-cathode vacuum-tube amplifier using a pentode is shown in Fig. 9-19*a* and its equivalent plate-circuit diagram is shown in Fig. 9-19*b*. Applying a signal voltage e_g to the input circuit of the tube may be considered the same as though the tube generated a current equal to $-g_m e_g$ that is made to flow through a parallel circuit formed by the plate resistance r_p and the load impedance Z_L. Thus an amplifier circuit may also be considered in the form shown in Fig. 9-19*b*, generally called the *constant-current-generator* form. This type of circuit is convenient for studying the operating characteristics of amplifier circuits using high-mu tubes.

Voltage Amplification. The maximum voltage amplification of a circuit using a low-mu triode can be made almost equal to its amplification factor because of the comparatively low values of plate resistance obtainable in low-mu triode amplifier tubes. In amplifier circuits using high-mu triodes, tetrodes, and pentodes the values of plate resistance are so high that the amount of voltage amplification obtainable is but a fraction of the amplification factor of the tube used. The large difference between the voltage amplification of the circuit and the amplification factor of these tubes is due to the high values of plate

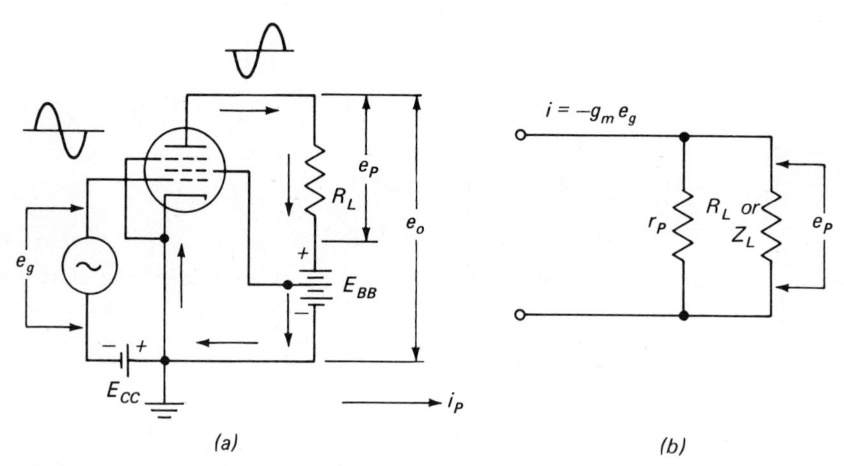

Fig. 9-19 Common-cathode vacuum-tube pentode amplifier circuit. (a) Basic circuit. (b) Equivalent plate circuit in the constant-current-generator form.

resistance and the comparatively low values of load impedance that must be used for practical amplifier circuits. When the ratio between the plate resistance and the load impedance becomes great, the transconductance of the output circuit approaches the value obtained when the load impedance is practically zero. The voltage amplification of the circuit will then be dependent on the value of the transconductance of the tube rather than on its amplification factor. This can be seen if Eqs. (9-37), (9-39), and (9-41) are expressed in terms of the transconductance instead of the amplification factor. Substituting $g_m r_p$ for μ [see Eq. (5-5)] in Eq. (9-37) and regrouping the terms,

$$i_p = g_m e_g \frac{r_p}{Z_L + r_p} \tag{9-46}$$

Substituting $g_m r_p$ for μ in Eq. (9-39) and regrouping, the voltage across the load impedance then becomes

$$e_p = g_m e_g \frac{r_p Z_L}{Z_L + r_p} \tag{9-47}$$

Substituting $g_m r_p$ for μ in Eq. (9-41) and regrouping, the voltage amplification of the circuit then becomes

$$A_v = g_m \frac{r_p Z_L}{Z_L + r_p} \tag{9-48}$$

When r_p is much larger than Z_L

$$A_v \cong g_m Z_L \tag{9-49}$$

When the terms in Eq. (9-47) are rearranged, the load impedance required to produce a definite value of voltage amplification may be found by

$$Z_L = \frac{A_v r_p}{g_m r_p - A_v} \tag{9-50}$$

When the load in the plate circuit is a resistance or a tuned circuit operated at its resonant frequency, the term R_L can be substituted for Z_L in Eqs. (9-46) to (9-50); also, R_L and r_p can then be added arithmetically.

Example 9-31 What is the voltage amplification of a vacuum-tube pentode amplifier circuit if the tube used has a plate resistance of 0.3 megohm, transconductance of 5,000 μmhos, and a load resistance of 100,000 ohms?

GIVEN: $r_p = 300{,}000$ ohms $\quad g_m = 5{,}000\ \mu$mhos $\quad R_L = 100{,}000$ ohms

FIND: A_v

SOLUTION:

$$A_v = g_m \frac{r_p R_L}{R_L + r_p} = 5{,}000 \times 10^{-6} \times \frac{300{,}000 \times 100{,}000}{100{,}000 + 300{,}000} = 375$$

Example 9-32 A certain pentode used in a voltage-amplifier circuit has a plate resistance of 300,000 ohms and a transconductance of 2,500 μmhos. If the plate load is a resistive load of 10,000 ohms, what is the voltage amplification of the circuit when determined by use of (*a*) Eq. (9-48)? (*b*) Eq. (9-49)?

GIVEN: $r_p = 300{,}000$ ohms $\quad g_m = 2{,}500$ μmhos $\quad R_L = 10{,}000$ ohms

FIND: (*a*) A_v [using Eq. (9-48)] $\quad$ (*b*) A_v [using Eq. (9-49)]

SOLUTION:

(*a*)
$$A_v = g_m \frac{r_p R_L}{R_L + r_p} = 2{,}500 \times 10^{-6} \times \frac{300{,}000 \times 10{,}000}{10{,}000 + 300{,}000} = 24.2$$

(*b*)
$$A_v \cong g_m R_L \cong 2{,}500 \times 10^{-6} \times 10{,}000 \cong 25$$

Example 9-33 A certain pentode is operated as a voltage amplifier. The transconductance of the tube is 1,575 μmhos and the plate resistance is 700,000 ohms when the tube is operated with 100 volts on the plate, 100 volts on the screen grid, and with a grid bias of 3 volts. (*a*) If the plate load is a tuned circuit, what must its effective impedance be in order to obtain a voltage amplification per stage of 90? (*b*) What is the a-c component of the plate current when a 20-mv a-c signal is applied to the grid circuit of the tube? (*c*) What amount of voltage will be developed across the load impedance?

GIVEN: $g_m = 1{,}575$ μmhos $\quad r_p = 700{,}000$ ohms $\quad A_v = 90 \quad e_g = 20$ mv

FIND: (*a*) Z_L $\quad$ (*b*) i_p $\quad$ (*c*) e_p

SOLUTION:

(*a*)
$$Z_L = \frac{A_v r_p}{g_m r_p - A_v} = \frac{90 \times 700{,}000}{(1{,}575 \times 10^{-6} \times 700{,}000) - 90} = 62{,}222 \text{ ohms}$$

(*b*)
$$i_p = g_m e_g \frac{r_p}{Z_L + r_p} = \frac{1{,}575 \times 10^{-6} \times 20 \times 10^{-3} \times 7 \times 10^5}{62{,}222 + 700{,}000} = 28.9 \ \mu\text{a}$$

(*c*)
$$e_p = i_p Z_L = 28.9 \times 10^{-6} \times 62{,}222 = 1.798 \text{ volts}$$

or
$$e_p = A_v e_g = 90 \times 20 \times 10^{-3} = 1.8 \text{ volts}$$

9-24 The Common-grid Vacuum-tube Amplifier

Circuit Operation. Figure 9-20 shows the basic circuit and a simplified equivalent circuit for a common-grid vacuum-tube amplifier using a triode. In this circuit, the input signal is injected into the grid-to-cathode circuit, and the output signal is taken from the plate-to-grid circuit, making the grid the common element to both the input and output circuits. The configuration of the basic common-grid vacuum-tube amplifier is similar to the basic common-base transistor amplifier (see Fig. 9-2).

In the common-grid vacuum-tube amplifier (Fig. 9-20) the power source E_{CC} makes the grid of the tube negative with respect to the cathode, and power source E_{BB} makes the plate of the tube positive with respect to the cathode. With zero input signal, power sources E_{CC} and E_{BB}, together with resistors R_k and R_L, establish the quiescent values of the grid voltage E_C and the plate

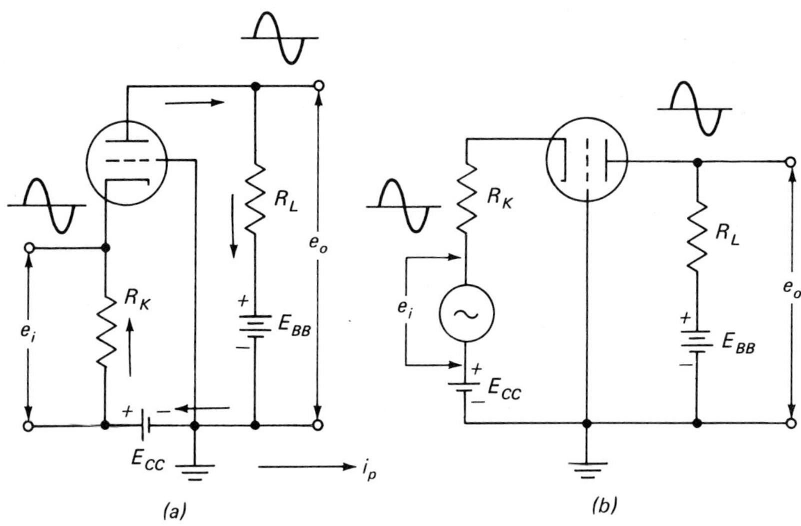

Fig. 9-20 Common-grid vacuum-tube triode amplifier circuit. (a) Basic circuit. (b) Simplified equivalent circuit of the basic circuit in (a).

current I_P. When the input signal is positive-going: (1) the effective grid voltage becomes less negative, (2) the plate current increases, (3) the voltage drop at R_L decreases, and (4) the output voltage increases. Thus a positive-going input signal produces a positive-going output signal and consequently the input and output signals are in phase. A similar step-by-step analysis will show that a negative-going input signal will result in a negative-going output signal. Because a small change in the input-signal voltage will produce a relatively large change in the output-signal voltage, amplification of the input signal takes place.

Characteristics. The common-grid vacuum-tube amplifier is used for very high frequency circuit applications such as in television and radar. Neutralization is not necessary with the common-grid amplifier circuit as the grid is grounded and thereby serves as a shield or screen between the plate and cathode. Regeneration can be caused only by the interelectrode capacitance between the plate and the cathode, since the input is to the cathode. Grounding the grid effectively shields the input circuit to the cathode and prevents it from receiving any feedback. Grounding the grid and introducing the signal between the cathode and ground produces about the same amount of amplification as applying the signal directly to the grid with the cathode grounded. The tube behaves as if the amplification factor was $\mu + 1$. However, the input resistance r_i is much lower and is approximately equal to

$$r_i \cong \frac{1}{g_m} \tag{9-51}$$

The voltage amplification of the common-grid circuit can be calculated in the same manner as with the common-cathode amplifier (Arts. 9-22 and 9-23); therefore, Eqs. (9-41) and (9-47) may be used.

The characteristics of the common-grid amplifier are (1) low input resistance, (2) low output resistance compared to the common-cathode amplifiers, (3) high gain when the source resistance is low, (4) plate current flows through the source of the input signal thereby producing a degenerative reaction which reduces the voltage amplification as compared with that of the common-cathode amplifier, (5) the output and input circuits are isolated insofar as stray capacitance is concerned, and (6) triodes can be used in low-noise, simple, inexpensive circuits.

Example 9-34 One section of a twin-triode tube being used as a common-grid high-frequency amplifier has a transconductance of 5,300 μmhos and a plate resistance of 7,100 ohms. What is the effective input resistance of the tube?

GIVEN: $g_m = 5{,}300$ μmhos

FIND: r_i

SOLUTION:

$$r_i \cong \frac{1}{g_m} \cong \frac{1}{5{,}300 \times 10^{-6}} \cong 188 \text{ ohms}$$

Example 9-35 What is the voltage amplification of the amplifier of Example 9-34 if the load resistance is 2,660 ohms?

GIVEN: $g_m = 5{,}300$ μmhos $r_p = 7{,}100$ ohms $R_L = 2{,}660$ ohms

FIND: A_v

SOLUTION:

$$A_v = g_m \frac{r_p R_L}{r_p + R_L} = 5{,}300 \times 10^{-6} \times \frac{7{,}100 \times 2{,}660}{7{,}100 + 2{,}660} = 10.27$$

9-25 The Common-plate Vacuum-tube Amplifier

Circuit Operation. Figure 9-21 shows the basic circuit and a simplified equivalent circuit for a common-plate vacuum-tube amplifier using a triode. This circuit is most commonly called a *cathode follower* but is also known as a *cathode-coupled amplifier*, and a *grounded-plate amplifier*. In this circuit, the input signal is injected into the grid-to-plate circuit and the output is taken from the load resistor R_L which is a-c (or signalwise) connected across the cathode-to-plate path by means of the low-impedance capacitor C. Thus the plate of the tube is the common element to both the input and output circuits. The configuration of the basic common-plate vacuum-tube amplifier is similar to the basic common-collector transistor amplifier (see Fig. 9-3).

In the common-plate vacuum-tube amplifier (Fig. 9-21) the grid-to-cathode voltage is the algebraic sum of e_o, E_{CC}, and e_i; the total voltage should always maintain the grid negative with respect to the cathode. The plate of the tube is made positive with respect to the cathode by the power source E_{BB}. With zero input signal, power sources E_{CC} and E_{BB}, together with the load resistor R_L, establish the quiescent values of the grid voltage E_C and the plate current I_P. When the input signal is positive-going: (1) the effective grid voltage becomes less negative, (2) the plate current increases, (3) the voltage drop at R_L increases, and (4) the output voltage, being taken off the load resistor R_L, therefore also increases. Thus a positive-going input signal produces a positive-going output signal and consequently the input and output signals are in phase. A similar step-by-step analysis will show that a negative-going input signal will result in a negative-going output signal. Capacitor C must be large enough to effectively ground the plate a-c signal variations. The cathode current is the sum of (1) the quiescent plate current produced by the power source E_{BB}, and (2) the a-c signal in the capacitor path. Because the output voltage is dependent on (1) the value of the resistor R_L in the cathode circuit and (2) the value of the cathode current, this circuit is often called a *cathode follower*.

In addition to the effect of a varying input signal just described above, applying a positive-going signal to the input circuit causes an increase in the voltage across R_L and makes the cathode more positive with respect to ground. This is equivalent to making the grid more negative with respect to the cathode

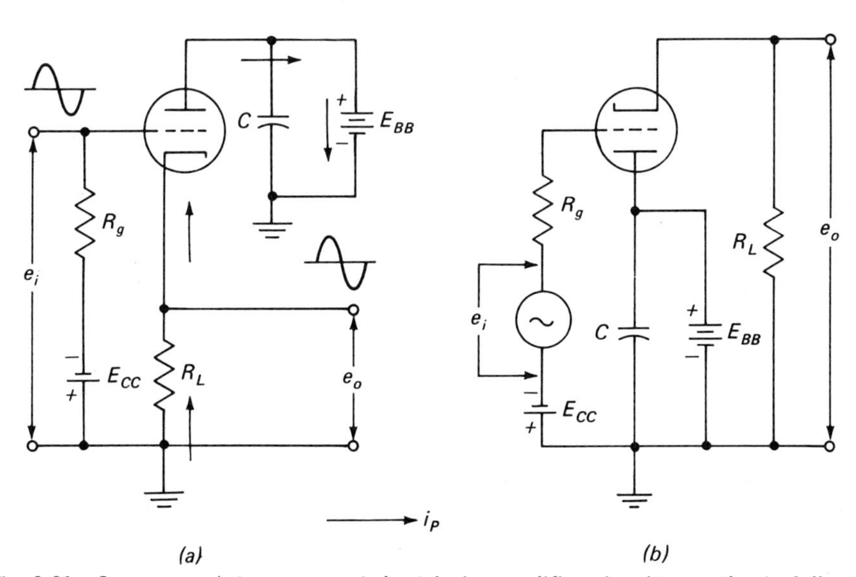

Fig. 9-21 Common-plate vacuum-tube triode amplifier circuit or cathode follower. (a) Basic circuit. (b) Simplified equivalent circuit of the basic circuit in (a).

which is opposite to the effect produced by the original positive-going signal. Thus, this circuit is a *degenerative-feedback amplifier*, also called an *inverse-feedback* or *negative-feedback amplifier*.

Voltage Amplification. In order that the signal applied to the grid may produce changes in the plate current, the voltage across R_L can never equal or exceed the signal voltage. The voltage amplification of a common-plate amplifier circuit is therefore always less than 1. However, the common-plate amplifier does provide a power gain and is capable of supplying relatively large values of signal current to low-impedance circuits. The voltage amplification of the common-plate amplifier circuit is expressed by

For triodes
$$A_v = \frac{\mu R_L}{r_p + R_L(\mu + 1)} \tag{9-52}$$

For pentodes
$$A_v = \frac{g_m R_L}{1 + g_m R_L} \tag{9-53}$$

In Eqs. (9-52) and (9-53) the denominator will always be greater than the numerator; hence the voltage amplification will always be less than 1. The higher the values are for μ and R_L with triodes, and g_m and R_L with pentodes, the closer the voltage amplification will approach 1.

Characteristics. The common-plate vacuum-tube amplifier is used as an impedance-matching or impedance-changing circuit. It is useful in converting a signal from a high-impedance load to a low-impedance load. Characteristics of the circuit are (1) high input impedance, (2) low output impedance, (3) low input capacitance, (4) voltage amplification of less than 1, (5) no phase inversion, and (6) low distortion.

Example 9-36 A certain triode having an amplification factor of 20 and a plate resistance of 6,500 ohms is used in a common-plate amplifier circuit. What is the voltage amplification of the circuit when the value of the load resistor is (*a*) 22,000 ohms? (*b*) 2,200 ohms? (*c*) 220 ohms?

GIVEN: $\mu = 20$ $r_p = 6{,}500$ ohms $R_L =$ (*a*) 22,000 ohms (*b*) 2,200 ohms (*c*) 220 ohms

FIND: (*a*) A_v (*b*) A_v (*c*) A_v

SOLUTION:

(*a*)
$$A_v = \frac{\mu R_L}{r_p + R_L(\mu + 1)} = \frac{20 \times 22{,}000}{6{,}500 + 22{,}000 \times (20 + 1)} \cong 0.94$$

(*b*)
$$A_v = \frac{\mu R_L}{r_p + R_L(\mu + 1)} = \frac{20 \times 2{,}200}{6{,}500 + 2{,}200 \times (20 + 1)} \cong 0.834$$

(*c*)
$$A_v = \frac{\mu R_L}{r_p + R_L(\mu + 1)} = \frac{20 \times 220}{6{,}500 + 220 \times (20 + 1)} \cong 0.396$$

Example 9-37 A certain pentode having a transconductance of 6,000 μmhos is used in a common-plate amplifier circuit. What is the voltage amplification of the circuit when the value of the load resistor is (*a*) 10,000 ohms? (*b*) 1,000 ohms? (*c*) 100 ohms?

GIVEN: $g_m = 6{,}000$ μmhos $R_L =$ (*a*) 10,000 ohms (*b*) 1,000 ohms (*c*) 100 ohms

FIND: (*a*) A_v (*b*) A_v (*c*) A_v

SOLUTION:

(*a*)
$$A_v = \frac{g_m R_L}{1 + g_m R_L} = \frac{6{,}000 \times 10^{-6} \times 10{,}000}{1 + 6{,}000 \times 10^{-6} \times 10{,}000} \cong 0.983$$

(*b*)
$$A_v = \frac{g_m R_L}{1 + g_m R_L} = \frac{6{,}000 \times 10^{-6} \times 1{,}000}{1 + 6{,}000 \times 10^{-6} \times 1{,}000} \cong 0.857$$

(*c*)
$$A_v = \frac{g_m R_L}{1 + g_m R_L} = \frac{6{,}000 \times 10^{-6} \times 100}{1 + 6{,}000 \times 10^{-6} \times 100} \cong 0.375$$

QUESTIONS

1. (*a*) What is the function of an amplifier? (*b*) Name two kinds of active components used in amplifiers.
2. Name six classifications of amplifiers.
3. (*a*) What are the input and output impedance characteristics of a transistor? (*b*) How is the gain of a transistor expressed?
4. (*a*) What are the input and output impedance characteristics of a vacuum tube (*b*) How is the gain of a vacuum tube expressed?
5. In terms of voltage and current as a controlling factor, how does a transistor amplifier differ from a vacuum-tube amplifier?
6. (*a*) Name three types of transistor amplifier circuits. (*b*) Name three types of vacuum-tube amplifier circuits.
7. Explain the use of *common* and *grounded* in describing a type of amplifier circuit.
8. What is the correlation among the emitter, base, and collector of a transistor and the grid, plate, and cathode of a vacuum tube?
9. Which vacuum-tube amplifier configuration is broadly taken as being comparable to (*a*) a common-emitter transistor amplifier, (*b*) a common-base transistor amplifier, (*c*) a common-collector transistor amplifier?
10. (*a*) Which transistor amplifier-circuit configuration is most frequently used? (*b*) Why?
11. Describe in detail the operation of a PNP common-emitter transistor amplifier circuit with a positive-going input signal.
12. Describe the following characteristics of the common-emitter transistor amplifier circuit: (*a*) current gain, (*b*) voltage gain, (*c*) power gain, (*d*) stability.
13. Describe in detail the operation of a PNP common-emitter transistor amplifier circuit with a negative-going input signal.
14. Describe in detail the operation of an NPN common-emitter transistor amplifier circuit with a positive-going input signal.
15. Describe the path of base-current flow for the circuit of Fig. 9-4*a*.

16. Describe the path of collector-current flow for the circuit of Fig. 9-4*a*.
17. Describe the composition of the emitter current for the circuit of Fig. 9-4*a*.
18. Describe in detail the operation of a PNP common-base transistor amplifier circuit with a positive-going input signal.
19. Describe the following characteristics of the common-base transistor amplifier circuit: (*a*) current gain, (*b*) voltage gain, (*c*) power gain, (*d*) stability.
20. Describe in detail the operation of a PNP common-base transistor amplifier circuit with a negative-going input signal.
21. Describe in detail the operation of an NPN common-base transistor amplifier circuit with a positive-going input signal.
22. Describe the path of the base-current flow for the circuit of Fig. 9-5*a*.
23. Describe the path of the collector-current flow for the circuit of Fig. 9-5*a*.
24. What is the composition of the emitter current for the circuit of Fig. 9-5*a*?
25. Describe in detail the operation of a PNP common-collector transistor amplifier circuit with a positive-going input signal.
26. Describe the following characteristics of the common-collector transistor amplifier circuit: (*a*) current gain, (*b*) voltage gain, (*c*) power gain, (*d*) stability.
27. Describe in detail the operation of a PNP common-collector transistor amplifier circuit with a negative-going input signal.
28. Describe in detail the operation of an NPN common-collector transistor amplifier circuit with a positive-going input signal.
29. Describe the path of the base-current flow for the circuit of Fig. 9-6*a*.
30. Describe the path of the collector-current flow for the circuit of Fig. 9-6*a*.
31. What is the composition of the emitter current for the circuit of Fig. 9-6*a*?
32. What is the relationship between a common-collector transistor amplifier and the emitter-follower amplifier?
33. Prepare a tabulation showing the relative values (low, medium, high) for the input and output impedances of the following amplifier configurations: (*a*) common-emitter, (*b*) common-base, (*c*) common-collector.
34. Compare the phase relation of the output signal with the input signal for the following amplifier configurations: (*a*) common-emitter, (*b*) common-base, (*c*) common-collector.
35. When using the *h*-parameters for analyzing the characteristics of an amplifier circuit, give (*a*) the advantages, (*b*) the disadvantages.
36. What are the advantages of determining the operating characteristics of a transistor amplifier circuit by using the beta value of the transistor and the values of the circuit components?
37. (*a*) Give a definition for a single-stage transistor amplifier. (*b*) What are the principal operating characteristics of this amplifier circuit?
38. (*a*) What is the purpose of setting up an a-c equivalent circuit for a transistor amplifier circuit? (*b*) What substitutions are made?
39. How are the *h*-parameters of a transistor determined?
40. Name five operating characteristics of a transistor amplifier circuit that can be determined with the *h*-parameters of the transistor.
41. What assumptions are made when determining the operating characteristics of a transistor amplifier circuit by using the beta value of the transistor and the values of the circuit components?

42. What is the effect on the current gain of a common-emitter transistor amplifier circuit when the value of the resistor in the base circuit is (*a*) increased? (*b*) Decreased?
43. What is the effect on the current gain of a common-emitter transistor amplifier circuit when the value of the load resistor is (*a*) increased? (*b*) Decreased?
44. What is the effect on the voltage gain of a common-emitter transistor amplifier circuit when the value of the signal source resistance is (*a*) increased? (*b*) Decreased?
45. (*a*) What is the purpose of plotting load lines on the transistor characteristics curves? (*b*) What characteristics of an amplifier circuit can be determined by use of load lines? (*c*) Why are both d-c and a-c load lines used?
46. To what extent do the a-c and d-c load lines (*a*) correspond? (*b*) Differ?
47. Where should the quiescent point be located on the load line in order to accommodate the maximum input-signal voltage without causing (*a*) cutoff? (*b*) Distortion?
48. Define (*a*) base bias, (*b*) emitter bias, (*c*) swamping resistor.
49. Explain the function and operation of a swamping resistor.
50. Explain the function and operation of voltage-divider-controlled base bias.
51. (*a*) Define collector bias. (*b*) How is the term *feedback* associated with collector bias?
52. Describe the operation of the common-base transistor amplifier circuit of Fig. 9-16*a*.
53. How can the operating characteristics of a common-base transistor amplifier circuit be determined when the *h*-parameters for only the common-emitter configuration are known?
54. Describe the operation of the common-collector transistor amplifier circuit of Fig. 9-17*a*.
55. How can the operating characteristics of a common-collector transistor amplifier circuit be determined when the *h*-parameters for only the common-emitter configuration are known?
56. Compare the basic operation of the common-cathode vacuum-tube amplifier with that of the common-emitter transistor amplifier.
57. Describe in detail the operation of the common-cathode vacuum-tube amplifier circuit with a positive-going input signal.
58. Describe in detail the operation of the common-cathode vacuum-tube amplifier circuit with a negative-going input signal.
59. Describe the following characteristics of the common-cathode vacuum-tube triode amplifier circuit: (*a*) output resistance, (*b*) input resistance, (*c*) voltage amplification, (*d*) power output, (*e*) power gain.
60. Describe the following characteristics of the common-cathode vacuum-tube pentode amplifier circuit: (*a*) output resistance, (*b*) input resistance, (*c*) voltage amplification, (*d*) power output, (*e*) power gain.
61. Describe some applications of the common-cathode vacuum-tube amplifier circuit.
62. (*a*) What is meant by a constant-voltage-generator form of equivalent circuit? (*b*) What is its use?
63. In the equivalent circuit vacuum-tube analysis, define (*a*) A_v, (*b*) μ, (*c*) r_p, (*d*) Z_L, (*e*) R_L, (*f*) e_g, (*g*) i_p, (*h*) e_p, (*i*) g_m, (*j*) e_i, (*k*) e_o.
64. How is the voltage gain of an amplifier using a triode vacuum tube affected by a change in the value of (*a*) r_p? (*b*) Z_L? (*c*) μ?
65. (*a*) What is meant by a constant-current-generator form of equivalent circuit? (*b*) What is its use?

66. How is the voltage gain of an amplifier using a pentode vacuum tube affected by a change in the value of (*a*) g_m? (*b*) r_p? (*c*) Z_L?
67. What two factors largely determine the voltage gain of a pentode vacuum-tube amplifier when the value of r_p is much higher than the value of Z_L?
68. Describe in detail the operation of the common-grid vacuum-tube amplifier circuit with a positive-going input signal.
69. Describe in detail the operation of the common-grid vacuum-tube amplifier circuit with a negative-going input signal.
70. Describe the operating characteristics of the common-grid vacuum-tube amplifier.
71. Compare the basic operation of the common-grid vacuum-tube amplifier with that of the common-base transistor amplifier.
72. Describe some applications of the common-grid vacuum-tube amplifier circuit.
73. Describe in detail the operation of the common-plate vacuum-tube amplifier circuit with a positive-going input signal.
74. Describe in detail the operation of the common-plate vacuum-tube amplifier circuit with a negative-going input signal.
75. Why is the common-plate vacuum-tube amplifier circuit frequently referred to as a *cathode follower*?
76. Describe the following characteristics of the common-plate vacuum-tube amplifier circuit: (*a*) input impedance, (*b*) output impedance, (*c*) input capacitance, (*d*) voltage amplification, (*e*) distortion.
77. Compare the basic operation of the common-plate vacuum-tube amplifier with that of the common-collector transistor amplifier.
78. Describe some applications of the common-plate vacuum-tube amplifier.
79. Compare the phase relation of the output signal with the input signal for the following amplifier configurations: (*a*) common cathode, (*b*) common grid, (*c*) common plate.
80. Why does the common-plate vacuum-tube amplifier circuit have a negative-feedback characteristic?

PROBLEMS

1. From the curves of Fig. 8-1, determine the value of h_{oe} when the base current is 20 μa.
2. From the curves of Fig. 8-1, determine the value of h_{oe} when the base current is 50 μa.
3. From the curves of Fig. 8-1, determine the value of h_{fe} when V_{CE} is 4 volts and the base current varies between 10 and 30 μa.
4. From the curves of Fig. 8-1, determine the value of h_{fe} when V_{CE} is 4 volts and the base current varies between 30 and 50 μa.

Table 9-2 Test Data for Input Static Characteristics with V_{CE} Constant at 4 Volts. The Transistor Represented Is the Same as Used for Fig. 8-1

I_B, μa	0	10	20	30	40	50	80	100
V_{BE}, mv	80	135	158	172	180	188	198	208

Table 9-3 Test Data for Input Static Characteristics with V_{CE} Constant at 8 Volts. The Transistor Represented Is the Same as Used for Fig. 8-1

I_B, μa.	0	10	20	30	40	50	80	100
V_{BE}, mv.	90	140	165	180	187	195	205	215

5. From the test data in Table 9-2, determine the value of h_{ie} for $V_{CE} = 4$ volts and I_B varying between 10 and 30 μa.
6. From the test data in Table 9-2, determine the value of h_{ie} for $V_{CE} = 4$ volts and I_B varying between 30 and 50 μa.
7. From the test data in Tables 9-2 and 9-3, determine the value of h_{re} for $I_B = 20$ μa and V_{CE} varying between 4 and 8 volts.
8. From the test data in Tables 9-2 and 9-3, determine the value of h_{re} for $I_B = 40$ μa and V_{CE} varying between 4 and 8 volts.
9. If the transistor whose h-parameters are found in Probs. 1, 3, 5, and 7 is substituted for the transistor shown in Fig. 9-8, determine the input resistance of the amplifier by use of (*a*) Eq. (9-5), (*b*) Eq. (9-5*a*). (*Note:* Use $h_{oe} = 2 \times 10^{-4}$ mho; $h_{fe} = 150$; $h_{ie} = 1{,}850$ ohms; $h_{re} = 175 \times 10^{-5}$.)
10. Find the input resistance of the common-emitter transistor amplifier circuit of Fig. 9-8 when the circuit parameters are $V_{CC} = 18$ volts, $R_B = 1$ megohm, $R_L = 10{,}000$ ohms, $R_o = 20{,}000$ ohms. The transistor parameters are $h_{ie} = 1{,}000$ ohms, $h_{fe} = 100$, $h_{oe} = 5 \times 10^{-5}$ mho, $h_{re} = 4 \times 10^{-4}$.
11. Find the output resistance r_o' of the transistor in the amplifier circuit of Prob. 9. (*Note:* $R_g = 0$.)
12. Find the output resistance r_o' of the transistor in the amplifier circuit of Prob. 10. (*Note:* $R_g = 0$.)
13. Find the current gain of the amplifier circuit of Prob. 9.
14. Find the current gain of the amplifier circuit of Prob. 10.
15. Find the voltage gain of the amplifier circuit of Prob. 9.
16. Find the voltage gain of the amplifier circuit of Prob. 10.
17. Find the power gain of the amplifier circuit of Prob. 9 in (*a*) power amplification, (*b*) decibels.
18. Find the power gain of the amplifier circuit of Prob. 10 in (*a*) power amplification. (*b*) decibels.
19. Find the a-c input resistance of the amplifier circuit of Fig. 9-8 if the beta of the transistor is 100.
20. Find the a-c input resistance of the amplifier circuit of Fig. 9-8 if the value of F_B is 500,000 ohms.
21. In the circuit of Fig. 9-8, what value would be required at R_B in order to obtain an a-c input resistance of 1,000 ohms?
22. Find the output resistance of the circuit in Fig. 9-8 if R_L is increased to 12,000 ohms.
23. In the circuit of Fig. 9-8, what value would be required at R_L in order to obtain an output resistance of 10,000 ohms?
24. In the circuit of Fig. 9-8, what value would be required at R_o in order to obtain an output resistance of 5,000 ohms?

25. What is the approximate current gain of the amplifier circuit of Fig. 9-8 for the circuit conditions of Prob. 19?

26. What is the approximate current gain of the amplifier circuit of Fig. 9-8 for the circuit conditions of Prob. 20?

27. What is the approximate current gain of the amplifier circuit of Fig. 9-8 for the circuit conditions of Prob. 21?

28. What is the approximate voltage gain of the amplifier circuit of Fig. 9-8 for the circuit conditions of Probs. 19 and 22?

29. What is the approximate voltage gain of the amplifier circuit of Fig. 9-8 for the circuit conditions of Probs. 20 and 23?

30. What is the approximate voltage gain of the amplifier circuit of Fig. 9-8 for the circuit conditions of Probs. 21 and 24?

31. What is the power-amplification ratio and the decibel gain for the amplifier of Probs. 25 and 28?

32. What is the power-amplification ratio and the decibel gain for the amplifier of Probs. 26 and 29?

33. To what per cent of the beta value is the current gain of the transistor amplifier circuit of Fig. 9-8 reduced if the value of R_B is ten times the value of βr_e, and β is 100?

34. To what per cent of the beta value is the current gain of the transistor amplifier circuit of Fig. 9-8 reduced if the value of R_B is ten times the value of βr_e, and β is 50?

35. To what per cent of the beta value is the current gain of the transistor amplifier circuit of Fig. 9-8 reduced if the value of R_B is double the value of βr_e?

36. To what per cent of the beta value is the current gain of the transistor amplifier circuit of Fig. 9-8 reduced if the value of R_B is 20 times the value of βr_e?

37. Using Eq. (9-21), to what per cent of the beta value is the current gain of an amplifier circuit reduced if the values of R_L and R_o are the same?

38. Using Eq. (9-21), to what per cent of the beta value is the current gain of an amplifier circuit reduced if the value of R_L is double the value of R_o?

39. Using Eq. (9-21), to what per cent of the beta value is the current gain of an amplifier circuit reduced if the value of R_o is double the value of R_L?

40. For Eq. (9-21), what ratio of values between R_L and R_o is required in order that the current gain will be 80 per cent of the beta value?

41. From the solution of Example 9-13, derive an equation that will express the ratio of A_v'/A_v in terms of βr_e and R_g.

42. Using the equation derived in Prob. 41, determine by what per cent the voltage gain of a transistor amplifier is reduced because of the effect of the signal-source resistance when R_g is equal to βr_e.

43. Using the equation derived in Prob. 41, determine by what per cent the voltage gain of a transistor amplifier is reduced because of the effect of the signal-source resistance when βr_e is ten times the value of R_g.

44. Using the equation derived in Prob. 41, determine by what per cent the voltage gain of a transistor amplifier is reduced because of the effect of the signal-source resistance when βr_e is double the value of R_g.

45. (*a*) Plot the d-c and a-c load lines for the amplifier circuit of Fig. 9-8 for the condition when the voltage of the power source is decreased to 20 volts. (*b*) Indicate the quiescent point on the load lines.

46. (*a*) Plot the d-c and a-c load lines for the amplifier circuit of Fig. 9-8 for the condition when R_L is increased to 12,000 ohms. (*b*) Indicate the quiescent point on the load lines.

47. (*a*) What is the maximum a-c output-signal voltage that can be accommodated by the amplifier of Prob. 45 without clipping off any of the signal? (*b*) What is the corresponding maximum a-c input-signal voltage?

48. (*a*) What is the maximum a-c output-signal voltage that can be accommodated by the amplifier of Prob. 46 without clipping off any of the signal? (*b*) What is the corresponding maximum a-c input-signal voltage?

49. (*a*) What value of base resistance should be substituted in the amplifier circuit of Prob. 45 in order that the amplifier will be capable of handling the maximum possible output signal? (*b*) What is the maximum a-c output signal that can be used without clipping off any of the signal? (*c*) What is the corresponding maximum a-c input-signal voltage?

50. (*a*) What value of base resistance should be substituted in the amplifier circuit of Prob. 46 in order that the amplifier will be capable of handling the maximum possible output signal? (*b*) What is the maximum a-c output signal that can be used without clipping off any of the signal? (*c*) What is the corresponding maximum a-c input-signal voltage?

51. If the value of the swamping resistor R_E in the circuit of Fig. 9-13 is reduced to 5,000 ohms, find (*a*) input resistance, (*b*) output resistance, (*c*) voltage gain of the circuit, (*d*) output voltage, (*e*) current gain of the circuit, (*f*) decibel power gain. (*Note:* Consider R_g to be negligible.)

52. If the value of the swamping resistor R_E in the circuit of Fig. 9-13 is reduced to 2,500 ohms, find (*a*) input resistance, (*b*) output resistance, (*c*) voltage gain of the circuit, (*d*) output voltage, (*e*) current gain of the circuit, (*f*) decibel power gain. (*Note:* Consider R_g to be negligible.)

53. If the values of R_B and R_S in the circuit of Fig. 9-14 are changed to 50,000 ohms and 100,000 ohms, respectively, find (*a*) emitter current, (*b*) input resistance, (*c*) output resistance, (*d*) voltage gain, (*e*) output voltage.

54. If the values of R_L, R_o, R_E, and β in the circuit of Fig. 9-14 are changed to 5,000 ohms, 2,000 ohms, 2,000 ohms, and 100 ohms, respectively, find (*a*) emitter current, (*b*) input resistance, (*c*) output resistance, (*d*) voltage gain, (*e*) output voltage.

55. If the parameters for the circuit of Fig. 9-15 are $V_{CC} = 22$ volts, $V_{BE} = 0.2$ volt, $\beta = 65$, $R_B = 100{,}000$ ohms, $R_L = 3{,}300$ ohms, and $R_o = 27{,}000$ ohms, find (*a*) emitter current, (*b*) output resistance, (*c*) input resistance, (*d*) voltage gain, (*e*) current gain using Eq. (9-21), (*f*) decibel power gain.

56. If the parameters for the circuit of Fig. 9-15 are $V_{CC} = 9$ volts, $V_{BE} = 0.2$ volt, $\beta = 150$, $R_B = 560{,}000$ ohms, $R_L = 4{,}700$ ohms, and $R_o = 15{,}000$ ohms, find (*a*) emitter current, (*b*) output resistance, (*c*) input resistance, (*d*) voltage gain, (*e*) current gain using Eq. (9-21), (*f*) decibel power gain.

57. If the parameters for the circuit of Fig. 9-16*a* are $V_{CC} = 21$ volts, $v_i = 25$ mv (peak), $\alpha = 0.98$, $R_B = 20{,}000$ ohms, $R_S = 40{,}000$ ohms, $R_E = 14{,}000$ ohms, $R_L = 7{,}000$ ohms, and $R_o = 7{,}000$ ohms, find (*a*) emitter current, (*b*) input resistance, (*c*) output resistance, (*d*) current gain, (*e*) voltage gain, (*f*) output voltage, (*g*) decibel power gain.

58. Derive an equation that will express the factor by which the voltage gain for the circuit of Fig. 9-16 will be reduced by the resistance of the signal source. (*Note:* A study of Fig. 9-16*c* should lead to the conclusion that the voltage gain of the amplifier circuit will be reduced substantially by any resistance at the signal source.)

59. When R_E is much greater than r_e for the circuit represented in Fig. 9-16*c*, the term R_E can be dropped in the solution of Example 9-21 and Prob. 58. Derive the simplified equation for the factor by which the voltage gain is reduced by the signal-source resistance when the effect of R_E is ignored.

60. If the signal source of the amplifier of Prob. 57 has a resistance of 100 ohms, find (*a*) the voltage-gain reduction factor by use of the equation derived in Prob. 58, (*b*) the voltage-gain reduction factor by use of the equation derived in Prob. 59, (*c*) the voltage gain of the circuit by use of the factor found in (*a*), (*d*) the voltage gain of the circuit by use of the factor found in (*b*).

61. From the common-emitter *h*-parameters found in Probs. 1, 3, 5, 7, determine the corresponding *h*-parameters for a common-base configuration.

62. From the common-emitter *h*-parameters found in Probs. 2, 4, 6, 8, determine the corresponding *h*-parameters for a common-base configuration.

63. Using a transistor having the *h*-parameters found in Prob. 61 in a circuit similar to Fig. 9-16*a* having a load resistor R_L of 7,000 ohms and an output resistor R_o of 7,000 ohms, find (*a*) input resistance, (*b*) current gain, (*c*) voltage gain, (*d*) decibel power gain.

64. Using a transistor having the *h*-parameters found in Prob. 62 in a circuit similar to Fig. 9-16*a* having a load resistor R_L of 7,000 ohms and an output resistor R_o of 7,000 ohms, find (*a*) input resistance, (*b*) current gain, (*c*) voltage gain, (*d*) decibel power gain.

65. If the parameters for the circuit of Fig. 9-17*a* are V_{EE} = 12 volts, β = 80, R_B = 80,000 ohms, R_E = 11,000 ohms, and R_o = 1,000 ohms, find (*a*) emitter current, (*b*) a-c output resistance r_o, (*c*) a-c input resistance of the amplifier circuit, (*d*) current gain, (*e*) voltage gain, (*f*) decibel power gain.

66. If the parameters of the circuit of Fig. 9-17*a* are V_{EE} = 30 volts, β = 100, R_B = 100,000 ohms, R_E = 20,000 ohms, and R_o = 500 ohms, find (*a*) emitter current, (*b*) a-c output resistance r_o, (*c*) a-c input resistance of the amplifier circuit, (*d*) current gain, (*e*) voltage gain, (*f*) decibel power gain.

67. From the common-emitter *h*-parameters found in Probs. 1, 3, 5, 7, determine the *h*-parameters for a common-collector configuration.

68. From the common-emitter *h*-parameters found in Probs. 2, 4, 6, 8, determine the *h*-parameters for a common-collector configuration.

69. Using a transistor having the *h*-parameters found in Prob. 67 in a circuit similar to Fig. 9-17*a* having an emitter-circuit resistor R_E of 11,000 ohms and an output-circuit resistor R_o of 1,000 ohms, find (*a*) a-c output resistance, (*b*) a-c input resistance, (*c*) current gain, (*d*) voltage gain, (*e*) decibel power gain.

70. For a transistor that has common-base *h*-parameters of h_{ib} = 12.2 ohms, h_{ob} = 1.32 μmhos, h_{fb} = -0.9934, and $h_{rb} = 70 \times 10^{-5}$, determine the *h*-parameters for a common-collector configuration. (*Note:* First convert to common-emitter *h*-parameters and then to common-collector *h*-parameters.)

71. A common-cathode vacuum-tube voltage amplifier using a triode has the following tube and circuit parameters: μ = 72, r_p = 50,000 ohms, R_L = 94,000 ohms. (*a*) What is the voltage amplification of the circuit? (*b*) What value of load resistance is required to obtain a voltage amplification of 40?

72. A common-cathode vacuum-tube power amplifier using a triode has the following tube and circuit parameters: $\mu = 4.2$, $r_p = 800$ ohms, $R_L = 2{,}500$ ohms. (*a*) What is the voltage amplification of the circuit? (*b*) What value of load resistance is required to obtain a voltage amplification of 3.6?

73. If a 0.2-volt a-c input signal is applied to the amplifier of Prob. 71, what is (*a*) the a-c component of the plate current, (*b*) the amount of signal voltage developed across the load resistor?

74. If a 5.5-volt a-c input signal is applied to the amplifier of Prob. 72, what is (*a*) the a-c component of the plate current, (*b*) the amount of signal voltage developed across the load resistor?

75. A common-cathode amplifier circuit using a medium-mu triode has an amplification factor of 45 and a plate resistance of 5,600 ohms. What is the voltage amplification of the circuit if the load resistor has a value of 12,400 ohms?

76. A common-cathode amplifier circuit using a high-mu triode has an amplification factor of 100 and a plate resistance of 31,500 ohms. What is the voltage amplification of the circuit if the load resistor has a value of 50,000 ohms?

77. When the a-c input signal applied to the amplifier of Prob. 75 is 0.6 volt, what is (*a*) the a-c signal voltage developed across the load resistor? (*b*) The a-c component of the plate current?

78. When the a-c input signal applied to the amplifier of Prob. 76 is 0.7 volt, what is (*a*) the a-c signal voltage developed across the load resistor? (*b*) The a-c component of the plate current?

79. What is the voltage amplification of the amplifier of Prob. 71 if the value of the load resistor is increased to 150,000 ohms?

80. What is the voltage amplification of the amplifier of Prob. 71 if the value of the load resistor is reduced to 30,000 ohms?

81. What is the voltage amplification of the amplifier of Prob. 75 if the value of the load resistor is reduced to 5,600 ohms?

82. What is the voltage amplification of the amplifier of Prob. 75 if the value of the load resistor is increased to 20,000 ohms?

83. What value of load resistance is required for the amplifier of Prob. 75 if a voltage amplification of 25 is desired?

84. What value of load resistance is required for the amplifier of Prob. 76 if a voltage amplification of 75 is desired?

85. What is the voltage amplification of a common-cathode vacuum-tube pentode voltage-amplifier circuit that has the following tube and circuit parameters: $r_p = 250{,}000$ ohms, $g_m = 4{,}000$ μmhos, $R_L = 200{,}000$ ohms?

86. What is the voltage amplification of a common-cathode vacuum-tube pentode power-amplifier circuit that has the following tube and circuit parameters: $r_p = 20{,}000$ ohms, $g_m = 6{,}400$ μmhos, $R_L = 10{,}000$ ohms?

87. If a 0.45-volt a-c input signal is applied to the amplifier of Prob. 85, what is (*a*) the a-c component of the plate current? (*b*) The amount of signal voltage developed across the load resistor?

88. If a 1-volt a-c input signal is applied to the amplifier of Prob. 86, what is (*a*) the a-c component of the plate current? (*b*) The amount of signal voltage developed across the load resistor?

89. What is the voltage amplification of the amplifier of Prob. 85 if the value of the load resistor is increased to 0.5 megohm?

90. What is the voltage amplification of the amplifier of Prob. 86 if the value of the load resistor is reduced to 5,000 ohms?

91. (*a*) What approximate value of voltage amplification is indicated by using Eq. (9-49) for the amplifier of Prob. 85? (*b*) What approximate value of voltage amplification is indicated by using Eq. (9-49) for the amplifier of Prob. 89? (*c*) Compare the values of (*a*) and (*b*) in this problem with the answers for Probs. 85 and 89. (*d*) Comparing the answers of this problem with the answers of Example 9-32, what conclusion can be made regarding the usefulness of Eq. (9-49)?

92. (*a*) What approximate value of voltage amplification is indicated by use of Eq. (9-49) for the amplifier of Prob. 86? (*b*) What approximate value of voltage amplification is indicated by use of Eq. (9-49) for the amplifier of Prob. 90? (*c*) Compare the values of (*a*) and (*b*) in this problem with the answers for Probs. 86 and 90.

93. A certain triode being operated as a common-grid r-f amplifier has a transconductance of 3,000 μmhos and a plate resistance of 6,000 ohms. (*a*) What is the approximate input resistance? (*b*) What is the voltage amplification of the circuit if the load resistance is 3,000 ohms?

94. One section of a certain twin triode being used as a common-grid r-f amplifier has a transconductance of 9,000 μmhos and a plate resistance of 4,500 ohms. (*a*) What is the approximate input resistance? (*b*) What is the voltage amplification of the circuit if the load resistance is 3,000 ohms?

95. What is the voltage amplification of a common-plate triode amplifier circuit if the tube and circuit parameters are $\mu = 19$, $r_p = 8{,}000$ ohms, $R_L = 24{,}000$ ohms?

96. One section of a twin triode is being used in a common-plate amplifier circuit to provide a low-impedance output. What is the voltage amplification of the circuit if the tube and circuit parameters are $\mu = 15$, $r_p = 6{,}400$ ohms, $R_L = 40{,}000$ ohms?

97. A certain pentode tube is being used in a common-plate configuration to obtain a low-impedance output. What is the voltage amplification of the circuit if the tube and circuit parameters are $g_m = 2{,}500$ μmhos, $R_L = 10{,}000$ ohms?

98. In order to obtain a low-impedance output, a certain pentode is being used in a common-plate configuration. If the transconductance of the tube is 4,000 μmhos, what is the voltage amplification of the circuit when the resistor in its cathode circuit has a value of (*a*) 5,000 ohms, (*b*) 25,000 ohms? (*c*) 100,000 ohms?

Chapter 10

Other Solid-state Devices

In addition to the types of diodes and transistors described in Chaps. 2 and 4, there are many variations of these basic types of solid-state devices. Each variation is designed for a specific type of application. Some of these variations are (1) tunnel diode, (2) varactor diode, (3) compensating diode, (4) junction field-effect transistor, (5) metal-oxide semiconductor field-effect transistor, (6) phototransistor, (7) silicon controlled rectifier, (8) triac, and (9) diac.

10-1 Tunnel Diode

Basic Construction. A small PN junction having a very high concentration of impurities in both the P and N sections is called a *tunnel diode.* Materials generally used in the fabrication of tunnel diodes are germanium, silicon, gallium antimonide, and gallium arsenide. Each section of the diode is doped with an impurity in increasing amounts until the critical voltage causing reverse breakdown has been reduced past the zero-voltage point and into the region of small forward bias voltage (see Fig. 10-1*b*). This process produces a relatively thin barrier region of less than one-millionth of an inch. In this condition, electrons, driven by the high field intensity across the junction, can *tunnel* through the barrier instantly.

Characteristics. A conventional diode will not conduct with a reverse bias until the breakdown voltage is reached, and will start conducting with a forward bias as low as 300 millivolts (Fig. 10-1*a*). The actions of a conventional diode are compared to those of a tunnel diode in Fig. 10-1. The current-voltage characteristics of a typical tunnel diode are shown in Fig. 10-1*b*. A small reverse bias, applied to a tunnel diode, causes the valence electrons near the junction to tunnel across the junction. As a result, the tunnel diode is highly conductive for all values of reverse bias. Similarly, with a small forward bias the valence electrons near the junction will tunnel across the junction. The diode current rises rapidly with small increases in forward bias until a maximum or *peak* value I_P is reached with a forward bias of V_P. As the forward bias is increased past this value, the diode current decreases rapidly to a *valley* having a relatively low value of I_V with a forward bias of V_V. Further increases in the forward bias produce a characteristic curve that resembles the shape of a conventional diode.

Negative Resistance. Negative resistance is defined as a condition producing a voltage-current relation in which an increase in voltage is associated with a decrease in current, and vice versa. The downward slope of the curve of Fig. 10-1*b* between V_P and V_V indicates that the current decreases as the applied voltage is increased within this range of bias voltage. The tunnel diode therefore possesses a negative-resistance characteristic within this region of its characteristic curve. The negative-resistance characteristic of the tunnel diode makes possible its use as an oscillator, amplifier, or switch.

Load Line. When the tunnel diode is used in amplifier or oscillator circuits, its operating point must be on the negative-resistance portion of its characteristic curve. The d-c load line must be very steep so that it will intersect the static characteristic curve at only one point. This is shown by the full-line d-c load line AB that intersects the curve at point G in Fig. 10-2*a*.

When a tunnel diode is used in an amplifier circuit, the a-c load line CD must be steep so that it will have only one intersection (point H) on any portion of its static characteristic curve. However, when the tunnel diode is used in oscillator circuits, the load line can be relatively flat, as shown by line EF, with as many as three operating points M, G, and N (Fig. 10-2*a*).

The location of the operating point is determined by (1) the signal swing, (2) the allowable signal-to-noise ratio, and (3) the operating temperature. To obtain the greatest amount of signal swing, the diode must be biased so that it operates around the center of the linear portion of the negative-resistance

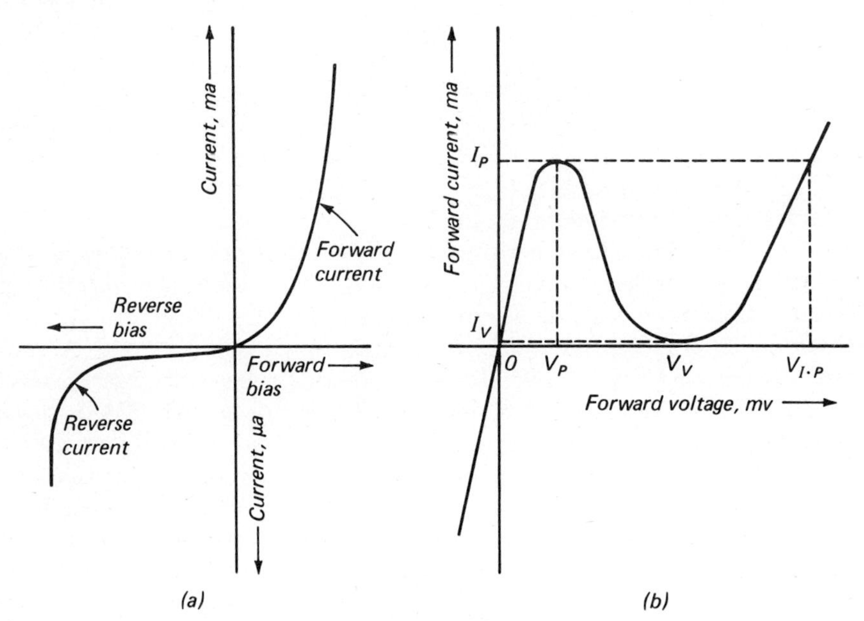

Fig. 10-1 Typical current-voltage characteristics of (a) a conventional diode, (b) a tunnel diode.

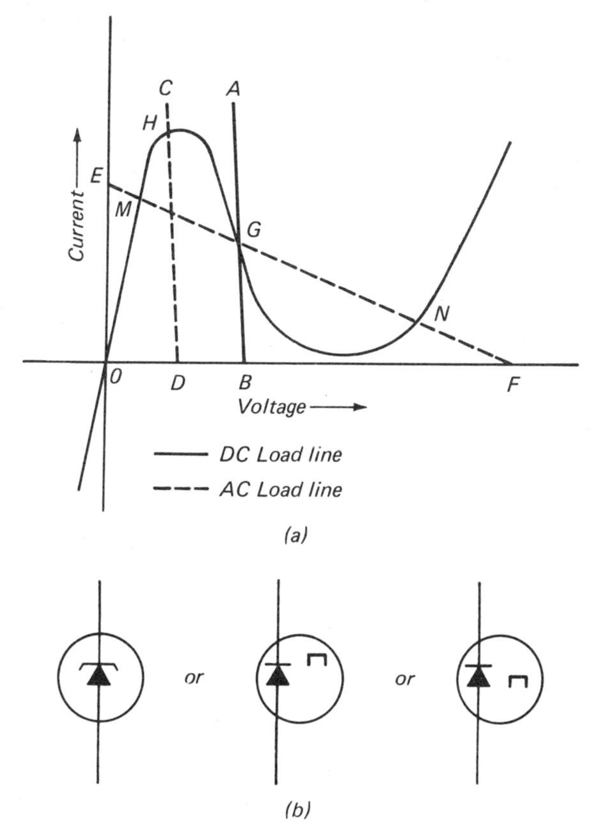

Fig. 10-2 (a) Typical load lines for tunnel-diode circuits. (b) Schematic symbols for a tunnel diode.

slope. To obtain the minimum amount of noise, the diode should be operated with the lowest amount of bias current possible. For high-temperature operation, a higher operating current should be selected.

Ratings. Tunnel-diode ratings are defined with reference to the static characteristic curves of Figs. 10-1 and 10-2.

Peak Point: The point on the forward characteristic curve corresponding to the lowest positive voltage at which $dI/dV = 0$.

Reverse Peak Point: The point on the reverse characteristic curve corresponding to the lowest reverse voltage at which $dI/dV = 0$.

Valley Point: The point on the forward characteristic curve which corresponds to the second lower positive voltage at which $dI/dV = 0$.

Peak-point Current I_P: The current at the peak point.

Reverse Peak-point Current I_P: The current at the reverse peak point.

Valley-point Current I_V: The current at the valley point.

Peak-to-valley-point Current Ratio I_P/I_V: The ratio of the peak-point current to the valley-point current.

Peak-point Voltage V_P: The voltage at which the peak point occurs.

Valley-point Voltage V_V: The voltage at the valley point.

Forward Voltage V_F: The voltage corresponding to a point on the forward characteristic curve at a specified current.

Forward Current I_F: The current in the first-quadrant conducting region.

Reverse Voltage V_R: The voltage corresponding to a point on the reverse characteristic curve at a specified current.

Reverse Current I_R: The current in the third-quadrant conducting region.

Negative Conductance G: The negative conductance of the intrinsic diode.

Capacity C: The barrier capacitance of the intrinsic diode.

Advantages. For certain applications, the tunnel diode has advantages over conventional solid-state diodes and electron-tube diodes.

1. Since electric charges move through the junction at speeds approaching the speed of light, instead of the relatively slow *drift rate* in conventional diodes and transistors, the tunnel diode can provide useful outputs at frequencies as high as 5,000 MHz. They are particularly suitable for use in microwave-amplifier or microwave-oscillator circuits.
2. In switching circuits for computers, industrial controls, etc., tunnel diodes can operate at speeds several hundred times faster than those available with circuits using transistors.
3. Compared to vacuum-tube microwave oscillators, tunnel-diode oscillators are inexpensive, require only a fraction of a volt for the d-c bias, and are rugged and reliable in severe environments.
4. Compared to transistor-driven varactor frequency-multiplier circuits, tunnel-diode oscillator circuits are simpler, more compact, and afford higher d-c to r-f conversion efficiencies.
5. Tunnel diodes operate at temperatures as high as 650°F, compared to 400°F for conventional silicon diodes and 200°F for conventional germanium diodes.
6. Tunnel diodes exhibit a negative-resistance characteristic which enables them to amplify and generate power at radio frequencies.
7. Tunnel diodes resist the damaging effects of a nuclear-radiation environment.
8. The simplicity of construction of tunnel diodes permits their fabrication in *micromodules* and in *integrated circuits* containing complete stages of amplifiers and oscillators that are formed on a single semiconductor that is not much larger than any one of the letters on this page.

10-2 Back Diode

Referring to Fig. 10-1*b*, it can be seen that when a reverse voltage is applied to a tunnel diode the current increases continuously with increases in voltage. When a negative voltage having a magnitude corresponding to the positive

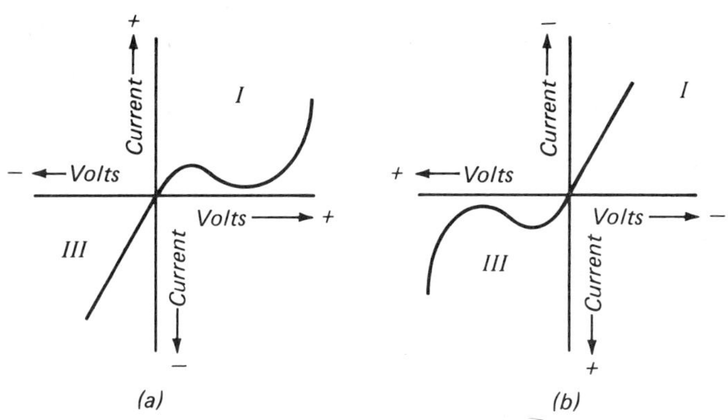

Fig. 10-3 Typical current-voltage characteristics of (a) a tunnel diode, (b) a back diode.

valley-point voltage V_V is applied to a tunnel diode, the resulting current will be many hundreds of times greater than the valley-point current I_V. A tunnel diode designed to take advantage of this large change in current with a polarity change of the applied voltage is called a *backward diode* or *back diode*. The term backward is used to indicate that the diode conducts current with negative applied voltage rather than with positive applied voltage.

The forward characteristics of a typical tunnel diode are shown in Fig. 10-3*a*. The characteristics of a back diode are obtained by interchanging the first and third quadrants as shown in Fig. 10-3*b*. From this curve it can be seen that the cut-in or threshold voltage is approximately zero volts. The new reverse characteristics (positive voltage) are not too different from that of a conventional diode. However, the back diode has a lower voltage drop at a given current than a conventional diode. This lower voltage drop is very advantageous in some tunnel-diode and transistor circuits. Current flow with a forward bias is undesirable in back diodes for the same reason that leakage current with reverse applied voltage is undesirable in conventional diodes. Back diodes are therefore designed to have a low peak-point current (usually less than 50 μa). Back diodes are well suited for rectifying very low-amplitude signals, and are therefore used for video detectors and microwave applications.

10-3 Varactor Diode

Basic Theory. A *varactor*, or *variable-capacitance diode*, is a variable-reactance PN-junction diode. When reverse bias is applied, the capacitance of the diode is a nonlinear function of the bias voltage. The variable-capacitance effect of a varactor is a result of the variation in the width of the depletion region at the junction of a diode with variations in bias voltage (Fig. 10-4). A high resistance exists at the junction, or depletion region, because of the lack of mobile

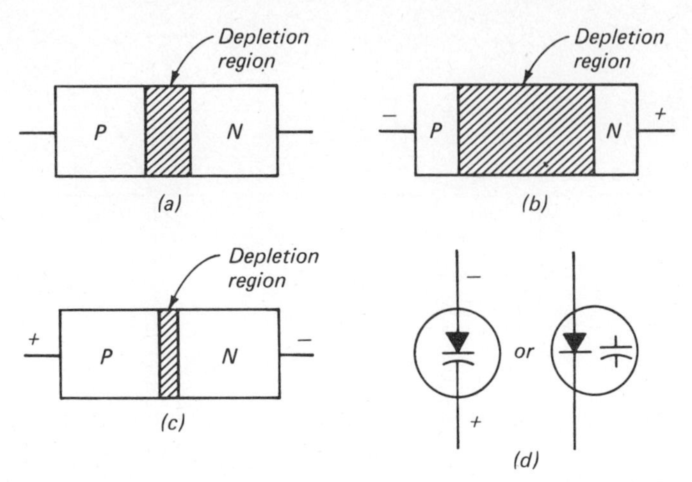

Fig. 10-4 Behavior of depletion region in a varactor under various bias conditions. (a) Zero bias. (b) Reverse bias. (c) A small forward bias. (d) Schematic symbol for a varactor.

carriers (electrons and/or holes). When a slight forward bias is applied, yet not high enough to cause forward conduction, (1) mobile carriers move toward the junction, (2) the resistance at the junction is decreased, and (3) the depletion region is narrowed. When a reverse bias is applied to the junction, (1) mobile carriers move away from the junction, (2) the resistance at the junction is increased, and (3) the depletion region is widened. The P and N sections of the diode act in the same manner as the conductive plates of a capacitor; that is, as the distance between the plates of the capacitor decreases the capacitance increases, and vice versa. The maximum value of capacitance of the varactor occurs at a very low value of forward bias, and the minimum value occurs at the reverse-breakdown voltage. Because the width of the depletion region changes as the voltage applied to the diode changes, the stored energy in this capacitor will also vary with changes in the applied voltage. This effect is called the *junction-transition capacitance.*

Characteristics. Varactor diodes are constructed with a very high impurity concentration outside the depletion region, and a relatively low impurity concentration at the junction. This type of construction permits this device to perform (1) oscillation, (2) frequency multiplication, and (3) switching functions at very high frequencies. Because the primary current at the junction is reactive, and shot-noise components are absent, the noise level of circuits using varactors is very low.

The gain and power output of a transistor circuit decrease rapidly as the frequency is increased and at ultrahigh frequencies a power amplifier may actually act as an attenuator. At these high frequencies higher power outputs can be obtained by using a varactor. Although the diode produces no gain, the losses are low since the varactor operates as a nonlinear reactance.

Applications. The applications of varactor diodes can basically be divided into two general-type categories: (1) tuning and (2) harmonic generation. When used as a tuning diode, the varactor is rugged, small, not affected by dust or moisture, and ideal for remote control and precision fine tuning. The uses of tuning diodes are many and span the range from a-m radio applications to microwave applications.

10-4 Compensating Diode

Stabilization of the output of a transistor for variations in both temperature and supply voltage can be obtained by the use of a compensating diode. This diode is connected to operate in the forward direction in the bias network of an amplifier or oscillator circuit. In a typical circuit, the diode is biased at the operating point represented by the broken line on the diode characteristic curves of Fig. 10-5*a*. The diode current, at this point on the curve, produces a bias voltage that determines the transistor idling current, represented by the broken line on the transistor characteristic curves of Fig. 10-5*b*. Since the bias voltage produced by the diode current will shift with varying temperatures in the same magnitude and direction as the transistor characteristic, the idling current becomes essentially independent of temperature change. The variations in diode current with changes in the supply voltage will vary in the same proportion as the changes in transistor bias voltage vary with changes in the supply voltage. Thus, the variations in the idling current with variations in supply voltage are minimized. The compensating diode therefore can also be used to stabilize the transistor output for variations in supply voltage.

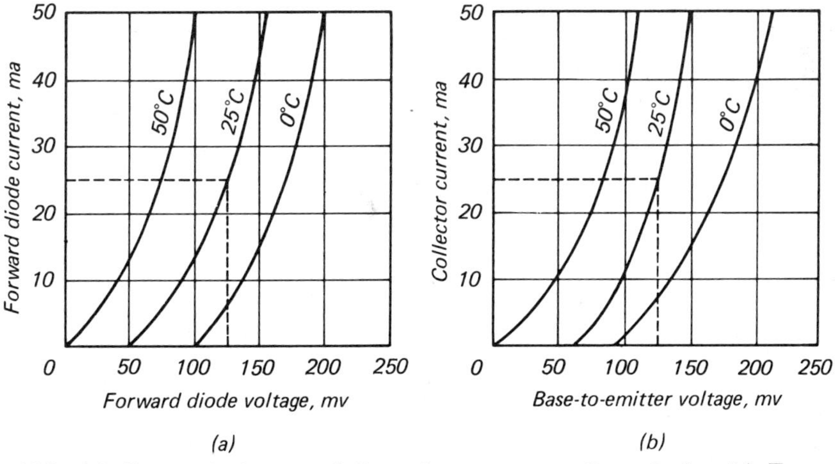

Fig. 10-5 (a) Forward characteristics of a compensating diode. (b) Transfer characteristics of a transistor.

10-5 Junction Field-effect Transistor

Basic Construction. The *junction field-effect transistor*, abbreviated JFET and the more commonly used FET, can be made with either an N channel or a P channel. In an N-channel FET (Fig. 10-6*a*) a narrow bar of an N-type semiconductor material is used to form the channel. Two junctions of a P-type material are diffused on opposite sides and in the middle of this bar. These two junctions form two PN diodes, or *gates*, at this section of the bar, and the area between the two gates is called the *channel*. Ohmic contacts (direct electrical connections) are made to the two P regions, which are then connected in parallel and a single lead brought out as the *gate terminal*. Leads are attached to the extreme ends of the N-type bar, and are identified as the *source terminal* and the *drain terminal*. In a P-channel FET, two N-type materials are diffused on opposite sides and in the middle of a P-type semiconductor material (Fig. 10-6*b*). Schematic symbols for N-channel and P-channel FETs are shown in Figs. 10-6*c* and 10-6*d*.

Theory of Operation. When a voltage V_{DD} is applied between the source and drain terminals of an N-channel FET, and a zero bias voltage is applied to the gate, electrons will flow from source to drain through the N-type material of the transistor (Fig. 10-7*a*). When a reverse bias V_{GG} is applied between the gate and source terminals, an electric field is produced about each PN junction. This electric field causes the electrons to move away from the

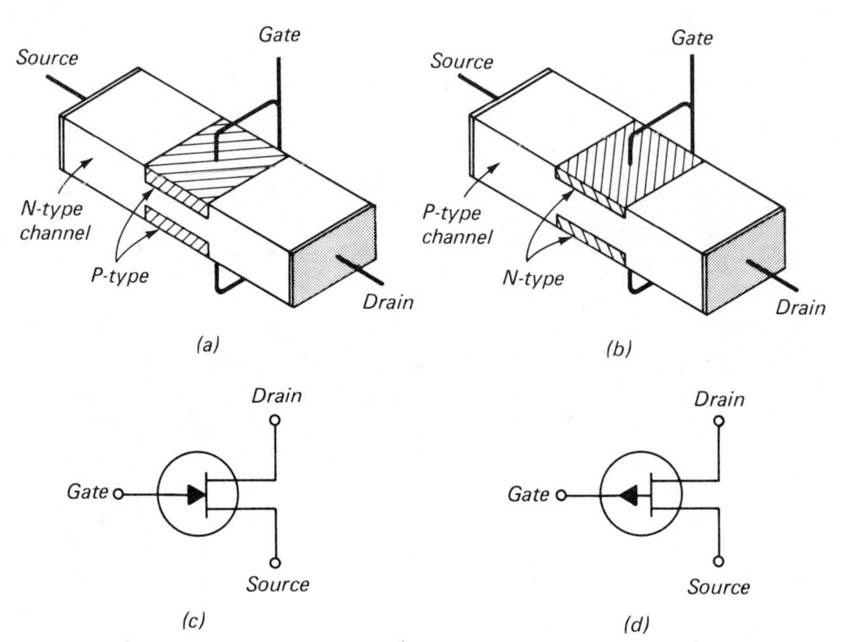

Fig. 10-6 Basic construction of a JFET: (a) N-channel. (b) P-channel. Schematic symbols for a JFET: (c) N-channel. (d) P-channel.

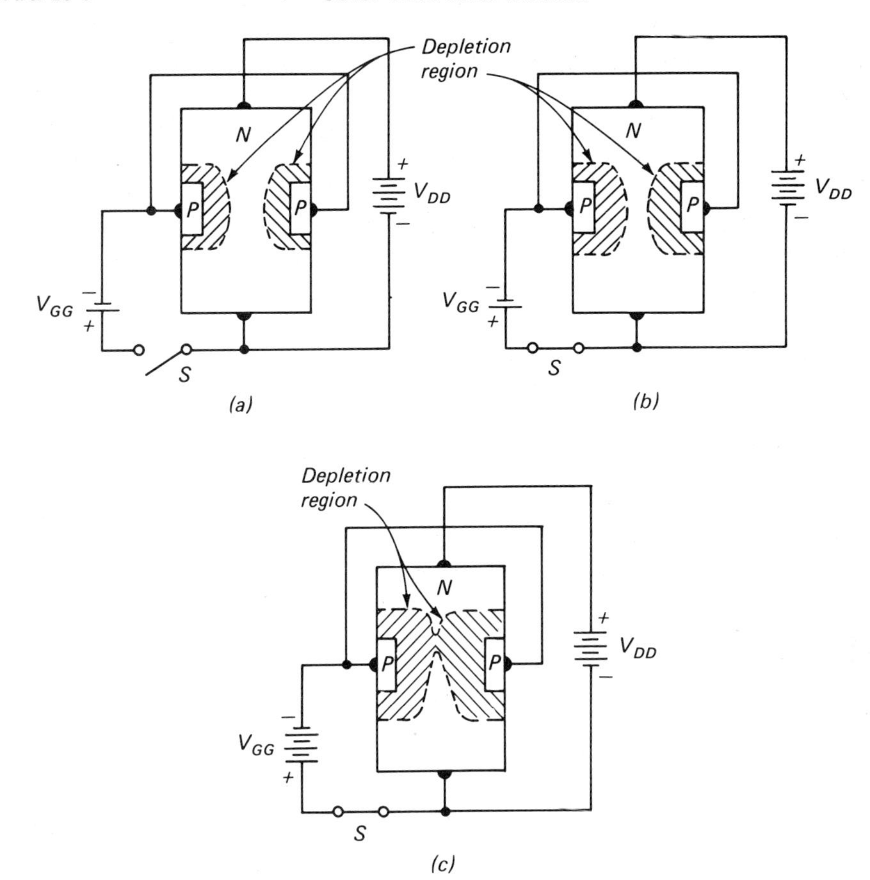

Fig. 10-7 Schematic representation of depletion-region variation in a JFET as the gate-bias voltage increases. (a) Zero gate bias. (b) Medium gate bias. (c) Gate biased to pinch-off.

junction, and the region near each junction becomes void of electrons. This region is called the *depletion region*. The effect of this depletion region is to reduce the cross-sectional area of the channel, thereby decreasing the conductance of the source-to-drain path. The penetration of the depletion regions into the channel follows a square-law characteristic, which is especially desirable for low cross modulation in r-f amplifiers. When the reverse bias is decreased to a point that causes the depletion areas to make contact with each other (Fig. 10-7*c*), the channel is referred to as being *pinched off*. The reverse bias V_{GG} applied between the gate and source terminals therefore controls the amount of electron flow between the source and drain. Since application of gate-bias voltage depletes the current flow, this device is sometimes referred to as a *depletion-type field-effect transistor*. A P-channel FET operates in the same manner as an N-channel FET, except that the channel current carriers will be holes instead of electrons, and the polarities of V_{DD} and V_{GG} are reversed.

Comparison of JFET, Conventional Transistor, and Vacuum Tube. Another method of explaining the operation of a FET is by comparing its operation with those of the conventional transistor and the vacuum tube. The symbols for all three devices, with each element labeled and the direction of current flow indicated by an arrow for a basic circuit connection, are shown in Fig. 10-8. With the NPN transistor connected in the common-emitter configuration (Fig. 10-8*a*) current flows from the emitter to the collector with the base controlling the amount of current flow. In a conventional transistor, both electrons and holes are used as current carriers; therefore, this type of transistor is referred to as a *bipolar device.* In the common-cathode vacuum-tube circuit (Fig. 10-8*b*) current flows from the cathode to the plate, and the grid is the controlling element. In a vacuum tube the plate current consists only of electrons; hence a vacuum tube is referred to as a *unipolar device.* When the FET is connected in the common-source configuration (Fig. 10-8*c*), current flows from the source to the drain with the gate controlling the amount of current flow. Since the channel current carrier is either only electrons for N-channel devices, or only holes for P-channel devices, a junction field-effect transistor is therefore a *unipolar device.*

Static Characteristic Curves. The static characteristic curves for an N-channel FET having a common-source mode of connection are shown in Fig. 10-9. This FET connection is comparable to the common-emitter connection of a junction transistor and the common-cathode connection of a vacuum-tube triode. In a manner similar to other three-terminal devices, the FET can be operated in three configurations: (1) common-source, (2) common-gate, and (3) common-drain. The static characteristic curves for an N-channel FET closely resemble those for a vacuum-tube pentode. The static characteristic curves for a P-channel FET with a common-source connection would be similar to those shown in Fig. 10-9; however, the polarities of the source-to-

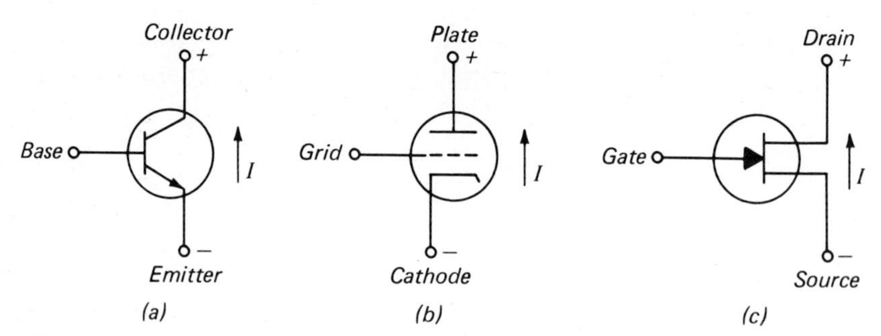

Fig. 10-8 Comparison of schematic symbols for (a) conventional transistor, (b) vacuum tube, (c) junction field-effect transistor.

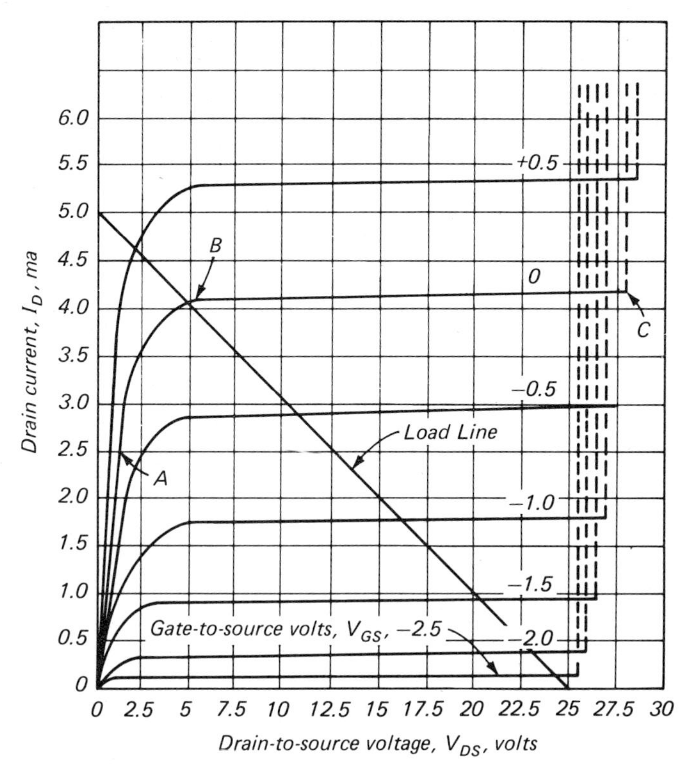

Fig. 10-9 Typical output characteristic curves for an N-channel junction field-effect transistor.

drain voltage, the gate-to-source bias voltage, and the drain current would all be reversed.

The shape of each of the static characteristic curves of Fig. 10-9 can be analyzed in the following manner: For example, the curve for $V_{GS} = 0$ shows that as the drain-to-source voltage is increased from zero to point A the curve varies as a straight line to point A, called the *knee*. Within this low drain-to-source voltage range, the resistance of the channel remains relatively constant and the current varies linearly with changes in voltage. This section is sometimes referred to as the *ohmic region* of operation. As the drain-to-source voltage is increased beyond point A the drain current increases at a reverse square-law rate to point B, called *pinch-off*. This point is comparable to cutoff in a vacuum tube. The progressive decrease in the rate of drain-current increase is caused by the square-law increases in the depletion areas at each gate up to point B, where the two regions make contact. This feature produces the second harmonic, which is essential for mixer operation of TV, f-m, and a-m tuners. Other devices, such as solid-state diodes, junction transistors, and vacuum tubes,

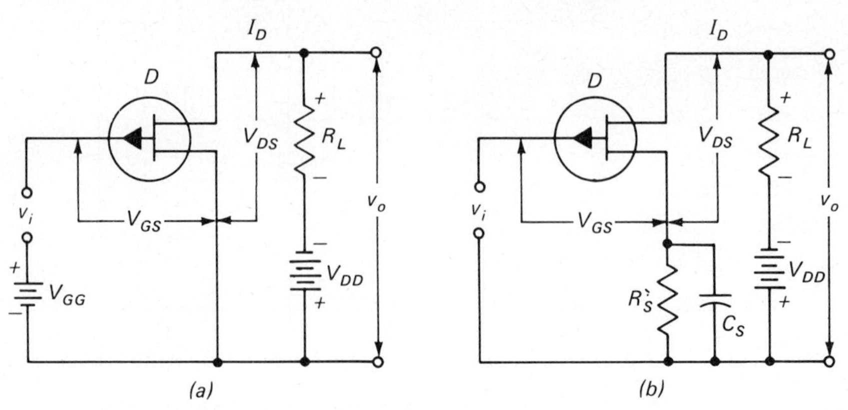

Fig. 10-10 Basic circuit bias arrangements in a common-source FET amplifier circuit. (a) Using a separate power source. (b) Using self-bias.

produce the third and other higher-order harmonics in addition to the second harmonic. Since these higher harmonics produce cross-modulation distortion, the FET, which produces only the second harmonic, is ideal for mixer-circuit applications in communications tuner circuits. The amount of the drain-to-source voltage required to produce pinch-off for a designated value of gate-bias voltage is called the *pinch-off voltage*, abbreviated V_p. Beyond the pinch-off point, B to C, the FET operates as a constant-current device. The drain current is relatively independent of drain-to-source voltage change, and its output is controlled only by variations in the grid-bias voltage. The section of the curve from B to C is used for normal operation of the FET as an amplifier, and it is called the *pinch-off* or *saturation region*. Increasing the drain-to-source voltage beyond point C causes the FET to enter the *breakdown region* where the drain current increases to excessive values.

Basic Biasing Circuit. The FET can be biased by using either (1) a separate power source V_{GG} (Fig. 10-10*a*) or (2) some form of self-bias as illustrated in Fig. 10-10*b*. The required amount of reverse bias for the common-source circuit of Fig. 10-10*b* is obtained by using the voltage drop produced across the biasing network R_SC_S. The d-c component (average value) of the drain current I_D flowing through the source-biasing resistor R_S produces the bias voltage V_{GS}. The a-c component (signal variations) of the drain current i_d flows through the source bypass capacitor C_S. The voltage drop across R_S is determined by the value of its resistance and the amount of current flowing through it. The value of R_S can be calculated by

$$R_S = \frac{V_{GS}}{I_D} \tag{10-1}$$

where R_S = source-bias resistance, ohms
V_{GS} = gate bias, volts
I_D = average or quiescent value of drain current, amp

Example 10-1 Determine the value of source-bias resistance required for the circuit of Fig. 10-10*b* to produce a gate bias of 1.2 volts with an average drain current of 0.6 ma.

GIVEN: $V_{GS} = 1.2$ volts $\quad I_D = 0.6$ ma

FIND: R_S

SOLUTION:

$$R_S = \frac{V_{GS}}{I_D} = \frac{1.2}{0.6 \times 10^{-3}} = 2{,}000 \text{ ohms}$$

Example 10-2 Determine the value of source-bias resistance required for the circuit of Fig. 10-10*b* using a FET whose characteristics are shown in Fig. 10-9. The transistor is to be operated on the load line indicated on the curve and with a quiescent value that produces a gate bias of −1.0 volt.

GIVEN: $V_{GS} = -1.0$ volt $\quad$ Characteristic curves, Fig. 10-9

FIND: R_S

SOLUTION: From Fig. 10-9, the quiescent point is located where the load line and the curve for $V_{GS} = -1.0$ volt intersect, and the drain current at this point has a value of 1.75 ma.

$$R_S = \frac{V_{GS}}{I_D} = \frac{1.0}{1.75 \times 10^{-3}} \cong 570 \text{ ohms}$$

The junction of the gate with the source and the drain of a FET forms an ohmic, or direct, electrical connection. Therefore, the input signal should not be permitted to drive the transistor into a forward bias in order to prevent having the gate circuit draw current. With a forward bias, the input impedance decreases and signal clipping will occur in a similar manner to vacuum tubes. In practical operation, the junction field-effect transistor performs in a similar manner to a pentode vacuum tube, and therefore it is sometimes referred to as a *solid-state tube*.

Parameters. The various parameters of a junction field-effect transistor can be obtained from its characteristic curves in a similar manner to those obtained for junction transistors (Arts. 4-12 to 4-17) and for vacuum tubes (Arts. 5-4 to 5-6). The two parameters usually specified for a FET having a common-source connection are (1) the *active channel resistance* (commonly referred to as the *drain resistance*) r_d, and (2) the *forward transconductance* g_{fs}. These parameters can be found by

$$r_d = \frac{dV_{DS}}{dI_D} \qquad (V_{GS}\text{—constant}) \tag{10-2}$$

$$g_{fs} = \frac{dI_D}{dV_{GS}} \qquad (V_{DS}\text{—constant}) \tag{10-3}$$

where r_d = drain resistance, ohms
g_{fs} = forward transconductance, mhos
V_{DS} = drain-to-source voltage, volts
dV_{DS} = change in drain-source voltage
dV_{GS} = change in gate-source voltage
dI_D = change in drain current

From the static characteristic curves shown in Fig. 10-9 it can be observed that the change in I_D over a wide range of V_{DS} is very small, and thus the drain resistance cannot be determined with any degree of accuracy from these curves. The value of drain resistance is very high, and is usually in the order of hundreds of thousand ohms.

Example 10-3 Determine the forward transconductance for the FET of Example 10-2 when V_{DS} is 10 volts.

GIVEN: $V_{GS} = 1.0$ volt $\quad I_D = 1.75$ ma $\quad V_{DS} = 10$ volts

FIND: g_{fs}

SOLUTION: From curves of Fig. 10-9, $dI_D = 0.9$ to 2.9 ma, and $dV_{GS} = -0.5$ to -1.5 volts.

$$g_{fs} = \frac{dI_D}{dV_{GS}} = \frac{0.0029 - 0.0009}{-0.5 - (-1.5)} = \frac{0.002}{1} = 0.002 \text{ mho} = 2{,}000\ \mu\text{mhos}$$

At high frequencies, the internal capacitances affect the operating parameters of a junction field-effect transistor. Because the resulting impedance is affected by the capacitive reactance at these frequencies, the forward transconductance is not usually specified by manufacturers. Instead the *forward transadmittance* for a definite frequency, such as 1 kHz, 100 MHz, and 200 MHz, is usually given. *Admittance* is a measure of how readily an alternating current will flow in a circuit. This value compares to conductance in a d-c circuit, and it is the reciprocal of impedance in the same manner that conductance is the reciprocal of resistance. Admittance is expressed in mhos and is designated by the symbol Y. The symbol for the forward transadmittance is Y_{fs}, and its value is expressed in mhos, or μmhos. The forward transconductance and/or the forward transadmittance, are important parameters of the FET. The gain of a FET amplifier circuit is directly proportional to the value of g_{fs} or Y_{fs}. Therefore, as the forward transconductance or transadmittance increases, the gain of the amplifier stage will also increase.

The gain of a FET is expressed as a power gain, rather than as a current gain in transistors or as a voltage gain in vacuum tubes. The power gain is expressed as a decibel and is abbreviated G_{ps}. This value is given for a definite frequency for the (1) maximum available power gain, (2) maximum usable power gain, unneutralized, and (3) maximum usable power gain, neutralized.

Voltage Amplification per Stage. The voltage gain of the amplifier circuit shown in Fig. 10-10 can be calculated by

$$A_v = \frac{\mu R_L}{r_d + R_L} \tag{10-4}$$

Since $\mu = g_{fs} r_d$

Then
$$A_v = \frac{g_{fs} r_d R_L}{r_d + R_L} \tag{10-5}$$

If $r_d \gg R_L$, the equation for the voltage gain at low frequencies can be expressed as

$$A_v = g_{fs} R_L \tag{10-6}$$

For high frequencies, this equation becomes

$$A_v = g_{fs} Z_L \tag{10-7}$$

Example 10-4 The forward transconductance of a FET used in a voltage-amplifier circuit is 2,000 μmhos, and the load resistance in this circuit is 10,000 ohms. Determine the voltage amplification of this circuit assuming that $r_d \gg$ than R_L.

GIVEN: $g_{fs} = 2{,}000$ μmhos $\quad R_L = 10{,}000$ ohms

FIND: A_v

SOLUTION:

$$A_v = g_{fs} R_L = 2{,}000 \times 10^{-6} \times 10{,}000 = 20$$

Example 10-5 The forward transadmittance of a FET used in a voltage-amplifier circuit is 7.5 mmhos at 100 MHz, and the output impedance of this circuit at 100 MHz is 4.2 kilohms. Determine the voltage amplification of this circuit.

GIVEN: $g_{fs} = 7.5$ mmhos $\quad Z_L = 4.2$ kilohms

FIND: A_v

SOLUTION:

$$A_v = g_{fs} Z_L = 7.5 \times 10^{-3} \times 4.2 \times 10^3 = 31.5$$

Advantages. The junction field-effect transistor combines the many advantages of both the bipolar junction transistor and the pentode vacuum tube without having any of the disadvantages inherent in these two circuit elements. FETs have (1) small size, (2) ruggedness, (3) long life, (4) geometry of a transistor, (5) high efficiency, (6) negative temperature coefficients, (7) high-frequency response, (8) high power gain, (9) low noise, (10) high input impedance, (11) high output impedance, and (12) square-law characteristics.

Since the FET can be fabricated by diffusion methods, it can be used in microminiaturized integrated circuits. Because its operation depends on bulk-material current carriers that do not cross junctions, the inherent noise sources of tubes (due to high-temperature operation and microphonics) and those of transistors (due to junction transition) are not present in the FET. Impedance matching is easily obtainable because of the high input and high output impedances. Its high power gain usually eliminates the necessity of using driver circuits. Because of its negative temperature coefficient, thermal runaway is impossible. Cross modulation in mixer and oscillator circuits is eliminated by virtue of its square-law characteristics. Because of these many superior operating characteristics, the FET is extensively used in the tuning, oscillator, and mixer sections of TV and f-m receivers.

The junction field-effect transistor is a voltage-controlled constant-current device, in a similar manner to the pentode vacuum tube, in which variations in the input voltage control the output current. The characteristics of the FET so closely resemble those of the pentode that the design and circuit analysis equations are similar for both devices.

Specifications. The various properties of FETs are listed under two general headings: (1) maximum ratings and (2) operating characteristics. Among the properties listed under maximum ratings are (1) drain-to-source voltage, (2) gate-to-source voltage, (3) drain current, (4) power dissipation at a specified temperature, and (5) operating and storage temperature range. The operating characteristics may include (1) drain-to-source voltage, (2) gate-to-source voltage, (3) drain current, (4) gate leakage current, (5) input resistance, (6) output resistance, (7) small-signal input capacitance, (8) small-signal reverse transfer capacitance, (9) forward transadmittance for a specified frequency, (10) noise figure (db), and (11) power gain (db) for usable, available neutralized, and available unneutralized power.

10-6 Metal-oxide Semiconductor Field-effect Transistor

The metal-oxide semiconductor field-effect transistor, abbreviated MOSFET or MOS transistor, can be made in either of two forms, (1) the depletion type and (2) the enhancement type. In the *depletion-type* MOSFET, drain current flows at zero gate-bias voltage, and this device operates in a similar manner to a FET, In the *enhancement-type* MOSFET no drain current flows at zero gate-bias voltage, and this device operates in a similar manner to a conventional junction transistor. In both types of MOSFETs, the gate is insulated from the conducting channel by a thin insulating film. Because of this insulating construction, a MOSFET is also called an *insulated-gate field-effect transistor*, abbreviated IGFET or IGT, which actually is a better descriptive designation of this type of device.

10-7 Depletion-type MOSFET

Basic Construction. The cross-sectional view of a depletion-type N-channel MOSFET is shown in Fig. 10-11. The MOSFET has three main terminals: (1) the source, (2) the drain, and (3) the gate; usually an auxiliary terminal (attached to the substrate) is also provided. The unit consists of (1) the *sub-*

strate of P-type material to which the auxiliary terminal is attached, (2) the *conducting channel* of N-type material which is diffused into the substrate and which has leads attached to each end—one being the source terminal and the other the drain terminal, (3) the *gate,* which is a thin metal plate or metal film to which the gate terminal is attached, and (4) an *insulator* in the form of an ultrathin film of insulating material such as silicon dioxide. The substrate forms the foundation of a MOSFET, and is sometimes called the *bulk, base,* or *body.* This type of construction is called an N-type MOSFET because the current flow from the source to the drain takes place in the N-type semiconductor material. A P-channel MOSFET is constructed in a similar manner to the N-channel MOSFET, except that the conducting channel is made of a P-type semiconductor and the substrate is made of an N-type semiconductor.

Theory of Operation. The basic difference in the operation of a JFET and a MOSFET is caused by the capacitor formed by the *metal gate,* the *silicon-dioxide insulator,* and the *semiconductor channel.* When the voltage between the gate and the source of a depletion-type N-channel MOSFET is zero (zero bias), electrons can flow freely from the source to the drain through the N-channel semiconductor (Fig. 10-11*a*). When a small amount of negative voltage is applied to the gate terminal, capacitor action will take place at the area where the gate and the channel are separated by the insulator. The gate takes on a surplus of electrons (a negative charge) and the electrostatic field produced at the insulator causes an equal number of electrons to be displaced from the area of the channel material that is adjacent to the gate, thereby causing a positive charge (or depletion of electrons) in this area of the channel (see Fig. 10-11*b*). The resultant decrease in the effective cross-sectional area of the N channel causes (1) a decrease in the conductance of the channel, and (2) a decrease in the amount of current that will flow from the source to the drain. If the gate is made more negative, the depletion area is increased and the source-to-drain current is decreased further. When the depletion area has been decreased to

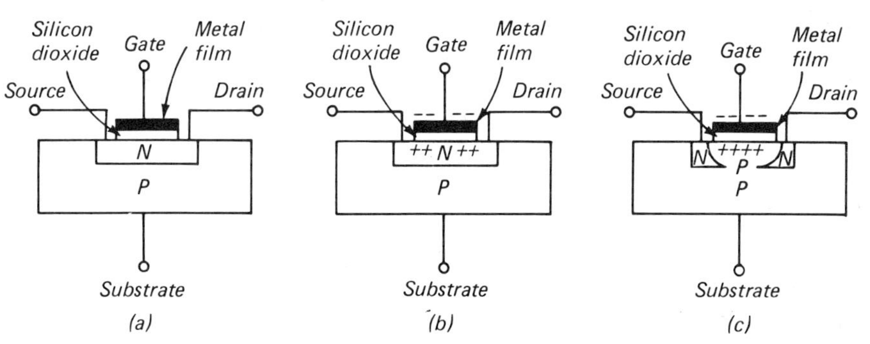

Fig. 10-11 Schematic representations of the depletion-region variation in a depletion-type N-channel MOSFET as the gate-bias voltage increases. (a) Zero gate bias. (b) Medium gate bias. (c) Gate-bias pinch-off.

the extent that it makes contact with the substrate, then (1) the conductance of the channel is reduced to zero, (2) the source-to-drain current is reduced to zero, and (3) the channel is said to be *pinched off* (see Fig. 10-11*c*). Under this condition two PN junctions are formed (1) between the substrate and source, and (2) between the substrate and drain; since a reverse bias will exist at one of these terminals for one polarity or the other of source-to-drain voltage, no current will flow through the channel at or beyond pinch-off. Although reversing the polarity of the voltage applied to the gate will theoretically have no harmful effect, the permissible magnitude of this voltage is severely restricted because of the ultrathin insulation used between the gate and the channel. A P-channel MOSFET operates in the same manner as an N-channel MOSFET, except that the channel carrier current will be holes instead of electrons, and the polarities at the source, drain, and gate are reversed.

Symbols. The schematic symbols for both the N- and P-channel depletion-type MOSFET shown in Fig. 10-12 are the ones most generally used. The letter symbols for the four terminals are S = source, D = drain, G = gate, and SS = substrate (or B = active bulk). Terminal B is connected to the substrate. In the schematic symbol for the depletion-type MOSFET, the channel bar is drawn with a solid line to identify the *normally-on characteristic* of this type of device.

Basic Circuit Operation. The basic circuit connections for the N- and P-channel depletion-type MOSFET are shown in Fig. 10-13. The substrate is usually connected internally to the source, although it may sometimes be connected externally to either the source or ground. This type of MOSFET operates in a similar manner to the JFET. A negative-going input signal applied to the gate input circuit of the N-channel MOSFET (Fig. 10-13*a*) introduces (or increases) a bias voltage on the gate, thereby depleting the channel area and reducing the source-to-drain current. The insulation between the gate and channel prevents the flow of any current in the gate circuit when a positive-going signal is high enough to drive the gate positive. Since the gate can be driven either negative or positive, no self-biasing source resistance is

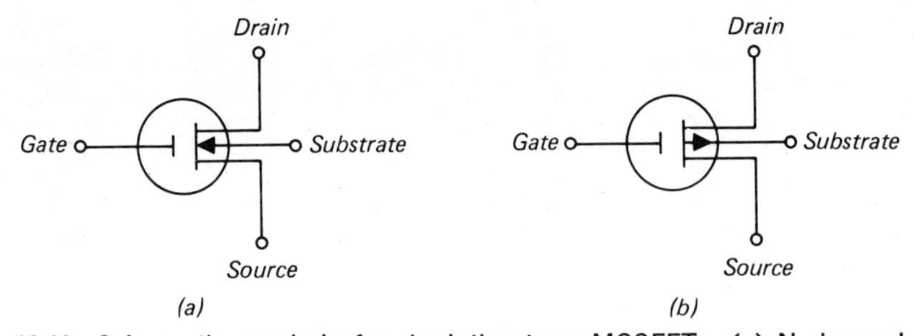

Fig. 10-12 Schematic symbols for depletion-type MOSFETs. (a) N-channel. (b) P-channel.

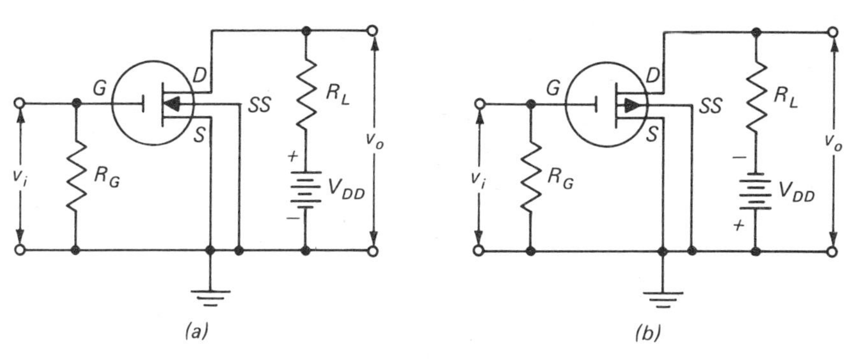

Fig. 10-13 Basic common-source circuit configuration for a depletion-type MOSFET. (a) N-channel. (b) P-channel.

required. Because of the relatively low value of capacitance of the gate-to-channel electrodes (in the order of 1 pf), the input impedance of this type of MOSFET is very high, as much as thousands of megohms. A wide range of values may be used for the gate resistance R_G; however, the value used must be considerably less than the input impedance of the MOSFET and high enough to prevent loading of the input-signal source. The P-channel MOSFET circuit of Fig. 10-13*b* operates in a similar manner to the N-channel circuit of Fig. 10-13*a*, except for (1) the reversal of the polarity of V_{DD}, and (2) the current carriers will be holes instead of electrons.

10-8 Enhancement-type MOSFET

Basic Construction. The cross-sectional view of an enhancement-type N-channel MOSFET is shown in Fig. 10-14. This device is formed by diffusing two N regions, to serve as source and drain, into the ends and on one side of a P-type semiconductor substrate. The substrate terminal is taken from the center of the opposite side of the semiconductor. A thin insulating film is formed on top of the semiconductor between the two *N* regions, and becomes the gate *insulator*. The gate consists of a thin metallic film that is deposited on top of the insulating film. A P-channel MOSFET is constructed in a similar manner to the N-channel MOSFET, except that the substrate is made of an N-type semiconductor with two P regions diffused into the ends of one side to serve as source and drain.

Theory of Operation. Comparison of Figs. 10-11 and 10-14 will indicate that the operation of an enhancement-type MOSFET is opposite to that of a depletion-type MOSFET. When the voltage between the gate and source of an enhancement-type MOSFET is zero (Fig. 10-14*a*) there is no channel to conduct current between the source and drain; therefore, no current will flow between the source and drain terminals at zero bias. When a positive voltage is applied to the gate, the input capacitor (formed by the gate, gate insulator,

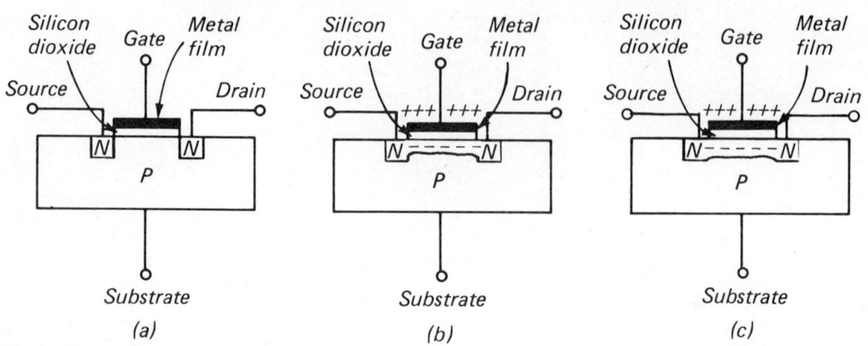

Fig. 10-14 Schematic representations of the enhancement-region variation in an enhancement-type N-channel MOSFET as the gate-bias voltage increases. (a) Zero gate bias. (b) Low gate bias. (c) High gate bias.

and channel) charges and the resulting electric field causes negative carriers to be drawn toward the positive gate. This action creates an N region in the substrate opposite the gate insulating film (Fig. 10-14*b*). Thus, a positive gate voltage in effect connects the N source and N drain by producing an N-N-N conducting channel from the source to the drain. The amount of forward-biasing gate voltage required to start current flow from source to drain is called the *threshold voltage*. As the positive voltage at the gate terminal is increased, the cross-sectional area of the created N region increases (Fig. 10-14*c*). This action increases the conductance of the channel between the source and drain, thus also increasing the amount of current flowing between these two terminals. A P-channel MOSFET operates in the same manner as an N-channel MOSFET, except that the channel carrier current will be holes instead of electrons, and the polarities at the source, drain, and gate are reversed.

Symbols. The schematic symbols for the enhancement-type MOSFET shown in Fig. 10-15 are similar to the schematic symbols for the depletion-type MOSFET except for the drawing of the channel bar. In this type of

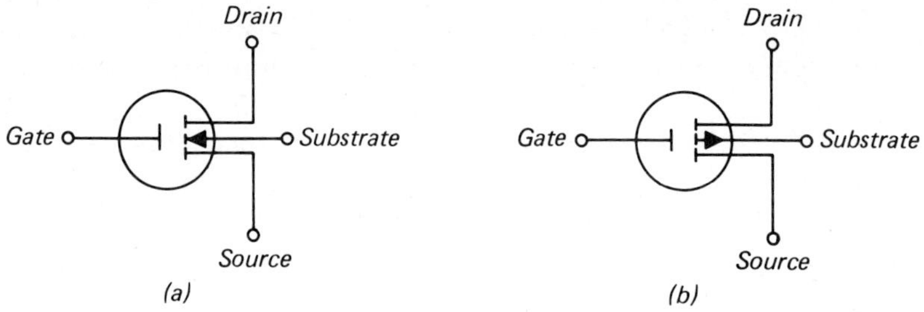

Fig. 10-15 Schematic symbols for enhancement-type MOSFETs. (a) N-channel. (b) P-channel.

device, the channel bar is drawn in three separate sections to identify the *normally-off characteristic* of the enhancement-type MOSFET.

Basic Circuit Operations. The basic characteristics of the enhancement-type MOSFET differ from the depletion type in that (1) with zero bias at the gate the enhancement-type MOSFET is *turned off* and the depletion-type MOSFET is *turned on;* (2) with the enhancement type, the gate is forward-biased to *enhance* the channel and turn the unit ON, while with the depletion type the gate is reverse-biased to *deplete* the channel and turn the unit OFF.

The basic circuit arrangements for a common-source connection of the N- and P-channel enhancement-type MOSFETs are shown in Fig. 10-16. With zero volts at the gate terminal, current cannot flow from the source to the drain (or vice versa) because a reverse bias will be present between either the source-to-channel or the channel-to-drain junctions depending on the polarities applied to the source and drain. With a positive voltage applied to the gate of an N-channel enhancement-type MOSFET (Fig. 10-16*a*) the gate takes on a positive charge and the P-type material takes on some negative charge, thereby changing a small area of the substrate channel material from P-type effect to N-type effect (Fig. 10-14*b*). Consequently the path from source to drain changes in effect from N-P-N material to N-N-N material, and current can now flow from source to drain with the polarity of V_{DD} as shown in Fig. 10-16*a*. Increasing the positive voltage at the gate results in further enhancement of the N-type character of the channel and thereby increases the amount of source-to-drain current. The increase in the source-to-drain current will cause an increase in the voltage drop at R_L and thus cause a decrease in the output voltage v_o. Thus, the output signal v_o will be 180 degrees out of phase with the input signal v_i.

When a varying signal is applied to the gate, any increase or decrease in voltage will increase or decrease the charge on the gate-to-channel capacitor and therefore cause an increase or decrease in the source-to-drain current. Curves illustrating the relation among the drain current, gate-source voltage, and drain-source voltage are shown in Figs. 10-20 and 10-21.

10-9 Biasing the MOSFET

Figure 10-16 illustrates several ways in which a gate-to-source bias voltage can be established. In the circuit of Fig. 10-16*a*, the gate-source bias is established soley by the input-signal voltage. A positive-going signal voltage will charge the gate-substrate capacitor with an increasing positive charge on the gate. After the signal reaches its maximum positive value and starts decreasing, some of the charge on the gate leaks off through R_G, thereby decreasing the gate-source bias and causing a decrease in the source-to-drain current. When the signal voltage decreases to a point where the gate-source bias is zero, the source-to-drain current will be cut off. Negative values of gate-to-source voltage will keep the source-to-drain current at zero. A varying signal

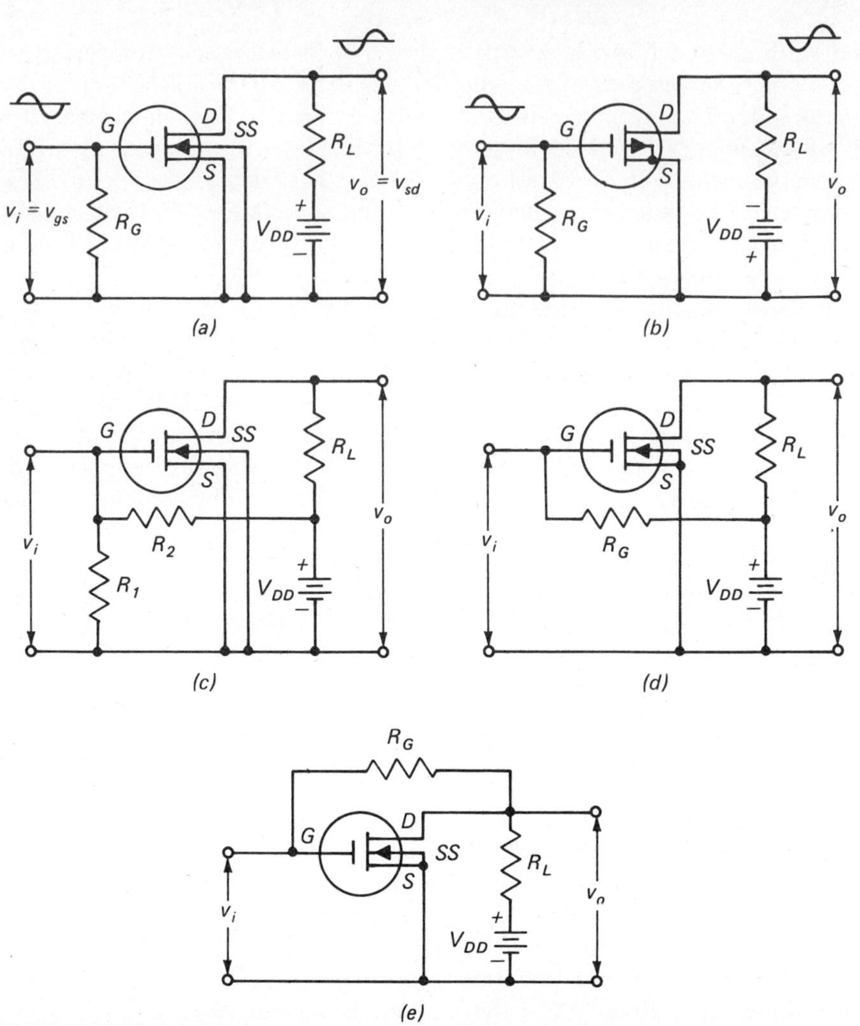

Fig. 10-16 Biasing circuits for MOSFETs using a common-source configuration. (a) Simple N-channel depletion-type MOSFET. (b) Simple P-channel enhancement-type MOSFET with source and substrate connected internally. (c) Fixed bias applied using a voltage-divider circuit. (d) Fixed bias applied using a single bias resistor. (e) Stabilized bias circuit.

voltage applied to the circuit of Fig. 10-16*b* will produce a somewhat similar pattern of changes in current and voltage values.

In the circuit of Fig. 10-16*c*, a fixed gate-to-source bias voltage is established by the voltage divider R_1R_2 and the power source V_{DD}. By making the fixed positive gate-source voltage equal to or greater than the negative swing of the input signal, the variations in the drain current and the output voltage will continuously follow changes in the input signal. The circuit of Fig. 10-16*d* is similar to that of Fig. 10-16*c* except that the gate-source voltage is greater than the drain-to-source voltage and is equal to the drain supply voltage V_{DD}. The circuit of Fig. 10-16*e* provides a stable operating bias that will not be affected appreciably by temperature changes. Resistor R_G can be in the order of 20 to 50 megohms, which is still small compared to the input impedance of the MOSFET gate-source impedance, which is in the order of 10^6 megohms. Under these conditions, R_G will not introduce any significant load to the input-signal source. Since the gate current of a MOSFET is always zero, no voltage drop can occur at R_G and the gate voltage v_{gs} will be equal to the source-drain voltage v_{sd}. As the signal voltage v_i varies, the action of the MOSFET causes a change in the source-to-drain current and also in the output voltage v_o.

10-10 Operating Modes of MOSFETs

Common-source Configuration. The common-source connection shown in Figs. 10-13 and 10-16 is the one most frequently used. The input signal is applied between the gate and source, and the output signal is taken off between the drain and source. This circuit arrangement provides (1) a high input impedance, (2) a medium to high output impedance, and (3) a voltage gain greater than 1. The voltage gain is expressed by

$$A_v = \frac{g_{fs} r_{os} R_L}{r_{os} + R_L} \tag{10-8}$$

where A_v = voltage gain, a dimensionless ratio
g_{fs} = gate-to-drain forward transconductance, mhos
r_{os} = common-source output resistance, ohms
R_L = effective load resistance, ohms

Example 10-6 Determine the voltage gain of the common-source MOSFET circuit of Fig. 10-13 if g_{fs} = 7,500 μmhos, r_{os} = 4,000 ohms, and R_L = 6,000 ohms.

GIVEN: g_{fs} = 7,500 μmhos r_{os} = 4,000 ohms R_L = 6,000 ohms

FIND: A_v

SOLUTION:

$$A_v = \frac{g_{fs} r_{os} R_L}{r_{os} + R_L} = \frac{7{,}500 \times 10^{-6} \times 4{,}000 \times 6{,}000}{4{,}000 + 6{,}000} = 18$$

Metal-oxide semiconductor field-effect transistors are usually rated as to their power gain rather than their voltage gain. Knowing the voltage gain, the power gain can be obtained by

$$A_p = \frac{A_v^2 R_i}{R_L} \tag{10-9}$$

where A_p = power gain, a dimensionless ratio
R_i = input-circuit resistance, ohms

Example 10-7 Determine the power gain in decibels of the MOSFET circuit of Example 10-6 if the input-circuit resistance is 2,000 ohms.

GIVEN: $A_v = 18 \qquad R_L = 6{,}000$ ohms $\qquad R_i = 2{,}000$ ohms

FIND: G_p

SOLUTION:

$$A_p = \frac{A_v^2 R_i}{R_L} = \frac{18 \times 18 \times 2{,}000}{6{,}000} = 108$$

$$G_p = 10 \log A_p = 10 \log 108 = 10 \times 2.033 = 20.33 \text{ db}$$

Common-gate Configuration. In the common-gate connection shown in Fig. 10-17 the input signal is applied between the source and the gate, and the output signal is taken off between the drain and gate. With this circuit arrangement (1) the input impedance is low and is approximately equal to the output impedance of the common-drain connection, (2) the output impedance is high, and (3) the voltage gain is relatively low. This type of circuit is used for impedance matching, transforming a low input impedance to a high output impedance. Because of its low voltage gain, neutralization is unnecessary, and this circuit arrangement is desirable for high-frequency applications. The voltage gain is expressed by

$$A_v = \frac{(g_{fs} r_{os} + 1) R_L}{(g_{fs} r_{os} + 1) R_G + r_{os} + R_L} \tag{10-10}$$

Example 10-8 Determine the voltage gain of the common-gate MOSFET circuit of Fig. 10-17 if $g_{fs} = 7{,}500$ μmhos, $r_{os} = 4{,}000$ ohms, $R_L = 2{,}000$ ohms, and $R_G = 800$ ohms.

GIVEN: $g_{fs} = 7{,}500$ μmhos $\qquad r_{os} = 4{,}000$ ohms $\qquad R_L = 2{,}000$ ohms
$R_G = 800$ ohms

FIND: A_v

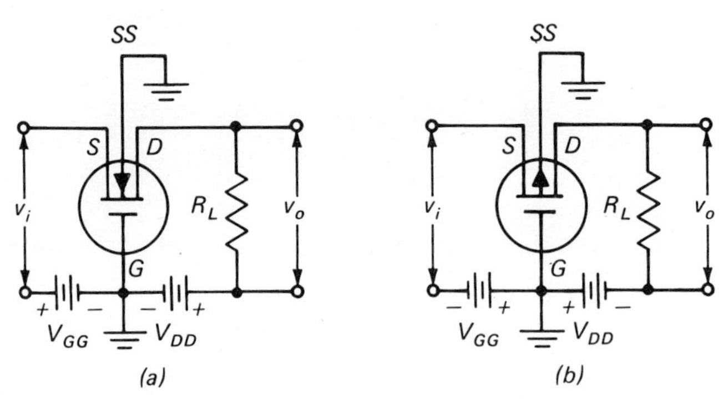

Fig. 10-17 Basic common-gate circuit configuration for a depletion-type MOSFET. (a) N-channel. (b) P-channel.

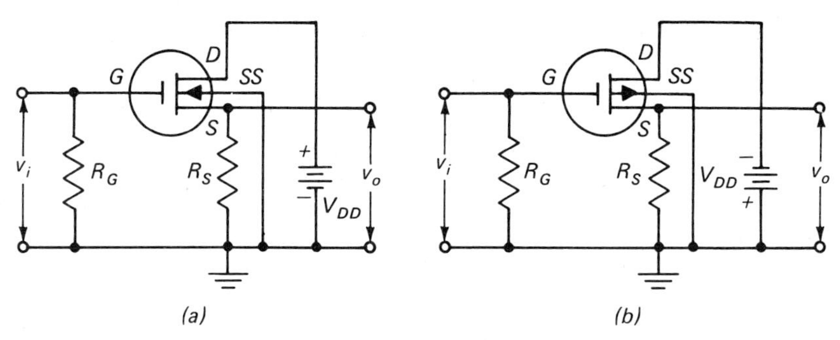

Fig. 10-18 Basic common-drain circuit arrangement for a depletion-type MOSFET. (a) N-channel. (b) P-channel.

SOLUTION:

$$A_v = \frac{(g_{fs}r_{os} + 1)R_L}{(g_{fs}r_{os} + 1)R_G + r_{os} + R_L}$$

$$= \frac{(7{,}500 \times 10^{-6} \times 4{,}000 + 1)2{,}000}{(7{,}500 \times 10^{-6} \times 4{,}000 + 1)800 + 4{,}000 + 2{,}000} = 2.01$$

Example 10-9 Determine the power gain in decibels of the MOSFET circuit of Example 10-8 if the input-circuit resistance is 800 ohms.

GIVEN: $A_v = 2.01$ $R_L = 2{,}000$ ohms $R_i = 800$ ohms

FIND: G_p

SOLUTION:

$$A_p = \frac{A_v^2 R_i}{R_L} = \frac{2.01 \times 2.01 \times 800}{2{,}000} = 1.61$$

$$G_p = 10 \log A_p = 10 \log 1.61 = 10 \times 0.2068 = 2.068 \text{ db}$$

Common-drain Configuration. The common-drain connection shown in Fig. 10-18 is also called a *source follower*. The input signal is effectively applied between the gate and drain, and the output signal is taken off between the source and drain. With this circuit arrangement (1) the input impedance is higher than with the common-source connection, (2) the output impedance is low, (3) there is no polarity reversal between the input and output signals, (4) the voltage gain is always less than 1, and (5) the distortion is low. This circuit inherently has 100 per cent negative feedback, and its voltage gain is expressed by

$$A_v' = \frac{g_{fs}R_s}{1 + g_{fs}R_s} \tag{10-11}$$

where A_v' = voltage amplification with feedback
R_s = source resistance, ohms

Example 10-10 Determine the voltage gain of the common-drain MOSFET circuit of Fig. 10-18 if $g_{fs} = 7{,}500$ μmhos and $R_s = 600$ ohms.

GIVEN: $g_{fs} = 7{,}500\ \mu$mhos $\qquad R_s = 600$ ohms

FIND: A_v'

SOLUTION:

$$A_v' = \frac{g_{fs}R_s}{1 + g_{fs}R_s} = \frac{7{,}500 \times 10^{-6} \times 600}{1 + 7{,}500 \times 10^{-6} \times 600} = 0.818$$

Example 10-11 Determine the power gain in decibels of the MOSFET circuit of Example 10-10 if the input-circuit resistance is 1 megohm and the load resistance is 5,000 ohms.

GIVEN: $A_v' = 0.818 \qquad R_i = 1{,}000{,}000$ ohms $\qquad R_L = 5{,}000$ ohms

FIND: G_p

SOLUTION:

$$A_p = \frac{A_v'^2 R_i}{R_L} = \frac{0.818 \times 0.818 \times 10^6}{5{,}000} = 134$$

$$G_p = 10 \log A_p = 10 \log 134 = 10 \times 2.127 = 21.27 \text{ db}$$

Characteristic Curves. Typical output characteristic curves for an N-channel depletion-type MOSFET are shown in Fig. 10-19, and those for an N-channel

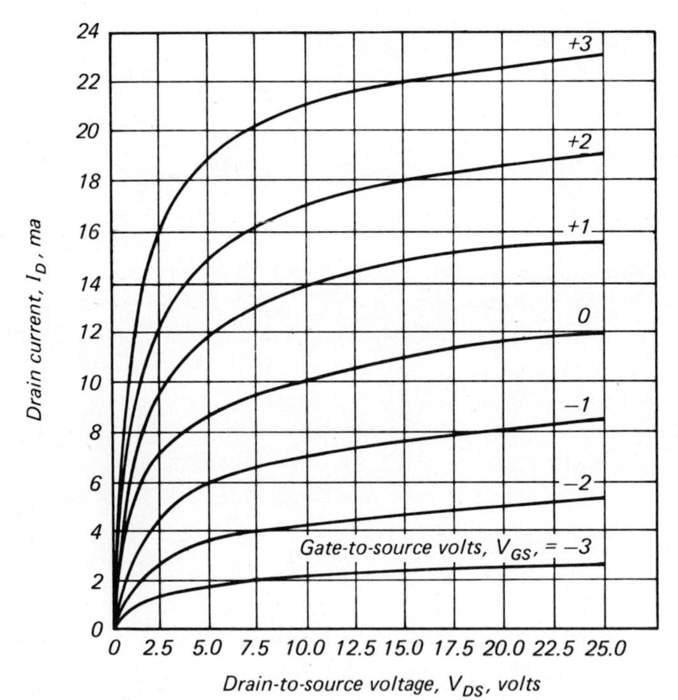

Fig. 10-19 Typical output characteristic curves for a common-source N-channel depletion-type MOSFET.

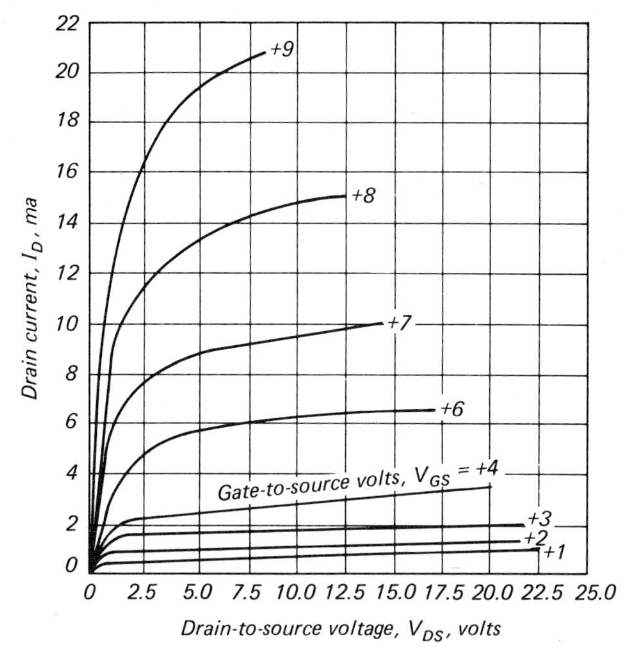

Fig. 10-20 Typical output characteristic curves for a common-source N-channel enhancement-type MOSFET.

enhancement-type MOSFET are shown in Fig. 10-20. Typical transfer characteristic curves for both N-channel depletion- and enhancement-type MOSFETs are shown in Fig. 10-21. The output and transfer characteristic curves for P-channel MOSFETs are identical to those for N-channel MOSFETs except that the polarities of voltages and current are reversed. It can be observed that the characteristic curves for MOSFETs are similar to those for junction bipolar transistors, electron tubes, and JFETS. Therefore, these curves can be used to determine the operating characteristics of MOSFETs in a similar manner as the comparable curves are used to determine the operating characteristics of junction bipolar transistors, electron tubes, and JFETs. Thus, the dynamic output resistance may be approximated from the slope of the output curve, and

$$r_{os} = \frac{dv_{ds}}{di_d} \qquad (V_{GS}\text{—constant}) \tag{10-12}$$

where r_{os} = dynamic output resistance, ohms
dv_{ds} = change in drain-to-source voltage, volts
di_d = change in drain current, amp
V_{GS} = gate-to-source voltage, volts

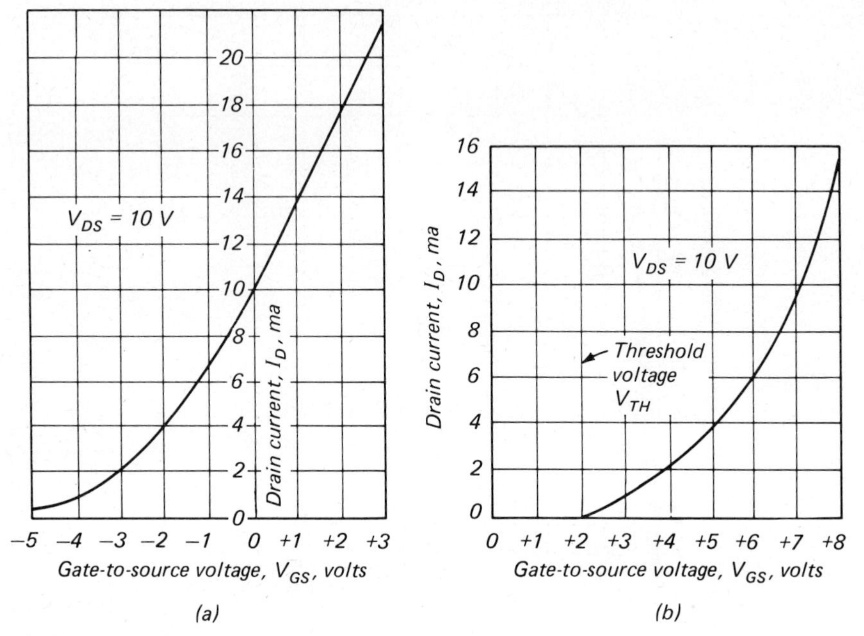

Fig. 10-21 Typical transfer characteristic curves for a common-source N-channel MOSFET. (a) Depletion type. (b) Enhancement type.

Example 10-12 Determine the approximate dynamic output resistance of a depletion-type MOSFET whose output characteristics are shown in Fig. 10-19 when $V_{DS} =$ 10 volts and $V_{GS} = -1.0$ volt.

GIVEN: $V_{DS} = 10$ volts $\quad V_{GS} = -1.0$ volt $\quad$ Characteristic curves, Fig. 10-19

FIND: r_{os}

SOLUTION: From Fig. 10-19

when $\quad V_{DS} = 12.5$ volts $\quad I_D = 7.2$ ma

$\quad V_{DS} = 7.5$ volts $\quad I_D = 6.4$ ma

$$r_{os} = \frac{dv_{ds}}{di_d} = \frac{12.5 - 7.5}{7.2 \times 10^{-3} - 6.4 \times 10^{-3}} = \frac{5}{0.8 \times 10^{-3}} = 6{,}250 \text{ ohms}$$

The approximate forward transconductance may be determined from the slope of the transfer characteristic curve, and

$$g_{fs} = \frac{di_d}{dv_{gs}} \qquad (V_{DS}\text{—constant}) \tag{10-13}$$

where g_{fs} = forward transconductance, mhos
dv_{gs} = change in gate-to-source voltage, volts
V_{DS} = drain-to-source voltage, volts

Example 10-13 Determine the approximate forward transconductance of a depletion-type MOSFET whose transfer characteristic curve is shown in Fig. 10-21*a* when $V_{DS} = 10$ volts and $V_{GS} = -1.0$ volt.

GIVEN: $V_{DS} = 10$ volts $\quad V_{GS} = -1.0$ volt $\quad$ Characteristic curves, Fig. 10-21*a*

FIND: g_{fs}

SOLUTION: From Fig. 10-21*a*

when $\quad V_{GS} = -0.5$ volt $\quad I_D = 8.2$ ma
$\quad V_{GS} = -1.5$ volts $\quad I_D = 5.2$ ma

$$g_{fs} = \frac{di_d}{dv_{gs}} = \frac{8.2 \times 10^{-3} - 5.2 \times 10^{-3}}{1.5 - 0.5} = \frac{3.0 \times 10^{-3}}{1.0} = 3{,}000\ \mu\text{mhos}$$

Advantages and Disadvantages. MOSFETs combine the inherent advantages of solid-state devices such as (1) small size, (2) low power consumption, (3) simplicity of construction, and (4) mechanical ruggedness; with the inherent advantage of electron tubes such as (1) very high input impedance and (2) square-law transfer characteristics.

Since a static electric charge may easily puncture the ultrathin silicon-dioxide dielectric insulation, extreme care must be exercised in handling MOSFETs. To minimize this type of failure, MOSFETs come with all four leads shorted. This short condition must not be removed until the device connections have been made to the circuit. To avoid (1) the problem of removing the short-circuiting connections, and/or (2) the troubles introduced by neglecting to remove the short-circuiting connections, some MOSFETs are provided with permanently connected Zener diodes (Fig. 10-22*g*) which safely shunt any static-voltage charges around the dielectric.

10-11 Dual-gate Field-effect Transistor

A multigate FET is made by diffusing two independent gates into the conducting channel of a JFET or a MOSFET. This type of device is called a *dual-gate FET* or a *tetrode FET*, and the symbols for this type of FET or MOSFET are shown in Fig. 10-22. Because each gate is independent of the other, the current-carrier streams they control are also independent of each other. This characteristic makes the dual-gate FET ideal for mixer applications in all types of communication receivers. Stable oscillator operation is obtained by the high isolation between the r-f amplifier and oscillator that is provided by the two independent gates. Cross modulation is minimized to a very low figure by the insulated-gate construction of MOSFETs. Dual-gate FETS are also ideal for use as color TV and stereo-f-m demodulators, and agc-driven i-f amplifiers.

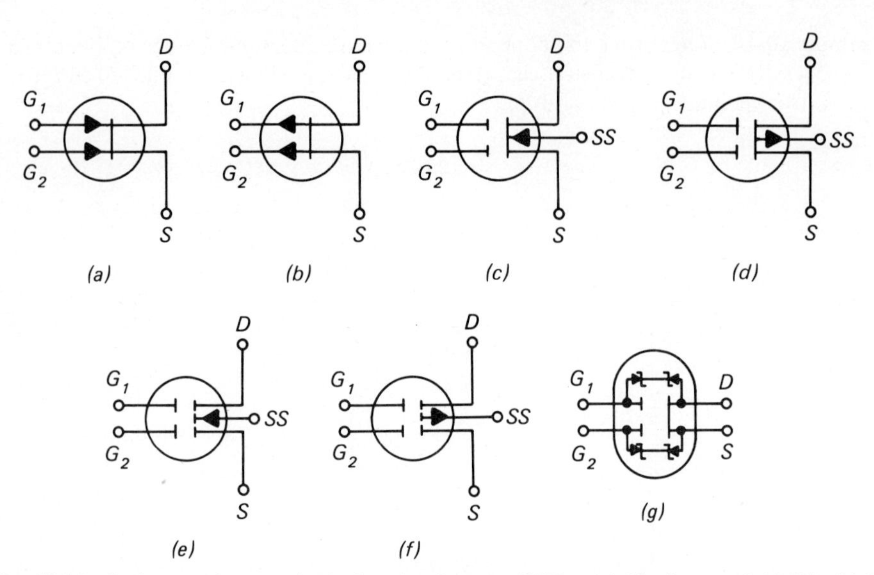

Fig. 10-22 Schematic symbols for dual-gate FETs. (a) N-channel JFET. (b) P-channel JFET. (c) N-channel depletion-type MOSFET. (d) P-channel depletion-type MOSFET. (e) N-channel enhancement-type MOSFET. (f) P-channel enhancement-type MOSFET. (g) With self-contained Zener diodes for protection against damage from static charges.

10-12 Light-sensitive Devices

Phototransistor. The construction of a phototransistor is similar to that of a junction transistor. However, only two external leads are used, one connected to the collector and the other connected to the emitter, with no connection being made to the base (Fig. 10-23). The transistor is positioned so that the base is illuminated through a small lens that is mounted at one end of the enclosing shell. A direct voltage is applied between the emitter and collector in the usual manner for a common-emitter amplifier. With no illumination the wafer is darkened and the collector current is extremely low, being the normal collector cutoff current for the applied voltage. Illumination of the wafer causes current carriers to be injected into the base region. A large collector current, which is proportional to the illumination and is equal to beta times the base current, will flow through the external circuit. The phototransistor has the advantage over the photodiode in that it will also provide amplification, because a relatively low light intensity will produce a comparatively high output current. Typical applications of the phototransistor are similar to those of the photodiode with the addition of optical coupling, light-beam reception, and electron-optical control service.

Miscellaneous Types of Light-sensitive Devices. The basic principles of the photocell, the photodiode, and the phototransistor are used to produce many other types of light-sensitive devices. Two of these are the light-activated switch and the Raysistor. The light-activated switch is a bistable component of the PNPN type. In a manner similar to a four-layer diode, once the unit is triggered to its ON state, it will continue to conduct current from its external power source until the power is interrupted. The Raysistor consists of a light source and a photoconductive cell that are enclosed in a lighttight housing. The light source may be either an incandescent or a glow lamp. The operation of the unit is similar to that of a lamp-photocell combination. The dark (OFF) resistance is of the order of 1,800 megohms, while the light (ON) resistance may drop to values as low as 300 ohms.

10-13 Thyristors

Types. A thyristor is a bistable semiconductor diode or triode, whose operations in the third quadrant of its voltage-current characteristic curve are (1) reverse blocking, (2) reverse conducting, and (3) bidirectional switching. The types of thyristor most commonly used are (1) the reverse-blocking triode, commonly called a *silicon controlled rectifier*, abbreviated SCR; and (2) the *bidirectional-switching triode*, usually referred to as a *triac.*

Uses. The SCR or the triac may be used either as a power control or as a power switch. As a power-control device, the thyristor is capable of furnishing power to a load that is variable from zero to full power without wasting any power. A rheostat, used in a similar power-control application, would waste power. A thyristor can be switched from a high-forward-resistance condition (OFF state) to a low-forward-resistance condition (ON state) in a relatively short period of time, sometimes as short as a few microseconds. Although the change in resistance is very great, involving high magnitudes of both voltage and current, this change can be achieved with very little control voltage, current, or power. The triac may also be used as a phase-control device in a-c circuit applications requiring phase control.

Because of the (1) relatively low cost, (2) high efficiency, (3) small size, and (4) wide range of current, voltage, power, and frequency ratings of the thyristor, the range of applications of this device is very wide. Thyristors are used in equipment ranging in power from the relatively small amount required by a high-fidelity radio receiver, to the very large amount required by a steel-mill drive. The frequency ratings range from the low power frequencies of approximately 60 hertz to the very high frequencies used in radar-pulse mod-

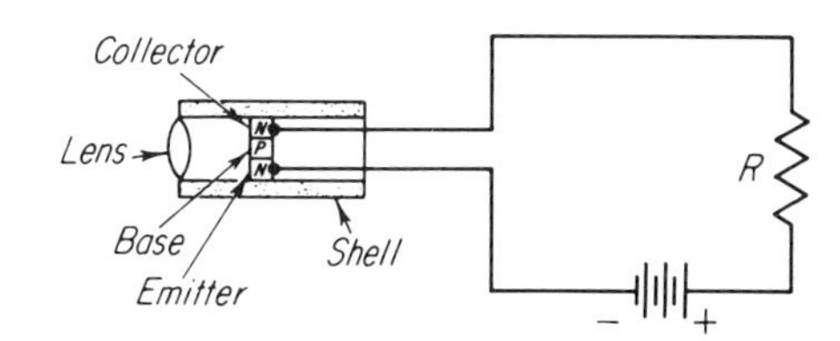

Fig. 10-23 A simple phototransistor and circuit.

ulators of hundreds of megahertz. Among the many applications of thyristors are (1) light dimming, (2) heat control, (3) d-c fluorescent lighting, (4) high-frequency induction heating, (5) a-c motor speed control, (6) ultrasonic cleaning, and (7) the replacement of electromechanical relays and switches in aircraft, computer, industrial, military, and business equipment.

10-14 Silicon Controlled Rectifier

Basic Theory. The silicon controlled rectifier consists of four alternating P and N layers of a semiconductor material having three external terminals, an *anode*, a *cathode*, and a *gate* (Fig. 10-24*e*). The schematic symbol for the SCR is shown in Fig. 10-24*f*. The effects on one another by the three junctions formed by the four layers of semiconductor material can best be explained by using the two-transistor analogy. The basic SCR structure of Fig. 10-24*a* can be considered as consisting of two transistors, a $P_1N_1P_2$ and an $N_2P_2N_1$, having a common-collector base junction N_1P_2 (Fig. 10-24*b* and *c*). According to the conventional manner of common-emitter operation, it can be observed

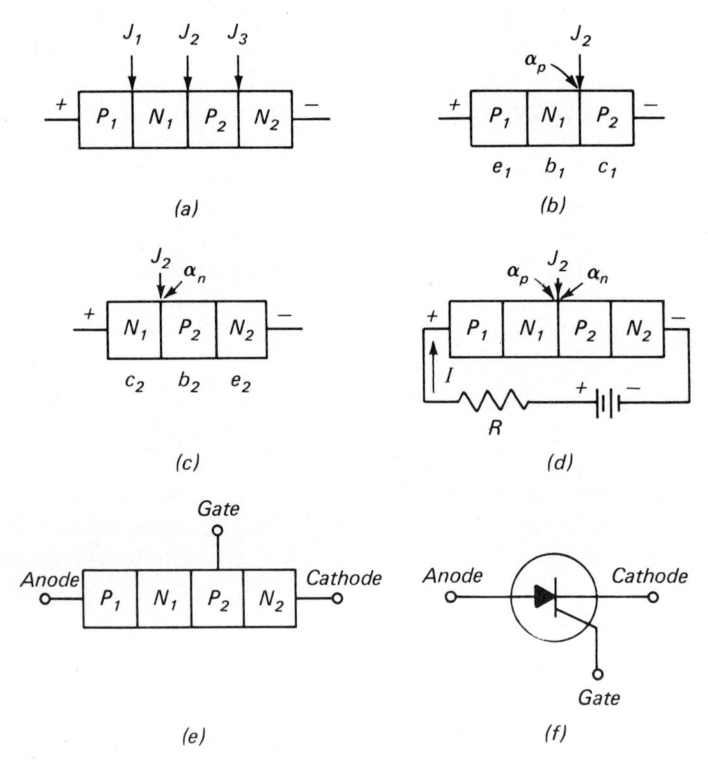

Fig. 10-24 Two-transistor SCR circuit analogy. (a) Basic four-layer structure. (b) PNP transistor section. (c) NPN transistor section. (d) Two-transistor analogy. (e) Terminal connections. (f) Schematic symbol.

that junctions J_1 and J_3 are forward-biased, and junction J_2 is reverse-biased. The current gain α_p of the $P_1N_1P_2$ transistor section, or the fraction of hole current injected in e_1 that reaches c_1, is equal to

$$\alpha_p = \frac{i_{c.1}}{i_{e.1}} \tag{10-14}$$

The current gain α_n of the $N_2P_2N_1$ transistor section, or the fraction of electron current injected in e_2 that reaches c_2, is equal to

$$\alpha_n = \frac{i_{c.2}}{i_{e.2}} \tag{10-15}$$

By combining the two transistor sections (Fig. 10-24*b* and *c*) as in Fig. 10-24*d*, the total current flowing in the $P_1N_1P_2N_2$ structure is equal to the sum of the currents for the individual transistor sections, or

$$I_{J.2} = I = I\alpha_n + I\alpha_p + I_{co} \tag{10-16}$$

where I = sum of the currents in all sections (line current), amp
$I_{J.2}$ = current at junction J_2, amp
$I\alpha_n$ = electron current from end N_1 region, amp
$I\alpha_p$ = hole current from end P_2 region, amp
I_{co} = leakage current, amp

Transposing Eq. (10-16),

$$I - I\alpha_n - I\alpha_p = I_{co} \tag{10-16a}$$

Combining like terms,

$$I(1 - \alpha_n - \alpha_p) = I_{co} \tag{10-16b}$$

Transposing to express I in terms of I_{co},

$$I = \frac{I_{co}}{1 - (\alpha_n + \alpha_p)} \tag{10-17}$$

From Eq. (10-17) it can be seen that the total current is dependent on $\alpha_n + \alpha_p$. When this sum is equal to 0.9, the total current will be ten times the leakage current I_{co}. However, the leakage current in a silicon PN junction is very small, and therefore the total current will also be very small. Hence, when $\alpha_n + \alpha_p$ is not greater than approximately 0.9, the SCR will be in its OFF condition. When $\alpha_n + \alpha_p$ is nearly unity, therefore, (1) the denominator is almost zero, (2) the total current is limited only by the resistance of the external circuit, and (3) the SCR will be in its ON condition. Physically, in the ON condition the two center regions N_1P_2 of the SCR are saturated with current carriers, and all three junctions have a forward bias. Hence, the entire voltage drop across the SCR is equivalent to that obtained from a forward-biased PN rectifier.

Methods Used to Turn an SCR ON. Two methods are employed for increasing the current gain in order to turn the SCR ON: (1) increasing the emitter

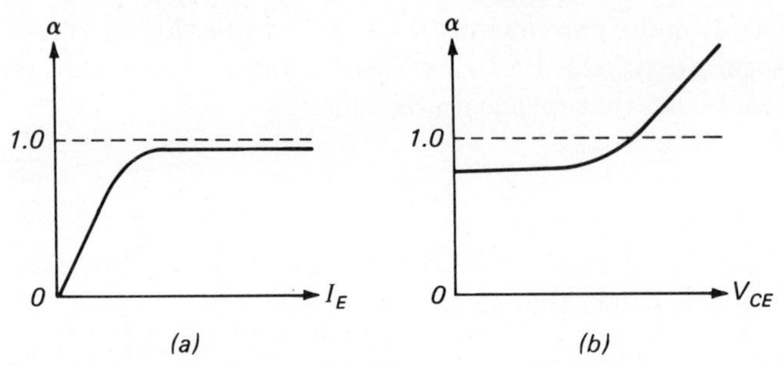

Fig. 10-25 Current and voltage variations of an SCR transistor. (a) Current gain versus emitter current. (b) Current gain versus collector-emitter voltage.

current and (2) increasing the collector-to-emitter voltage. In most silicon transistors, the current gain is quite low at low values of emitter current and increases quite rapidly as the emitter current is increased (Fig. 10-25*a*). This increase in emitter current can be obtained by the introduction of current at either of the two base sections N_1 and P_2 (Fig. 10-24*b* and *c*). The current gain of a silicon transistor increases slightly as the collector-to-emitter voltage is increased (Fig. 10-25*b*). This slight increase continues until a voltage is reached at which the energy of the current carriers arriving at the collector P_2N_1 junction (J_2, Fig. 10-24*b* and *c*) is sufficient to dislodge additional carriers. This action produces an *avalanche breakdown* at J_2, resulting in a large reverse current passing through it, which in turn increases both α_n and α_p.

Characteristic Curves. A typical forward- and reverse-bias characteristic curve for an SCR having a zero gate bias is shown in Fig. 10-26. Increasing the forward voltage does not increase the value of current until a point is reached where avalanche breakdown begins to occur. The voltage required to produce this breakdown is called the *breakover voltage*, abbreviated BV_F. Increasing the voltage past this point causes the current to increase quite rapidly until the total current is sufficient to maintain $\alpha_n + \alpha_p$ greater than 1. At this point, called the *pickup current*, switchback occurs and the SCR will go into the high-conduction region. In this region the SCR operates in the same manner as a low-resistance, single-junction, forward-biased silicon diode. The device will remain in this ON state as long as the current through J_2 (external-circuit current) is large enough to maintain $\alpha_n + \alpha_p$ equal to unity. The minimum value of current required to maintain the SCR in its ON state is called the *holding current*, abbreviated I_H. Should the current drop below this minimum value the device will revert to the forward blocking region, or OFF state.

When the polarity of the external voltage is reversed, it causes junctions J_1 and J_3 to be reverse-biased, and junction J_2 to be forward-biased (Fig. 10-24*a*). Thus, the SCR is then a device that is essentially made up of two reverse-biased PN junctions in series. It will therefore behave in the same manner as a single-junction reverse-biased silicon diode, and the SCR will be in the OFF state.

Gate Control. The purpose of the gate electrode is to control the voltage at which breakover occurs. A typical family of SCR forward- and reverse-bias characteristic curves for increasing positive values of gate currents is shown in Fig. 10-27. From this set of curves it can be observed that for increasing values of positive gate current (1) the region between the pickup current and the holding current decreases, and (2) the value of the breakover voltage also decreases. For relatively high values of positive gate current, the entire forward-blocking region is removed and the voltage-current characteristics of the SCR are essentially the same as those of a single-junction silicon rectifier. The SCR is usually biased well below the forward breakdown voltage. Triggering is obtained by injecting current into the gate terminal. This mode of operation has the advantage of (1) using a forward breakdown voltage that is much higher than any voltage that will normally occur in the circuit, and (2) using only a moderate amount of trigger power to start high conduction. Once the gate has triggered the SCR into its ON state, it no longer has any control over the device. The only method of turning the SCR OFF is to reduce the circuit current to a value that is lower than the holding current. Since the increase of $\alpha_n + \alpha_p$ is obtained by increasing the circuit current, the SCR is a *current-triggered device* as opposed to the thyratron, which is a *voltage-triggered device.*

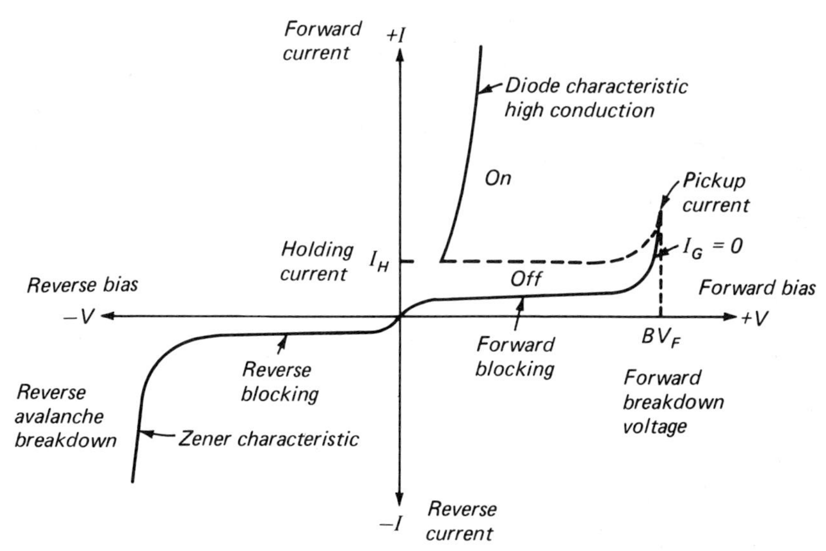

Fig. 10-26 Forward- and reverse-bias characteristics for an SCR with zero gate current.

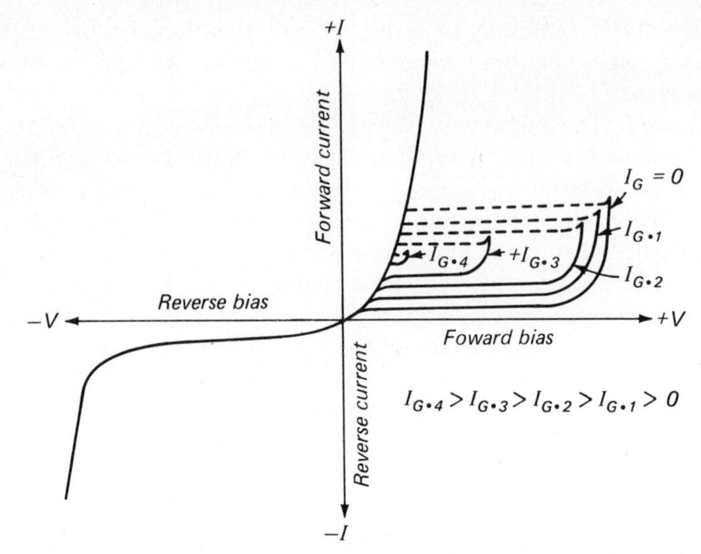

Fig. 10-27 Forward- and reverse-bias characteristics for an SCR with various values of positive gate current.

10-15 Triacs

Basic Structure. A disadvantage of the SCR is its inability to conduct current in both directions. The *triac* (triode a-c operated) can conduct current in both directions and is sometimes referred to as a *bilateral triode switch.* The basic structure of a triac is shown in Fig. 10-28*a*, and its schematic symbol in Fig. 10-28*b*. This device has three terminals, two main terminals and one gate terminal. The fabrication and internal behavior of a triac are beyond the scope of this text. Basically, however, this device can be considered as consisting of two parallel PNPN structures that are oriented in opposite directions in order to provide bidirectional characteristics.

Operating Characteristics. The triac basically operates as two SCRs connected in parallel, with the anode of each connected to the cathode of the other

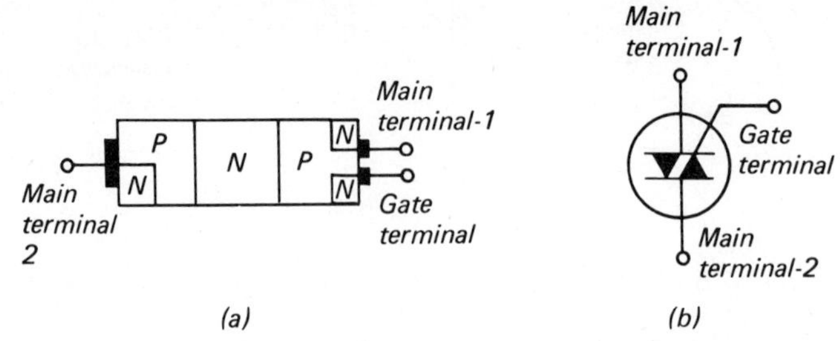

Fig. 10-28 (a) Basic structure of a triac. (b) Schematic symbol of a triac.

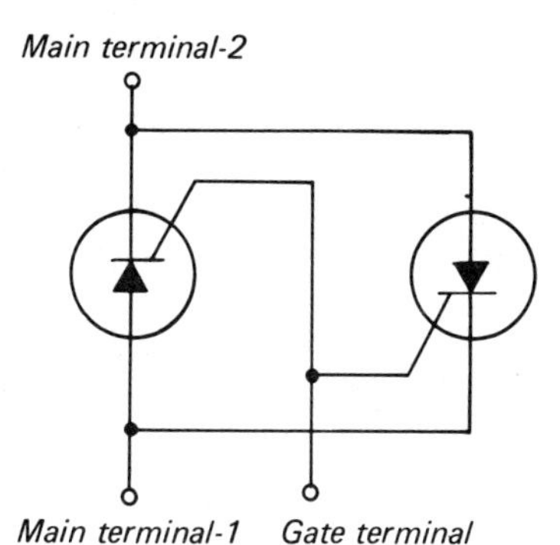

Fig. 10-29 Equivalent circuit of a triac using two SCRs connected in parallel and oriented in opposite directions.

(Fig. 10-29). This type of connection causes the operating characteristics of the triac in the first and third quadrants of its voltage-current characteristic curves (Fig. 10-30) to be the same, except for the direction of current flow and applied voltage. The voltage-current characteristics for the triac in these two quadrants are essentially identical to those of an SCR in its first quadrant. Because of the symmetrical construction of the triac, the terms forward and reverse are not used in reference to this device. Hence, the high-impedance state of the first and third quadrants is referred to as the OFF *state* rather than the forward- or reverse-blocking state; and the low-impedance state of the first and third quadrants is referred to as the ON *state* rather than the forward- or reverse-conducting state.

Gate Control. The point at which the triac switches from an OFF state to an ON state is the breakover voltage. In a similar manner to the SCR, the break-

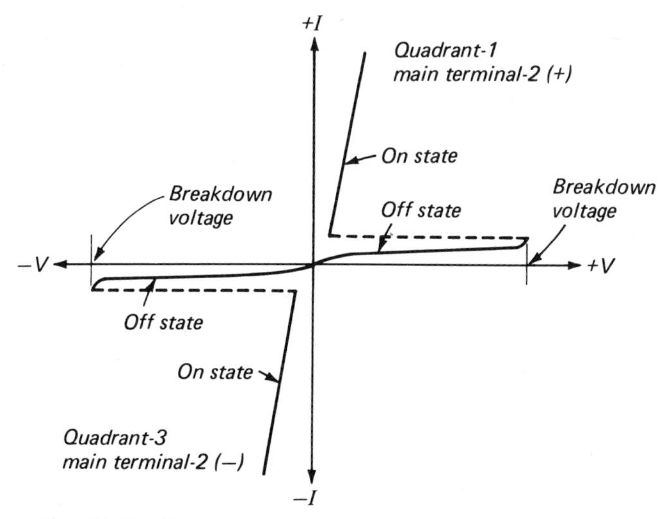

Fig. 10-30 Voltage-current characteristics of a triac.

over voltage of the triac can be controlled or varied by the application of a positive or negative current to its gate terminal. It is also possible to introduce a reactive component in the gate circuit that will cause a phase shift in the gate-cathode circuit. This reactive component can be used to shift the gate-trigger waveform up to 180 degrees relative to the anode-cathode voltage, and the leading edge of this waveform can thus be made to attain its trigger level at any time during the anode-cathode half-cycle. With this type of arrangement, the triac can be used to control power in a load from zero to full power in a smooth and continuous manner with no loss of power in the controlling device.

10-16 Diacs

A diac is a two-electrode, three-layer bidirectional avalanche diode, which can be switched from its OFF state to its ON state for either polarity of applied voltage. The structure of the diac (Fig. 10-31*a*) is similar to that of a bipolar transistor, and its schematic symbol is shown in Fig. 10-31*b*. The basic differences in the structure of a diac and a bipolar junction transistor are (1) the doping concentration at the two junctions is made approximately the same,

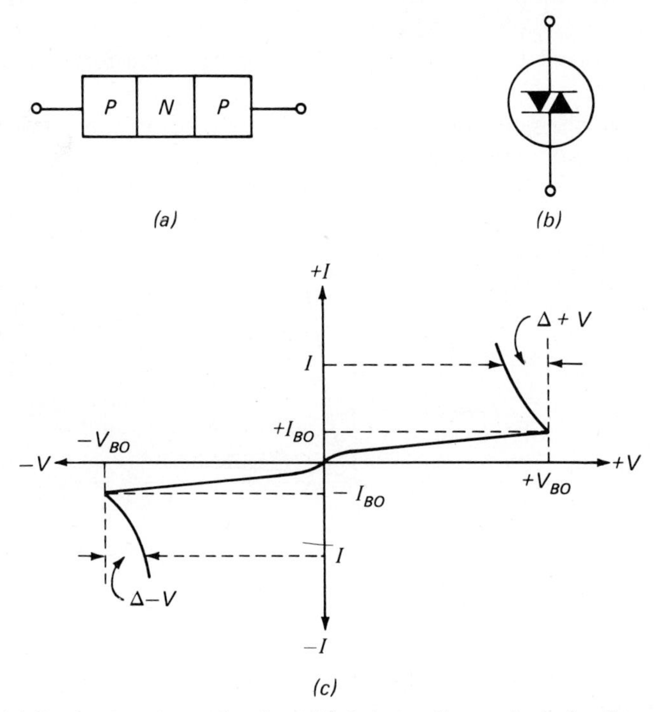

Fig. 10-31 (a) Basic structure of a diac. (b) Schematic symbol of a diac. (c) Voltage-current characteristics of a diac.

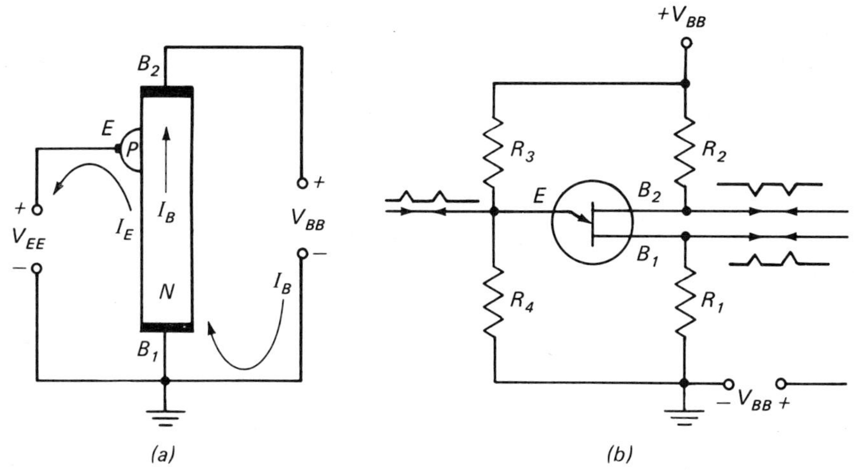

Fig. 10-32 Unijunction transistor. (a) Basic circuit analysis. (b) Basic circuit operation.

and (2) there is no terminal attached to the base layer. Because of the equal doping levels, the bidirectional switching characteristics are symmetrical (Fig. 10-31*c*).

When a positive or negative voltage is applied across the terminals of a diac, a minimum leakage current I_{BO} will flow through the device. As the applied voltage is increased, this minimum current continues to flow until the voltage reaches the breakover point V_{BO}. At this point (1) avalanche breakdown occurs at the reverse-biased junction, (2) the device exhibits a negative-resistance characteristic, and (3) current through the device increases substantially with decreasing values of applied voltage.

Diacs are used primarily as triggering devices in triac phase-control circuits. Some of the circuit applications are (1) light dimming, (2) heat control, and (3) universal-motor speed control.

10-17 Unijunction Transistors

Characteristics. The unijunction transistor is basically a three-terminal semiconductor diode. This device has a unique characteristic in that it can be *triggered by*, or an *output can be taken from*, any one of the three terminals. When this transistor is triggered, the emitter current increases regeneratively until it reaches a limiting point that is determined by the external power supply. Because of this characteristic the unijunction transistor can be employed in a variety of circuit applications, some of which are (1) switching, (2) pulse generator, (3) sawtooth generator, (4) sine-wave generator, (5) monocycle multivibrator, and (6) square-wave multivibrator.

The silicon-base bar is temperature-sensitive, and both the interbase resistance and temperature increase at a practically constant rate. The unijunction transistor can therefore be used in temperature-sensitive control applications.

Basic Operation. The basic construction and bias connections of a unijunction transistor are shown in Fig. 10-32*a*. This device consists of an N-type silicon bar having two base terminals, one at each end, B_1 and B_2; and a P-type emitter whose junction along the N-type bar is made closer to one base terminal than to the other. Because the two base terminals are taken from one section of the semiconductor diode, this device is also called a *double-based diode*. A voltage gradient is established along the N-type bar by the base-biasing potential V_{BB}. Since the emitter is located nearer to B_2, more than half of the base-bias voltage appears between the emitter and B_1. When an external potential V_{EE} is applied between the emitter terminal and B_1 that is greater than the internal voltage gradient between these two points, the junction is forward-biased. When the external voltage V_{EE} is less than the internal voltage gradient, a reverse bias is produced.

Normally, a reverse bias is applied between the emitter and B_1 to produce emitter-current cutoff in its OFF condition. When a positive trigger-pulse voltage is applied to the emitter, or a negative trigger-pulse voltage to B_1 or B_2, the emitter is forward-biased. An increase in the hole current causes (1) a reduction in the internal resistance between the emitter and B_1, and (2) a reduction in the internal voltage drop between the emitter and B_1. These actions cause the emitter current to increase regeneratively until it is limited by the external power source and is referred to as the *conductivity modulation of the interbase current*. The device is returned to its OFF state by applying a negative trigger-pulse voltage at the emitter.

Basic Circuit Operation. A typical unijunction transistor circuit with the input and output points indicated is shown in Fig. 10-32*b*. The circuit is normally held at cutoff by the emitter-bias voltage divider R_3R_4. An input voltage applied across R_4 produces a relatively high output voltage across R_1 and a relatively low output voltage across R_2. In a similar manner an input voltage applied across R_2 produces output voltages across R_1 and R_4; and an input voltage applied across R_1 produces output voltages across R_2 and R_4.

QUESTIONS

1. Describe (*a*) the basic construction of a tunnel diode, (*b*) the basic principle of operation of a tunnel diode.

2. Explain the operating characteristics of a tunnel diode for (*a*) a small reverse bias, (*b*) a small forward bias, (*c*) small increases in forward bias.

3. Explain the operating characteristics of a tunnel diode for bias values of (*a*) peak point I_P, (*b*) valley point I_V, (*c*) greater than I_V.

4. (*a*) Define the term negative resistance. (*b*) Describe the negative-resistance characteristic of a tunnel diode.

5. Describe the d-c load-line operating-point characteristics for a tunnel-diode (*a*) amplifier circuit, (*b*) oscillator circuit.

6. Describe the a-c load-line operating-point characteristics for a tunnel-diode (*a*) amplifier circuit, (*b*) oscillator circuit.
7. Define the following terms: (*a*) peak point, (*b*) reverse peak point, (*c*) valley point, (*d*) peak-point current, (*e*) reverse peak-point current.
8. Define the following terms: (*a*) valley-point current, (*b*) peak-to-valley-point current ratio, (*c*) peak-point voltage, (*d*) valley-point voltage, (*e*) forward voltage.
9. Define the following terms: (*a*) forward current, (*b*) reverse current, (*c*) reverse voltage, (*d*) negative conductance, (*e*) capacity.
10. Describe five advantages of a tunnel diode.
11. (*a*) Explain the basic principle of operation of a back diode. (*b*) Name two applications of back diodes.
12. Explain the operating characteristics of a back diode.
13. Describe the operating characteristics at the PN junction of a varactor diode with (*a*) a slight forward bias, (*b*) a reverse bias.
14. (*a*) Describe the basic construction of a varactor diode. (*b*) What is meant by the term junction-transition capacitance?
15. (*a*) Describe three advantageous operating characteristics of varactor diodes. (*b*) Describe two general applications of varactors.
16. Explain how a compensating diode can be used to stabilize the output of a transistor for variations in temperature and/or supply voltage.
17. Describe the basic construction of a junction field-effect transistor having (*a*) an N channel, (*b*) a P channel.
18. Explain the theory of operation of a JFET.
19. Compare the circuit operations of a FET with (*a*) a conventional transistor, (*b*) a vacuum tube.
20. What is meant by (*a*) a bipolar device? (*b*) A unipolar device?
21. (*a*) Name the three circuit configurations of a FET. How do these configurations compare with those of (*b*) a conventional transistor? (*c*) A vacuum tube?
22. Explain the circuit analysis of the static characteristic curve of a FET for (*a*) the ohmic region, (*b*) the region between the knee and pinch-off, (*c*) the saturation region, (*d*) the breakdown region.
23. Describe the advantages of using a FET as a mixer in TV, a-m, and f-m tuner circuits as compared to other electronic devices.
24. In the common-source mode of operation, why should the bias circuit be of such values as to prevent the input signal from driving the FET into a forward bias?
25. Define the following terms: (*a*) drain resistance, (*b*) forward transconductance, (*c*) admittance, (*d*) forward transadmittance.
26. (*a*) Why is it not possible to obtain an accurate value of the drain resistance from the static characteristic curves? (*b*) What is the general value of drain resistance?
27. (*a*) Why is the forward transadmittance usually specified by manufacturers rather than the forward transconductance? (*b*) Why is the transconductance or transadmittance an important parameter rating of a FET?
28. Explain the various power-gain ratings of a FET.
29. Name 10 advantages of the FET.
30. (*a*) Explain the impedance-matching characteristics of a FET. (*b*) Explain the power-gain characteristics of a FET.
31. (*a*) Why are the inherent noise sources of other electronic devices not present in a FET? (*b*) Why is it impossible to have thermal runaway in a FET?

32. Name the following FET specifications usually listed by manufacturers: (*a*) five maximum ratings, (*b*) ten operating characteristics.

33. (*a*) What does the abbreviation MOSFET represent? (*b*) What is another name for a MOSFET?

34. (*a*) Which type of MOSFET operates in a similar manner to a FET, and why? (*b*) Which type of MOSFET operates in a similar manner to a conventional junction transistor, and why?

35. Describe the basic construction of a depletion-type (*a*) N-channel MOSFET, (*b*) P-channel MOSFET.

36. Explain the theory of operation of an N-channel depletion-type MOSFET (*a*) at zero bias, (*b*) with a small amount of negative gate bias.

37. Explain the theory of operation of a P-channel depletion-type MOSFET (*a*) at zero bias, (*b*) with a small amount of positive gate bias.

38. Explain the theory of operation of an N-channel depletion-type MOSFET when the negative bias causes the depletion area to make contact with the substrate.

39. Explain the theory of operation of a P-channel depletion-type MOSFET when the positive bias causes the depletion area to make contact with the substrate.

40. What factor limits the magnitude of reverse bias that can be applied to the gate terminal of a MOSFET?

41. Describe the basic circuit operation and relative values of the circuit components for a common-source depletion-type basic MOSFET circuit.

42. Describe the construction of an enhancement-type (*a*) N-channel MOSFET, (*b*) P-channel MOSFET.

43. Explain the theory of operation of an N-channel enhancement-type MOSFET (*a*) at zero bias, (*b*) with a small amount of positive gate bias, (*c*) with increasing amounts of positive gate bias.

44. Explain the theory of operation of a P-channel enhancement-type MOSFET (*a*) at zero bias, (*b*) with a small amount of negative gate bias, (*c*) with increasing amounts of negative gate bias.

45. Define the following MOSFET terms: (*a*) threshold voltage, (*b*) turned on, (*c*) turned off.

46. Describe the circuit operation of a common-source enhancement-type basic MOS-FET circuit.

47. Explain the operation of a common-source depletion-type MOSFET circuit using the input signal to establish the gate-to-source bias, for (*a*) an N-channel transistor, (*b*) a P-channel transistor.

48. Explain the operation of a common-source depletion-type MOSFET circuit using (*a*) a voltage divider to obtain the gate bias, (*b*) a single resistor to obtain the gate bias, (*c*) a stabilized bias circuit.

49. What are the basic characteristics of the following modes of MOSFET operation: (*a*) the common-source configuration? (*b*) The common-gate configuration? (*c*) The common-drain configuration?

50. Explain the advantages of MOSFETs as compared to other solid-state devices.

51. (*a*) Why must extreme care be exercised in handling MOSFETs? (*b*) What precaution is used to prevent this type of failure?

52. (*a*) What is meant by a tetrode FET, or a tetrode MOSFET? (*b*) What are the advantages of a dual-gate FET or MOSFET?

53. Describe the basic (*a*) construction of a phototransistor, (*b*) circuit operation of a phototransistor.
54. What are the advantages of the phototransistor as compared to those of a photodiode?
55. Describe the operation of (*a*) a light-sensitive switch, (*b*) a Raysistor.
56. What is (*a*) a thyristor? (*b*) A silicon controlled rectifier? (*c*) A bidirectional-switching triode?
57. What are the advantages of using a thyristor as (*a*) a power-control device? (*b*) A power switch?
58. Name four general advantages of the thyristor.
59. Name seven general applications of the thyristor.
60. Describe the basic structure of an SCR.
61. Explain the theory of operation of an SCR using the two-transistor analogy.
62. Explain the relation of $\alpha_n + \alpha_p$ in an SCR for (*a*) the OFF state, (*b*) the ON state.
63. Explain the two methods generally used to turn an SCR ON.
64. Define the following SCR terms: (*a*) breakover voltage, (*b*) pickup current, (*c*) holding current.
65. Explain the operation of an SCR with (*a*) a small amount of forward voltage, (*b*) a reverse bias.
66. (*a*) What is the purpose of the gate electrode in an SCR? (*b*) What effects are produced by increasing amounts of positive gate current?
67. (*a*) Describe the method generally used to trigger an SCR. (*b*) What are the advantages of using this mode of operation? (*c*) What method is used to turn an SCR OFF?
68. (*a*) Describe the basic structure of a triac. (*b*) What is another name for the triac?
69. Explain the theory of operation of a triac.
70. Explain why the terms forward and reverse are not used in reference to the operation of a triac.
71. What terms are used in a triac to describe (*a*) the OFF state? (*b*) The ON state?
72. Explain how the breakover voltage in a triac is controlled by (*a*) using a positive or negative current, (*b*) using a reactive component.
73. Describe the basic (*a*) structure of a diac, (*b*) application of a diac.
74. Explain the theory of operation of a diac.
75. (*a*) What are the characteristics of a unijunction transistor? (*b*) Name six applications of the unijunction transistor.
76. Describe (*a*) the basic structure of a unijunction transistor, (*b*) the theory of operation of a unijunction transistor.
77. What is meant by conductivity modulation of the interbase current of a unijunction transistor?
78. Describe the operation of a typical unijunction transistor circuit.

PROBLEMS

1. Determine the value of source-bias resistance required for the circuit of Fig. 10-10*b* to produce a gate bias of 0.6 volt with an average drain current of (*a*) 3 ma, (*b*) 2 ma.
2. A source-bias resistance of 600 ohms is used in the FET circuit of Fig. 10-10*b*. Determine the drain current when the gate-bias voltage is (*a*) 1.8 volts, (*b*) 2.1 volts.
3. Determine the approximate value of source-bias resistance required for the circuit

of Fig. 10-10*b* using a FET whose characteristics are shown in Fig. 10-9. The transistor is to be operated on the load line indicated on the curves and with a quiescent value that produces a gate bias of (*a*) −1.5 volts, (*b*) −0.5 volt.

4. Determine the approximate value of source-bias resistance required for the circuit of Fig. 10-10*b* using a FET whose characteristics are shown in Fig. 10-9. The transistor is to be operated on the load line indicated on the curves and with a quiescent value that produces a gate bias of (*a*) −1.75 volts, (*b*) −0.25 volt.
5. Determine the forward transconductance for the FET used in Prob. 3, for each of the gate biases used.
6. Determine the forward transconductance for the FET used in Prob. 4, for each of the gate biases used.
7. The forward transconductance of a FET used in a voltage-amplifier circuit is 4.5 mmhos. Determine the voltage amplification of this circuit when the load resistance is (*a*) 6,000 ohms, (*b*) 8,000 ohms.
8. The forward transconductance of a FET used in a voltage-amplifier circuit is 6.3 mmhos. Determine the load resistance required to produce a voltage amplification of (*a*) 35, (*b*) 47.25.
9. The forward transadmittance of a FET used in a voltage-amplifier circuit is 8.3 mmhos at 100 MHz. Determine the voltage amplification of this circuit when the output impedance of this circuit at 100 MHz is (*a*) 4,000 ohms, (*b*) 5,400 ohms.
10. The forward transadmittance of a FET used in a voltage-amplifier circuit is 6.5 mmhos at 100 MHz. Determine the output impedance required of this circuit at 100 MHz to produce a voltage amplification of (*a*) 27.3, (*b*) 44.2.
11. Determine the voltage gain of the common-source MOSFET circuit of Fig. 10-13 if g_{fs} = 7 mmhos, r_{os} = 3,800 ohms, and R_L = 5,000 ohms.
12. Determine the voltage gain of the common-source MOSFET circuit of Fig. 10-13 if g_{fs} = 10,000 μmhos, r_{os} = 2,200 ohms, and R_L = 4,000 ohms.
13. Determine the power gain in decibels of the MOSFET circuit of Prob. 11 if the input-circuit resistance is 3,000 ohms.
14. Determine the power gain in decibels of the MOSFET circuit of Prob. 12 if the input-circuit resistance is 2,000 ohms.
15. Determine the voltage gain of the common-gate MOSFET circuit of Fig. 10-17 if g_{fs} = 7 mmhos, r_{os} = 3,800 ohms, R_L = 1,500 ohms, and R_G = 800 ohms.
16. Determine the voltage gain of the common-gate MOSFET circuit of Fig. 10-17 if g_{fs} = 10,000 μmhos, r_{os} = 2,200 ohms, R_L = 1,800 ohms, and R_G = 600 ohms.
17. Determine the power gain in decibels of the MOSFET circuit of Prob. 15 if the input-circuit resistance is 750 ohms.
18. Determine the power gain in decibels of the MOSFET circuit of Prob. 16 if the input-circuit resistance is 900 ohms.
19. Determine the voltage gain of the common-drain MOSFET circuit of Fig. 10-18 if g_{fs} = 7 mmhos and R_s = 700 ohms.
20. Determine the voltage gain of the common-drain MOSFET circuit of Fig. 10-18 if g_{fs} = 10,000 μmhos and R_s = 800 ohms.
21. Determine the power gain in decibels of the MOSFET circuit of Prob. 19 if its input resistance is 0.8 megohm, and the load resistance is 4,000 ohms.
22. Determine the power gain in decibels of the MOSFET circuit of Prob. 20 if its input resistance is 1.2 megohms, and the load resistance is 6,000 ohms.
23. Determine the approximate dynamic output resistance of a depletion-type MOS-

FET whose output characteristics are shown in Fig. 10-19 when V_{DS} = 12.5 volts, and V_{GS} = −2.0 volts.

24. Determine the approximate dynamic output resistance of a depletion-type MOSFET whose output characteristics are shown in Fig. 10-19 when V_{DS} = 7.5 volts and V_{GS} = +1.0 volt.

25. Determine the approximate forward transconductance of a depletion-type MOSFET whose transfer characteristic curve is shown in Fig. 10-21*a* when V_{DS} = 10.0 volts and V_{GS} = −2.0 volts.

26. Determine the approximate forward transconductance of a depletion-type MOSFET whose transfer characteristic curve is shown in Fig. 10-21*a* when V_{DS} = 10.0 volts and V_{GS} = +1.0 volt.

27. Determine the approximate dynamic output resistance of an enhancement-type MOSFET whose output characteristics are shown in Fig. 10-20 when V_{DS} = 10.0 volts and V_{GS} = +4.0 volts.

28. Determine the approximate dynamic output resistance of an enhancement-type MOSFET whose output characteristics are shown in Fig. 10-20 when V_{DS} = 7.5 volts and V_{GS} = +7.0 volts.

29. Determine the approximate forward transconductance of an enhancement-type MOSFET whose transfer characteristic curve is shown in Fig. 10-21*b* when V_{DS} = 10.0 volts and V_{GS} = +4.0 volts.

30. Determine the approximate forward transconductance of an enhancement-type MOSFET whose transfer characteristic curve is shown in Fig. 10-21*b* when V_{DS} = 10.0 volts and V_{GS} = +7.0 volts.

Chapter 11

Tuning Circuits

A *tuning circuit* is one in which the circuit parameters can be varied in order to tune the circuit to a desired value of frequency, and hence make it resonant to that specific frequency. A *tuned circuit* is one in which the circuit parameters have values that will make the circuit resonant to a specific frequency. A knowledge of resonant circuits is important in the study of electronics. In the operation of a radio or television receiver, the desired program is obtained by tuning the receiver. Actually, when a desired station is being selected, the tuning circuit of the receiver is being adjusted so that it is in resonance with the carrier frequency of the station transmitting the desired program. Tuned circuits are also used in all other forms of communications, such as radar, sonar, and telemetry. In other fields of electronics tuned circuits have many applications, some of which are quality control, thickness control, capacitor-relay alarms, and data processing.

11-1 Tuning

Operating Characteristics. Selecting a desired signal is only one of three important functions performed by the tuning circuit. In addition, it must reject all undesired signals, and in many instances it also increases the voltage of the desired signal before passing it on to the following circuit. The ability of a radio receiver to accomplish each of its three functions is referred to as its sensitivity, selectivity, and fidelity. Although the terms and definitions presented in this article refer to radio receivers, they apply to tuning circuits in general as well as to all other forms of communication systems.

Sensitivity is a measure of the ability of a radio receiver to reproduce, with satisfactory volume, weak signals received by the antenna. It may further be defined as the minimum strength of signal input required to produce a specified audio-frequency power output at the loudspeaker; it is generally expressed either in microvolts or in decibels below 1 volt.

Selectivity is a measure of the ability of a radio receiver to reproduce the signal of one desired station and to exclude the signals from all others. The selectivity of a receiver (or a tuning circuit) is generally expressed in the form of a graph, also called a *response curve,* showing the signal strength at its resonant frequency and the variation in signal current when the frequency is varied a specified amount above and below the frequency of resonance.

Fidelity is a measure of the ability of a radio receiver to reproduce faithfully all the frequencies present in the original signal. The fidelity of a receiver is generally expressed in the form of a graph showing the ratio of the actual output to the output of a standard audio frequency of 400 hertz. For good fidelity of reproduction, the bandwidth as shown by the selectivity graphs should be great enough to accommodate all the frequencies of the signal to be reproduced.

The Response Curve. The process of selecting the carrier wave of a desired station is called *tuning*. This may be accomplished by adjusting one or more components of a series tuned circuit so that its resonant frequency will be equal to that of the desired carrier wave. The impedance of the series tuned circuit at resonance will be at its minimum value; therefore, the current in the series tuned circuit produced by the desired station adjustment will be at its maximum value. As the resonant frequency of the tuning circuit is varied, either above or below the frequency of the desired station, the impedance of the circuit will increase and the signal current of the desired station will therefore decrease. A graph showing the current in a tuned circuit at resonance and the decrease in current at frequencies off resonance is called a *response curve* (Fig. 11-1*b*).

Ideal Response Curve. A transmitted wave is made up of the carrier wave modulated by two sidebands whose frequencies are equal to the frequency of the carrier wave plus or minus the frequency of the audio signal. For general commercial a-m broadcast transmission, the maximum frequency of the audio signal is 5 kHz. The sidebands of a modulated carrier wave will therefore vary up to 5 kHz above and below its carrier frequency. For example, a broadcasting station operating on a carrier frequency of 1,000 kHz will have sideband frequencies ranging from 995 to 1,005 kHz.

In order to reproduce the signals as transmitted, the ideal response curve should have a flat top and straight sides, so that it may pass a 10-kHz band for commercial a-m broadcast signals. This ideal can be closely approximated by proper use of resonant circuits. Three methods of increasing the fidelity, sensitivity, and selectivity of a radio receiver are (1) increasing the circuit Q, (2) using two or more tuned circuits, and (3) using a bandpass amplifier.

A-M Radio Channels. Each transmitting station requires a band 10 kHz wide for commercial a-m broadcasting stations; this band is called a *radio channel*. In order to prevent interference between two stations operating on adjacent channels there should be a difference of at least 10 kHz in their carrier frequencies. For the same reason, the tuning circuit of a radio receiver should be capable of selecting signals from stations approximately 10 kHz apart without any interference. The commercial a-m broadcasting range extending from 535 to 1,605 kHz could accommodate 107 channels. More than 1,000 stations are assigned to these channels. Under this crowded condition more than one station may be operating on the same frequency; also the sidebands from adjacent channels may overlap the signals of more than one station. This interference between stations transmitting on the same frequency or operating partly within the same channel may produce a hum, whistle, or cross talk in the receiver.

In order to prevent interference between stations, the FCC has set up a zoning system and assigns the carrier frequency to broadcasting stations so that the interference is reduced to a minimum. Furthermore, stations are licensed as to the amount of power they may use, and in some cases they are also limited as to the hours during which they may broadcast. For example, 13 stations in the United States are assigned to the 710-kHz channel. Four of these, located at New York, Los Angeles, Miami, and Seattle, employ high-power transmitters; but because they are located great distances from one another, they do not interfere with each other and may therefore operate on a continuous time schedule. On the other hand, over 150 stations in the United States are assigned to the 1,240-kHz channel, with as many as 10 stations in the same state assigned to the same channel. The licensed power of these stations is kept low, and some of the stations are restricted to operating at only certain hours of the day.

The ability of a radio receiver to minimize interference depends on the selectivity of the tuned circuits. In order to eliminate interference from adjacent stations, the selectivity has to be increased. This can reduce the width of the response curve and also decrease the fidelity of the receiver, as the high-frequency notes are not reproduced. For example, if the width of the response curve is reduced to 8 kHz, only 4 kHz of the 5-kHz sidebands as transmitted will be reproduced. Under this condition all audio signals between 4,000 and 5,000 hertz will not be reproduced by the radio receiver.

F-M Radio Channels. F-m radio stations operate on assigned frequencies ranging from 88 to 108 MHz and the FCC-assigned channels are kept at least 200 kHz apart. The channel assignments permit f-m transmitters to use a signal having a 150-kHz carrier swing, thereby enabling the transmission of a-f signals up to 15 kHz.

Television Channels. Television stations are assigned two frequency bands: (1) the *vhf band* with a frequency range of 55 to 88 MHz (channels 2 to 6) and 174 to 216 MHz (channels 7 to 13); and (2) the *uhf band* with a frequency range of 470 to 890 MHz (channels 14 to 83). Each channel is allotted a 6-MHz band.

11-2 Resonance

Resonance is a condition that exists when the inductive reactance and the capacitive reactance of a circuit are equal. For a given value of inductance and capacitance resonance can occur at only one value of frequency, called the *resonant frequency of the circuit.* When the inductive reactance and the capacitive reactance are equal, then (1) the impedance of a series resonant circuit will be at its lowest possible value, and (2) the impedance of a parallel resonant circuit will be at its highest possible value.

When an alternating current flows through a circuit that contains resistance, inductance, and capacitance, the flow of current is opposed by the combined effects of resistance, inductive reactance, and capacitive reactance. The current due to the inductive reactance lags the voltage by 90°, and the current due to the capacitive reactance leads the voltage by 90°. Because the effects of inductive reactance and capacitive reactance are 180° out of phase with each other, they neutralize one another and make the phenomenon of tuning possible. The principles of resonance are presented in great detail in the authors' "Essentials of Electricity—Electronics."

11-3 Series Resonant Circuit

Characteristics at Resonance. At resonance, the impedance of a series resonant circuit is at its minimum, and its value is equal to the resistance of the circuit. The circuit acts the same as a resistor. The current through all parts of the circuit is the same and is equal to the line current; the current is at its maximum value and is in phase with the applied voltage. The power factor of the circuit will therefore be unity. The voltages across the reactances are approximately equal and nearly 180° out of phase with each other, and the voltage across the resistance is equal to the applied voltage. Increasing the value of the resistance will decrease both the current in the line and the voltage across each reactance. For frequencies below resonance, the capacitive reactance is greatest and the current is leading. For frequencies above resonance, the inductive reactance is greatest and the current is lagging.

Impedance and Current Characteristics. When the inductance and the capacitance of a series circuit are of such values that the inductive reactance is equal to the capacitive reactance at the frequency of the applied voltage, the circuit is called a *series resonant circuit* (Fig. 11-1*a*). The impedance of a series circuit is expressed by Eq. (11-1):

$$Z = \sqrt{R^2 + (X_L - X_C)^2} \tag{11-1}$$

However, at resonance $(X_L - X_C) = 0$ and the impedance of the resonant series circuit becomes equal to the resistance of the circuit, or

$$Z_{s \cdot r} = R \tag{11-2}$$

The current of a series resonant circuit may then be expressed as

$$I_{s \cdot r} = \frac{E}{R} \tag{11-3}$$

If the frequency is either increased or decreased from its resonant value, the quantity $(X_L - X_C)$ will no longer be zero. Also, if the frequency is kept constant and either the inductance or the capacitance is changed, the quantity $(X_L - X_C)$ will no longer be zero. As $(X_L - X_C)$ is zero only when the circuit

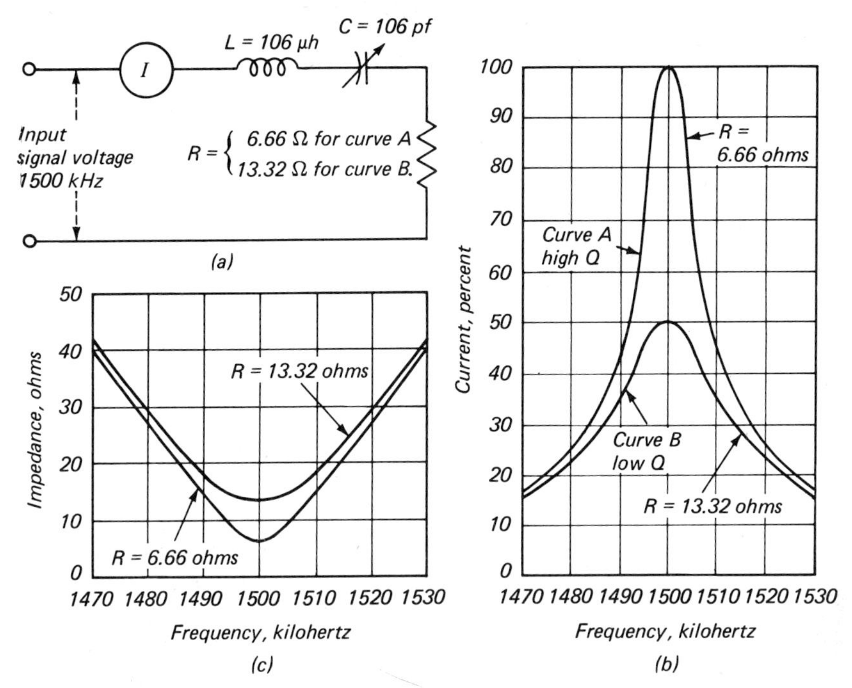

Fig. 11-1 Series resonance. (a) A series resonant circuit. (b) Current resonance curves. (c) Impedance resonance curves.

is at resonance, any change in frequency, inductance, or capacitance will result in a significant value for $(X_L - X_C)$ and the impedance of the circuit must increase. Therefore, *in a series resonant circuit, the impedance will be at its minimum value and the current will be at its maximum value.*

Frequency of Resonance. The frequency at which the inductive reactance and the capacitive reactance are equal is called the *resonant frequency* and is designated as f_r. The following equations are used to find the resonant frequency, the inductance, or the capacitance of a resonant circuit.

$$f_r = \frac{159}{\sqrt{LC}} \tag{11-4}$$

$$L = \frac{25{,}300}{f_r^2 C} \tag{11-5}$$

$$C = \frac{25{,}300}{f_r^2 L} \tag{11-6}$$

where f_r = resonant frequency, kHz
L = inductance, μh
C = capacitance, μf

Example 11-1 What is the resonant frequency of a series circuit having an inductance of 250 μh if the capacitor is adjusted to 350 pf?

GIVEN: $C = 350$ pf $\quad L = 250\ \mu$h

FIND: f_r

SOLUTION:

$$f_r = \frac{159}{\sqrt{LC}} = \frac{159}{\sqrt{250 \times 350 \times 10^{-6}}} = \frac{159}{0.296} = 538 \text{ kHz}$$

Example 11-2 How much inductance is required in a series circuit having a capacitance of 250 pf to produce resonance with a 500-kHz signal input?

GIVEN: $C = 250$ pf $\quad f_r = 500$ kHz

FIND: L

SOLUTION:

$$L = \frac{25{,}300}{f_r^2 C} = \frac{25{,}300}{500 \times 500 \times 250 \times 10^{-6}} = 405\ \mu\text{h}$$

Example 11-3 To what value of capacitance must the variable capacitor of a series circuit be adjusted to produce resonance at 600 kHz if the inductance of the circuit is 300 μh?

GIVEN: $L = 300\ \mu$h $\quad f_r = 600$ kHz

FIND: C

SOLUTION:

$$C = \frac{25{,}300}{f_r^2 L} = \frac{25{,}300}{600 \times 600 \times 300} = 0.000234\ \mu\text{f} = 234 \text{ pf}$$

11-4 Circuit Q, Bandwidth, LC Product, LC Ratio

Circuit Q. The ratio of the inductive reactance of a tuning circuit to the resistance of the circuit is called the *circuit Q;* it is also called the *figure of merit.* Basically Q is the ratio of two amounts of energy: (1) the energy stored in an inductor during the time that a magnetic field is being built up, and (2) the energy lost in the inductor (due to its resistance) during the same period of time. Mathematically

$$Q = \frac{X_L}{R} \tag{11-7}$$

where Q = quality, or figure of merit (a dimensionless ratio)
X_L = inductive reactance of the circuit, ohms
R = total effective resistance of the circuit, ohms

Equation (11-7) can be used to determine the Q of a coil (or inductor) as well as for a tuning circuit. Similarly, the Q of a capacitor can be determined by

$$Q = \frac{X_C}{R} \tag{11-7a}$$

where X_C = capacitive reactance of the capacitor, ohms
R = total effective resistance of the capacitor, ohms

For a resonant circuit

$$Q = \frac{1}{R}\sqrt{\frac{L}{C}} \tag{11-8}$$

where R = total effective resistance of the circuit, ohms
L = inductance of the circuit, henrys
C = capacitance of the circuit, farads

Bandwidth. The current in a series tuning circuit is at its maximum value when the frequency of the applied voltage corresponds to the resonant frequency of the circuit. At frequencies above or below the resonant frequency value, the impedance of the circuit increases and the current decreases. Because the frequency values of the signal in communication systems vary above and below the resonant frequency, owing to the modulating signals being transmitted, the amount by which the current decreases for frequencies off resonance is of great importance. A commonly used measure is the bandwidth of a tuned circuit, which expresses the total spread of frequency values above and below the resonant frequency for which the current does not decrease below 0.707 times the current at the resonant frequency (Fig. 11-2). The bandwidth of a tuned circuit is expressed by

$$BW = f_2 - f_1 \tag{11-9}$$

where BW = bandwidth of circuit, hertz
f_1 = lower frequency at which $I = 0.707\ I_{max}$, hertz
f_2 = upper frequency at which $I = 0.707\ I_{max}$, hertz

The bandwidth of a tuning circuit is directly proportional to its circuit resistance and inversely proportional to its inductance. In order to obtain high selectivity in tuning circuits it is essential that the width of the resonance curve be as narrow as practically possible. This means that the resistance of the circuit must be low and the inductance of the coil high; in other words the circuit Q must be high. The relationship among the bandwidth of a tuned circuit and its parameters is expressed by

$$f_2 - f_1 = \frac{f_r}{Q} \tag{11-10}$$

$$f_2 - f_1 = \frac{R}{2\pi L} \tag{11-11}$$

Example 11-4 A series tuned circuit has a resistance of 5 ohms and an inductance of 225 μh. (*a*) What is the width in hertz of its resonance curve at a point where the current in the circuit is equal to 0.707 times the current at resonance? (*b*) If the resistance of the circuit is increased to 10 ohms, how does the change affect the width of the resonance curve?

GIVEN: (*a*) $R = 5$ ohms $L = 225\ \mu$h (*b*) $R = 10$ ohms

FIND: (*a*) $f_2 - f_1$ (*b*) $f_2 - f_1$

SOLUTION:

(*a*)
$$f_2 - f_1 = \frac{R}{2\pi L} = \frac{5}{6.28 \times 225 \times 10^{-6}} = 3{,}538 \text{ hertz}$$

(*b*)
$$f_2 - f_1 = \frac{R}{2\pi L} = \frac{10}{6.28 \times 225 \times 10^{-6}} = 7{,}076 \text{ hertz}$$

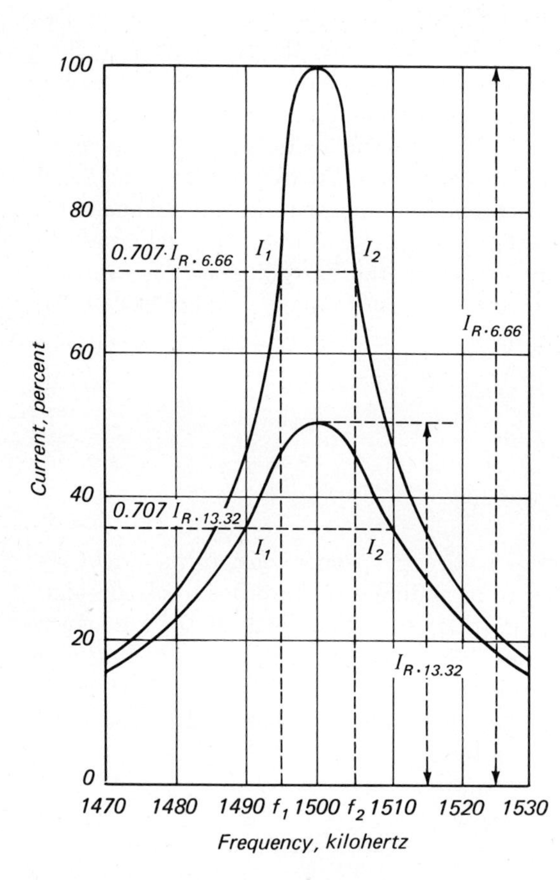

Fig. 11-2 Series resonance curves with the width of the band ($f_2 - f_1$) indicated.

***LC* Product.** Equation (11-4) shows that the frequency of resonance is dependent on the product of the inductance and capacitance of the circuit. For each value of L times C there can be only one frequency at which resonance occurs and, conversely, there can be only one LC product for each resonant frequency. If the LC product for a desired frequency is known, the capacitance required can be found by dividing this product by the value of inductance used, or the inductance required can be found by dividing the LC product by the capacitance used. Values of L times C for commonly used frequencies are often listed in tabular or graphical form. These charts or tables can be found in reference books.

***LC* Ratio.** As the frequency of resonance of a circuit is dependent only on its LC product, any number of combinations of L and C can be used to obtain the same resonant frequency. The numerical value of inductance can be made greater than, less than, or equal to that of the capacitance. The resonant-circuit application determines to a large extent the LC ratio that should be used.

Voltage Ratios at Resonance. In a series resonant circuit, the impedance of the circuit at resonance is very low, thus allowing a comparatively high current to flow. This high current flowing through the capacitor and inductor causes a voltage to be developed across these reactances which is higher than the voltage applied to the circuit. The voltages across the various portions of a series tuned circuit are

$$E_R = E_{\text{line}} \tag{11-12}$$

$$E_{X_L} = E_{\text{line}}\, Q \tag{11-13}$$

$$E_{X_C} = E_{\text{line}}\, Q \tag{11-14}$$

Example 11-5 The resistance of a series resonant circuit is 10 ohms, and its inductive reactance is equal to 500 ohms at resonant frequency. What voltage is developed across the inductor, capacitor, and resistor when the applied voltage is 5 volts?

GIVEN: $E = 5$ volts $\quad R = 10$ ohms $\quad X_L = 500$ ohms

FIND: $E_{X_L} \quad E_{X_C} \quad E_R$

SOLUTION:

$$I_r = \frac{E}{R} = \frac{5}{10} = 0.5 \text{ amp}$$

$$E_{X_L} = I_r X_L = 0.5 \times 500 = 250 \text{ volts}$$

At resonance, $\quad X_L = X_C$

Therefore $\quad E_{X_C} = E_{X_L} = 250$ volts

$$E_R = I_r R = 0.5 \times 10 = 5 \text{ volts}$$

11-5 Resonance Curves

Basic Principle. The action of a resonant circuit is best explained by reference to graphs illustrating the variation in the circuit conditions at or near resonance. Figure 11-1*b* shows two resonance curves for the series tuning circuit of Fig. 11-1*a*; curve *A* shows the circuit characteristics when the resistance of the circuit is 6.66 ohms and curve *B* when the resistance is increased to 13.32 ohms.

Effect of Resistance on the Current in a Series Resonant Circuit. At resonance, the current in a series tuned circuit (with a constant voltage) is dependent entirely upon the resistance of the circuit. Figure 11-1*b* shows the resonance curves for two circuits tuned to the same frequency having different values of resistance. At resonance, the current flowing through the circuit having 6.66 ohms resistance is twice the amount flowing through the circuit having 13.32 ohms resistance. Therefore, to produce maximum current, it is best that the resistance of a series tuned circuit be as low as possible.

Effect of Resistance on the Bandwidth in a Series Resonant Circuit. The resonance curves of Fig. 11-2 are the same as those of Fig. 11-1*b* and are repeated to show the bandwidth characteristics of the circuit. The curves show that with an increase in the circuit resistance from 6.66 to 13.32 ohms, the bandwidth of the circuit increases from approximately 10 to 20 kHz. The conclusion indicated here is that a lower circuit resistance produces (1) a narrower bandwidth and (2) sharper tuning characteristics.

Example 11-6 What is the bandwidth of the circuit of Fig. 11-1*a* when the circuit resistance is (*a*) 6.66 ohms? (*b*) 13.32 ohms?

GIVEN: $L = 106\ \mu h$ (*a*) $R = 6.66$ ohms (*b*) $R = 13.32$ ohms

FIND: (*a*) BW (*b*) BW

SOLUTION:

$$(a) \qquad BW = f_2 - f_1 = \frac{R}{2\pi L} = \frac{6.66}{6.28 \times 106 \times 10^{-6}} \cong 10{,}000 \text{ hertz} \cong 10 \text{ kHz}$$

$$(b) \qquad BW = f_2 - f_1 = \frac{R}{2\pi L} = \frac{13.32}{6.28 \times 106 \times 10^{-6}} \cong 20{,}000 \text{ hertz} \cong 20 \text{ kHz}$$

11-6 Parallel Resonant Circuit

Characteristics at Resonance. At resonance, the impedance of parallel resonant circuits is at its maximum, and its value is equal to the product of either reactance and the circuit Q. The circuit acts the same as a resistor, and the current and voltage are in phase; the power factor of the circuit will therefore be unity. The line current is at a minimum and is equal to the applied voltage divided by the impedance of the circuit. The voltages across the inductance and the capacitance are the same and are equal to the applied voltage. At resonance, the currents in the inductor and capacitor are approximately equal and nearly 180° out of phase with each other. Increasing the resistance of the circuit decreases the circuit impedance, thereby increasing the line current. For frequencies below resonance, the current in the inductor increases and the line current is lagging. For frequencies above resonance, the current in the capacitor increases and the line current is leading.

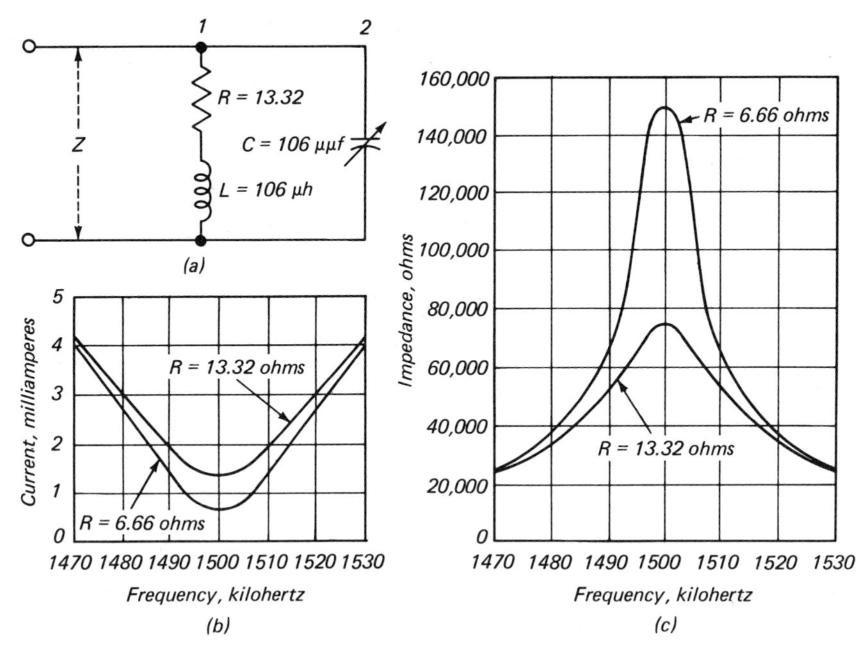

Fig. 11-3 Parallel resonance. (a) A parallel resonant circuit. (b) Current resonance curves. (c) Impedance resonance curves.

Impedance and Current Characteristics. The impedance of a parallel circuit having two paths, such as Fig. 11-3*a*, is expressed by

$$Z_p = \frac{Z_1 Z_2}{Z_1 + Z_2} \tag{11-15}$$

When $X_L \gg R$

$$Z_P \cong \frac{X_L X_C}{\sqrt{(X_L - X_C)^2 + R^2}} \tag{11-15a}$$

The impedance of a parallel circuit at its resonant frequency can be expressed in a simplified form as

$$Z_{pr} = \frac{{X_L}^2}{R} \tag{11-16}$$

substituting Q for $\dfrac{X_L}{R}$

$$Z_{pr} = X_L Q \tag{11-17}$$

or

$$Z_{pr} = \frac{L}{CR} \tag{11-18}$$

The current of a parallel resonant circuit can then be expressed as

$$I_{pr} = \frac{E}{Z_{pr}} \tag{11-19}$$

also

$$I_{pr} = \frac{ER}{X_L^2} \tag{11-20}$$

or

$$I_{pr} = \frac{ECR}{L} \tag{11-21}$$

Example 11-7 What are the impedance and current at the resonant frequency of the tuning circuit of Fig. 11-3*a* when the voltage of the power source is 100 volts and the value of R is (*a*) 6.66 ohms? (*b*) 13.32 ohms?

GIVEN: $L = 106\ \mu h$ $C = 106$ pf $E = 100$ volts (*a*) $R = 6.66$ ohms (*b*) $R = 13.32$ ohms

FIND: (*a*) Z_{pr} and I_{pr} (*b*) Z_{pr} and I_{pr}

SOLUTION:

(*a*)

$$Z_{pr} = \frac{L}{CR} = \frac{106 \times 10^{-6}}{106 \times 10^{-12} \times 6.66} = 150{,}000 \text{ ohms}$$

$$I_{pr} = \frac{E}{Z_{pr}} = \frac{100}{150{,}000} = 0.666 \text{ ma}$$

(*b*)

$$Z_{pr} = \frac{L}{CR} = \frac{106 \times 10^{-6}}{106 \times 10^{-12} \times 13.32} = 75{,}000 \text{ ohms}$$

$$I_{pr} = \frac{E}{Z_{pr}} = \frac{100}{75{,}000} = 1.33 \text{ ma}$$

Frequency of Resonance. Equations (11-4), (11-5), and (11-6) established for the series tuning circuit also apply to the parallel tuning circuit.

Circuit Q, Bandwidth, LC Product, LC Ratio. Equations (11-7) through (11-11) established for the series tuning circuit also apply to the parallel tuning circuit.

Current Ratios at Resonance. In a parallel resonant circuit, the impedance of the circuit is at its maximum value and therefore the line current will be at its minimum value. Because the line current is equal to the vector sum of the separate branch currents (which are nearly 180° out of phase) it is possible that the branch currents will be higher than the line current. At resonance,

$$I_{X_L} = I_{\text{line}}\, Q \tag{11-22}$$

$$I_{X_C} \cong I_{\text{line}}\, Q \tag{11-23}$$

Example 11-8 How much current flows in the separate branches of the parallel tuning circuit of Example 11-7 at its resonant frequency when the value of R is 13.32 ohms?

GIVEN: $I_{\text{line}} = 1.33$ ma $R = 13.32$ ohms $L = 106\ \mu h$ $C = 106$ pf

FIND: I_{X_L} I_{X_C}

SOLUTION:

$$I_{X_L} = I_{X_C} = I_{\text{line}}\, Q$$

where
$$Q = \frac{1}{R}\sqrt{\frac{L}{C}} = \frac{1}{13.32}\sqrt{\frac{106 \times 10^{-6}}{106 \times 10^{-12}}} = \frac{1}{13.32}\sqrt{10^6} = 75$$

then
$$I_{X_L} \text{ and } I_{X_C} = I_{\text{line}}\, Q = 1.33 \times 75 = 100 \text{ ma}$$

Resonance Curves. The resonance curves for a parallel tuning circuit are shown in Fig. 11-3. The curves are generally plotted with frequency versus impedance as in Fig. 11-3*c*, although occasionally curves may be plotted with frequency versus current as in Fig. 11-3*b*.

11-7 Comparison of Series and Parallel Resonant Circuits

In both types of circuits, resonance occurs when the capacitive reactance equals the inductive reactance. The resonant frequency of both circuits can be calculated by Eq. (11-4). The circuit Q and the bandwidth of both circuits can be calculated by Eqs. (11-7) and (11-10), respectively. The impedance versus frequency resonance curves of the parallel tuning circuit are like those of the series tuning circuit when inverted; the same is true of their respective current versus frequency curves. Increasing the circuit Q will increase the slope of the curve and decrease the bandwidth in both the series and parallel tuning circuits. The characteristics of series tuning circuits and parallel tuning circuits may be readily compared with each other by reference to the tabular listing of their characteristics in Table 11-1.

11-8 Classification of Circuits as Series or Parallel

Series tuned circuits are used wherever maximum current is desired for a definite frequency or band of frequencies. Parallel tuned circuits are used wherever the signal strength of any one frequency or band of frequencies is to be reduced to a minimum. Frequently tuned circuits form only a part of the complete circuit. Because of this, it may be difficult to ascertain whether the tuned circuit is of the series or parallel type. In some instances, circuits that have all the appearances of a parallel tuned circuit are actually classed as series tuned circuits. This is usually the case when the inductor is the secondary of a transformer and the voltage of the circuit is an induced voltage rather than a voltage applied from an entirely separate source. For example, Fig. 11-4 shows two tuned circuits (drawn in heavy lines), each of which has the general appearance of a parallel circuit, but in analyzing the complete circuit the one at the left is considered a parallel tuned circuit and the one at the right is considered a series tuned circuit. The circuit at the left is considered a parallel tuned circuit because it receives its electrical energy from the plate circuit of tube 1. The circuit at the right is considered a series tuned circuit because no separate voltage is applied to the inductor and capacitor, but in-

Table 11-1 Characteristics of Series and Parallel Tuning Circuits

QUANTITY	SERIES CIRCUIT	PARALLEL CIRCUIT
At resonance:		
Reactance $(X_L - X_C)$	Zero (because $X_L = X_C$)	Zero (because $X_L = X_C$)
Frequency of resonance	$\frac{159}{\sqrt{LC}}$	$\frac{159}{\sqrt{LC}}$
Impedance	Minimum value; $Z = R$	Maximum value; $Z = QX_L$
I_{line}	Maximum value	Minimum value
I_L	I_{line}	Q times I_{line}
I_C	I_{line}	Q times I_{line}
E_L	Q times E_{line}	E_{line}
E_C	Q times E_{line}	E_{line}
Phase angle between I_{line} and E_{line}	0°	0°
Angle between E_L and E_C	180°	0°
Angle between I_L and I_C	0°	180°
Desired value of Q	High	High
Desired value of R	Low	Low
Highest selectivity	High Q; low R; high LC ratio	High Q; low R; high CL ratio
When f is greater than f_r:		
Reactance $(X_L - X_C)$	Inductive	Capacitive
Phase angle between I_{line} and E_{line}	Lagging current	Leading current
When f is less than f_r:		
Reactance $(X_L - X_C)$	Capacitive	Inductive
Phase angle between I_{line} and E_{line}	Leading current	Lagging current

stead a voltage is induced in the inductor (secondary of an r-f transformer) which is considered as a voltage connected in series with the inductor and the capacitor. The following procedure will help to determine whether a circuit should be classed as parallel or series: (1) locate the inductive and capacitive components forming the tuned circuit, (2) locate the source of a-c voltage for

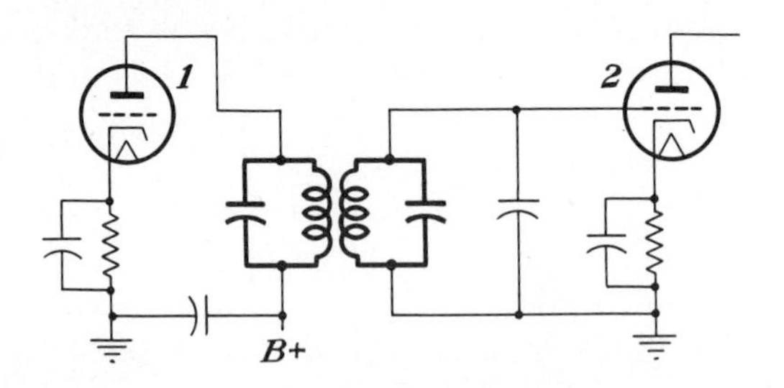

Fig. 11-4 An application of a series and parallel resonant circuit in a radio receiver.

these components, (3) determine whether the components are in series or parallel with the source of a-c voltage. In electronic circuits, the source of a-c voltage may be obtained from any one of the following circuits: antenna, output of transistor or vacuum tube, or the induced voltage from other circuits.

11-9 Circuit Elements

Variable Capacitors. Tuning can be accomplished by varying either the capacitance or the inductance of a tuning circuit. The method commonly used is varying the capacitance. Three types of variable capacitors used in tuning are the *straight-line-capacity*, *straight-line-wavelength*, and *straight-line-frequency;* abbreviated as slc, slw, and slf.

With an slc-type capacitor, the capacitance increases directly with the amount of rotation of its movable plates (Fig. 11-5*a*). For example, if the rotor plates are one-quarter in mesh with the stator, its capacitance will be one-quarter of its total value; if the rotor plates are one-half in mesh with the stator, its capacitance will be one-half of its total value, etc. If an slc capacitor is used in the tuning circuit of an a-m radio receiver, most of the stations will appear on one-half of the dial (see Fig. 11-5*a*), because the resonant frequency of a tuned circuit does not vary in a direct ratio with changes in its capacitance. The upper half of the frequency band (1,075 to 1,600 kHz) will appear on approximately only one-eighth of the dial; hence when this type of capacitor is used it is difficult to separate the signals from adjacent stations in the upper half of the frequency band.

In the slw capacitor the area of the rotor plates is reduced on the side that first enters into mesh with the stator plates, so that the wavelength of the tuned circuit increases in a direct ratio with the amount of rotation of the

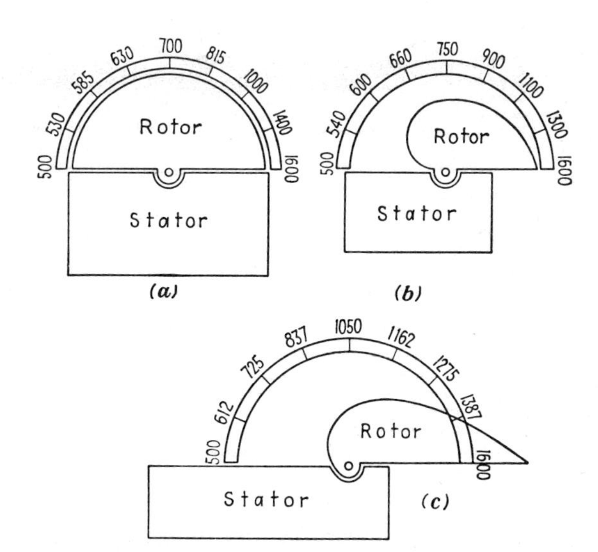

Fig. 11-5 Frequency distribution obtained with three types of variable capacitors. (a) Straight-line capacity. (b) Straight-line wavelength. (c) Straight-line frequency.

movable plates. If an slw capacitor is used in the tuning circuit of a broadcast receiver, most of the stations will appear on approximately three-fourths of the dial (Fig. 11-5*b*).

With the slf capacitor, the resonant frequency of the tuned circuit will vary directly with the amount of rotation of the rotor plates (Fig. 11-5*c*).

In some tuning-circuit applications varactors are used in place of variable capacitors (see Art. 10-3).

Inductors. The secondary winding of an r-f transformer provides the fixed inductance that is used with a variable capacitor to form the tuning circuit in many types of communication receivers. In this type of tuning circuit the input signal flows through the primary winding of the r-f transformer, and by means of the mutual inductance between the two windings the signal is transferred to the secondary. The variable capacitor and the secondary winding form a series resonant circuit which must be tuned to the frequency of the desired signal. The amount of inductance required in the secondary winding will depend on the frequency range desired and the capacitance of the variable capacitor that is used. Increasing the inductance of the secondary coil increases the amount of voltage developed across the winding, thus increasing the gain of the tuned circuit. The selectivity of the tuned circuit is also increased, as increasing the inductance increases both the L/C ratio and the circuit Q. Several types of r-f coils are shown in Fig. 11-7.

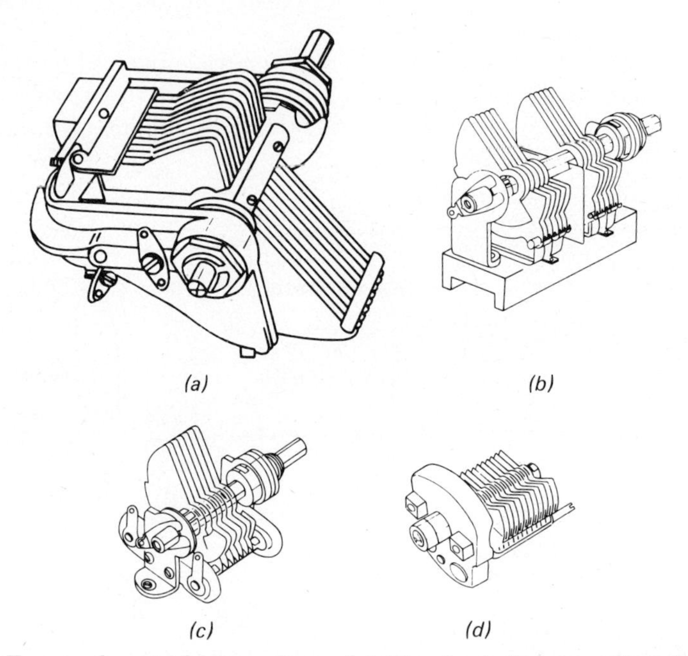

Fig. 11-6 Types of variable capacitors. (a) Standard-size capacitor. (b) Midget split-stator capacitor. (c) Midget single-unit capacitor. (d) Micro capacitor.

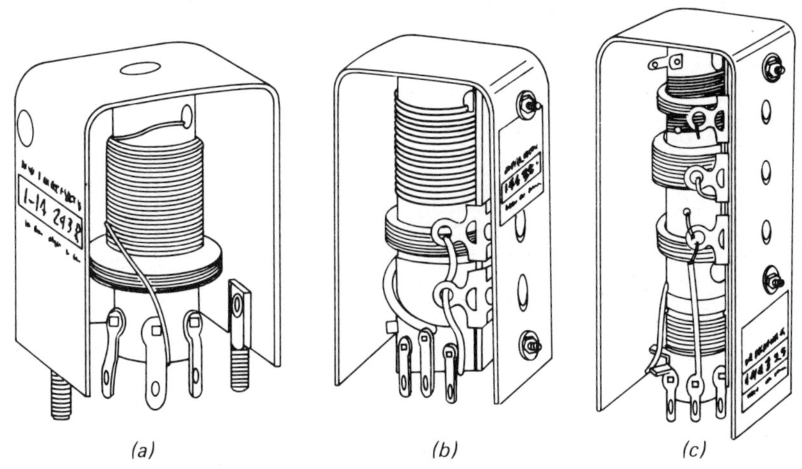

(a) (b) (c)

Fig. 11-7 Types of r-f coils. (a) Single-band coil. (b) Dual-band coil. (c) Triple-band coil.

The amount of inductance that should be used is such a value that when combined with the distributed capacitance of the secondary winding and the minimum capacitance of the tuning capacitor it will form a resonant circuit whose frequency of resonance is equal to or greater than the highest frequency desired. For broadcast reception this frequency is approximately 1,600 kHz. The maximum value of the variable capacitor must be equal to the amount of capacitance required to adjust the frequency of resonance of the tuned circuit to the minimum desired frequency, generally 550 kHz.

Example 11-9 It is desired to determine the inductance required in the secondary winding of an r-f transformer, and it is assumed that the distributed capacitance of the winding and the circuit wiring is 23 pf (considered acting in parallel with the tuning capacitor). (*a*) What value of inductance is required in order to obtain resonance at 1,600 kHz if the minimum value of the variable capacitor is 17 pf? (*b*) What value of capacitance must the capacitor have if it is desired to tune in stations as low as 550 kHz with the coil used in part (*a*)?

GIVEN: $C_D = 23$ pf $\quad C_V = 17$ pf (min) $\quad f = 1{,}600$ kHz (*a*) $\quad f = 550$ kHz (*b*)

FIND: (*a*) L $\quad$ (*b*) C_V

SOLUTION:

(*a*)
$$L = \frac{25{,}300}{f_r^2 C} = \frac{25{,}300}{1{,}600^2 \times (17 + 23)10^{-6}} = 246\ \mu\text{h}$$

(*b*)
$$C_T = \frac{25{,}300}{f_r^2 L} = \frac{25{,}300}{550^2 \times 246} = 0.0003399\ \mu\text{f or } 340\ \text{pf}$$

$$C_V = C_T - C_D = 340 - 23 = 317\ \text{pf}$$

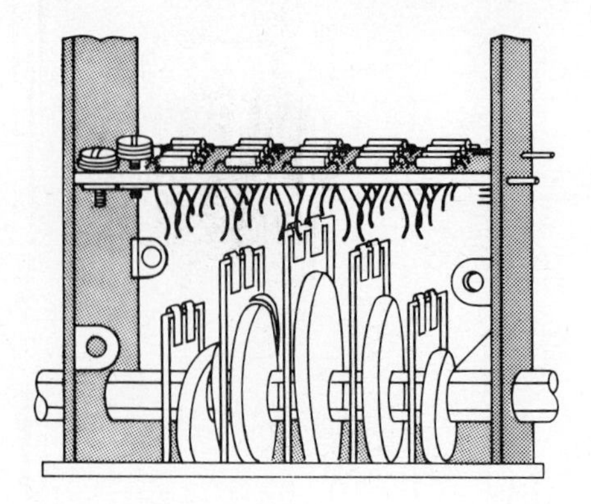

Fig. 11-8 Band switch.

Variable Inductors. In radio receivers made to tune the broadcast band and one or more short-wave bands, the capacitor and inductor used in the tuning circuit are constructed so that either one or both can be adjusted to the value required for the band desired. One method used for changing from one frequency band to another is to have the primary and secondary windings for each band wound on a single coil form (Fig. 11-7). By means of a band-selector switch (Fig. 11-8), any set of primary and secondary windings may be connected into or out of the tuning circuit. The switch contacts are usually arranged so that the unused coils are short-circuited. If the coils not actually in use are not short-circuited, the distributed capacitance of these coils may cause them to become resonant at some frequency, each within its own band, thereby coupling impedance into the coils that are being used. The switch should provide (1) a low-impedance contact and (2) some means whereby the connections between the coils and the switch are made as short as possible.

UHF Tuning Elements. Some radical construction features are used in the uhf circuit elements because of the high operating frequencies (Figs. 11-9 and 11-10). Coils L_1 and L_5 are small two-turn loops of wire. Coil L_2 has the appearance of a small pin pressed into a recessed hole in the shield S between coils L_3 and L_4. Coils L_3 and L_4 are connected to the other end of this pin, which becomes the mutual inductance between L_3 and L_4. Coupling of the antenna signal to these circuits is produced by placing L_1 close to the low-potential end of L_3. Inductors L_3 and L_4 are wound with 0.015-inch-diameter silver-plated steel wire to form coils of approximately $\frac{1}{8}$ inch diameter and $\frac{1}{2}$ inch length. Coils L_3, L_4, and L_6 are shielded from one another by being mounted in a small metal container having three separate cavities. The tuning capacitors C_1, C_2, and C_3 are formed by small machine screws (1-72) mounted at the top of the cavities in the metal shielding container. The tops of the coils, the tips of the machine screws, and the small space between them form the capacitances, and no connections are made to the top ends of the coils. The minimum values of these capacitances is in the order of 0.25 pf.

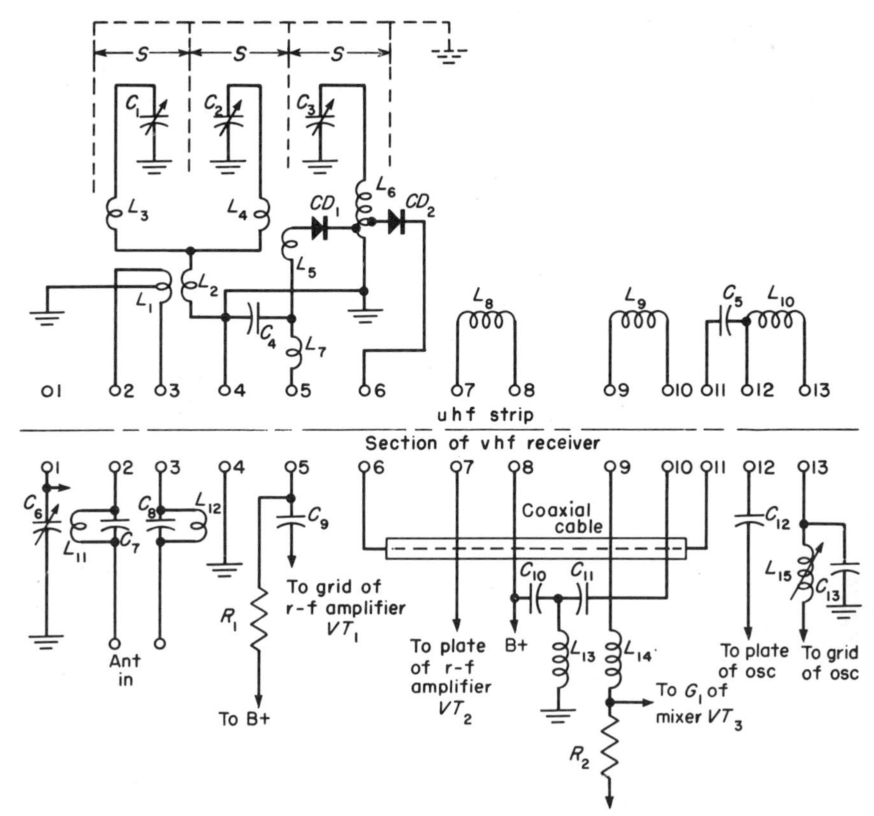

Fig. 11-9 Schematic diagram of a uhf strip and the corresponding points of connection in the vhf circuits. (*Zenith Radio Corp.*)

11-10 Bandspread

Frequency Range of the Individual Bands. The use of separate coils or fixed capacitors provides a means of covering the broadcast band and a number of short-wave bands with one receiver. To eliminate the possibility of a gap between two adjacent bands, the coils are usually designed to overlap the extreme frequencies of the adjacent bands.

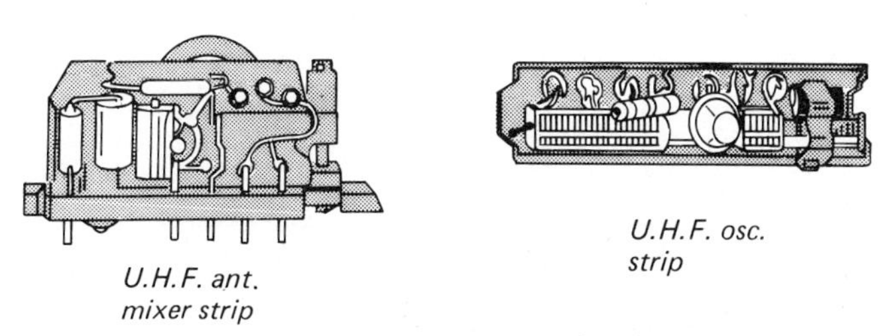

Fig. 11-10 UHF channel strips.

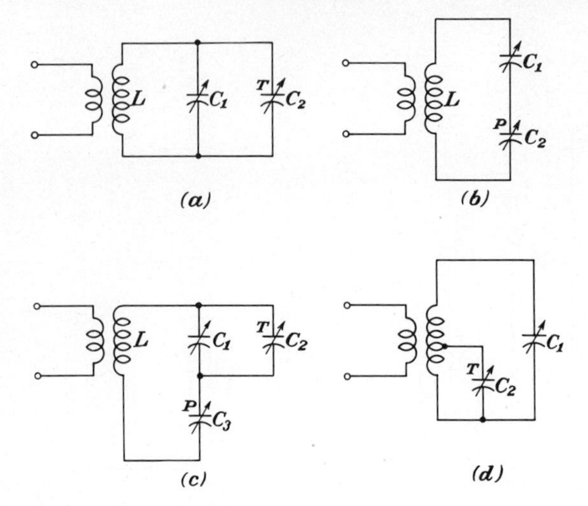

Fig. 11-11 Trimmer and padder capacitors connected to produce bandspread and band compression.

To facilitate tuning, the tuning range for each band should occupy practically the entire scale of the dial. Because of the varying widths of these bands, special tuning circuits are used to obtain the correct maximum-minimum capacitance ratio for each band. This process, called *bandspreading*, is accomplished by connecting a small adjustable capacitor in series, parallel, or combination with the main tuning capacitor (Fig. 11-11).

Trimmers. The operation of these auxiliary capacitors is based on the principle that the capacitance of a circuit is increased by connecting capacitors in parallel and decreased by connecting them in series.

Example 11-10 A 20-pf auxiliary capacitor, C_2 of Fig. 11-11*a*, is connected in parallel with a tuning capacitor C_1 having a range of 10 to 100 pf. What is the range of capacitance of the combined circuit?

GIVEN: $C_2 = 20$ pf $\quad C_{1\text{-min}} = 10$ pf $\quad C_{1\text{-max}} = 100$ pf

FIND: $C_{\min}$ to $C_{\max}$

SOLUTION:

$$\begin{aligned} C_{\min} &= C_{1\text{-min}} + C_2 = 10 + 20 = 30 \text{ pf} \\ C_{\max} &= C_{1\text{-max}} + C_2 = 100 + 20 = 120 \text{ pf} \\ \text{Range} &= 30 \text{ to } 120 \text{ pf} \end{aligned}$$

Example 11-11 What is the range of the combined circuit of Example 11-10 if a 100-pf auxiliary capacitor is used in place of the 20-pf capacitor?

GIVEN: $C_2 = 100$ pf $\quad C_{1\text{-min}} = 10$ pf $\quad C_{1\text{-max}} = 100$ pf

FIND: $C_{\min}$ to $C_{\max}$

SOLUTION:

$$\begin{aligned} C_{\min} &= C_{1\text{-min}} + C_2 = 10 + 100 = 110 \text{ pf} \\ C_{\max} &= C_{1\text{-max}} + C_2 = 100 + 100 = 200 \text{ pf} \\ \text{Range} &= 110 \text{ to } 200 \text{ pf} \end{aligned}$$

In Example 11-10, adding a 20-pf auxiliary capacitor in parallel with the 10- to 100-pf tuning capacitor increases its range to 30 to 120 pf. Example 11-11 shows that adding a 100-pf auxiliary capacitor in parallel with the tuning capacitor increases the range to 110 to 200 pf. In each case the greatest per cent of increase occurs at the minimum value of capacitance. If the auxiliary capacitor is made adjustable between the values of 20 and 100 pf, then a large number of minimum and maximum values of capacitance can be obtained to produce a corresponding bandspread. Such an adjustable capacitor is called a *trimmer*. Trimmers have the same effect on a tuned circuit as does distributed capacitance.

Padders. In order that a desired frequency range may be obtained, it is sometimes desirable to restrict the maximum capacitance of the tuning circuit without greatly changing its minimum value. To accomplish this, an auxiliary capacitor, C_2 of Fig. 11-11*b*, is connected in series with the secondary coil L and the tuning capacitor C_1. This capacitor, called a *padder*, is also made adjustable so that the minimum capacitance of the tuning circuit can be kept fairly constant.

Example 11-12 A 20- to 200-pf adjustable capacitor (C_2 of Fig. 11-11*b*) is connected in series with the tuning capacitor C_1, having a range of 10 to 100 pf. What is the range of capacitance of the combined circuit if the adjustable capacitor is set at 20 pf?

GIVEN: $C_2 = 20$ pf $\quad C_{1\text{-min}} = 10$ pf $\quad C_{1\text{-max}} = 100$ pf

FIND: C_{min} to C_{max}

SOLUTION:

$$C_{min} = \frac{C_{1\text{-min}}C_2}{C_{1\text{-min}} + C_2} = \frac{10 \times 20}{10 + 20} = 6.66 \text{ pf}$$

$$C_{max} = \frac{C_{1\text{-max}}C_2}{C_{1\text{-max}} + C_2} = \frac{100 \times 20}{100 + 20} = 16.66 \text{ pf}$$

$$\text{Range} = 6.66 \text{ to } 16.66 \text{ pf}$$

Example 11-13 What is the range of the combined circuit of Example 11-12 if the padder capacitor is set at 200 pf?

GIVEN: $C_2 = 200$ pf $\quad C_{1\text{-min}} = 10$ pf $\quad C_{1\text{-max}} = 100$ pf

FIND: C_{min} to C_{max}

SOLUTION:

$$C_{min} = \frac{C_{1\text{-min}}C_2}{C_{1\text{-min}} + C_2} = \frac{10 \times 200}{10 + 200} = 9.52 \text{ pf}$$

$$C_{max} = \frac{C_{1\text{-max}}C_2}{C_{1\text{-max}} + C_2} = \frac{100 \times 200}{100 + 200} = 66.66 \text{ pf}$$

$$\text{Range} = 9.52 \text{ to } 66.66 \text{ pf}$$

Example 11-12 shows that adding a 20-pf capacitance in series with the 10- to 100-pf tuning capacitor decreases its range to 6.66 to 16.66 pf. Example

11-13 shows that adding a 200-pf capacitance in series with the tuning capacitor decreases the range to 9.52 to 66.66 pf. In each case the greatest per cent of decrease occurs at the maximum value of capacitance; hence the padding capacitor has comparatively little effect on the minimum capacitance of the tuning circuit but definitely controls its maximum capacitance.

Radio receivers may use both trimmers and padders (Fig. 11-11*c*) in order to obtain the desired bandspread. The desired frequency limits are obtained by adjusting these two capacitors to the correct values. Sometimes the trimmer is connected across only part of the secondary coil L, as shown in Fig. 11-11*d*. The amount of bandspread increases as the tap is made closer to the bottom of the coil.

Example 11-14 The tuning circuit of the oscillator section of a superheterodyne receiver is to be adjusted so that it will always tune 465 kHz higher than the tuning circuit of the receiver, whose frequency limits are to be 530 kHz and 1,550 kHz. A circuit similar to that shown in Fig. 11-11*c* is to be used, and the circuit elements are to have the following values: $L = 200\ \mu h$, $C_{1\cdot min} = 12.5$ pf, $C_{1\cdot max} = 250$ pf, and $C_2 = 6.5$ pf. The distributed capacitance of the circuit is 15 pf. (*a*) What value of padder capacitor C_3 is required when the tuning circuit is adjusted to its minimum frequency? (*b*) Using the padder capacitance as calculated in part (*a*), what is the resonant frequency of the oscillator section when the tuning circuit is adjusted to its maximum frequency?

GIVEN: $f = 530$ kHz to 1,550 kHz $\quad C_1 = 12.5$ to 250 pf $\quad C_2 = 6.5$ pf
$C_D = 15$ pf $\quad L = 200\ \mu h$

FIND: (*a*) C_3 $\quad$ (*b*) $f_{osc\cdot max}$

SOLUTION:

(*a*)

$$f_{osc\cdot min} = f_{min} + 465 = 530 + 465 = 995 \text{ kHz}$$

$$C_{f\cdot min} = \frac{25{,}300}{(f_{osc\cdot min})^2 L} = \frac{25{,}300}{995^2 \times 200} = 127.7 \text{ pf}$$

$$C_3 = \frac{(C_{1\cdot max} + C_2)(C_{f\cdot min} - C_D)}{(C_{1\cdot max} + C_2) - (C_{f\cdot min} - C_D)} = \frac{256.5 \times 112.7}{256.5 - 112.7} = 201 \text{ pf}$$

(*b*)

$$C_{min} = \frac{(C_{1\cdot min} + C_2)C_3}{(C_{1\cdot min} + C_2) + C_3} + C_D = \frac{19 \times 201}{19 + 201} + 15 = 32.35 \text{ pf}$$

$$f_{osc\cdot max} = \frac{159}{\sqrt{LC_{min}}} = \frac{159}{\sqrt{200 \times 32.35 \times 10^{-6}}} = 1{,}976 \text{ kHz}$$

The maximum frequency of the oscillator as calculated in Example 11-14 is equal to 1,976 kHz instead of 2,015 kHz (1,550 + 465). By adjusting the trimmer capacitor C_2 the correct value of maximum frequency can be obtained. As any change in the value of trimmer capacitance will have only a slight effect on the maximum value of capacitance of the entire circuit, the minimum frequency will remain practically the same.

Bandspread Tuning. Dials used for tuning generally have a rotation of 180 mechanical degrees. With such a dial, the amount of change in kHz for

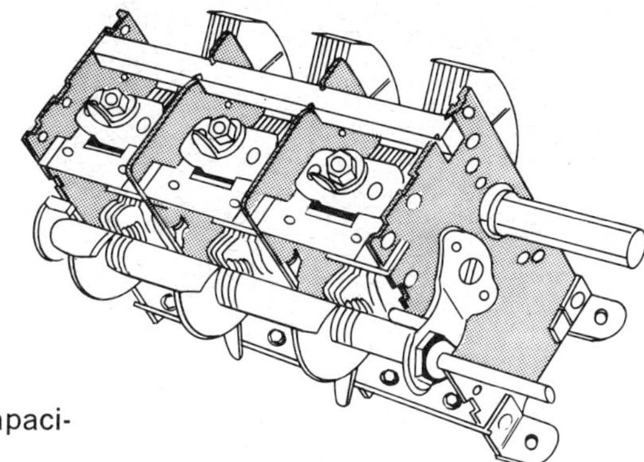

Fig. 11-12 Bandspread tuning capacitor.

each degree of rotation will vary with the amount of bandspread. On the broadcast band, 535 to 1,605 kHz, each degree would represent a change of 1,070 divided by 180, or approximately 6 kHz. As there should be at least a 10-kHz difference between adjacent stations, there is no problem in tuning a desired station on the broadcast band.

On the short-wave bands, tuning a desired station increases in difficulty as the frequency of the band is increased. Two methods are used to spread the amount of kilohertz change over a greater portion of the dial, one mechanical and the other electrical. The mechanical method employs a geared dial that causes the variable capacitor to move at a slower rate than the dial. The electrical method utilizes the principle of the trimmer capacitor to obtain bandspread. A midget or micro variable capacitor, depending on the frequency band, is substituted for C_2 (Fig. 11-11). This capacitor is called the *bandspread capacitor,* and is used to obtain the exact frequency, as a considerable movement of this capacitor will change the resonant frequency by only a fraction of a kilohertz.

11-11 Multiple Stages

Selectivity of Multiple Stages. A single tuning stage does not always provide sufficient selectivity for satisfactory reception; hence in order to obtain the desired selectivity a number of similar tuning stages are sometimes used. The selectivity of the receiver will then be dependent on the selectivity of each stage and the number of stages used. The extent to which the selectivity is improved by increasing the number of tuned circuits is shown in Fig. 11-13. The frequency of resonance for each tuned circuit is 1,000 kHz. The signal output at the resonant frequency is taken as 100 per cent. The signal output

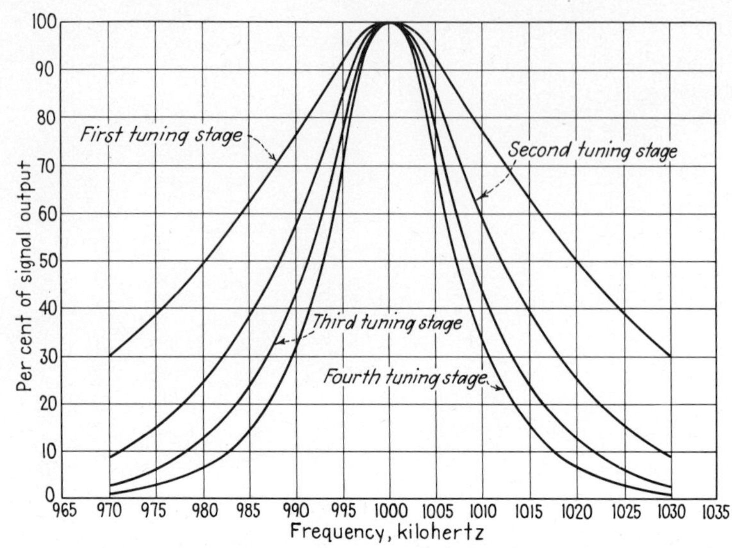

Fig. 11-13 Graph illustrating the manner in which the selectivity of a radio receiver increases with additional stages of tuning.

at all other frequencies will be a definite percentage of this output. If the characteristics of each tuning circuit are the same, the percentage of the input signal, for all frequencies above and below resonance, can be represented by the response curve for the first stage of tuning. For example, at 5 kHz off resonance, that is, 995 or 1,005 kHz, the input signal is reduced to 92 per cent, as represented by the response curve for the first stage of tuning. The output for the second stage for this frequency would be equal to 92 per cent of 92 per cent, or 84.6 per cent. The output from the third stage for the same frequency would be 92 per cent of 84.6 per cent, or 77.8 per cent; for the fourth stage the output would be 92 per cent of 77.8 per cent, or 71.6 per cent. This procedure can be followed for any value of frequency.

Number of Tuning Stages to Be Used. Theoretically, it would seem that any number of tuning stages could be added, but practically this is not so, since other factors must be taken into consideration. In order to obtain good fidelity from commercial a-m radio stations, the width of the band passed (measured at 0.707 of the maximum value) should not be less than 10 kHz. Adding too many stages may decrease the width of the band passed below this value, thus decreasing the fidelity of reception.

In modern receivers, the tuning circuit forms a definite part of each stage of r-f amplification and the complete unit is generally called a *stage of r-f amplification*. A number of methods are used to couple one stage of r-f amplification to another, and the choice of method used will depend on (1) the type of amplifier circuit used, (2) the amount of amplification required, (3) the amount of amplification produced by each stage, (4) the coupling device used, (5) the tuning circuit used, and (6) the type of composite circuit of which the amplifier stages are a part.

11-12 Single-control Tuning

When a device employs two or more tuning circuits for which simultaneous adjustments are desirable (such as a radio receiver), the variable capacitors may be gang-controlled by building them into a single frame so that all rotors can be operated by a single shaft (Figs. 11-6*b* and 11-12). In order to reduce the capacitance between adjacent stator sections, a flat metal grounded plate is mounted between each section.

11-13 Equalizing the Tuning Circuits

Tracking. Although the single-control tuning system rotates each tuning capacitor by the same amount, the electrical adjustment, while approximately equal, is not the same for each tuning circuit. In order to obtain the maximum fidelity, selectivity, and sensitivity for a communication receiver, all the tuning circuits must track together over the entire range of the receiver. Therefore, it is necessary to employ some means of adjusting each circuit. Usually, it is sufficient to make adjustments only for differences in capacitance to ensure satisfactory tracking of two or more tuning circuits. One method used to compensate for differences in tuning-circuit capacitance is to connect a small adjustable trimmer capacitor in parallel with the main tuning capacitor (Fig. 11-12). The trimmer capacitors can be adjusted for only general capacitance differences between each tuning circuit but cannot be used to align each section for all positions of the tuning control. It is general practice to adjust the capacitor sections for three positions, namely, the approximate center and each end of the frequency band. A more accurate method is to cut slots in the end rotor plates, as shown in Fig. 11-14. These plates generally have four or more sections, and the capacitance for each section can be changed by bending the plate at that section. If the plate is bent toward the stator, the capacitance is increased; and if it is bent away from the stator, the capacitance is decreased. With this system, the tuning circuits can be made to track for all positions of the tuning control. Some communication receivers use a combination of both of these methods. In these receivers the trimmers are used to compensate for general differences in capacitance and the slotted rotor

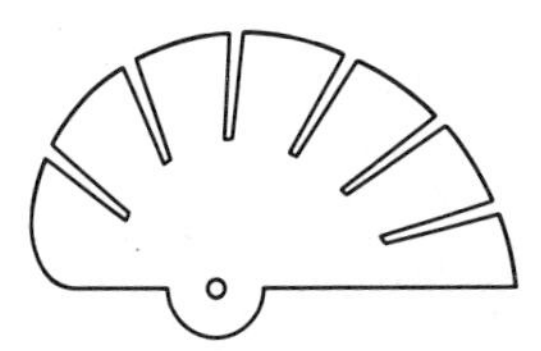

Fig. 11-14 A slotted rotor plate.

plates are used to align the capacitor sections for various positions of the tuning control.

11-14 Automatic Tuning

Methods of Obtaining Automatic Tuning Control. In the methods of tuning described in the previous articles, the selecting of stations on a receiver is accomplished by rotating the tuning dial to the desired position by hand. This is sometimes called *manual tuning*. Another method used to tune a radio or television receiver is by merely pushing a button or a lever. With this system, the receiver may automatically be tuned to the station or channel predetermined for each tuning button. This method is called *automatic tuning*, and the number of tuning buttons on a receiver varies with the type of receiver, the manufacturer, and the model.

Although numerous systems have been devised for obtaining automatic tuning, they may be classified into three general types: (1) mechanically operated manual types, (2) tuned-circuit substitution types, and (3) motor-operated types. In the *mechanically operated manual types*, the shaft of the tuning capacitor is turned to the preset desired position by pressing a button, key, or some type of lever. This type of tuning may be further subdivided into five common methods of operation, namely, linear, rocker bar, rotary, indent, and flash. The *linear* method employs a series of cams and levers to obtain the station selection. The *rocker-bar* method uses a series of pushbuttons to translate the motion to preset positions of the tuning capacitor. The *rotary* type uses a dial similar in appearance to the dial on a telephone. The *indent* method, also known as *spot tuning*, uses a steel ball that is pressed in a groove of a soft metal cylinder at preset positions and thereby provides indents to aid the manual tuning. The *flash* method, also known as *light-indicator tuning*, does not use any button arrangement but is tuned by manual tuning. When the tuning dial is in a position corresponding to one of the preset flash-tuned stations, a dial lamp is caused to light up a marker indicating the station to which the receiver is tuned. In the *tuned-circuit substitution types*, a number of precalibrated tuned circuits are connected to a pushbutton or rotary type of selector switch. The precalibrated tuned circuits are generally tuned by means of mica trimmer capacitors, permeability tuned coils, or a combination of both. In most cases where the tuned-circuit substitution method is used, manual tuning is also provided. In the *motor-operated types*, the shaft of the tuning capacitor is turned to the position required to tune a station by means of a small electric motor. The selection of desired stations is obtained by a station-selector switch or a number of pushbuttons, and generally requires the use of a selecting commutator or some other device for stopping the motor at the desired point.

Auxiliary Controls Required with Automatic Tuning. Adding automatic tuning to a radio or television receiver requires also including a number of extra auxiliary circuits or controls, the most important of which are (1) a transfer circuit or mechanism to change from manual tuning to automatic tuning, (2) an audio silencing or muting provision that will silence the receiver when the automatic tuning mechanism is changing the receiver from one station to another, (3) a station-selecting commutator mechanism for stopping the motor at the correct position for station reception with motor-operated types.

QUESTIONS

1. What is meant by (*a*) a tuning circuit? (*b*) A tuned circuit?
2. Why are resonant circuits considered important to the study of electronics?
3. Name and explain three functions generally performed by tuning circuits.
4. Define the following terms: (*a*) sensitivity, (*b*) selectivity, (*c*) fidelity.
5. (*a*) What is meant by tuning? (*b*) How is tuning accomplished?
6. Explain the following terms: (*a*) carrier wave, (*b*) sidebands, (*c*) response curve, (*d*) ideal response curve.
7. Name three methods generally used to approximate the ideal response curve.
8. (*a*) What factors in radio transmission may produce interference such as a hum, whistle, or cross talk in a radio receiver? (*b*) How can this interference be minimized?
9. Explain why increasing the selectivity beyond a certain point decreases the fidelity of a receiver.
10. What is the assigned frequency range of (*a*) a-m radio channels? (*b*) f-m radio channels? (*c*) vhf television channels? (*d*) uhf television channels?
11. What is the assigned bandwidth of (*a*) a-m radio channels? (*b*) f-m radio channels? (*c*) Television channels?
12. What is meant by the following terms as applied to electronics: (*a*) Resonance? (*b*) Resonant frequency?
13. In a series circuit containing inductance and capacitance, how do the inductive and capacitive reactances vary when the frequency of the input voltage is (*a*) increased? (*b*) Decreased?
14. Why should the resistance of a series resonant circuit be kept at a minimum?
15. For a series resonant circuit operating at its resonant frequency, describe the characteristics of the circuit in terms of (*a*) impedances, (*b*) currents, (*c*) voltages, (*d*) power factor.
16. Explain what happens to the currents in the inductor, capacitor, and line of a series tuned circuit for frequencies (*a*) below the resonant frequency, (*b*) above the resonant frequency.
17. (*a*) What is meant by the term figure of merit? (*b*) How does the circuit Q of a tuned circuit compare with the coil Q?
18. For a tuning circuit operating at its resonant frequency, what is the relation between its bandwidth and (*a*) circuit Q? (*b*) Resistance? (*c*) Inductance?
19. What is meant by (*a*) LC product? (*b*) L/C ratio?
20. Explain how the voltage across either reactance in a series tuned circuit can be greater than the applied voltage.
21. For a series tuned circuit, explain the effect of resistance on (*a*) the current, (*b*) the bandwidth.

22. For a parallel resonant circuit operating at its resonant frequency, describe the characteristics of the circuit in terms of (*a*) impedances, (*b*) currents, (*c*) voltages, (*d*) power factor.
23. How does a decrease in the circuit resistance affect the impedance of a parallel resonant circuit?
24. How does the impedance of each branch of a parallel resonant circuit compare with the impedance of the circuit (*a*) at the resonant frequency? (*b*) For frequencies below resonance? (*c*) For frequencies above resonance?
25. How does the current flowing in each branch of a parallel resonant circuit compare with the line current (*a*) at the resonant frequency? (*b*) For frequencies above resonance? (*c*) For frequencies below resonance?
26. How does the power factor of the inductor and capacitor of a parallel or series tuned circuit compare with the power factor of the line (*a*) at the resonant frequency? (*b*) For frequencies below resonance? (*c*) For frequencies above resonance?
27. (*a*) How do the rules for the interpretation of parallel resonance curves compare with those for a series circuit? (*b*) Why?
28. In what respects are parallel tuned circuits (*a*) similar to series tuned circuits? (*b*) Different from series tuned circuits?
29. What factor determines whether a resonant circuit should be classed as a series or parallel type?
30. What procedure should be followed in determining whether a resonant circuit is of the series or parallel type?
31. How does the frequency vary with the amount of rotation for each of the following types of variable capacitors: (*a*) slc? (*b*) slw? (*c*) slf?
32. What factors determine the amount of inductance required of the secondary winding of a tuned r-f transformer?
33. Explain one of the methods used to tune a radio receiver having a broadcast band and one or more short-wave bands.
34. (*a*) Explain two requirements of a band-selector switch. (*b*) Why is it necessary to short-circuit the unused coils?
35. Explain the construction features of a uhf tuning circuit for (*a*) inductors, (*b*) capacitors, (*c*) coupling.
36. (*a*) What is meant by bandspreading? (*b*) Why is it used? (*c*) How may it be accomplished?
37. (*a*) How is a trimmer capacitor connected in a circuit? (*b*) What is the basic principle in the application of trimmer capacitors?
38. If trimming means to cut down or decrease, does the addition of a trimmer capacitor trim (*a*) the capacitance of the circuit or (*b*) the resonant frequency of the circuit?
39. (*a*) How is a padder capacitor connected in a circuit? (*b*) What is the basic principle in the application of padder capacitors?
40. If padding means to build up or increase, does the addition of a padder capacitor pad (*a*) the capacitance of the circuit or (*b*) the resonant frequency of the circuit?
41. When used in conjunction with bandspread coils, what is the purpose of (*a*) trimmers? (*b*) padders?
42. (*a*) What is meant by bandspread tuning? (*b*) Where is it used? (*c*) Why is it desirable?
43. Describe the mechanical method of obtaining bandspread tuning.

44. Describe the electrical method of obtaining bandspread tuning.
45. Explain how the selectivity of a receiver is increased by using multiple-stage tuning.
46. What factors determine the number of stages of tuning to be used?
47. What are the advantages of single-control tuning?
48. (*a*) What is a multiple-variable capacitor? (*b*) What are its advantages?
49. What is meant by tracking?
50. (*a*) What are equalizing capacitors? (*b*) Where are they used? (*c*) Why are they used?
51. (*a*) How are slotted rotor plates used to align the tuning circuit? (*b*) What are their advantages?
52. What is meant by automatic tuning?
53. Describe the various types of the mechanically operated method of automatic tuning.
54. Describe the tuned-circuit substitution method of automatic tuning.
55. Describe the motor-operated method of automatic tuning.
56. Describe the three important auxiliary controls or circuits that should be included when automatic tuning is used.

PROBLEMS

1. A series tuned circuit has an inductance of 316 μh. To what value of capacitance must its variable capacitor be adjusted in order to obtain resonance for the following frequencies: (*a*) 500 kHz? (*b*) 1,500 kHz?
2. A series tuned circuit has a capacitance of 9.2 pf. Determine the value of inductance required to obtain resonance for the following frequencies: (*a*) 54 MHz, (*b*) 198 MHz.
3. A variable capacitor having a maximum capacitance of 350 pf is used for tuning a broadcast radio receiver. (*a*) What inductance is required to make the circuit resonant at the lowest frequency 500 kHz? (*b*) If the minimum capacitance of the capacitor is 15 pf, what is the highest frequency that can be obtained with the inductance determined in part (*a*)?
4. Determine the value of capacitance required for a tuned circuit having the following values: (*a*) $L = 0.0755$ μh, $f_r = 473$ MHz, (*b*) $L = 0.08$ μh, $f_r = 749$ MHz.
5. A 0.25-mh coil and an adjustable capacitor are connected to form the secondary side of an i-f transformer whose resonant frequency is 460 kHz (Fig. 11-15). (*a*) What is the capacitance of the adjustable capacitor? (*b*) What is the Q of the secondary winding if its resistance is 15.75 ohms? (*c*) What is the bandwidth of this circuit?

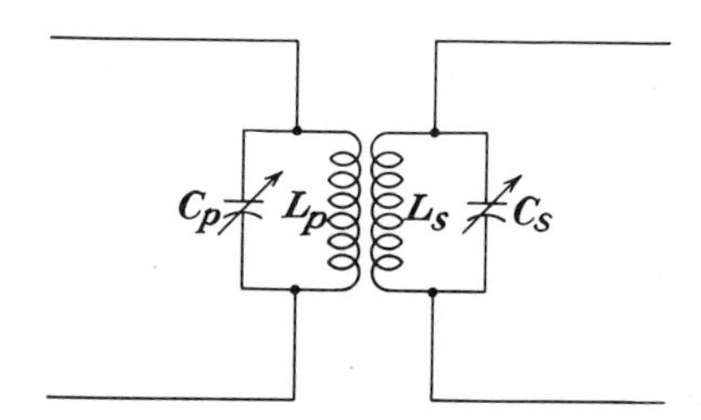

Fig. 11-15

6. An adjustable inductance coil that is set at 10 μh and a fixed capacitor are connected to form the secondary side of an i-f transformer whose resonant frequency is 10.7 MHz. (*a*) What is the capacitance of the fixed capacitor? (*b*) What is the Q of the secondary winding if its resistance is 12.6 ohms? (*c*) What is the bandwidth of this circuit?
7. A series tuned circuit has a resistance of 6 ohms and an inductance of 191 μh. (*a*) What is the width in hertz of its resonance curve at a point where the current in the circuit is equal to 0.707 times the current at resonance? (*b*) If the resistance of the circuit is increased to 12 ohms, how does the change affect the width of the resonance curve?
8. A series tuned circuit has an inductance of 12 μh. What circuit resistance is required to produce a bandwidth at 0.707 times the current at resonance of (*a*) 100 kHz? (*b*) 200 KHz?
9. A 10-mv signal is applied to a series tuned circuit having an inductance of 316 μh, a capacitance of 80 pf, and a resistance of 10 ohms. (*a*) Determine the voltage developed across the resistance, inductance, and capacitance of this circuit at resonance. (*b*) What is the circuit Q?
10. A 10-mv signal is applied to a series tuned circuit having an inductance of 0.775 μh, a capacitance of 7.5 pf, and a resistance of 9.72 ohms. (*a*) Determine the voltage developed across the resistance, inductance, and capacitance of this circuit at resonance. (*b*) What is the circuit Q?
11. A resonant circuit has an inductance of 316 μh and a capacitance of 80 pf. (*a*) Plot a curve showing how the inductive reactance changes when the frequency is varied from 975 to 1,025 kHz. (*b*) On the same paper, using the same reference abscissa, plot another curve showing how the capacitive reactance changes for the same frequency range. In plotting these curves, obtain values for every 5 kHz.
12. A resonant circuit has an inductance of 0.775 μh and a capacitance of 7.5 pf. (*a*) Plot a curve showing how the inductive reactance changes when the frequency is varied from 56 to 76 MHz. (*b*) On the same paper, using the same reference abscissa, plot another curve showing how the capacitive reactance changes for the same frequency range. In plotting these curves, obtain values for every 2 MHz.
13. Using the curves obtained in Prob. 11, determine the resultant circuit reactance at the following frequencies: (*a*) 980, (*b*) 990, (*c*) 1,000, (*d*) 1,010, (*e*) 1,020 kHz.
14. Using the curves obtained in Prob. 12, determine the resultant circuit reactance at the following frequencies: (*a*) 57, (*b*) 63, (*c*) 66, (*d*) 69, (*e*) 75 MHz.
15. Determine the resonant frequency of the circuit used in Prob. 11 (*a*) by referring to the curves obtained in Prob. 11, (*b*) by substituting the values of inductance and capacitance in the equation for finding the frequency of resonance.
16. Determine the resonant frequency of the circuit used in Prob. 12 (*a*) by referring to the curves obtained in Prob. 12, (*b*) by substituting the values of inductance and capacitance in the equation for finding the frequency of resonance.
17. A 10-mv signal is applied to a series resonant circuit having an inductance of 316 μh, a capacitance of 80 pf (same as Prob. 11), and a resistance of 10 ohms. Plot the resonance curve for this circuit.
18. A 10-mv signal is applied to a series resonant circuit having an inductance of 0.775 μh, a capacitance of 7.5 pf (same as Prob. 12), and a resistance of 9.72 ohms. Plot the resonance curve for this circuit.

19. Determine the width of the frequency band of the circuit used in Prob. 17 (*a*) from the resonance curve as plotted, (*b*) by substituting the values of inductance and resistance in the equation for finding the width of the frequency band.
20. Determine the width of the frequency band of the circuit used in Prob. 18 (*a*) from the resonance curve as plotted, (*b*) by substituting the values of inductance and resistance in the equation for finding the width of the frequency band.
21. A 120-pf adjustable capacitor and an inductance are connected in parallel to form the primary side of an i-f transformer whose resonant frequency is 460 kHz (Fig. 11-15). (*a*) What is the inductance of the primary winding? (*b*) What is the Q of the primary winding if its resistance is 9.85 ohms?
22. A 4.5-pf capacitor and an adjustable inductance coil are connected in parallel to form the primary side of an i-f transformer whose resonant frequency is 41.25 MHz. (*a*) What is the inductance of the primary winding? (*b*) What is the Q of the primary winding if its resistance is 4.62 ohms?
23. What is the impedance of the parallel resonant circuit used in Prob. 21?
24. What is the impedance of the parallel resonant circuit used in Prob. 22?
25. From the curves of Fig. 11-3*c*, what is the approximate impedance of the circuit with a resistance of 6.66 ohms at (*a*) 1,492.5 kHz? (*b*) 1,502.5 kHz? Using Eq. (11-15*a*) what is the impedance at (*c*) 1,492.5 kHz? (*d*) 1,502.5 kHz?
26. From the curves of Fig. 11-3*c*, what is the approximate impedance of the circuit with a resistance of 13.32 ohms at (*a*) 1,497.5 kHz? (*b*) 1,507.5 kHz? Using Eq. (11-15*a*) what is the impedance at (*c*) 1,497.5 kHz? (*d*) 1,507.5 kHz?
27. Plot a parallel resonance impedance curve for a circuit having a capacitance of 80 pf, an inductance of 316 μh, and a resistance of 10 ohms [use Eq. (11-15)]. *Note:* These are the same values used in Prob. 11.
28. Plot a parallel resonance impedance curve for a circuit having a capacitance of 7.5 pf, an inductance of 0.775 μh, and a resistance of 0.1 ohm [use Eq. (11-15)]. *Note:* These are the same values used in Prob. 12.
29. What are the impedance and current at the resonant frequency of the tuning circuit of Prob. 27 when the voltage of the power source is 100 volts and the value of R is (*a*) 5 ohms? (*b*) 10 ohms?
30. What are the impedance and current at the resonant frequency of the tuning circuit of Prob. 28 when the voltage of the power source is 100 volts and the value of R is (*a*) 0.05 ohm? (*b*) 0.1 ohm?
31. How much current flows in the separate branches of the parallel tuning circuit of Prob. 29 at the resonant frequency when the value of R is (*a*) 5 ohms? (*b*) 10 ohms?
32. How much current flows in the separate branches of the parallel tuning circuit of Prob. 30 at the resonant frequency when the value of R is (*a*) 0.05 ohm? (*b*) 0.1 ohm?
33. The coil used in Prob. 3 plus the circuit wiring has a distributed capacitance of 15 pf, thus increasing the circuit capacitance by this amount. What is the frequency range of the circuit?
34. The coils used in Prob. 4 plus the circuit wiring have a distributed capacitance of 0.05 pf, thus increasing the circuit capacitance by this amount. Determine the resonant frequency when the effect of this distributed capacitance is considered for part (*a*), for part (*b*).
35. The inductance of a tuned circuit is 0.0304 μh, the capacitance is 20 pf, and its effective distributed capacitance is 2 pf. (*a*) What is the resonant frequency of the

circuit? (*b*) What is the apparent resonant frequency if the effect of the distributed capacitance is ignored?

36. A variable capacitor having a maximum capacitance of 140 pf and a minimum capacitance of 10 pf is used to tune a frequency band whose lowest frequency is 6.5 MHz. (*a*) Find the inductance of the coil. (*b*) What is the highest frequency that can be tuned?

37. A series resonant circuit is to be used as a wave trap to eliminate the effect of a 1,200-kHz signal. What value of capacitance must be used if the coil has an inductance of 80 μh and a distributed capacitance of 10 pf?

38. A coil having a distributed capacitance of 10 pf is connected in series with a 350-pf capacitor in order to pass a 750-kHz signal. Find the inductance of this coil.

39. A parallel resonant circuit is to be used as a wave trap to eliminate the effects of a 1,300-kHz signal. The circuit has a resistance of 1.5 ohms and a capacitance (distributed and wiring) of 10 pf. (*a*) What value of inductance must be used with a capacitor whose value is 65 pf? (*b*) What is the circuit Q? (*c*) What is the width of the band being eliminated?

40. A parallel resonant circuit is to be used as a wave trap to eliminate the effects of a 100-MHz signal. The circuit has a resistance of 1.25 ohms and a capacitance (distributed and wiring) of 5.3 pf. (*a*) What value of inductance must be used with a capacitor of 20 pf? (*b*) What is the circuit Q? (*c*) What is the width of the band being eliminated?

41. A coil having an inductance of 320 μh is connected in parallel with an adjustable capacitor in order to bypass a band of frequencies between 999.25 and 1,000.75 kHz. The distributed capacitance of the coil and circuit is 9 pf. (*a*) What is the value of the adjustable capacitor? (*b*) What is the resistance of the circuit?

42. A coil having an inductance of 0.9 μh is connected in parallel with an adjustable capacitor in order to bypass a band of frequencies between 39.75 and 40.25 MHz. The distributed capacitance of the coil and circuit is 5.5 pf. (*a*) What is the value of the adjustable capacitor? (*b*) What is the resistance of the circuit?

43. A 15-pf auxiliary capacitor, C_2 of Fig. 11-11*a*, is connected in parallel with the tuning capacitor C_1 and the secondary winding L. The capacitor C_1 has a range of 15 to 320 pf, and the inductance of L is 316 μh. Determine (*a*) the capacitance range of the entire circuit, (*b*) the frequency range of the entire circuit.

44. A series tuned circuit has a tuning capacitor of 6 pf with a 3.2-pf auxiliary capacitor in parallel with it. What is the resonant frequency when this parallel combination is connected in series with a secondary winding having an inductance of (*a*) 0.944 μh? (*b*) 0.07 μh?

45. Determine the frequency range of the tuned circuit of Prob. 43 if the circuit and elements have a total distributed capacitance of 10 pf.

46. Determine the resonant frequency of the two tuned circuits in Prob. 44 if each circuit has a distributed capacitance of 1.2 pf.

47. A tuning circuit using a variable capacitor that has a maximum capacitance of 250 pf and a minimum capacitance of 10 pf has a frequency range of 500 to 1,800 kHz. The distributed capacitance of the circuit is 10 pf. (*a*) What is the value of the trimmer capacitor required to decrease the higher limit of the frequency range to 1,600 kHz? (*b*) What effect does the use of this trimmer capacitor have on the lower limit of the frequency range?

48. The upper and lower limits of a frequency band are 11.8 and 3.88 MHz, respectively. The maximum and minimum capacitances of the tuning capacitor are 140 and 6 pf. The distributed capacitance of the circuit is 10 pf. (*a*) What is the value of the trimmer capacitor required to decrease the higher limit of the frequency range to 10 MHz? (*b*) What effect does the use of this trimmer capacitor have on the lower limit of the frequency range?

49. A 10- to 100-pf adjustable capacitor is connected in series with a tuning capacitor having a range of 15 to 320 pf (Fig. 11-11*b*). The inductance of the secondary winding is 316 μh and the padder capacitor is set at 20 pf. Determine (*a*) the capacitance range of this circuit, (*b*) the frequency range of this circuit.

50. The padder capacitor of Prob. 49 is set at 80 pf. Determine (*a*) the capacitance range of this circuit, (*b*) the frequency range of this circuit.

51. The tuning circuit of the oscillator section of a superheterodyne receiver is to be adjusted so that it will always tune 455 kHz higher than the tuning circuit of the receiver whose frequency range is 550 to 1,630 kHz. A circuit similar to Fig. 11-11*c* is used, and the circuit elements have the following values: $L = 100\ \mu$h, $C_{1\cdot max} = 320$ pf, $C_{1\cdot min} = 20$ pf, and $C_2 = 20$ pf. The distributed capacitance of the circuit is 20 pf. (*a*) What value is required of the padder capacitor when the tuning circuit is adjusted to its minimum frequency? (*b*) Using the padder capacitor as calculated in part (*a*), what is the resonant frequency of the oscillator section when the tuning circuit is adjusted to its maximum frequency?

52. If the distributed capacitance of the circuit of Prob. 51 could be reduced to zero, (*a*) what value of padder capacitance would be required when the tuning circuit is adjusted to its minimum frequency? (*b*) Using the padder capacitor as calculated in part (*a*), what is the resonant frequency of the oscillator section when the tuning circuit is adjusted for its maximum frequency?

53. The tuning circuit of the oscillator section of a superheterodyne receiver is to be adjusted so that it will always tune 265 kHz higher than the tuning circuit of the receiver, whose frequency limits are 550 and 1,625 kHz. A circuit similar to that shown in Fig. 11-11*c* is used and the circuit elements have the following values: $L = 135\ \mu$h, $C_{1\cdot max} = 320$ pf, $C_{1\cdot min} = 13.5$ pf, and $C_2 = 20$ pf. The distributed capacitance of the circuit is 20 pf. (*a*) What value is required of the padder capacitor when the tuning circuit is adjusted to its minimum frequency? (*b*) Using the padder capacitor as calculated in part (*a*), what is the resonant frequency of the oscillator section when the tuning circuit is adjusted to its maximum frequency?

54. The tuning circuit of the oscillator section of a superheterodyne receiver is to be adjusted so that it will always tune 455 kHz higher than the tuning circuit of the receiver, whose frequency limits are 550 and 1,650 kHz. A circuit similar to that shown in Fig. 11-11*c* is used, and the circuit elements have the following values: $L = 130\ \mu$h, $C_{1\cdot max} = 260$ pf, $C_{1\cdot min} = 10$ pf, and $C_2 = 20$ pf. The distributed capacitance of the circuit is 15 pf. (*a*) What value is required of the padder capacitor when the tuning circuit is adjusted to its minimum frequency? (*b*) Using the padder capacitor as calculated in part (*a*), what is the resonant frequency of the oscillator section when the tuning circuit is adjusted to its maximum frequency?

55. It is desired to use the variable capacitor (350 to 15 pf) of Prob. 3 to tune a frequency band whose lowest frequency is 1,700 kHz by substituting a different coil for the one used in Prob. 3. (*a*) Find the inductance of the coil. (*b*) What is the highest frequency that can be tuned with this coil and capacitor?

56. A variable capacitor having a maximum capacitance of 140 pf and a minimum

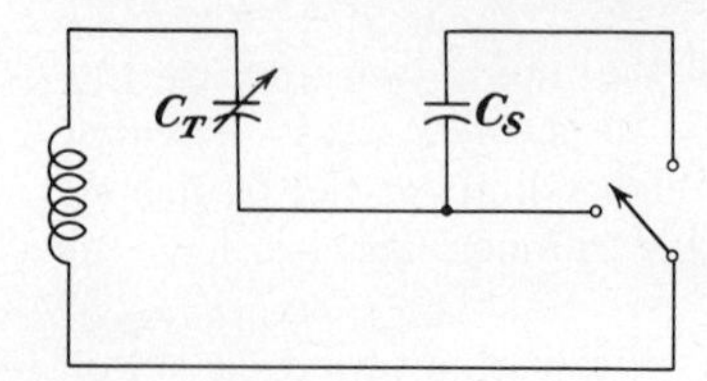

Fig. 11-16

capacitance of 10 pf is used to tune a frequency band whose lowest frequency is 1,700 kHz. (*a*) Find the inductance of the coil. (*b*) What is the highest frequency that can be tuned with this coil and capacitor?

57. In Prob. 55, the presence of distributed capacitance in the circuit was ignored. If the distributed capacitance of the circuit is 15 pf, find (*a*) the inductance of the coil, (*b*) the highest frequency that can be tuned with this coil and capacitor.

58. In Prob. 56, the presence of distributed capacitance in the circuit was ignored. If the distributed capacitance of the circuit is 10 pf, find (*a*) the inductance of the coil, (*b*) the highest frequency that can be tuned with this coil and capacitor.

59. It is desired to use the coil (290 μh) and the capacitor (350 to 15 pf) of Prob. 3 and add a fixed capacitor C_S, shown in Fig. 11-16, to tune a frequency band whose lowest frequency is 1,700 kHz. The distributed capacitance of the circuit is 15 pf. (*a*) What value of capacitance is required at C_S? (*b*) What is the highest frequency to which this circuit may be tuned?

60. A variable capacitor having a maximum capacitance of 140 pf and a minimum capacitance of 10 pf is used to tune a frequency band whose lowest frequency is 4.63 MHz. The distributed capacitance of the circuit is 5 pf, and the inductance of the secondary winding is 58.5 μh. (*a*) What value of capacitance must be connected in series with the tuning capacitor in order to obtain this minimum frequency when all the plates are in mesh? (*b*) What is the highest frequency to which the circuit may be tuned?

Chapter 12

Coupling and Filter Circuits

In order to transfer electrical energy from one *circuit element* to another, a conducting material, usually a wire, is used. In order to transfer electrical energy from one circuit to another, a coupling impedance is used. There are various ways in which two circuits may be coupled with each other, and each method has a definite use and produces different effects. When two or more individual electronic circuits are combined, they form a complex circuit through which the following kinds of currents may flow: (1) direct, (2) power frequency (60 hertz), (3) audio frequencies (20 to 20,000 hertz), and (4) radio frequencies (100 kHz to 30,000 MHz). One side of any number of these circuits may be completed through a common wire or through the chassis. It is the purpose of a filter to separate these currents at any desired point and to direct each of them into the conductor or circuit through which it is desired to have them flow.

12-1 Coupling of Circuits

Principles of Coupling. Two circuits are *coupled* when they have a common impedance that permits the transfer of electrical energy from one circuit to another. This common impedance, called the *coupling element,* may be a resistor, an inductor, a capacitor, a transformer, or a combination of two or more of these elements.

Coupling elements are usually required to perform some filter action in addition to transferring energy from one circuit to another. Conversely, every filter circuit contains a section that acts as a coupling device. Coupling circuits and filter circuits are so much alike that it is sometimes difficult to state whether they should be called filters or coupling units. The choice of name may be governed by that function which is considered of major importance. The type of impedance used is determined by the kinds of currents flowing in the input circuit and the kind desired in the output circuit.

Simple-coupled Circuits. A simple-coupled circuit is one in which the common impedance consists of a single element. A group of simple-coupled circuits is shown in Fig. 12-1.

The resistance-, inductive-, and capacitive-coupled circuits are also called *direct-coupled* circuits. In these circuits the coupling is accomplished by having the current of the input circuit flow through the common impedance where it

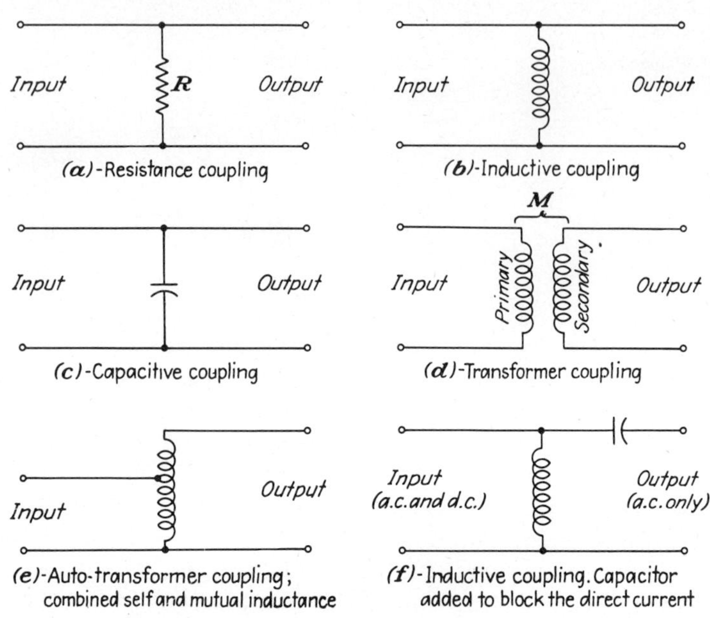

Fig. 12-1 Types of simple-coupled circuits.

produces a voltage drop. This voltage is applied to the output circuit, thus resulting in a transfer of electrical energy from the input to the output circuit. The output voltage is equal to the product of the current in the coupling element and its impedance.

The transformer-coupled circuit shown in Fig. 12-1*d* is also referred to as *indirect* coupling, *magnetic* coupling, or *mutual-inductive* coupling. In this type of coupling, the transfer of energy is accomplished by having the alternating current of the input circuit flow through the primary winding and setting up an alternating magnetic field. The magnetic lines of this field link the turns of the secondary winding and induce the voltage that supplies the energy for the output circuit.

In some applications of coupling devices the input circuit may have both alternating and direct current flowing, and it is desired that the coupling unit transfer only the alternating current to the output circuit. The transformer-coupled unit will serve this purpose satisfactorily, as it will pass only the alternating current. The other simple-coupled circuits (Fig. 12-1*a*, *b*, and *c*) can be modified so that no direct current can reach the load. This is accomplished by placing a capacitor in the output side of the coupling element as shown in Fig. 12-1*f*.

Complex-coupled Circuits. A complex-coupled circuit is one in which the common impedance consists of two or more circuit elements. A few of the numerous types of complex coupling are shown in Fig. 12-2.

The proportion of energy transferred in a simple inductive-coupled circuit increases as the frequency increases, while with simple capacitive coupling, the proportion of energy transferred decreases as the frequency increases. Using combinations of two or more elements in the coupling unit makes it possible to obtain various proportions of energy transfer for inputs of varying frequency. For example, the coupling element of Fig. 12-2*a* is a series tuned circuit and hence will have a minimum impedance at its resonant frequency. The proportion of energy transferred will be lowest when the frequency of the input circuit is equal to the resonant frequency of the coupling unit. At frequencies above resonance the proportion of energy transfer will increase and will be inductive. At frequencies below resonance the proportion of energy transfer will also increase but will be capacitive. The fact that the energy transfer is minimum at the resonant frequency may be more clearly understood when the input side of the filter is shown to be a part of a series circuit, for example, the plate circuit of a tube (Fig. 12-2*e*). At resonance, when the impedance of the coupling unit is minimum, its voltage drop will be at its minimum and the proportion of energy transfer must also be at its minimum.

In general, the amount of energy transferred will be proportional to the current flowing through the coupling unit and to the impedance of the unit. For purpose of analysis, complex-coupled circuits may be represented by a simple equivalent circuit, as shown in Fig. 12-2*f*.

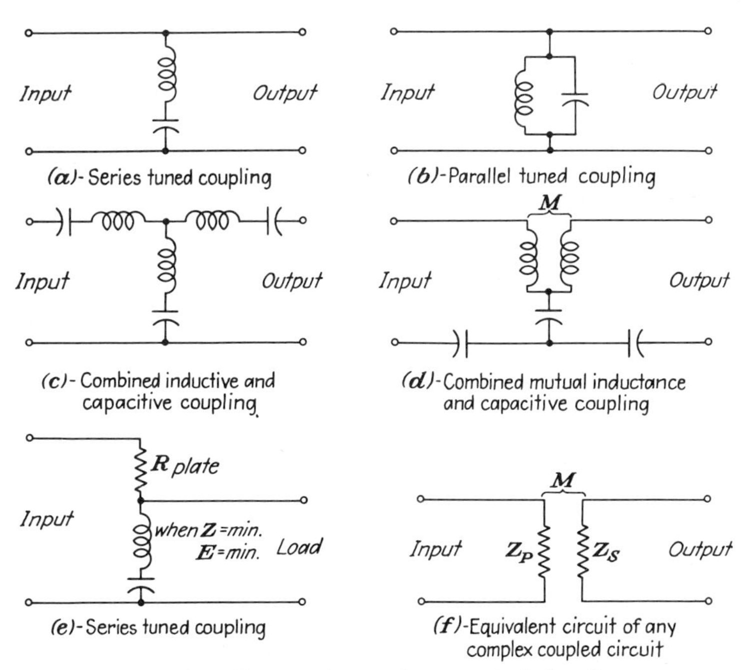

Fig. 12-2 Types of complex-coupled circuits.

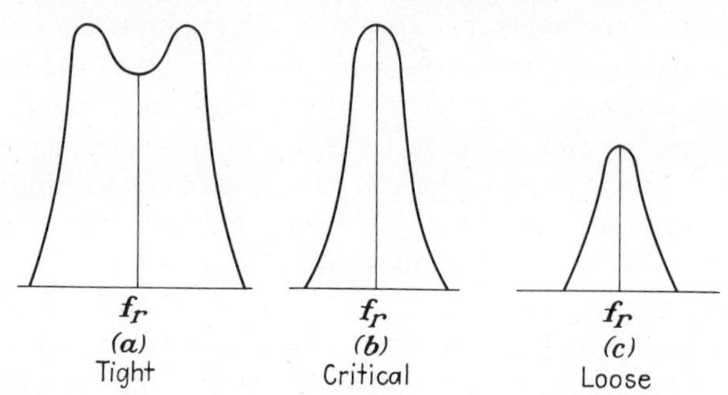

Fig. 12-3 Response curves showing the effect of various amounts of coupling.

Coefficient of Coupling. The ratio of the energy of the output circuit to the energy of the input circuit is called the *coefficient of coupling*. *Critical*, *tight*, and *loose coupling* are terms used to express the relative value of the coefficient of coupling for mutual-inductive-coupled circuits.

The response curves for tight, critical, and loose coupling are shown in Fig. 12-3. When the maximum amount of energy is transferred from one circuit to another, the circuits possess *critical coupling*, also referred to as *optimum coupling*. If the coefficient of coupling is higher than that required to produce critical coupling, the coils are *tightly coupled;* if it is less than that required for critical coupling, the coils are *loosely coupled.*

The effect of varying the coupling between two circuits may be seen from the response curves of Figs. 12-3 and 12-29. When two circuits are very tightly coupled, resonance will be obtained at two new frequencies, one below and the other above the normal frequency of resonance for the capacitor and inductor used. As the coupling is decreased, the two peaks come closer together until critical coupling is reached and a single peak of maximum height is obtained. If the coupling is decreased below the critical value, a single peak of reduced height is obtained.

Air-core transformers illustrate the importance of the amount of coupling between the primary and the secondary windings. As it is difficult to design an air-core transformer in which a large portion of the magnetic lines set up by the primary winding will link the turns of the secondary winding, the coefficient of coupling is generally low. A low value for the coefficient of coupling is not objectionable in some circuits, as it provides certain desirable characteristics which are presented in the following article.

12-2 Characteristics of Mutual-inductive-coupled Circuits

Mutual-inductive coupling, as provided by a transformer, is a means commonly used to transfer energy from one circuit to another. The characteristics of these circuits depend upon (1) the type of circuit, that is, whether a capacitor is connected to the primary, to the secondary, or to both, and (2) the amount of coupling between the two circuits.

Coupled Impedance. The primary and the secondary circuits of a transformer are separate electric circuits that are magnetically coupled. Each circuit has an impedance of its own, generally designated as Z_p and Z_s. The impedance of the primary winding, when no load is applied to the secondary, consists of the resistance and inductance of the primary winding. The impedance of the secondary circuit consists of the resistance and inductance of the secondary winding plus the impedance of any load connected to the circuit. When the secondary circuit is left open, that is, when no load is applied to its terminals, the impedance of the secondary will be infinity, or so large that it is immeasurable. Under this condition the presence of the secondary will have no effect upon the primary circuit.

When a load is applied to the secondary, the impedance will have a significant value and a current will flow in the secondary circuit. The amount of energy in this circuit will depend upon the secondary voltage and impedance. The secondary voltage, however, is dependent upon the number of magnetic lines linking the two circuits. The number of linkages is proportional to the coefficient of coupling; therefore, the amount of energy transferred is also dependent upon the coefficient of coupling. As the energy in the secondary circuit must come from the primary, it is evident that the primary impedance will be affected by the impedance of the secondary circuit. The effect of the secondary circuit upon the primary is equivalent to adding an impedance in series with the primary. This added impedance is generally referred to as the *coupled impedance.*

The numerical value of the coupled impedance of a mutual-inductive-coupled circuit may be found by the equation

$$Z_{p-s} = \frac{(2\pi f M)^2}{Z_s} \tag{12-1}$$

where Z_{p-s} = impedance coupled into the primary by the secondary, ohms
f = frequency of the power source, hertz
M = mutual inductance, henrys
Z_s = secondary impedance, ohms

The derivation of this equation is explained in the following steps:

1. From the definition, two circuits have a mutual inductance of 1 henry when a current in one circuit, changing at the rate of 1 amp per sec, induces an average emf of 1 volt in the second circuit; the induced voltage in the second circuit may be expressed as

$$e_s = M \frac{I_{p2} - I_{p1}}{t_2 - t_1} \tag{12-2}$$

This equation indicates that, when the mutual inductance M is 1 henry and the rate of current change in the primary $\frac{I_{p2} - I_{p1}}{t_2 - t_1}$ is 1 amp per sec, the average value of the voltage e_s induced in the secondary will be 1 volt. Thus, this equation is derived from the definition of the unit of mutual inductance.

2. When an alternating current I_p is flowing, the current is continually changing from a maximum value to zero in a positive and negative direction and at a rate proportional to the frequency. As the alternating current I_p is an effective value, the maximum current will be I_p divided by 0.707. Also, a change in current from the maximum value to zero occurs in a period of time corresponding to one-quarter of a cycle. Therefore

$$\frac{I_{p2} - I_{p1}}{t_2 - t_1} = \frac{I_{\max} - I_o}{\frac{1}{4f}} = \frac{I_{\max}}{\frac{1}{4f}} = \frac{\frac{I_p}{0.707}}{\frac{1}{4f}} = \frac{4fI_p}{0.707} \tag{12-3}$$

Substituting Eq. (12-3) in Eq. (12-2),

$$e_s = M\frac{4fI_p}{0.707} \tag{12-4}$$

3. The induced secondary voltage e_s is expressed as an average value, and in practical work it is desired to have it expressed as the effective value E. As the average value is equal to $2/\pi$ (or 0.637) times the maximum value and the effective value is equal to 0.707 times the maximum value, then the effective value may be expressed as

$$E_s = \frac{e_s}{\frac{2}{\pi}} \times 0.707 = \frac{0.707\pi e_s}{2} \tag{12-5}$$

or

$$e_s = \frac{2E_s}{0.707\pi} \tag{12-6}$$

Substituting Eq. (12-6) in Eq. (12-4),

$$\frac{2E_s}{0.707\pi} = M\frac{4fI_p}{0.707} \tag{12-7}$$

or

$$E_s = 2\pi fMI_p \tag{12-8}$$

4. The secondary current I_s will therefore be

$$I_s = \frac{E_s}{Z_s} = \frac{2\pi fMI_p}{Z_s} \tag{12-9}$$

5. This secondary current upon flowing through the secondary winding sets up a magnetic field of its own that induces a voltage in the primary. This in-

duced voltage will be 180° out of phase with the primary impressed voltage and is referred to as a counter, or back, voltage. By the same reasoning as was used to derive the secondary induced voltage, it may be shown that this counter voltage induced in the primary will be

$$E_{\text{counter}} = 2\pi f M I_s \tag{12-10}$$

Substituting Eq. (12-9) for I_s in Eq. (12-10),

$$E_{\text{counter}} = (2\pi f M)\frac{(2\pi f M) I_p}{Z_s} \tag{12-11}$$

or

$$E_{\text{counter}} = \frac{(2\pi f M)^2}{Z_s} I_p \tag{12-12}$$

6. As this voltage represents the effect that the secondary has upon the primary and as an alternating voltage is equal to the product of impedance and current, it may be stated [from Eq. (12-12)] that the effect of the secondary impedance upon the primary is

$$Z_{p-s} = \frac{(2\pi f M)^2}{Z_s} \tag{12-1}$$

The coupled impedance expressed by Eq. (12-1) may be represented by an equivalent resistance and an equivalent reactance connected in series with the primary circuit. The numerical values of the equivalent resistance and equivalent reactance are expressed by the following equations:

$$R_{p-s} = \frac{(2\pi f M)^2 R_s}{Z_s^2} \tag{12-13}$$

$$X_{p-s} = -\frac{(2\pi f M^2) X_s}{Z_s^2} \tag{12-14}$$

where R_{p-s} = resistance coupled into the primary by the secondary, ohms
X_{p-s} = reactance coupled into the primary by the secondary, ohms
f = frequency of the power source, hertz
M = mutual inductance, henrys
R_s = resistance of the secondary circuit, ohms
Z_s = impedance of the secondary circuit, ohms
X_s = reactance of the secondary circuit, ohms

Note: When X_s is inductive, then X_{p-s} has a negative sign, and when X_s is capacitive, X_{p-s} has a positive sign.

Example 12-1 A mutual-inductance-coupled circuit is shown in Fig. 12-4 together with the circuit values. Find (*a*) the inductance reactance of the primary, (*b*) the inductive reactance of the secondary, (*c*) the impedance coupled into the primary by the secondary, (*d*) the equivalent resistance component of the coupled impedance, (*e*) the equivalent reactance component of the coupled impedance, (*f*) the equivalent circuit

M=0.08 mh
10 volts
456 kHz
L_P=3mh
R_P=80Ω
L_S=0.8mh
R_S=40Ω
R_L=160Ω

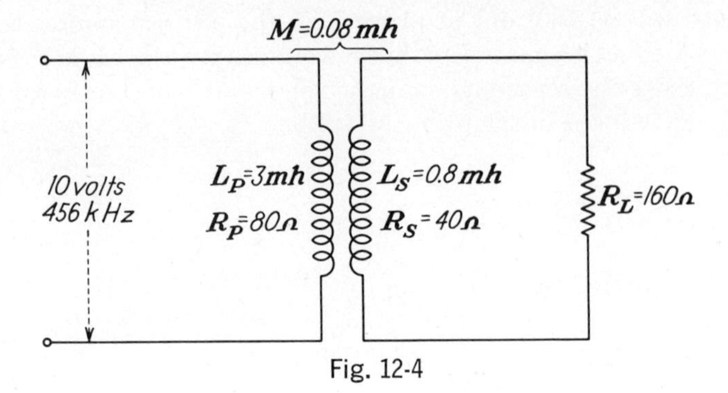

Fig. 12-4

diagram, (*g*) the effective impedance of the primary, (*h*) the primary current, (*i*) the secondary voltage, (*j*) the secondary current.

GIVEN: Fig. 12-4

FIND: (*a*) X_{LP} (*b*) X_{LS} (*c*) Z_{p-s} (*d*) R_{p-s} (*e*) X_{p-s} (*f*) Diagram
(*g*) Z_{pT} (*h*) I_p (*i*) E_s (*j*) I_s

SOLUTION:

(*a*) $$X_{LP} = 2\pi f L_P = 6.28 \times 456 \times 10^3 \times 3 \times 10^{-3} = 8{,}590 \text{ ohms}$$

(*b*) $$X_{LS} = 2\pi f L_S = 6.28 \times 456 \times 10^3 \times 0.8 \times 10^{-3} = 2{,}290 \text{ ohms}$$

(*c*) $$Z_{p-s} = \frac{(2\pi fM)^2}{Z_s} = \frac{(2\pi fM)^2}{\sqrt{(R_s + R_L)^2 + X_{LS}^2}}$$

$$= \frac{(6.28 \times 456 \times 10^3 \times 0.08 \times 10^{-3})^2}{\sqrt{(40 + 160)^2 + (2{,}290)^2}} = \frac{52{,}441}{2{,}298} = 22.8 \text{ ohms}$$

(*d*) $$R_{p-s} = \frac{(2\pi fM)^2(R_s + R_L)}{Z_s^2} = \frac{52{,}441 \times 200}{(2{,}298)^2} = 1.98 \text{ ohms}$$

(*e*) $$X_{p-s} = \frac{(2\pi fM)^2 X_s}{Z_s^2} = \frac{52{,}441 \times 2{,}290}{(2{,}298)^2} = 22.7 \text{ ohms}$$

(*f*) Fig. 12-5

(*g*) $$Z_{pT} = \sqrt{(R_p + R_{p-s})^2 + (X_{LP} - X_{p-s})^2}$$

$$= \sqrt{(80 + 1.98)^2 + (8{,}590 - 22.7)^2} = 8{,}567 \text{ ohms}$$

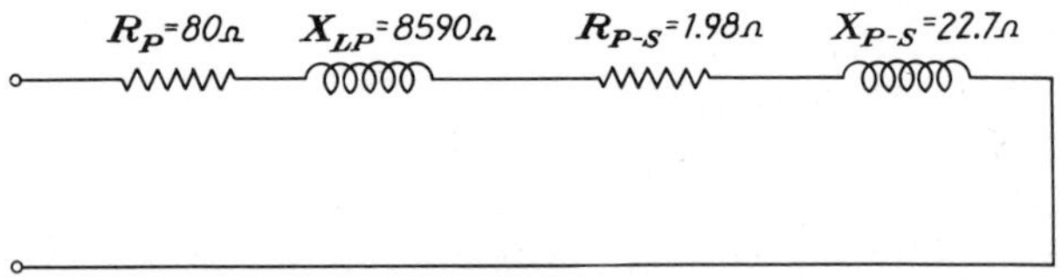

R_P=80Ω
X_{LP}=8590Ω
R_{P-S}=1.98Ω
X_{P-S}=22.7Ω

Fig. 12-5

(h) $I_p = \frac{E}{Z_{pT}} = \frac{10}{8{,}567} = 0.00116 \text{ amp} = 1.16 \text{ ma}$

(i) $E_s = 2\pi f M I_p = 6.28 \times 456 \times 10^3 \times 0.08 \times 10^{-3} \times 1.16 \times 10^{-3} = 0.266 \text{ volt}$

(j) $I_s = \frac{E_s}{Z_s} = \frac{0.266}{2{,}298} = 0.000115 \text{ amp} = 115\ \mu\text{a}$

The results of this example indicate that the effect of the coupled equivalent resistance is to increase the effective resistance of the primary circuit. The equivalent reactance that is coupled into the primary by a secondary whose reactance is inductive is opposite in phase to the primary reactance and hence reduces the effective reactance of the primary circuit. The net result is a reduction in the effective primary impedance, more current thereby being allowed to flow in the primary circuit, thus making possible the transfer of more energy to the secondary circuit.

Many of the important characteristics of coupled circuits are explained by the effects of coupled impedance. Examination of Eq. (12-1) indicates that the coupled impedance will be low when the coefficient of coupling is low because the value of M decreases when the coefficient of coupling is decreased. Also, the coupled impedance will be low when the secondary impedance is high. Thus, when the coefficient of coupling is low or when very little load is applied to the secondary (high secondary impedance), the coupled impedance will be low and the effect of the secondary upon the primary will be negligible. However, when the coefficient of coupling is high or when the secondary carries considerable amounts of load (low secondary impedance), the coupled impedance will be high and the secondary will produce considerable effect upon the primary circuit.

Circuit with Untuned Primary and Untuned Secondary. The simplest type of transformer coupling is a circuit having an untuned primary and an untuned secondary with a resistance or inductance load. Such a circuit is shown in Fig. 12-6*a*. This circuit is often used as an equivalent circuit to represent the effects produced by a shield, metal panel, or other metal object located near a coil. The effect of the shield or panel upon the coil would be the same as that of a secondary winding consisting of inductance and resistance in series. The coupled impedance of such a circuit will increase the effective resistance of the primary and reduce its effective reactance. It also indicates that losses of the

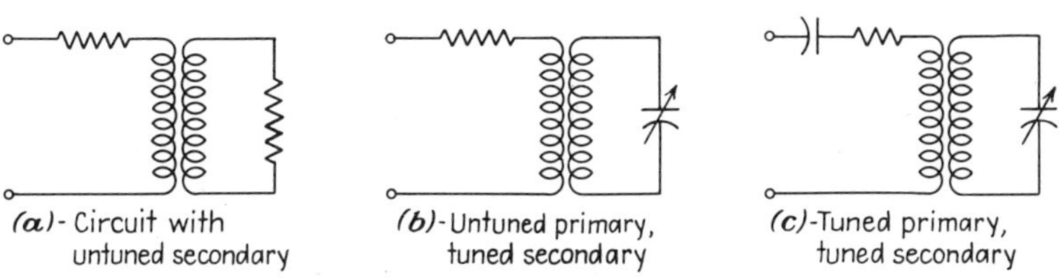

Fig. 12-6 Fundamental transformer-coupled circuits.

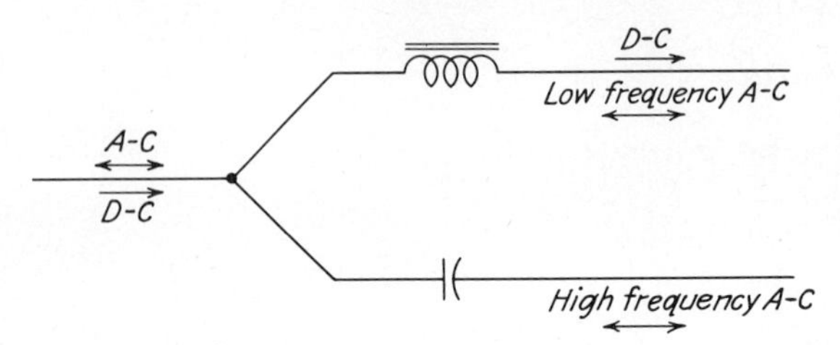

Fig. 12-7 A capacitor and an inductor used to separate alternating current from direct current and low-frequency currents from high-frequency currents.

coil circuit are increased by an amount proportional to the resistance coupled into the primary by the secondary, which is actually the shield or nearby metal panel.

Circuit with Untuned Primary and Tuned Secondary. The circuit shown in Fig. 12-6*b* differs from Fig. 12-6*a* in that a capacitor is used in place of the resistor in the secondary circuit. The tuned r-f amplifier circuit is similar to this fundamental circuit.

The secondary circuit is similar to the series tuned circuit, and its characteristics will be the same as those of the series tuned circuit. At the resonant frequency the impedance will be at its minimum and the current at its maximum. The impedance coupled into the primary will be high and will have a critical effect upon the primary current. At frequencies above or below resonance the secondary impedance increases and its current decreases. The impedance coupled into the primary decreases, and its effect on the primary circuit is decreased.

Circuit with Tuned Primary and Tuned Secondary. This type of circuit (Fig. 12-6*c*) is used extensively in radio and television receivers. A common application of this circuit is the i-f amplifier of the superheterodyne receiver. This type of coupling is very useful for amplifiers because it can be designed to provide an approximately uniform secondary current response over the band of frequencies that are normally applied to the primary circuit.

12-3 Filter Action

A filter circuit consists of a combination of capacitors, inductors, and resistors connected so that it will separate alternating currents from direct currents or alternating currents within a band of frequencies from those alternating currents outside of this band. Filter circuits may range from a very simple circuit to a very intricate circuit, depending upon their application. However, no matter how simple or involved a filter circuit may be, its action must depend upon the following principles of a-c circuits:

1. The opposition offered to the flow of alternating currents by a circuit containing only inductance will increase with frequency increase. Such a circuit will offer comparatively little opposition to the flow of direct, pulsating, or low-frequency alternating currents. The opposition offered to r-f currents by such a circuit will be comparatively high (Figs. 12-7 and 12-8).

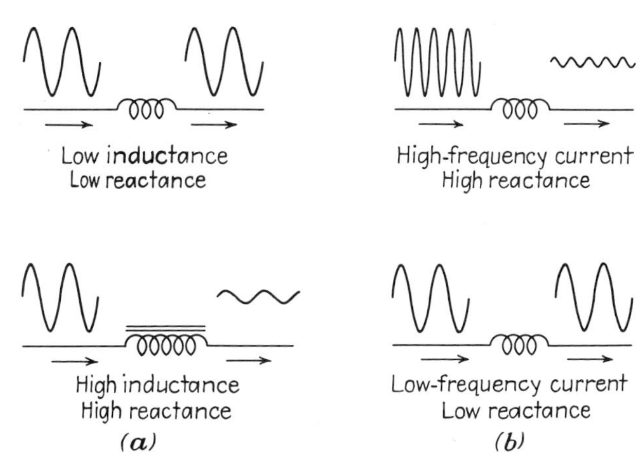

Fig. 12-8 Effects of inductance and frequency on current flow. (a) Variable inductance, constant frequency. (b) Variable frequency, constant inductance.

2. The opposition offered to the flow of alternating currents by a circuit containing only capacitance will decrease with frequency increase. Such a circuit will offer a comparatively high opposition to low-frequency currents, little opposition to r-f currents, and will block the flow of direct currents (Figs. 12-7 and 12-9).

3. A series resonant circuit has a low impedance at resonance and will offer little opposition to the flow of all currents whose frequencies lie within a narrow band above and below the resonant frequency. Such a circuit will offer a comparatively high opposition to the flow of currents of all other frequencies (Fig. 12-10*a*).

4. A parallel resonant circuit has a high impedance at resonance and will

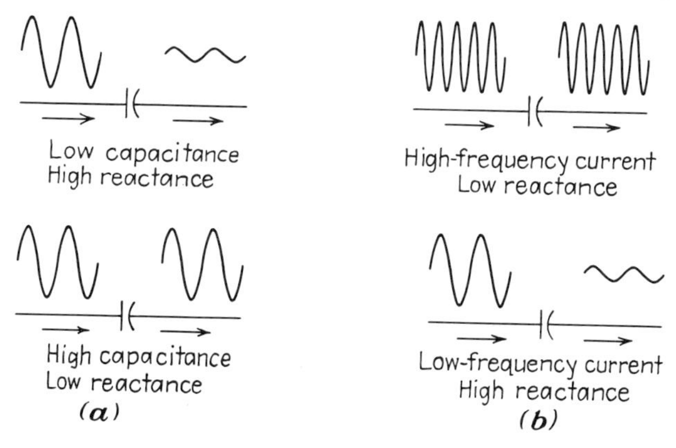

Fig. 12-9 Effects of capacitance and frequency on current flow. (a) Variable capacitance, constant frequency. (b) Variable frequency, constant capacitance.

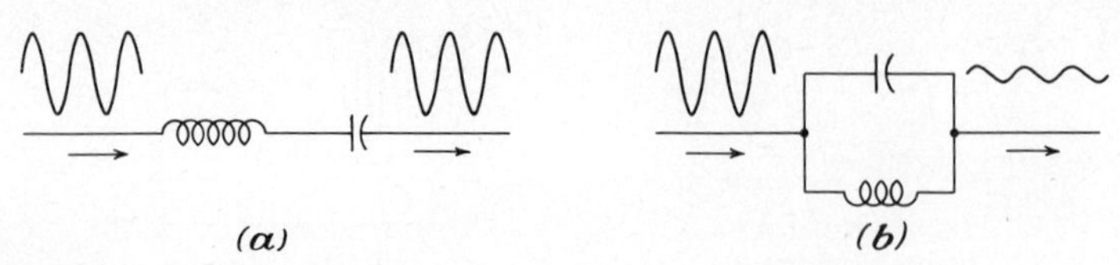

Fig. 12-10 Effect of resonant circuits on the amount of current flow at or near the resonant frequency. (a) Series resonant circuit. (b) Parallel resonant circuit.

offer a comparatively high opposition to the flow of all currents whose frequencies lie within a narrow band above and below the resonant frequency. Such a circuit will offer little opposition to the flow of currents of all other frequencies (Fig. 12-10*b*).

5. Resistors do not provide any filtering action when used alone, as they oppose the flow of all currents regardless of their frequency. When connected in series with a capacitor or inductor or both, it increases the impedance of the circuit, thus decreasing the sharpness of the filter circuit. Increasing the resistance of the series resonant circuit decreases the current at the resonant frequency, thus decreasing the slope of the curve on either side of the resonant frequency (Fig. 11-1*b*).

Example 12-2 A 10-henry filter choke has a resistance of 475 ohms. Determine its opposition (*a*) to direct current, (*b*) to 60-hertz alternating current.

GIVEN: $R = 475$ ohms $\quad L = 10$ henrys

FIND: (*a*) R $\quad$ (*b*) Z

SOLUTION:

(*a*) $$R = 475 \text{ ohms}$$

(*b*) $$X_L = 2\pi fL = 6.28 \times 60 \times 10 = 3{,}768 \text{ ohms}$$

$$Z = \sqrt{R^2 + X_L^2} = \sqrt{475^2 + 3{,}768^2} = 3{,}797 \text{ ohms}$$

Example 12-2 indicates that the opposition offered to the flow of 60-hertz alternating current is approximately eight times that offered to the d-c flow. This type of coil can therefore be used to pass direct current and block the flow of low-frequency alternating currents.

Example 12-3 (*a*) To which type of current will a 0.2-μf capacitor offer the greater opposition, a 5,000-hertz audio signal or a 5,000-kHz r-f signal? (*b*) How many times greater is the larger impedance than the smaller impedance?

GIVEN: $C = 0.2\ \mu\text{f}$ $\quad f_{af} = 5$ kHz $\quad f_{rf} = 5{,}000$ kHz

FIND: (*a*) $X_{C \cdot af}$ or $X_{C \cdot rf}$ $\quad$ (*b*) Ratio $X_{C \cdot af} - X_{C \cdot rf}$

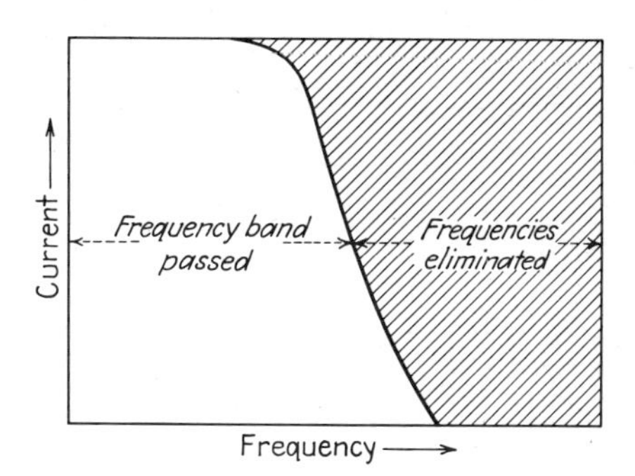

Fig. 12-11 Characteristic curve for a simple low-pass filter circuit.

SOLUTION:

(*a*) At 5,000 hertz

$$X_C = \frac{159{,}000}{f_{af}C} = \frac{159{,}000}{5 \times 10^3 \times 2 \times 10^{-1}} = 159 \text{ ohms}$$

At 5,000 kHz

$$X_C = \frac{159{,}000}{f_{rf}C} = \frac{159{,}000}{5 \times 10^6 \times 2 \times 10^{-1}} = 0.159 \text{ ohm}$$

Greater opposition is offered to the a-f current.

(*b*)

$$\text{Ratio of opposition} = \frac{X_{C\cdot af}}{X_{C\cdot rf}} = \frac{159}{0.159} = 1{,}000 \text{ to } 1$$

12-4 Types of Filter Circuits

Filter circuits are composed of inductors and capacitors having losses as low as commercially and economically practical. In the elementary consideration of filter circuits, it is assumed that the inductors and capacitors have no internal effective resistance. There are four general types of filter circuits, namely, low-pass filter, high-pass filter, bandpass filter, bandstop filter.

Low-pass Filter. A low-pass filter circuit allows all currents having a frequency below a certain value to pass into a desired circuit and diverts the flow of all currents having a frequency above this value (Fig. 12-11). An inductance coil inserted in the line (Fig. 12-12*a*) will offer little opposition to the flow of low-frequency currents and a large amount of opposition to the flow of high-frequency currents. In order to divert the undesired high-frequency currents

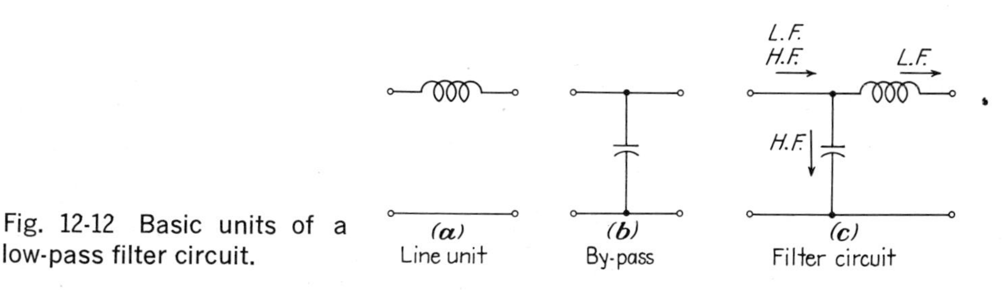

Fig. 12-12 Basic units of a low-pass filter circuit.

back to the source, a capacitor is used as a bypass path (Fig. 12-12*b*). The value of this bypass capacitor should be such that it will offer little opposition to the flow of current for all frequencies above a definite value and greatly oppose the flow of current for all frequencies below this value. When the coil and capacitor are connected as shown in Fig. 12-12*c*, the simplest type of low-pass filter circuit is obtained.

Example 12-4 A 5-mh choke coil and a 0.001-μf capacitor are connected as shown in Fig. 12-12*c* to form a low-pass filter circuit. Determine the opposition offered to a 5-kHz a-f signal (*a*) by the capacitor, (*b*) by the inductor. Determine the opposition offered to a 500-kHz r-f signal (*c*) by the capacitor, (*d*) by the inductor.

GIVEN: $C = 0.001\ \mu f \quad L = 5\ mh \quad f_{af} = 5\ kHz \quad f_{rf} = 500\ kHz$

FIND: (*a*) X_C at 5 kHz (*b*) X_L at 5 kHz (*c*) X_C at 500 kHz (*d*) X_L at 500 kHz

SOLUTION:

(*a*) $$X_C = \frac{159{,}000}{f_{af}C} = \frac{159{,}000}{5 \times 10^3 \times 1 \times 10^{-3}} = 31{,}800 \text{ ohms}$$

(*b*) $$X_L = 2\pi f_{af}L = 6.28 \times 5 \times 10^3 \times 5 \times 10^{-3} = 157 \text{ ohms}$$

(*c*) $$X_C = \frac{159{,}000}{f_{rf}C} = \frac{159{,}000}{500 \times 10^3 \times 1 \times 10^{-3}} = 318 \text{ ohms}$$

(*d*) $$X_L = 2\pi f_{rf}L = 6.28 \times 5 \times 10^5 \times 5 \times 10^{-3} = 15{,}700 \text{ ohms}$$

Analyzing the values obtained in this example, it can be seen that the r-f currents will find the path of least opposition through the capacitor and hence will take that path. The a-f currents will find the path of least opposition through the inductor and hence will take that path. A measure of the ability of a filter circuit to bypass undesired currents may be expressed in terms of the ratio of the impedances at the frequency limits.

High-pass Filter. A high-pass filter circuit allows all currents having a frequency above a certain value to pass into a desired circuit and diverts the flow of all currents having a frequency below this value (Fig. 12-13). A capacitor inserted in the line (Fig. 12-14*a*) will offer little opposition to the flow of high-

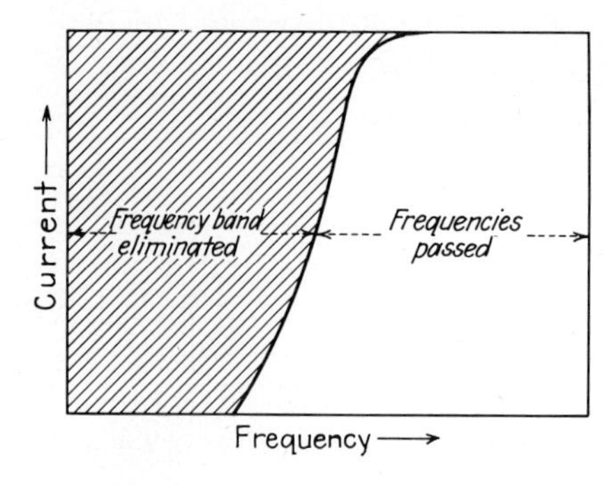

Fig. 12-13 Characteristic curve for a simple high-pass filter circuit.

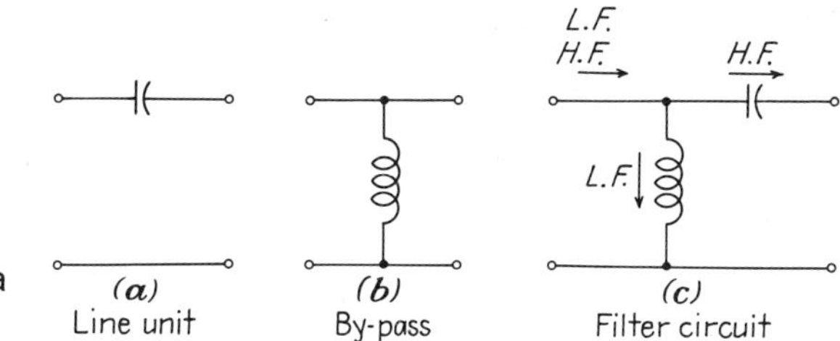

Fig. 12-14 Basic units of a high-pass filter circuit.

frequency currents, a large amount of opposition to the flow of low-frequency currents, and will block the flow of direct currents. The capacitor used should be of such a value that it will allow the passage of all currents whose frequencies are above a definite value and greatly oppose the flow of current for all frequencies below this value. In order to divert the undesired low-frequency currents back to the source, an inductance coil is used as a bypass path (Fig. 12-14*b*). The inductance of this coil should be of such a value that it will carry off the currents whose frequencies are below the cutoff point and reject the currents whose frequencies are above this value, thus forcing them to pass on through the circuit. When the coil and capacitor are connected as shown in Fig. 12-14*c*, the simplest type of high-pass filter circuit is obtained.

Example 12-5 A 2-henry choke coil and a 0.5-μf capacitor are connected as shown in Fig. 12-14*c* to form a high-pass filter circuit. Determine the opposition offered to a 60-hertz power interference signal (*a*) by the capacitor, (*b*) by the inductor. Determine the opposition offered to a 1,200-hertz a-f signal (*c*) by the capacitor, (*d*) by the inductor.

GIVEN: $C = 0.5\ \mu\text{f}$ $\quad L = 2$ henrys $\quad f_{pf} = 60$ hertz $\quad f_{af} = 1{,}200$ hertz

FIND: (*a*) X_C at 60 hertz (*b*) X_L at 60 hertz (*c*) X_C at 1,200 hertz (*d*) X_L at 1,200 hertz

SOLUTION:

(*a*) $$X_C = \frac{159{,}000}{f_{pf}C} = \frac{159{,}000}{60 \times 0.5} = 5{,}300 \text{ ohms}$$

(*b*) $$X_L = 2\pi f_{pf}L = 6.28 \times 60 \times 2 = 753.6 \text{ ohms}$$

(*c*) $$X_C = \frac{159{,}000}{f_{af}C} = \frac{159{,}000}{1{,}200 \times 0.5} = 265 \text{ ohms}$$

(*d*) $$X_L = 2\pi f_{af}L = 6.28 \times 1{,}200 \times 2 = 15{,}072 \text{ ohms}$$

Analyzing the values obtained in this example, it can be seen that the high-frequency currents will take the path of the capacitor and the low-frequency currents will be bypassed through the inductor.

Bandpass Filter. A bandpass filter allows the current of a narrow band of frequencies to pass through a circuit and excludes all currents whose frequencies are either greater or less than the extreme limits of the band (Fig. 12-15).

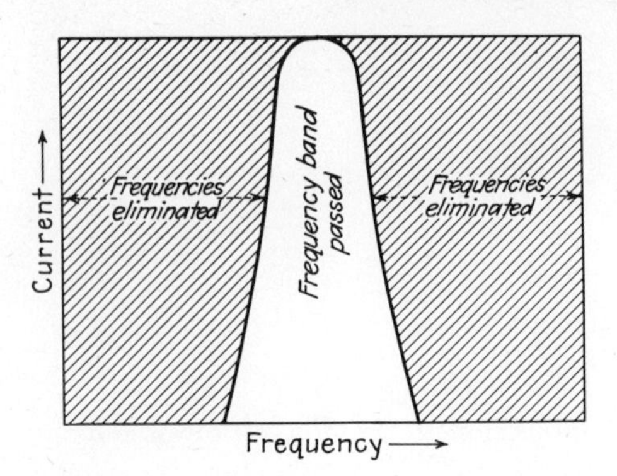

Fig. 12-15 Characteristic curve for a simple bandpass filter circuit.

Resonant circuits can serve as filters in a manner similar to the action of individual capacitors and inductors. The series resonant circuit (Fig. 12-16*a*) replacing the inductor of Fig. 12-12*a* would act as a bandpass filter, passing currents whose frequencies are at or near the resonant frequency and blocking the passage of all currents whose frequencies are outside this narrow band. The parallel resonant circuit (Fig. 12-16*b*) replacing the capacitor of Fig. 12-12*b*, if tuned to the same frequency as the series resonant circuit, will provide a path for all currents whose frequencies are outside the limits of the frequency band passed by the series resonant circuit. When the two resonant circuits are connected as shown in Fig. 12-16*c*, the simplest type of bandpass filter circuit is obtained.

Bandstop Filter. A bandstop filter opposes the flow of current for a narrow band of frequencies while allowing the current to flow for all frequencies above or below this band (Fig. 12-17). Bandstop filters are also known as *band-suppression* and *band-elimination filters*. Their purpose is opposite to that of a bandpass filter, and the relative positions of the resonant circuits in the filter circuit are interchanged.

The parallel resonant circuit of Fig. 12-18*a*, which replaces the capacitor

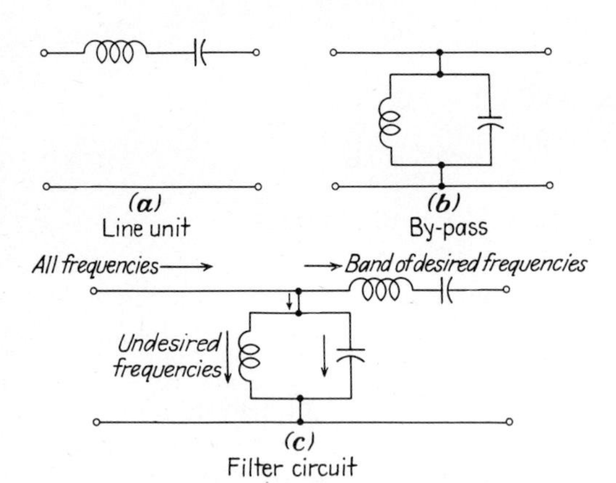

Fig. 12-16 Basic units of a bandpass filter circuit.

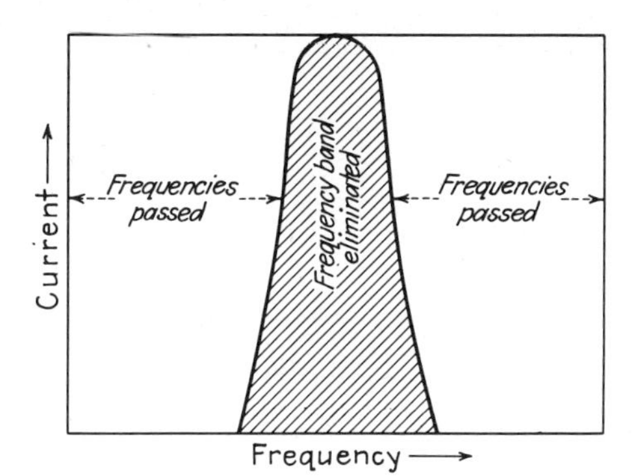

Fig. 12-17 Characteristic curve for a simple bandstop filter circuit.

of Fig. 12-14*a*, acts as a bandstop filter circuit and blocks the passage of all currents whose frequencies are at or near its resonant frequency, passing all currents whose frequencies are outside this band. The series resonant circuit of Fig. 12-18*b*, which replaces the inductor of Fig. 12-14*b*, when tuned to the same resonant frequency as the parallel resonant circuit provides a bypass path for the undesired band of frequencies. When the two resonant circuits are connected as shown in Fig. 12-18*c*, the simplest type of bandstop filter circuit is obtained.

12-5 Multisection Filter Circuits

Need for Multisection Filter Circuits. All the filter circuits explained in the previous article have only one section. None of these single-section filter circuits provide a sharp reduction of current at the cutoff frequency. Adding a capacitor, inductor, or resonant circuit in series or parallel (depending on the circuit) with the filter circuit will improve its filtering action, thus sharpening the reduction of current at the desired frequency. When an additional unit is added to a single-section filter circuit, the form of the resulting circuit will resemble the letter T or the symbol π. They are therefore called T- or π-type filters, depending on which symbol they resemble. Two or more sections of either the T- or π-type filters may be joined to produce a sharper cutoff.

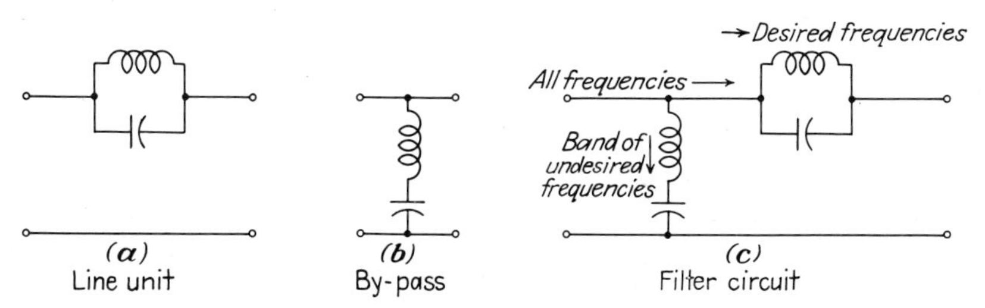

Fig. 12-18 Basic units of a bandstop filter circuit.

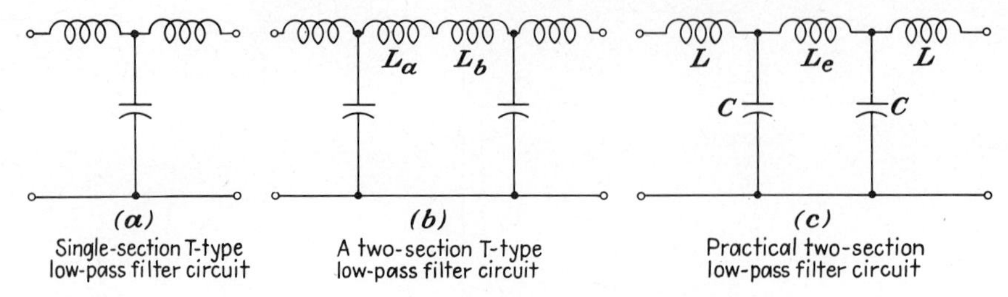

Fig. 12-19 T-type low-pass filter circuits.

T-type Filter Circuits. When an inductor is connected in series with the simple low-pass filter of Fig. 12-12*c*, a T-type low-pass filter circuit is formed (Fig. 12-19*a*). When two of these filter circuits are connected as shown in Fig. 12-19*b*, the inductors L_a and L_b can be replaced by a single inductor L_e (Fig. 12-19*c*) whose value is equal to $L_a + L_b$. As all the inductors originally are of the same size, the center inductor L_e should have a value of twice the inductance of either of the end inductors.

When a capacitor is connected in series with the simple high-pass filter of Fig. 12-14*c*, a T-type high-pass filter circuit is formed (Fig. 12-20*a*). When two of these filter circuits are connected as shown in Fig. 12-20*b*, the capacitors C_a and C_b can be replaced by a single capacitor C_e (Fig. 12-20*c*). As the capacitors C_a and C_b are of equal value and connected in series, the value of C_e should be one-half the value of either C_a or C_b.

When a series resonant circuit is connected in series with the simple bandpass filter circuit of Fig. 12-16*c*, a T-type bandpass filter is formed (Fig. 12-21*a*). When a parallel resonant circuit is connected in series with the simple bandstop filter circuit of Fig, 12-18*c*, a T-type bandstop filter is formed (Fig. 12-21*b*).

Pi-type Filter Circuits. When a capacitor is connected to the simple low-pass filter of Fig. 12-12*c*, so that the line is shunted at both ends of the inductor by a capacitor, a π-type low-pass filter is formed (Fig. 12-22*a*). When two of these filter circuits are connected as shown in Fig. 12-22*b*, the capacitors C_a and C_b can be replaced by a single capacitor C_e (Fig. 12-22*c*) whose value is equal to $C_a + C_b$. As all the capacitors originally are of the same size, the cen-

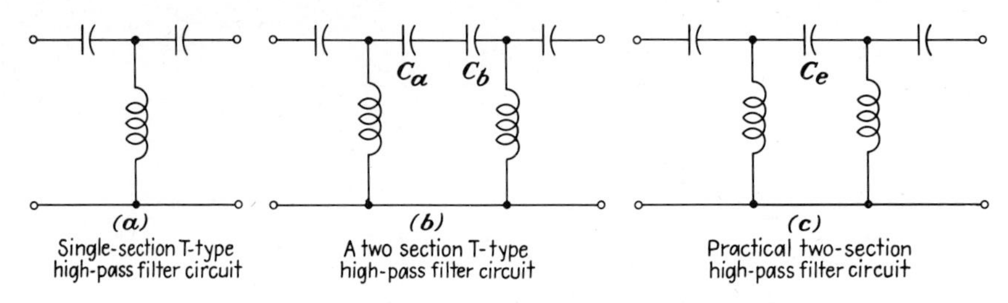

Fig. 12-20 T-type high-pass filter circuits.

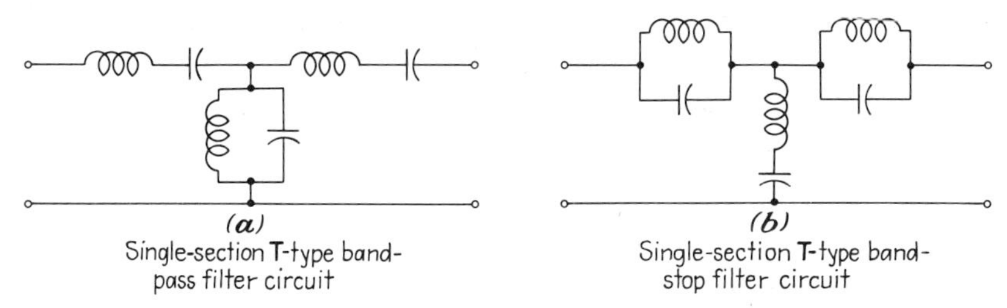

Fig. 12-21 T-type bandpass and bandstop filter circuits.

ter capacitor should have a value of twice the capacitance of either of the end capacitors.

When an inductor is connected to the simple high-pass filter of Fig. 12-14*c*, so that the line is shunted at both ends of the capacitor by an inductor, a π-type high-pass filter is formed (Fig. 12-23*a*). When two of these filter circuits are connected as shown in Fig. 12-23*b*, the inductors L_a and L_b can be replaced by a single inductor L_e (Fig. 12-23*c*). As the inductors L_a and L_b are of equal value and connected in parallel, the value of L_e should be one-half the value of either L_a or L_b.

When a parallel resonant circuit is connected to the simple bandpass filter circuit of Fig. 12-16*c*, so that the line is shunted at both ends of the series resonant circuit by a parallel resonant circuit, a π-type bandpass filter circuit is formed (Fig. 12-24*a*). When a series resonant circuit is connected to the simple bandstop filter circuit of Fig. 12-18*c*, so that the line is shunted at both ends of the parallel resonant circuit by a series resonant circuit, a π-type bandstop filter circuit is formed (Fig. 12-24*b*).

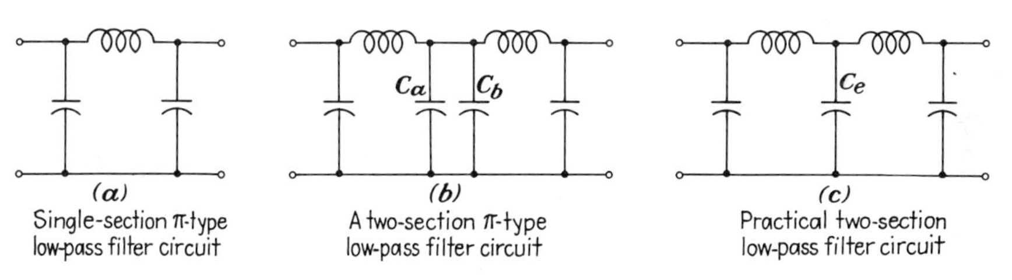

Fig. 12-22 π-type low-pass filter circuits.

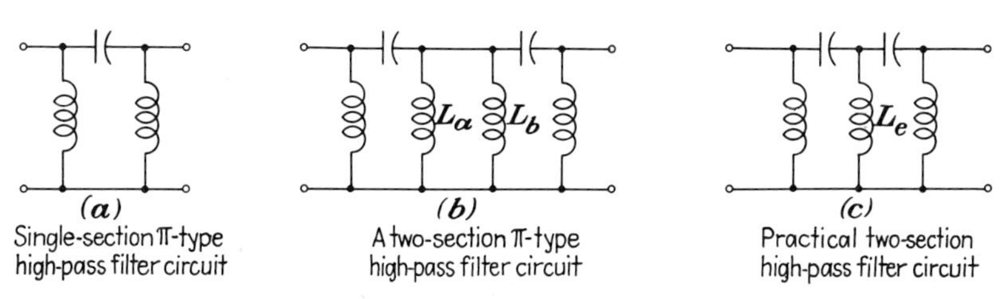

Fig. 12-23 π-type high-pass filter circuits.

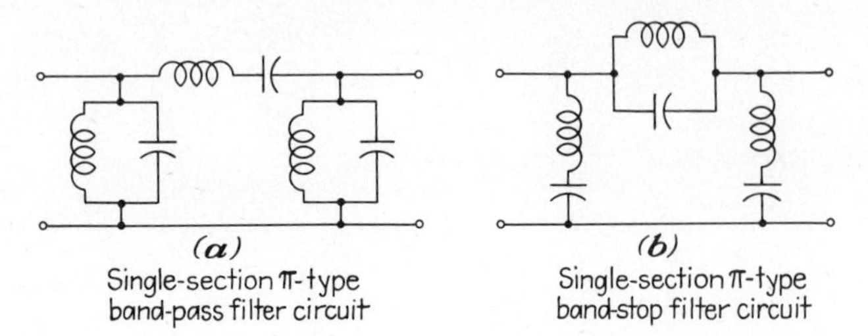

Fig. 12-24 π-type bandpass and bandstop filter circuits.

12-6 Filter Circuits as a Whole

The design of filter circuits is a specialized field of electronics, and the calculations for the component parts is left to specialized texts. The choice of the type of circuit to be used is a matter for the designing engineer to decide. However, the following terms, generally used in connection with filter circuits, should be understood.

Source Impedance. The impedance of the circuit leading into the filter circuit is called the *source impedance.* This may be the plate circuit of a tube, a high resistance, etc.

Load Impedance. The impedance of the circuit into which the filter circuit feeds is called the *load impedance.* This may be a voltage divider, the plate resistor of a resistance coupled amplifier unit, etc.

Image Impedances. These are the impedances at each end of the filter. In order that there will be no reflection loss, the image impedance at the load end should equal the load impedance and the image impedance at the source end should equal the source impedance.

Characteristic Impedance. This is sometimes called the *iterative impedance* and is equal to the impedance that the filter circuit offers the source. Filter circuits are generally designed so that the load impedance equals the input impedance. The image impedances for this condition are equal and also equal to the characteristic impedance. The image impedance and characteristic impedance generally vary with frequency; therefore, the two measurements must be taken at the same frequency.

Constant-k Filter. The filter circuits described up to now are of the *constant-k* type. This means that the product of the impedance of the series arm or arms and the impedance of the shunt arm or arms is constant for all frequencies and is equal to k^2. This value k is also equal to the characteristic impedance of the filter circuit over the greater portion of the passband, and

$$k = \sqrt{Z_{se}Z_{sh}} = Z_{ch} \tag{12-15}$$

where Z_{ch} = characteristic impedance, ohms
Z_{se} = impedance of series arm, ohms
Z_{sh} = impedance of shunt arm, ohms

The ideal properly terminated constant-k filter circuit acts as a resistance load throughout the passband. At the cutoff frequency the load becomes either zero or infinite, and thereafter it is imaginary. For example, in the attenuation band the ideal filter acts as a reactive load, does not take any energy from the source, and does not transmit any energy to its terminating impedance. In the ideal filter, (1) the frequencies within the passband would have zero attenuation, (2) the frequencies outside the passband would have infinite attenuation, and (3) the frequency band between passing and attenuation would be very narrow, thereby producing very sharp cutoff. Because of the resistance present in the circuit components, and because the impedances do not remain constant over the frequency range of the filter, the characteristics of the constant-k type of filter may vary considerably from the ideal filter.

When the performance of the constant-k type of filter does not fulfill the requirements of a particular application, the constant-k filter circuit may be modified by the use of additional inductors and/or capacitors in order to produce the desired operating characteristics. If the arrangement of the circuit components is varied, an infinite number of possible types of filters may be obtained; the m-derived filter is a commonly used example. Because the constant-k type of filter is the basic or elementary filter circuit, it is often referred to as being the *prototype filter*.

12-7 Other Filter Circuits

***m*-Derived Filter.** The m-derived filter circuits are variations of the basic prototype filters, and their behavior depends on a factor that is a function of a constant, m. Additional impedances are inserted into the basic circuit to form either a shunt-derived or a series-derived type of filter. If the additional impedances are added to the *shunt arm* of the section, the filter circuit is *series-derived* (Fig. 12-25). If the additional impedances are added to the *series arm* of the section, the filter circuit is *shunt-derived* (Fig. 12-26).

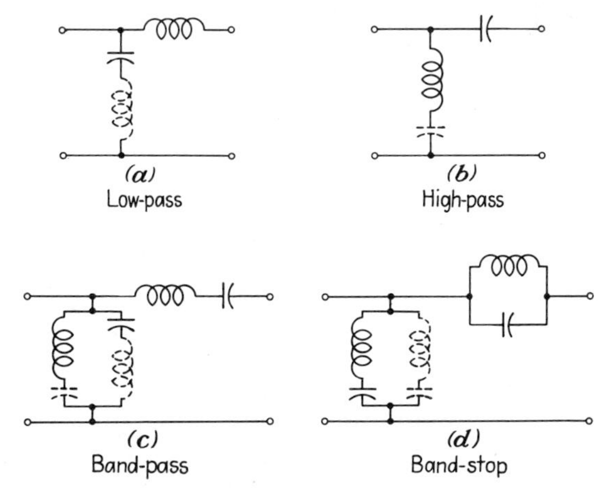

Fig. 12-25 Series-derived m-type filter circuits. The impedances shown in broken line are those added to the shunt unit to form the m-derived filter circuit.

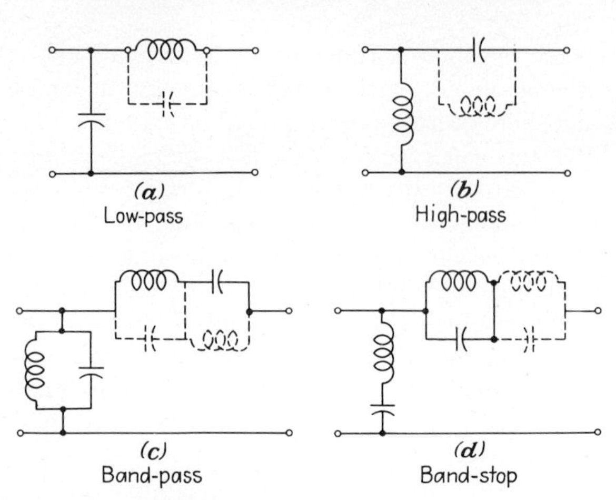

Fig. 12-26 Shunt-derived m-type filter circuits. The impedances shown in broken line are those added to the series unit to form the m-derived filter circuit.

The addition of an inductor or a capacitor in the shunt arm of the m-derived filter (Fig. 12-25a and b) introduces a series resonant circuit in the bypass path which results in higher attenuation at the resonant frequency of the shunt arm and produces a sharper cutoff characteristic for the filter. The extent of the change in the characteristics of the filter circuit depends on the value of m, which may range between zero and unity; a commonly used value is in the order of 0.6. The value of m is determined by the components of the filter circuit and is expressed by the following equations:

For low-pass filters:

$$m = \sqrt{1 - \left(\frac{f_o}{f_{rs}}\right)^2} \tag{12-16}$$

For high-pass filters:

$$m = \sqrt{1 - \left(\frac{f_{rs}}{f_o}\right)^2} \tag{12-17}$$

where f_o = cutoff frequency of the prototype filter
f_{rs} = resonant frequency of the shunt arm

From these equations, it can be seen that the lower values of m can be obtained by making the values of f_o and f_{rs} closer to each other.

The addition of an inductor or a capacitor in the series arm of the m-derived filter (Fig. 12-26a and b) introduces a parallel resonant circuit in the line path which results in higher attenuation at the resonant frequency of the series arm and produces a sharper cutoff characteristic for the filter. The value of m may

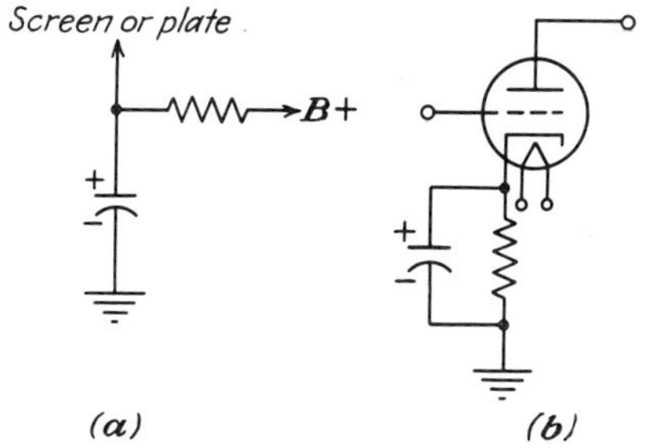

Fig. 12-27 Resistor-capacitor filter circuits. (a) Filter action in the plate or screen-grid circuit. (b) Cathode grid-bias circuit.

be determined by using the equations for obtaining m for series-derived filter circuits [Eqs. (12-16) and (12-17)].

Resistor-Capacitor Circuits. Many electronic circuits carry both alternating and direct current. A vacuum-tube circuit may carry direct current for the plate supply and an a-c signal at the same time. It is often necessary to separate the direct current and alternating signal current and to provide a path for the signal currents so that the voltage produced may be applied only to certain portions of the circuit.

One method of accomplishing this separation is to use a capacitor to provide a path for the alternating current and a resistor to provide a path for the direct current (Fig. 12-27). The circuit of Fig. 12-27*a* uses a capacitor to allow the passage of the alternating signal current from the screen-grid circuit of a tube to ground. The resistor keeps the alternating signal current from getting into the B supply, where it might cause trouble. The resistor is also used to provide the correct voltage for the screen grid by acting as a dropping resistor.

In Fig. 12-27*b*, the resistor connected between cathode and ground is used to make the cathode positive with respect to ground, which in turn makes the grid negative with respect to the cathode. This resistor offers an impedance to the signal current and may reduce it to a critical value. When this reduction in signal current is large, it may introduce degeneration, an action that should be avoided. If a capacitor is connected across the resistor as shown in Fig. 12-27*b*, it will provide a path for the alternating signal current. The diversion of the signal current from the resistor will aid in fulfilling the purpose of the resistor, namely, to provide a fixed amount of bias for the grid of the tube.

12-8 Bandpass Amplifier Circuits

Ideal Response Curve. In order to obtain high fidelity and high selectivity, the ideal response curve should have a flat top and straight sides. This ideal can be closely approximated by using two resonant circuits tuned to the same frequency and coupled to each other (Fig. 12-28). This circuit is called a *bandpass amplifier*, *bandpass filter*, or *bandpass circuit*.

The width of the frequency band that will be passed depends upon the circuit application. In a-m radio receivers this band is usually 10 kHz wide and represents a 5-kHz sideband above and below the resonant frequency. These sidebands are ptat of an a-m wave and represent the amount that the frequency varies from the resonant frequency. The frequency deviation will depend upon the frequency of the audio signal that is being received. Although the frequency of an audio signal may vary from 20 to 20,000 hertz, most commercial a-m receivers reproduce sounds only up to 5,000 hertz. Amplifiers are designed to provide specific amounts of bandwidth depending upon the application. For example, (1) a 200-kHz bandwidth is used in the i-f amplifier of an f-m radio receiver, (2) a 50-kHz bandwidth is used in the sound section of a television receiver, and (3) a 4.5-MHz bandwidth is used in the picture section of a television receiver.

The important characteristic of the bandpass amplifier circuit is the manner in which the current in the secondary circuit varies with the frequency when a constant voltage is applied to the primary circuit. The amount of current in the secondary is directly proportional to the amount of coupling between the two windings. The shape of the resonance curve will therefore be dependent on the coefficient of coupling and may be a narrow peaked curve, a flat-top curve, or a curve having two separate peaks with a valley between them.

Figure 12-29 illustrates the manner in which the current in the secondary of such a circuit varies with the amount of coupling for frequencies above and below its resonant frequency. When the coefficient of coupling is low, $K = 0.01$ for resonance at 500 kHz, the secondary current will be low and the curve will be peaked. As the amount of coupling between the two circuits is increased, the amount of secondary current will increase and there will be a reduction in the sharpness of the peak, $K = 0.015$. With critical coupling $K = 0.02$, the maximum amount of secondary current is obtained and the resonance curve will be comparatively flat at the top and will have steep sides. With tight coupling, the coupled impedance at resonance is high, thus reducing the primary current, which in turn reduces the amount of voltage induced into the secondary, thus reducing the secondary current. This accounts for the decrease in secondary currents at resonance for coefficients of coupling greater than the critical value as indicated by the curves in Fig. 12-29. The reactance coupled into the primary is inductive for frequencies below resonance and capacitive for frequencies above resonance. This reactance is opposite to that of the primary circuit and will therefore reduce the equivalent impedance offered to the applied voltage. The primary current, and therefore the voltage

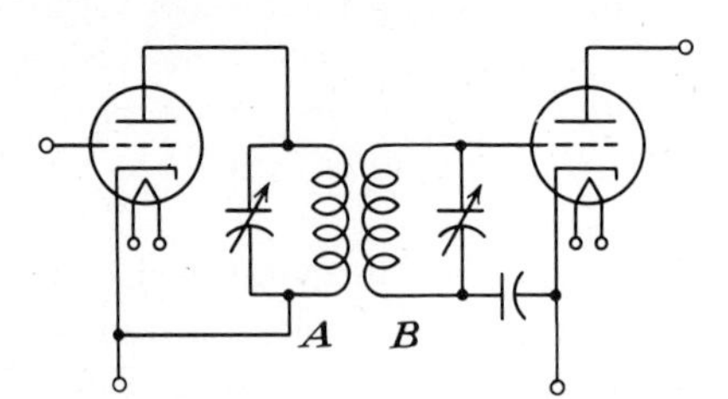

Fig. 12-28 Bandpass amplifier circuit.

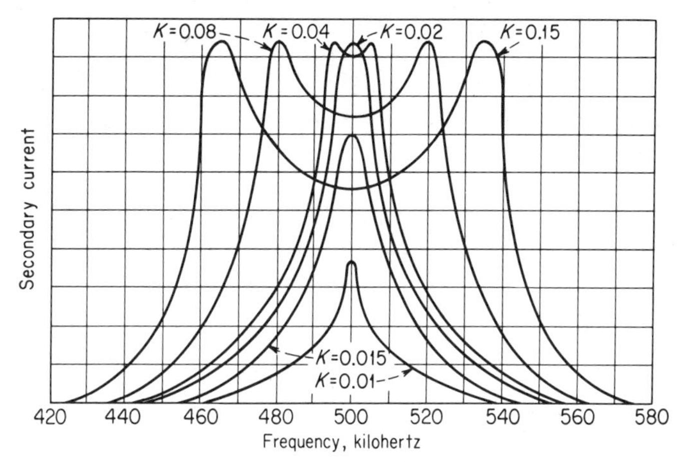

Fig. 12-29 Curves for a bandpass amplifier circuit showing the variation in secondary current with frequency.

induced into the secondary, will increase for frequencies off resonance. When the coupling is tight, this action will introduce new resonant frequencies above and below resonance corresponding to the amount of secondary current. This accounts for the peaks in the resonance curves when the coupling is greater than the critical value. The current at these peaks is practically the same as the peak current with critical coupling. The spacing between these peaks is directly proportional to the coefficient of coupling.

Width of Bandpass. The maximum possible transfer of energy to the secondary at the resonant frequency is obtained with critical coupling. With tight coupling, a fairly constant secondary current and voltage can be obtained for a narrow band of frequencies. The width of this band, measured at 0.707 of the maximum response, is directly proportional to the coefficient of coupling and the resonant frequency of the tuned circuits. An approximate value of the bandwidth can be obtained by

$$\text{Width of bandpass} = K \times f_r \tag{12-18}$$

where K = coefficient of coupling
f_r = resonant frequency of the tuned circuits

This equation indicates that the larger the value of K, the wider the bandpass will be. The coupled impedance will also increase with an increase in the amount of coupling, thus causing a decrease in the output current (Fig. 12-29).

Example 12-6 What is the approximate bandwidth of a bandpass filter circuit having a resonant frequency of 456 kHz and a coefficient of coupling of 0.02?

GIVEN: $K = 0.02 \qquad f_r = 456$ kHz

FIND: Width of bandpass

SOLUTION:

$$\text{Width of bandpass} = K \times f_r = 0.02 \times 456 = 9.12 \text{ kc} = 9{,}120 \text{ hertz}$$

The most important properties of a bandpass circuit are the width of the band of frequencies it allows to pass and the uniformity of response within this band. Referring to Fig. 12-29 it can be seen that, when $K = 0.02$, the secondary current is fairly constant for a band of frequencies between 495 and 505 kHz. The response will therefore be uniform for this band of frequencies with a coefficient of coupling of 0.02 at 500 kHz. As the coefficient of coupling is increased ($K = 0.04$; $K = 0.08$; $K = 0.15$), the band becomes wider and less uniform.

The coefficient of coupling of bandpass circuits is usually of such a value that uniform response is obtained for the band of frequencies to be passed. The response should decrease rapidly for frequencies beyond these limits (Fig. 12-29).

The uniformity of response within the band of frequencies to be passed is dependent on the circuit Q and the value of the coefficient of coupling. The equation for finding the amount of coupling required to produce maximum transfer of energy is usually expressed in terms of the primary and secondary Q as follows:

$$K_c = \frac{1}{\sqrt{Q_p Q_s}} \tag{12-19}$$

where K_c = critical coupling
Q_p = Q of the primary circuit
Q_s = Q of the secondary circuit
$\sqrt{Q_p Q_s}$ = Q of the complete circuit

Figure 12-30 shows that, if the circuit Q is too high, pronounced double peaks occur and, if too low, the response curve is round instead of flat. Experiments have shown that the best value of Q is approximately 50 per cent more than that required to produce critical coupling. Substituting this value in Eq. (12-19) and solving for the circuit Q,

$$\sqrt{Q_p Q_s} = \frac{1.5}{K_c} \tag{12-20}$$

or

$$Q_p Q_s = \frac{2.25}{K_c^2} \tag{12-21}$$

The coefficient of coupling required may be obtained by transposing Eq. (12-18), as

$$K = \frac{\text{width of bandpass}}{f_r} \tag{12-22}$$

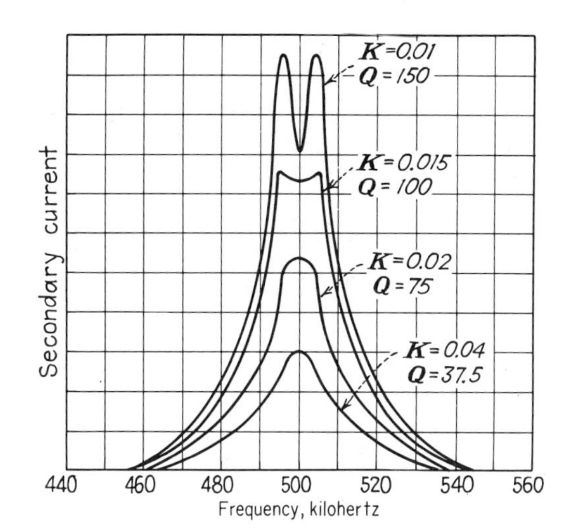

Fig. 12-30 Characteristics of a bandpass amplifier circuit showing the effect of circuit Q on uniformity of response within the band being passed.

The resonant frequency of the i-f bandpass amplifier is approximately 460 kHz for a-m radio receivers, 10.7 MHz for f-m radio receivers, and 44 MHz for television receivers.

Example 12-7 A bandpass amplifier circuit used in an a-m radio receiver is tuned to a resonant frequency of 465 kHz. What values of circuit Qs are necessary to produce uniform response for a 10-kHz band?

GIVEN: Width of bandpass = 10 kHz $\quad f_r = 465$ kHz

FIND: Q_p, Q_s

SOLUTION:

$$K = \frac{\text{width of bandpass}}{f_r} = \frac{10}{465} = 0.0215$$

$$Q_pQ_s = \frac{2.25}{K_c^2} = \frac{2.25}{(0.0215)^2} = 4{,}870$$

If $Q_p = Q_s$, then

$$Q_p = \sqrt{4{,}870} \cong 70$$

Substituting values of 200 kHz for the width of the bandpass and 10.7 MHz for the frequency of resonance, it will be evident that the coupling required for the i-f bandpass amplifiers of f-m radio receivers is approximately 0.0187 and the Q of each coil is approximately 80.

12-9 Wide-bandpass Amplifier Circuits

Methods Used to Obtain a Wide Bandpass. The width of the band of frequencies to be passed in the r-f and i-f amplifier circuits of f-m radio receivers and television receivers is much greater than that which can normally be obtained by the use of a tuned circuit consisting only of a capacitor and an inductor. In order to obtain a wide bandpass, one or more of the following methods are generally used: (1) decreasing the value of the circuit Q, (2) increasing the value of the coefficient of coupling, (3) stagger tuning.

The value of the circuit Q for a series-resonant circuit [Eq. (11-8)] can be decreased by (1) adding resistance to the external circuit thereby increasing the value of R, or (2) changing the value of the inductance, capacitance, or both in such a manner that will decrease the L/C ratio. The effect of various values of Q on the width of the bandpass was presented in Arts. 11-4 and 11-5 and is also illustrated by Figs. 11-1 and 11-2.

Obtaining a wide bandpass by increasing the value of the coefficient of coupling has been discussed in the previous article. The relation between the coefficient of coupling and the width of the bandpass is illustrated in Fig. 12-29 and can be seen by a study of Eq. (12-18).

A wide bandpass can also be obtained by stagger-tuning several stages of tuned amplifier circuits. In this method, each tuned circuit in the various stages of amplification involved is tuned to a different value of frequency, but all are within the desired range of the band to be passed. In order to obtain the overall characteristics desired, the Q of each circuit is adjusted to produce a certain amount of overlap.

Bandpass in F-M Receivers. A wide bandpass (Fig. 12-31) may be obtained by using either stagger tuning or overcoupled i-f transformers. When overcoupling is employed to obtain a wide bandpass, three i-f transformers tuned to the same frequency are generally used. The first and third transformers are

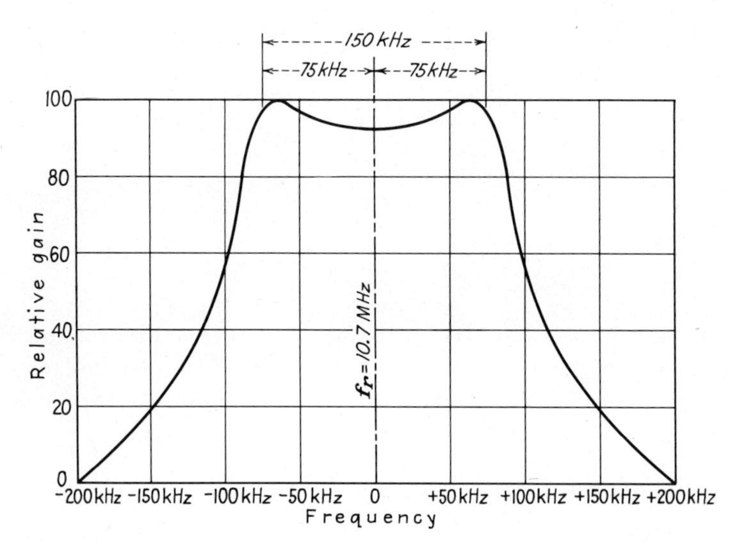

Fig. 12-31 A double-peak i-f response curve for an f-m radio receiver produced by overcoupling.

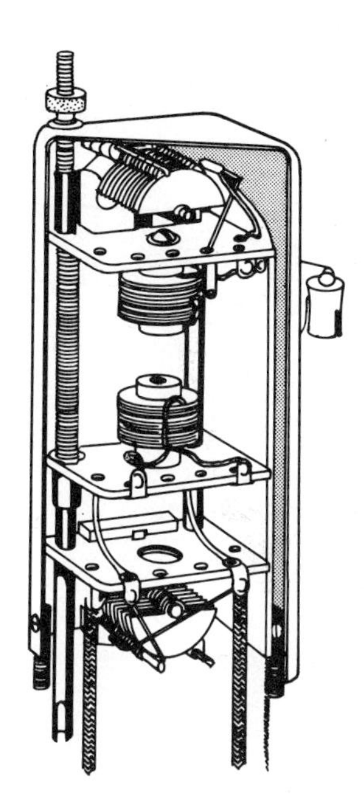

Fig. 12-32 Variable-coupling i-f transformer used to adjust the width of bandpass.

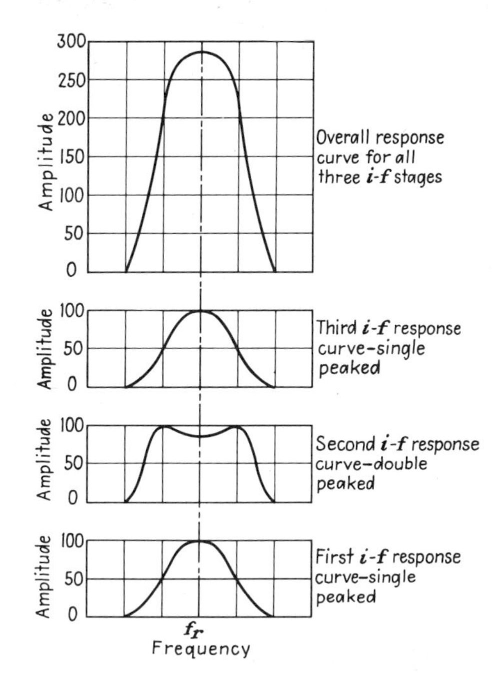

Fig. 12-33 Wide passband for the i-f reponse curve in an f-m radio receiver obtained by use of three stages with different values of coupling.

single-peaked and operate just under critical coupling. The second transformer is overcoupled to produce a double-peaked response curve (Fig. 12-33).

Bandpass in Television Receivers. The overall bandpass of the i-f section of a television receiver is not made symmetrical because of the vestigial-sideband method of transmission. The ideal response curve for reception of this type of transmission is shown in Fig. 12-34. The operating frequency of i-f transformers is approximately 40 MHz, and the width of its unsymmetrical bandpass approximates 4 MHz. The ideal response curve is difficult to obtain; however, if various combinations of stagger tuning are used, a close approximation can be achieved. The combination used will vary with the designer. In general, three or more stages of tuned amplification are used. Each circuit may be tuned to a single frequency or may be overcoupled to produce peaks at two (or more) frequencies within the desired range of the bandpass. It is beyond the scope of this text to analyze each of the variations that may be used. An example of one of the combinations that may be used is illustrated by the response curves for a four-stage stagger-tuned i-f amplifier circuit shown in Fig. 12-35. The second and third i-f stages are tuned to the upper end of the desired passband. The first and fourth i-f stages and the mixer are tuned to the lower end of the desired

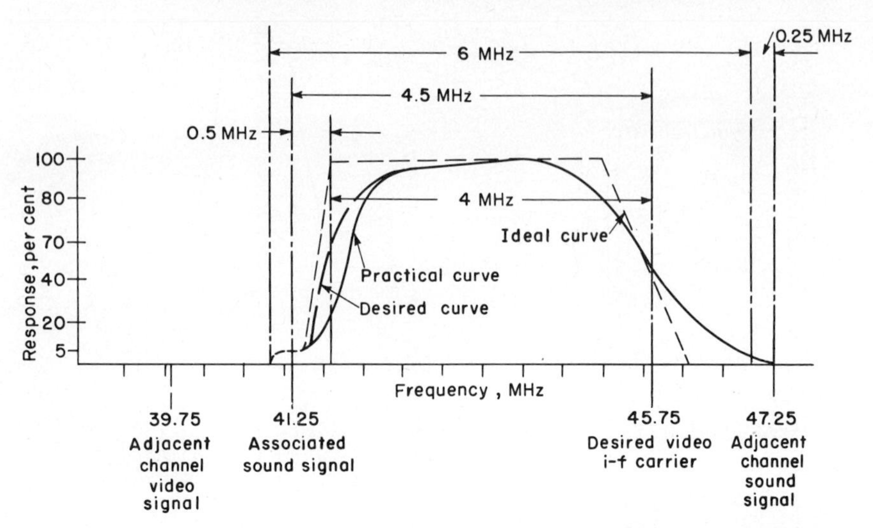

Fig. 12-34 The ideal, desired, and practical i-f response curves for the reception of vestigial-side-band transmission of television signals.

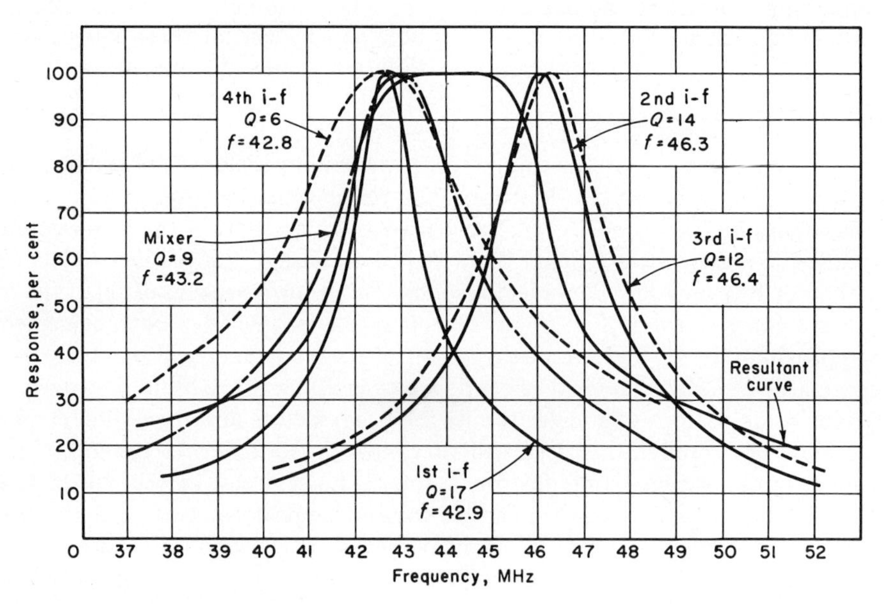

Fig. 12-35 Wide passband for the i-f response curve of a television receiver obtained by use of stagger tuning.

passband. The resultant overall response curve approximates the desired curve of Fig. 12-34.

12-10 Ceramic Filters

Use. When space, weight, reliability, and a relatively high sensitivity are important requirements, a ceramic filter may be used in place of an LC tuned circuit. Ceramic filters constructed to have bandpass characteristics similar to an LC tuned circuit can be used in the amplifier circuits described in Arts. 12-8 and 12-9. In some communication equipment, ceramic filters (often having the appearance of a capacitor) have replaced the r-f and i-f transformers.

Principle. When the material of a ceramic filter is placed under a mechanical strain, such as compression, expansion, bending, and/or twisting, the mechanical energy expended is converted to electrical energy. This action is called the *piezoelectric effect,* and as the process is reversible the ceramic material will be set in vibration when an alternating voltage is applied. The frequency of vibration is determined by the structural characteristics of the ceramic material used. Thus, a properly designed piece of ceramic material (lead-zirconate and lead-titanate) can be used to simulate the characteristics of a tuned circuit (see also Art. 17-22).

Electrical Characteristics. Figure 12-36*c* illustrates the equivalent electric circuit of a ceramic filter. The bandpass characteristics of this simple filter are approximately the same as the curves of Fig. 12-29 with values of K from 0.01 to 0.02. By using two or more simple filters in series and/or parallel combinations, a wide variety of bandpass characteristics can be obtained. For example,

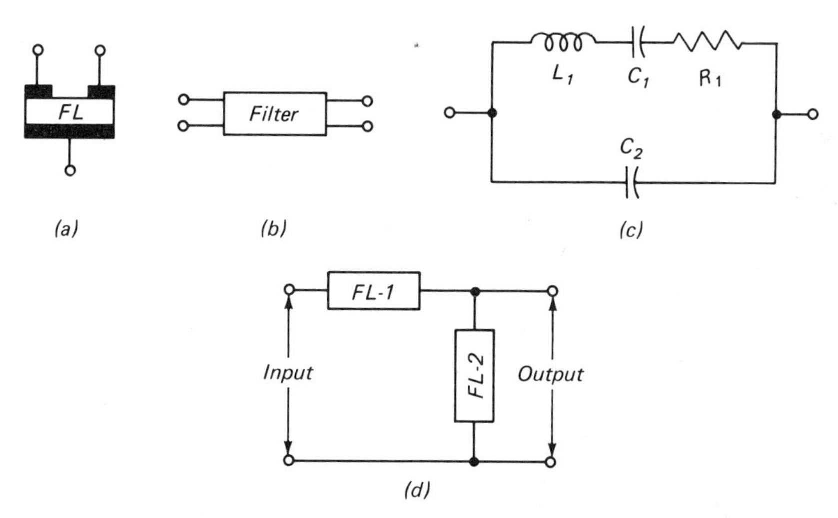

Fig. 12-36 Ceramic filters. (a) Drawing symbol. (b) Simplified schematic symbol. (c) Equivalent electric circuit. (d) Two filters used to improve the bandpass characteristics.

the two filters connected as in Fig. 12-36*d* will produce bandpass characteristics somewhat similar to the curves of Fig. 12-29 with values of K from 0.04 to 0.08. The high electromechanical coupling of the ceramic filter circuit produces values of Q from 450 to 1,600, which makes it possible to construct a wide-bandpass filter with low insertion loss. The higher the Q the wider the bandpass and the lower the insertion loss.

Ceramic filter circuits can be designed for center frequency values of from 50 Hz to many megahertz and with various bandpass characteristics. A 455-kHz ceramic filter can be made in the form of a disk approximately 0.25 inch in diameter and 0.02 inch thick. As the frequency characteristics of these filters are fixed and have a stability in the order of 0.2 per cent, care must be exercised in selecting ceramic filters for original design and replacement.

12-11 Attenuators

Types of Attenuators. A resistance network that is used to reduce voltage, current, or power in controllable and known amounts that are independent of frequency is called an *attenuator*. The resistors in an attenuator circuit may be connected in various ways: (1) so that the input and output impedances are equal, (2) so that the input and output impedances are unequal, and (3) to provide different amounts of attenuation. Some of the ways in which resistors are connected to form an attenuator are the L, T, H, π, O, ladder, and bridged-T types of circuits (Fig. 12-37). When only fixed resistors are used in the network, the attenuation is a fixed amount that is determined by the type of network and the values of the resistors used. This type of network is sometimes called a *fixed attenuator pad*. When some or all of the resistors used in a network are variable, the amount of attenuation can be controlled by varying one or more of the variable resistors. This type of network is sometimes called a *variable attenuator pad*.

The amount of attenuation produced by a network may be expressed by

$$\alpha = \frac{I_L}{I_{L\cdot a}} \tag{12-23}$$

where α = image transfer constant
I_L = load current without attenuator
$I_{L\cdot a}$ = load current with attenuator

L Attenuator. The L attenuator is an unbalanced network that consists of two resistors, one placed in either of the series arms and the other in the shunt arm (Fig. 12-37*a* and *b*). These resistors have such values that one of the image impedances of the attenuator remains constant regardless of the amount of attenuation. The L attenuator will therefore maintain an impedance that is independent of attenuation at either the input or output terminals. Because of this characteristic, the L attenuator is generally used when several different loads are connected to a common power source and the amount of power delivered to each load must be varied without affecting the impedance offered to the signal source.

T Attenuator. The T attenuator is an unbalanced network in which three resistors are arranged to form the letter T (Fig. 12-37*c*). In this type of attenuator, resistors in the series arms, R_1 and R_3, are generally identical in value in order to make the input image impedance R_i equal to the output image impedance R_o. The T attenuator is generally used where the amount of attenuation or the presence of the network has no effect upon the impedance relations of the circuit. This is accomplished by making the image impedance of the attenuator equal to R_i or R_o. This type of attenuator will therefore maintain a constant impedance at both the input and output terminals irrespective of the amount of attenuation.

H Attenuator. The H attenuator is a balanced T network in which five resistors are arranged to form the letter H when viewed sidewise (Fig. 12-37*d*). The balanced H network can be made to produce the same operating characteristics as the unbalanced T network. This is accomplished by making the resistance of all the resistors in the series arms equal to each other and also equal to one-half the value required for a resistor to be used in the series arm of a T attenuator. In the circuit shown in Fig. 12-37*d*, $R_1 = R_3 = R_4 = R_5 =$ one-half the resistance of either R_1 or R_3 of Fig. 12-37*c*.

π Attenuator. The π attenuator is an unbalanced network containing three resistors connected to form the Greek letter π (Fig. 12-37*e*). In this type of attenuator, the resistors in the shunt arms, R_1 and R_3, are usually identical in value in order to make the input image impedance equal to the output image impedance. The π attenuator can be designed to produce the same operating characteristics as the T attenuator, namely, that the impedance relations of the circuit are not affected by any changes in attenuation or the presence of the network.

O Attenuator. The O attenuator is a balanced π network containing four resistors connected to form the letter O (Fig. 12-37*f*). The balanced O network can be made to produce the same operating characteristics as the unbalanced π network. This is accomplished by making the two resistors in the series arms identical in value and equal to one-half the value required of a resistor to be used in the series arm of a π attenuator. In the circuit shown in Fig. 12-37*f*, $R_4 = R_5 =$ one-half the resistance of R_2 of Fig. 12-37*e*.

Ladder Attenuator. A ladder attenuator consists of a series of two or more symmetrical π-section networks. The arrangement shown in Fig. 12-37*g* consists of three cascaded π sections. The values of resistance used for the shunt and series arms of each section are such that they will produce image impedances at each end that are equal to the input and output image impedances. It should be noted that R_1, R_2, and R_3 represent the series arms of π sections 1, 2, and 3, respectively; R_4 and R_7 the shunt arms of sections 1 and 3, respec-

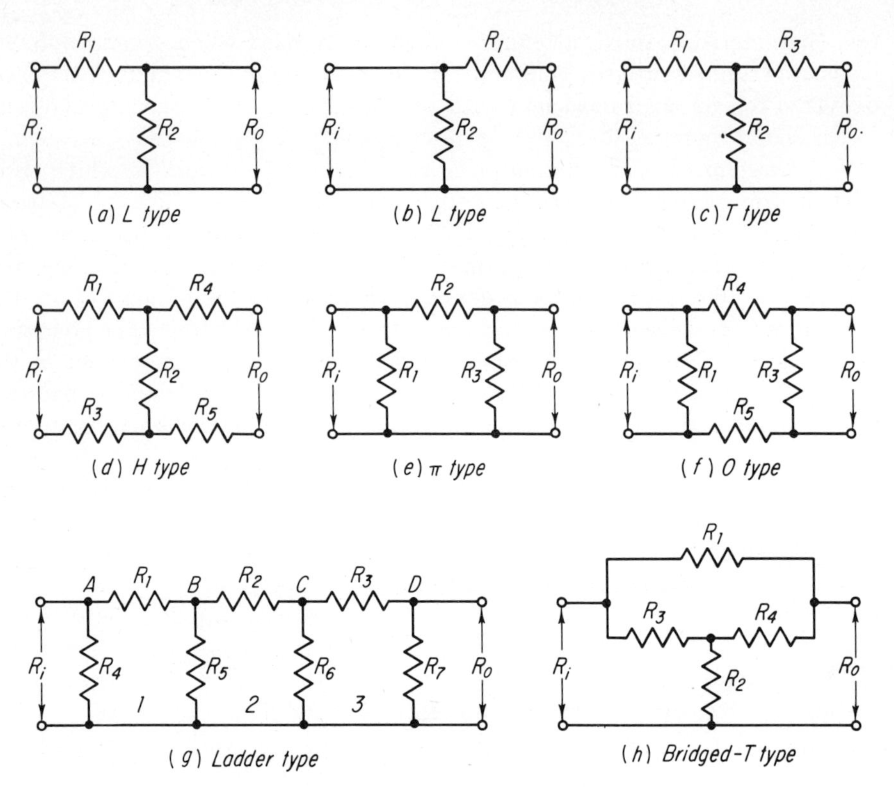

Fig. 12-37 Attenuator networks.

tively; R_5 and R_6 the equivalent resistance of the two adjacent shunt arms for sections 2 and 3, respectively. The impedance between any junction point and the common side of a ladder attenuator is equal to one-half the image impedance.

The total attenuation of a ladder network is equal to the product of the attenuation of each section. Thus, if the image transfer constant of each section is 10, then each section reduces the load current by $\frac{1}{10}$ and the total attenuation of the three-section attenuator would be $\frac{1}{10} \times \frac{1}{10} \times \frac{1}{10}$, or $\frac{1}{1,000}$. Ladder networks are generally used in electronic equipment, such as signal generators, where it is required that the voltage and current be reduced in known ratios. When the various junction terminals of the ladder attenuator of Fig. 12-37*g* are connected to the terminals of a rotary deck switch and values are used to produce an α of 10, attenuations of 0, $\frac{1}{10}$, $\frac{1}{100}$, and $\frac{1}{1,000}$ can be obtained.

Bridged-T Attenuator. A bridged-T attenuator consists of four resistors connected as shown in Fig. 12-37*h*. To obtain a constant value of image impedance that is independent of attenuation, the four resistors should have values of $R_3 = R_4 = R_i$ and $R_1R_2 = R_i^2$. This type of attenuator is equivalent to the simple T attenuator. However, to obtain variable attenuation with constant values of image impedance the bridged-T network requires the variation of only two resistors R_1 and R_2 while the simple T network requires that all three resistors be varied.

Attenuator Circuit Calculations. The design of attenuators is a specialized area in electronics. The following equations are for only the L-, T-, and π-type attenuators; data for the other types of attenuators may be obtained from specialized reference books featuring attenuators.

For the simple resistor-type L attenuator circuit of Fig. 12-37*a*, supplying power to a load of R_L,

$$R_1 = \frac{R_L(K-1)}{K} \tag{12-24}$$

$$R_2 = \frac{R_L}{K-1} \tag{12-25}$$

$$K = \frac{e_i}{e_o} \tag{12-26}$$

Example 12-8 A certain 1,000-ohm load is supplied by a 100-volt d-c power source. It is desired to reduce the load voltage to 10 volts by means of an L-type attenuator similar to Fig. 12-37*a* and still retain a 1,000-ohm load condition at the power source. (*a*) What are the values of K, R_1, and R_2? Using the values obtained in part (*a*), (*b*) calculate the resistance at the power source terminals, (*c*) check the output voltage and the value of K.

GIVEN: $R_L = 1{,}000$ ohms $e_i = 100$ volts $e_o = 10$ volts

FIND: (*a*) K, R_1, R_2 (*b*) R_i (*c*) e_o, K

SOLUTION:

(*a*)
$$K = \frac{e_i}{e_o} = \frac{100}{10} = 10$$

$$R_1 = \frac{R_L(K-1)}{K} = \frac{1{,}000(10-1)}{10} = 900 \text{ ohms}$$

$$R_2 = \frac{R_L}{K-1} = \frac{1{,}000}{10-1} = 111.1 \text{ ohms}$$

(*b*)
$$R_i = R_1 + R = 900 + 100 = 1{,}000 \text{ ohms}$$

where
$$R = \frac{R_2R_L}{R_2 + R_L} = \frac{111.1 \times 1{,}000}{111.1 + 1{,}000} = 100 \text{ ohms}$$

(*c*)
$$e_o = \frac{e_iR}{R_1 + R} = \frac{100 \times 100}{900 + 100} = 10 \text{ volts}$$

$$K = \frac{e_i}{e_o} = \frac{100}{10} = 10$$

For the simple resistor-type T attenuator circuit of Fig. 12-37*c*, supplying power to a load of R_L,

$$R_1 = R_3 = R_L \frac{K-1}{K+1} \tag{12-27}$$

$$R_2 = \frac{2R_LK}{K^2-1} \tag{12-28}$$

$$K = \frac{e_i}{e_o} \tag{12-26}$$

Example 12-9 A certain 2,000-ohm load is supplied by a 50-volt d-c power source. It is desired to reduce the load voltage to 25 volts by means of a T-type attenuator similar to Fig. 12-37*c* and still retain a 2,000-ohm load condition at the power source. (*a*) What are the values of K, R_1, R_2, and R_3? (*b*) By using the values obtained in part (*a*) calculate the resistance at the power source terminals.

GIVEN: $R_L = 2{,}000$ ohms $\quad e_i = 50$ volts $\quad e_o = 25$ volts

FIND: (*a*) K, R_1, R_2, R_3 $\quad$ (*b*) R_i

SOLUTION:

(*a*)
$$K = \frac{e_i}{e_o} = \frac{50}{25} = 2$$

$$R_1 = R_3 = R_L\frac{K-1}{K+1} = 2{,}000 \times \frac{2-1}{2+1} = 666.6 \text{ ohms}$$

$$R_2 = \frac{2R_LK}{K^2-1} = \frac{2 \times 2{,}000 \times 2}{(2 \times 2) - 1} = 2{,}666.6 \text{ ohms}$$

(*b*)
$$R_i = R_1 + \frac{R_2(R_3 + R_L)}{R_2 + (R_3 + R_L)} = 666.6 + \frac{2{,}666.6(666.6 + 2{,}000)}{2{,}666.6 + 666.6 + 2{,}000}$$
$$= 666.6 + 1{,}333.3 \cong 2{,}000 \text{ ohms}$$

For the simple resistor-type π attenuator circuit of Fig. 12-37*e*, supplying power to a load of R_L,

$$R_1 = R_3 = R_L \frac{K+1}{K-1} \tag{12-29}$$

$$R_2 = R_L \frac{K^2-1}{2K} \tag{12-30}$$

$$K = \frac{e_i}{e_o} \tag{12-26}$$

Example 12-10 A certain 12,000-ohm load is supplied by a 250-volt d-c power source. It is desired to reduce the load voltage to 50 volts by means of a π-type attenuator similar to Fig. 12-37*e* and still retain a 12,000-ohm load condition at the power source. (*a*) What are the values of K, R_1, R_2, and R_3? (*b*) By using the values obtained in part (*a*) calculate the resistance at the power source terminals.

GIVEN: $R_L = 12{,}000$ ohms $\quad e_i = 250$ volts $\quad e_o = 50$ volts

FIND: (a) K, R_1, R_2, R_3 $\quad$ (b) R_i

SOLUTION:

(a)
$$K = \frac{e_i}{e_o} = \frac{250}{50} = 5$$

$$R_1 = R_3 = R_L \frac{K+1}{K-1} = 12{,}000 \times \frac{5+1}{5-1} = 18{,}000 \text{ ohms}$$

$$R_2 = R_L \frac{K^2-1}{2K} = 12{,}000 \times \frac{(5 \times 5) - 1}{2 \times 5} = 12{,}000 \times 2.4 = 28{,}800 \text{ ohms}$$

(b)
$$R_i = \frac{R_1(R_2 + R)}{R_1 - (R_2 + R)} = \frac{18{,}000(28{,}800 + 7{,}200)}{18{,}000 + 28{,}800 + 7{,}200} = 12{,}000 \text{ ohms}$$

where
$$R = \frac{R_L R_3}{R_L + R_3} = \frac{12{,}000 \times 18{,}000}{12{,}000 + 18{,}000} = 7{,}200 \text{ ohms}$$

12-12 Delayed-action Circuits

Inductors or capacitors may be used in electric and electronic circuits to control the time required for the current or voltage to reach a predetermined value. The operation of these circuits is based on the time constant of the resistance-inductance or the resistance-capacitance combination. These circuits are generally referred to as *RL* and *RC* circuits.

Time Constant of Resistance-inductance Circuits. If an inductor, which may be considered as a resistance and inductance in series, is connected to a d-c power source, a current will flow in the circuit. The amount of current that will flow will be its Ohm's law value, namely, the voltage applied to the circuit divided by the resistance of the circuit. In a circuit having only resistance (Fig. 12-38*a*), the current will rise to its Ohm's law value practically

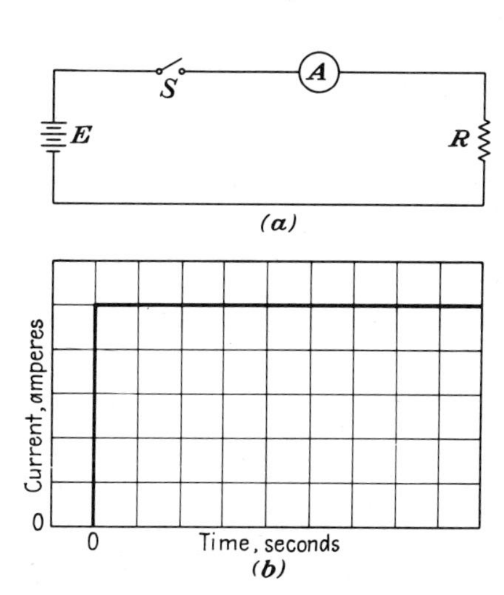

Fig. 12-38 Characteristics of current versus time for a circuit containing only resistance. (a) The circuit. (b) Current versus time characteristics.

instantaneously, as indicated in Fig. 12-38*b*. However, as the inductor has the effect of a resistance and inductance connected in series (Fig. 12-39*a*), the current will require an appreciable amount of time to reach its Ohm's law value, as is shown in Fig. 12-39*b*. This is explained by the fact that, for the current to reach its final value of 5 amp, it must progressively pass through its lesser values such as 1, 2, 3, and 4 amp. Under these conditions, the current is changing in amount, and the circuit will have an emf induced in it owing to the self-inductance of the circuit. This induced emf will oppose the impressed voltage and thus will prevent the current from reaching its Ohm's law value as long as the induced emf is present. The current will, however, eventually reach its Ohm's law value, the time required to accomplish this depending upon the relative values of the inductance and resistance. The current increases in a manner indicated by the graph shown in Fig. 12-39*b* and will rise to 63.2 per cent of its final value in a period of time, expressed in seconds, equal to the inductance of the circuit divided by the resistance of the circuit. This time interval is called the *time constant* of the circuit and is expressed mathematically as

$$t = \frac{L}{R} \tag{12-31}$$

Example 12-11 An RL circuit is used to control the time of closing a relay. The relay closes when the current reaches 63.2 per cent of its final value and the circuit resistance and inductance are 12 ohms and 2.4 henrys, respectively. What is the time interval between the closing of the line switch and the operation of the relay?

GIVEN: $R = 12$ ohms $\quad L = 2.4$ henrys

FIND: t

SOLUTION:

$$t = \frac{L}{R} = \frac{2.4}{12} = 0.2 \text{ sec}$$

The time required for the current to reach values other than 63.2 per cent of the final value follows a curve known mathematically as an *exponential curve*. The universal time-constant curves of Fig. 12-43 provide a simple means of finding the current at any instant of time.

Further analysis of the RL circuit will show that, when the current is increasing, the voltage drop across the resistance will increase at the same time rate as the current. This is so because the voltage drop across the resistance at any instant of time is equal to the product of the current and the resistance. Furthermore, as the sum of the voltages around the circuit must be equal to the applied voltage, the induced emf at any instant of time due to the inductance must be equal to the applied voltage less the IR drop. Figure 12-39*c* shows the voltage characteristics of the circuit when the current is building up.

The circuit shown in Fig. 12-40*a* is arranged so that the RL circuit may

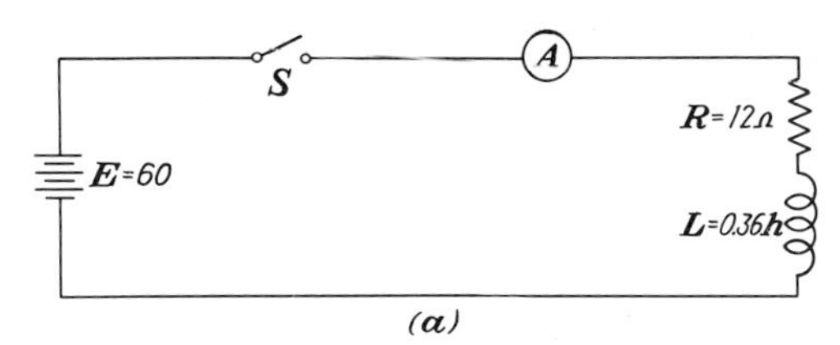

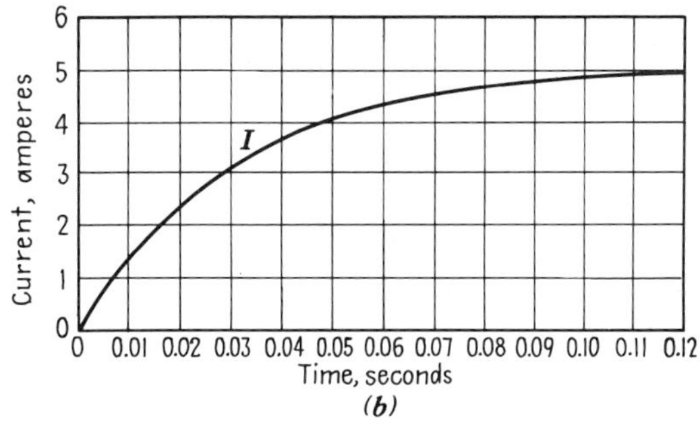

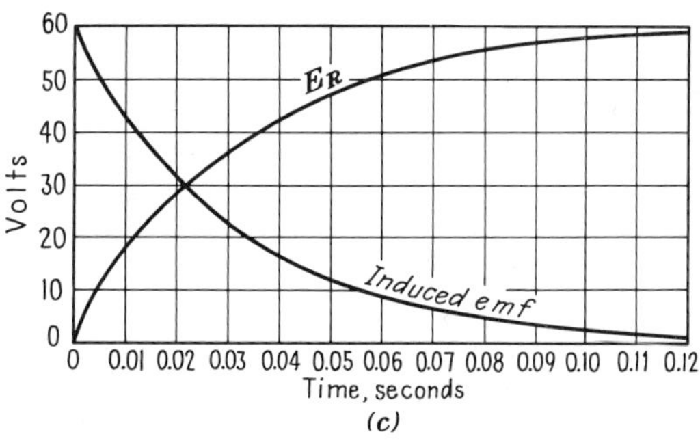

Fig. 12-39 Characteristics of current and voltage versus time for a circuit containing resistance and inductance. (a) The circuit. (b) Current versus time characteristics immediately after closing switch S. (c) Voltage versus time characteristics immediately after closing switch S.

either be connected to the d-c power source or be connected so that the inductance will be short-circuited through the resistance R. If the circuit is connected to the d-c power source, a current will flow in the circuit and will rise to its Ohm's law value according to the current-time curve previously described. When a current is flowing in the circuit, energy is transferred to the magnetic field. If the switch S is changed from position 1 to position 2, so that the inductance is disconnected from the power source and then instantaneously short-circuited across the resistance, the energy in the collapsing magnetic field will induce a voltage in the turns of the coil and cause a current to flow in the circuit. The current will decrease as the energy is dissipated in the resistance. The rate at which the current decreases will depend upon the relative values of the inductance and the resistance. The current-time changes will also follow an exponential curve but will be a descending curve. As the inductance is now actually in parallel with the resistance, the resistance voltage drop and the induced emf will be equal in value and will decrease according to an exponential curve. The current-time characteristics are shown in Fig. 12-40*b*, and the voltage-time characteristics are shown in Fig. 12-40*c*. The time in seconds as determined by dividing L by R represents the time in which the current (and voltage) decreases 63.2 per cent; hence the current and voltage will drop to 36.8 per cent of their maximum values in L/R sec. The time required for the current and voltage to decrease to values other than 36.8 per cent of their maximum values can be found by use of the universal exponential curves presented at the end of this article.

Time Constant of Resistance-capacitance Circuits. If a perfect capacitor, that is, one having no resistance, is connected to a d-c power source, a high current surge will flow instantly and will charge the capacitor. As the capacitor becomes charged almost instantaneously, the amount of current flow will decrease rapidly. The capacitor will charge to the value of the impressed voltage, and the current flow will diminish to zero practically instantaneously.

If the circuit contains resistance in addition to the capacitance (Fig. 12-41*a*),

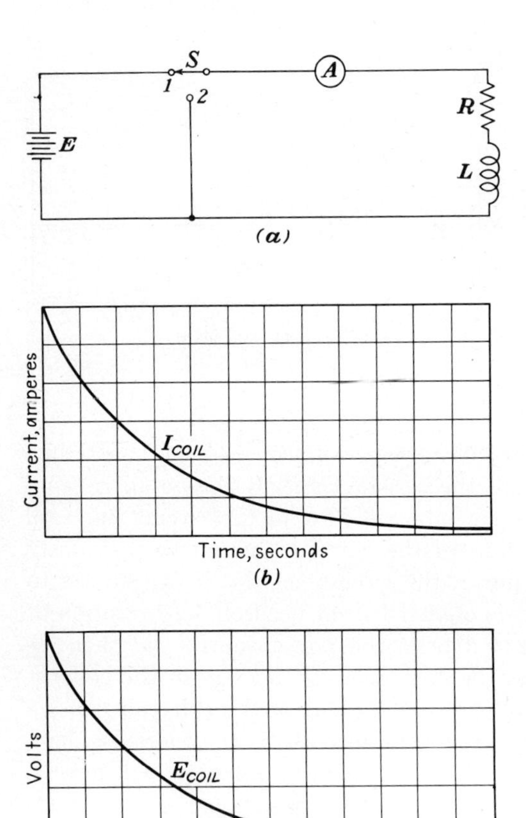

Fig. 12-40 Characteristics of current and voltage versus time. (a) The circuit. (b) Current versus time characteristics immediately after switch *S* is placed in position 2. (c) Voltage versus time characteristics immediately after switch *S* is placed in position 2.

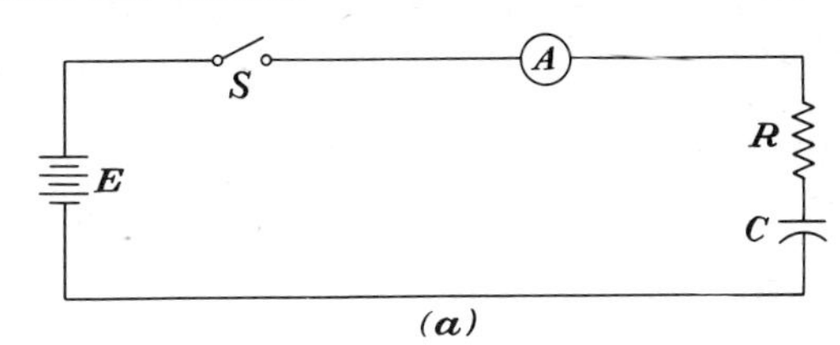

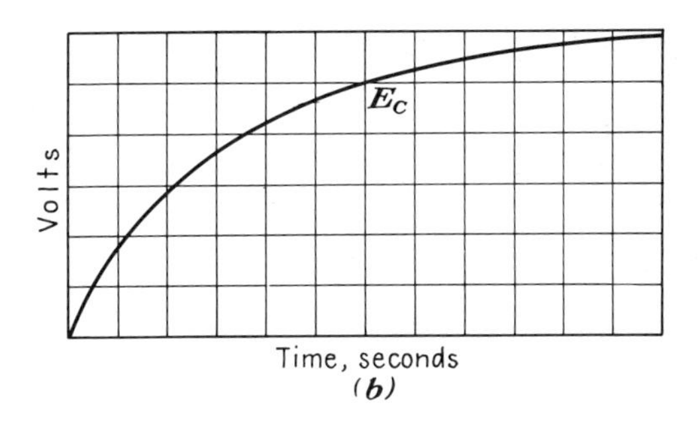

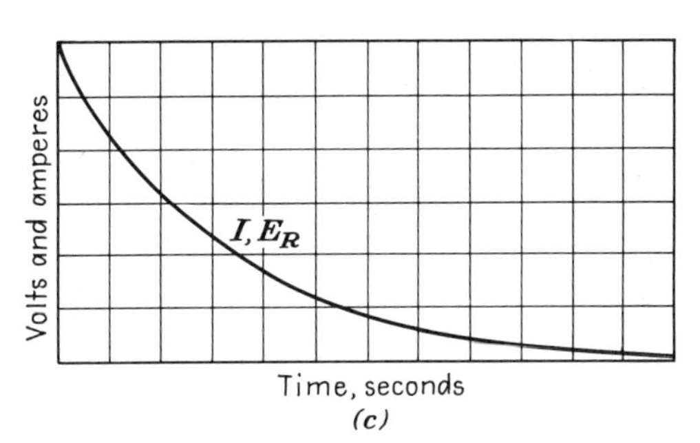

Fig. 12-41 Characteristics of current and voltage versus time for a circuit containing resistance and capacitance. (a) The circuit. (b) Capacitor volts versus time characteristics immediately after closing switch *S*. (c) Current and resistor volts versus time characteristics immediately after closing switch *S*.

the capacitor will become charged to the same value of voltage but will require a longer period of time to reach its final value. The voltage increases in a manner indicated by the graph shown in Fig. 12-41*b* and will rise to 63.2 per cent of its final value in a period of time, expressed in seconds, equal to the product of the capacitance and resistance of the circuit. This is called the *time constant* of the circuit and is expressed mathematically as

$$t = CR \qquad (12\text{-}32)$$

Example 12-12 What is the time constant of an automatic-volume-control filter circuit that uses a 1.25-megohm resistor and a 0.25-μf capacitor?

GIVEN: $R = 1.25$ megohms $\qquad C = 0.25\ \mu$f

FIND: t

SOLUTION:

$$t = CR = 0.25 \times 10^{-6} \times 1.25 \times 10^{6} = 0.3125 \text{ sec}$$

The time required for the voltage to reach values other than 63.2 per cent of the final value follows an exponential curve. The universal time-constant curves of Fig. 12-43 provide a simple means of finding the voltage at any instant of time.

If the switch S of Fig. 12-42*a* is closed to position 1, the voltage and current characteristics of the circuit will conform to the voltage-time and current-time curves shown in Fig. 12-41. While a current is flowing in the circuit, energy is being stored in the capacitor. If the switch S is changed from position 1 to position 2, the energy stored in the capacitor will cause a current to flow and the capacitor will discharge through the resistor. At the instant of closing the switch the current will be at its highest value (Ohm's law value) and will decrease exponentially, as shown in Fig. 12-42*b*. The voltage across the capacitor and resistor will be equal in amount and will also decrease exponentially with time, as shown in Fig. 12-42*c*.

Universal Time-constant Curves. The time required for the current of an *RL* circuit or the voltage across the capacitor of an *RC* circuit to reach values other than 63.2 per cent of their final values may be determined mathemati-

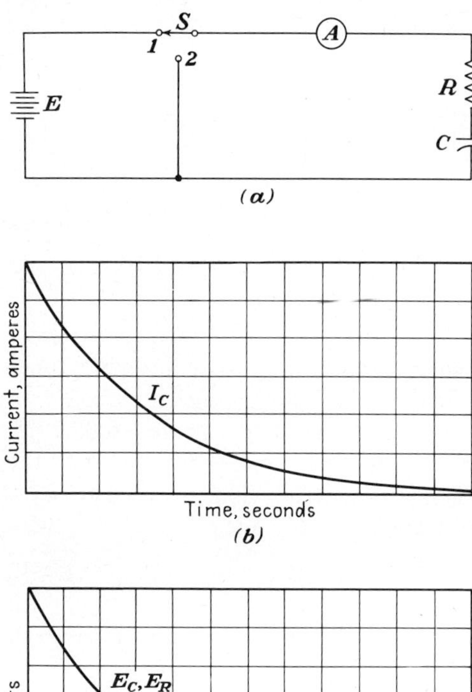

Fig. 12-42 Characteristics of current and voltage versus time. (a) The circuit. (b) Current versus time characteristics immediately after switch *S* is placed in position 2. (c) Voltage versus time characteristics immediately after switch *S* is placed in position 2.

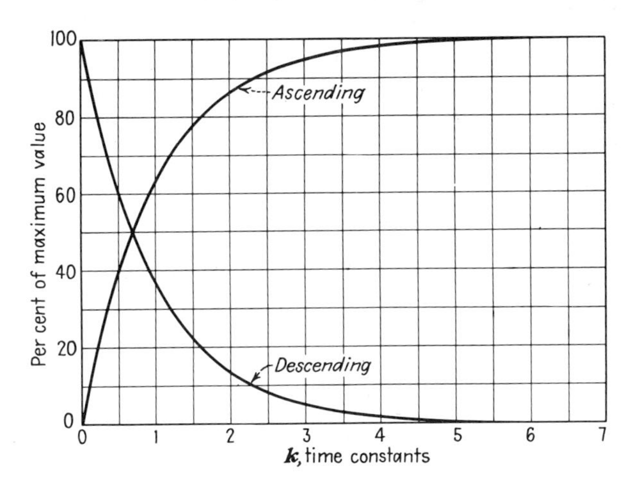

Fig. 12-43 Universal time-constant curves.

cally by use of suitable equations. The mathematics involved is beyond the scope of this text. A shorter and more convenient method of determining the time required to attain any percentage of the final value is by use of time-constant curves. As all the current-time and voltage-time relations vary exponentially, it is possible to represent these variations by the two general exponential curves shown in Fig. 12-43. These curves are plotted from values obtained mathematically and listed in Tables 12-1 and 12-2.

Example 12-13 An *RL* circuit used to control the action of a switch has a resistance of 12 ohms and an inductance of 0.5 henry and is connected to a 6-volt battery. (*a*) If the switch operates when the current attains 63.2 per cent of its final value, what time is required to operate the switch? (*b*) If the switch requires 400 ma to operate, what is the time between the start of current flow and the closing of the switch?

GIVEN: $R = 12$ ohms $L = 0.5$ henry $I = 400$ ma $E = 6$ volts

Table 12-1 Ascending Curve

k TIME CONSTANTS	PER CENT OF MAXIMUM VALUE	k TIME CONSTANTS	PER CENT OF MAXIMUM VALUE	k TIME CONSTANTS	PER CENT OF MAXIMUM VALUE
0.00	0.000	0.70	50.3	2.50	91.8
0.05	4.9	0.80	55.1	3.00	95.0
0.10	9.5	0.90	59.3	3.50	97.0
0.15	14.0	1.00	63.2	4.00	98.2
0.20	18.1	1.20	69.9	4.50	98.9
0.30	25.9	1.40	75.3	5.00	99.3
0.40	33.0	1.60	79.8	5.50	99.6
0.50	39.3	1.80	83.5	6.00	99.8
0.60	45.1	2.00	86.5	7.00	99.9

Table 12-2 Descending Curve

k TIME CONSTANTS	PER CENT OF MAXIMUM VALUE	k TIME CONSTANTS	PER CENT OF MAXIMUM VALUE	k TIME CONSTANTS	PER CENT OF MAXIMUM VALUE
0.00	100	0.70	49.7	2.50	8.2
0.05	95.1	0.80	44.9	3.00	5.0
0.10	90.5	0.90	40.7	3.50	3.0
0.15	86.0	1.00	36.8	4.00	1.8
0.20	81.9	1.20	30.1	4.50	1.1
0.30	74.1	1.40	24.7	5.00	0.7
0.40	67.0	1.60	20.2	5.50	0.4
0.50	60.7	1.80	16.5	6.00	0.2
0.60	54.9	2.00	13.5	7.00	0.1

FIND: (*a*) t (*b*) t

SOLUTION:

(*a*)
$$t = \frac{L}{R} = \frac{0.5}{12} = 0.0416 \text{ sec}$$

(*b*)
$$I_{\max} = \frac{E}{R} = \frac{6}{12} = 0.5 \text{ amp}$$

$$\% \text{ of } I_{\max} \text{ required to operate the switch} = \frac{I}{I_{\max}} \times 100 = \frac{400}{500} \times 100 = 80$$

From Fig. 12-43 or Table 12-1, $k = 1.6$

$$t = k\frac{L}{R} = \frac{1.6 \times 0.5}{12} = 0.0666 \text{ sec}$$

Example 12-14 A 0.005-μf capacitor and a 2-megohm resistor are connected to form an RC circuit. If the RC combination is connected to a 300-volt source of d-c power, what time is required for the voltage across the capacitor to reach (*a*) 100 volts? (*b*) 200 volts? (*c*) 270 volts? If the capacitor becomes fully charged (300 volts) and is then discharged through the 2-megohm resistor, what time is required to discharge the capacitor to (*d*) 250 volts? (*e*) 200 volts? (*f*) 50 volts?

GIVEN: $R = 2$ megohms $C = 0.005\ \mu$f $E_{\max} = 300$ volts

FIND: t

SOLUTION:

(*a*)
$$\% \text{ of } E_{\max} = \frac{100}{300} \times 100 = 33.3$$

$k = 0.40$ from Fig. 12-43

$$t = kCR = 0.40 \times 0.005 \times 10^{-6} \times 2 \times 10^{6} = 0.004 \text{ sec}$$

(*b*)
$$\% \text{ of } E_{\text{max}} = \frac{200}{300} \times 100 = 66.6$$
$$k = 1.1 \qquad \text{from Fig. 12-43}$$
$$t = kCR = 1.1 \times 0.005 \times 10^{-6} \times 2 \times 10^{6} = 0.011 \text{ sec}$$

(*c*)
$$\% \text{ of } E_{\text{max}} = \frac{270}{300} \times 100 = 90$$
$$k = 2.27 \qquad \text{from Fig. 12-43}$$
$$t = kCR = 2.27 \times 0.005 \times 10^{-6} \times 2 \times 10^{6} = 0.0227 \text{ sec}$$

(*d*)
$$\% \text{ of } E_{\text{max}} = \frac{250}{300} \times 100 = 83.3$$
$$k = 0.19 \qquad \text{from Fig. 12-43}$$
$$t = kCR = 0.19 \times 0.005 \times 10^{-6} \times 2 \times 10^{6} = 0.0019 \text{ sec}$$

(*e*)
$$\% \text{ of } E_{\text{max}} = \frac{200}{300} \times 100 = 66.6$$
$$k = 0.40 \qquad \text{from Fig. 12-43}$$
$$t = kCR = 0.40 \times 0.005 \times 10^{-6} \times 2 \times 10^{6} = 0.004 \text{ sec}$$

(*f*)
$$\% \text{ of } E_{\text{max}} = \frac{50}{300} \times 100 = 16.6$$
$$k = 1.8 \qquad \text{from Fig. 12-43}$$
$$t = kCR = 1.8 \times 0.005 \times 10^{-6} \times 2 \times 10^{6} = 0.018 \text{ sec}$$

Uses of Delayed-action Circuits. There are numerous applications of *RC* and *RL* circuits in all branches of electronics. A few of the many applications are as follows.

One type of detector circuit, found in radio receivers, uses a capacitor and a resistor connected in parallel in the plate circuit of a diode. Actually this *RC* combination is a time-constant circuit, and the values of *R* and *C* are chosen to produce a time constant of sufficient duration so that the charge on the capacitor gained during the positive half-cycles does not have time to dissipate any appreciable amount through the resistor during the negative half-cycles. Other examples of *RC* circuits in radio apparatus include grid-leak detection, automatic-volume-control circuits, automatic-frequency-control circuits, and relaxation oscillators.

RC and/or *RL* circuits are used in television receivers and transmitters for producing timing, pulsing, triggering, and synchronizing circuits. The names of some of these circuits are differentiator, integrator, horizontal sweep, horizontal blanking, vertical blanking, automatic gain control, and automatic frequency control.

Industrial applications of time-constant circuits include controlling the length of time for a specific manufacturing operation, timing of electric welders, timing the exposures of photoflashes and photofinishing processes, timing of instruments, producing repeated action for life tests, timing and triggering circuits for computers and data-process equipment, and motor control.

QUESTIONS

1. (*a*) Why is it necessary to couple circuits? (*b*) What is meant by the coupling element?
2. What is meant by (*a*) simple-coupled circuit? (*b*) Complex-coupled circuit?
3. Draw the circuit diagram and explain the action of four simple coupling circuits using a different coupling element for each.
4. Draw the circuit diagram and explain the action of three complex coupling circuits using a different coupling element for each.
5. What is the general purpose for using complex-coupled circuits?
6. Explain what is meant by (*a*) critical coupling, (*b*) loose coupling, (*c*) tight coupling.
7. What type of coupling element is commonly used to provide mutual-inductive coupling in electronic circuits?
8. Explain what is meant by coupled impedance.
9. How does the amount of coupled impedance affect the primary impedance?
10. How does the coupled impedance vary with the amount of coupling?
11. How is electrical energy transferred from (*a*) one circuit element to another? (*b*) One circuit to another?
12. Why is it necessary to use filters in electronic circuits?
13. (*a*) What are the essential parts of a filter circuit? (*b*) Explain the action of each part.
14. Explain the purpose of using each of the following units in a filter circuit: (*a*) capacitor, (*b*) inductor, (*c*) resistor, (*d*) series resonant circuit, (*e*) parallel resonant circuit.
15. (*a*) What is meant by a low-pass filter circuit? (*b*) Draw a simple low-pass circuit, and explain its action.
16. (*a*) What is meant by a high-pass filter circuit? (*b*) Draw a simple high-pass circuit, and explain its action.
17. (*a*) What is meant by a bandpass filter circuit? (*b*) Draw a simple bandpass circuit, and explain its action.
18. (*a*) What is meant by a bandstop filter circuit? (*b*) Draw a simple bandstop circuit, and explain its action.
19. Why is it necessary to use multisection filter circuits?
20. What is meant by (*a*) T-type filter? (*b*) π-type filter?
21. Draw a circuit diagram of one section of a T-type filter circuit for (*a*) low pass, (*b*) high pass, (*c*) bandpass, (*d*) bandstop.
22. Draw a circuit diagram of a two-section T-type filter circuit for (*a*) low pass, (*b*) high pass.
23. When two or more T-type low-pass filter circuits are joined, why should the connecting inductor have a value of twice the inductance of either of the end inductors?
24. When two or more T-type high-pass filter circuits are joined, why should the connecting capacitor have a value of one-half the capacitance of either of the end capacitors?
25. Draw a circuit diagram of one section of a π-type filter circuit for (*a*) low pass, (*b*) high pass, (*c*) bandpass, (*d*) bandstop.
26. Draw a circuit diagram of a two-section π-type filter circuit for (*a*) low pass, (*b*) high pass.

27. When two or more π-type low-pass filter circuits are joined, why should the connecting capacitor have a value of twice the capacitance of either end capacitor?
28. When two or more π-type high-pass filter circuits are joined, why should the connecting inductor have a value of one-half the inductance of either end inductor?
29. What is meant by the following terms: (*a*) Source impedance? (*b*) Load impedance? (*c*) Image impedance? (*d*) Characteristic impedance?
30. What is meant by (*a*) constant-k filter? (*b*) Imaginary load?
31. Describe the frequency requirements of an ideal filter.
32. What is meant by (*a*) m-derived filter? (*b*) Shunt-derived? (*c*) Series-derived?
33. Describe the circuit characteristics of (*a*) a shunt-derived low-pass filter, (*b*) a series-derived high-pass filter.
34. Explain how a resistor and a capacitor combine to form a filter circuit.
35. What is the bandwidth of the bandpass circuit used in (*a*) a-m radio receivers? (*b*) F-m radio receivers? (*c*) The picture section of television receivers?
36. Explain the relation between the width of band passed and the amount of coupling.
37. Explain the effect on secondary current of two resonant circuits coupled by means of a transformer with (*a*) critical coupling, (*b*) tight coupling, (*c*) loose coupling.
38. What are the important properties of a bandpass circuit?
39. How does the amount of circuit Q affect the uniformity of response?
40. Name and explain three methods of obtaining a wide bandpass.
41. Describe a method for obtaining a wide bandpass in the i-f section of an f-m receiver.
42. Describe how the unsymmetrical wide bandpass required by the i-f video section of a television receiver is obtained.
43. Describe four characteristics of ceramic filters that make them desirable for use as bandpass filter circuits.
44. (*a*) What is meant by the piezoelectric effect? (*b*) How is this effect used in ceramic filters?
45. Describe how the bandpass characteristics of an amplifier circuit using a ceramic filter compare with those of a similar amplifier circuit using i-f transformers.
46. What are the factors of a ceramic filter that determine (*a*) its insertion loss? (*b*) The width of its bandpass?
47. Define (*a*) attenuator, (*b*) fixed-attenuator pad, (*c*) variable-attenuator pad.
48. (*a*) Describe an L-attenuator circuit. (*b*) What are its circuit characteristics?
49. (*a*) Describe a T-attenuator circuit. (*b*) What are its circuit characteristics?
50. Describe (*a*) an H-attenuator circuit, (*b*) an O-attenuator circuit.
51. (*a*) Describe a π-attenuator circuit. (*b*) What are its circuit characteristics?
52. (*a*) Describe a ladder-attenuator circuit. (*b*) What are its circuit characteristics?
53. (*a*) Describe a bridged-T attenuator circuit. (*b*) How does this circuit compare with the simple-T attenuator circuit?
54. What is the essential purpose of RL and RC circuits?
55. Explain the operation of a delayed-action (*a*) RL circuit, (*b*) RC circuit.
56. What is meant by the time constant of a delayed-action (*a*) RL circuit? (*b*) RC circuit?
57. (*a*) What is the purpose of the universal time-constant curve? (*b*) How is it used in the solution of time-delay circuits?
58. (*a*) Name four applications of time-delay circuits. (*b*) Explain the circuit actions of one of these applications.

PROBLEMS

1. A mutual-inductance-coupled circuit similar to Fig. 12-4 has the following circuit parameters: $L_P = 1.15$ mh, $L_S = 0.575$ mh, $R_P = 72$ ohms, $R_S = 36$ ohms, $M = 0.08$ mh, $e_i = 10$ volts, $f_r = 456$ kHz, $R_L = 150$ ohms. Find (*a*) the inductive reactance of the primary, (*b*) the inductive reactance of the secondary, (*c*) the impedance coupled into the primary by the secondary, (*d*) the equivalent resistance component of the coupled impedance, (*e*) the equivalent reactance component of the coupled impedance, (*f*) the equivalent circuit diagram, (*g*) the effective impedance of the primary, (*h*) the primary current, (*i*) the secondary voltage, (*j*) the secondary current.
2. A mutual-inductance-coupled circuit similar to Fig. 12-4 has the following circuit parameters: $L_P = 8.8$ μh, $L_S = 6.6$ μh, $R_P = 11$ ohms, $R_S = 8.3$ ohms, $M = 0.3$ μh, $e_i = 10$ volts, $f_r = 10.7$ MHz, $R_L = 100$ ohms. Find (*a*) the inductive reactance of the primary, (*b*) the inductive reactance of the secondary, (*c*) the impedance coupled into the primary by the secondary, (*d*) the equivalent resistance component of the coupled impedance, (*e*) the equivalent reactance component of the coupled impedance, (*f*) the equivalent circuit diagram, (*g*) the effective impedance of the primary, (*h*) the primary current, (*i*) the secondary voltage, (*j*) the secondary current.
3. It is desired that a filter choke, having a resistance of 45 ohms, oppose the flow of a 60-hertz current with ten times the opposition that it offers to direct current. What is the inductance of the coil?
4. A 15-henry filter choke has a resistance of 400 ohms. Determine (*a*) its opposition to direct current, (*b*) its opposition to 60-hertz alternating current, (*c*) its ratio of opposition, alternating to direct current.
5. (*a*) To which type of current will a 0.04-μf capacitor offer the greater opposition, a 4,000-hertz a-f signal or a 1,500-kHz r-f signal? (*b*) How many times greater is the larger impedance than the smaller impedance? (*c*) Which type of current is blocked by this capacitor?
6. (*a*) To which type of current will a 33-pf capacitor offer the greater opposition, a 4.5-MHz signal or a 45.75-MHz signal? (*b*) How many times greater is the larger impedance than the smaller impedance? (*c*) Which type of current is blocked by this capacitor?
7. A 10-mh inductor and a 250-pf capacitor are connected as shown in Fig. 12-12*c* to form a low-pass filter circuit. Determine the opposition offered to a 10-kHz signal (*a*) by the capacitor, (*b*) by the inductor. Determine the opposition offered to a 1,500-kHz signal (*c*) by the capacitor, (*d*) by the inductor.
8. A 1.8-μh inductor and a 68-pf capacitor are connected as shown in Fig. 12-12*c* to form a low-pass filter circuit. Determine the opposition offered to a 4.5-MHz signal (*a*) by the capacitor, (*b*) by the inductor. Determine the opposition offered to a 45.75-MHz signal (*c*) by the capacitor, (*d*) by the inductor.
9. A 4-henry inductor and a 0.4-μf capacitor are connected as shown in Fig. 12-14*c* to form a high-pass filter circuit. Determine the opposition offered to a 60-hertz power disturbance (*a*) by the capacitor, (*b*) by the inductor. Determine the opposition offered to a 1,200-hertz a-f signal (*c*) by the capacitor, (*d*) by the inductor.
10. A 1.2-μh inductor and a 20-pf capacitor are connected as shown in Fig. 12-14*c* to form a high-pass filter circuit. Determine the opposition offered to a 10.7-MHz i-f

signal (*a*) by the capacitor, (*b*) by the inductor. Determine the opposition offered to an 88-MHz f-m signal (*c*) by the capacitor, (*d*) by the inductor.

11. Determine the value of m for a low-pass filter circuit whose resonant frequency is 40 MHz and whose cutoff frequency is 32 MHz.

12. Determine the value of m for a high-pass filter circuit whose resonant frequency is 4.5 MHz and whose cutoff frequency is 5.0 MHz.

13. A circuit similar to the one shown in Fig. 12-27*a* is to be used in the r-f stage of a receiver. It is desired that the capacitor offer an impedance of 100 times that of the resistor whose value is 6,500 ohms. What size capacitor is required if the signal is 1,500 kHz.

14. A circuit similar to the one shown in Fig. 12-27*b* has the following values: The resistor has 7,200 ohms, and the capacitor has a value of 20 μf. (*a*) What impedance does the capacitor offer to a 5,000-hertz current? (*b*) Will the a-f signal flow through the resistor or the capacitor?

15. What is the approximate width of the frequency band passed by a bandpass amplifier circuit having a resonant frequency of 465 kHz and a coefficient of coupling of 0.02?

16. What coefficient of coupling is required to produce a 150-kHz bandpass at an operating frequency of (*a*) 10.7 MHz? (*b*) 9 MHz?

17. It is necessary that the critical coupling of an inductance-coupled bandpass circuit be equal to 0.008. If the circuit Q of the primary and secondary circuit are equal to each other, what is their value?

18. It is necessary that the critical coupling of an inductance-coupled bandpass circuit be equal to 0.02. If the circuit Q of the primary and secondary circuit are equal to each other, what is their value?

19. It is desired that a circuit pass a band 10-kHz wide. The circuit Q is equal to 68.5. (*a*) What value of coefficient of coupling is required? (*b*) What are the extreme limits of the frequency band passed?

20. It is desired that a circuit pass a band 10-kHz wide. The circuit Q is equal to 40. (*a*) What value of coefficient of coupling is required? (*b*) What are the extreme limits of the frequency band passed?

21. A certain 1,800-ohm load is supplied by a 60-volt d-c power source. It is desired to reduce the load voltage to 10 volts by means of an L-type attenuator similar to Fig. 12-37*a* and still retain a 1,800-ohm load condition at the power source. (*a*) What are the values of K, R_1, and R_2? Using the values obtained in part (*a*), (*b*) calculate the resistance at the power source terminals, (*c*) check the output voltage and the value of K.

22. A certain 48,000-ohm load is supplied by a 4.8-volt signal. It is desired that the signal be reduced to one-eighth its input value by an L-type attenuator similar to Fig. 12-37*a* and still retain a 48,000-ohm load condition at the signal source. (*a*) What are the values of K, R_1, and R_2? Using the values obtained in part (*a*), (*b*) calculate the resistance at the signal-input terminals, (*c*) check the output voltage and the value of K.

23. A certain 120,000-ohm load is supplied by a 150-volt d-c power source. It is desired to reduce the load voltage by means of a T-type attenuator similar to Fig. 12-37*c* with an attenuation factor of 5 and still retain a 120,000-ohm load condition at the power source. (*a*) What are the values of e_o, R_1, R_2, and R_3? (*b*) By using the values obtained in part (*a*) calculate the resistance at the power source terminals.

24. A certain 8,000-ohm load is supplied by a 3.0-volt signal. It is desired that the signal be reduced to 0.2 volt by using a T-type attenuator similar to Fig. 12-37*c* and still retain an 8,000-ohm load condition at the signal source. (*a*) What are the values of K, R_1, R_2, and R_3? (*b*) By using the values obtained in part (*a*) calculate the resistance at the signal-input terminals.
25. A certain 12,000-ohm load is supplied by an 18-volt signal. It is desired that the signal be reduced to 12 volts by using a π-type attenuator similar to Fig. 12-37*e* and still retain a 12,000-ohm load condition at the signal source. (*a*) What are the values of K, R_1, R_2, and R_3? (*b*) By using the values obtained in part (*a*) calculate the resistance at the signal-input terminals.
26. A certain 60,000-ohm load is supplied by a 200-volt d-c power source. It is desired to reduce the load voltage to 150 volts by means of a π-type attenuator similar to Fig. 12-37*e* and still retain a 60,000-ohm load condition at the power source. (*a*) What are the values of K, R_1, R_2, and R_3? By using the values obtained in part (*a*) calculate the resistance at the power source terminals.
27. A low-current d-c relay that has an inductance of 25 henrys is connected in series with a 1,000-ohm resistor to form an RL time-delay control circuit operated on a 110-volt d-c circuit. (*a*) What is the time constant of the circuit? (*b*) If the relay closes when the current is 88 ma, what time elapses between closing the line switch and operation of the relay?
28. A low-current d-c relay having an inductance of 10 henrys is to close 0.02 sec after the line switch is closed. What value of resistor should be connected in series with the relay if it closes when the current reaches (*a*) 63.2 per cent of its final value? (*b*) 80 per cent of its final value?
29. A radio receiver is to have an RC circuit with a time constant of 0.2 sec for its automatic-volume-control circuit. (*a*) What value of resistor is required if a 0.1-μf capacitor is used? (*b*) What value of resistor is required if a 0.15-μf capacitor is used? (*c*) What value of capacitor is required if a 1-megohm resistor is used?
30. A 1-megohm grid resistor is shunted by a 250-pf capacitor. (*a*) What is the time constant of this circuit? (*b*) If the highest a-f signal to be applied to the circuit is 5,000 cycles, what is the time required to complete one of these cycles? (*c*) Under the conditions of (*a*) and (*b*) will the capacitor ever become completely discharged? (*d*) Explain your answer to (*c*).
31. A 0.05-μf capacitor and a 0.5-megohm resistor are connected to form an RC circuit. The RC combination is connected to a 250-volt d-c source. (*a*) What time is required for the voltage across the capacitor to reach 50, 100, 200 volts? (*b*) What current flows when the switch is closed? (*c*) What current flows when the voltage across the capacitor reaches 200 volts? (*d*) If the capacitor is fully charged and is then discharged through the 0.5-megohm resistor, what is the current at the instant it starts to discharge? (*e*) What is the value of the voltage RC sec after the capacitor starts to discharge? (*f*) At what time will the capacitor be discharged to half voltage?
32. What is the time constant of an RC circuit as used in a television receiver when (*a*) $R = 2.2$ megohms, $C = 0.05$ μf? (*b*) $R = 470{,}000$ ohms, $C = 0.1$ μf? (*c*) $R = 680{,}000$ ohms, $C = 820$ pf?

Chapter 13

Basic Voltage-amplifier Circuits

The discussion of amplifier circuits in the preceding chapters described the operation and characteristics of amplifier circuits using only a single transistor or vacuum tube. The voltage amplification or the power gain obtained with a single stage of amplification is frequently insufficient to meet the requirements of a composite electronic circuit or device, and therefore multiple stages of amplification are commonly used. This chapter presents the operation and the characteristics of the basic types of multistage amplifier circuits used for obtaining voltage amplification. Amplifiers used to obtain power gains are presented in a later chapter.

13-1 Methods of Coupling

When two or more single stages of amplification are needed to obtain the desired gain, the output of the first stage becomes the input of the second stage. The composite amplifier is called a *cascade amplifier*. The manner in which the output of the first stage is coupled into the second stage is important, as it determines the characteristics of the cascade amplifier. The basic methods of coupling two stages to one another are (1) resistance-capacitance (RC) coupling, (2) impedance coupling, (3) transformer coupling, and (4) direct coupling.

13-2 Resistance-Capacitance Coupling

Basic Circuit. RC coupling, a widely used method, uses two resistors and a capacitor to couple one circuit to another. In the circuit of Fig. 13-1, the RC network consists of (1) the input resistor R_i, (2) the coupling capacitor C_c, and (3) the output resistor R_o. In the circuits of Figs. 13-2 and 13-3, the input resistor is R_L, the coupling capacitor is C_2, and the output resistor is R_o.

Circuit Operation. The basic function of the RC coupling network is (1) to pass any a-c signal from one circuit to another and (2) to prevent any d-c voltage present at its input from reaching the output. With this circuit, the voltage v_i from a signal source is applied to R_i; frequently, the input signal also contains a steady d-c voltage component.

When the signal voltage at R_i in Fig. 13-1 increases (making T_1 more positive), electrons flow through R_o to the capacitor and increase the charge on the capacitor. While a current is flowing through R_o, a voltage drop IR is

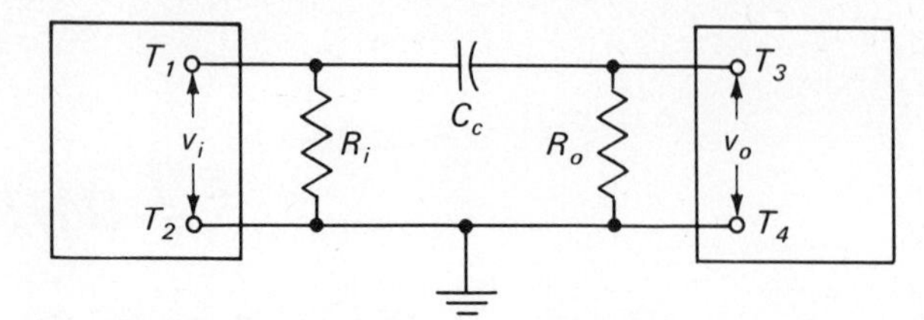

Fig. 13-1 Basic resistance-capacitance coupling circuit.

being developed across R_o with T_3 positive. When the input-signal voltage decreases, electrons flow through R_o to ground. The IR voltage drop across R_o now makes T_3 negative. Any d-c voltage present in the input at R_i will be blocked from reaching the output terminals by the normal action of the capacitor to a d-c voltage. Thus, the coupling capacitor permits a-c signals to pass from the input circuit to the output circuit but prevents any d-c voltage present at the input from reaching the output. Because of its dual role, the capacitor is referred to as either a *coupling capacitor* or a *blocking capacitor* and is represented by the symbol C_c or C_b. Also, because of the important functions performed by the capacitor, this circuit is also called *capacitive coupling.*

Characteristics. All the gain in the RC-coupled amplifier is provided by the active element in the circuit. The associated circuit elements do not add to the gain but rather reduce the effective gain of the circuit so that the resultant amplification is less than the gain provided by the active element. However, the overall amplification of an RC-coupled amplifier circuit is generally higher than can be obtained with other coupling methods. This is possible because of the high values of resistance that can be used to match the high values of output resistance of the active circuit element. Other advantages of this type of amplifier circuit are: (1) there are few parts that are low in cost and require very little space; (2) as there are no coils or transformers in the circuit, there is

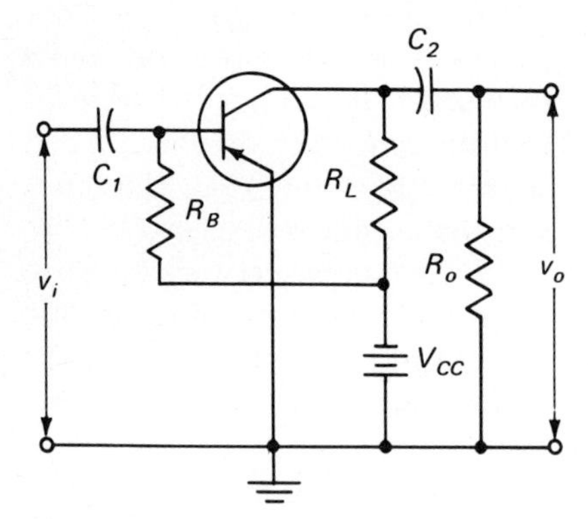

Fig. 13-2 Basic single-stage transistor amplifier.

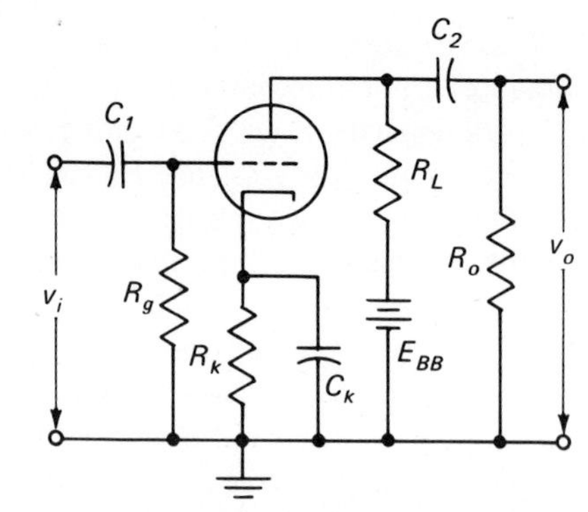

Fig. 13-3 Basic single-stage vacuum-tube amplifier.

very little pickup of undesirable currents from any a-c leads; thus the amount of nonlinear distortion is minimized. A disadvantage of the RC-coupled amplifier is that a higher supply voltage must be used to compensate for the voltage drop across the coupling resistor.

Frequency Characteristics. Although the gain of an amplifier is not uniform for the entire range of a particular frequency band (a-f, r-f, video, etc.), it is practically constant over a fairly wide range of frequencies within its band, decreasing rapidly at both the very low and very high frequencies at the ends of the band (Fig. 14-6). The effect of frequency on the gain is of sufficient importance that it is discussed fully in Chap. 14. Accordingly, the characteristics of amplifiers presented in this chapter are for the values in the middle of its normal frequency range.

13-3 Basic Multistage *RC*-coupled Amplifier

Transistor Amplifier. Figure 13-4 illustrates the basic circuit of an RC-coupled two-stage transistor amplifier.

Vacuum-tube Amplifier. Figure 13-8 illustrates the basic circuit of an RC-coupled two-stage vacuum-tube amplifier.

Comparison. These two circuits show that the RC-coupling elements (R_2, C_2, R_3) have the same circuit configuration for adaptation to either transistors or vacuum tubes. However, in actual use, the values of resistance and capacitance for transistor applications are much different from those for vacuum tubes.

13-4 *RC*-coupled Two-stage Transistor Amplifier

The Circuit. The basic RC-coupling network of Fig. 13-1 is used in the circuit of Fig. 13-4 to produce a two-stage amplifier. In order to determine the operating characteristics of this circuit it is necessary to consider the effects of the transistors and all of its circuit components.

Circuit Operation. The circuit of Fig. 13-4 consists of two single-stage transistor amplifiers using the common-emitter configuration. A two-stage cascade amplifier is formed by joining the two single stages with the RC-coupling network R_2, C_2, R_3. The operation and the characteristics of each single stage are the same as described in Arts. 9-5 and 9-8.

In the circuit of Fig. 13-4, (1) the input signal v_i is amplified by Q_1 and its phase is reversed, (2) the amplified output of Q_1 appears at R_2, (3) the output of the first stage (at R_2) is coupled to the input of the second stage (at R_3) by the coupling capacitor C_2, (4) the signal at the base of Q_2 is amplified by Q_2 and its phase is reversed, (5) the output of Q_2 appears at R_4, and (6) capacitor C_3 couples the a-c signals at R_4 to the load resistor R_5 but blocks the d-c voltage of V_{CC} from the load resistor. Thus, the output signal v_o is a twice-amplified replica of the input signal v_i. The phase of the output signal is the same as that of the input signal because undergoing phase reversal twice restores the phase of the signal to its original condition.

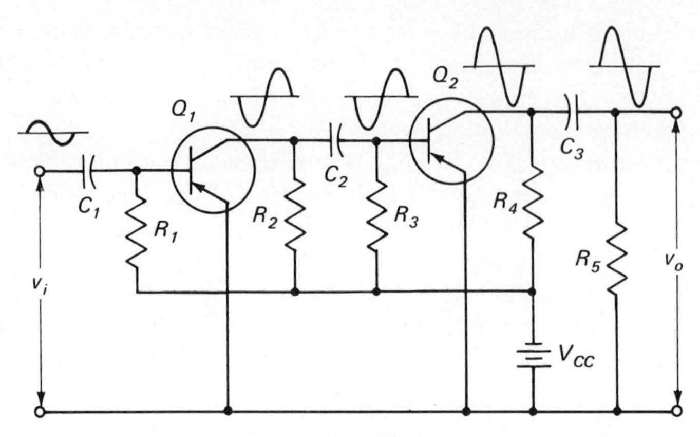

$C_1, C_2, C_3 = 1\ \mu f$; $R_1, R_3 = 750{,}000$ ohms; $R_2, R_4 = 4{,}000$ ohms; $R_5 = 16{,}000$ ohms; $V_{CC} = 15$ volts; $\beta_1, \beta_2 = 100$.

Fig. 13-4 Two-stage *RC*-coupled transistor amplifier.

Voltage Gain. The overall voltage gain of the multistage amplifier is equal to the product of the gains of its component stages, thus

$$A_v = A_{v\cdot 1}A_{v\cdot 2}A_{v\cdot 3} \qquad \text{etc.} \tag{13-1}$$

It is therefore necessary first to determine the gain of each stage individually.

Because the input of the second stage forms a part of the output of the first stage, it is desirable first to determine the gain of the second stage. From Eq. (9-17),

$$A_{v\cdot 2} = \beta_2 \frac{r_{o\cdot 2}}{r_{i\cdot 2}} \tag{13-2}$$

However, for the ideal transistor approximation $r_{i\cdot 2} = \beta_2 r_{e\cdot 2}$. Substituting $\beta_2 r_{e\cdot 2}$ for $r_{i\cdot 2}$ in Eq. (13-2) produces

$$A_{v\cdot 2} = \beta_2 \frac{r_{o\cdot 2}}{\beta_2 r_{e\cdot 2}} \tag{13-2a}$$

Factoring out β_2, then

$$A_{v\cdot 2} = \frac{r_{o\cdot 2}}{r_{e\cdot 2}} \tag{13-2b}$$

The voltage gain for the first stage is found with the same basic equation, modified to

$$A_{v\cdot 1} = \frac{r_{o\cdot 1}}{r_{e\cdot 1}} \tag{13-3}$$

Example 13-1 (*a*) Draw the a-c equivalent circuit for the second stage of the amplifier of Fig. 13-4. (*b*) What is the voltage gain of this stage?

GIVEN: Fig. 13-4

FIND: (*a*) Circuit diagram (*b*) $A_{v \cdot 2}$

SOLUTION:

(*a*) Fig. 13-5

(*b*)
$$A_{v \cdot 2} = \frac{r_{o \cdot 2}}{r_{e \cdot 2}} = \frac{3{,}200}{12.5} = 256$$

where
$$r_{o \cdot 2} = \frac{R_4 R_5}{R_4 + R_5} = \frac{4{,}000 \times 16{,}000}{4{,}000 + 16{,}000} = 3{,}200 \text{ ohms}$$

$$r_{e \cdot 2} = \frac{25}{I_{E \cdot 2}} = \frac{25}{2} = 12.5 \text{ ohms}$$

$$I_{E \cdot 2} = \beta_2 I_{B \cdot 2} = \frac{\beta_2 V_{CC}}{R_3} = \frac{100 \times 15}{750{,}000} = 2 \text{ ma}$$

Example 13-2 (*a*) Draw the a-c equivalent circuit for the first stage of the amplifier of Fig. 13-4. (*b*) What is the voltage gain for this stage?

GIVEN: Fig. 13-4

FIND: (*a*) Circuit diagram (*b*) $A_{v \cdot 1}$

SOLUTION:

(*a*) Fig. 13-6

(*b*)
$$A_{v \cdot 1} = \frac{r_{o \cdot 1}}{r_{e \cdot 1}} = \frac{952}{12.5} \cong 76.2$$

where
$$r_{o \cdot 1} = \frac{R_2 r_{i \cdot 2}}{R_2 + r_{i \cdot 2}} = \frac{4{,}000 \times 1{,}250}{4{,}000 + 1{,}250} \cong 952$$

$$r_{i \cdot 2} = \frac{R_3 \beta_2 r_{e \cdot 2}}{R_3 + \beta_2 r_{e \cdot 2}} = \frac{750{,}000 \times (100 \times 12.5)}{750{,}000 + (100 \times 12.5)} \cong 1{,}250 \text{ ohms}$$

$$r_{e \cdot 1} = \frac{25}{I_{E \cdot 1}} = \frac{25}{2} = 12.5 \text{ ohms}$$

$$I_{E \cdot 1} = \beta_1 I_{B \cdot 1} = \frac{\beta_1 V_{CC}}{R_1} = \frac{100 \times 15}{750{,}000} = 2 \text{ ma}$$

Example 13-3 (*a*) What is the voltage gain of the amplifier of Fig. 13-4? (*b*) What is the decibel voltage gain of the amplifier? (*c*) What is the a-c input resistance of this amplifier?

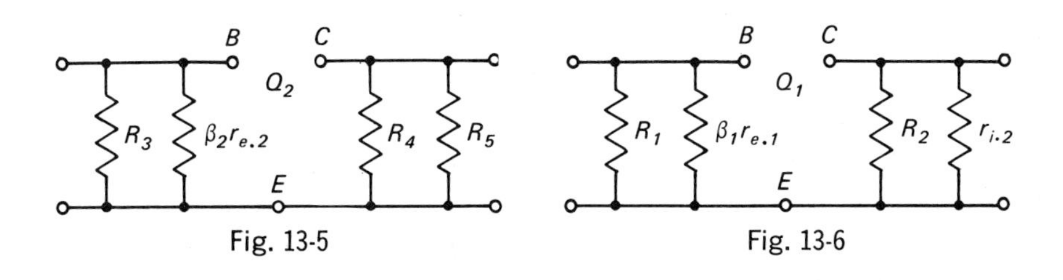

Fig. 13-5 Fig. 13-6

GIVEN: Figs. 13-4, 13-5, 13-6 $\quad A_{v\cdot1} = 76.2 \quad A_{v\cdot2} = 256$

FIND: (*a*) A_v (*b*) db (*c*) r_i

SOLUTION:

(*a*) $$A_v = A_{v\cdot1}A_{v\cdot2} = 76.2 \times 256 \cong 19{,}500$$

(*b*) $$\text{db} = 20 \log A_v = 20 \log 19{,}500 = 20 \times 4.29 = 85.8 \text{ db}$$

(*c*) $$r_i = \frac{R_1\beta_1 r_{e\cdot1}}{R_1 + \beta_1 r_{e\cdot1}} = \frac{750{,}000 \times (100 \times 12.5)}{750{,}000 + (100 \times 12.5)} \cong 1{,}250 \text{ ohms}$$

13-5 Effect of Beta on the Gain

The beta value of a transistor may change because of (1) aging of the transistor, (2) changes in the operating current, voltage, or frequency, (3) changes in the values of associated circuit components, and (4) replacement of a transistor. In a multistage amplifier, the effect of a change in beta is much greater than with a single-stage amplifier, as is indicated by the following example.

Example 13-4 If the values of beta for the transistors used in Examples 13-1, 13-2, and 13-3 decrease to 75, what is (*a*) the voltage gain of the second stage? (*b*) The voltage gain of the first stage? (*c*) The voltage gain of the amplifier? (*d*) The decibel gain of the amplifier? (*e*) The a-c input resistance of the amplifier?

GIVEN: Examples 13-1, 13-2, 13-3 $\quad$ New value of beta = 75

FIND: (*a*) $A_{v\cdot2}$ (*b*) $A_{v\cdot1}$ (*c*) A_v (*d*) db (*e*) r_i

SOLUTION:

(*a*) $$A_{v\cdot2} = \frac{r_{o\cdot2}}{r_{e\cdot2}} = \frac{3{,}200}{16.7} \cong 192$$

where $$r_{o\cdot2} = \frac{R_4R_5}{R_4 + R_5} = \frac{4{,}000 \times 16{,}000}{4{,}000 + 16{,}000} = 3{,}200 \text{ ohms}$$

$$r_{e\cdot2} = \frac{25}{I_{E\cdot2}} = \frac{25}{1.5} \cong 16.7 \text{ ohms}$$

$$I_{E\cdot2} = \beta_2 I_{B\cdot2} = \frac{\beta_2 V_{CC}}{R_3} = \frac{75 \times 15}{750{,}000} = 1.5 \text{ ma}$$

(*b*) $$A_{v\cdot1} = \frac{r_{o\cdot1}}{r_{e\cdot1}} = \frac{952}{16.7} \cong 57.2$$

where $$r_{o\cdot1} = \frac{R_2 r_{i\cdot2}}{R_2 + r_{i\cdot2}} = \frac{4{,}000 \times 1{,}250}{4{,}000 + 1{,}250} \cong 952$$

$$r_{i\cdot2} = \frac{R_3\beta_2 r_{e\cdot2}}{R_3 + \beta_2 r_{e\cdot2}} = \frac{750{,}000 \times (75 \times 16.7)}{750{,}000 + (75 \times 16.7)} \cong 1{,}250 \text{ ohms}$$

$$r_{e\cdot1} = \frac{25}{I_{E\cdot1}} = \frac{25}{1.5} \cong 16.7 \text{ ohms}$$

$$I_{E\cdot1} = \beta_1 I_{B\cdot1} = \frac{\beta_1 V_{CC}}{R_1} = \frac{75 \times 15}{750{,}000} = 1.5 \text{ ma}$$

(c) $A_v = A_{v\cdot 1}A_{v\cdot 2} = 57.2 \times 192 \cong 11{,}000$

(d) $\text{db} = 20 \log A_v = 20 \log 11{,}000 = 20 \times 4.041 \cong 80.8 \text{ db}$

(e) $$r_i = \frac{R_1\beta_1 r_{e\cdot 1}}{R_1 + \beta_1 r_{e\cdot 1}} = \frac{750{,}000 \times (75 \times 16.7)}{750{,}000 + (75 \times 16.7)} \cong 1{,}250 \text{ ohms}$$

Comparing the results of Examples 13-1, 13-2, and 13-3 with the results of Example 13-4 shows that a 25 per cent decrease in beta produces a 25 per cent decrease in the emitter current and approximately a 43 per cent decrease in the voltage gain of the amplifier, a condition which cannot be tolerated in most applications. This condition can be corrected to some extent by (1) use of other bias arrangements (Chaps. 8 and 9), (2) adding swamping resistors (Chap. 9), and (3) using feedback arrangements (Chap. 15).

13-6 Amplifier Using Base Bias and Emitter Bias

The circuit of Fig. 13-7 shows two resistors (R_6 and R_7) and two bypass capacitors (C_4 and C_5) added to the emitter circuits of Fig. 13-4. Also, to compensate for the voltage drop at the emitter resistors, the supply voltage V_{CC} is increased from 15 to 22 volts. The capacitors C_4 and C_5 keep the emitters at a-c ground potential, and the resistors R_6 and R_7 control the d-c bias of the transistors in a manner that will reduce the effects of variations in beta on (1) the emitter current, and (2) the amount of decrease in the voltage gain.

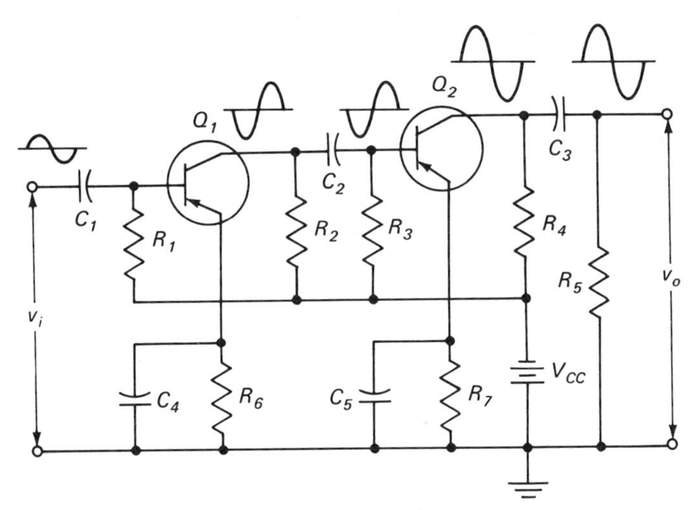

$C_1, C_2, C_3 = 1\ \mu f$; $C_4, C_5 = 50\ \mu f$; $R_1, R_3 = 1{,}500{,}000$ ohms; $R_2, R_4 = 4{,}000$ ohms; $R_5 = 16{,}000$ ohms; $R_6, R_7 = 7{,}000$ ohms; $V_{CC} = 22$ volts; $\beta_1, \beta_2 = 100$.

Fig. 13-7 Two-stage *RC*-coupled transistor amplifier with emitter bias added to the circuit of Fig. 13-4.

Example 13-5 For the amplifier of Fig. 13-7, what is the voltage gain of (*a*) the second stage? (*b*) The first stage? (*c*) The amplifier? (*d*) What is the gain of the amplifier in decibels? (*e*) What is the a-c input resistance of the amplifier?

GIVEN: Fig. 13-7

FIND: (*a*) $A_{v \cdot 2}$ (*b*) $A_{v \cdot 1}$ (*c*) A_v (*d*) db (*e*) r_i

SOLUTION:

(*a*) $$A_{v \cdot 2} = \frac{r_{o \cdot 2}}{r_{e \cdot 2}} = \frac{3{,}200}{25} = 128$$

where $$r_{o \cdot 2} = \frac{R_4 R_5}{R_4 + R_5} = \frac{4{,}000 \times 16{,}000}{4{,}000 + 16{,}000} = 3{,}200 \text{ ohms}$$

$$r_{e \cdot 2} = \frac{25}{I_{E \cdot 2}} = \frac{25}{1} = 25 \text{ ohms}$$

$$I_{E \cdot 2} = \frac{V_{CC}}{R_7 + \dfrac{R_3}{\beta_2}} = \frac{22}{7{,}000 + \dfrac{1{,}500{,}000}{100}} = 1 \text{ ma}$$

(*b*) $$A_{v \cdot 1} = \frac{r_{o \cdot 1}}{r_{e \cdot 1}} = \frac{1{,}540}{25} \cong 61.6$$

where $$r_{o \cdot 1} = \frac{R_2 r_{i \cdot 2}}{R_2 + r_{i \cdot 2}} = \frac{4{,}000 \times 2{,}500}{4{,}000 + 2{,}500} \cong 1{,}540 \text{ ohms}$$

$$r_{i \cdot 2} = \frac{R_3 \beta_2 r_{e \cdot 2}}{R_3 + \beta_2 r_{e \cdot 2}} = \frac{1{,}500{,}000 \times (100 \times 25)}{1{,}500{,}000 + (100 \times 25)} \cong 2{,}500 \text{ ohms}$$

$$r_{e \cdot 1} = \frac{25}{I_{E \cdot 1}} = \frac{25}{1} = 25 \text{ ohms}$$

$$I_{E \cdot 1} = \frac{V_{CC}}{R_6 + \dfrac{R_1}{\beta_1}} = \frac{22}{7{,}000 + \dfrac{1{,}500{,}000}{100}} = 1 \text{ ma}$$

(*c*) $$A_v = A_{v \cdot 1} A_{v \cdot 2} = 61.6 \times 128 \cong 7{,}900$$

(*d*) $$\text{db} = 20 \log A_v = 20 \log 7{,}900 = 20 \times 3.897 \cong 78 \text{ db}$$

(*e*) $$r_i = \frac{R_1 \beta_1 r_{e \cdot 1}}{R_1 + \beta_1 r_{e \cdot 1}} = \frac{1{,}500{,}000 \times (100 \times 25)}{1{,}500{,}000 + (100 \times 25)} \cong 2{,}500 \text{ ohms}$$

Example 13-6 If the values of beta for the transistors used in Example 13-5 decrease to 75, what is (*a*) the voltage gain of the second stage? (*b*) The voltage gain of the first stage? (*c*) The voltage gain of the amplifier? (*d*) The gain of the amplifier in decibels? (*e*) The a-c input resistance of the amplifier?

GIVEN: Example 13-5 New value of beta = 75

FIND: (*a*) $A_{v \cdot 2}$ (*b*) $A_{v \cdot 1}$ (*c*) A_v (*d*) db (*e*) r_i

SOLUTION:

(*a*) $$A_{v \cdot 2} = \frac{r_{o \cdot 2}}{r_{e \cdot 2}} = \frac{3{,}200}{30.7} \cong 104$$

where $$r_{o\cdot 2} = \frac{R_4R_5}{R_4 + R_5} = \frac{4{,}000 \times 16{,}000}{4{,}000 + 16{,}000} = 3{,}200 \text{ ohms}$$

$$r_{e\cdot 2} = \frac{25}{I_{E\cdot 2}} = \frac{25}{0.815} \cong 30.7 \text{ ohms}$$

$$I_{E\cdot 2} = \frac{V_{CC}}{R_7 + \dfrac{R_3}{\beta_2}} = \frac{22}{7{,}000 + \dfrac{1{,}500{,}000}{75}} \cong 0.815 \text{ ma}$$

(*b*) $$A_{v\cdot 1} = \frac{r_{o\cdot 1}}{r_{e\cdot 1}} = \frac{1{,}460}{30.7} \cong 47.5$$

where $$r_{o\cdot 1} = \frac{R_2r_{i\cdot 2}}{R_2 + r_{i\cdot 2}} = \frac{4{,}000 \times 2{,}300}{4{,}000 + 2{,}300} \cong 1{,}460 \text{ ohms}$$

$$r_{i\cdot 2} = \frac{R_3\beta_2r_{e\cdot 2}}{R_3 + \beta_2r_{e\cdot 2}} = \frac{1{,}500{,}000 \times (75 \times 30.7)}{1{,}500{,}000 + (75 \times 30.7)} \cong 2{,}300 \text{ ohms}$$

$$r_{e\cdot 1} = \frac{25}{I_{E\cdot 1}} = \frac{25}{0.815} \cong 30.7 \text{ ohms}$$

$$I_{E\cdot 1} = \frac{V_{CC}}{R_6 + \dfrac{R_1}{\beta_1}} = \frac{22}{7{,}000 + \dfrac{1{,}500{,}000}{75}} \cong 0.815 \text{ ma}$$

(*c*) $$A_v = A_{v\cdot 1}A_{v\cdot 2} = 47.5 \times 104 \cong 4{,}950$$

(*d*) $$\text{db} = 20 \log A_v = 20 \log 4{,}950 = 20 \times 3.695 \cong 74 \text{ db}$$

(*e*) $$r_i = \frac{R_1\beta_1r_{e\cdot 1}}{R_1 + \beta_1r_{e\cdot 1}} = \frac{1{,}500{,}000 \times (75 \times 30.7)}{1{,}500{,}000 + (75 \times 30.7)} \cong 2{,}300 \text{ ohms}$$

The results of Examples 13-1 through 13-6 are summarized in Table 13-1 to show the relative effects on the emitter current and the voltage gain caused by a 25 per cent decrease in beta.

13-7 *RC*-coupled Two-stage Vacuum-tube Amplifier Using Triodes

The Circuit. Although the numerical values of the components in the vacuum-tube amplifiers differ considerably from those in transistor amplifiers, the basic principles relating to determining their characteristics are essentially the same. A typical two-stage vacuum-tube amplifier using triodes is

Table 13-1 Effect of a Decrease in Beta for the Circuits of Figs. 13-4 and 13-7

CIRCUIT OF	DECREASE IN		
	BETA	EMITTER CURRENT	VOLTAGE GAIN
Fig. 13-4	25%	25%	43.5%
Fig. 13-7	25%	18.5%	37.3%

shown in Fig. 13-8. Comparing this circuit with those of Figs. 13-4 and 13-7 will show similarities in the circuit configuration but a wide difference in values of comparable circuit components.

The circuit of Fig. 13-8 consists of two single-stage vacuum-tube amplifiers using the common-cathode configuration and coupled together to form a cascade amplifier. The two stages are joined by the coupling capacitor C_2 which together with resistors R_2 and R_3 forms the RC coupling network.

Circuit Operation. The operation and the characteristics of each single stage are the same as described in Arts. 9-21 and 9-22.

In the cascade amplifier of Fig. 13-8, (1) the input signal v_i is amplified by VT_1 and its phase is reversed, (2) the amplified output of VT_1 appears at R_2, (3) the output of the first stage (at R_2) is coupled to the input of the second stage (at R_3) by the coupling capacitor C_2, (4) the signal at the grid of VT_2 is

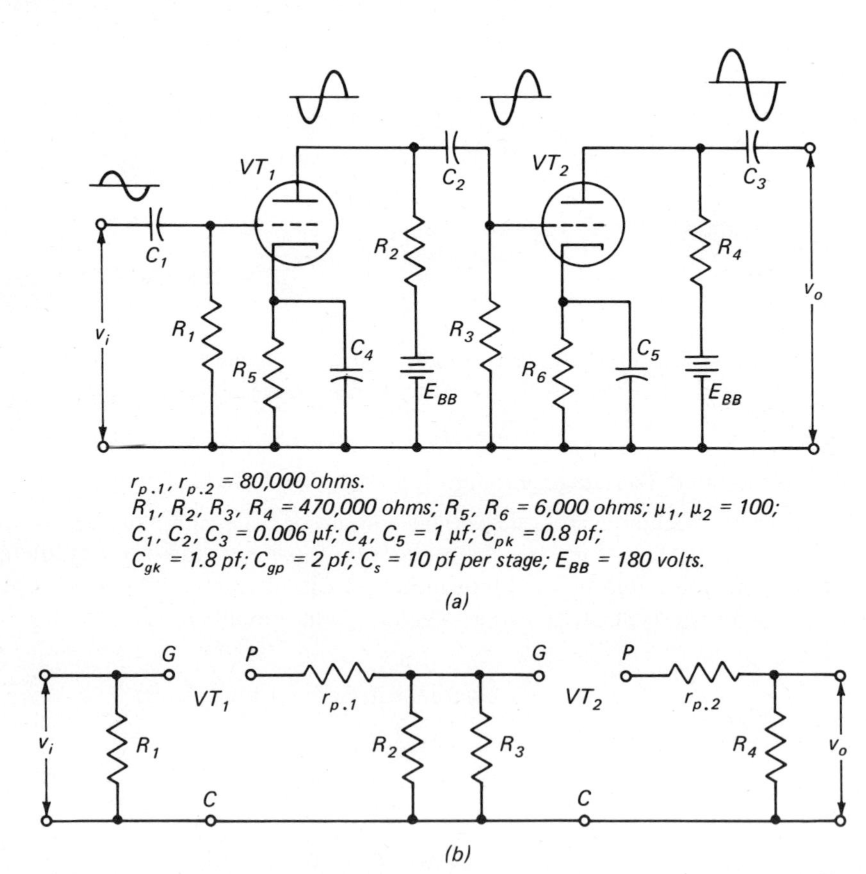

Fig. 13-8 Two-stage RC-coupled vacuum-tube amplifier using triodes. (a) Circuit diagram. (b) Simplified a-c equivalent circuit at the mid-frequency range where effects of all capacitances are negligible.

amplified by VT_2 and its phase is reversed, (5) the output of VT_2 appears at R_4, and (6) capacitor C_3 couples the a-c signals at R_4 to the load while also blocking the d-c voltage of E_{BB} from the load. Thus, the output signal v_o is a twice-amplified replica of the input signal v_i. The phase of the output signal is the same as that of the input signal because undergoing phase reversal twice restores the phase of the signal to its original condition.

Voltage Gain. The overall voltage gain of the multistage amplifier is equal to the product of the gains of its component stages, thus

$$A_v = A_{v\cdot 1}A_{v\cdot 2}A_{v\cdot 3} \qquad \text{etc.} \tag{13-1}$$

From Fig. 13-8*b* and Eq. (9-44) the gains of the individual stages are

$$A_{v\cdot 1} = \frac{\mu_1 r_o}{r_{p\cdot 1} + r_o} \tag{13-4}$$

where

$$r_o = \frac{R_2R_3}{R_2 + R_3}$$

$$A_{v\cdot 2} = \frac{\mu_2 R_4}{r_{p\cdot 2} + R_4} \tag{13-5}$$

Example 13-7 For the amplifier of Fig. 13-8, what is the voltage gain of (*a*) the first stage? (*b*) The second stage? (*c*) The amplifier? (*d*) The amplifier expressed in decibels?

GIVEN: Fig. 13-8

FIND: (*a*) $A_{v\cdot 1}$ (*b*) $A_{v\cdot 2}$ (*c*) A_v (*d*) db

SOLUTION:

(*a*) $$A_{v\cdot 1} = \frac{\mu_1 r_o}{r_{p\cdot 1} + r_o} = \frac{100 \times 235{,}000}{80{,}000 + 235{,}000} \cong 74.5$$

where $$r_o = \frac{R_2R_3}{R_2 + R_3} = \frac{470{,}000 \times 470{,}000}{470{,}000 + 470{,}000} = 235{,}000 \text{ ohms}$$

(*b*) $$A_{v\cdot 2} = \frac{\mu_2 R_4}{r_{p\cdot 2} + R_4} = \frac{100 \times 470{,}000}{80{,}000 + 470{,}000} \cong 85.5$$

(*c*) $$A_v = A_{v\cdot 1}A_{v\cdot 2} = 74.5 \times 85.5 \cong 6{,}400$$

(*d*) $$\text{db} = 20 \log A_v = 20 \times \log 6{,}400 = 20 \times 3.806 \cong 76 \text{ db}$$

13-8 *RC*-coupled Two-stage Vacuum-tube Amplifier Using Pentodes

The Circuit. A typical two-stage vacuum-tube amplifier using pentodes is shown in Fig. 13-9. Comparing this circuit with Fig. 13-8 shows that the circuit configuration of the pentode amplifier differs from the triode amplifier only by the addition of a resistor and bypass capacitor in the screen-grid circuit. The values of the resistors used with pentodes are generally much higher than those used with triodes while the values of the coupling capacitors are about the same in both cases. The plate resistance of pentode tubes is higher

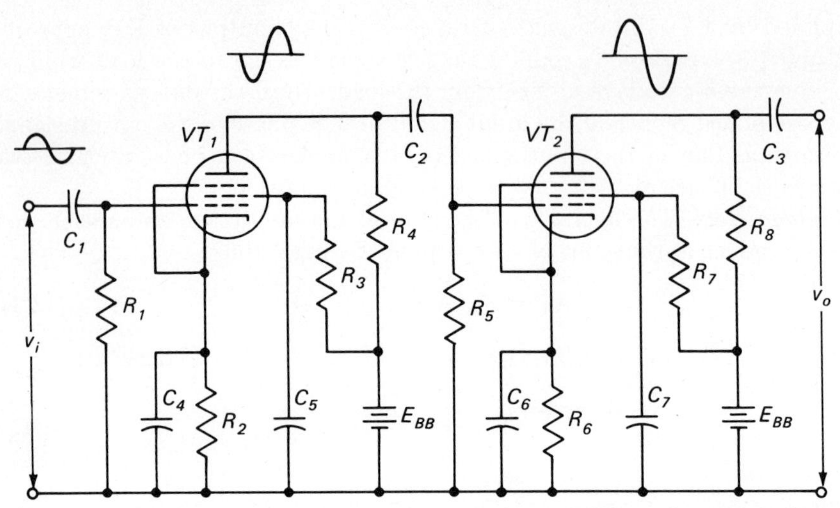

Fig. 13-9 Two-stage *RC*-coupled vacuum-tube amplifier using pentodes.

than for triodes, which contributes to the potentially higher voltage gains obtained with pentode amplifiers.

Circuit Operation. The operation and the characteristics of each stage are the same as described in Art. 9-23. The operation of the cascade amplifier is the same as described for the triode amplifier in Art. 13-7.

Voltage Gain. The voltage gain for the multistage pentode amplifier is the same as expressed by Eq. (13-1). The gains for the individual stages can be derived by (1) preparing an a-c equivalent circuit diagram (which will be the same as Fig. 13-8*b* since the screen-grid resistors do not carry any of the signal current), and (2) modifying Eq. (9-48) to satisfy the conditions of this circuit. Thus,

$$A_{v\cdot 1} = g_{m\cdot 1} \frac{r_{p\cdot 1} r_{o\cdot 1}}{r_{p\cdot 1} + r_{o\cdot 1}} \tag{13-6}$$

where

$$r_{o\cdot 1} = \frac{R_4 R_5}{R_4 + R_5}$$

$$A_{v\cdot 2} = g_{m\cdot 2} \frac{r_{p\cdot 2} R_8}{r_{p\cdot 2} + R_8} \tag{13-7}$$

13-9 Impedance Coupling

Basic Circuit. An impedance-coupling network (Fig. 13-10) is similar to the basic *RC*-coupling network (Fig. 13-1) except that R_i is replaced with an inductor L, called the *coupling coil.*

Basic Circuit Action. The action of the impedance-coupling network is similar to that of the *RC*-coupling network described in Art. 13-2. It differs mainly

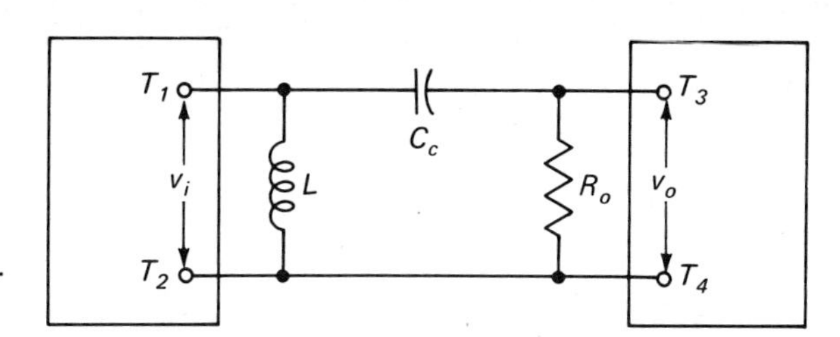

Fig. 13-10 Basic impedance-coupling circuit.

in that the impedance of the coupling coil to a-c signals is dependent on (1) the inductance of the coil, and (2) the frequency of the signal.

Circuit Characteristics. The characteristics of impedance coupling are similar to those of RC coupling. An advantage of impedance coupling over RC coupling is that the resistance of the input element L is much lower than the resistance of the RC-coupling input element R_i and therefore lower power-supply voltages can be used.

Frequency Characteristics. Because the impedance of the input inductor L varies directly with the frequency while the impedance of the input resistor R_i is constant, the frequency characteristics of impedance coupling are not so good as those obtained with RC coupling.

13-10 Impedance-coupled Two-stage Transistor Amplifier

The Circuit. The basic impedance-coupling network of Fig. 13-10 is used in the circuit of Fig. 13-11 to couple the two stages of the amplifier by means of the coupling network L, C_2, R_3. The output of the second stage is coupled to the load by the RC-coupling network R_4, C_3, R_5.

Circuit Operation. The operation of the circuit of Fig. 13-11 is the same as that of the RC-coupled amplifier of Fig. 13-4 described in Art. 13-4.

Voltage Gain. The overall voltage gain of the amplifier is equal to the product of the gains of its component stages, thus

$$A_v = A_{v\cdot 1}A_{v\cdot 2}A_{v\cdot 3} \quad \text{etc.} \tag{13-1}$$

From Fig. 13-11, it can be shown that the voltage gain of the first stage (impedance-coupled) is

$$A_{v\cdot 1} = \frac{Z_{o\cdot 1}}{r_{e\cdot 1}} \tag{13-8}$$

where $Z_{o\cdot 1}$ = impedance of X_L and $r_{i\cdot 2}$ in parallel
$r_{i\cdot 2}$ = resistance of R_3 and $\beta_2 r_{e\cdot 2}$ in parallel
When $X_L \gg r_{i\cdot 2}$, $Z_{o\cdot 1} \cong r_{i\cdot 2}$

and

$$A_{v\cdot 1} = \frac{r_{i\cdot 2}}{r_{e\cdot 1}} \tag{13-8a}$$

The voltage gain of the second stage (RC-coupled) can be found in the manner used in Example 13-1.

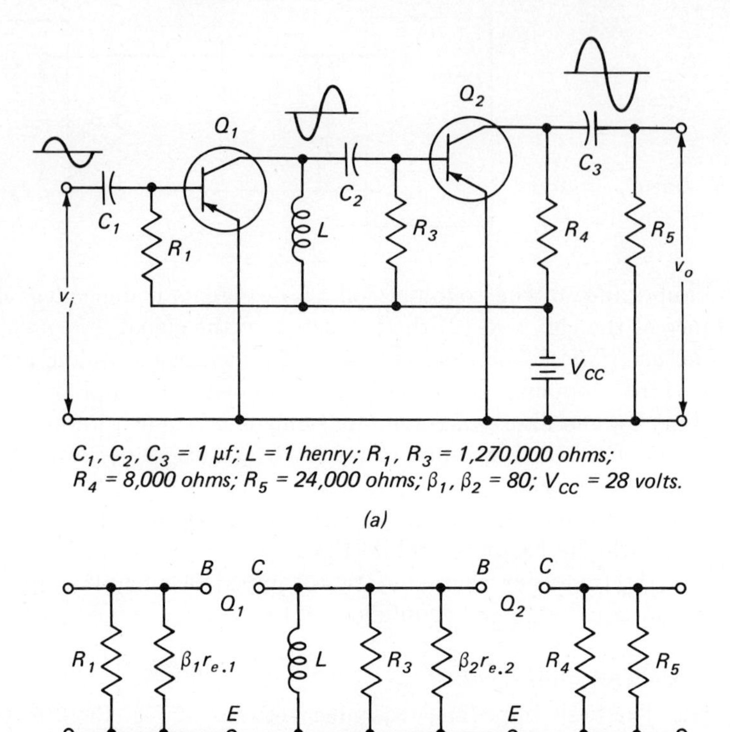

Fig. 13-11 Two-stage impedance-coupled transistor amplifier. (a) Circuit diagram. (b) Simplified a-c equivalent circuit at the mid-frequency range where effects of all capacitances are negligible.

Example 13-8 For the circuit of Fig. 13-11, what is (*a*) the voltage gain of the second stage? (*b*) The voltage gain of the first stage at a frequency of 2,000 Hz? (*c*) The voltage gain of the amplifier? (*d*) The decibel gain of the amplifier?

GIVEN: Fig. 13-11

FIND: (*a*) $A_{v \cdot 2}$ (*b*) $A_{v \cdot 1}$ (*c*) A_v (*d*) db

SOLUTION:

(*a*)
$$A_{v \cdot 2} = \frac{r_{o \cdot 2}}{r_{e \cdot 2}} = \frac{6{,}000}{14.2} \cong 420$$

where
$$r_{o \cdot 2} = \frac{R_4 R_5}{R_4 + R_5} = \frac{8{,}000 \times 24{,}000}{8{,}000 + 24{,}000} = 6{,}000 \text{ ohms}$$

$$r_{e \cdot 2} = \frac{25}{I_{E \cdot 2}} = \frac{25}{1.76} \cong 14.2 \text{ ohms}$$

$$I_{E \cdot 2} = \beta_2 I_{B \cdot 2} = \frac{\beta_2 V_{CC}}{R_3} = \frac{80 \times 28}{1{,}270{,}000} \cong 1.76 \text{ ma}$$

(b) $$A_{v \cdot 1} = \frac{Z_{o \cdot 1}}{r_{e \cdot 1}} \cong \frac{r_{i \cdot 2}}{r_{e \cdot 1}} \cong \frac{1{,}140}{14.2} \cong 80$$

where $$X_L = 2\pi fL = 6.28 \times 2{,}000 \times 1 = 12{,}560 \text{ ohms}$$

$$r_{i \cdot 2} = \frac{R_3 \beta_2 r_{e \cdot 2}}{R_3 + \beta_2 r_{e \cdot 2}} = \frac{1{,}270{,}000 \times (80 \times 14.2)}{1{,}270{,}000 + (80 \times 14.2)} \cong 1{,}140 \text{ ohms}$$

$$r_{e \cdot 1} = \frac{25}{I_{E \cdot 1}} = \frac{25}{1.76} \cong 14.2 \text{ ohms}$$

$$I_{E \cdot 1} = \beta_1 I_{B \cdot 1} = \frac{\beta_1 V_{CC}}{R_1} = \frac{80 \times 28}{1{,}270{,}000} \cong 1.76 \text{ ma}$$

(c) $$A_v = A_{v \cdot 1} A_{v \cdot 2} = 80 \times 420 = 33{,}600$$

(d) $$\text{db} = 20 \log A_v = 20 \log 33{,}600 = 20 \times 4.526 \cong 90.5 \text{ db}$$

13-11 Impedance-coupled Two-stage Vacuum-tube Amplifier

The Circuit. The basic impedance-coupled network of Fig. 13-10 is used in the circuit of Fig. 13-12 to couple two single-stage triode amplifiers by means of the coupling network L, C_2, R_3. This circuit differs from that of Fig. 13-8 only in that (1) an inductor L is substituted for the resistor R_2, and (2) the output load resistor R_5 is included. The output of the second stage is RC-coupled to the load on the amplifier.

Circuit Operation. The operation of the circuit of Fig. 13-12 is the same as that of the RC-coupled amplifier of Fig. 13-8 described in Art. 13-7.

Voltage Gain. The overall voltage gain of the amplifier is equal to the product of the gains of its component stages, thus

$$A_v = A_{v \cdot 1} A_{v \cdot 2} A_{v \cdot 3} \qquad \text{etc.} \tag{13-1}$$

By substituting L for R_2 in Fig. 13-8b it can be shown that voltage gain for the first stage is

$$A_{v \cdot 1} = \frac{\mu Z_{\text{eq}}}{r_{p \cdot 1} + Z_{\text{eq}}} \tag{13-9}$$

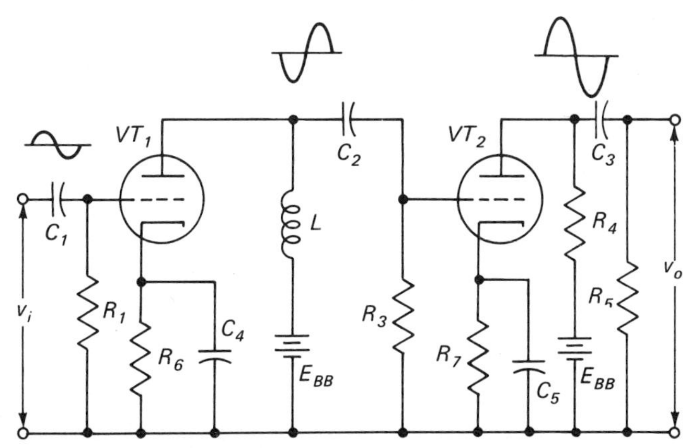

Fig. 13-12 Two-stage impedance-coupled vacuum-tube amplifier.

Note: · under $\underset{\cdot}{r}$ and $\underset{\cdot}{Z}$ indicates vector addition.

where

$$Z_{\text{eq}} = \text{vector sum of } \underset{\cdot}{X}_L \text{ and } \underset{\cdot}{R}_3 \text{ in parallel}$$

When $R_3 \gg X_L$, $Z_{\text{eq}} \cong X_L$

and
$$A_{v\cdot1} \cong \frac{\mu X_L}{\sqrt{r_{p\cdot1}^2 + X_L^2}} \tag{13-9a}$$

When $X_L \gg R_3$, $Z_{\text{eq}} \cong R_3$

and
$$A_{v\cdot1} \cong \frac{\mu R_3}{r_{p\cdot1} + R_3} \tag{13-9b}$$

The voltage gain of the second stage is equal to

$$A_{v\cdot2} = \frac{\mu r_{o\cdot2}}{r_{p\cdot2} + r_{o\cdot2}} \tag{13-10}$$

where
$$r_{o\cdot2} = \frac{R_4 R_5}{R_4 + R_5}$$

13-12 Parallel Plate Feed for Impedance-coupled Amplifier

In the impedance-coupled circuit of Fig. 13-12, direct current flows through the inductor in addition to the desired a-c signal currents. If the inductor has an iron core, any change in the direct current can change appreciably the magnetic strength in the iron core and therefore also the inductance of the coil. This condition can be corrected by the circuit of Fig. 13-13, called a *parallel-plate-feed amplifier*. In this circuit, the d-c portion of the plate current is prevented from reaching the inductor by the capacitor C_6 and returns to the power source through resistor R_2. The capacitor allows only the a-c signal to pass through the inductor. The voltage gain will have a high peak

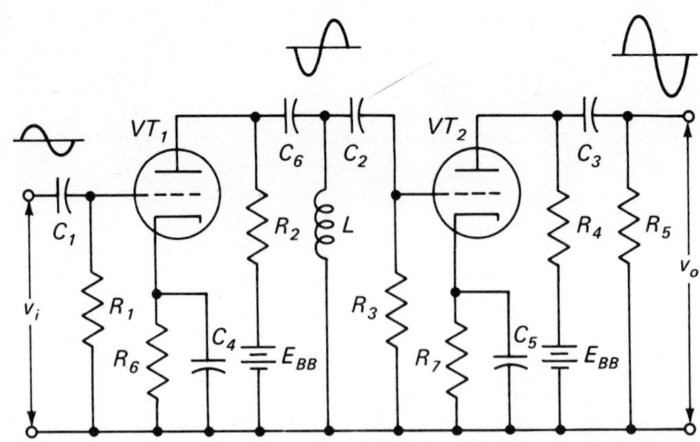

Fig. 13-13 Impedance-coupled vacuum-tube amplifier with parallel plate feed.

(Fig. 14-15) at the resonant frequency of the series-tuned circuit formed by L and C_6. Selecting appropriate values for L and C_6 can place this peak at the preferred frequency value.

13-13 Transformer Coupling

Basic Circuit. The transformer-coupling network shown in Fig. 13-14 is the simplest type of coupling arrangement, as only four terminals are required. The transformer may have (1) an iron core as with low-frequency applications, (2) an air core as with high-frequency applications, or (3) a powdered-iron core and/or highly magnetic alloys for special applications.

Circuit Operation. The input signal is applied to the primary winding of the transformer and is coupled to the secondary winding (and the output load) by magnetic induction. As there is no conductive coupling between the input and output circuits, any d-c voltage on the input side is isolated from the output side. The a-c input signals are coupled by magnetic induction to the output circuit.

Characteristics. The basic principles of transformer coupling were presented in Arts. 12-2 and 12-8. Transformer-coupled amplifiers (1) provide a higher voltage gain, (2) accomplish impedance matching between the input and the output load, and (3) produce a combination of higher voltage gain and impedance matching.

The transformer is a very efficient device, and iron-core transformers have efficiencies of over 90 per cent. Among the characteristics of transformer coupling are (1) it introduces inductance into both the input and output circuits, (2) leakage inductance exists between the primary and secondary windings and vice versa, (3) both the primary and secondary windings introduce shunting (distributed) capacitances to the circuit. Also, a power loss is introduced by the iron core of the transformer, but since this is usually very small its effect is disregarded in the following discussions on transformer coupling.

Voltage Conversion. In addition to coupling a-c signals from one circuit to another, transformer coupling can produce greater voltage gains than the gain produced by the active element in the amplifier circuit. Thus, higher voltage gains are possible with transformer coupling than with other methods of coupling.

When the secondary winding has more turns than the primary winding, the voltage at the output of the transformer will be higher than the voltage at

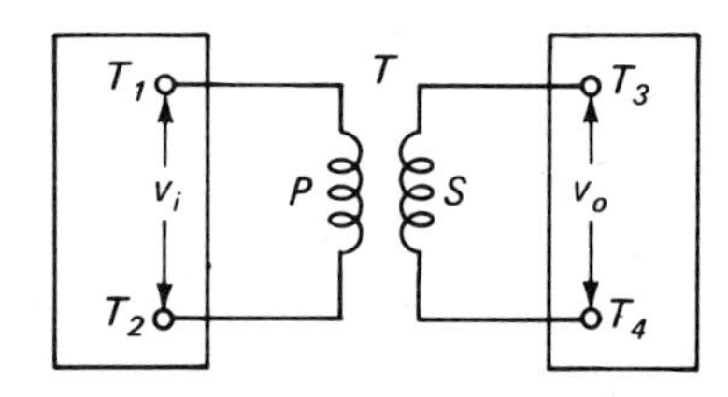

Fig. 13-14 Basic transformer-coupling circuit.

the input to the transformer by the ratio of the secondary to the primary turns. Mathematically,

$$\frac{v_o}{v_i} = \frac{N_S}{N_P} \tag{13-11}$$

and

$$v_o = v_i \frac{N_S}{N_P} \tag{13-11a}$$

where N_S = number of turns on secondary winding
N_P = number of turns on primary winding

Impedance Conversion. The transformer also changes the effective value of the output load impedance so that the impedance presented by the primary-winding terminals will be either higher or lower than the actual load impedance, depending on the ratio of the primary to secondary turns, as is indicated by

$$\frac{Z_i}{Z_o} = \left(\frac{N_P}{N_S}\right)^2 \tag{13-12}$$

and

$$Z_i = Z_o \left(\frac{N_P}{N_S}\right)^2 \tag{13-12a}$$

Frequency Characteristics. At those frequencies in the middle of its normal frequency range, the impedances of the inductances, capacitances, and resistances introduced by the transformer usually have so little effect on the circuit operation that an approximate value of the voltage gain can be obtained even though these parameters are disregarded. In this case, the transformer is considered to be an ideal circuit element. The voltage gain at those frequencies for which the impedances introduced by the transformer cannot be disregarded is discussed in Art. 14-10. The gain versus frequency characteristics for transformer coupling are not so good as with impedance coupling and *RC* coupling.

13-14 Transformer-coupled Amplifier Using Transistors

A basic transistor amplifier circuit using a transformer to couple the output signal of the amplifier to a low-resistance load is shown in Fig. 13-15*a*. Figure 13-15*b* shows a circuit with the output signal being fed directly to the low-resistance load. An important advantage of transformer coupling is illustrated by the results of Examples 13-9 and 13-10.

Example 13-9 (*a*) Draw the a-c equivalent circuit for the amplifier of Fig. 13-15*a*, assuming an ideal transformer. (*b*) What is the voltage gain up to the primary of the transformer? (*c*) What is the signal voltage at the primary of the transformer? (*d*) What is the signal voltage at the load? (*e*) What is the overall voltage gain of the amplifier? (*f*) What is the maximum input-signal voltage the amplifier can accommodate without clipping? (*g*) Draw the d-c and a-c load lines.

GIVEN: Fig. 13-15*a*

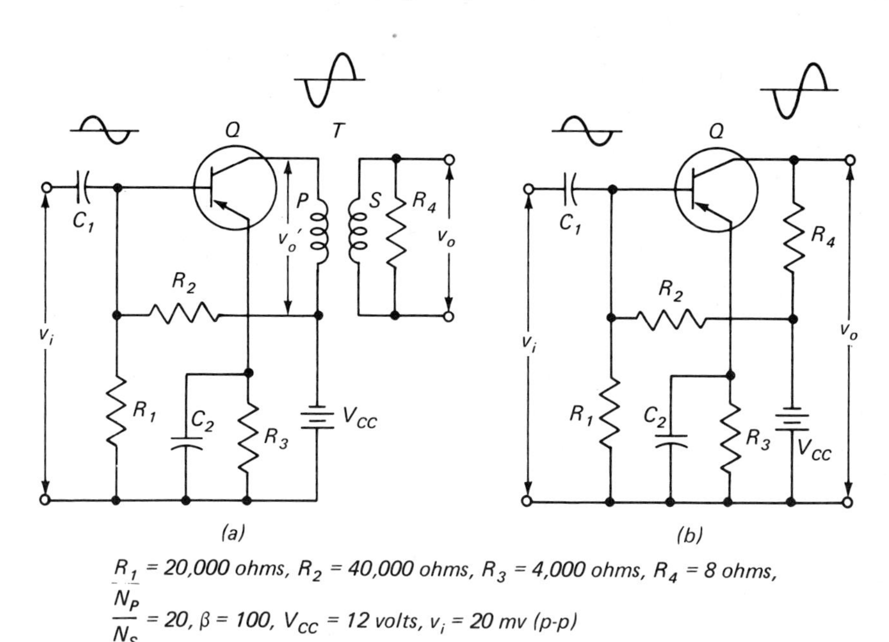

Fig. 13-15 (a) Transistor amplifier with transformer coupling to low-impedance load. (b) Amplifier of (a) without the transformer.

FIND: (*a*) Diagram (*b*) A_v' (*c*) v_o' (*d*) v_o (*e*) A_v (*f*) $v_{i\cdot\max}$
(*g*) d-c and a-c load lines

SOLUTION:

(*a*) Fig. 13-16*a*

(*b*) $$A_v' = \frac{r_o'}{r_e} = \frac{3{,}200}{25} = 128$$

where $$r_o' = R_4\left(\frac{N_P}{N_S}\right)^2 = 8 \times 20^2 = 3{,}200 \text{ ohms}$$

$$r_e = \frac{25}{I_E} = \frac{25}{1} = 25 \text{ ohms}$$

$$I_E = \frac{E_B}{R_3} = \frac{V_{CC}R_1}{(R_1 + R_2)R_3} = \frac{12 \times 20{,}000}{(20{,}000 + 40{,}000) \times 4{,}000} = 1 \text{ ma}$$

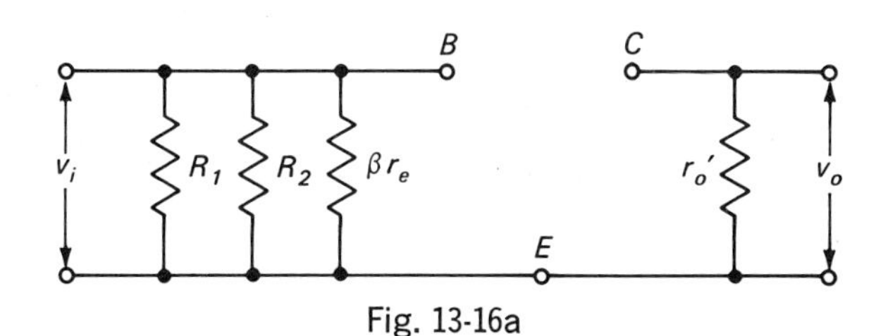

Fig. 13-16a

(c) $$v_o' = A_v'v_i = 128 \times 20 \times 10^{-3} = 2.56 \text{ volts}$$

(d) $$v_o = v_o' \frac{N_S}{N_P} = 2.56 \times \frac{1}{20} = 128 \text{ mv}$$

(e) $$A_v = \frac{v_o}{v_i} = \frac{128}{20} = 6.4$$

(f) The lower of $2V_{CE}$ or $2I_C r_o'$

$$2V_{CE} = 2(V_{CC} - I_C R_3) = 2(12 - 1 \times 10^{-3} \times 4{,}000) = 16 \text{ volts (p-p)}$$

$$2I_C r_o' = 2 \times 1 \times 10^{-3} \times 3{,}200 = 6.4 \text{ volts (p-p)}$$

$$v_{i \cdot \max} = \frac{2I_C r_o'}{A_v'} = \frac{6.4}{128} = 50 \text{ mv (p-p)}$$

(g) For d-c load line

$$V_{CE} = V_{CC} = 12 \text{ volts}$$

$$I_{C \cdot \text{sat}} = \frac{V_{CC}}{R_E + R_C} = \frac{V_{CC}}{R_3} = \frac{12}{4{,}000} = 3 \text{ ma}$$

For a-c load line

$$V_{CE} = V_{CE \cdot Q} + I_C r_o' = 8 + (1 \times 10^{-3} \times 3{,}200) = 11.2 \text{ volts}$$

$$V_{CE \cdot Q} = V_{CC} - I_C R_3 = 12 - (1 \times 10^{-3} \times 4{,}000) = 8 \text{ volts}$$

$$I_{C \cdot \text{sat}} = I_C + \frac{V_{CE \cdot Q}}{r_o'} = 1 + \frac{8}{3{,}200} = 3.5 \text{ ma}$$

For graph, see Fig. 13-16b

Example 13-10 (a) Draw the a-c equivalent circuit for the amplifier of Fig. 13-15b. (b) What is the voltage gain of this amplifier? (c) What is the output voltage of the

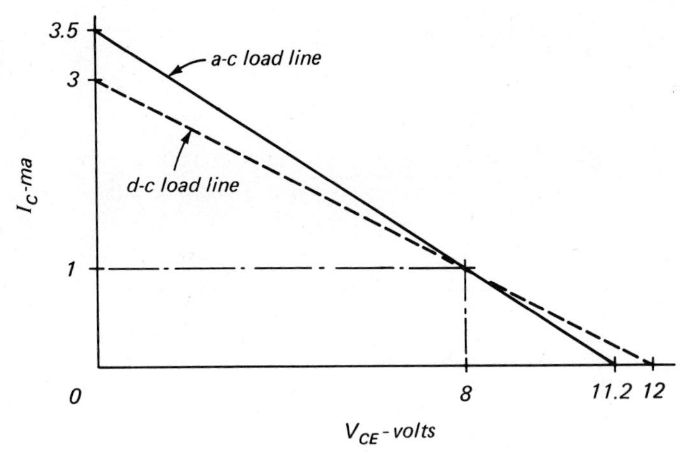

Fig. 13-16b

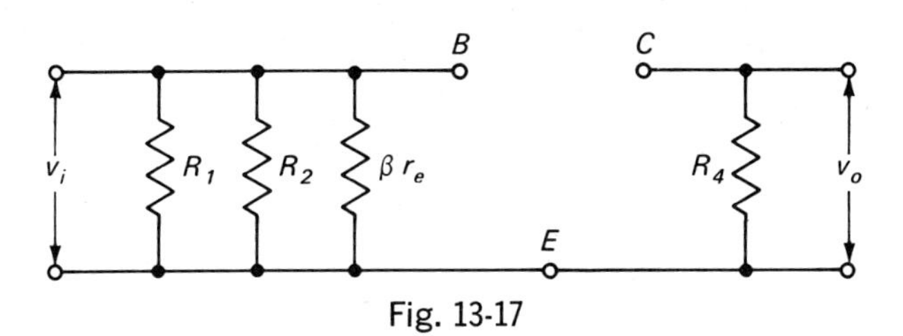

Fig. 13-17

amplifier? (*d*) What is the maximum input-signal voltage the amplifier can accommodate without clipping?

GIVEN: Fig. 13-15*b*

FIND: (*a*) Diagram (*b*) A_v (*c*) v_o (*d*) $v_{i \cdot \max}$

SOLUTION:

(*a*) Fig. 13-17

(*b*) $$A_v = \frac{r_o}{r_e} = \frac{8}{25} = 0.32$$

where $r_e = 25$ see Solution 13-9 (*b*)

(*c*) $$v_o = A_v v_i = 0.32 \times 20 = 6.4 \text{ mv}$$

(*d*) The lower of $2V_{CE}$ or $2I_C r_o$

$$2V_{CE} = 2(V_{CC} - I_C R_3) = 2(12 - 1 \times 10^{-3} \times 4{,}000) = 16 \text{ volts (p-p)}$$
$$2I_C r_o = 2 \times 1 \times 10^{-3} \times 8 = 16 \text{ mv (p-p)}$$
$$v_{i \cdot \max} = \frac{2I_C r_o}{A_v} = \frac{16}{0.32} = 50 \text{ mv (p-p)}$$

Comparing the results of Examples 13-9 and 13-10 shows that with transformer coupling (1) the output voltage increased from 6.4 to 128 mv, (2) the load seen by the collector of the transistor increased from 8 to 3,200 ohms, and (3) the peak-to-peak signal-handling capability at the collector of the transistor increased from 16 mv to 6.4 volts.

13-15 Transformer-coupled Amplifier Using Vacuum Tubes

The Circuit. Figure 13-18 illustrates two applications of transformer coupling. Transformer T_1 is used as an interstage coupling element to (1) couple the output of VT_1 to the input of VT_2, and (2) step up the voltage of the a-c signal. Transformer T_2 is used as an output transformer to (1) couple the output of VT_2 to the load R_4, and (2) increase the effective load impedance in the plate circuit of VT_2.

Circuit Operation. In the cascade amplifier of Fig. 13-18, (1) the input signal v_i is amplified at VT_1 by the factor μ_1 and its phase is reversed, (2) the output of VT_1 is coupled to the input of VT_2 by the step-up transformer T_1

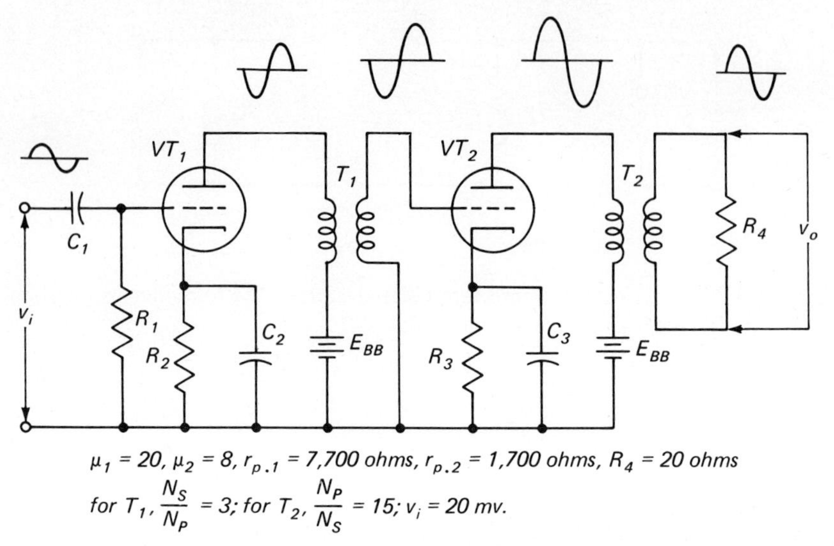

Fig. 13-18 Two-stage transformer-coupled amplifier using triode vacuum tubes.

which also increases the signal voltage threefold because of its 1 to 3 turns ratio, (3) the signal at the grid of VT_2 is amplified at VT_2 by the factor μ_2 and its phase is reversed, (4) the output of VT_2 is coupled to the load R_4 by the step-down transformer T_2 which (*a*) increases the effective load resistance seen by the plate of VT_2 and (*b*) decreases the signal voltage by the 15 to 1 turns ratio of T_2. Because the signal voltage was reversed at VT_1 and again at VT_2, the phase of the output signal v_o is the same as that of the input signal v_i.

Voltage Gain. The two equations presented here are for the frequencies in the mid-frequency range where the effects of the transformer constants introduced into the circuit are considered to be negligible. The voltage gains for frequencies above and below the mid-frequencies are discussed in Art. 14-11.

The voltage gain of the first stage is dependent chiefly on (1) the amplification factor of VT_1, and (2) the turns ratio of T_1. Because the secondary of T_1 is connected directly to the grid-to-cathode circuit of VT_2, the load impedance seen by the plate of VT_1 is very high compared to $r_{p\cdot 1}$ and therefore the quantity $Z_{o\cdot 1}/r_{p\cdot 1} + Z_{o\cdot 1}$ [as appears in Eq. (13-9)] is approximately 1 and thus can be omitted. Therefore,

$$A_{v\cdot 1} \cong \mu_1 \frac{N_{S\cdot 1}}{N_{P\cdot 1}} \tag{13-13}$$

The voltage gain of the second stage is dependent on (1) the amplification factor of VT_2, (2) the turns ratio of T_2, and (3) the load resistance seen by the plate of VT_2 and the plate resistance of VT_2. Because the secondary of T_2 is connected to a low-impedance load (R_4) and the quantity $Z_{o\cdot 2}/r_{p\cdot 2} + Z_{o\cdot 2}$ is substantially less than 1, this quantity cannot be disregarded as was the case in the first stage. Therefore,

$$A_{v\cdot 2} = \mu_2 \times \frac{r_{o\cdot 2}'}{r_{p\cdot 2} + r_{o\cdot 2}'} \times \frac{N_{S\cdot 2}}{N_{P\cdot 2}} \tag{13-14}$$

where $$r_{o\cdot 2}' = R_4 \left(\frac{N_{P\cdot 2}}{N_{S\cdot 2}}\right)^2$$

Example 13-11 For the amplifier circuit of Fig. 13-18 assume that (1) T_1 and T_2 are ideal circuit elements, and (2) the voltage gains are to be determined for those frequencies in the mid-frequency range. Find (*a*) the voltage gain of the first stage, (*b*) the voltage gain of the second stage, (*c*) the voltage gain of the cascade amplifier, (*d*) the a-c output-signal voltage, (*e*) the load impedance seen by the plate of VT_2.

GIVEN: Fig. 13-18

FIND: (*a*) $A_{v\cdot 1}$ (*b*) $A_{v\cdot 2}$ (*c*) A_v (*d*) v_o (*e*) $r_{o\cdot 2}'$

SOLUTION:

(*a*) $$A_{v\cdot 1} = \mu_1 \frac{N_{S\cdot 1}}{N_{P\cdot 1}} = 20 \times 3 = 60$$

(*b*) $$A_{v\cdot 2} = \mu_2 \times \frac{r_{o\cdot 2}'}{r_{p\cdot 2} + r_{o\cdot 2}'} \times \frac{N_{S\cdot 2}}{N_{P\cdot 2}} = 8 \times \frac{4{,}500}{1{,}700 + 4{,}500} \times \frac{1}{15} \times 0.387$$

where $$r_{o\cdot 2}' = R_4 \left(\frac{N_{P\cdot 2}}{N_{S\cdot 2}}\right)^2 = 20 \times 15^2 = 4{,}500 \text{ ohms}$$

(*c*) $$A_v = A_{v\cdot 1}A_{v\cdot 2} = 60 \times 0.387 = 23.2$$

(*d*) $$v_o = A_v v_i = 23.2 \times 20 \times 10^{-3} = 0.464 \text{ volt}$$

(*e*) $$r_{o\cdot 2}' = R_4 \left(\frac{N_{P\cdot 2}}{N_{S\cdot 2}}\right)^2 = 20 \times 15^2 = 4{,}500 \text{ ohms}$$

13-16 Direct Coupling

In some applications it becomes desirable to connect a load device directly in series with the output electrode of the active circuit element; such an arrangement is called a *direct-coupled circuit*. The load device may be (1) a d-c relay, (2) a d-c meter, (3) headphones, (4) a loudspeaker, (5) the input circuit of a transistor, etc. This method of connection is permissible when (1) the d-c component of the output current does not interfere with the normal operation of the device, and (2) the resistance of the device is low enough to avoid appreciable reduction of the voltages at the electrodes.

Because circuits can be designed to operate without the use of frequency-sensitive components such as capacitors, inductors, or transformers, direct coupling can be used for amplifiers required to amplify (1) a-c signals with frequencies as low as a fraction of a hertz, and (2) changes in d-c voltages. Because of its ability to amplify d-c voltage changes, this circuit is also referred to as a *d-c* or *direct-current amplifier*. Transistors are more readily adaptable than vacuum tubes for use in direct-coupled amplifiers. Some applications of direct-coupled amplifiers are (1) regulator circuits of power supplies, (2) pulse amplifiers, (3) differential amplifiers, (4) impedance matching, (5) computer circuitry, (6) electronic instruments.

13-17 Direct-coupled Two-stage Amplifier Using Two Similar Transistors

The circuit of Fig. 13-19*a* illustrates a basic cascade direct-coupled amplifier using similar transistors (i.e., both are NPN types), each connected in the common-emitter configuration. Both stages employ direct coupling as (1) the collector of the transistor in the first stage is connected directly to the base of the transistor in the second stage, and (2) the load (R_2) is connected directly to the collector of the transistor in the second stage. Resistor R_1 establishes the forward bias for Q_1 and also indirectly for Q_2. Observing that the collector current of Q_1 is also the base current of Q_2, any signal current at the base of Q_1 is amplified β_1 times at Q_1 and again amplified β_2 times at Q_2. Thus, the signal-current gain of the amplifier is

$$A_i = \beta_1\beta_2 \tag{13-15}$$

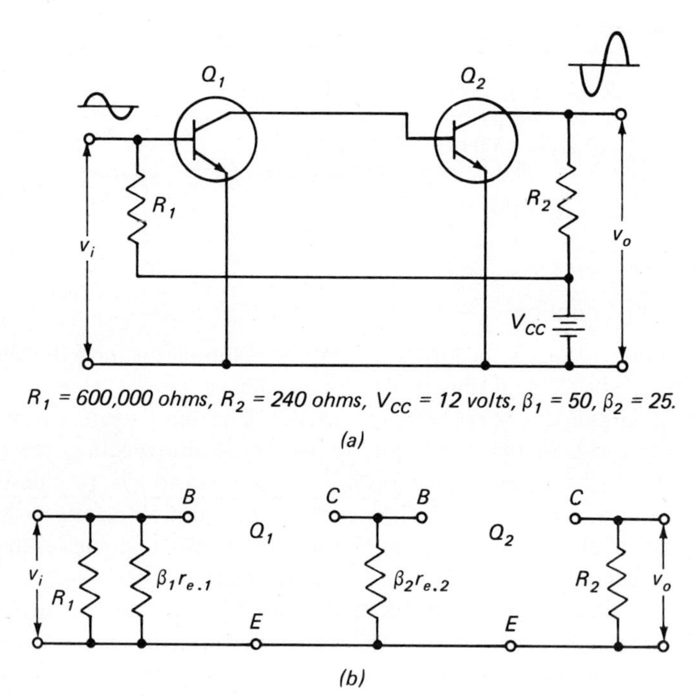

Fig. 13-19 Basic direct-coupled amplifier using similar transistors. (a) Circuit diagram. (b) A-c equivalent circuit.

The first stage does not produce any voltage gain as is shown in the following analysis of the a-c equivalent circuit (Fig. 13-19*b*).

$$A_{v\cdot 1} = \frac{r_{o\cdot 1}}{r_{e\cdot 1}} \tag{13-3}$$

From Fig. 13-19*b*, it is seen that $r_{o\cdot 1} = \beta_2 r_{e\cdot 2}$ which when substituted in Eq. (13-3) results in

$$A_{v\cdot 1} = \frac{\beta_2 r_{e\cdot 2}}{r_{e\cdot 1}} \tag{13-16}$$

also,

$$I_{E\cdot 2} = \beta_2 I_{B\cdot 2} \tag{13-17}$$

However, $I_{B\cdot 2} = I_{E\cdot 1}$ which when substituted in Eq. (13-17) results in

$$I_{E\cdot 2} = \beta_2 I_{E\cdot 1} \tag{13-17a}$$

and

$$\beta_2 = \frac{I_{E\cdot 2}}{I_{E\cdot 1}} \tag{13-17b}$$

also,

$$r_{e\cdot 1} = \frac{25}{I_{E\cdot 1}} \tag{13-18}$$

$$r_{e\cdot 2} = \frac{25}{I_{E\cdot 2}} \tag{13-19}$$

Substituting Eqs. (13-17*b*), (13-18), and (13-19) in Eq. (13-16), then

$$A_{v\cdot 1} = \frac{I_{E\cdot 2}}{I_{E\cdot 1}} \times \frac{25}{I_{E\cdot 2}} \times \frac{I_{E\cdot 1}}{25} \tag{13-16a}$$

after factoring,

$$A_{v\cdot 1} = 1 \tag{13-16b}$$

The voltage gain of the second stage can be determined directly with Eq. (13-2*b*).

Example 13-12 For the circuit of Fig. 13-19*a*, find (*a*) the current gain of the amplifier, (*b*) the voltage gain of the first stage, (*c*) the voltage gain of the second stage, (*d*) the voltage gain of the amplifier, (*e*) the power gain (in decibels) for the amplifier, (*f*) the input resistance of the amplifier.

GIVEN: Fig. 13-19

FIND: (*a*) A_i (*b*) $A_{v\cdot 1}$ (*c*) $A_{v\cdot 2}$ (*d*) A_v (*e*) G_p (*f*) r_i

SOLUTION:

(*a*) $A_i = \beta_1 \beta_2 = 50 \times 25 = 1{,}250$

(*b*) $A_{v\cdot 1} = 1$

(*c*) $A_{v\cdot 2} = \dfrac{r_{o\cdot 2}}{r_{e\cdot 2}} = \dfrac{240}{1} = 240$

where $$r_{e \cdot 2} = \frac{25}{I_{E \cdot 2}} = \frac{25}{25} = 1 \text{ ohm}$$

$$I_{E \cdot 2} = \beta_2 I_{E \cdot 1} = 25 \times 1 = 25 \text{ ma}$$

$$I_{E \cdot 1} = \beta_1 I_{B \cdot 1} = \beta_1 \frac{V_{CC}}{R_1} = 50 \times \frac{12}{600{,}000} = 1 \text{ ma}$$

(*d*) $$A_v = A_{v \cdot 1} A_{v \cdot 2} = 1 \times 240 = 240$$

(*e*) $$A_p = A_i A_v = 1{,}250 \times 240 = 300{,}000$$

$$G_p = 10 \log A_p = 10 \log 300{,}000 = 10 \times 5.476 \cong 55 \text{ db}$$

(*f*) $$r_i = \frac{R_1 \beta_1 r_{e \cdot 1}}{R_1 + \beta_1 r_{e \cdot 1}} = \frac{600{,}000 \times (50 \times 25)}{600{,}000 + (50 \times 25)} \cong 1{,}250 \text{ ohms}$$

The circuit of Fig. 13-19*a* uses a minimum number of circuit components. A disadvantage of this circuit is that any variation in the bias current, due to a change in temperature, in one stage is amplified in the following stage (or stages); thus this circuit has poor temperature-stability characteristics. For this reason, only a few stages of direct-coupled amplification are practicable.

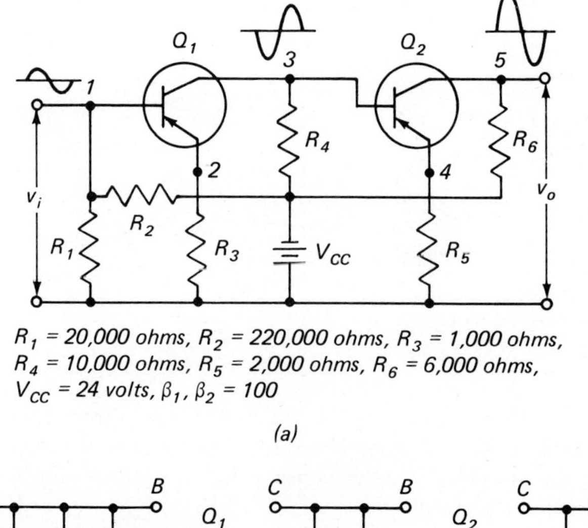

$R_1 = 20{,}000$ ohms, $R_2 = 220{,}000$ ohms, $R_3 = 1{,}000$ ohms,
$R_4 = 10{,}000$ ohms, $R_5 = 2{,}000$ ohms, $R_6 = 6{,}000$ ohms,
$V_{CC} = 24$ volts, $\beta_1, \beta_2 = 100$

(a)

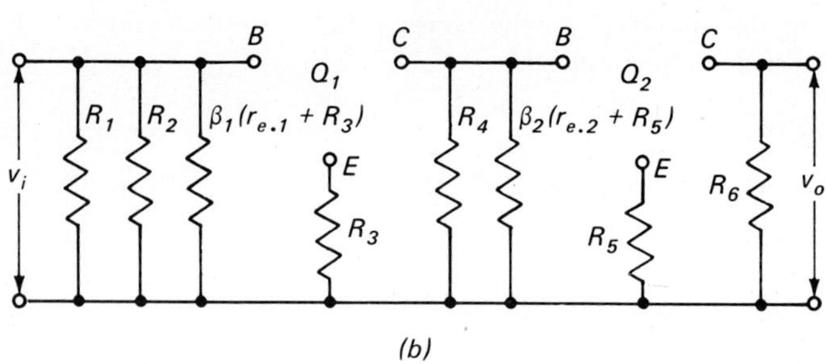

(b)

Fig. 13-20 Two-stage direct-coupled amplifier using two PNP transistors. (a) Circuit diagram. (b) A-c equivalent circuit.

In the circuit of Fig. 13-20*a*, the emitter (swamping) resistors R_3 and R_5 reduce the effect of temperature instability. However, the emitter resistors introduce degenerative feedback and consequently reduce the gain of the amplifier. In this circuit, the voltage divider R_1R_2 together with the emitter resistor R_3 determine the emitter current of Q_1. Resistor R_4 performs two functions: (1) it is the load resistor for Q_1, and (2) it establishes the bias voltage for Q_2.

Example 13-13 For the amplifier of Fig. 13-20*a*, find (*a*) the voltage gain of the first stage, (*b*) the voltage gain of the second stage, (*c*) the voltage gain of the amplifier, (*d*) the input resistance of the amplifier.

GIVEN: Fig. 13-20*a*

FIND: (*a*) $A_{v\cdot 1}$ (*b*) $A_{v\cdot 2}$ (*c*) A_v (*d*) r_i

SOLUTION:

(*a*)
$$A_{v\cdot 1} = \frac{r_{o\cdot 1}}{r_{e\cdot 1} + R_3} = \frac{9{,}500}{12.5 + 1{,}000} \cong 9.4$$

where
$$r_{o\cdot 1} = \frac{R_4\beta_2(r_{e\cdot 2} + R_5)}{R_4 + \beta_2(r_{e\cdot 2} + R_5)} = \frac{10{,}000 \times [100(12.5 + 2000)]}{10{,}000 + [100(12.5 + 2{,}000)]} \cong 9{,}500 \text{ ohms}$$

$$r_{e\cdot 1} = \frac{25}{I_{E\cdot 1}} = \frac{25}{2} = 12.5 \text{ ohms}$$

$$I_{E\cdot 1} = \frac{V_{CC}R_1}{(R_1 + R_2)R_3} = \frac{24 \times 20{,}000}{(20{,}000 + 220{,}000)1{,}000} = 2 \text{ ma}$$

(*b*)
$$A_{v\cdot 2} = \frac{R_6}{r_{e\cdot 2} + R_5} = \frac{6{,}000}{12.5 + 2{,}000} \cong 3$$

where
$$r_{e\cdot 2} = \frac{25}{I_{E\cdot 2}} = \frac{25}{2} = 12.5 \text{ ohms}$$

$$I_{E\cdot 2} = \frac{V_{CC} - I_{E\cdot 1}R_4}{R_5} = \frac{24 - (2 \times 10^{-3} \times 10{,}000)}{2{,}000} = 2 \text{ ma}$$

(*c*)
$$A_v = A_{v\cdot 1}A_{v\cdot 2} = 9.4 \times 3 = 28.2$$

(*d*)
$$r_i = \frac{1}{\dfrac{1}{R_1} + \dfrac{1}{R_2} + \dfrac{1}{\beta_1(r_{e\cdot 1} + R_3)}} = \frac{1}{\dfrac{1}{20{,}000} + \dfrac{1}{220{,}000} + \dfrac{1}{100(12.5 + 1{,}000)}}$$
$$\cong 15{,}500 \text{ ohms}$$

Example 13-14 (*a*) For the circuit of Fig. 13.20*a*, what are the d-c voltages at points 1, 2, 3, 4, and 5? (Assume V_{BE} to be negligible.) (*b*) If a 2-volt d-c input signal increases by 0.05 volt, what is the amount of voltage variation at the output of the amplifier?

GIVEN: Fig. 13-20*a* $v_i = 2$ volts $\Delta v_i = 0.05$ volt $A_v = 28.2$

FIND: (*a*) V_1, V_2, V_3, V_4, V_5 (*b*) Δv_o

SOLUTION:

(a)
$$V_1 = \frac{V_{CC}R_1}{R_1 + R_2} = \frac{24 \times 20{,}000}{20{,}000 + 220{,}000} = -2 \text{ volts}$$

$$V_2 \cong V_1 \cong -2 \text{ volts}$$

$$V_3 = V_{CC} - I_{C\cdot 1}R_4 = 24 - (2 \times 10^{-3} \times 10{,}000) = -4 \text{ volts}$$

where
$$I_{C\cdot 1} \cong I_{E\cdot 1} = \frac{V_2}{R_3} = \frac{2}{1{,}000} = 2 \text{ ma}$$

$$V_4 \cong V_3 \cong -4 \text{ volts}$$

$$V_5 = V_{CC} - I_{C\cdot 2}R_6 = 24 - (2 \times 10^{-3} \times 6{,}000) = -12 \text{ volts}$$

where
$$I_{C\cdot 2} \cong I_{E\cdot 2} = \frac{V_4}{R_5} = \frac{4}{2{,}000} = 2 \text{ ma}$$

(b)
$$\Delta v_o = A_v \, \Delta v_i = 28.2 \times 0.05 \cong 1.4 \text{ volts}$$

13-18 Direct-coupled Amplifier Using the Complementary Symmetry of Two Transistors

Basic Principle. An NPN transistor may be coupled to a similar PNP transistor (or vice versa) by means of direct coupling (Fig. 13-21). This method of coupling uses the principle that the two types of transistors are symmetrical counterparts of each other; that is, they have the same beta ratings and the same operating voltages but of the opposite polarity. For each operating point of an NPN transistor there is an equivalent operating point for a similar PNP transistor. The polarity of the input signal required to increase conduction in an NPN transistor is opposite to that required to increase conduction in a PNP transistor. A circuit using this method of coupling is called a *complementary symmetrical circuit.*

Basic Circuit. The circuit of Fig. 13-21 illustrates a two-stage cascade amplifier using one NPN transistor and one PNP transistor, each connected in

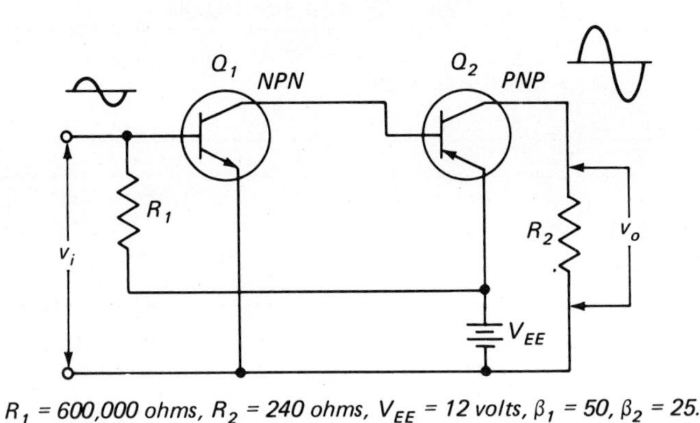

R_1 = 600,000 ohms, R_2 = 240 ohms, V_{EE} = 12 volts, β_1 = 50, β_2 = 25.

Fig. 13-21 Basic direct-coupled amplifier using complementary symmetry of two transistors.

the common-emitter configuration. This circuit differs from that of Fig. 13-19 in that (1) complementary transistors are used in place of two similar transistors, (2) the power source is relocated to provide the proper bias voltages for complementary symmetry operation, and (3) the output voltage is taken directly from across the load-resistor terminals as compared with the algebraic sum of the load-resistor voltage and the power-source voltage. The gains of the two circuits are the same.

Typical Circuit. In the circuit of Fig. 13-22, the bias current of Q_1 is determined by the voltage divider R_1R_2 and the emitter resistor R_3. Resistor R_4 performs two functions: (1) it is the load for Q_1, and (2) it establishes the bias voltage for Q_2. The emitter resistors R_3 and R_5 improve the stability of the amplifier.

When a positive-going signal is applied to the base of the NPN transistor Q_1, the circuit actions are (1) the base current increases, (2) the collector current increases, (3) the voltage drop at R_4 increases, (4) the voltage at the collector of Q_1 and the base of Q_2 becomes less positive—hence more negative, and (5) a negative-going signal is applied to the input of Q_2. The negative-going signal at the base of the PNP transistor Q_2 causes (1) an increase in the forward bias, (2) an increase in the collector current, (3) an increase in the voltage developed across R_6, and thus (4) an amplified positive-going output signal at R_6. Therefore, a signal applied to the input of a two-stage complementary symmetrical amplifier will appear at the output of the amplifier in an amplified form and of the same polarity.

Example 13-15 For the amplifier of Fig. 13-22, find (*a*) the approximate voltage gain of the amplifier, (*b*) the d-c voltages at points 1, 2, 3, 4, and 5. (Assume R_3, $R_5 \gg r_e$, and $V_E \gg V_{BE}$.)

GIVEN: Fig. 13-22

FIND: (*a*) A_v (*b*) V_1, V_2, V_3, V_4, V_5

SOLUTION:

(*a*)
$$A_v = A_{v\cdot 1}A_{v\cdot 2} \cong \frac{R_4}{R_3} \times \frac{R_6}{R_5} \cong \frac{6{,}000}{2{,}000} \times \frac{12{,}000}{6{,}000} \cong 6$$

(*b*)
$$V_1 = \frac{V_{EE}R_1}{R_1 + R_2} = \frac{24 \times 20{,}000}{20{,}000 + 220{,}000} = 2 \text{ volts}$$

$$V_2 \cong V_1 \cong 2 \text{ volts}$$

$$V_3 = V_{EE} - I_{C\cdot 1}R_4 = 24 - (1 \times 10^{-3} \times 6{,}000) = 18 \text{ volts}$$

where
$$I_{c\cdot 1} = \frac{V_2}{R_3} = \frac{2}{2{,}000} = 1 \text{ ma}$$

$$V_4 \cong V_3 \cong 18 \text{ volts}$$

$$V_5 = I_{C\cdot 2}R_6 = 1 \times 10^{-3} \times 12{,}000 = 12 \text{ volts}$$

where
$$I_{C\cdot 2} = \frac{V_{EE} - V_4}{R_5} = \frac{24 - 18}{6{,}000} = 1 \text{ ma}$$

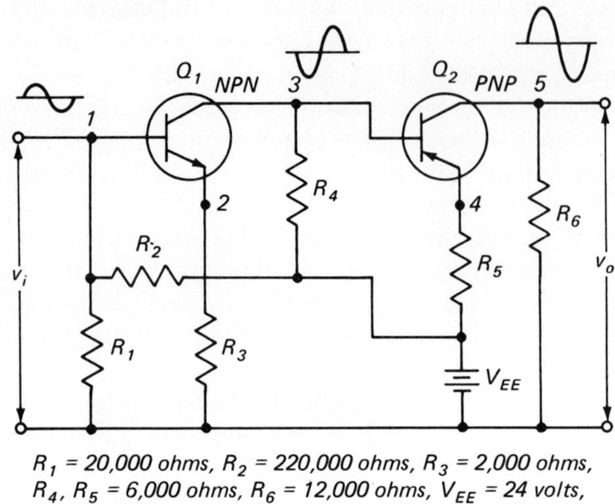

R_1 = 20,000 ohms, R_2 = 220,000 ohms, R_3 = 2,000 ohms, R_4, R_5 = 6,000 ohms, R_6 = 12,000 ohms, V_{EE} = 24 volts, β_1, β_2 = 100.

Fig. 13-22 Two-stage direct-coupled amplifier using complementary symmetry of two transistors.

13-19 Direct-coupled Cascade Amplifiers Using Two Different Configurations

Principle. Amplifiers can be designed with different configurations for the transistors in the individual stages. The use of such amplifiers is determined by the characteristics desired of the amplifier. Table 13-2 shows the relative impedance match of the output circuit of a transistor with the input circuit of another transistor for the various common electrode connections. A good match is indicated when two impedances are of the same order of magnitude.

Common-base Coupled to Common-emitter. Figure 13-23*a* shows a two-stage direct-coupled cascade amplifier employing complementary symmetry in which the first stage uses an NPN transistor connected in the common-base configuration and the second stage uses a PNP transistor connected in the

Table 13-2 Relative Impedance Matching of Transistors

FROM	TO: COMMON EMITTER	COMMON BASE	COMMON COLLECTOR
Common emitter	Fair	Poor	Good
Common base	Poor to fair	Poor	Good
Common collector	Good	Good	Poor

common-emitter configuration. Two characteristics of this circuit are (1) the input resistance is low, and (2) the phase of the output voltage v_o is opposite that of the input voltage v_i since phase reversal occurs at the common-emitter stage but not at the common-base stage (Art. 9-6).

Because the two transistors are directly coupled, (1) the collector current of Q_1 returns to the power source through the emitter of Q_2, and (2) the collector current of Q_2 is equal to beta times the collector current of Q_1. The emitter current of Q_2 is equal to the sum of the collector currents of Q_1 and Q_2. However, as the collector current of Q_2 is much greater than that of Q_1, it can be assumed that the emitter current of Q_2 is approximately equal to the collector current of Q_2.

The current amplification of the amplifier is

$$A_i = \alpha_1\beta_2 \qquad (13\text{-}20)$$

By the same reasoning as presented in Art. 13-17

$$A_{v\cdot 1} = 1 \qquad (13\text{-}16b)$$

Example 13-16 For the circuit of Fig. 13-23*a*, find (*a*) emitter current and emitter-diode resistance of Q_1, (*b*) emitter current and emitter-diode resistance of Q_2, (*c*) voltage gain of the first stage, (*d*) voltage gain of the second stage, (*e*) voltage gain of the amplifier.

GIVEN: Fig. 13-23*a*

FIND: (*a*) $I_{E\cdot 1}$, $r_{e\cdot 1}$ (*b*) $I_{E\cdot 2}$, $r_{e\cdot 2}$ (*c*) $A_{v\cdot 1}$ (*d*) $A_{v\cdot 2}$ (*e*) A_v

SOLUTION:

(*a*)
$$I_{E\cdot 1} = \frac{V_{CC}}{R_1} = \frac{12}{12{,}000} = 1 \text{ ma}$$
$$r_{e\cdot 1} = \frac{25}{I_{E\cdot 1}} = \frac{25}{1} = 25 \text{ ohms}$$

(*b*)
$$I_{E\cdot 2} \cong I_{C\cdot 2} \cong \beta_2 I_{E\cdot 1} = 25 \times 1 = 25 \text{ ma}$$
$$r_{e\cdot 2} = \frac{25}{I_{E\cdot 2}} = \frac{25}{25} = 1 \text{ ohm}$$

(*c*)
$$A_{v\cdot 1} = 1$$

(*d*)
$$A_{v\cdot 2} = \frac{R_2}{r_{e\cdot 2}} = \frac{240}{1} = 240$$

(*e*)
$$A_v = A_{v\cdot 1}A_{v\cdot 2} = 1 \times 240 = 240$$

Example 13-17 For the circuit of Example 13-16, find (*a*) current gain of the amplifier, (*b*) decibel power gain of the amplifier, (*c*) input resistance of the amplifier.

GIVEN: Example 13-16

FIND: (*a*) A_i (*b*) G_p (*c*) r_i

SOLUTION:

(*a*) $A_i = \alpha_1\beta_2 = 0.98 \times 25 \cong 25$

(*b*) $G_p = 10 \log A_p = 10 \log 6{,}000 = 10 \times 3.778 \cong 37.8 \text{ db}$

where $A_p = A_iA_v = 25 \times 240 = 6{,}000$

(*c*) $$r_i = \frac{R_1 r_{e\cdot 1}}{R_1 + r_{e\cdot 1}} = \frac{12{,}000 \times 25}{12{,}000 + 25} \cong 25 \text{ ohms}$$

13-20 Three-stage Direct-coupled Amplifier

Figure 13-24 illustrates the basic principle of a three-stage direct-coupled cascade amplifier. In this circuit, the collector current of Q_1 makes the base of Q_2 negative with respect to its emitter, and the collector current of Q_2 makes the base of Q_3 positive with respect to its emitter. It should be noted that the base current of Q_2 is the collector current of Q_1, and the base current of Q_3 is the collector current of Q_2. Thus it can be observed that successive transistors change types, that is, from NPN to PNP to NPN, and the amplifier uses a complementary symmetrical circuit. The resistors R_1, R_2, and R_3 are of low values and act as swamping resistors to attain good temperature stability. Resistor R_4 is the load of the complete amplifier.

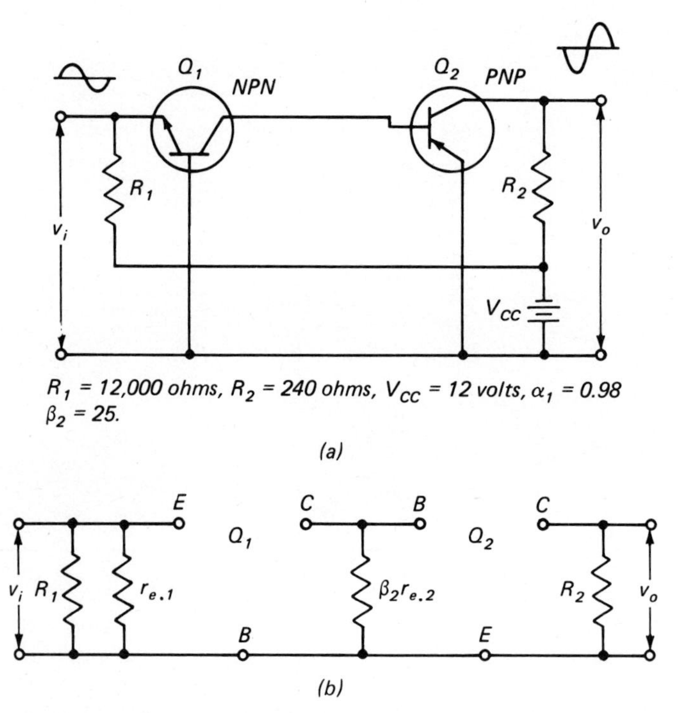

Fig. 13-23 Direct-coupled amplifier with common-base to common-emitter configurations. (a) Circuit diagram. (b) A-c equivalent circuit.

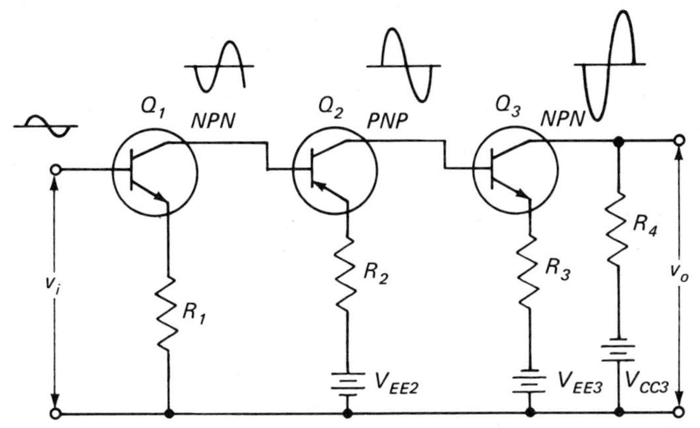

Fig. 13-24 Three-stage direct-coupled transistor amplifier circuit.

13-21 Direct-coupled Amplifier Using Vacuum Tubes

Use of the principle of direct coupling with vacuum-tube amplifiers requires that the direct current (or low-frequency a-c) signal be applied directly to the grid of the amplifier tube. Figure 13-25 shows the application of a direct-coupled amplifier in a basic vacuum-tube-voltmeter circuit. The variable resistor R_4 provides for adjustment of the circuit in order to set the meter M at its zero signal ($e_s = 0$) reference. When the meter leads (T_1-T_2) are connected to a d-c voltage, the grid-to-cathode voltage at the tube is increased (or decreased) and causes a change in the plate current of the tube. The change in plate current causes a change in the amount of current flowing through the meter and changes the meter indication.

13-22 Decoupling Circuits

Need for Decoupling Circuits. In many amplifier circuits it is common practice to supply the d-c operating voltages for the active components from a single source of d-c power. This power source then acts as a common imped-

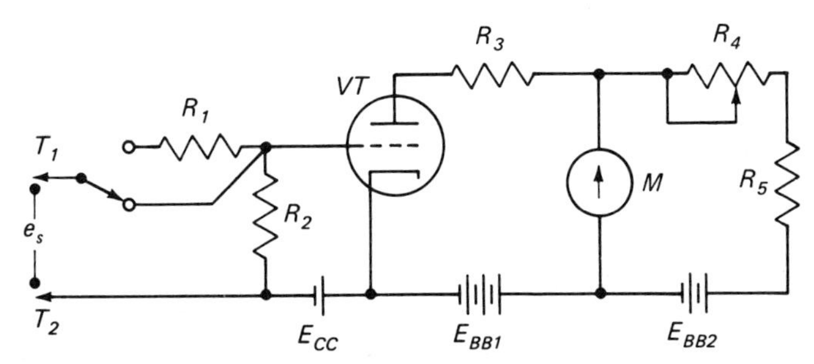

Fig. 13-25 Application of direct coupling to a vacuum-tube voltmeter.

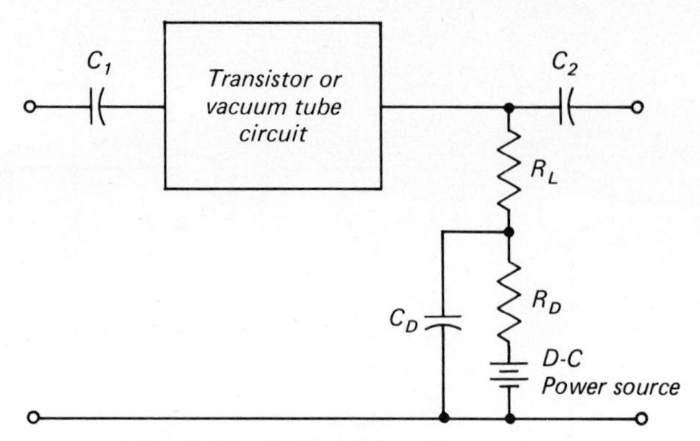

Fig. 13-26 Basic decoupling circuit.

ance for all the circuits drawing power from it. Undesired coupling may exist between circuits operating at the same frequency and having a common impedance. This coupling may be either regenerative or degenerative depending upon the phase relation of the original signal current in a circuit and the signal current coupled back by way of the power source. If the current coupled into a circuit through its power source is sufficient to cause undesirable reactions upon the operation of the amplifier, it becomes necessary to decouple each stage of the amplifier from the other stages.

Basic Decoupling Circuit. The basic decoupling circuit (Fig. 13-26) is formed by the series-connected resistor R_D and the bypass capacitor C_D. Satisfactory decoupling is usually obtained when the time constant of R_DC_D is equal to or greater than the reciprocal of the lowest frequency of the input signal, or

$$R_DC_D = \frac{10^6}{f_l} \tag{13-21}$$

where R_D = resistance of the decoupling resistor, ohms
C_D = capacitance of the decoupling capacitor, μf
f_l = lowest frequency of the input signal, Hz

From this equation, it can be seen that many combinations of R_D and C_D will produce the desired time constant. However, the effectiveness of the decoupling circuit is dependent upon the ratio of the reactance X_D to the resistance R_D, or

$$K_D \cong \frac{X_D}{R_D} \tag{13-22}$$

Also,

$$K_D \cong \frac{10^6}{2\pi f_l C_D R_D} \tag{13-22a}$$

$$K_D \cong \frac{159{,}000}{f_l C_D R_D} \tag{13-22b}$$

where K_D = decoupling factor, dimensionless ratio
X_D = reactance of the capacitor at frequency of f_l, ohms

The undesired coupling through the common power source is reduced by the factor K_D. Actually, the value of resistance used in Eqs. (13-21) and (13-22) should be the sum of R_D and the internal resistance of the power source. As the resistance of the power source is usually very low compared with R_D, it is omitted in this presentation.

Equation (13-22) suggests that more effective decoupling is achieved when R_D is much greater than X_D. However, the value of R_D is limited by the voltage drop at this resistor as it reduces the voltage available at the elements of the transistors or vacuum tubes in the circuits.

Decoupling in Transistor Circuits. Figure 13-27 illustrates two applications of decoupling circuits used with transistors. In the circuit of Fig. 13-27*a*, the decoupling network R_DC_D reduces the a-c component of the base current that passes through the power source to a negligible amount. In the circuit of Fig. 13-27*b*, the decoupling network R_DC_D keeps the a-c component of the collector current entering the power source at a negligible amount.

Example 13-18 If for the circuit of Fig. 13-27*a*, the maximum loss of voltage that can be tolerated at the base of the transistor is 0.1 volt, the lowest signal frequency is 30 Hz, and a decoupling factor of 0.106 is desired, what is the approximate value for (*a*) R_D? (*b*) C_D?

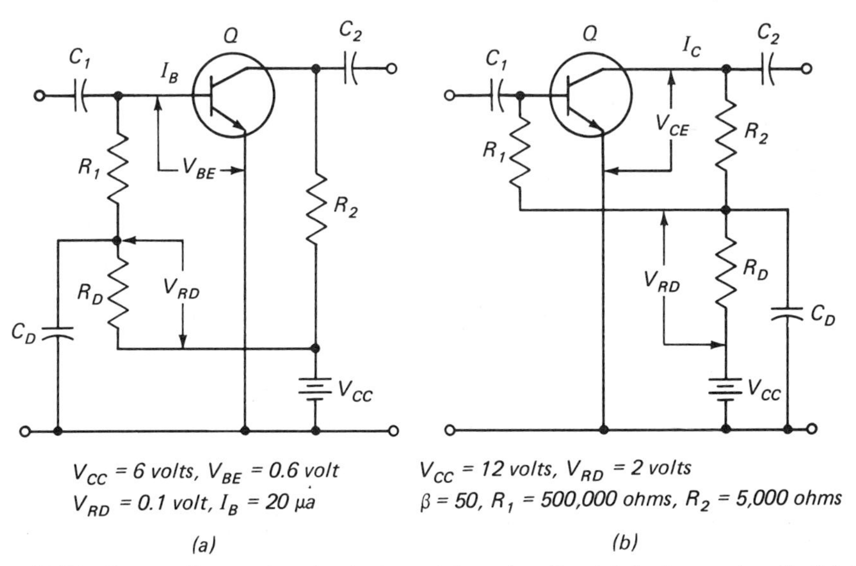

Fig. 13-27 Decoupling networks in transistor circuits. (a) In base circuit. (b) In collector circuit.

GIVEN: Fig. 13-27*a* $V_{R\cdot D} = 0.1$ volt $I_B = 20\ \mu a$ $f_l = 30$ Hz $K_D = 0.106$

FIND: (*a*) R_D (*b*) C_D

SOLUTION:

(*a*)
$$R_D = \frac{V_{R\cdot D}}{I_B} = \frac{0.1}{20 \times 10^{-6}} = 5{,}000 \text{ ohms}$$

(*b*)
$$C_D = \frac{159{,}000}{f_l R_D K_D} = \frac{159{,}000}{30 \times 5{,}000 \times 0.106} = 10\ \mu\text{f}$$

Decoupling in Vacuum-tube Circuits. Figure 13-28 illustrates the application of a decoupling network to a vacuum-tube amplifier circuit. Because of the higher working voltages used with tubes as compared with transistors, a higher voltage drop can be tolerated at the decoupling resistor.

Example 13-19 If for the circuit of Fig. 13-28 a 10-volt drop can be tolerated at R_D, the normal plate current is 4 ma, the lowest signal frequency is 50 Hz, and a decoupling factor of 0.125 is desired, what is the approximate value for (*a*) R_D? (*b*) C_D?

GIVEN: Fig. 13-28 $E_{R\cdot D} = 10$ volts $I_P = 4$ ma $f_l = 50$ Hz $K_D = 0.125$

FIND: (*a*) R_D (*b*) C_D

SOLUTION:

(*a*)
$$R_D = \frac{E_{R\cdot D}}{I_P} = \frac{10}{4 \times 10^{-3}} = 2{,}500 \text{ ohms}$$

(*b*)
$$C_D = \frac{159{,}000}{f_l R_D K_D} = \frac{159{,}000}{50 \times 2{,}500 \times 0.125} \cong 10\ \mu\text{f}$$

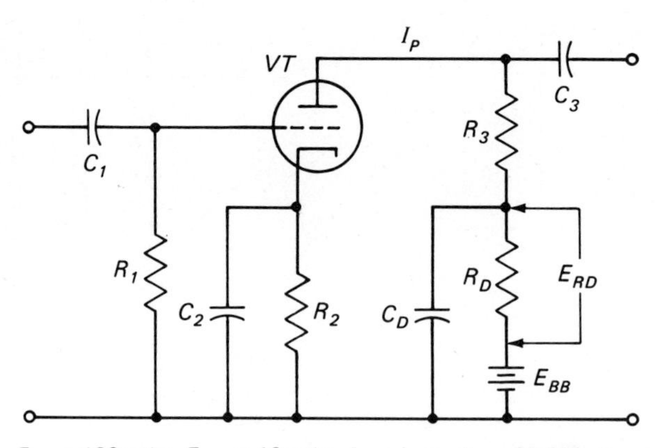

Fig. 13-28 Decoupling network in a vacuum-tube circuit.

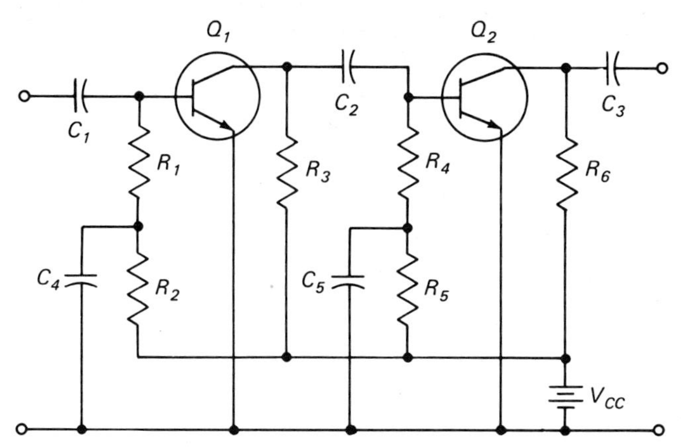

Fig. 13-29 Two-stage transistor amplifier using two decoupling networks R_2C_4 and R_5C_5.

Multistage Decoupling. When a multistage amplifier uses two or more decoupling circuits, the overall decoupling factor is equal to the product of the individual decoupling factors, or

$$K_D = K_{D\cdot1}K_{D\cdot2}K_{D\cdot3} \qquad \text{etc.} \tag{13-23}$$

Example 13-20 If for the circuit of Fig. 13-29, R_2 and R_5 are 5,000 ohms, C_4 and C_5 are 10 μf, and the lowest signal frequency is 30 Hz, find (*a*) $K_{D\cdot1}$, (*b*) $K_{D\cdot2}$, (*c*) K_D.

GIVEN: Fig. 13-29 $\quad R_2, R_5 = 5{,}000$ ohms $\quad C_4, C_5 = 10\ \mu$f $\quad f_l = 30$ Hz

FIND: (*a*) $K_{D\cdot1}$ $\quad$ (*b*) $K_{D\cdot2}$ $\quad$ (*c*) K_D

SOLUTION:

(*a*)
$$K_{D\cdot1} = \frac{159{,}000}{f_lC_4R_2} = \frac{159{,}000}{30 \times 10 \times 5{,}000} \cong 0.106$$

(*b*)
$$K_{D\cdot2} = \frac{159{,}000}{f_lC_5R_5} = \frac{159{,}000}{30 \times 10 \times 5{,}000} \cong 0.106$$

(*c*)
$$K_D = K_{D\cdot1}K_{D\cdot2} \cong 0.106 \times 0.106 \cong 0.0112$$

QUESTIONS

1. Name four basic methods of coupling.
2. What is a cascade amplifier?
3. Describe the network components of an *RC*-coupling circuit.
4. Describe the operation of the basic *RC*-coupling circuit.
5. Describe the function of (*a*) the coupling capacitor, (*b*) a blocking capacitor. (*c*) What is the relationship of the coupling capacitor to a blocking capacitor?
6. Describe the characteristics of the *RC*-coupling network.

7. Describe the operation of the two-stage *RC*-coupled amplifier of Fig. 13-4.
8. (*a*) How is the overall voltage gain of a multistage amplifier determined? (*b*) Why is it desirable to determine the gain of the second stage first?
9. (*a*) What is an a-c equivalent circuit? (*b*) What is its purpose?
10. In preparing an a-c equivalent circuit diagram, what assumptions are made in regard to (*a*) series-connected capacitors? (*b*) Parallel-connected capacitors? (*c*) Power supplies?
11. (*a*) How does a change in the beta value affect the gain of the amplifier of Fig. 13-4? (*b*) What are some of the causes of a change in the value of beta?
12. In the circuit of Fig. 13-7, what purpose is served by (*a*) R_6 and R_7? (*b*) C_4 and C_5?
13. (*a*) In the circuit of Fig. 13-7, can R_6 and R_7 be classed as swamping resistors? (*b*) Why?
14. Describe the operation of the circuit of Fig. 13-8.
15. Upon comparing the circuits of Figs. 13-7 and 13-8, in what manner are they (*a*) alike? (*b*) Different?
16. Describe the operation of the circuit of Fig. 13-9.
17. (*a*) Describe the basic impedance-coupled circuit. (*b*) How does it differ from the *RC*-coupled circuit?
18. Comparing impedance coupling with *RC* coupling, what are its (*a*) advantages? (*b*) Disadvantages?
19. Describe the operation of the circuit of Fig. 13-11.
20. Describe the operation of the circuit of Fig. 13-12.
21. Why is vector addition required in the denominator of Eq. (13-9)?
22. (*a*) What is the purpose of parallel plate feed in a vacuum-tube amplifier? (*b*) What new characteristic does it introduce?
23. (*a*) Describe transformer coupling. (*b*) What are its advantages? (*c*) Describe three types of construction features.
24. Describe the manner in which transformer coupling achieves (*a*) voltage conversion, (*b*) impedance conversion.
25. How is the improvement in gain with the circuit of Fig. 13-15*a* achieved over the gain with the circuit of Fig. 13-15*b*? (*b*) Is the amount of gain significant?
26. (*a*) What is the primary function of the first stage of the amplifier of Fig. 13-18? (*b*) Describe the operation of this stage.
27. (*a*) What is the primary function of the second stage of the amplifier of Fig. 13-18? (*b*) Describe the operation of this stage.
28. (*a*) What is the principle of direct coupling? (*b*) Where can it be used? (*c*) What are some of the restrictions for its use?
29. What is the relationship between a direct-coupled amplifier and a direct-current amplifier?
30. For the circuit of Fig. 13-19*a*, how is the forward bias established for (*a*) Q_1? (*b*) Q_2?
31. For the circuit of Fig. 13-19*a*, how is the collector bias established for (*a*) Q_1? (*b*) Q_2?
32. For the circuit of Fig. 13-19*a*, describe the path (or paths) for the emitter current of (*a*) Q_1, (*b*) Q_2.
33. For the circuit of Fig. 13-20*a*, how is the forward bias established for (*a*) Q_1? (*b*) Q_2?
34. For the circuit of Fig. 13-20*a*, how is the collector bias established for (*a*) Q_1? (*b*) Q_2?
35. For the circuit of Fig. 13-20*a*, describe the path (or paths) for the emitter current of (*a*) Q_1, (*b*) Q_2.

36. For the circuit of Fig. 13-20*a*, describe how R_3 and R_4 affect (*a*) temperature stability of the amplifier, (*b*) voltage gain of the amplifier.
37. What is the basic principle of complementary symmetry for transistors?
38. Describe the operation of the circuit of Fig. 13-21.
39. Upon comparing the circuits of Figs. 13-19*a* and 13-21, in what manner are they (*a*) alike? (*b*) Different?
40. Describe the operation of the circuit of Fig. 13-22.
41. Upon comparing the circuits of Figs. 13-20 and 13-22, in what manner are they (*a*) alike? (*b*) Different?
42. Upon comparing the circuits of Figs. 13-19*a* and 13-23*a*, in what manner are they (*a*) alike? (*b*) Different?
43. For the circuit of Fig. 13-23*a*, how is the forward bias established for (*a*) Q_1? (*b*) Q_2?
44. For the circuit of Fig. 13-23*a*, describe the path (or paths) for the emitter current of (*a*) Q_1, (*b*) Q_2.
45. For the circuit of Fig. 13-23*a*, describe the path (or paths) for the collector current of (*a*) Q_1, (*b*) Q_2.
46. Describe the operation of the circuit of Fig. 13-24.
47. For the circuit of Fig. 13-24, describe the path (or paths) for the emitter current of (*a*) Q_1, (*b*) Q_2, (*c*) Q_3.
48. For the circuit of Fig. 13-24, describe the path (or paths) for the collector current of (*a*) Q_1, (*b*) Q_2, (*c*) Q_3.
49. For the circuit of Fig. 13-25, explain how the variable resistor R_4 controls the adjustment of the meter for zero signal input.
50. For the circuit of Fig. 13-25, explain how a change in the value of e_s causes a movement of the needle of the meter.
51. What purpose do decoupling circuits serve?
52. Name the components of a basic decoupling circuit.
53. Describe the operation of a basic decoupling circuit.
54. Under what condition is it necessary to include the resistance of the power source in analyzing a decoupling circuit?
55. (*a*) How does the ratio of reactance to resistance affect the decoupling factor? (*b*) What usually determines this ratio?
56. (*a*) In the circuit of Fig. 13-15, can the combination of R_3 and C_2 be classed as a decoupling circuit? (*b*) Why?
57. (*a*) In the circuit of Fig. 13-28, can the combination of R_2 and C_2 be classed as a decoupling circuit? (*b*) Why?

PROBLEMS

1. Draw the a-c equivalent circuit for the amplifier of Fig. 13-2.
2. Draw the a-c equivalent circuit for the amplifier of Fig. 13-3.
3. If the parameters for the circuit of Fig. 13-2 are $R_B = 1{,}000{,}000$ ohms, $R_L = 7{,}500$ ohms, $R_o = 5{,}000$ ohms, $V_{CC} = 20$ volts, $v_i = 15$ mv, and $\beta = 75$, find (*a*) voltage gain of the amplifier, (*b*) input resistance, (*c*) output voltage, (*d*) collector-to-emitter volts.
4. Repeat Prob. 3 for the condition with R_B changed to 600,000 ohms.
5. What is the voltage gain of the amplifier of Fig. 13-3 if the circuit parameters are $R_L = 250{,}000$ ohms, $R_o = 500{,}000$ ohms, $r_p = 50{,}000$ ohms, and $\mu = 80$?

6. What is the voltage gain of the amplifier of Prob. 5 after substituting a tube that has an amplification factor of 50 and a plate resistance of 25,000 ohms?

7. If the circuit parameters for Fig. 13-4 are R_1 = 1,000,000 ohms, R_2 = 6,000 ohms, R_3 = 800,000 ohms, R_4 = 5,000 ohms, R_5 = 20,000 ohms, V_{CC} = 20 volts, $\beta_1 = 75$, $\beta_2 = 60$, find (*a*) voltage gain of the second stage, (*b*) voltage gain of the first stage, (*c*) decibel gain of the amplifier, (*d*) collector-to-emitter voltage at Q_1, (*e*) collector-to-emitter voltage at Q_2. (*Note:* Use $30/I_E$ for determining r_e.)

8. If when wiring the amplifier of Prob. 7, resistors R_4 and R_5 are inadvertently interchanged, how would the performance of the amplifier be affected?

9. If the voltage of V_{CC} in Prob. 7 decreases to 15 volts, how would the performance of the amplifier be affected in terms of (*a*) voltage gain of the second stage? (*b*) Voltage gain of the first stage? (*c*) Decibel gain of the amplifier? (*d*) Collector-to-emitter voltage at Q_1? (*e*) Collector-to-emitter voltage at Q_2?

10. If the voltage of V_{CC} in Prob. 7 decreases to 10 volts, how would the performance of the amplifier be affected in terms of (*a*) voltage gain of the second stage? (*b*) Voltage gain of the first stage? (*c*) Decibel gain of the amplifier? (*d*) Collector-to-emitter voltage at Q_1? (*e*) Collector-to-emitter voltage at Q_2?

11. If in the amplifier of Example 13-5, C_5 becomes open-circuited, how will the performance of the amplifier be affected in terms of (*a*) voltage gain of the second stage? (*b*) Voltage gain of the first stage? (*c*) Decibel gain of the amplifier? (*d*) Collector-to-emitter voltage at Q_1? (*e*) Collector-to-emitter voltage at Q_2?

12. If in the amplifier of Example 13-5, C_5 becomes short-circuited, how will the performance of the amplifier be affected in terms of (*a*) voltage gain of the second stage? (*b*) Voltage gain of the first stage? (*c*) Decibel gain of the amplifier? (*d*) Collector-to-emitter voltage at Q_1? (*e*) Collector-to-emitter voltage at Q_2?

13. If the parameters for the circuit of Fig. 13-8 are R_2, R_3 = 250,000 ohms, R_4 = 5,000 ohms, $r_{p\cdot 1}$ = 66,000 ohms, $r_{p\cdot 2}$ = 1,700 ohms, $\mu_1 = 100$, $\mu_2 = 8$, find (*a*) voltage gain of the first stage, (*b*) voltage gain of the second stage, (*c*) voltage gain of the amplifier, (*d*) decibel gain of the amplifier.

14. If the parameters for the circuit of Fig. 13-8 are R_2, R_3, R_4 = 250,000 ohms, $r_{p\cdot 1}$, $r_{p\cdot 2}$ = 66,000 ohms, μ_1, μ_2 = 100, find (*a*) voltage gain of the first stage, (*b*) voltage gain of the second stage, (*c*) voltage gain of the amplifier, (*d*) decibel gain of the amplifier.

15. If the parameters for the circuit of Fig. 13-9 are R_4 = 220,000 ohms, R_5 = 330,000 ohms, R_8 = 3,000 ohms, $r_{p\cdot 1}$ = 1,000,000 ohms, $r_{p\cdot 2}$ = 25,000 ohms, $g_{m\cdot 1}$ = 1,200 μmhos, $g_{m\cdot 2}$ = 6,000 μmhos, find (*a*) voltage gain of the first stage, (*b*) voltage gain of the second stage, (*c*) voltage gain of the amplifier, (*d*) decibel gain of the amplifier.

16. If the parameters for the circuit of Fig. 13-9 are R_5 = 250,000 ohms, R_4, R_8 = 100,000 ohms, R_L (connected to C_3) = 100,000 ohms, $r_{p\cdot 1}$, $r_{p\cdot 2}$ = 1,000,000 ohms, $g_{m\cdot 1}$, $g_{m\cdot 2}$ = 1,200 μmhos, find (*a*) voltage gain of the first stage, (*b*) voltage gain of the second stage, (*c*) voltage gain of the amplifier, (*d*) decibel gain of the amplifier.

17. For the circuit of Fig. 13-11, at 200 Hz find (*a*) voltage gain of the second stage, (*b*) voltage gain of the first stage, (*c*) voltage gain of the amplifier, (*d*) decibel gain of the amplifier.

18. If for the amplifier of Example 13-8, the output resistor R_5 is made equal to the output-stage collector resistor R_4, find the new values of (*a*) voltage gain of the second stage, (*b*) voltage gain of the first stage, (*c*) voltage gain of the amplifier, (*d*) decibel gain of the amplifier.

19. The amplifier circuit of Fig. 13-12 has the following parameters: $L = 300$ henrys, $R_3 = 500{,}000$ ohms, R_4, $R_5 = 100{,}000$ ohms, $r_{p\cdot1} = 7{,}700$ ohms, $r_{p\cdot2} = 50{,}000$ ohms, $\mu_1 = 20$, and $\mu_2 = 70$. For a 1,000-Hz signal, find (*a*) voltage gain at the second stage, (*b*) voltage gain at the first stage, (*c*) voltage gain of the amplifier, (*d*) decibel gain of the amplifier.

20. For the amplifier of Prob. 19, (*a*) what voltage is required for the B power supply of VT_1 if the plate current is 10 ma, the resistance of the inductor L is 5,000 ohms, and the voltage at the plate of the tube is to be 250 volts? (*b*) What voltage is required for the B power supply of VT_2 if the plate current is 1 ma and the voltage at the plate of the tube is to be 250 volts?

21. If the basic transformer coupling circuit of Fig. 13-14 has a 60-mv input signal, what turns ratio is required of the transformer in order to produce a 1.5-volt output signal?

22. If a 200-mv input signal is applied to the transformer coupling circuit of Fig. 13-14, what is the output-signal voltage if the transformer has 50 turns on the primary winding and 300 turns on the secondary winding?

23. The transformer in a simple coupling circuit has 400 primary winding turns and 20 secondary winding turns, the load on its secondary terminals is 8 ohms, and the input signal is 2.4 volts. (*a*) What is the impedance seen by the primary circuit? (*b*) What is the voltage at the 8-ohm load terminals?

24. What turns ratio is required for the coupling transformer if it is desired to match a 10-ohm load to a 6,250-ohm circuit?

25. Assuming an ideal transformer for the circuit of Fig. 13-30*a*, find (*a*) voltage gain up to the primary of the transformer, (*b*) signal voltage at the primary of the transformer, (*c*) signal voltage at the load resistor R_3, (*d*) overall voltage gain, (*e*) maximum voltage at v_o' without clipping, (*f*) maximum input-signal voltage without clipping, (*g*) draw the d-c and a-c load lines for the amplifier.

26. To illustrate the advantage of using the transformer in the circuit of Fig. 13-30*a*,

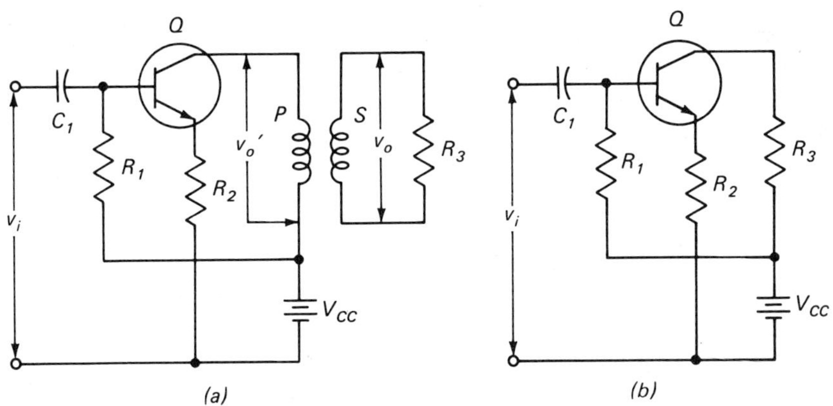

$R_1 = 300{,}000$ ohms, $R_2 = 4{,}000$ ohms, $R_3 = 40$ ohms, $V_{CC} = 10$ volts,

$v_i = 40$ mv (p-p), $\beta = 50$, $\dfrac{N_p}{N_s} = 10$

Fig. 13-30

find for the circuit of Fig. 13-30*b*: (*a*) voltage gain of the circuit, (*b*) signal voltage at the load resistor, (*c*) maximum output voltage without clipping, (*d*) maximum input-signal voltage without clipping.

27. What new value of R_1 will provide the maximum possible voltage gain without clipping for the amplifier of Prob. 25? (*Note:* This condition will be fulfilled when the quiescent value of collector current is one-half of the collector saturation current.)

28. What new value of R_2 will provide the maximum possible voltage gain without clipping for the amplifier of Prob. 25? (*Note:* This will occur when $R_2 = R_1/\beta$.)

29. For the amplifier of Prob. 27, find (*a*) overall voltage gain, (*b*) maximum signal voltage at v_o' without clipping, (*c*) maximum input-signal voltage without clipping. (*d*) Draw the d-c and a-c load lines for the amplifier.

30. If the voltage source V_{CC} is increased to 12 volts for the amplifier of Prob. 28, find (*a*) voltage gain of the amplifier, (*b*) maximum voltage at v_o' without clipping, (*c*) maximum input-signal voltage without clipping. (*d*) Draw the d-c and a-c load lines.

31. For the amplifier circuit of Fig. 13-31 assume (1) T is an ideal transformer, and (2) the voltage gains are to be determined for the mid-frequency range. Find (*a*) voltage gain for the first stage, (*b*) voltage gain for the second stage, (*c*) voltage gain for the amplifier, (*d*) the output-signal voltage.

32. What is the overall voltage gain for the amplifier of Prob. 31 if the value of R_4 is decreased to 100,000 ohms?

33. If for the amplifier circuit of Fig. 13-19*a*, the load resistance is decreased to 120 ohms, what are the values of (*a*) current gain? (*b*) Voltage gain of the first stage? (*c*) Voltage gain of the second stage? (*d*) Voltage gain of the amplifier? (*e*) Decibel gain (power) of the amplifier?

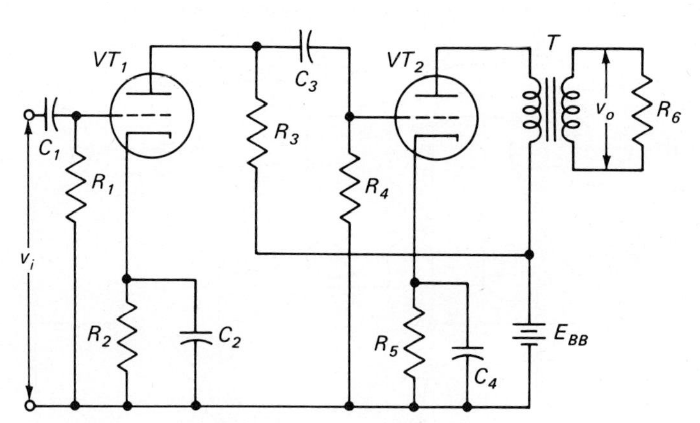

$R_1, R_3, R_4 = 220{,}000$ ohms, $R_2 = 3{,}000$ ohms, $R_5 = 5{,}000$ ohms, $R_6 = 6$ ohms, $r_{p.1} = 60{,}000$ ohms, $r_{p.2} = 7{,}700$ ohms, $\mu_1 = 100$, $\mu_2 = 17$, $E_{BB} = 180$ volts, $C_1, C_3 = 0.012\ \mu f$, $C_2 = 2\ \mu f$,

$C_4 = 1\ \mu f$, $T - \dfrac{N_p}{N_S} = 30$, $v_i = 50$ mv

Fig. 13-31

34. If for the amplifier circuit of Fig. 13-19*a*, the load resistance is increased to 360 ohms, what are the values of (*a*) current gain? (*b*) Voltage gain of the first stage? (*c*) Voltage gain of the second stage? (*d*) Voltage gain of the amplifier? (*e*) Decibel gain (power) of the amplifier?

35. If Q_1 in the circuit of Fig. 13-20*a* is replaced with a transistor having a beta value of 50, find (*a*) voltage gain of the first stage, (*b*) voltage gain of the second stage, (*c*) voltage gain of the amplifier, (*d*) input resistance of the amplifier.

36. If Q_2 in the circuit of Fig. 13-20*a* is replaced with a transistor having a beta value of 50, find (*a*) voltage gain of the first stage, (*b*) voltage gain of the second stage, (*c*) voltage gain of the amplifier, (*d*) input resistance of the amplifier.

37. For the circuit of Fig. 13-21, find (*a*) emitter current at Q_1, (*b*) emitter current at Q_2, (*c*) current gain of the amplifier, (*d*) voltage gain of the amplifier, (*e*) power gain of the amplifier.

38. If Q_2 in the circuit of Fig. 13-21 is replaced with a transistor having a beta value of 50, find (*a*) emitter current at Q_1, (*b*) emitter current at Q_2, (*c*) current gain of the amplifier, (*d*) voltage gain of the amplifier, (*e*) power gain of the amplifier.

39. It is desired to add a capacitor across R_3 in the circuit of Fig. 13-22 to bypass any a-c signals down to 10 Hz. (*a*) What value of capacitance is required if a ratio of R to X_C of 10 to 1 is considered satisfactory? Also, find (*b*) voltage gain of the first stage, (*c*) voltage gain of the second stage, (*d*) voltage gain of the amplifier.

40. A capacitor is to be added across R_5 in the circuit of Fig. 13-22 to bypass any a-c signals down to 10 Hz, and the ratio of R to X_C is to be approximately 10 to 1. Find (*a*) the value of the added capacitor, (*b*) voltage gain of the first stage, (*c*) voltage gain of the second stage, (*d*) voltage gain of the amplifier.

41. For the circuit of Example 13-15, (*a*) what is the collector-to-emitter voltage at Q_1? (*b*) What is the collector-to-emitter voltage at Q_2? (*c*) What new resistance value would be required at R_6 in order to produce 12 volts from collector to emitter at Q_2?

42. For the circuit of Prob. 41, what is the maximum signal-handling capability at the (*a*) output of Q_2? (*b*) Output of Q_1? (*c*) Input of Q_1?

43. If the components in the circuit of Fig. 13-23*a* are changed to R_1 = 1,600 ohms, R_2 = 16 ohms, and β_2 = 50, find (*a*) emitter current and base-emitter resistance at Q_1, (*b*) emitter current and base-emitter resistance at Q_2, (*c*) voltage gain of the first stage, (*d*) voltage gain of the second stage, (*e*) voltage gain of the amplifier.

44. For the circuit of Prob. 43, find (*a*) current gain of the amplifier, (*b*) power gain of the amplifier, (*c*) input resistance of the amplifier.

45. The circuit of Fig. 13-32 illustrates a cascade amplifier with a common-base first stage and a common-collector second stage. Find (*a*) collector current at Q_1, (*b*) collector current at Q_2, (*c*) voltage gain of the second stage, (*d*) voltage gain of the first stage, (*e*) voltage gain of the amplifier, (*f*) input resistance of the amplifier.

46. For the amplifier of Prob. 45, find (*a*) maximum output of the second stage without clipping, (*b*) maximum output of the first stage without clipping, (*c*) maximum input-signal voltage without causing clipping.

47. For the circuit of Fig. 13-33, find (*a*) collector current at Q_1, (*b*) collector current at Q_2, (*c*) voltage gain of the second stage, (*d*) voltage gain at the first stage, (*e*) voltage gain of the amplifier, (*f*) input resistance of the amplifier.

48. For the amplifier of Prob. 47, find (*a*) maximum output of the second stage with-

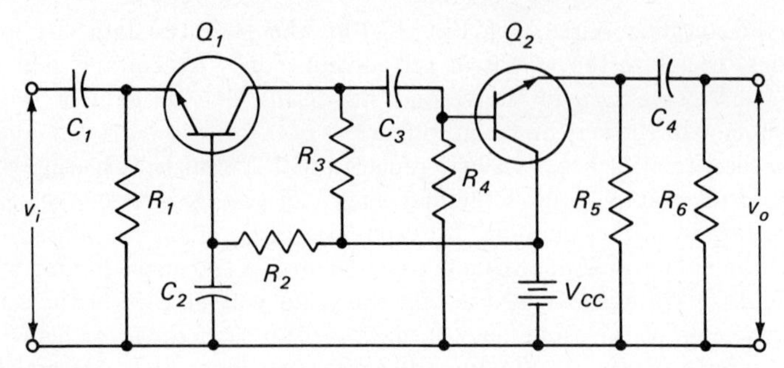

R_1 = 6,000 ohms, R_2 = 1,400,000 ohms, R_3 = 4,000 ohms, R_4 = 1,000,000 ohms, R_5, R_6 = 10,000 ohms, V_{CC} = 20 volts, $\alpha_1 = 0.99$, β_1, β_2 = 100

Fig. 13-32

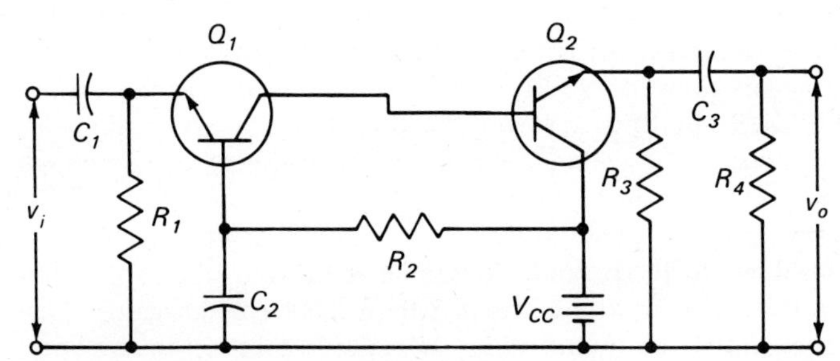

R_1 = 10,000 ohms, R_2 = 1,000,000 ohms, R_3 = 200 ohms, R_4 = 20 ohms, V_{CC} = 10 volts, $\alpha_1 = 0.99$, $\beta_1 = 100$, $\beta_2 = 50$

Fig. 13-33

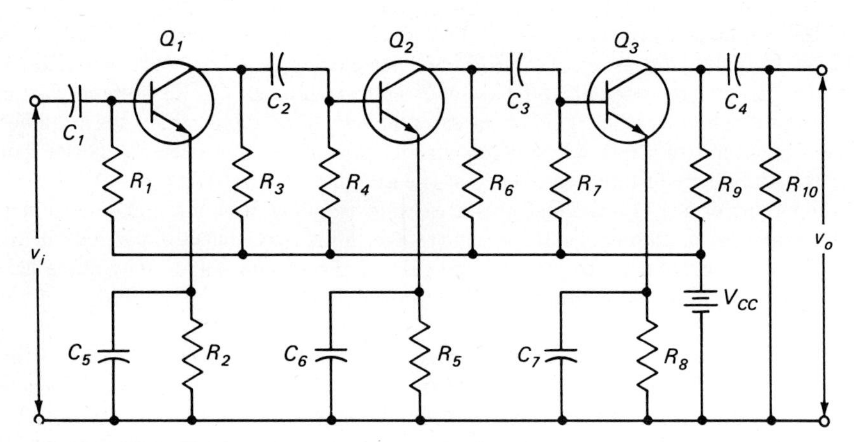

R_1, R_4, R_7 = 1,250,000 ohms, R_2, R_5, R_8 = 2,500 ohms, R_3, R_6, R_9, R_{10} = 5,000 ohms V_{CC} = 30 volts, β_1, β_2, β_3 = 100

Fig. 13-34

out clipping, (*b*) maximum output of the first stage without clipping, (*c*) maximum input-signal voltage without causing clipping, (*d*) current gain of the amplifier, (*e*) power gain of the amplifier.

49. For the circuit of Fig. 13-34, find (*a*) emitter current for each transistor, (*b*) base-emitter resistance for each transistor, (*c*) voltage gain of the third stage, (*d*) voltage gain of the second stage, (*e*) voltage gain of the first stage, (*f*) voltage gain of the amplifier.

50. Repeat Prob. 49 for the condition when the beta value of each transistor is 50.

51. Repeat Prob. 49 for the condition produced by changing each emitter-circuit resistor to a 500-ohm unbypassed resistor in series with a 2,000-ohm bypassed resistor.

52. Repeat Prob. 51 for the condition when the beta value of each transistor is 50.

53. It is desired that adding the decoupling network to the circuit of Example 13-18 should not cause a decrease in the base current or base-emitter voltage. This can be accomplished by replacing R_1 when adding R_D and C_D. What is the value for R_1: (*a*) before adding R_D and C_D? (*b*) After adding R_D and C_D?

54. It is desired that the decoupling factor for the circuit of Fig. 13-27*a* should have a value of 0.1 at the minimum signal frequency of 10 Hz and that the maximum voltage drop at R_D should not exceed 0.15 volt. What values are required for (*a*) R_D? (*b*) C_D?

55. If for the circuit of Fig. 13-27*b*, the voltage drop at R_D is limited to 2 volts, the lowest frequency of the input signal is 30 Hz, and a decoupling factor of 0.05 is desired, find (*a*) R_D, (*b*) C_D, (*c*) V_{CE}.

56. If the decoupling components for the circuit of Fig. 13-27*b* are removed, what will be the values of (*a*) I_C? (*b*) V_{CE}?

57. It is desired that the circuit of Fig. 13-28 should have a decoupling factor of 0.08 at its lowest frequency of 40 Hz. What are the values of (*a*) R_D? (*b*) C_D? (*c*) The voltage at the plate of the tube?

58. A vacuum-tube amplifier circuit using a 250,000-ohm resistor as its plate load also has a 30,000-ohm decoupling resistor and a 0.25-μf decoupling capacitor. (*a*) What is the decoupling factor for a 50-Hz signal? (*b*) What value of capacitance is required to make the decoupling factor approximately 0.1?

59. It is desired that the overall decoupling factor for the circuit of Example 13-20 should be approximately 0.025. (*a*) What is the decoupling factor for each stage? (*b*) What new value of C_D will satisfy this condition?

60. Two stages of a cascade transistor amplifier have identical decoupling circuits similar to Fig. 13-27*b*, and each produces a decoupling factor of 0.16. (*a*) What is the overall decoupling factor of the amplifier? (*b*) By what per cent could the capacitance of C_D be reduced if an overall decoupling factor of 0.1 is sufficient?

Chapter 14

Frequency Response of Amplifiers

An important characteristic of an amplifier is the manner in which it performs when the frequency of the input signal varies over a wide range. The preceding chapter presented the characteristics of four basic types of amplifier circuits for frequencies at or near the middle of its normal operational frequency range. In this chapter, the characteristics of the basic amplifiers will be studied for the mid-frequency range and the practical limits for the low- and high-frequency ends.

14-1 Variation in Amplification with Frequency

Principle. If the input to an amplifier is kept at a constant voltage while its frequency is being varied, it will be found that the voltage gain of the amplifier will (1) remain practically constant over a moderate range of mid-frequencies and (2) decrease at both the low- and high-frequency ends of the frequency band.

Frequency Range. When analyzing the gain versus frequency characteristics of amplifiers, three values of frequency are important, namely, (1) the mid-frequency range, (2) the low-frequency cutoff, and (3) the high-frequency cutoff. For practical reasons, the low-frequency and high-frequency cutoff values represent the respective frequencies at which the gain has decreased to 0.707 times the gain at the mid-frequencies. The frequency span from low-frequency cutoff to high-frequency cutoff is called the *passband* of the amplifier. The frequency at which the gain of a circuit is 0.707 times the mid-frequency gain is called the *cutoff* or *rolloff frequency*. A decrease by the factor 0.707 can be expressed in decibels as (1) −3 db, (2) down 3 db, or (3) a 3-db loss.

14-2 Variation in the Voltage Gain

Causes. The gain of an amplifier is fairly uniform over a moderately wide band of frequencies but decreases quite rapidly at both the low- and high-frequency ends of the passband. The variation in gain is due to changes in the impedance of any capacitance or inductance present in the circuit with any changes in the frequency of the signal current being amplified.

Capacitances in Amplifier Circuits. Capacitances present in amplifier circuits may be classified as (1) capacitances due to components included in the circuit to perform a specific function such as coupling and bypass capacitors,

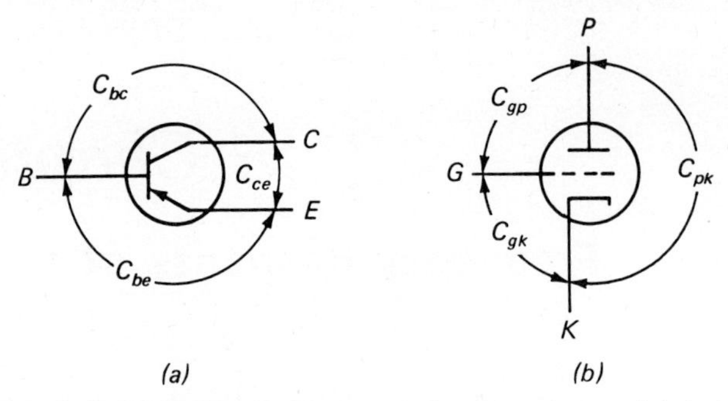

Fig. 14-1 (a) Interelement capacitances of a transistor. (b) Interelectrode capacitances of a vacuum tube.

and (2) the undesired inherent capacitances such as the interelement capacitances of active devices and the stray capacitance of the wiring. The interelement capacitances of a transistor and the interelectrode capacitances of a vacuum tube are shown in Fig. 14-1.

It will be helpful also to classify the capacitances as being (1) effectively series-connected or (2) effectively parallel-connected.

Inductances in Amplifier Circuits. Inductances present in amplifier circuits are usually in the form of (1) inductors introduced into the circuit to perform a specific function, (2) the inductance of transformer windings, (3) inductance of wirewound resistors, and (4) stray inductance of the wiring, etc.

Effect of Capacitances on the Voltage Gain. Because the impedance of a capacitance changes with frequency, the gain versus frequency characteristics of an amplifier are affected by all the capacitances present in a circuit. The impedance of a capacitance varies inversely with the frequency; hence (1) at low frequencies the impedance will be high, and (2) at high frequencies the impedance will be low. At the lower frequencies, the effect of a series-connected capacitance is very pronounced, and the effect of a parallel-connected capacitance is negligible. Conversely, at the higher frequencies, the effect of a series-connected capacitance is negligible, and the effect of a parallel-connected capacitance is very pronounced. The effects of both series-connected and parallel-connected capacitances usually can be neglected for the mid-frequency range (see Chap. 13).

14-3 Frequency Response with *RC* Coupling

The Circuit. *RC* coupling is often used to couple the output signal of one active circuit device to the input of another active device. The characteristics of the *RC*-coupled amplifier are best understood by studying the circuit actions at the low-, intermediate-, and high-frequency ranges of the amplifier.

Taking into account the interelement capacitances of the transistors or vacuum tubes, the basic *RC*-coupling circuit of Fig. 13-1 is modified as shown

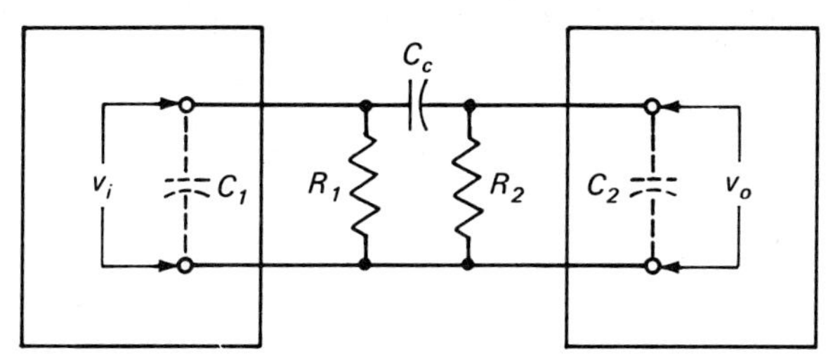

Fig. 14-2 Basic *RC*-coupling circuit.

in Fig. 14-2. This figure shows that the *RC*-coupling network is affected by three capacitances: (1) the coupling capacitor C_c, (2) the input capacitance C_1 representing the combined effect of the interelectrode capacitances of the active device in the input circuit, and (3) the output capacitance C_2 representing the combined effect of the interelectrode capacitances of the active device in the output circuit. Stray capacitances such as those produced by the wiring, etc., are included in C_1 and C_2.

Circuit Values. The values of R_1, R_2, and C_c are generally selected so that maximum gain occurs in the desired mid-frequency range. For example, in a circuit designed to amplify signals from 20 to 20,000 Hz, maximum gain should be obtained in the range of 100 to 15,000 Hz. For satisfactory mid-frequency operation, it is desirable that the value of C_c is high enough to make its impedance low compared with R_2 so that the voltage drop across C_c will be negligible. Also, the values of C_1 and C_2 should be low enough so that their impedances will be high compared with R_1 and R_2 and thus make it possible to disregard these capacitances at the mid-frequencies. By considering C_c as a short circuit, and C_1 and C_2 as open circuits, the net effect is that of having only R_1 and R_2 in the circuit (Fig. 14-3). With these two resistors effectively connected in parallel, the output voltage is equal to the input voltage and maximum voltage gain is achieved.

Gain at Mid-frequencies. At mid-frequencies, the impedance of the coupling capacitor C_c is so low in comparison with R_1 and R_2 that this capacitor may be disregarded. The capacitances C_1 and C_2 are so low that their impedances at the mid-frequencies are very high compared with R_1 and R_2, and their effect in the circuit is so slight that these capacitances too may be disregarded. Figure 14-3 is an a-c equivalent circuit for the mid-frequencies. From this circuit it can be shown that the output voltage at the mid-frequencies is

$$v_{o\cdot m} = \frac{v_i' R_2}{R_1 + R_2} \qquad (14\text{-}1$$

The voltage gain is

$$A_{v\cdot m} = \frac{v_{o\cdot m}}{v_i'} \qquad (14\text{-}2)$$

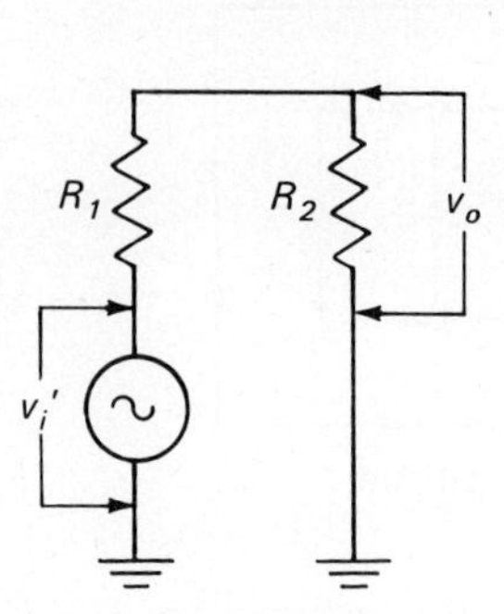

Fig. 14-3 A-c equivalent circuit of *RC*-coupling network at mid-frequencies.

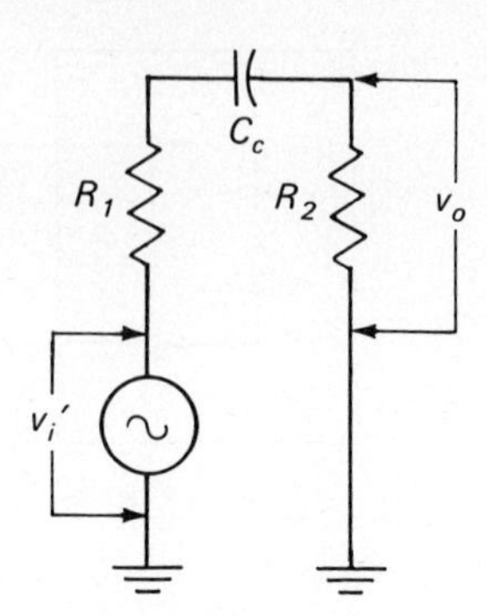

Fig. 14-4 A-c equivalent circuit of *RC*-coupling network at low frequencies.

Substituting Eq. (14-1) in (14-2)

$$A_{v\cdot m} = \frac{R_2}{R_1 + R_2} \tag{14-2a}$$

Example 14-1 What is the mid-frequency gain for the circuit of Fig. 14-2, if $R_1 =$ 6,000 ohms, $R_2 = 30{,}000$ ohms, and $C_c = 0.1$ μf?

GIVEN: Fig. 14-2 $R_1 = 6{,}000$ ohms $R_2 = 30{,}000$ ohms $C_c = 0.1$ μf

FIND: $A_{v\cdot m}$

SOLUTION:

$$A_{v\cdot m} = \frac{R_2}{R_1 + R_2} = \frac{30{,}000}{6{,}000 + 30{,}000} = 0.833$$

Gain at Low Frequencies. Because the effect of the coupling capacitor is significant at the low frequencies, capacitor C_c is included in the low-frequency equivalent circuit (Fig. 14-4). From this circuit it can be shown that the output voltage at the low frequencies is

$$v_{o\cdot l} = \frac{v_i' R_2}{\sqrt{(R_1 + R_2)^2 + X_c^2}} \tag{14-3}$$

The voltage gain at the low frequencies is

$$A_{v\cdot l} = \frac{R_2}{\sqrt{(R_1 + R_2)^2 + X_c^2}} \tag{14-4}$$

The gain at the low frequencies will be less than the gain at the mid-frequencies by a factor K_l, obtained by dividing Eq. (14-4) by Eq. (14-2*a*); thus,

$$K_l = \frac{R_2}{\sqrt{(R_1 + R_2)^2 + X_c^2}} \div \frac{R_2}{R_1 + R_2} \tag{14-5}$$

after factoring,

$$K_l = \frac{R_1 + R_2}{\sqrt{(R_1 + R_2)^2 + X_c^2}} \tag{14-5a}$$

Squaring both sides of Eq. (14-5*a*),

$$K_l^2 = \frac{(R_1 + R_2)^2}{(R_1 + R_2)^2 + X_c^2} \tag{14-5b}$$

Dividing the numerator and denominator of the right-hand member by $(R_1 + R_2)^2$, then

$$K_l^2 = \frac{1}{1 + \dfrac{X_c^2}{(R_1 + R_2)^2}} \tag{14-5c}$$

Taking the square root of both sides of Eq. (14-5*c*),

$$K_l = \frac{1}{\sqrt{1 + \left(\dfrac{X_c}{R_1 + R_2}\right)^2}} \tag{14-5d}$$

When K_l and $A_{v \cdot m}$ are known, the gain at the low frequencies also can be determined by

$$A_{v \cdot l} = K_l A_{v \cdot m} \tag{14-6}$$

Low-frequency Cutoff. The low-frequency cutoff value is dependent on the values of C_c, R_1, and R_2. The low-frequency cutoff is that frequency value for which the voltage gain is 0.707 times the mid-frequency gain and therefore can be obtained by substituting 0.707 for K_1 in Eq. (14-5*d*) and solving for frequency. Then

$$0.707 = \frac{1}{\sqrt{1 + \left(\dfrac{X_c}{R_1 + R_2}\right)^2}} \tag{14-7}$$

Squaring both sides of Eq. (14-7),

$$0.5 = \frac{1}{1 + \dfrac{X_c^2}{(R_1 + R_2)^2}} \tag{14-7a}$$

Transposing,

$$0.5 + \frac{0.5X_c^2}{(R_1 + R_2)^2} = 1 \tag{14-7b}$$

Rearranging the terms,

$$0.5X_c^2 = (1 - 0.5)(R_1 + R_2)^2 \tag{14-7c}$$

and

$$X_c^2 = (R_1 + R_2)^2 \tag{14-7d}$$

Taking the square root of both sides,

$$X_c = R_1 + R_2 \tag{14-7e}$$

Stating X_c in terms of f and C,

$$\frac{10^6}{2\pi f C_c} = R_1 + R_2 \tag{14-7f}$$

Rearranging the terms, the low-frequency cutoff becomes

$$f_{l\cdot co} = \frac{10^6}{2\pi C_c(R_1 + R_2)} \tag{14-7g}$$

$$f_{l\cdot co} = \frac{159{,}000}{C_c(R_1 + R_2)} \tag{14-7h}$$

Example 14-2 What is the low-frequency cutoff for the *RC*-coupling circuit of Fig. 14-2 if $R_1 = 6{,}000$ ohms, $R_2 = 30{,}000$ ohms, and $C_c = 0.1$ μf?

GIVEN: $R_1 = 6{,}000$ ohms $\quad R_2 = 30{,}000$ ohms $\quad C_c = 0.1\ \mu$f

FIND: $f_{l\cdot co}$

SOLUTION:

$$f_{l\cdot co} = \frac{159{,}000}{C_c(R_1 + R_2)} = \frac{159{,}000}{0.1 \times (6{,}000 + 30{,}000)} \cong 44 \text{ Hz}$$

Example 14-3 (*a*) What is the gain of the circuit of Example 14-2 at the cutoff frequency found in Example 14-2? (*b*) Using the low-frequency gain found in (*a*) and the mid-frequency gain found in Example 14-1, what is the low-frequency gain factor?

GIVEN: $R_1 = 6{,}000$ ohms $\quad R_2 = 30{,}000$ ohms $\quad C_c = 0.1\ \mu$f $\quad A_{v\cdot m} = 0.833$

FIND: (*a*) $A_{v\cdot l}$ $\quad$ (*b*) K_l

SOLUTION:

(*a*)
$$A_{v\cdot l} = \frac{R_2}{\sqrt{(R_1 + R_2)^2 + X_c^2}} = \frac{30{,}000}{\sqrt{(6{,}000 + 30{,}000)^2 + 36{,}000^2}} \cong 0.59$$

$$X_c = \frac{159{,}000}{fC} = \frac{159{,}000}{44 \times 0.1} \cong 36{,}000 \text{ ohms}$$

(*b*)
$$K_l = \frac{A_{v\cdot l}}{A_{v\cdot m}} = \frac{0.59}{0.833} \cong 0.707$$

Example 14-4 What is the decibel loss at the low-frequency cutoff?

GIVEN: $K_l = 0.707$

FIND: db

SOLUTION:

$$\text{db} = 20 \log K_l = 20 \log 0.707 = 20(9.85 - 10) = -3 \text{ db}$$

Gain at High Frequencies. The equivalent circuit of Fig. 14-5*a* reflects the facts that at the high frequencies (1) the impedances of the parallel-connected capacitances of the active devices are significant, and (2) the impedance of the series-connected coupling capacitor is insignificant and can be disregarded. From Fig. 14-5*b*, which is the series circuit equivalent of Fig. 14-5*a*, it can be shown that the output voltage at the high frequencies is

$$v_{o \cdot h} = \frac{v_i' R X_{C \cdot T}}{R_1 \sqrt{R^2 + X_{C \cdot T}^2}} \tag{14-8}$$

The voltage gain at the high frequencies is

$$A_{v \cdot h} = \frac{R X_{C \cdot T}}{R_1 \sqrt{R^2 + X_{C \cdot T}^2}} \tag{14-9}$$

The gain at the high frequencies will be less than the gain at the mid-frequencies by a factor K_h, obtained by dividing Eq. (14-9) by Eq. (14-2*a*); thus,

$$K_h = \frac{R X_{C \cdot T}}{R_1 \sqrt{R^2 + X_{C \cdot T}^2}} \div \frac{R_2}{R_1 + R_2} \tag{14-10}$$

From $R = \dfrac{R_1 R_2}{R_1 + R_2}$, it can be shown that $\dfrac{R_2}{R_1 + R_2} = \dfrac{R}{R_1}$

Substituting $\dfrac{R}{R_1}$ for $\dfrac{R_2}{R_1 + R_2}$ in Eq. (14-10),

$$K_h = \frac{R X_{C \cdot T}}{R_1 \sqrt{R^2 + C_{C \cdot T}^2}} \div \frac{R}{R_1} \tag{14-10a}$$

After performing the division

$$K_h = \frac{X_{C \cdot T}}{\sqrt{R^2 + X_{C \cdot T}^2}} \tag{14-10b}$$

Dividing the numerator and denominator of the right-hand member by $X_{C \cdot T}$, then

$$K_h = \frac{1}{\sqrt{1 + \left(\dfrac{R}{X_{C \cdot T}}\right)^2}} \tag{14-10c}$$

When K_h and $A_{v \cdot m}$ are known, the gain at the high frequencies also can be determined by

$$A_{v \cdot h} = K_h A_{v \cdot m} \tag{14-11}$$

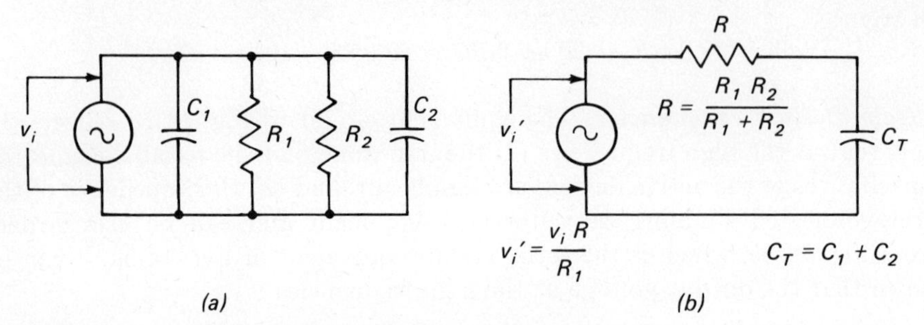

Fig. 14-5 A-c equivalent circuit of *RC*-coupling network at high frequencies. (a) Basic circuit. (b) Series circuit equivalent of (a).

High-frequency Cutoff. The high-frequency cutoff value is dependent on the values of R_1, R_2, C_1, and C_2. The high-frequency cutoff is that frequency value for which the voltage gain is 0.707 times the mid-frequency gain and therefore can be obtained by substituting 0.707 for K_h in Eq. (14-10*c*) and solving for frequency. Then,

$$0.707 = \frac{1}{\sqrt{1 + \left(\frac{R}{X_{C \cdot T}}\right)^2}} \tag{14-12}$$

Squaring both sides of Eq. (14-12),

$$0.5 = \frac{1}{1 + \frac{R^2}{X_{C \cdot T}^2}} \tag{14-12a}$$

Transposing,

$$0.5 + \frac{0.5R^2}{X_{C \cdot T}^2} = 1 \tag{14-12b}$$

Rearranging the terms,

$$0.5R^2 = (1 - 0.5)X_{C \cdot T}^2 \tag{14-12c}$$

and

$$R^2 = X_{C \cdot T}^2 \tag{14-12d}$$

Taking the square root,

$$R = X_{C \cdot T} \tag{14-12e}$$

Stating $X_{C \cdot T}$ in terms of f and C_T,

$$R = \frac{10^6}{2\pi f C_T} \tag{14-12f}$$

Rearranging the terms, the high-frequency cutoff becomes

$$f_{h \cdot co} = \frac{10^6}{2\pi C_T R} \tag{14-12g}$$

and

$$f_{h \cdot co} = \frac{159{,}000}{C_T R} \tag{14-12h}$$

Example 14-5 What is the high-frequency cutoff for the *RC*-coupling circuit of Fig. 14-2 if $R_1 = 6{,}000$ ohms, $R_2 = 30{,}000$ ohms, $C_1 = 120$ pf, and $C_2 = 120$ pf?

GIVEN: $R_1 = 6{,}000$ ohms $\quad R_2 = 30{,}000$ ohms $\quad C_1 = 120$ pf $\quad C_2 = 120$ pf

FIND: $f_{h\cdot co}$

SOLUTION:

$$f_{h\cdot co} = \frac{159{,}000}{C_T R} = \frac{159{,}000}{240 \times 10^{-6} \times 5{,}000} \cong 132 \text{ kHz}$$

$$C_T = C_1 + C_2 = 120 + 120 = 240 \text{ pf}$$

$$R = \frac{R_1 R_2}{R_1 + R_2} = \frac{6{,}000 \times 30{,}000}{6{,}000 + 30{,}000} = 5{,}000 \text{ ohms}$$

Example 14-6 What is the gain of the circuit of Example 14-5 at the cutoff frequency found in Example 14-5?

GIVEN: $R_1 = 6{,}000$ ohms $\quad R = 5{,}000$ ohms $\quad C_T = 240$ pf $\quad f_{h\cdot co} = 132$ kHz

FIND: $A_{v\cdot h}$

SOLUTION:

$$A_{v\cdot h} = \frac{R X_{C\cdot T}}{R_1 \sqrt{R^2 + X_{C\cdot T}^2}} = \frac{5{,}000 \times 5{,}000}{6{,}000 \sqrt{5{,}000^2 + 5{,}000^2}} \cong 0.59$$

$$X_{C\cdot T} = \frac{159{,}000}{f_{h\cdot co} C_T} = \frac{159{,}000}{132 \times 10^3 \times 240 \times 10^{-6}} \cong 5{,}000 \text{ ohms}$$

Example 14-7 (*a*) Find the high-frequency cutoff factor for the circuit of Examples 14-5 and 14-6 by using Eq. (14-10*c*). (*b*) What is the corresponding decibel loss?

GIVEN: $R = 5{,}000$ ohms $\quad X_{C\cdot T} = 5{,}000$ ohms

FIND: (*a*) K_h $\quad$ (*b*) db

SOLUTION:

(*a*)
$$K_h = \frac{1}{\sqrt{1 + \left(\dfrac{R}{X_{C\cdot T}}\right)^2}} = \frac{1}{\sqrt{1 + \left(\dfrac{5{,}000}{5{,}000}\right)^2}} = 0.707$$

(*b*)
$$\text{db} = 20 \log K_h = 20 \log 0.707 = 20(9.85 - 10) = -3 \text{ db}$$

Example 14-8 If for the circuit of Fig. 14-2, $R_1 = 20{,}000$ ohms, $R_2 = 80{,}000$ ohms, $C_c = 0.016$ μf, $C_1 = 120$ pf, and $C_2 = 120$ pf, find (*a*) low-frequency cutoff, (*b*) high-frequency cutoff.

GIVEN: $R_1 = 20{,}000$ ohms $\quad R_2 = 80{,}000$ ohms $\quad C_c = 0.016$ μf $\quad C_1, C_2 = 120$ pf

FIND: (*a*) $f_{l\cdot co}$ $\quad$ (*b*) $f_{h\cdot co}$

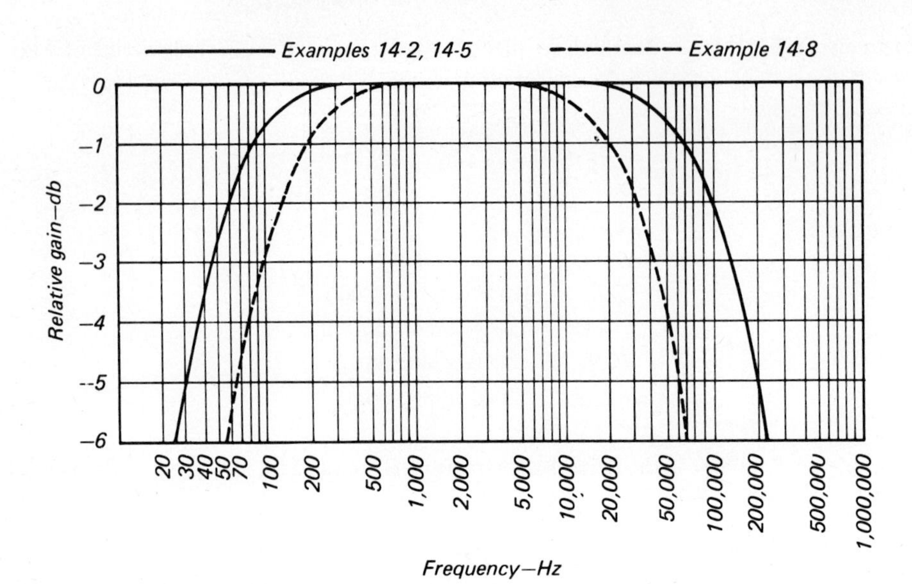

Fig. 14-6 Gain versus frequency characteristic curves for *RC*-coupled circuits of Examples 14-2, 14-5, and 14-8.

SOLUTION:

(*a*)
$$f_{l\text{-co}} = \frac{159{,}000}{C_c(R_1 + R_2)} = \frac{159{,}000}{0.016(20{,}000 + 80{,}000)} \cong 100 \text{ Hz}$$

(*b*)
$$f_{h\text{-co}} = \frac{159{,}000}{C_T R} = \frac{159{,}000}{240 \times 10^{-6} \times 16{,}000} \cong 41.4 \text{ kHz}$$

$$C_T = C_1 + C_2 = 120 + 120 = 240 \text{ pf}$$

$$R = \frac{R_1 R_2}{R_1 + R_2} = \frac{20{,}000 \times 80{,}000}{20{,}000 + 80{,}000} = 16{,}000 \text{ ohms}$$

Response Curves. The gain versus frequency characteristics of an *RC*-coupled amplifier are frequently expressed graphically by *frequency response curves*. Because of the wide range of frequency values involved, the frequency scale of the curve is made logarithmic. Figure 14-6 illustrates the frequency response curves for the circuit and conditions of Examples 14-2, 14-5, and 14-8.

14-4 Frequency Response of Cascaded *RC*-coupled Circuits

Principle. The gain versus frequency characteristics presented in Art. 14-3 apply to a single stage of *RC* coupling. When two or more coupled circuits are cascaded, the combined effects on the response characteristics must be taken into consideration.

Low-frequency Cutoff. When two or more circuits affect the low-frequency cutoff, the highest value is the predominant one because its action voids the others. If two circuits have approximately the same cutoff value, a reasonable approximation of the cutoff frequency for the complete circuit is obtained by using the lower value increased by 50 per cent.

High-frequency Cutoff. When two or more circuits affect the high-frequency cutoff, the lowest value is the predominant one because its action voids the others. If two circuits have approximately the same cutoff value, a reasonable approximation of the cutoff frequency for the complete circuit is obtained by using the higher value decreased by 50 per cent.

Cascaded Similar Stages. When two or more similar stages of coupling are cascaded, the approximate value of the composite circuit frequency responses can be found by

$$f_{l\cdot co}\text{ total} = 1.1\sqrt{n}\, f_{l\cdot co}\text{ per stage} \tag{14-13}$$

$$f_{h\cdot co}\text{ total} = \frac{f_{h\cdot co}\text{ per stage}}{1.1\sqrt{n}} \tag{14-14}$$

where n = number of stages

Example 14-9 If an amplifier consists of three similar stages each having a low-frequency cutoff of 30 Hz and a high-frequency cutoff of 1.5 MHz, what is the overall (*a*) low-frequency cutoff? (*b*) High-frequency cutoff?

GIVEN: $n = 3$ $\quad f_{l\cdot co}/\text{stage} = 30$ Hz $\quad f_{h\cdot co}/\text{stage} = 1.5$ MHz

FIND: (*a*) $f_{l\cdot co}$ $\quad$ (*b*) $f_{h\cdot co}$

SOLUTION:

(*a*) $$f_{l\cdot co} = 1.1\sqrt{n}\, f_{l\cdot co} = 1.1\sqrt{3} \times 30 \cong 57\text{ Hz}$$

(*b*) $$f_{h\cdot co} = \frac{f_{h\cdot co}}{1.1\sqrt{n}} = \frac{1.5 \times 10^6}{1.1\sqrt{3}} = 790\text{ kHz}$$

14-5 Frequency Response of Multistage *RC*-coupled Transistor Amplifiers

The Circuit. Figure 14-7 illustrates a typical two-stage transistor amplifier containing a number of modifications to the basic circuit.

Low-frequency Cutoff. In the circuit of Fig. 14-7, the rolloff in gain at the lower frequencies is caused by the series-connected capacitors C_1, C_2, and C_3, and the bypass capacitors C_4 and C_5. For the circuits involving C_1, C_2, and C_3, the low-frequency cutoff occurs when the reactance of the capacitor is equal to the resistance in series with the capacitor because the input signal to the stage then will be down 3 db. For these circuits, the low-frequency cutoff can be found with Eq. (14-7*h*).

Low-frequency Cutoff Due to Bypass Capacitors. The low-frequency cutoff of the amplifier is also dependent on the bypass capacitors C_4 and C_5 because at the lower frequencies the reactances of these capacitors become greater than their associated resistances R_6 and R_7. Under these conditions (1) a significant

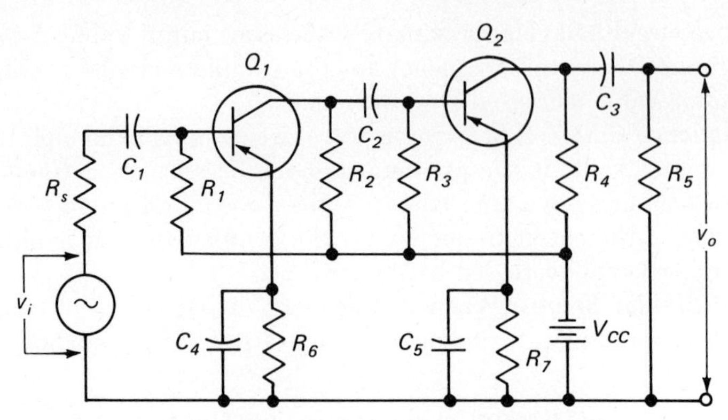

C_1, C_2, C_3 = 1 μf; C_4, C_5 = 50 μf; R_1, R_3 = 1,500,000 ohms;
R_2, R_4 = 4,000 ohms; R_5 = 16,000 ohms; R_6, R_7 = 7,000 ohms;
V_{CC} = 22 volts; C_s for each stage = 10 pf; R_s = 1,000 ohms;
For each transistor: β = 100; f_T = 120 MHz; C_{eb} = 20 pf; C_{cb} = 2 pf.

Fig. 14-7 Two-stage *RC*-coupled transistor amplifier.

amount of signal current flows through these resistors, (2) degeneration takes place in the emitter circuit, and (3) the voltage gain of the amplifier stage decreases. Cutoff (−3 db) occurs when $X_C = R_{eq}$; and for the circuit involving C_4

$$f_{l\text{-}co} = \frac{159{,}000}{C_4 R_{eq}} \tag{14-15}$$

where $R_{eq} = R_6 \| r_x$

$$r_x = r_{e\text{-}1} + \frac{R_1 \| R_s}{\beta_1}$$

For the general case,

$$R_{eq} = R_x \| (R_y + r_x) \tag{14-16}$$

where R_{eq} = resistance seen by the bypass capacitor, ohms
R_x = bypassed portion of the emitter resistance, ohms
R_y = unbypassed portion of the emitter resistance, ohms

Example 14-10 For the amplifier circuit of Fig. 14-7, what is the low-frequency cutoff (*a*) at C_1? (*b*) At C_2? (*c*) At C_3? (*d*) At C_4? (*e*) At C_5? (*f*) For the complete circuit?

GIVEN: Fig. 14-7

FIND: (*a*) $f_{C\text{-}1}$ (*b*) $f_{C\text{-}2}$ (*c*) $f_{C\text{-}3}$ (*d*) $f_{C\text{-}4}$ (*e*) $f_{C\text{-}5}$ (*f*) $f_{l\text{-}co}$

SOLUTION:

(*a*) At C_1

$$f_{l\text{-}co} = \frac{159{,}000}{C_1 R_{eq}} = \frac{159{,}000}{1 \times 3{,}500} \cong 45.5 \text{ Hz}$$

where, assuming C_4 holds the emitter of Q_1 at a-c ground

$$R_{eq} = R_s + (R_1 \| \beta_1 r_{e\cdot 1}) = 1{,}000 + (1{,}500{,}000 \| 100 \times 25) \cong 3{,}500 \text{ ohms}$$
$$r_{e\cdot 1} = 25/I_{E\cdot 1} = 25/1 = 25 \text{ ohms}$$
$$I_{E\cdot 1} = \frac{V_{CC}}{R_6 + R_1/\beta_1} = \frac{22}{7{,}000 + 1{,}500{,}000/100} = 1 \text{ ma}$$

(*b*) At C_2

$$f_{l\cdot co} = \frac{159{,}000}{C_2 R_{eq}} = \frac{159{,}000}{1 \times 6{,}500} \cong 24.5 \text{ Hz}$$

where, assuming C_5 holds the emitter of Q_2 at a-c ground

$$R_{eq} = R_2 + r_{i\cdot 2} = 4{,}000 + 2{,}500 = 6{,}500 \text{ ohms}$$
$$r_{i\cdot 2} = R_3 \| \beta_2 r_{e\cdot 2} = 1{,}500{,}000 \| 100 \times 25 \cong 2{,}500 \text{ ohms}$$
$$r_{e\cdot 2} = 25/I_{E\cdot 2} = 25/1 = 25 \text{ ohms}$$
$$I_{E\cdot 2} = \frac{V_{CC}}{R_7 + R_3/\beta_2} = \frac{22}{7{,}000 + 1{,}500{,}000/100} = 1 \text{ ma}$$

(*c*) At C_3

$$f_{l\cdot co} = \frac{159{,}000}{C_3 R_{eq}} = \frac{159{,}000}{1 \times 20{,}000} \cong 8 \text{ Hz}$$

where $R_{eq} = R_4 + R_5 = 4{,}000 + 16{,}000 = 20{,}000$ ohms

(*d*) At C_4

$$f_{l\cdot co} = \frac{159{,}000}{C_4 R_{eq}} = \frac{159{,}000}{50 \times 35} \cong 91 \text{ Hz}$$

where $R_{eq} = R_6 \| \left(r_{e\cdot 1} + \dfrac{R_1 \| R_s}{\beta_1} \right) = 7{,}000 \| \left(25 + \dfrac{1{,}500{,}000 \| 1{,}000}{100} \right) \cong 35$ ohms

(*e*) At C_5

$$f_{l\cdot co} = \frac{159{,}000}{C_5 R_{eq}} = \frac{159{,}000}{50 \times 65} \cong 49 \text{ Hz}$$

where $R_{eq} = R_7 \| \left(r_{e\cdot 2} + \dfrac{R_3 \| R_2}{\beta_2} \right) = 7{,}000 \| \left(25 + \dfrac{1{,}500{,}000 \| 4{,}000}{100} \right) \cong 65$ ohms

(*f*) For the circuit

$$f_{l\cdot co} \cong \text{highest value among } (a) \text{ to } (e) \cong 91 \text{ Hz}$$

Low-frequency Gain. At the low frequencies, the voltage gain of a stage will be lower than the gain at the mid-frequencies due to the voltage drop at the coupling capacitor. The gain at the low frequencies can be determined by use of Eqs. (14-6), (14-5*d*), and (14-2*a*).

Example 14-11 What is the voltage gain at 30 Hz for the first stage of the amplifier of Fig. 14-7? [*Note:* (1) Assume that the voltage drops at R_s and C_1 are negligible; (2) from Example 13-5*b*, $A_{v\cdot m} = 61.6$.]

GIVEN: Fig. 14-7 $A_{v \cdot m} = 61.6$ $f = 30$ Hz

FIND: A_v

SOLUTION:

$$A_v = K_l A_{v \cdot m} = 0.775 \times 61.6 \cong 47.8$$

$$K_l = \frac{1}{\sqrt{1 + \left(\frac{X_c}{R_{eq}}\right)^2}} = \frac{1}{\sqrt{1 + \left(\frac{5{,}300}{6{,}500}\right)^2}} = 0.775$$

$$X_{C \cdot 2} = \frac{159{,}000}{fC_2} = \frac{159{,}000}{30 \times 1} = 5{,}300 \text{ ohms}$$

$$R_{eq} = 6{,}500 \text{ ohms} \qquad \text{(from Example 14-10}b\text{)}$$

High-frequency Cutoff. The high-frequency cutoff value is dependent chiefly upon (1) the interelement capacitances of the transistors, and (2) the stray capacitances of the wiring, etc. At the high frequencies, the impedances of C_1, C_2, C_3, C_4, and C_5 are so low that these capacitors do not affect the high-frequency cutoff value.

Determining the high-frequency cutoff value is more difficult than determining the low-frequency cutoff value because (1) the input and output capacitances of the transistors have interrelated actions called the *Miller effect,* and (2) the interelement capacitances frequently are not available.

Simple Method of Determining the Approximate High-frequency Cutoff. The value generally given for the beta of a transistor is for the band of frequencies at which beta is constant, sometimes called the *low-frequency beta.* For frequencies above and below this band, the value of beta decreases. At high frequencies, beta decreases rapidly, and at some frequency value it decreases to unity. The frequency at which $\beta = 1$ is called the gain-bandwidth product (Art. 4-19) and is represented by the symbol f_T. (The value of f_T is usually given with the manufacturers' technical data.)

The frequency at which beta has decreased to 0.707 times the low-frequency beta is called the *beta cutoff frequency* and is represented by the symbol f_β. The beta cutoff frequency occurs at

$$f_\beta = \frac{f_T}{\beta} \tag{4-23}$$

The beta cutoff frequency is sometimes used as the approximate high-frequency cutoff value for the composite circuit, thus

$$f_{h \cdot co} \cong \frac{f_T}{\beta} \tag{14-17}$$

Because this value does not include the effects of the (1) interelement capacitances, (2) stray capacitances, and (3) other circuit components, it is only an approximate value for the high-frequency cutoff for the circuit.

Comprehensive Method of Determining the High-frequency Cutoff. A more accurate value for the high-frequency cutoff of a transistor amplifier circuit is obtained by determining the cutoff frequency: (1) due to the transit time of the transistor by use of Eq. (14-17), and (2) at the input and output of each stage when including the effects of all the circuit parameters involved—by use of Eqs. (14-18) and (14-20). The lowest of these values more accurately represents the high-frequency cutoff of the circuit.

The high-frequency cutoff at the input side of an amplifier stage may be found by

$$f_{h\cdot co\cdot i} = \frac{1}{2\pi r_i C_i} \tag{14-18}$$

and when C_i is expressed in microfarads

$$f_{h\cdot co\cdot i} = \frac{159{,}000}{r_i C_i} \tag{14-18a}$$

where

$$C_i = C_{eb} + C_{cb}(1 + A_v) \tag{14-19}$$

The high-frequency cutoff at the output side of an amplifier stage may be found by

$$f_{h\cdot co\cdot o} = \frac{1}{2\pi r_o C_o} \tag{14-20}$$

and when C_o is expressed in microfarads

$$f_{h\cdot co\cdot o} = \frac{159{,}000}{r_o C_o} \tag{14-20a}$$

where

$$C_o = C_s + C_{cb}\left(1 + \frac{1}{A_v}\right) \tag{14-21}$$

In Eqs. (14-19) and (14-21), the terms $C_{cb}(1 + A_v)$ and $C_{cb}(1 + 1/A_v)$ represent the Miller effect of the capacitance C_{cb}. The Miller effect takes into account the feedback from the collector to the base, and vice versa, due to the collector-to-base capacitance.

Example 14-12 For the circuit of Fig. 14-7, what is the high-frequency cutoff at Q_1 (*a*) by Eq. (14-17)? (*b*) By Eq. (14-18*a*)? (*c*) By Eq. (14-20*a*)?

GIVEN: Fig. 14-7

FIND: (*a*) $f_{h\cdot co}$ (*b*) $f_{h\cdot co\cdot i}$ (*c*) $f_{h\cdot co\cdot o}$

SOLUTION:

(*a*) $$f_{h\cdot co} = \frac{f_{T\cdot 1}}{\beta_1} = \frac{120}{100} = 1.2 \text{ MHz}$$

(*b*) $$f_{h\cdot co\cdot i} = \frac{159{,}000}{r_i C_i} = \frac{159{,}000}{715 \times 145 \times 10^{-6}} \cong 1.53 \text{ MHz}$$

where $$r_i = R_s \| R_1 \| \beta_1 r_{e\cdot 1} \cong R_s \| \beta_1 r_{e\cdot 1} = \frac{1{,}000 \times 2{,}500}{1{,}000 + 2{,}500} \cong 715 \text{ ohms}$$

$$r_{e\cdot 1} = \frac{25}{I_{E\cdot 1}} = \frac{25}{1} = 25 \text{ ohms}$$

$$I_{E\cdot 1} = \frac{V_{CC}}{R_6 + \dfrac{R_1}{\beta_1}} = \frac{22}{7{,}000 + \dfrac{1{,}500{,}000}{100}} = 1 \text{ ma}$$

$$C_i = C_{eb\cdot 1} + C_{cb\cdot 1}(1 + A_{v\cdot 1}) = 20 + 2(1 + 61.5) = 145 \text{ pf}$$

$$A_{v\cdot 1} = \frac{r_{o\cdot 1}}{r_{e\cdot 1}} = \frac{1{,}540}{25} \cong 61.5$$

$$r_{o\cdot 1} = R_2 \| R_3 \| \beta_2 r_{e\cdot 2} \cong R_2 \| \beta_2 r_{e\cdot 2} = \frac{4{,}000 \times (100 \times 25)}{4{,}000 + (100 \times 25)} \cong 1{,}540 \text{ ohms}$$

$$r_{e\cdot 2} = \frac{25}{I_{E\cdot 2}} = \frac{25}{1} = 25 \text{ ohms}$$

$$I_{E\cdot 2} = \frac{V_{CC}}{R_7 + \dfrac{R_3}{\beta_2}} = \frac{22}{7{,}000 + \dfrac{1{,}500{,}000}{100}} = 1 \text{ ma}$$

(*c*) $$f_{h\cdot co\cdot o} = \frac{159{,}000}{r_o C_o} = \frac{159{,}000}{1{,}540 \times 12 \times 10^{-6}} \cong 8.6 \text{ MHz}$$

$$r_o = r_{o\cdot 1} = 1{,}540 \text{ ohms} \qquad \text{[see part (}b\text{)]}$$

$$C_o = C_s + C_{cb}\left(1 + \frac{1}{A_v}\right) = 10 + 2\left(1 + \frac{1}{61.5}\right) \cong 12 \text{ pf}$$

High-frequency Gain. At the high frequencies, the voltage gain of a stage will be lower than the gain at the mid-frequencies due to parallel-connected capacitances. The gain at the high frequencies can be determined by use of Eqs. (14-11), (14-10*c*), and (14-2*a*).

Example 14-13 What is the voltage gain at 1 MHz for the first stage of the amplifier of Fig. 14-7? [*Note:* (1) Assume that the voltage drops at R_s and C_1 are negligible; (2) from Example 13-5*b*, $A_{v\cdot m} = 61.6$.]

GIVEN: Fig. 14-7 $\quad A_{v\cdot m} = 61.6 \quad f = 1$ MHz

FIND: A_v

SOLUTION:

$$A_v = K_h A_{v\cdot m} = 0.84 \times 61.6 \cong 51.8$$

At the output side

$$K_h = \frac{1}{\sqrt{1 + \left(\dfrac{R}{X_c}\right)^2}} = \frac{1}{\sqrt{1 + \left(\dfrac{1{,}540}{13{,}250}\right)^2}} \cong 1$$

$$R = 1{,}540 \text{ ohms} \qquad \text{(from Example 14-12}b\text{)}$$

$$X_c = \frac{159{,}000}{fC_o} = \frac{159{,}000}{10^6 \times 12 \times 10^{-6}} \cong 13{,}250 \text{ ohms}$$

$$C_o = 12 \text{ pf} \qquad \text{(from Example 14-12}c\text{)}$$

At the input side

$$K_h = \frac{1}{\sqrt{1 + \left(\frac{R}{X_c}\right)^2}} = \frac{1}{\sqrt{1 + \left(\frac{715}{1,100}\right)^2}} \cong 0.84$$

$$R = 715 \text{ ohms} \qquad \text{(from Example 14-12}b)$$

$$X_c = \frac{159,000}{fC_i} = \frac{159,000}{10^6 \times 145 \times 10^{-6}} \cong 1,100 \text{ ohms}$$

$$C_i = 145 \text{ pf} \qquad \text{(from Example 14-12}b)$$

14-6 Frequency Response of *RC*-coupled Triode Vacuum-tube Amplifier

The Circuit. Figure 14-8 illustrates the circuit of a typical vacuum-tube amplifier using a triode; the circuit parameters are similar to those of Fig. 13-8. Comparing this circuit with the first stage of the transistor amplifier of Fig. 14-7 will show similarities in the circuit configuration and a wide difference in the values of comparable circuit components.

Frequency-response Characteristics. The gain versus frequency response characteristics of the vacuum-tube amplifier are similar to those of the transistor amplifier. Again, the maximum gain occurs in the mid-frequency range with some rolloff at both the lower and higher frequencies.

Low-frequency Cutoff. In the circuit of Fig. 14-8, the rolloff in voltage gain at the low frequencies is caused by the series-connected capacitors C_1 and C_2, and the bypass capacitor C_3. As with the transistor amplifier, the cutoff frequency can be found with Eqs. (14-7*h*) and (14-15).

Example 14-14 For the circuit of Fig. 14-8, find the cutoff frequency due to (*a*) C_1, (*b*) C_2, (*c*) C_3.

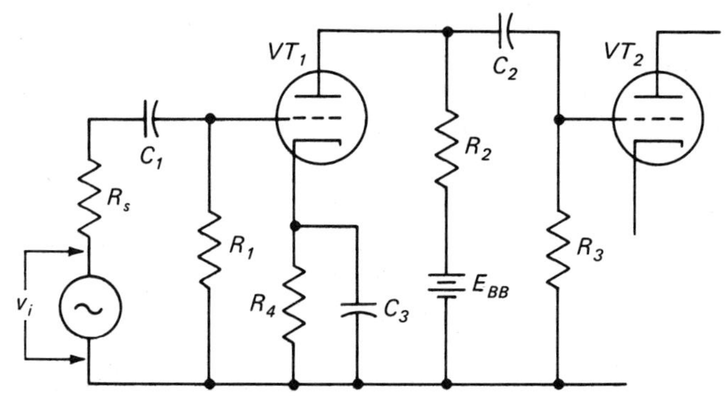

R_1, R_2, R_3 = 470,000 ohms, R_4 = 6,000 ohms, R_s = 2,000 ohms, r_p = 80,000 ohms, C_1, C_2 = 0.006 μf, C_3 = 1 μf, E_{BB} = 180 volts, μ = 100, C_s = 10 pf, C_{gk} = 1.8 pf, C_{pk} = 0.8 pf, C_{gp} = 2 pf; VT_1 and VT_2 are similar tubes.

Fig. 14-8 Typical *RC*-coupled amplifier using triodes.

GIVEN: Fig. 14-8

FIND: (a) $f_{l\cdot co}$ (b) $f_{l\cdot co}$ (c) $f_{l\cdot co}$

SOLUTION:

(a) $$f_{l\cdot co} = \frac{159{,}000}{C_1(R_s + R_1)} = \frac{159{,}000}{0.006(2{,}000 + 470{,}000)} \cong 56 \text{ Hz}$$

(b) $$f_{l\cdot co} = \frac{159{,}000}{C_2 R} = \frac{159{,}000}{0.006 \times 538{,}000} \cong 49 \text{ Hz}$$

$$R = R_3 + (r_p \| R_2) = 470{,}000 + \frac{80{,}000 \times 470{,}000}{80{,}000 + 470{,}000} \cong 538{,}000 \text{ ohms}$$

(c) $$f_{l\cdot co} = \frac{159{,}000}{C_3 R_4} = \frac{159{,}000}{1 \times 6{,}000} \cong 26.5 \text{ Hz}$$

Low-frequency Gain. The voltage gain at the low frequencies is less than the mid-frequency gain and can be found with Eqs. (14-6), (14-5*d*), and (13-4).

Example 14-15 What is the approximate voltage gain at 30 Hz for the amplifier of Fig. 14-8? (*Note:* Assume that the voltage drops at R_s and C_1 are negligible.)

GIVEN: Fig. 14-8 $f = 30$ Hz

FIND: A_v

SOLUTION:

$$A_v = K_l A_{v\cdot m} = 0.47 \times 74.5 = 35$$

$$K_l = \frac{1}{\sqrt{1 + \left(\frac{X_c}{R}\right)^2}} = \frac{1}{\sqrt{1 + \left(\frac{884{,}000}{470{,}000}\right)^2}} \cong 0.47$$

$$X_c = \frac{159{,}000}{fC_2} = \frac{159{,}000}{30 \times 0.006} \cong 884{,}000 \text{ ohms}$$

$$R = R_3 = 470{,}000 \text{ ohms}$$

$$A_{v\cdot m} = \frac{\mu r_o}{r_p + r_o} = \frac{100 \times 235{,}000}{80{,}000 + 235{,}000} \cong 74.5$$

$$r_o = \frac{R_2 R_3}{R_2 + R_3} = \frac{470{,}000 \times 470{,}000}{470{,}000 + 470{,}000} = 235{,}000 \text{ ohms}$$

High-frequency Cutoff. In the circuit of Fig. 14-8, the rolloff in voltage gain at the high frequencies is caused by the parallel-connected capacitances C_s, C_{pk}, C_{gk}, and C_{gp}. As with the transistor amplifier, the cutoff frequency can be found with Eqs. (14-12*h*) and (14-10*c*). In using Eq. (14-10*c*),

$$C_T \text{ (at input side of } VT_1) = C_{gk} + C_{gp}(1 + A_v) \tag{14-22}$$

$$C_T \text{ (at output side of } VT_1) = C_s + C_{pk} + C_{gp}\left(1 + \frac{1}{A_v}\right) \tag{14-23}$$

Example 14-16 For the circuit of Fig. 14-8, find the high-frequency cutoff at (*a*) the input of VT_1, (*b*) the output of VT_1. (*Note:* From Example 14-15, $A_v = 74.5$.)

GIVEN: Fig. 14-8 $A_v = 74.5$

FIND: (*a*) $f_{h \cdot co}$ (*b*) $f_{h \cdot co}$

SOLUTION:

$$(a) \quad f_{h \cdot co} = \frac{159{,}000}{C_T R} = \frac{159{,}000}{152.8 \times 10^{-6} \times 2{,}000} \cong 520 \text{ kHz}$$

$$C_T = C_{gk} + C_{gp}(1 + A_v) = 1.8 + 2(1 + 74.5) = 152.8 \text{ pf}$$

$$R = R_s \| R_1 = \frac{R_s R_1}{R_s + R_1} = \frac{2{,}000 \times 470{,}000}{2{,}000 + 470{,}000} \cong 2{,}000 \text{ ohms}$$

$$(b) \quad f_{h \cdot co} = \frac{159{,}000}{C_T R} = \frac{159{,}000}{12.8 \times 10^{-6} \times 60{,}000} \cong 207 \text{ kHz}$$

$$C_T = C_s + C_{pk} + C_{gp}\left(1 + \frac{1}{A_v}\right) = 10 + 0.8 + 2\left(1 + \frac{1}{74.5}\right) \cong 12.8 \text{ pf}$$

$$R = r_p \| R_2 \| R_3 = 80{,}000 \| 470{,}000 \| 470{,}000 \cong 60{,}000 \text{ ohms}$$

High-frequency Gain. The voltage gain at the high frequencies is less than the mid-frequency gain and can be found with Eqs. (14-11), (14-10*c*), (14-22), (14-23), and (13-4).

Example 14-17 What is the approximate voltage gain at 100 kHz for the amplifier of Fig. 14-8? (*Note:* Use $A_{v \cdot m} = 74.5$.)

GIVEN: Fig. 14-8 $A_{v \cdot m} = 74.5$

FIND: A_v

SOLUTION:

$$A_v = K_h A_{v \cdot m} = 0.9 \times 74.5 \cong 67$$

At the input to VT_1

$$K_h = \frac{1}{\sqrt{1 + \left(\dfrac{R}{X_{C \cdot T}}\right)^2}} = \frac{1}{\sqrt{1 + \left(\dfrac{2{,}000}{10{,}400}\right)^2}} \cong 0.985$$

$$R = 2{,}000 \text{ ohms} \qquad \text{(from Example 14-16}a\text{)}$$

$$X_{C \cdot T} = \frac{159{,}000}{f C_T} = \frac{159{,}000}{10^5 \times 152.8 \times 10^{-6}} \cong 10{,}400 \text{ ohms}$$

$$C_T = 152.8 \text{ pf} \qquad \text{(from Example 14-16}a\text{)}$$

At the output of VT_1

$$K_h = \frac{1}{\sqrt{1 + \left(\dfrac{R}{X_{C \cdot T}}\right)^2}} = \frac{1}{\sqrt{1 + \left(\dfrac{60{,}000}{124{,}000}\right)^2}} \cong 0.90$$

$$R = 60{,}000 \text{ ohms} \qquad \text{(from Example 14-16}b\text{)}$$

$$X_{C \cdot T} = \frac{159{,}000}{f C_T} = \frac{159{,}000}{10^5 \times 12.8 \times 10^{-6}} \cong 124{,}000 \text{ ohms}$$

$$C_T = 12.8 \text{ pf} \qquad \text{(from Example 14-16}b\text{)}$$

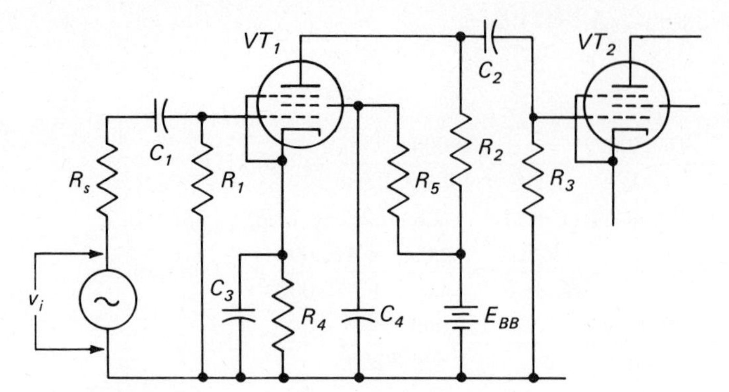

$R_s = 2{,}000$ ohms, $R_1, R_2, R_3 = 220{,}000$ ohms, $R_4 = 2{,}700$ ohms, $R_5 = 340{,}000$ ohms, $C_1, C_2 = 0.008\ \mu f$, $C_3 = 10\ \mu f$, $C_4 = 0.06\ \mu f$, $r_p = 500{,}000$ ohms, $g_m = 4{,}000\ \mu$mhos, $E_{BB} = 90$ volts, $C_s = 10$ pf, $C_{gk} = 5$ pf, $C_{pk} = 5$ pf, $C_{gp} = 0.0035$ pf; VT_1, VT_2 are similar tubes.

Fig. 14-9 Typical *RC*-coupled amplifier using pentodes.

14-7 Frequency Response of *RC*-coupled Pentode Vacuum-tube Amplifier

The Circuit. Figure 14-9 illustrates the circuit of a typical vacuum-tube amplifier using a pentode. The circuit differs from the triode amplifier (Fig. 14-8) only by the addition of the screen-grid resistor R_5 and its bypass capacitor C_4. The major differences in the circuit parameters are (1) the pentode tube has a much higher plate resistance than the triode, and (2) the grid-to-plate capacitance of the pentode is much lower than for the triode.

Frequency Response Characteristics. The gain versus frequency response of the *RC*-coupled pentode amplifier characteristics are similar to the corresponding transistor and triode amplifiers.

Voltage Gain at the Mid-frequencies. The gain for the *RC*-coupled pentode amplifier at the mid-frequencies can be found with Eq. (13-6).

Cutoff Frequencies. The low-frequency and high-frequency cutoff values can be found in the same manner used with the triode amplifiers in Examples 14-14 and 14-16.

Voltage Gain at the Low and High Frequencies. As with the amplifier using a triode, the voltage gain of the pentode amplifier at the low and high frequencies will be lower than the gain at the mid-frequencies. The reduction in the gains will be by the same factors K_l and K_h as for the triode amplifiers. The gains at the low and high frequencies can be found in the same manner as used for the triode amplifier in Examples 14-15 and 14-17.

14-8 Frequency Response with Impedance Coupling

Circuit Characteristics. The circuit for the basic impedance-coupled network is shown in Fig. 14-10. The inductance of the coil L should be high enough so that its impedance at the higher frequencies is much greater than the value of the output resistor R. At the lower frequencies, the impedance of the coil may become so low that the amount of current bypassed by the coil will in effect short-circuit the signal source and drastically reduce the output-signal voltage v_o.

Equivalent Circuits. The a-c equivalent circuits for the impedance-coupled network (Figs. 14-11 to 14-13) reflect the same circuit characteristics as with the RC-coupled network. The main difference between the two systems is that with impedance coupling the impedance of the input varies with the frequency of the input signal while with RC coupling the impedance of the input remains constant.

Frequency Response Characteristics. The frequency response of impedance coupling is not as good as that obtained with RC coupling, especially at the low frequencies. Impedance coupling provides more efficient use of the power source since the resistance of the coupling coil is much lower than the resistance it replaces in the RC-coupling network and hence less d-c power is dissipated by the coupling element.

Gain at Mid-frequencies. From the a-c equivalent circuit (Fig. 14-11) it can be shown that the voltage gain at the mid-frequencies is

$$A_{v \cdot m} = \frac{X_L}{\sqrt{R^2 + X_L{}^2}} \tag{14-24}$$

Dividing the numerator and denominator by X_L

$$A_{v \cdot m} = \frac{1}{\sqrt{1 + \left(\dfrac{R}{X_L}\right)^2}} \tag{14-24a}$$

Gain at Low Frequencies. The voltage gain at the low frequencies will decrease as the frequency of the input signal decreases. This is because (1) the

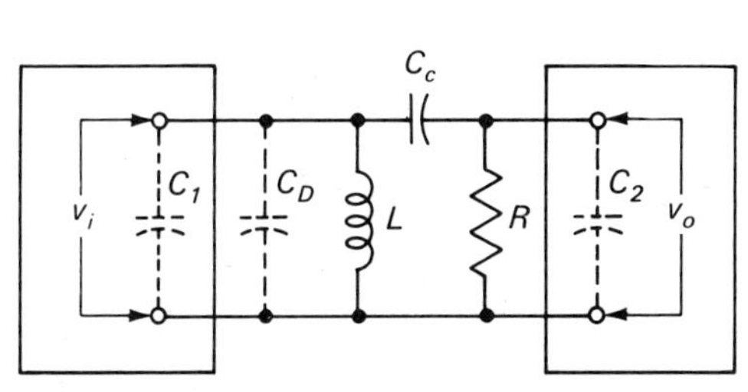

Fig. 14-10 Basic impedance-coupling circuit.

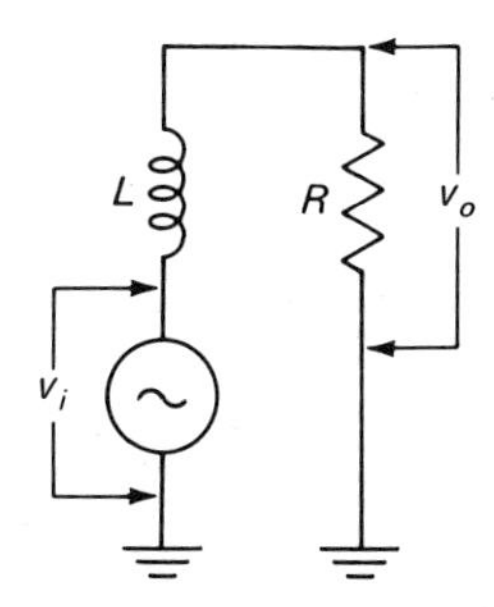

Fig. 14-11 A-c equivalent circuit of impedance-coupling network at mid-frequencies.

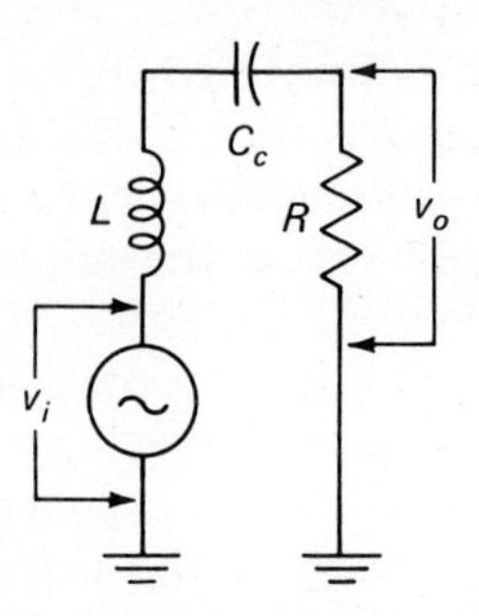

Fig. 14-12 A-c equivalent circuit of impedance-coupling network at low frequencies.

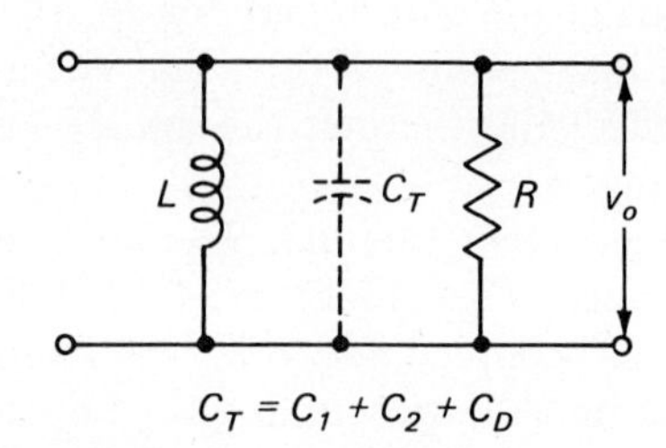

Fig. 14-13 A-c equivalent circuit of impedance-coupling network at high frequencies.

lower value of X_L causes a greater portion of the signal current to flow through the coil and hence a smaller portion of signal current will flow through the output resistor, and (2) the higher value of X_C causes an increase in the voltage across the coupling capacitor and a decrease in the voltage across the output resistor. The dual effect on the low-frequency gain is indicated in Eq. (14-25). From the a-c equivalent circuit (Fig. 14-12) it can be shown that the low-frequency gain factor is

$$K_l = \frac{1}{\sqrt{1 + \left(\frac{R}{X_L}\right)^2 + \left(\frac{X_C}{R}\right)^2}} \tag{14-25}$$

The voltage gain at the low frequencies is expressed by

$$A_{v\cdot l} = K_l A_{v\cdot m} \tag{14-6}$$

Low-frequency Cutoff. Equation (14-25) indicates that the low-frequency cutoff is dependent on (1) the ratio of R to X_L, and (2) the ratio on X_C to R. A simple way to determine the low-frequency cutoff is to find the separate cutoff values due to (1) the coupling coil, and (2) the coupling capacitor. The higher of these two values represents the approximate low-frequency cutoff for the circuit. The separate cutoff values can be found by (1) using -3 db as the cutoff level, and (2) the fact that a 3-db loss occurs when the reactance is equal to the resistance. Thus, for the coil when $2\pi fL = R$, solving for f produces

$$f_{l\cdot co} = \frac{R}{2\pi L} \tag{14-26}$$

For the coupling capacitor, when $\frac{159{,}000}{fC} = R$ solving for f produces

$$f_{l\cdot co} = \frac{159{,}000}{CR} \tag{14-27}$$

Gain at High Frequencies. At the high frequencies, the effects of the coupling coil and the coupling capacitor can be disregarded (as with the RC-coupling system) and only the shunting capacitances C_D, C_1, and C_2 appreciably affect the high-frequency gain; this is indicated in Eq. (14-28). From the a-c equivalent circuit (Fig. 14-13) it can be shown that the high-frequency gain factor is

$$K_h = \frac{1}{\sqrt{1 + (2\pi f C_T R)^2}} \tag{14-28}$$

The voltage gain at the high frequencies is expressed by

$$A_{v \cdot h} = K_h A_{v \cdot m} \tag{14-11}$$

High-frequency Cutoff. The equation for the high-frequency cutoff can be derived by substituting 0.707 for K_h in Eq. (14-28) and solving for f. Then

$$f_{h \cdot co} = \frac{1}{2\pi C_T R} \tag{14-29}$$

Where C is expressed in microfarads

$$f_{h \cdot co} = \frac{159{,}000}{C_T R} \tag{14-29a}$$

Example 14-18 If the parameters for the circuit of Fig. 14-10 are $L = 8$ henrys, $R = 5{,}000$ ohms, $C_c = 0.3\ \mu$f, $C_D = 100$ pf, $C_1 = 50$ pf, and $C_2 = 150$ pf, what is the (*a*) mid-frequency gain (use 1,000 Hz)? (*b*) Low-frequency cutoff? (*c*) High-frequency cutoff?

GIVEN: Fig. 14-10 $L = 8$ henrys $R = 5{,}000$ ohms $C_c = 0.3\ \mu$f
$C_D = 100$ pf $C_1 = 50$ pf $C_2 = 150$ pf $f_m = 1{,}000$ Hz

FIND: (*a*) $A_{v \cdot m}$ (*b*) $f_{l \cdot co}$ (*c*) $f_{h \cdot co}$

SOLUTION:

(*a*)

$$A_{v \cdot m} = \frac{X_L}{\sqrt{R^2 + X_L^2}} = \frac{50{,}200}{\sqrt{5{,}000^2 + 50{,}200^2}} \cong 0.995$$

$$X_L = 2\pi f L = 6.28 \times 1{,}000 \times 8 \cong 50{,}200 \text{ ohms}$$

(*b*) For the coil

$$f_{l \cdot co} = \frac{R}{2\pi L} = \frac{5{,}000}{6.28 \times 8} \cong 100 \text{ Hz}$$

For the capacitor

$$f_{l \cdot co} = \frac{159{,}000}{CR} = \frac{159{,}000}{0.3 \times 5{,}000} \cong 106 \text{ Hz}$$

For the circuit

$$f_{l \cdot co} = 1.1 \sqrt{n}\, f_{l \cdot co} = 1.1 \sqrt{2} \times 106 \cong 165 \text{ Hz}$$

(*c*)

$$f_{h \cdot co} = \frac{159{,}000}{C_T R} = \frac{159{,}000}{300 \times 10^{-6} \times 5{,}000} \cong 106 \text{ kHz}$$

$$C_T = C_D + C_1 + C_2 = 100 + 50 + 150 = 300 \text{ pf}$$

Example 14-19 For the circuit of Example 14-18, what is (*a*) the gain factor at 50 Hz? (*b*) The voltage gain at 50 Hz? (*c*) The gain factor at 50 kHz? (*d*) The voltage gain at 50 kHz?

GIVEN: Example 14-18 $A_{v\cdot m} = 0.995$

FIND: (*a*) K_l (*b*) $A_{v\cdot l}$ (*c*) K_h (*d*) $A_{v\cdot h}$

SOLUTION:

(*a*) $$K_l = \frac{1}{\sqrt{1 + \left(\frac{R}{X_L}\right)^2 + \left(\frac{X_C}{R}\right)^2}} = \frac{1}{\sqrt{1 + \left(\frac{5{,}000}{2{,}510}\right)^2 + \left(\frac{10{,}600}{5{,}000}\right)^2}} \cong 0.325$$

$$X_L = 2\pi fL = 6.28 \times 50 \times 8 \cong 2{,}510 \text{ ohms}$$

$$X_C = \frac{159{,}000}{fC_c} = \frac{159{,}000}{50 \times 0.3} \cong 10{,}600 \text{ ohms}$$

(*b*) $$A_{v\cdot l} = K_l A_{v\cdot m} = 0.325 \times 0.995 \cong 0.324$$

(*c*) $$K_h = \frac{1}{\sqrt{1 + (2\pi fC_TR)^2}} = \frac{1}{\sqrt{1 + (6.28 \times 50 \times 10^3 \times 300 \times 10^{-12} \times 5{,}000)^2}} \cong 0.91$$

$$C_T = C_D + C_1 + C_2 = 100 + 50 + 150 = 300 \text{ pf}$$

(*d*) $$A_{v\cdot h} = K_h A_{v\cdot m} = 0.91 \times 0.995 \cong 0.905$$

Modified Impedance-coupled Circuits. The voltage gain of the impedance-coupled amplifier at the low frequencies can be increased by adding an *LC*-

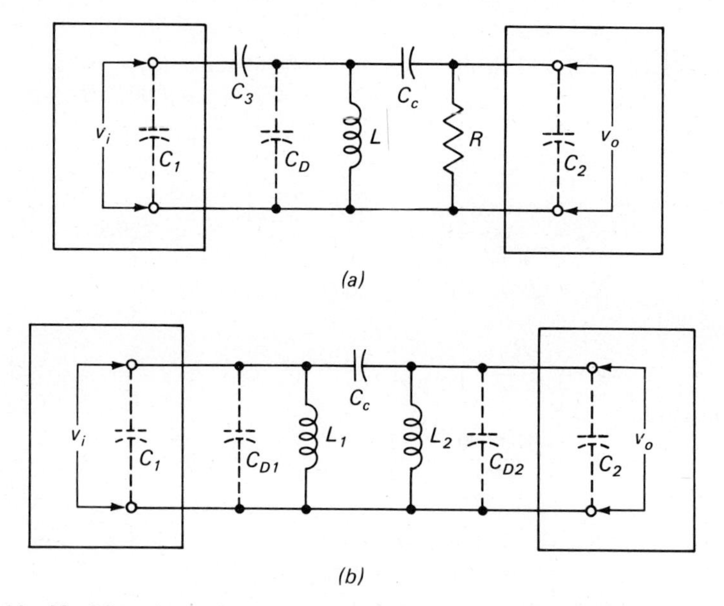

Fig. 14-14 Modified impedance-coupling circuits. (a) With tuned circuit C_3L. (b) With tuned circuit C_cL_2.

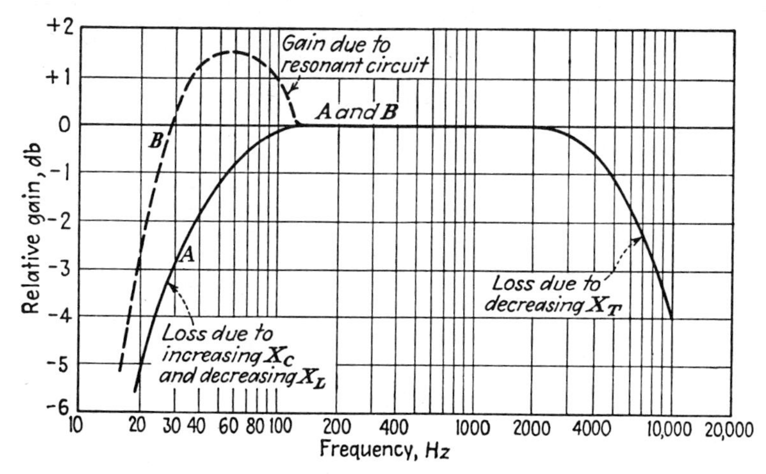

Fig. 14-15 Gain versus frequency characteristics for a typical low-frequency impedance-coupled amplifier. Curve *A* for the basic impedance-coupled circuit of Fig. 14-10. Curve *B* for a circuit with a resonant section as in Fig. 14-14.

tuned circuit to the basic circuit of Fig. 14-10. The inductance and capacitance form a series-tuned circuit that is resonant at a low frequency value and provides voltage amplification at this frequency (Fig. 14-15). In the circuit of Fig. 14-14*a*, C_3 and L form the desired tuned circuit. In the circuit of Fig. 14-14*b*, C_c and L_2 form the desired tuned circuit; this circuit is referred to as being *double-impedance-coupled*. The series resonant circuit also results in a reduction in the value of the low-frequency cutoff.

A disadvantage of double-impedance coupling is that the use of two coils in parallel increases the amount of the distributed capacitance, thus further increasing the shunting capacitance of the circuit. The added shunting capacitance (C_{D2}) results in a reduction in the value of the high-frequency cutoff.

14-9 Frequency Response of Impedance-coupled Amplifiers

Characteristics. The characteristics established in Art. 14-8 for the impedance-coupling network can be applied to amplifier circuits.

Impedance-coupled Transistor Amplifiers. A two-stage impedance-coupled transistor amplifier is described in Art. 13-10 including its characteristics for the mid-frequencies. The low- and high-frequency characteristics presented in Art. 14-8 can be applied to the two-stage amplifier of Fig. 14-16, which is similar to the circuit of Fig. 13-11.

Example 14-20 For the circuit of Fig. 14-16, what is the (*a*) low-frequency cutoff? (*b*) The high-frequency cutoff?

GIVEN: Fig. 14-16 From Example 13-8, $r_{i \cdot 2} = 1{,}140$ ohms and $r_{o \cdot 2} = 6{,}000$ ohms

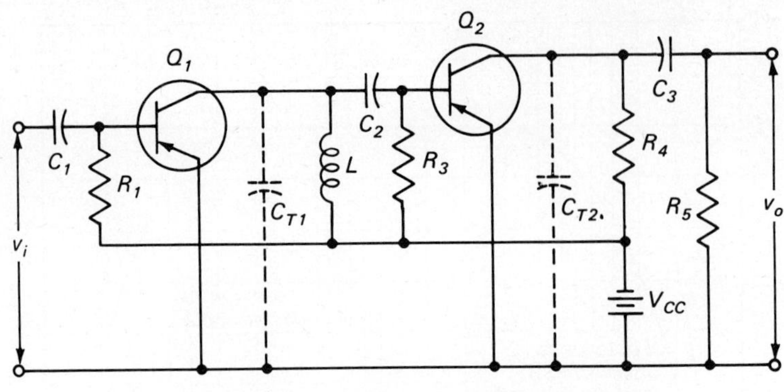

$C_1, C_2, C_3 = 1\ \mu f$; $C_{T1} = 200\ pf$; $C_{T2} = 44\ pf$; $L = 1$ henry;
$R_1, R_3 = 1{,}270{,}000$ ohms; $R_4 = 8{,}000$ ohms; $R_5 = 24{,}000$ ohms;
$V_{CC} = 28$ volts; $\beta_1, \beta_2 = 80$.

Fig. 14-16 Two-stage impedance-coupled transistor amplifier.

FIND: (a) $f_{l\cdot co}$ (b) $f_{h\cdot co}$

SOLUTION:
(a) For R_4, C_3, R_5

$$f_{l\cdot co} = \frac{159{,}000}{C_3(R_4 + R_5)} = \frac{159{,}000}{1(8{,}000 + 24{,}000)} \cong 5 \text{ Hz}$$

For L, C_2, R_3,
Due to the coupling coil

$$f_{l\cdot co} = \frac{R}{2\pi L} = \frac{r_{i\cdot 2}}{2\pi L} = \frac{1{,}140}{6.28 \times 1} \cong 182 \text{ Hz}$$

Due to the coupling capacitor

$$f_{l\cdot co} = \frac{159{,}000}{RC_2} = \frac{159{,}000}{r_{i\cdot 2}C_2} = \frac{159{,}000}{1{,}140 \times 1} \cong 140 \text{ Hz}$$

For the amplifier

$$f_{l\cdot co} \cong 1.1\sqrt{n}\, f_{l\cdot co} \cong 1.1\sqrt{2} \times 182 \cong 284 \text{ Hz}$$

(b) At Q_1

$$f_{h\cdot co} = \frac{159{,}000}{C_{T\cdot 1}r_{i\cdot 2}} = \frac{159{,}000}{200 \times 10^{-6} \times 1{,}140} \cong 700 \text{ kHz}$$

At Q_2

$$f_{h\cdot co} = \frac{159{,}000}{C_{T\cdot 2}r_{o\cdot 2}} = \frac{159{,}000}{44 \times 10^{-6} \times 6{,}000} \cong 600 \text{ kHz}$$

For the amplifier

$$f_{h\cdot co} \cong \frac{f_{h\cdot co}}{1.1\sqrt{n}} \cong \frac{600}{1.1\sqrt{2}} \cong 385 \text{ kHz}$$

Impedance-coupled Vacuum-tube Amplifiers. A two-stage impedance-coupled vacuum-tube amplifier is described in Art. 13-11 including its characteristics for the mid-frequencies. The low- and high-frequency characteristics presented in Art. 14-8 can also be applied to this vacuum-tube amplifier.

14-10 Frequency Response with Transformer Coupling

Circuit Characteristics. The circuit for the basic transformer-coupled network is illustrated in Fig. 14-17*a*. The transformer introduces inductance into both the input and output circuits, as is indicated by L_P and L_S in Fig. 14-17*b*. Also introduced into the circuit are (1) leakage inductances L_P' and L_S', (2) resistances of the transformer windings R_P and R_S, and (3) shunting (distributed) capacitances C_P and C_S. The load R_L is reflected to the primary side as described in Art. 13-13.

Figure 14-18 illustrates typical gain versus frequency characteristic curves for transformer coupling showing (1) a decrease in the gain at the low- and high-frequency ends of the response curves, and (2) a peak gain at the resonant frequency due to the inductances and capacitances in the circuit.

Equivalent Circuits. The equivalent circuits for the low-, mid-, and high frequencies are shown in Figs. 14-19 and 14-20. At the low frequencies: (1) the impedances of the shunting capacitances are very high and therefore can be ignored; (2) the leakage inductances are generally so low that their impedances can be ignored; (3) the total effective reactance of the primary side acts as a parallel-connected inductor and must be considered; and (4) the total

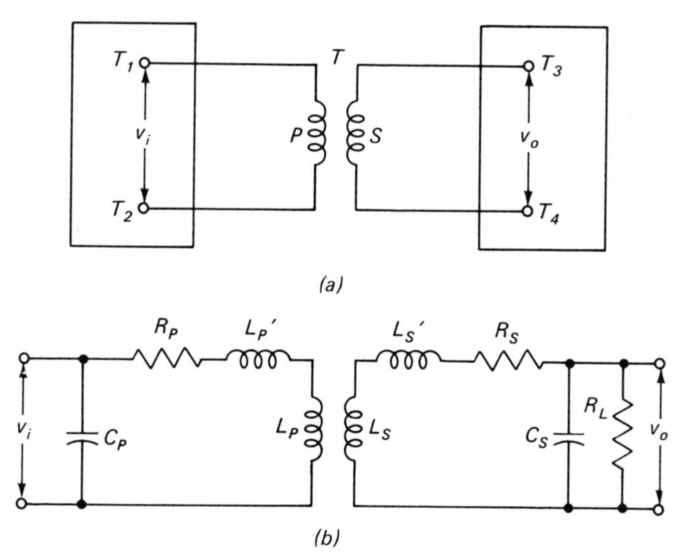

Fig. 14-17 Transformer-coupling network. (a) Basic circuit. (b) Equivalent circuit.

equivalent series-connected resistance in the primary circuit must be considered (see Fig. 14-19).

At the mid-frequencies, the impedances of the inductances, capacitances, and resistances usually have so little effect on the circuit that they can be disregarded. In this case, the transformer is considered to be an ideal circuit element (Fig. 14-17). When the impedances introduced by the transformer have a significant effect at the mid-frequencies, the circuit is treated in the same manner as for the low frequencies.

At the high frequencies: (1) the impedance due to the inductance of the primary winding is so high that it can be ignored; (2) the impedances of all the shunting capacitances are so low that they must be included; (3) the effect of the leakage inductances must be included; and (4) the resistances of the secondary circuit must be included (see Fig. 14-20).

Gain at Mid-frequencies. When the transformer is considered to be an ideal circuit element, the voltage gain of the coupling network is equal to the turns ratio of the transformer, or

$$A_{v\cdot m} = n \tag{14-30}$$

where $n = \dfrac{N_S}{N_P}$

When the parameters of the transformer have a significant effect at the mid-frequencies, the voltage gain at the mid-frequencies can be found with Eq. (14-33).

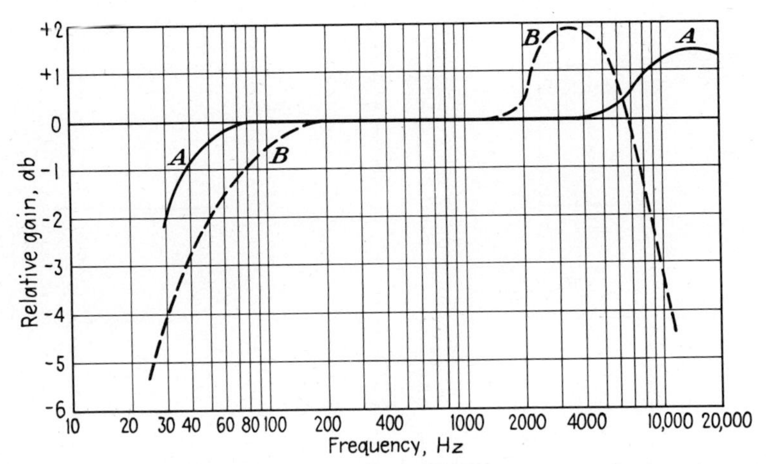

Fig. 14-18 Gain versus frequency characteristics for a transformer-coupled a-f amplifier. Curve *A*—with high-grade transformer. Curve *B*—with low-grade transformer.

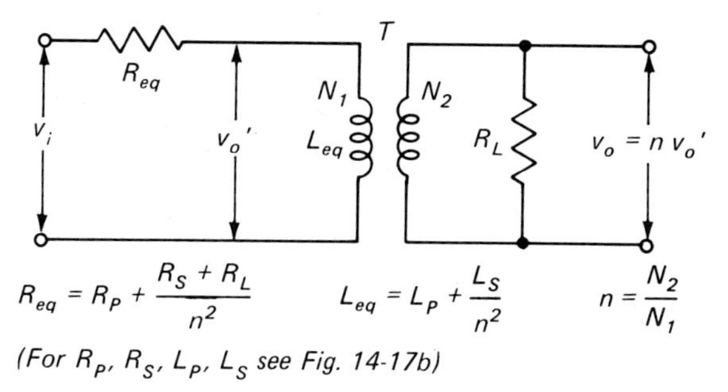

Fig. 14-19 Equivalent circuit of transformer coupling for low frequencies and mid-frequencies.

Gain at Low Frequencies. At the low frequencies, the impedances of the transformer windings decrease and tend to short-circuit the input signal. Thus, the gain of the circuit at low frequencies decreases. From Fig. 14-19, it can be shown that

$$v_{o \cdot l} = \frac{v_i n X_{L \cdot \mathrm{eq}}}{\sqrt{R_{\mathrm{eq}}{}^2 + X_{L \cdot \mathrm{eq}}{}^2}} \tag{14-31}$$

The low-frequency gain factor is

$$K_l = \frac{X_{L \cdot \mathrm{eq}}}{\sqrt{R_{\mathrm{eq}}{}^2 + X_{L \cdot \mathrm{eq}}{}^2}} \tag{14-32}$$

or

$$K_l = \frac{1}{\sqrt{1 + \left(\dfrac{R_{\mathrm{eq}}}{X_{L \cdot \mathrm{eq}}}\right)^2}} \tag{14-32a}$$

The gain at the low frequencies is

$$A_{v \cdot l} = K_l A_{v \cdot m} \tag{14-6}$$

and

$$A_{v \cdot l} = \frac{n}{\sqrt{1 + \left(\dfrac{R_{\mathrm{eq}}}{X_{L \cdot \mathrm{eq}}}\right)^2}} \tag{14-33}$$

where R_{eq} and L_{eq} are as indicated on Fig. 14-19.

Low-frequency Cutoff. The low-frequency cutoff occurs at -3 db, or when R_{eq} is equal to $X_{L \cdot \mathrm{eq}}$. Thus,

$$2\pi f_{l \cdot \mathrm{co}} L_{\mathrm{eq}} = R_{\mathrm{eq}} \tag{14-34}$$

and

$$f_{l \cdot \mathrm{co}} = \frac{R_{\mathrm{eq}}}{2\pi L_{\mathrm{eq}}} \tag{14-34a}$$

Gain at High Frequencies. From the equivalent circuit for the high frequencies (Fig. 14-20) it can be shown that

$$v_{o \cdot h} = \frac{v_i n X_{C \cdot T}}{\sqrt{R_{eq}^2 + X_{eq}^2}} \tag{14-35}$$

The high-frequency gain factor is

$$K_h = \frac{X_{C \cdot T}}{\sqrt{R_{eq}^2 + X_{eq}^2}} \tag{14-36}$$

The gain at the high frequencies is

$$A_{v \cdot h} = K_h A_{v \cdot m} \tag{14-11}$$

and

$$A_{v \cdot h} = \frac{n X_{C \cdot T}}{\sqrt{R_{eq}^2 + X_{eq}^2}} \tag{14-37}$$

where $X_{C \cdot T}$ = reactance of the total shunting capacitance referred to the primary, ohms
R_{eq} = total equivalent resistance referred to the primary, ohms
$X_{eq} = X_{L \cdot T}' - X_{C \cdot T}$ = total equivalent reactance referred to the primary, ohms
$X_{L \cdot T}'$ = total equivalent leakage reactance referred to the primary, ohms

High-frequency Cutoff. The high-frequency cutoff occurs at -3 db and can be found by substituting 0.707 for K_h in Eq. (14-36). Because the cutoff frequency is dependent on both inductance and capacitance, the equation for the cutoff frequency becomes complex. However, an approximate value can be obtained with a simplified form of the complex equation

$$f_{h \cdot co} \cong \frac{0.175 \sqrt{(R_{eq} C_T)^2 - 2L_T' C_T}}{L_T' C_T} \tag{14-38}$$

where R_{eq}, L_T', and C_T are as indicated on Fig. 14-20.

Peak High-frequency Gain. Because the leakage inductance L_T' and the shunting capacitances C_T form a tuned circuit, a peak voltage gain (see Fig.

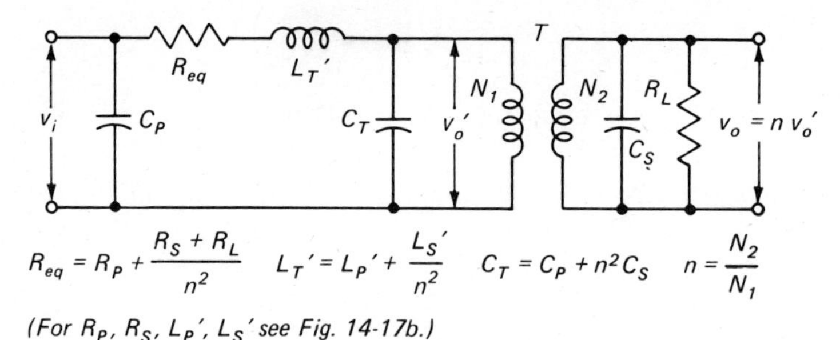

Fig. 14-20 Equivalent circuit of transformer coupling for high frequencies.

14-18) occurs at the resonant frequency of this tuned circuit. The frequency at which this peak occurs is

$$f_r = \frac{1}{2\pi\sqrt{L_T'C_T}} \tag{14-39}$$

The voltage gain factor at the resonant peak is

$$K_{h\cdot r} = \frac{\sqrt{\dfrac{L_T'}{C_T}}}{R_{eq}} \tag{14-40}$$

The voltage gain at the resonant peak is

$$A_{v\cdot r} = K_{h\cdot r}A_{v\cdot m} \tag{14-41}$$

Example 14-21 A transformer-coupling network similar to Fig. 14-17*b* has the following parameters: $n = 3$, $L_P = 20$ henrys, $L_S = 180$ henrys, $L_P' = 0.125$ henry, $L_S' = 1.125$ henrys, $R_P = 200$ ohms, $R_S = 1{,}200$ ohms, $R_L = 60{,}000$ ohms, $C_P = 101$ pf, $C_S = 11$ pf. Find (*a*) mid-frequency gain assuming an ideal transformer, (*b*) mid-frequency gain at 500 Hz using Eq. (14-33), (*c*) low-frequency cutoff, (*d*) gain at the low-frequency cutoff, (*e*) high-frequency cutoff, (*f*) gain at the high-frequency cutoff.

GIVEN: $n = 3$ $L_P = 20$ henrys $L_S = 180$ henrys $L_P' = 0.125$ henry
$L_S' = 1.125$ henrys $R_P = 200$ ohms $R_S = 1{,}200$ ohms
$R_L = 60{,}000$ ohms $C_P = 101$ pf $C_S = 11$ pf.

FIND: (*a*) $A_{v\cdot m}$ (*b*) $A_{v\cdot m}$ (*c*) $f_{l\cdot co}$ (*d*) $A_{v\cdot l}$ (*e*) $f_{h\cdot co}$ (*f*) $A_{v\cdot h}$

SOLUTION:

(*a*) $A_{v\cdot m} \cong n \cong 3$

(*b*)
$$A_{v\cdot m} = \frac{n}{\sqrt{1 + \left(\dfrac{R_{eq}}{X_{L\cdot eq}}\right)^2}} = \frac{3}{\sqrt{1 + \left(\dfrac{7{,}000}{125{,}600}\right)^2}} \cong 3$$

where
$$R_{eq} = R_P + \frac{R_S + R_L}{n^2} = 200 + \frac{1{,}200 + 60{,}000}{3^2} = 7{,}000 \text{ ohms}$$

$$X_{L\cdot eq} = 2\pi f L_{eq} = 6.28 \times 500 \times 40 = 125{,}600 \text{ ohms}$$

$$L_{eq} = L_P + \frac{L_S}{n^2} = 20 + \frac{180}{3^2} = 40 \text{ henrys}$$

(*c*)
$$f_{l\cdot co} = \frac{R_{eq}}{2\pi L_{eq}} = \frac{7{,}000}{6.28 \times 40} \cong 28 \text{ Hz}$$

(*d*)
$$A_{v\cdot l} = K_l A_{v\cdot m} = 0.707 \times 3 \cong 2.12$$

where
$$K_l = \frac{1}{\sqrt{1 + \left(\dfrac{R_{eq}}{X_{L\cdot eq}}\right)^2}} = \frac{1}{\sqrt{1 + \left(\dfrac{7{,}000}{7{,}000}\right)^2}} = 0.707$$

$$X_{L\cdot eq} = 2\pi f L_{eq} = 6.28 \times 28 \times 40 \cong 7{,}000 \text{ ohms}$$

$$(e)\qquad f_{h\cdot co} \cong \frac{0.175\sqrt{(R_{eq}C_T)^2 - 2L_T'C_T}}{L_T'C_T}$$

$$\cong \frac{0.175\sqrt{(7{,}000 \times 200 \times 10^{-12})^2 - 2 \times 0.25 \times 200 \times 10^{-12}}}{0.25 \times 200 \times 10^{-12}} \cong 34.6 \text{ kHz}$$

where $C_T = C_P + n^2C_s = 101 + (3^2 \times 11) = 200$ pf

$$L_T' = L_P' + \frac{L_S'}{n^2} = 0.125 + \frac{1.125}{3^2} = 0.25 \text{ henry}$$

$$(f)\qquad A_{v\cdot h} = K_h A_{v\cdot m} = 0.71 \times 3 \cong 2.13$$

where $K_h = \dfrac{X_{C\cdot T}}{\sqrt{R_{eq}^2 + X_{eq}^2}} = \dfrac{23{,}000}{\sqrt{7{,}000^2 + 31{,}300^2}} \cong 0.71$

$$X_{eq} = X_{L\cdot T'} - X_{C\cdot T} = 54{,}300 - 23{,}000 = 31{,}300 \text{ ohms}$$

$$X_{C\cdot T} = \frac{159{,}000}{f_{h\cdot co}C_T} = \frac{159{,}000}{34.6 \times 10^3 \times 200 \times 10^{-6}} \cong 23{,}000 \text{ ohms}$$

$$X_{L\cdot T'} = 2\pi f_{h\cdot co}L_T' = 6.28 \times 34.6 \times 10^3 \times 0.25 \cong 54{,}300 \text{ ohms}$$

Example 14-22 For the circuit and values of Example 14-21, what is (*a*) the frequency at which the peak voltage gain occurs? (*b*) The peak voltage gain?

GIVEN: Example 14-21 $L_T' = 0.25$ henry $C_T = 200$ pf $R_{eq} = 7{,}000$ ohms

FIND: (*a*) f_r (*b*) $A_{v\cdot r}$

SOLUTION:

$$(a)\qquad f_r = \frac{1}{2\pi\sqrt{L_T'C_T}} = \frac{1}{6.28\sqrt{0.25 \times 200 \times 10^{-12}}} \cong 22.5 \text{ kHz}$$

$$(b)\qquad A_{v\cdot r} = K_{h\cdot r}A_{v\cdot m} = 5 \times 3 = 15$$

where $K_{h\cdot r} = \dfrac{\sqrt{\dfrac{L_T'}{C_T}}}{R_{eq}} = \dfrac{\sqrt{\dfrac{0.25}{200 \times 10^{-12}}}}{7{,}000} \cong 5$

14-11 Frequency Response of Transformer-coupled Amplifiers

The characteristics established in Art. 14-10 for the transformer-coupled network can be applied to transistor and vacuum-tube amplifier circuits.

Transformer-coupled Transistor Amplifiers. A basic transistor amplifier circuit using a transformer to couple the output signal of the amplifier to the load is described in Art. 13-14. The voltage gain at the mid-frequencies also was developed in Art. 13-14. The low- and high-frequency characteristics presented in Art. 14-10 can be applied to the transformer-coupled transistor amplifier of Fig. 14-21.

Example 14-23 The circuit of Fig. 14-21 consists of (1) a transistor amplifier, and (2) the transformer-coupling network of Example 14-21. Find (*a*) the voltage gain at

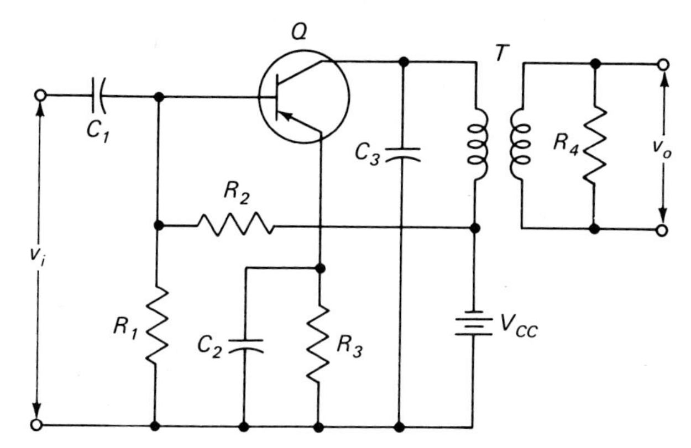

R_1 = 20,000 ohms, R_2 = 40,000 ohms, R_3 = 4,000 ohms, R_4 = 60,000 ohms,
$C_3 = C_s + C_{ce}$ = 20 pf, n = 3, β = 100, V_{CC} = 12 volts,
Transformer T = same as in Example 14-21.

Fig. 14-21 Transistor amplifier with transformer coupling.

the mid-frequencies, (*b*) the low-frequency cutoff, (*c*) the gain at the low-frequency cutoff, (*d*) the high-frequency cutoff, (*e*) the gain at the high-frequency cutoff.

GIVEN: Fig. 14-21 Example 14-21

FIND: (*a*) $A_{v \cdot m}$ (*b*) $f_{l \cdot co}$ (*c*) $A_{v \cdot l}$ (*d*) $f_{h \cdot co}$ (*e*) $A_{v \cdot h}$

SOLUTION:

(*a*) $$A_{v \cdot m} \cong n \frac{r_0}{r_e} \cong \frac{3 \times 7{,}000}{25} \cong 840$$

where $r_0 = R_{eq}$ of Example 14-21 = 7,000 ohms

$$r_e = \frac{25}{I_E} = \frac{25}{1} = 25 \text{ ohms}$$

$$I_E = \frac{V_{CC}R_1}{(R_1 + R_2)R_3} = \frac{12 \times 20{,}000}{(20{,}000 + 40{,}000) \times 4{,}000} = 1 \text{ ma}$$

(*b*) $f_{l \cdot co}$ = same as $f_{l \cdot co}$ of Example 14-21 $\cong$ 28 Hz

(*c*) $A_{v \cdot l} \cong 0.707 A_{v \cdot m} \cong 0.707 \times 840 \cong 595$

(*d*) $$f_{h \cdot co} \cong \frac{0.175 \sqrt{(R_{eq}C_T)^2 - 2L_T'C_T}}{L_T'C_T}$$

$$\cong \frac{0.175 \sqrt{(7{,}000 \times 220 \times 10^{-12})^2 - 2 \times 0.25 \times 220 \times 10^{-12}}}{0.25 \times 220 \times 10^{-12}} \cong 33 \text{ kHz}$$

where R_{eq} = same as in Example 14-21 = 7,000 ohms
L_T' = same as in Example 14-21 = 0.25 henry
$C_T = C_3 + C_P + n^2C_S = 20 + 101 + (3^2 \times 11) = 220$ pf

(*e*) $A_{v \cdot h} \cong 0.707 A_{v \cdot m} \cong 0.707 \times 840 \cong 595$

Transformer-coupled Vacuum-tube Amplifiers. The basic vacuum-tube amplifier circuit using a transformer to couple the signal from one circuit or stage to another is described in Art. 13-15. The voltage gain for the mid-frequencies also was developed in Art. 13-15. The low- and high-frequency characteristics presented in Art. 14-10 can be applied to the transformer-coupled vacuum-tube circuits.

14-12 Frequency Response of Direct-coupled Amplifiers

Because direct-coupled amplifier circuits do not use capacitors, inductors, or transformers, the gain is practically constant from zero hertz to relatively high values of frequency at which the impedances due to the stray capacitance of the wiring and the interelectrode capacitances of the active devices have a significant effect.

QUESTIONS

1. What is meant by the *passband* of an amplifier?
2. What is meant by the cutoff or rolloff frequency of a circuit?
3. What is the significance of a 3-db decrease in the gain of an amplifier circuit?
4. Describe (*a*) mid-frequency range, (*b*) low-frequency cutoff, (*c*) high-frequency cutoff.
5. What causes the variation in the voltage gain of an amplifier when the signal frequency changes?
6. In amplifier circuits, what accounts for the presence of (*a*) capacitance? (*b*) Inductance?
7. Describe the capacitive effects introduced by (*a*) transistors, (*b*) vacuum tubes.
8. Describe the effect of series-connected capacitances in a coupling circuit at (*a*) mid-frequencies, (*b*) low frequencies, (*c*) high frequencies.
9. Describe the effect of parallel-connected capacitances in a coupling circuit at (*a*) mid-frequencies, (*b*) low frequencies, (*c*) high frequencies.
10. Describe the effect of series-connected inductances in a coupling circuit at (*a*) mid-frequencies, (*b*) low frequencies, (*c*) high frequencies.
11. Describe the effect of parallel-connected inductances in a coupling circuit at (*a*) mid-frequencies, (*b*) low frequencies, (*c*) high frequencies.
12. Why is it desirable to prepare a-c equivalent circuits to represent the mid-, low-, and high-frequency signal conditions?
13. Describe the a-c characteristics of the *RC*-coupling network at the mid-frequencies.
14. (*a*) Describe the a-c characteristics of the *RC*-coupling network at the low frequencies. (*b*) What factors largely determine the low-frequency cutoff value?
15. What does the low-frequency gain factor K_l represent?
16. (*a*) Describe the a-c characteristics of the *RC*-coupling network at the high frequencies. (*b*) What factors largely determine the high-frequency cutoff value?
17. What does the high-frequency gain factor K_h represent?
18. (*a*) What is a frequency response curve? (*b*) What purpose does this type of curve serve?
19. What is the effect on the low-frequency cutoff when two similar stages are cascaded?

20. What is the effect on the high-frequency cutoff when two similar stages are cascaded?
21. In a circuit containing resistance and reactance, what relative values of resistance and reactance will result in a 3-db loss in voltage gain?
22. What is meant by (*a*) low-frequency beta? (*b*) Gain-bandwidth product? (*c*) Beta cutoff frequency?
23. (*a*) Why is calculating for the high-frequency cutoff more difficult than for the low-frequency cutoff? (*b*) What simple method is available for calculating an approximate value of the high-frequency cutoff?
24. (*a*) What is meant by the Miller effect of interelectrode capacitances? (*b*) How is the Miller effect accounted for in the high-frequency cutoff calculations?
25. How does impedance coupling differ from *RC* coupling with respect to the circuit characteristics?
26. How does impedance coupling differ from *RC* coupling with respect to frequency response characteristics?
27. Describe two modifications of the basic impedance-coupling circuit.
28. Name eight circuit parameters that are introduced into an amplifier circuit by the transformer-coupling network.
29. What causes a peak in the voltage-gain characteristics of a transformer-coupled circuit?
30. Describe the frequency characteristics of direct-coupled amplifier circuits.

PROBLEMS

1. What is the mid-frequency voltage gain of the *RC*-coupling network of Fig. 14-2 when the parameters of the circuit are R_1 = 4,000 ohms, R_2 = 2,000 ohms, C_c = 0.7 μf, C_1 = 60 pf, C_2 = 20 pf?
2. What is the mid-frequency voltage gain of the *RC*-coupling network of Fig. 14-2 when the parameters of the circuit are R_1 = 10,000 ohms, R_2 = 10,000 ohms, C_c = 0.5 μf, C_1 = 200 pf, C_2 = 25 pf?
3. What is the low-frequency cutoff for the circuit of Prob. 1?
4. What is the low-frequency cutoff for the circuit of Prob. 2?
5. What is the low-frequency gain factor for the circuit of Prob. 1 at the frequency found in Prob. 3?
6. What is the low-frequency gain factor for the circuit of Prob. 2 at the frequency found in Prob. 4?
7. What is the voltage gain for the circuit of Prob. 1 at the frequency found in Prob. 3?
8. What is the voltage gain for the circuit of Prob. 2 at the frequency found in Prob. 4?
9. What is the voltage gain for the circuit of Prob. 1 at a frequency of 100 Hz?
10. What is the voltage gain for the circuit of Prob. 2 at a frequency of 100 Hz?
11. What is the decibel change for the circuit of Prob. 1 when the signal frequency is 100 Hz?
12. What is the decibel change for the circuit of Prob. 2 when the signal frequency is 100 Hz?
13. What is the high-frequency cutoff for the circuit of Prob. 1?
14. What is the high-frequency cutoff for the circuit of Prob. 2?
15. What is the voltage gain for the circuit of Prob. 1 at the frequency found in Prob. 13?

16. What is the voltage gain for the circuit of Prob. 2 at the frequency found in Prob. 14?
17. What is the voltage gain for the circuit of Prob. 1 when the signal frequency is 500 kHz?
18. What is the voltage gain for the circuit of Prob. 2 when the signal frequency is 200 kHz?
19. What is the decibel change for the circuit of Prob. 1 when the signal frequency is 500 kHz?
20. What is the decibel change for the circuit of Prob. 2 when the signal frequency is 200 kHz?
21. What are the low- and high-frequency cutoffs when two stages of coupling similar to Prob. 1 are cascaded? (*Note:* Use the single-stage values found in Probs. 3 and 13.)
22. What are the low- and high-frequency cutoffs when two stages of coupling similar to Prob. 2 are cascaded? (*Note:* Use the single-stage values found in Probs. 4 and 14.)
23. What change of a single component in the circuit of Example 14-10 can reduce the low-frequency cutoff to approximately 50 Hz?
24. What would be the effect on the low-frequency cutoff for the circuit of Example 14-10 if when servicing the unit a 0.1-μf capacitor is inadvertently substituted for C_1?
25. If a 100-ohm resistor is added to the circuit of Fig. 14-7 between the emitter of Q_1 and the junction of C_4R_6, what is the low-frequency cutoff for this circuit (*a*) at C_4? (*b*) For the circuit?
26. What is the decibel change in the voltage gain of the first stage for the circuit of Prob. 25?
27. It is desired to lower the low-frequency cutoff of the circuit of Fig. 14-7 by adding an unbypassed resistor in the emitter circuit of Q_1. What value of resistance is needed to reduce the cutoff frequency at C_4 to approximately 50 Hz?
28. What is the decibel change in the voltage gain by adding the resistor required in Prob. 27?
29. For the circuit of Fig. 14-7, what is the high-frequency cutoff at Q_2: (*a*) by Eq. (14-17)? (*b*) By Eq. (14-18*a*)? (*c*) By Eq. (14-20*a*)?
30. For the amplifier of Prob. 29 (and Example 14-12), what is the approximate high-frequency cutoff for the composite circuit?
31. What is the effect on the high-frequency cutoff at the input side of Q_1 in Example 14-12 (*b*) when a 28.5-ohm unbypassed resistor is added to the emitter circuit of Q_1?
32. What is the effect on the high-frequency cutoff at the output side of Q_1 in Example 14-12(*c*) when a 28.5-ohm unbypassed resistor is added to the emitter circuit of Q_1?
33. What is the high-frequency cutoff at the input side of Q_2 if a 25-ohm unbypassed resistor is added to the emitter circuit of Q_2 in Prob. 29?
34. What is the decibel change in the mid-frequency voltage gain at Q_2 due to adding the 25-ohm unbypassed resistor in Prob. 33?
35. What is the high-frequency cutoff at the input side of Q_2 if a 50-ohm unbypassed resistor is added to the emitter circuit of Q_2 in Prob. 29?
36. What is the decibel change in the mid-frequency voltage gain at Q_2 due to adding the 50-ohm unbypassed resistor in Prob. 35?
37. What is the mid-frequency voltage gain for the amplifier of Fig. 14-8 if the cir-

cuit parameters are changed to $R_s = 1{,}000$ ohms, $R_1 = 100{,}000$ ohms, $R_2 = 47{,}000$ ohms, $R_3 = 22{,}000$ ohms, $R_4 = 1{,}200$ ohms, $r_p = 7{,}700$ ohms, $\mu = 20$, $E_{BB} = 180$ volts, C_1, $C_2 = 0.1\ \mu f$, $C_3 = 3\ \mu f$, $C_s = 10$ pf, $C_{gp} = 1.5$ pf, $C_{gk} = 1.8$ pf, $C_{pk} = 1.3$ pf?

38. What is the mid-frequency voltage gain for the amplifier of Fig. 14-8 if the circuit parameters are: R_s is a variable resistor, $R_1 = 6{,}800{,}000$ ohms, $R_2 = 220{,}000$ ohms, $R_3 = 470{,}000$ ohms, R_4 is omitted, $r_p = 60{,}000$ ohms, C_1, $C_2 = 0.002\ \mu f$, C_3 is omitted, $C_s = 10$ pf, $C_{gk} = 2.2$ pf, $C_{pk} = 0.5$ pf, $C_{gp} = 1.5$ pf, $\mu = 70$, $E_{BB} = 250$ volts?

39. What is the low-frequency cutoff of the amplifier of Prob. 37 due to (*a*) C_1? (*b*) C_2? (*c*) C_3?

40. What is the low-frequency cutoff of the amplifier of Prob. 38 due to C_2?

41. What is the high-frequency cutoff of the amplifier of Prob. 37 (*a*) at the input of VT_1? (*b*) At the output of VT_1?

42. What is the high-frequency cutoff of the amplifier of Prob. 38 at the output side of VT_1?

43. For the amplifier stage of Prob. 37, what is (*a*) the low-frequency gain factor for a signal frequency of 30 Hz? (*b*) The voltage gain at this frequency? (*c*) The decibel change? (*Note:* Consider the stage to be from the grid of VT_1 to the grid of VT_2.)

44. For the amplifier stage of Prob. 38, what is (*a*) the low-frequency gain factor for a signal frequency of 100 Hz? (*b*) The voltage gain at this frequency? (*c*) The decibel change? (*Note:* Consider the stage to be from the grid of VT_1 to the grid of VT_2.)

45. For the amplifier of Prob. 37, what is (*a*) the high-frequency gain factor for a signal frequency of 300 kHz? (*b*) The voltage gain at this frequency? (*c*) The decibel change?

46. The input to the amplifier of Prob. 38 is controlled by a potentiometer set so that $R_s = 60{,}000$ ohms. With a 50-kHz input signal, what is (*a*) the high-frequency gain factor at the input to VT_1? (*b*) The high-frequency gain factor at the output of VT_1? (*c*) The approximate voltage gain of the amplifier? (*d*) The decibel loss?

47. For the amplifier circuit of Fig. 14-9, find (*a*) the mid-frequency voltage gain, (*b*) the low-frequency cutoff, (*c*) the high-frequency cutoff.

48. The parameters for the circuit of Prob. 47 are changed to $E_{BB} = 180$ volts, R_1, $R_3 = 1{,}000{,}000$ ohms, $R_2 = 470{,}000$ ohms, $R_4 = 2{,}900$ ohms, $R_5 = 1{,}150{,}000$ ohms, C_1, $C_2 = 0.003\ \mu f$, $C_3 = 5\ \mu f$, $C_4 = 0.04\ \mu f$. The tube parameters are $r_p = 1{,}000{,}000$ ohms, $g_m = 5{,}000\ \mu$mhos; the tube capacitances are unchanged. Find (*a*) the mid-frequency voltage gain, (*b*) the low-frequency cutoff, (*c*) the high-frequency cutoff.

49. It is desired to reduce the low-frequency cutoff for the circuit of Example 14-18 from 165 to 100 Hz by replacing the coupling coil and the coupling capacitor. Also, it is desired to keep the cutoff frequency for the coil and the capacitor about equal. What value is recommended for (*a*) the new inductor? (*b*) The new capacitor? (*c*) What is the approximate voltage gain for the circuit at 64 Hz?

50. It is desired to reduce the low-frequency cutoff for the circuit of Example 14-18 from 165 to 100 Hz by replacing the load resistor and the coupling capacitor. Also, it is desired to keep the cutoff frequency for the coil and the capacitor about equal. What value is recommended for (*a*) the new resistor? (*b*) The new capacitor? (*c*) What is the approximate voltage gain for the circuit at 64 Hz?

51. For the circuit of Fig. 14-16, find the voltage gain at the mid-frequency of 2,000 Hz for (*a*) the first stage, (*b*) the second stage, (*c*) the amplifier.

52. When the signal frequency applied to the amplifier of Fig. 14-16 is 100 Hz, find (*a*) the voltage-gain factor for the first stage, (*b*) the voltage-gain factor for the second stage, (*c*) the voltage gain of the amplifier.

53. When the signal frequency applied to the amplifier of Fig. 14-16 is 500 kHz, find (*a*) the voltage-gain factor for the first stage, (*b*) the voltage-gain factor for the second stage.

54. What is the voltage gain of the amplifier of Fig. 14-16 when the signal frequency is 500 kHz?

55. For the circuit of Fig. 14-22, find (*a*) the low-frequency cutoff, (*b*) the high-frequency cutoff.

56. For the amplifier of Fig. 14-22, what is the voltage gain when the signal frequency is (*a*) 500 Hz? (*b*) 30 Hz? (*c*) 100 kHz?

57. What is the resonant frequency for L and C_2 in the circuit of Fig. 14-16?

58. What is the resonant frequency for L and C_3 in the circuit of Fig. 14-22?

59. A transformer-coupled network similar to Fig. 14-17*b* has the following circuit parameters: $n = 2.5$, $L_P = 40$ henrys, $L_S = 250$ henrys, $L_P' = 0.1$ henry, $L_S' = 0.625$ henry, $R_P = 80$ ohms, $R_S = 500$ ohms, $R_L = 100{,}000$ ohms, $C_P = 27.5$ pf, $C_S = 10$ pf, $C_s = 10$ pf. Find (*a*) low-frequency cutoff, (*b*) high-frequency cutoff, (*c*) frequency at which the peak gain occurs.

60. Repeat Prob. 59 for the condition when the load resistance is reduced to 10,000 ohms.

61. For the circuit and values of Prob. 59, find (*a*) voltage gain at the mid-frequencies, (*b*) voltage gain at the low-frequency cutoff, (*c*) voltage gain at the high-frequency cutoff, (*d*) peak voltage gain.

62. For the circuit and values of Prob. 60, find (*a*) voltage gain at the mid-frequencies, (*b*) voltage gain at the low-frequency cutoff, (*c*) voltage gain at the high-frequency cutoff, (*d*) peak voltage gain.

63. For the circuit and values of Prob. 59, what is the voltage gain when the signal frequency is (*a*) 25 Hz? (*b*) 40 kHz?

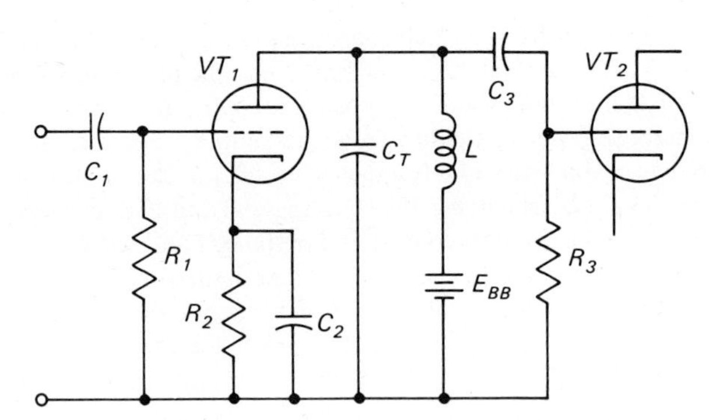

R_1, R_3 = 100,000 ohms, R_2 = 1,400 ohms, L = 150 henrys, C_1, C_3 = 0.03 μf, C_4 = 4 μf, C_T = 175 pf, r_p = 15,000 ohms, μ = 60, E_{BB} = 180 volts, I_P = 5 ma, R_{coil} = 3,000 ohms.

Fig. 14-22

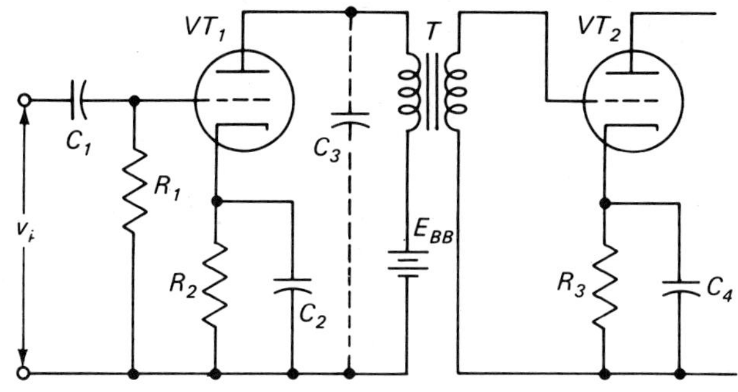

VT_1, VT_2: $\mu = 20$, $r_p = 7{,}800$ ohms, $C_{gk} = 2.4$ pf, $C_{pk} = 3.6$ pf,
$C_{gp} = 1.5$ pf, $C_3 = C_D + C_s = 62.5$ pf.
Transformer: $n = 3$, $R_P = 100$ ohms, $R_S = 900$ ohms, $L_P = 100$ henrys,
$L_S = 900$ henrys, $L_P' = 0.15$ henry, $L_S' = 1.35$ henrys.

Fig. 14-23

64. For the circuit and values of Prob. 59, what is the voltage gain when the signal frequency is (*a*) 50 Hz? (*b*) 60 kHz?

65. From Eq. (14-36), prove that the high-frequency cutoff of -3 db occurs when $R_{eq}^2 = 2X_{C \cdot T}^2 - X_{eq}^2$.

66. From the equation $R_{eq}^2 = 2X_{C \cdot T}^2 - X_{eq}^2$, derive an equation for the high-frequency cutoff in terms of L, C, and R. (*Note:* For convenience, rewrite the equation as $R^2 = 2X_C^2 - X^2$.)

67. Using the equation

$$f^2 = \frac{\pm\sqrt{4L^2C^2 + (R^2C^2 - 2LC)^2} - (R^2C^2 - 2LC)}{78.8L^2C^2}$$

found in Prob. 66, (*a*) determine the high-frequency cutoff for Prob. 59, and (*b*) compare the answer with the approximate value found in Prob. 59 (*b*).

68. For the circuit of Fig. 14-23, (*a*) draw the complete a-c equivalent circuit of the coupling network from the plate of VT_1 to the plate of VT_2. Also, find the values of (*b*) R_{eq}, (*c*) L_{eq}, (*d*) L_T', (*e*) C_T, including the Miller effect of VT_2. (*Note:* Assign the stray capacitance C_s to the primary side of the transformer.)

69. For the circuit of Fig. 14-23, find (*a*) low-frequency cutoff, (*b*) high-frequency cutoff, (*c*) frequency at which the peak gain occurs, (*d*) gain factor at the frequency corresponding to the peak gain. (Use values from Prob. 68, namely, $R_{eq} = 8{,}000$ ohms, $L_{eq} = 200$ henrys, $L_T' = 0.3$ henry, and $C_T = 100$ pf.)

Chapter 15

Other Amplifier Circuits

There are many modifications of the basic amplifier circuits presented in the preceding chapters. Special features may be added to the basic circuits to (1) improve their performance and (2) adapt them for specific applications. Among the types of amplifiers discussed in this chapter are feedback, tuned, wideband, emitter-follower, cathode-follower, limiter, differential, and Darlington-pair.

15-1 Feedback Amplifiers

Feedback. When a signal passing through a circuit produces a secondary signal effect, either by design or unintentionally, the effect on the input signal of the circuit is called *feedback*. An amplifier designed to utilize the effects of feedback is called a *feedback amplifier*.

Positive Feedback. When the effect of the feedback signal on the action of the circuit is the same as that produced by the input signal, the feedback is called *positive, regenerative*, or *direct*.

Negative Feedback. When the effect of the feedback signal on the action of the circuit is opposite to that produced by the input signal, the feedback is called *negative, degenerative*, or *inverse*.

Uses of Feedback. Negative feedback is frequently used in amplifier systems. Although it may seem undesirable to operate an amplifier in a manner that does not produce the maximum possible gain for the circuit, the advantages of negative-feedback amplifiers outweigh the disadvantage of reduced gain. Among the advantages are (1) higher fidelity, (2) improved stability, (3) less amplitude distortion, (4) less harmonic distortion, (5) less frequency distortion, (6) less phase distortion, (7) lower ratio of noise level. Furthermore, by using an active device capable of producing a relatively high gain, or by using an additional stage of amplification, it is possible to obtain the desired overall gain for the amplifier circuit.

Positive feedback generally produces excessive distortion of the input signal and therefore is seldom used with circuits intended chiefly for amplification. However, because positive feedback increases the power of the original signal, it is useful in oscillator circuits.

Principle of Feedback Amplifiers. Figure 15-1*b* illustrates the principle of feedback amplifiers. In this type of amplifier, a portion of the output voltage is returned to the input circuit either (1) in phase with the input voltage for positive feedback, or (2) 180 degrees out of phase with the input voltage for negative feedback. The amount of voltage that is fed back is generally expressed as a percentage of the output voltage. The decimal equivalent of this percentage is represented by the symbol B.

The output voltage of the amplifier circuit without feedback is

$$e_o = Ae_i \tag{15-1}$$

With feedback, the output voltage of the circuit is

$$e_o' = A(e_i + Be_o') \tag{15-2}$$

where e_o' = output-signal voltage with feedback, volts
A = voltage amplification of the circuit without feedback
B = decimal equivalent of output voltage being fed back

Note: In this discussion on feedback amplifiers: (1) the mark ′ indicates values when the effect of feedback is included, and (2) A is used in place of A_v previously used for voltage amplification.

Rearranging the terms of Eq. (15-2)

$$e_o' = e_i \frac{A}{1 - AB} \tag{15-2a}$$

and

$$Ae_i = e_o'(1 - AB) \tag{15-2b}$$

The gain of the amplifier with feedback is

$$A' = \frac{e_o'}{e_i} \tag{15-3}$$

Substituting Eq. (15-2*a*) in Eq. (15-3), then

$$A' = \frac{A}{1 - AB} \tag{15-3a}$$

With a single-stage common-emitter amplifier, and also a single-stage vacuum-tube amplifier, phase reversal takes place between the output and input signals and the gain of the amplifier is indicated as $-A$. In the feedback loop of Fig. 15-1*b*, no phase reversal takes place and the feedback decimal value of B is positive. When A and B have opposite signs, negative feedback takes place, and when A and B have the same signs, positive feedback occurs. The product AB is called the *feedback factor*. Mathematically, the feedback factor represents the ratio of the feedback voltage e_f to the input voltage when the feedback effect is included, or

$$AB = \frac{e_f}{e_i + e_f} \tag{15-4}$$

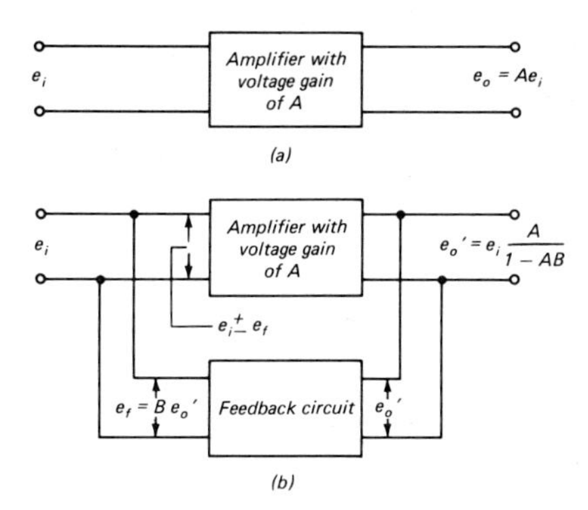

Fig. 15-1 Block diagram illustrating the principle of feedback. (a) Amplifier with out feedback. (b) Amplifier with feedback.

Rearranging the terms in Eq. (15-4)

$$e_f = e_i \frac{AB}{1 - AB} \tag{15-5}$$

Also, from Fig. 15-1*b*,

$$e_f = Be_o' \tag{15-5a}$$

Example 15-1 The amplifier represented by Fig. 15-1*b* has the following parameters: $A = -100$, $B = 0.04$, $e_i = 20$ mv. Find (*a*) voltage gain with feedback included, (*b*) output voltage, (*c*) feedback factor, (*d*) feedback voltage.

GIVEN: $A = -100 \qquad B = 0.04 \qquad e_i = 20$ mv

FIND: (*a*) A' (*b*) e_o' (*c*) AB (*d*) e_f

SOLUTION:

(*a*) $$A' = \frac{A}{1 - AB} = \frac{-100}{1 - (-100 \times 0.04)} = -20$$

(*b*) $$e_o' = e_iA' = 20 \times 10^{-3} \times -20 = -0.4 \text{ volt}$$

(*c*) $$AB = -100 \times 0.04 = -4$$

(*d*) $$e_f = Be_o' = 0.04 \times (-0.4) = -0.016 \text{ volt} = -16 \text{ mv}$$

Limitations of Feedback Amplifiers. Any increase in the feedback factor will cause an increase in the proportion of the output voltage that is returned. The overall voltage amplification of the circuit will then be lower than the amplification of the same amplifier without feedback. In order to produce the same amplification as was obtained without feedback, it becomes necessary to increase the input voltage e_i by an amount AB times the voltage of the input signal. Thus, to make $e_o' = e_o$

$$e_i' = e_i(1 - AB) \tag{15-6}$$

Example 15-2 What value of input-signal voltage is required for the amplifier of Example 15-1 in order that the output-signal voltage with feedback will be equal to the output-signal voltage without feedback?

GIVEN: $e_i = 20$ mv $\quad A = -100 \quad B = 0.04$

FIND: e_i'

SOLUTION:

$$e_i' = e_i(1 - AB) = 20 \times 10^{-3}[1 - (-100 \times 0.04)] = 100 \text{ mv}$$

The preceding discussion relating to negative feedback has assumed that the feedback voltage was exactly 180 degrees out of phase with the input signal. The actual angle of lead or lag is dependent upon the resistances and reactances of the coupling units, the interelement capacitances of the active device, and the frequency of the input signal. If the angle of lead or lag differs greatly from the desired 180 degrees, the circuit operation may become unstable and thus change the feedback characteristics from negative to positive.

15-2 Effect of Negative Feedback on Distortion

Kinds of Distortion. A common fault in amplifier circuits is that the waveform of the output signal is not a true replica of the input-signal waveform and hence the output is distorted. Four common types of distortion are (1) linearity, (2) phase, (3) frequency, and (4) noise. The output of an amplifier having distortion can be considered as consisting of the amplified input signal plus an added new signal, as is indicated in Eq. (15-8*a*).

Nonlinear distortion is caused by operating the active device of an amplifier over a nonlinear portion of its characteristic curve. *Phase distortion* occurs when the input- and output-signal voltages of an amplifier are not either exactly in phase or 180 degrees out of phase with each other.

Frequency distortion occurs when the gain of an amplifier varies with changes in the frequency of the input signal. Because B (the portion of the output signal being fed back) is generally independent of the frequency, the variation in the gain of an amplifier due to changes in frequency will be reduced when negative feedback is used. The relative effect of the frequency on the gain of an amplifier with and without feedback is shown in Fig. 15-2. By increasing the value of B, the gain of an amplifier can be made fairly uniform over a wide range of frequency. Because of this feature, negative-feedback amplifiers are used in high-fidelity audio amplifiers and in video-amplifier circuits where wideband amplification is required.

Noise and hum distortion may be due to (1) actions in the amplifier circuit itself, or (2) external causes such as static, man-made interferences, or circuit defects. The noises produced within the amplifier itself are due to (1) thermal

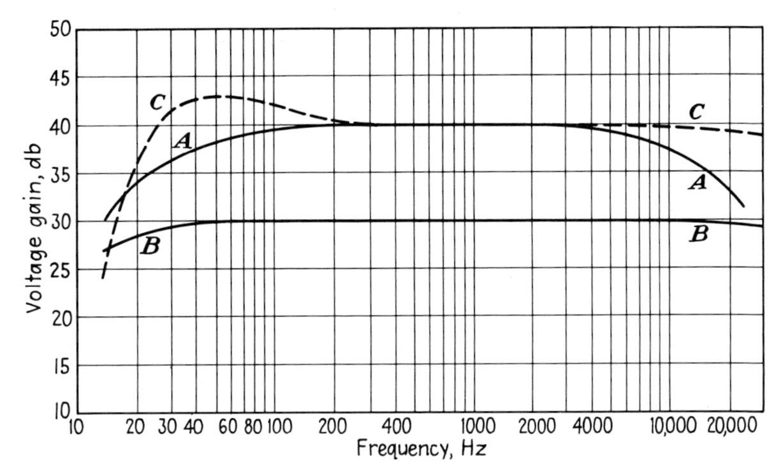

Fig. 15-2 Variation of voltage gain with frequency. Curve *A* for an amplifier without feedback, curve *B* for a similar amplifier with negative feedback, curve *C* for an amplifier with balanced feedback.

agitation, (2) shot effect, (3) microphonics, and (4) hum from an a-c power source. *Thermal noise* is caused by irregular random movement of the free electrons which produces minute currents in the circuit components, particularly active devices and resistors. Noise generated in transistors is described in Art. 4-22. The magnitude of these noise currents increases with increases in temperature. *Shot effect* is caused by small irregularities in current introduced by the currents present in an active device. *Microphonic* noises are caused by mechanical vibration at one or more points in the circuit. *Hum* may be caused by (1) operating an active device with voltages obtained from a rectified a-c power source having insufficient filtering, or (2) stray magnetic and electrostatic fields near the amplifier or its circuit components.

Effect of Distortion on the Output Signal. Distortion in an amplifier causes a new false signal to appear in the output of the amplifier. The magnitude of this spurious signal is dependent on the operating parameters of the circuit and generally is independent of the magnitude of the input-signal voltage. The distortion in an amplifier is specified as a decimal portion of the output-signal voltage for a specified input-signal voltage and that would exist without feedback. The relationship among the input, output, and distortion signals (without feedback) is shown in Fig. 15-3*a*, and is expressed by

$$D = de_o \tag{15-7}$$

$$e_o = e_{o \cdot s} + D \tag{15-8}$$

and

$$e_o = Ae_i + D \tag{15-8a}$$

where D = amount of distortion without feedback, volts
d = decimal portion of the output signal without feedback for a specified value of input signal
$e_{o \cdot s}$ = output due to input signal, volts

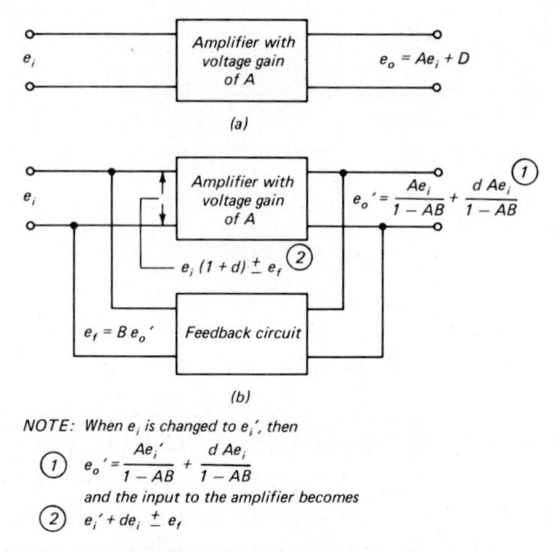

Fig. 15-3 Block diagram illustrating the effect of feedback on distortion. (a) Amplifier without feedback. (b) Amplifier with feedback.

Example 15-3 An input signal of 1 volt is applied to an amplifier that has a gain of -80 and 5 per cent distortion when the input signal is 1 volt. What is (*a*) the output-signal voltage? (*b*) The distortion voltage? (*c*) The output voltage?

GIVEN: $e_i = 1$ volt $\quad A = -80 \quad d = 0.05$

FIND: (*a*) $e_{o \cdot s}$ $\quad$ (*b*) D $\quad$ (*c*) e_o

SOLUTION:

(*a*) $e_{o \cdot s} = Ae_i = -80 \times 1 = -80$ volts

(*b*) $D = de_o = 0.05 \times -80 = -4$ volts

(*c*) $e_o = e_{o \cdot s} + D = -80 + (-4) = -84$ volts

Effect of Negative Feedback on Distortion. Introducing negative feedback into an amplifier circuit improves its operating characteristics but at a sacrifice in the voltage gain. Negative feedback reduces the effect of distortion, as is indicated by

$$D' = \frac{de_o}{1 - AB} \tag{15-9}$$

Substituting Eq. (15-1) in Eq. (15-9),

$$D' = \frac{dAe_i}{1 - AB} \tag{15-9a}$$

where D' = distortion with feedback, volts

An important point about distortion with negative feedback is that the distortion voltage de_o is independent of the input-signal voltage e_i. Therefore, if the input-signal voltage is increased in order to restore the output-signal voltage to its former value without feedback, the distortion component of the output voltage remains unchanged. The relationship among the input, output, and distortion voltages with feedback is shown in Fig. 15-3*b*, and is expressed mathematically as

$$e_o' = e_{o \cdot s}' + D' \tag{15-10}$$

Substituting Eqs. (15-2*a*) and (15-9*a*) in Eq. (15-10),

$$e_o' = \frac{Ae_i}{1 - AB} + \frac{dAe_i}{1 - AB} \tag{15-10a}$$

When the input-signal voltage is increased to e_i', as in the case of Eq. (15-6), only the signal portion $e_{o \cdot s}'$ of Eq. (15-10) will be affected, and therefore

$$e_o' = \frac{Ae_i'}{1 - AB} + \frac{dAe_i}{1 - AB} \tag{15-10b}$$

Example 15-4 If a feedback loop (Fig. 15-3*b*) whose value of B is 0.01 is added to the amplifier of Example 15-3, what is (*a*) the signal component of the output voltage? (*b*) The distortion component of the output voltage? (*c*) The output voltage?

GIVEN: $e_i = 1$ volt $A = -80$ $d = 0.05$ $B = 0.01$

FIND: (*a*) $e_{o \cdot s}'$ (*b*) D' (*c*) e_o'

SOLUTION:

(*a*) $$e_{o \cdot s}' = e_i \frac{A}{1 - AB} = \frac{1 \times (-80)}{1.8} = -44.44 \text{ volts}$$

$$(1 - AB) = 1 - (-80) \times 0.01 = 1.8$$

(*b*) $$D' = \frac{dAe_i}{1 - AB} = \frac{0.05 \times (-80) \times 1}{1.8} = -2.22 \text{ volts}$$

(*c*) $$e_o' = e_{o \cdot s}' + D' = -44.44 + (-2.22) = -46.66 \text{ volts}$$

Example 15-5 If for the amplifier of Example 15-4, the input-signal voltage is raised in order to increase the signal component of the output voltage to the same value it would have without feedback, what is (*a*) the magnitude of the new input signal?

(*b*) The signal component of the output voltage? (*c*) The distortion component of the output voltage? (*d*) The output voltage?

GIVEN: $e_i = 1$ volt $A = -80$ $d = 0.05$ $B = 0.01$ $(1 - AB) = 1.8$

FIND: (*a*) e_i' (*b*) $e_{o \cdot s}'$ (*c*) D' (*d*) e_o'

SOLUTION:

(*a*) $$e_i' = e_i(1 - AB) = 1 \times 1.8 = 1.8 \text{ volts}$$

(*b*) $$e_{o \cdot s}' = e_i' \frac{A}{1 - AB} = \frac{1.8 \times (-80)}{1.8} = -80 \text{ volts}$$

(*c*) $$D' = \frac{dAe_i}{1 - AB} = \frac{0.05 \times (-80) \times 1}{1.8} = -2.22 \text{ volts}$$

(*d*) $$e_o' = e_{o \cdot s}' + D' = -80 + (-2.22) = -82.22 \text{ volts}$$

The values of output voltage and distortion voltage obtained in Examples 15-3 and 15-5 indicate that, although the same output voltage is obtained with feedback as without it, the distortion voltage with feedback is reduced to approximately 55 per cent of its former value.

By increasing the value of the feedback factor, the proportion of the output voltage that is fed back is increased, thus further decreasing the distortion voltage. It is therefore possible to reduce the distortion voltage by controlling the amount of the feedback voltage.

15-3 Stability of Negative-feedback Amplifiers

The gain of an amplifier circuit may vary with (1) the active device used, (2) the operating voltages, and (3) the load impedance. A change in one or more of these variables may cause a change in the gain of the amplifier, thus affecting its stability. Negative feedback can be used to improve the stability of the amplifier. The manner in which negative feedback affects stability can be shown mathematically by rearranging the terms of Eq. (15-3*a*) as

$$A' = \frac{A}{1 - AB} \tag{15-3a}$$

then

$$A' = -\frac{1}{B}\left(\frac{1}{1 - \dfrac{1}{AB}}\right) \tag{15-3b}$$

When the feedback factor AB is many times greater than 1, the gain of the amplifier will then be approximately equal to

$$A' \cong -\frac{1}{B} \tag{15-3c}$$

Equations (15-3*b*) and (15-3*c*) show that when the feedback factor AB is much greater than 1, the voltage amplification of the circuit is dependent on the percentage of the output voltage returned B rather than on the voltage gain of the amplifier. The value of B is generally dependent upon the resistance network whose component values are independent of the frequency, the voltages applied to the circuit, and the characteristics of the active device. The stability of negative-feedback amplifiers can be made to be comparatively high by adjusting the percentage of the feedback to produce a high feedback factor.

15-4 Negative-feedback Amplifier Circuits

Classifications. Negative feedback can be applied to a single-stage or multistage amplifier in a number of ways. Basically, feedback circuits may be divided into three general classes: (1) voltage-feedback circuits, (2) current-feedback circuits, and (3) a combination of both voltage and current feedback.

Voltage-controlled Feedback. The basic principle of voltage-controlled feedback is illustrated in Fig. 15-4. From this figure it can be seen that the feedback voltage is dependent on the output voltage e_o' and is not directly affected by the current in the output impedance R_o, and therefore the feedback is said to be *voltage-controlled.* The feedback voltage is obtained from the voltage-divider network R_1R_f and is dependent on the output voltage e_o' and the values of R_1 and R_f; thus

$$D' = e_o' \frac{R_f}{R_1 + R_f} \tag{15-11}$$

The decimal portion of the feedback voltage in relation to the output voltage is expressed by

$$B = \frac{R_f}{R_1 + R_f} \tag{15-12}$$

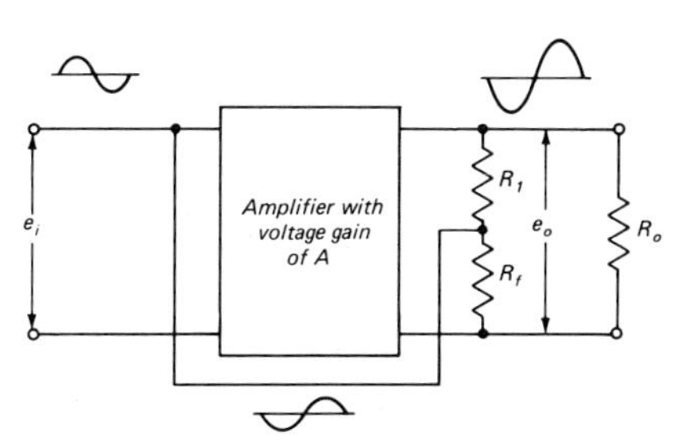

Fig. 15-4 Basic principle of voltage-controlled feedback.

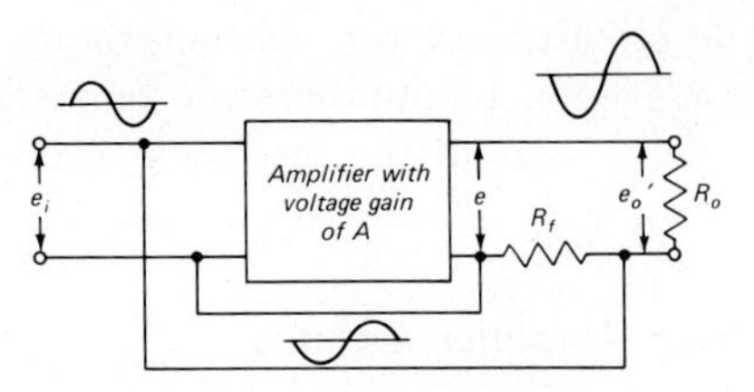

Fig. 15-5 Basic principle of current-controlled feedback.

Current-controlled Feedback. The basic principle of current-controlled feedback is illustrated in Fig. 15-5. From this figure it can be seen that the feedback voltage is dependent on the signal current in the output circuit, and therefore the feedback is said to be *current-controlled*. The feedback voltage is determined by the amount of current flowing through the feedback resistor R_f and is equal to

$$D' = e \frac{R_f}{R_o + R_f} \tag{15-13}$$

and

$$B = \frac{R_f}{R_o + R_f} \tag{15-14}$$

15-5 Transistor Amplifiers with Negative Feedback

Voltage-controlled Negative Feedback. Figure 15-6*a* shows a basic common-emitter amplifier using fixed bias and with negative feedback added by means of C_1 and R_2. The capacitor C_1 is used to block the d-c component of the output voltage e_o from the base of the transistor Q. For the mid-frequency range of the amplifier, the effect of the impedance of C_1 is generally negligible, and

$$B = \frac{R_1}{R_1 + R_2} \tag{15-12}$$

for low frequencies

$$B = \frac{R_1}{\sqrt{(R_1 + R_2)^2 + X_C^2}} \tag{15-15}$$

Example 15-6 If the parameters for the amplifier circuit of Fig. 15-6*a* are R_1 = 1,600,000 ohms, R_2 = 10,000,000 ohms, R_3 = 8,000 ohms, β = 100, and V_{CC} = 20 volts, find (*a*) the voltage gain without feedback, (*b*) the feedback factor, (*c*) the voltage gain with feedback.

GIVEN: R_1 = 1,600,000 ohms $\quad R_2$ = 10,000,000 ohms $\quad R_3$ = 8,000 ohms
β = 100 $\quad V_{CC}$ = 20 volts

FIND: (*a*) A (*b*) AB (*c*) A'

SOLUTION:

(*a*)
$$A = \frac{r_o}{r_e} = \frac{R_3}{r_e} = \frac{8{,}000}{20} = -400$$

$$r_e = \frac{25}{I_E} = \frac{25}{1.25} = 20 \text{ ohms}$$

$$I_E = \beta \frac{V_{CC}}{R_1} = \frac{100 \times 20}{1{,}600{,}000} = 1.25 \text{ ma}$$

(*b*)
$$AB = -400 \times 0.138 \cong -55$$

$$B = \frac{R_1}{R_1 + R_2} = \frac{1.6 \times 10^6}{(1.6 + 10) \times 10^6} \cong 0.138$$

(*c*)
$$A' = \frac{A}{1 - AB} = \frac{-400}{1 - (-55)} \cong -7.15$$

Figure 15-6*b* shows the basic common-emitter amplifier with collector feedback (Art. 8-16). The amount of voltage feedback is determined by the voltage divider formed by R_1 and the a-c base-emitter resistance as expressed by

$$B = \frac{\beta r_e}{R_1 + \beta r_e} \tag{15-16}$$

Example 15-7 If the parameters for the amplifier circuit of Fig. 15-6*b* are R_1 = 800,000 ohms, R_2 = 8,000 ohms, β = 100, and V_{CC} = 20 volts, find (*a*) the voltage gain without feedback, (*b*) the feedback factor, (*c*) the voltage gain with feedback.

GIVEN: R_1 = 800,000 ohms R_2 = 8,000 ohms β = 100 V_{CC} = 20 volts

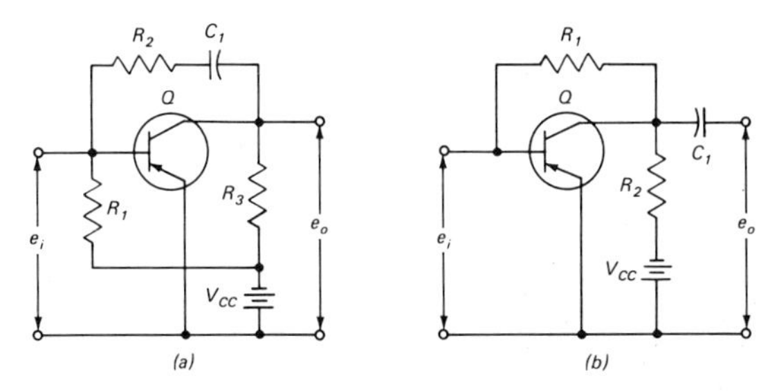

Fig. 15-6 Transistor amplifier with voltage-controlled feedback. (a) Common-emitter amplifier with fixed bias and negative feedback. (b) Common-emitter amplifier with collector bias.

FIND: (*a*) A (*b*) AB (*c*) A'

SOLUTION:

(*a*)
$$A = \frac{r_o}{r_e} = \frac{R_2}{r_e} = \frac{8{,}000}{20} = -400$$

$$r_e = \frac{25}{I_E} = \frac{25}{1.25} = 20 \text{ ohms}$$

$$I_E = \frac{V_{CC}}{R_2 + \dfrac{R_1}{\beta}} = \frac{20}{8{,}000 + \dfrac{800{,}000}{100}} = 1.25 \text{ ma}$$

(*b*)
$$AB = -400 \times 0.0025 = -1$$

$$B = \frac{\beta r_e}{R_1 + \beta r_e} = \frac{100 \times 20}{800{,}000 + (100 \times 20)} \cong 0.0025$$

(*c*)
$$A' = \frac{A}{1 - AB} = \frac{-400}{1 - (-1)} = -200$$

Current-controlled Negative Feedback. An example of current-controlled negative feedback is illustrated by Fig. 15-7 in which the unbypassed resistor R_E in the emitter circuit provides the source of the feedback. Any signal current flowing through this resistor develops a voltage across it that will (1) vary directly with the signal current, and (2) be opposite in phase to the input-signal voltage. The effect of this negative-feedback voltage on the circuit is to (1) improve the stability, (2) increase the input resistance, and (3) decrease the voltage gain. The basic equation for voltage gain [Eq. (9-17*a*)] is therefore modified to

$$A = \frac{r_o}{R_E + r_e} \tag{15-17}$$

When $R_E \gg r_e$, then for the circuit of Fig. 15-7

$$B = \frac{R_E}{R_E + R_2} \tag{15-18}$$

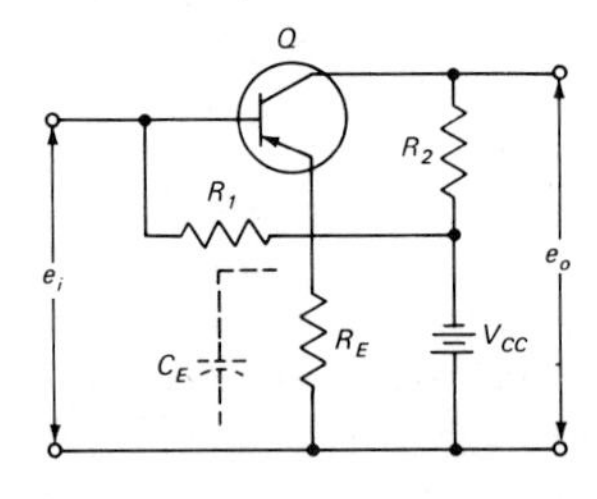

Fig. 15-7 Basic common-emitter amplifier circuit with current-controlled (emitter) feedback.

Example 15-8 If the parameters for the amplifier circuit of Fig. 15-7 are $R_1 =$ 1,500,000 ohms, $R_E = 1{,}000$ ohms, $R_2 = 7{,}000$ ohms, $\beta = 100$, and $V_{CC} = 20$ volts, find (*a*) the normal voltage gain of the circuit as if R_E was bypassed by a capacitor, (*b*) the feedback factor, (*c*) the voltage gain with feedback.

GIVEN: $R_1 = 1{,}500{,}000$ ohms $\quad R_E = 1{,}000$ ohms $\quad R_2 = 7{,}000$ ohms $\quad \beta = 100$
$V_{CC} = 20$ volts

FIND: (*a*) A (*b*) AB (*c*) A'

SOLUTION:

(*a*)
$$A = \frac{r_o}{r_e} = \frac{R_2}{r_e} = \frac{7{,}000}{20} = -350$$

$$r_e = \frac{25}{I_E} = \frac{25}{1.25} = 20 \text{ ohms}$$

$$I_E = \frac{V_{CC}}{R_E + \dfrac{R_1}{\beta}} = \frac{20}{1{,}000 + \dfrac{1{,}500{,}000}{100}} = 1.25 \text{ ma}$$

(*b*)
$$AB = -350 \times 0.125 \cong -43.8$$

$$B = \frac{R_E}{R_E + R_2} = \frac{1{,}000}{1{,}000 + 7{,}000} = 0.125$$

(*c*)
$$A' = \frac{A}{1 - AB} = \frac{-350}{1 - (-43.8)} \cong -7.8$$

Effect of Negative Feedback on Input and Output Impedances. When a negative-feedback signal is applied to the input voltage the input impedance increases, and when applied to the input current the input impedance decreases. Voltage-controlled negative feedback reduces the output impedance and current-controlled negative feedback increases the output impedance.

15-6 Vacuum-tube Amplifiers with Negative Feedback

Voltage-controlled Negative Feedback. Figure 15-8*a* shows a basic triode amplifier with a feedback circuit consisting of C_3, R_2, and R_1. Capacitor C_3 blocks the d-c component of the output voltage e_o from reaching the grid of the tube, and resistors R_1 and R_2 form a voltage divider to control the amount of output voltage that is fed back to the input of the tube. The value of C_3 is chosen so that its impedance is negligible (compared with $R_1 + R_2$) at the frequencies to be handled by the amplifier. With a positive-going input signal, the voltages at points A and B will be negative-going. The input signal from grid to cathode of the tube is the sum of the positive-going input signal e_i and the negative-going feedback signal e_f; therefore, negative feedback is accomplished.

Current-controlled Negative Feedback. An example of current-controlled negative feedback is illustrated by Fig. 15-8*b*. By leaving the cathode resistor R_1 unbypassed, negative feedback is produced in a manner similar to the

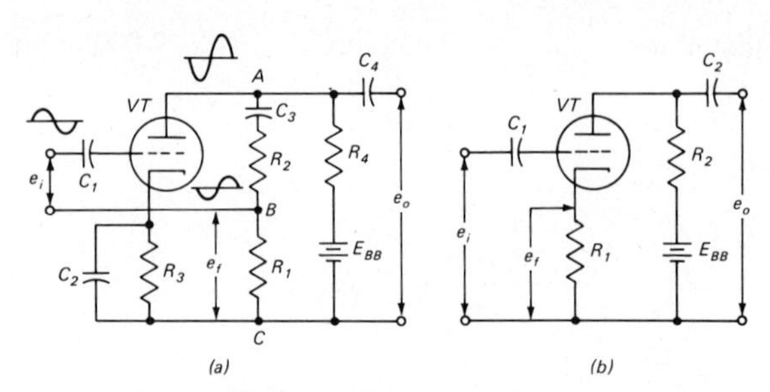

Fig. 15-8 Vacuum-tube amplifier with negative feedback. (a) Voltage-controlled feedback. (b) Current-controlled feedback.

unbypassed resistor in the emitter circuit of a transistor. The portion of feedback for the circuit of Fig. 15-8*b* is

$$B = \frac{R_1}{R_1 + r_p} \tag{15-19}$$

15-7 Feedback for Multistage Amplifiers

The feedback principle, both voltage-controlled and current-controlled, can be applied to multistage as well as to single-stage amplifiers. Care must be exercised in determining and selecting the phase of the feedback signal at the takeoff point in relation to the phase of the signal at the point of feedback.

15-8 Balanced Feedback Amplifier

If the feedback factor AB is large and the phase shift approaches 180 degrees, positive feedback will occur, and the frequency response characteristics will be affected as is illustrated in Fig. 15-9. These representative curves show that while an increase in the amount of feedback makes the voltage gain practically constant over a wider range of frequencies, it also introduces peaks of voltage gain at the lower and upper ends of the frequency range. When the effects of these peaks of amplification are great enough to warrant special consideration, two methods of correction are available.

One method of reducing the effect of the peaks is by means of filter circuits as shown in Fig. 15-10. The RC filter network at the input side will neutralize the peaks that occur at the low frequency, and the LR network at the output side will neutralize the high-frequency peaks. This method, called an *equalized circuit*, is useful only when the peaks are not too large.

A second method of reducing the effects of the peaks is by providing the

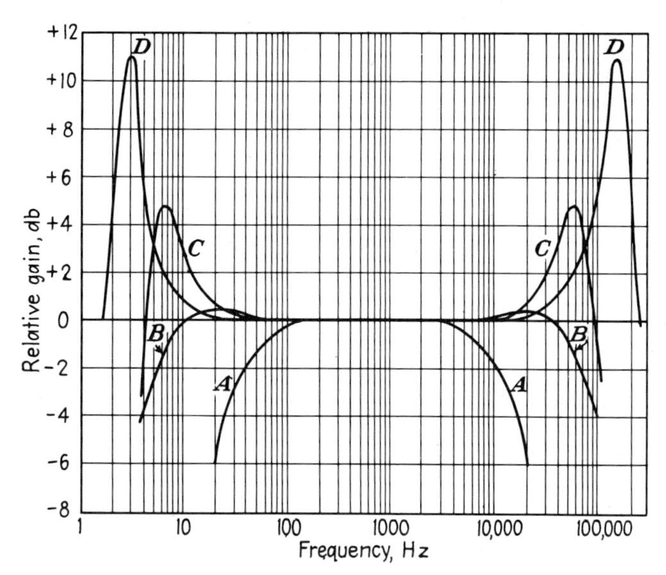

Fig. 15-9 Relative gain versus frequency characteristics of a two-stage *RC*-coupled a-f amplifier with various amounts of feedback. Curve *A*, amplifier without feedback. Curve *B*, *AB* equal to —2 at mid-frequency. Curve *C*, *AB* equal to —10 at mid-frequency. Curve *D*, *AB* equal to —50 at mid-frequency.

amplifier with both positive and negative feedback. This type of amplifier is called a *balanced-feedback amplifier*. With this type of circuit a more nearly constant response is obtained for a multistage amplifier by designing the first portion of the amplifier to provide practically uniform response to all frequencies and then applying both positive- and negative-feedback voltages. The feedback voltages are generally of such proportions that the amount of positive feedback at the mid-frequency will cancel the negative feedback. At frequencies above and below this point, the amounts of positive and negative feedback will vary in a manner tending to neutralize any variations from the optimum response of the first portion of the amplifier and thus providing practically uniform response over a wider range of frequency.

An advantage of the balanced-feedback amplifier is that it provides prac-

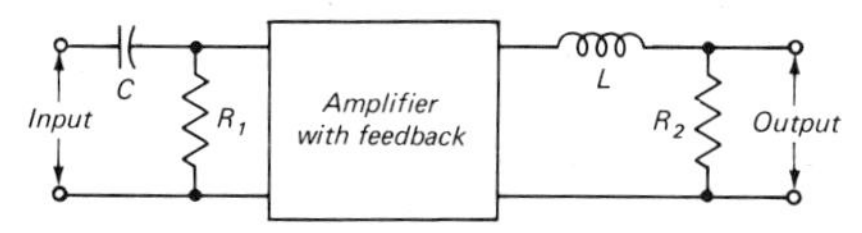

Fig. 15-10 A feedback amplifier with equalizing filter networks for neutralizing the amplification peaks.

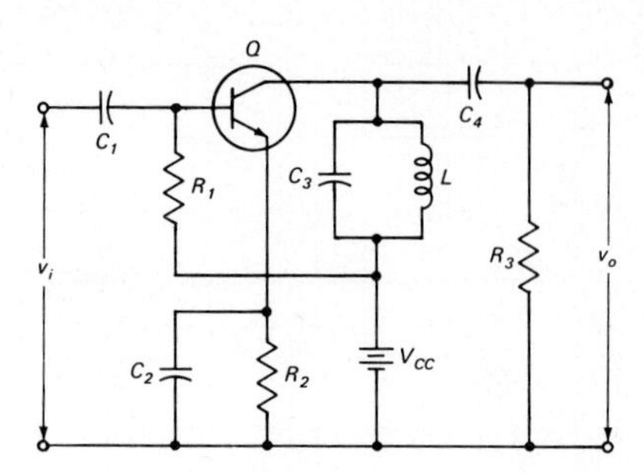

Fig. 15-11 Basic tuned amplifier.

tically uniform amplification over a wider range of frequency than an amplifier employing only negative feedback. A higher overall gain is also obtained with the balanced-feedback amplifier, as is illustrated in Fig. 15-2.

15-9 Tuned Amplifier

Some amplifier applications require that only a small portion of the mid-frequency range need be amplified. Combining an active device and a tuned circuit as its load (Fig. 15-11) is the method commonly used for this type of application; this circuit is generally called a *tuned amplifier*. The characteristics of the tuned amplifier are determined largely by the tuned circuit and have been fully described in Chaps. 11 and 12.

The characteristics of major concern for the tuned amplifier are (1) the resonant frequency, (2) the bandpass at -3 db, and (3) the voltage gain of the amplifier. These characteristics have been discussed in preceding chapters and are expressed by

$$f_r = \frac{159}{\sqrt{LC}} \tag{11-4}$$

$$BW = f_2 - f_1 \tag{11-9}$$

$$f_2 - f_1 = \frac{f_r}{Q} \tag{11-10}$$

The value of Q for the circuit of Fig. 15-11 is expressed by

$$Q = \frac{r_o}{X_L} \tag{15-20}$$

For the circuit of Fig. 15-11, the impedance of the parallel tuned circuit formed by C_3 and L is very high at (and near) its resonant frequency. When the impedance of this tuned circuit is much greater than the output resistance R_3, the voltage gain of the amplifier approaches the value of

$$A_v = \frac{r_o}{r_e} \tag{9-17a}$$

Example 15-9 If the parameters for the circuit of Fig. 15-11 are R_1 = 1,270,000 ohms, R_2 = 1,500 ohms, R_3 = 6,000 ohms, C_3 = 1,000 pf, L = 120 μh, β = 100, and V_{CC} = 15 volts, find (*a*) the resonant frequency, (*b*) the bandwidth, (*c*) the voltage gain.

GIVEN: $R_1 = 1{,}270{,}000$ ohms $\quad R_2 = 1{,}500$ ohms $\quad R_3 = 6{,}000$ ohms
$C_3 = 1{,}000$ pf $\quad L = 120\ \mu$h $\quad \beta = 100 \quad V_{CC} = 15$ volts

FIND: (*a*) f_r (*b*) BW (*c*) A_v

SOLUTION:

(*a*)
$$f_r = \frac{159}{\sqrt{LC}} = \frac{159}{\sqrt{120 \times 1{,}000 \times 10^{-6}}} \cong 460 \text{ kHz}$$

(*b*)
$$BW = \frac{f_r}{Q} = \frac{460}{17.3} \cong 26.6 \text{ kHz}$$

$$Q = \frac{r_o}{X_L} = \frac{6{,}000}{346} \cong 17.3$$

$$r_o = R_3 = 6{,}000 \text{ ohms}$$

$$X_L = 2\pi f L = 6.28 \times 460 \times 10^3 \times 120 \times 10^{-6} \cong 346 \text{ ohms}$$

(*c*)
$$A_v = \frac{r_o}{r_e} = \frac{6{,}000}{23.6} \cong 254$$

$$r_e = \frac{25}{I_E} = \frac{25}{1.06} \cong 23.6 \text{ ohms}$$

$$I_E = \frac{V_{CC}}{R_2 + \dfrac{R_1}{\beta}} = \frac{15}{1{,}500 + \dfrac{1{,}270{,}000}{100}} \cong 1.06 \text{ ma}$$

A tuned amplifier using a vacuum tube as the active device can be treated in the same manner as in the preceding paragraphs for the tuned amplifier using a transistor.

15-10 Wideband Amplifier

Need for a Wide Passband. The operation of cathode-ray tubes, whether for television viewing, radar, electronic instruments, or industrial controls, etc., requires the use of voltages having a sawtooth waveform. Any nonsinusoidal waveform such as sawtooth, square, or pulse consists of a fundamental frequency and a number of harmonics (Chap. 16). The number of harmonics, and whether odd or even, determine the shape of the composite waveform. In order to amplify nonsinusoidal signals without introducing distortion of the waveform, an amplifier must provide uniform amplification to signals ranging from very low frequencies (10 Hz or lower) to very high frequencies (4 MHz or higher). An amplifier capable of handling such a range of frequencies is called a *wideband amplifier*. When a wideband amplifier is used with a cathode-ray tube to provide a visible signal display, it is called a *video amplifier*. A wideband amplifier used in circuits to amplify pulses is called a *pulse amplifier*.

Methods of Obtaining a Wide Passband. It was shown in Chap. 14 that the passband of an ordinary amplifier, whether *RC*-, impedance-, or transformer-coupled, suffers a 3-db loss of voltage gain at the low- and high-frequency ends of the mid-frequency range long before reaching the respective values required for a wideband amplifier.

For an *RC*-coupled amplifier, the 3-db frequency response range can be extended by adding (1) a high-frequency compensating circuit, (2) a low-frequency compensating circuit, or (3) both high- and low-frequency compensating circuits. The same principle can be applied to impedance-coupled amplifiers.

For a transformer-coupled amplifier, the frequency range can be extended by use of two or more stages of tuned circuits with each stage made resonant to a slightly different frequency. Wideband amplification using tuned primary and tuned secondary circuits is described in Art. 12-9.

The following discussion will introduce the methods of achieving a wide passband for the *RC*-coupled transistor amplifier. Similar principles can be applied to other methods of coupling and to vacuum-tube amplifier circuits.

High-frequency Compensation. The rolloff of voltage gain at the high frequencies in an amplifier is due largely to the shunting capacitances in the amplifier circuit (Fig. 15-12), namely, (1) the capacitance C_1 consisting of the output capacitance of Q_1 and any stray capacitances at the input side of the coupling network, and (2) the capacitance C_3 consisting of the input capacitance of Q_2 and any stray capacitances at the output side of the coupling network. At the high frequencies, the impedance of the coupling capacitor C_2 is so low that this series-connected capacitance can be disregarded.

The value of frequency at which the gain decreases by 3 db can be increased by adding inductance to the circuit and thereby reducing the effect of the

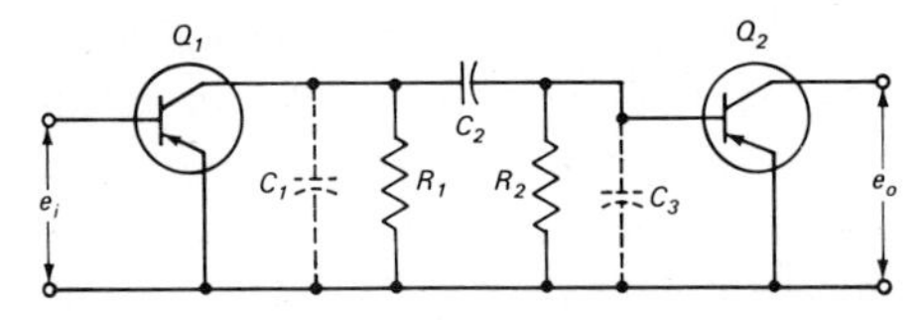

Fig. 15-12 Basic *RC*-coupled transistor amplifier with high-frequency shunting capacitances C_1 and C_3 indicated.

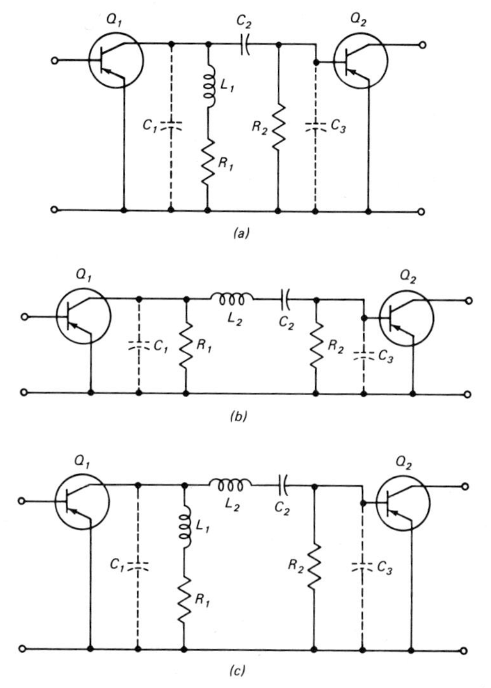

Fig. 15-13 Types of high-frequency compensation for wideband amplifiers. (a) Shunt compensation. (b) Series compensation. (c) Series-shunt compensation.

shunting capacitances C_1 and C_3. The inductance can be added to the circuit in the three ways shown in Fig. 15-13: (1) *shunt compensation* by adding an inductor L_1 in series with the resistor R_1; (2) *series compensation* by adding an inductor L_2 in series with the coupling capacitor C_2; and (3) *series-shunt compensation* by using both (1) and (2) methods.

With *shunt compensation*, also called *shunt peaking*, the inductor L_1 in series with R_1 is connected in parallel with capacitances C_1 and C_3 to form a parallel tuned circuit. With optimum values of $L_1C_1C_3$, the impedance of the circuit

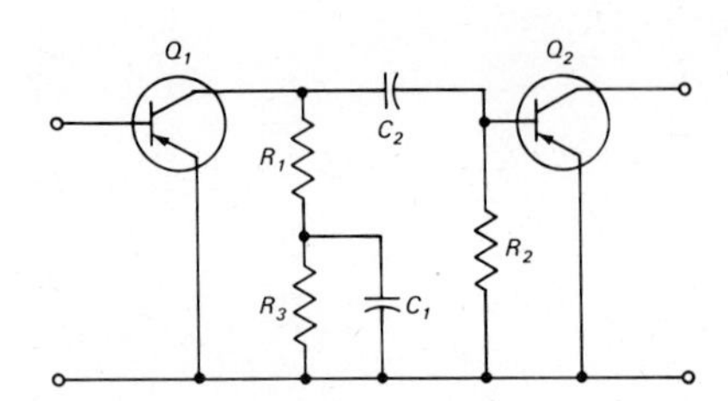

Fig. 15-14 Basic *RC*-coupled transistor amplifier with low-frequency compensation network R_3C_1.

remains nearly constant over a wide range of high frequencies as decreases in X_C are offset by increases in X_L and vice versa. Consequently, satisfactory high-frequency response is extended considerably before the 3-db rolloff point is reached.

With *series compensation*, also called *series peaking*, the inductor L_2 together with capacitance C_3 forms a series tuned circuit; the capacitance of C_2 is usually many times greater than that of C_3, and thus C_2 can be ignored. As the frequency increases, the impedance of C_1 decreases, causing a decrease in the load impedance on Q_1 and a corresponding decrease in the voltage gain. However, the impedance of the series tuned circuit L_2C_3 also decreases as the frequency increases toward the resonant frequency of the series tuned circuit, thereby increasing the voltage across C_3 and R_2. Consequently, the drop in voltage at R_1 is offset by an increase in voltage at R_2, thereby extending the frequency range before the 3-db rolloff point is reached.

With *series-shunt compensation*, also called *series-shunt peaking*, the characteristics of both shunt and series compensation are combined.

Low-frequency Compensation. The rolloff of voltage gain at the low frequencies is due chiefly to the impedance of the coupling capacitor C_2 (Figs. 15-12 and 15-13). At low frequencies, the input and output capacitances C_1 and C_3 have no appreciable effect on the frequency response and therefore can be disregarded. In the basic *RC*-coupled amplifier, at the lower frequencies the voltage drop across C_2 increases and the voltage drop across R_2 decreases and hence results in a decrease in the voltage gain of the amplifier.

Adding a low-frequency compensating network (C_1C_3 of Fig. 15-14) to the basic circuit extends the low-frequency range of the amplifier. With optimum values for C_1 and R_3, the impedance of C_1 at the mid- and high frequencies is much lower than the value of R_3 and in effect short-circuits R_3. When $R_2 \gg R_1$, the value of r_o for the circuit is then approximately equal to R_1. At very low frequencies, the impedance of C_1 increases to a value much greater than R_3 and acts as an open circuit. When $R_2 \gg R_1$, the value of r_o for the circuit is then approximately equal to $R_1 + R_3$, and the voltage gain of the circuit is increased. Thus, the increase in the voltage gain provided by C_1R_3 compensates for the decrease in the voltage gain due to the voltage-divider action of C_2R_2.

15-11 High-fidelity Amplifier

High-fidelity sound systems for reproducing musical selections recorded on disk and tapes require carefully designed amplifiers. The important characteristics of a high-fidelity amplifier system are (1) frequency response, (2) total harmonic distortion, (3) noise level, and (4) maximum power output.

The *frequency response* of a high-fidelity amplifier should not vary more than 1 db for a range of frequencies from below the lowest to well above the highest frequencies to be reproduced. A commonly used rating is 20 to 20,000 Hz with a variation in response of not more than 1 db.

Harmonic distortion affects the quality of the reproduced sound and is expressed as a percentage of the output power. This form of distortion in an amplifier system should not exceed 0.5 per cent at the desired listening level.

The *level of noise* produced by the amplifier system relative to the signal strength also expresses the quality of a high-fidelity amplifier. The noise level determines the range in volume of an amplifier between the loudest and softest sounds reproduced. If the loudest volume is considered to be 60 db, then the noise level of the amplifier should be at least 60 db below the signal level at the desired listening level.

The *maximum power output* required for a high-fidelity amplifier depends on (1) size of the listening area, (2) volume of sound desired, and (3) efficiency of the loudspeaker system. The acoustic power required to reproduce a concert-orchestra program properly in the average home at a volume corresponding to the sound level in a concert hall is in the order of 0.5 watt. The efficiency of loudspeakers used in the home high-fidelity systems is in the order of 5 per cent or lower. In order to produce 0.5 watt of acoustic power with a loudspeaker system having only 5 per cent efficiency a power-output rating of at least 10 watts is required.

15-12 Emitter-follower Amplifier

The emitter-follower amplifier is basically the same as the common-collector amplifier described in Arts. 9-7 and 9-19. Advantages of the emitter follower are (1) high input resistance, (2) low output resistance, (3) low distortion, and (4) no phase inversion from input to output signals. Because of these characteristics, the emitter follower is useful in coupling a high-impedance signal source to a low-impedance load. For example, an emitter follower can be used to couple the output signal of a voltage-amplifier stage having a high output impedance to the input of a power-output stage having a low input impedance.

The equations presented in Art. 9-19 apply also to the emitter follower of Fig. 15-15. However, it should be noted that Eq. (9-33) represents the a-c out-

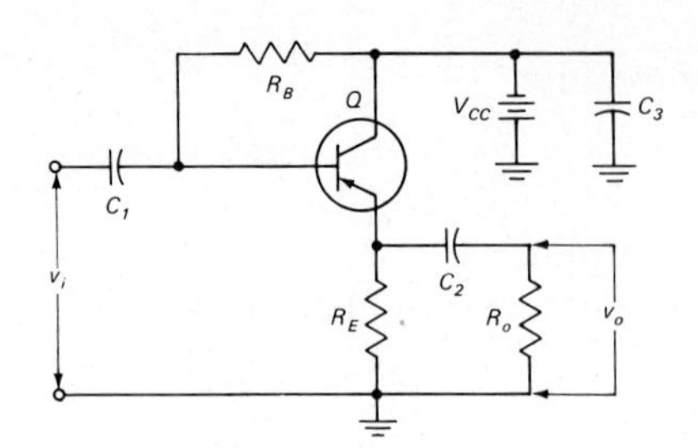

Fig. 15-15 Emitter-follower amplifier.

put resistance seen by the emitter. The output resistance of the amplifier seen by the load R_o is

$$r_o' = R_E \| r_{o \cdot Q}$$

where

$$r_{o \cdot Q} = \frac{R_B}{\beta} + r_e = \text{output resistance of the transistor, ohms} \qquad (15\text{-}21)$$

Example 15-10 If the parameters for the circuit of Fig. 15-15 are $R_B = 50{,}000$ ohms, $R_E = 500$ ohms, $R_o = 50$ ohms, $\beta = 100$, and $V_{CC} = 10$ volts, find (*a*) input resistance seen by the base of the transistor, (*b*) input resistance of the circuit, (*c*) voltage gain of the circuit, (*d*) output resistance of the amplifier seen by the load resistance R_o, (*e*) current gain of the circuit, (*f*) power gain of the circuit.

GIVEN: $R_B = 50{,}000$ ohms $\quad R_E = 500$ ohms $\quad R_o = 50$ ohms $\quad \beta = 100$
$V_{CC} = 10$ volts

FIND: (*a*) $r_{i \cdot B}$ $\quad$ (*b*) $r_{i \cdot cct}$ $\quad$ (*c*) A_v $\quad$ (*d*) r_o' $\quad$ (*e*) A_i $\quad$ (*f*) A_p

SOLUTION:

(*a*)

$$r_{i \cdot B} = \beta(r_e + r_o) = 100(2.5 + 45.5) \cong 4{,}800 \text{ ohms}$$

$$r_e = \frac{25}{I_E} = \frac{25}{10} = 2.5 \text{ ohms}$$

$$I_E = \frac{V_{CC}}{R_E + \dfrac{R_B}{\beta}} = \frac{10}{500 + \dfrac{50{,}000}{100}} = 10 \text{ ma}$$

$$r_o = R_E \| R_o = 500 \| 50 \cong 45.5 \text{ ohms}$$

(*b*)

$$r_{i \cdot cct} = R_B \| r_{i \cdot B} = 50{,}000 \| 4{,}800 \cong 4{,}400 \text{ ohms}$$

(*c*)

$$A_v = \frac{r_o}{r_o + r_e} = \frac{45.5}{45.5 + 2.5} \cong 0.95$$

(*d*)

$$r_o' = R_E \| r_{o \cdot Q} = 500 \| 500 = 250 \text{ ohms}$$

$$r_{o \cdot Q} = \frac{R_B}{\beta} + r_e = \frac{50{,}000}{100} + 2.5 \cong 500 \text{ ohms}$$

(*e*) $A_i \cong \beta \cong 100$
(*f*) $A_p = A_v A_i = 0.95 \times 100 \cong 95$

Example 15-10 illustrates two important characteristics of the emitter-follower amplifier: (1) a low-impedance load (50 ohms) is transformed to a high-impedance value (4,400 ohms), and (2) a power gain is achieved even though the voltage gain is less than 1.

15-13 Cathode-follower Amplifier

Principle. The cathode-follower amplifier is basically the same as the common-plate amplifier described in Art. 9-25. Like its transistor counterpart, the emitter follower, its advantages are (1) high input resistance, (2) low output resistance, (3) low distortion, and (4) no phase inversion from input to output signals. The voltage gain of the cathode follower is less than 1, as has been indicated in Art. 9-25.

Output Resistance. The a-c output resistance of the cathode-follower circuit (Fig. 15-16*b*) consists of the effect of the plate resistance r_p in parallel with the cathode (load) resistance R_k and is expressed by

$$r_o = \frac{r_p}{\mu + 1} \Bigg\| R_k \tag{15-22}$$

and, when $\mu \gg 1$

$$r_o \cong \frac{r_p R_k}{r_p + \mu R_k} \tag{15-22a}$$

In terms of mutual conductance,

$$r_o \cong \frac{R_k}{1 + g_m R_k} \tag{15-22b}$$

Input Resistance. The cathode-follower amplifier has a high input resistance. When the grid resistor is returned to ground, as in Fig. 15-16*a*, the input resistance is equal to R_g. However, when R_g is returned directly to the cathode, the input resistance increases, as is indicated by

$$r_i = \frac{R_g}{1 - A_v} \tag{15-23}$$

Input Capacitance. In the cathode-follower circuit, the large increase in the input capacitance due to the Miller effect (Arts. 14-5 and 14-6) is eliminated by grounding the plate to a-c signals. The input capacitance then is determined by

$$C_i = C_{gp} + C_{gk}(1 - A_v) \tag{15-24}$$

Because of the degenerative action of this type of circuit, the Miller-effect factor has a negative sign as compared with the positive effect in the grounded-cathode amplifier; see Eq. (14-22). Consequently, both the input capacitance

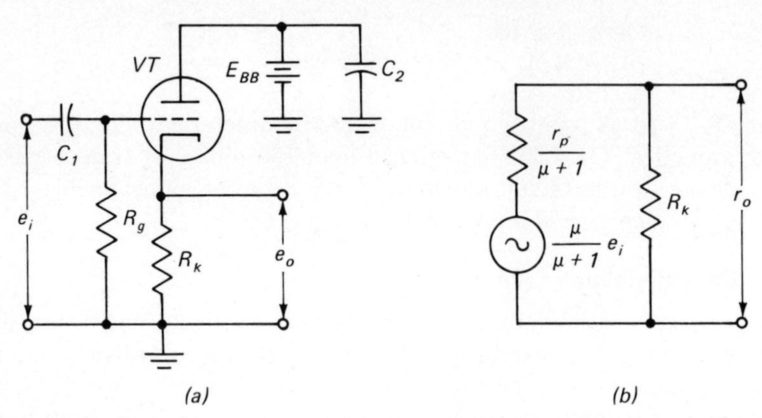

Fig. 15-16 Cathode-follower amplifier. (a) Basic circuit. (b) A-c equivalent of the basic circuit.

C_i and the total shunting capacitance C_T are much lower for the cathode-follower circuit.

Grid-bias Compensation. Grid bias is established by the average value of the plate current flowing through R_k. When the voltage developed across R_k is greater than the grid bias desired, the cathode resistance is made up of two resistors, R_2 and R_3 in Fig. 15-17, and the grid is returned to the junction of R_2 and R_3 in order to provide the desired grid bias. For the circuit of Fig. 15-17,

$$r_i = \frac{R_g}{1 - \dfrac{A_v R_3}{R_2 + R_3}} \tag{15-25}$$

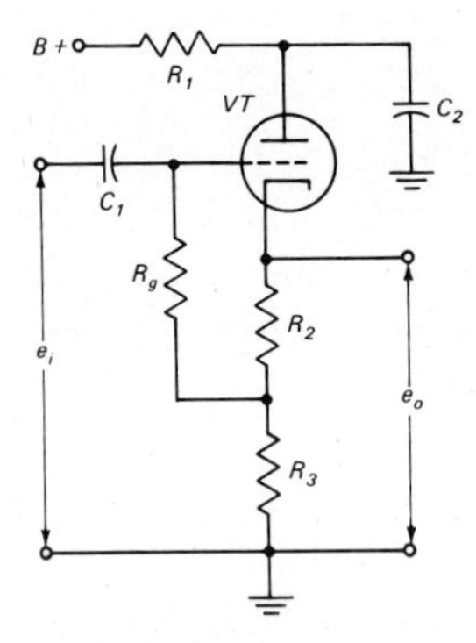

Fig. 15-17 Cathode-follower amplifier with grid-bias compensation.

15-14 Limiting Amplifier

Principle. The limiter, sometimes called a *clipper* or *chopper,* is essentially an amplifier operated with lower than normal voltages at the terminals of its active device so that saturation or cutoff is reached for specific values of input-signal voltages. Limiting circuits can be designed to use diodes, transistors, or vacuum tubes. A limiting circuit using a transistor or a vacuum tube can provide both clipping action and amplification, whereas a diode can provide only clipping action. Limiting circuits are classified as (1) a *series limiting circuit* in which the active device is connected in series with the load, and (2) a *shunt limiting circuit* in which the active device is connected in parallel with the load.

Series Limiting. Figure 15-18*a* illustrates the basic series limiting circuit and its input and output waveforms. In the circuit of Fig. 15-18*a*, when T_1 is positive, (1) the diode D will have a low resistance and will therefore be conducting, and (2) the current flowing through the load resistor R will establish the output voltage indicated by the waveform T_3T_4. Because the negative half-cycles of the input voltage have been eliminated from the output, the circuit is called a *negative clipper*. In the circuit of Fig. 15-18*b*, the terminals of the diode have been reversed and the circuit becomes a *positive clipper.*

Shunt Limiting. Figure 15-19*a* illustrates the basic shunt limiting circuit and its input and output waveforms. In the circuit of Fig. 15-19*a* when T_1 is positive (1) the diode D will conduct, and (2) because the resistance of the diode is very low the output voltage will be practically zero. When T_1 is negative, (1) the diode will not conduct, and (2) the output voltage will vary in the same manner as the input voltage. Consequently, positive clipping takes place. In the circuit of Fig. 15-19*b*, the terminals of the diode have been reversed and the circuit becomes a negative clipper.

Partial Half-cycle Limiting. The circuit of Fig. 15-20*a* is similar to that of Fig. 15-19*a* with an added fixed voltage source. During the negative half-cycle when the diode is not conducting, (1) the diode acts as an open switch and thereby disconnects the power source E_B, (2) the output voltage will vary directly with the input voltage as with the circuit of Fig. 15-19*a*, and (3) the output voltage will be as shown in Fig. 15-20*a*. In the circuit of Fig. 15-20*b*, the

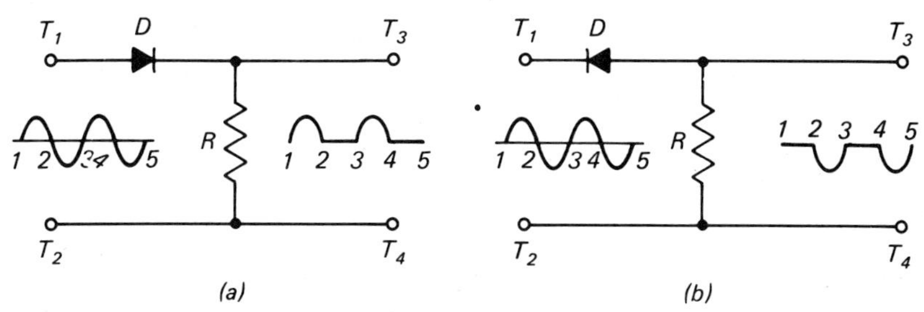

Fig. 15-18 Basic series limiter circuit for (a) negative clipping, (b) positive clipping.

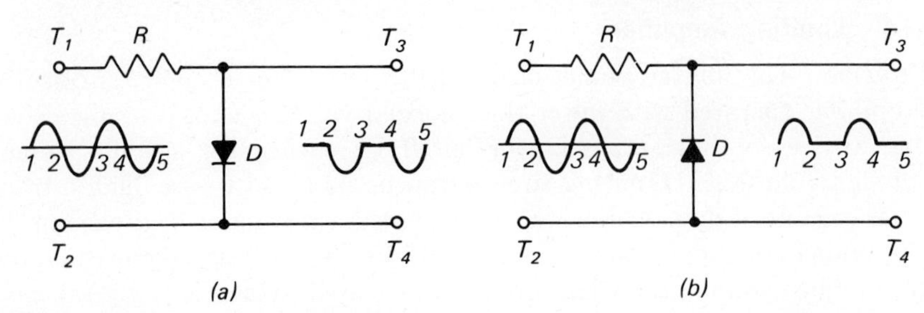

Fig. 15-19 Basic shunt limiter circuit for (a) positive clipping, (b) negative clipping.

diode and the fixed power source are reversed from that in Fig. 15-20*a* and only partial clipping of the negative half-cycles occurs as is shown in Fig. 15-20*b*.

Double Limiting Circuit. Figure 15-21 illustrates the use of partial limiting for both the positive and negative half-cycles. This circuit uses the principles of both Fig. 15-20*a* and *b* to form a *double limiting circuit*. The action of each circuit individually is the same as previously described. The waveform of the output voltage is the combined results of both the positive and negative clipping action, as is indicated at T_3T_4 of Fig. 15-21.

Limiting Amplifier. By selecting suitable circuit parameters, operating voltages, and active device, a limiter can be designed to achieve both amplifica-

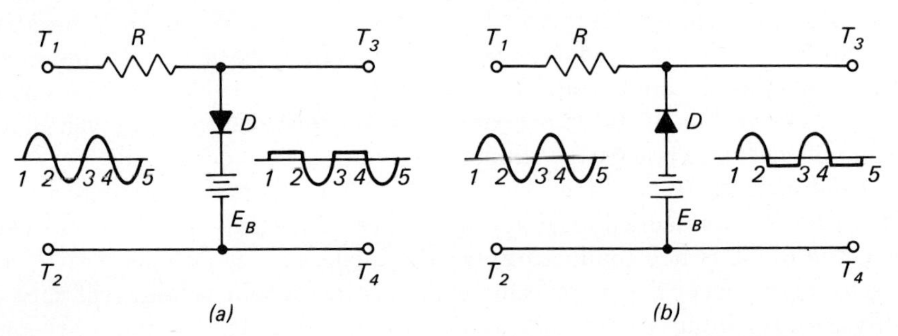

Fig. 15-20 Partial-cycle limiting. (a) Partial positive clipping. (b) Partial negative clipping.

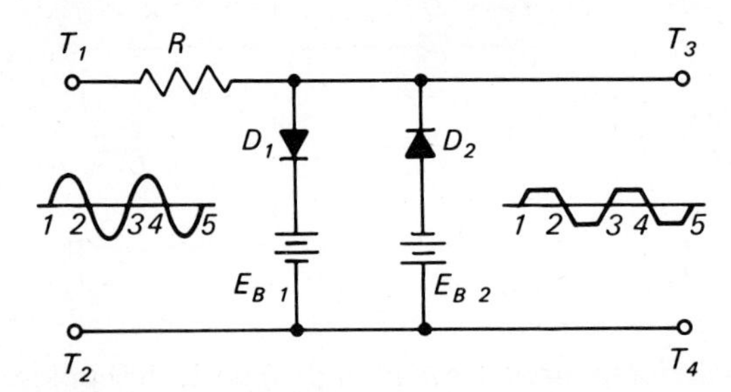

Fig. 15-21 Double limiting.

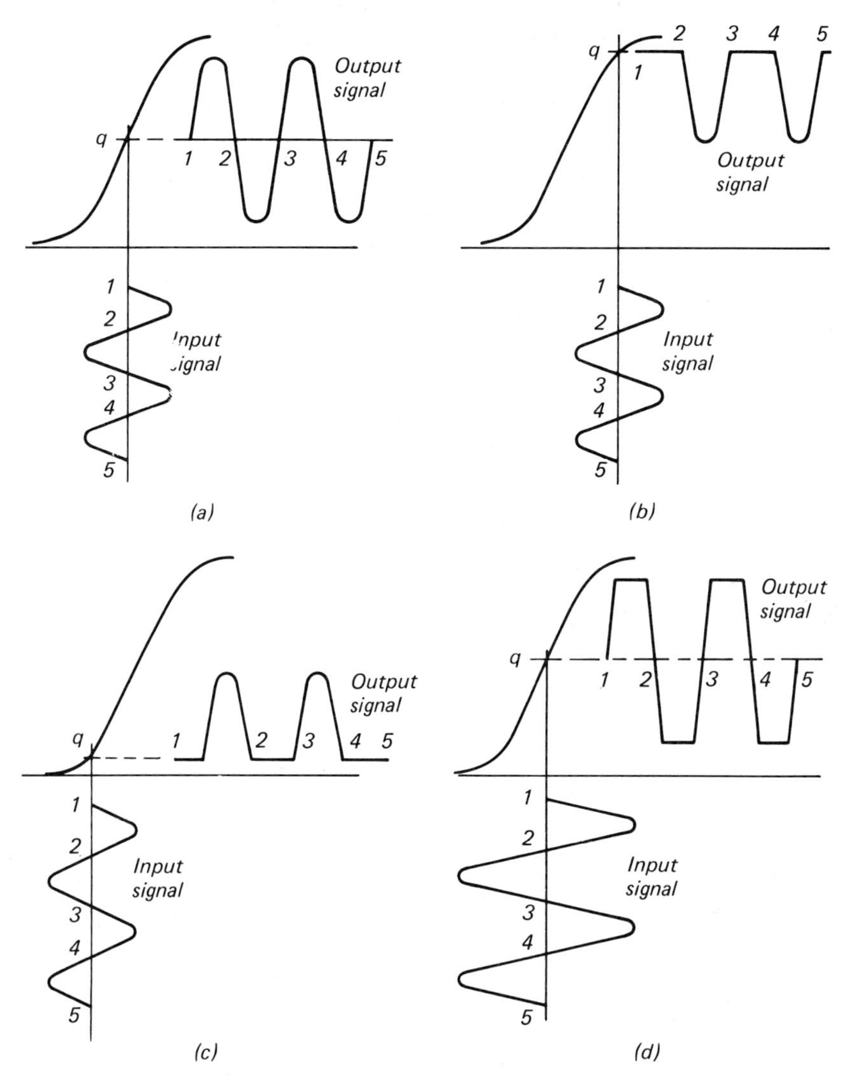

Fig. 15-22 Characteristics of limiting amplifiers. (a) Class A amplifier, no clipping. (b) Class B amplifier with positive clipping. (c) Class B amplifier with negative clipping. (d) Class AB amplifier with both positive and negative clipping.

tion and clipping of the output signal. The curves of Fig. 15-22 illustrate the principles involved in such limiting circuits.

The curve of Fig. 15-22*a* shows the input and output waveforms for the condition of Class A operation with no clipping action. By shifting the quiescent operating point q and controlling the magnitude of the input signal, positive, negative, or combined clipping can be achieved. In Fig. 15-22*b*, the quiescent point is located at the saturation area of the characteristic curve and the active

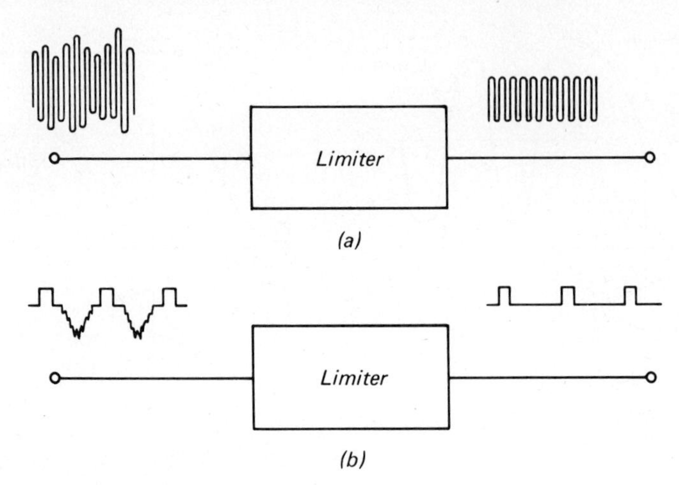

Fig. 15-23 Uses of limiter circuits. (a) Converting an f-m signal of varying amplitude to a signal with constant amplitude. (b) Recovering the sync signal from a composite television signal.

device operates as a Class B amplifier with resultant positive clipping action. In Fig. 15-22*c*, the quiescent point is located at the cutoff region of the active device and operates as a Class B amplifier with resultant negative clipping action. In Fig. 15-22*d*, the quiescent point is located at the middle of the straight portion of the characteristic curve, the same as in Fig. 15-22*a*. However, in Fig. 15-22*d*, the input-signal voltage has been increased to the extent that the active device is driven beyond saturation during a part of the positive portion of the input signal and beyond cutoff during a part of the negative portion of the input signal. Consequently, portions of both the positive and negative halves of the input signal are clipped.

Uses of Limiters. Limiter circuits have many applications in the various fields of communications, two of which are illustrated in Fig. 15-23. The waveform of the input of Fig. 15-23*a* represents the varying amplitude of an f-m signal in the input stages of an f-m radio receiver. Passing these signals through a limiter circuit before detection of the a-f signals eliminates much of the noise that is characteristic of the a-m radio system.

One of the functions in a television receiver is to isolate and recover the synchronizing pulses present in the composite television signal. This function is accomplished by limiting action. The waveforms in Fig. 15-23*b* show the composite television signal at the input to the limiter circuit and the synchronizing pulses at the output of the limiter.

15-15 Differential Amplifier

The Circuit. Figure 15-24 illustrates a circuit called a *differential* (or *difference*) *amplifier*. Features of this amplifier are (1) it contains two basic common-emitter amplifier circuits, (2) only a small number of components (such as resistors and capacitors) are required, (3) it is a directly coupled amplifier (emitter-to-emitter), (4) it can accommodate two inputs by means of T_1 and ground and T_5 and ground, and (5) it can provide two separate outputs by means of terminals T_3 and ground and T_4 and ground, or a single output between terminals T_3 and T_4. A number of modifications can be made to the basic circuit such as (1) adding emitter bias, (2) adding emitter feedback, (3) using a single input by eliminating R_4 and returning the base of Q_2 to ground, (4) using a single output by eliminating R_2 and taking the output between T_4 and ground, etc. (see Probs. 49 to 52).

Advantages. Among the advantages of the differential amplifier are (1) a higher gain for a direct-coupled amplifier than for two cascaded stages of ordinary direct coupling; (2) fairly uniform amplification of signals from d-c up into the very high frequencies; (3) isolation between the input and output signals; (4) a wide variety of applications such as amplification, limiting, mixing, amplitude modulation, signal generation, frequency multiplication, and temperature compensation; (5) a minimum number of capacitors and resistors; (6) high adaptability to monolithic integrated-circuit construction.

Circuit Operation. In the circuit of Fig. 15-24, when the parameters of Q_1 and Q_2 are identical and the values of their associated circuit components are matched, the circuit is called a *balanced differential amplifier*. Under these conditions, each stage of the amplifier produces the same voltage gain, and

$$A_v = \frac{r_o}{r_e} \tag{9-17a}$$

where in Fig. 15-24, $r_o = R_2 = R_3$

The output voltage between terminals T_3 and T_4 is

$$v_o(T_3T_4) = A_v(v_{i\cdot 1} - v_{i\cdot 2}) \tag{15-26}$$

where A_v = voltage gain of each section

Example 15-11 For the amplifier of Fig. 15-24, the voltage gain of each section is 200, $v_{i\cdot 1}$ is 25 mv, and $v_{i\cdot 2}$ is 10 mv. Find the voltage between terminals (*a*) T_3 and ground, (*b*) T_4 and ground, (*c*) T_3 and T_4. (*d*) What are the polarities at terminals T_3 and T_4?

GIVEN: $A_{v\cdot 1}$, $A_{v\cdot 2} = 200$ $v_{i\cdot 1} = 25$ mv $v_{i\cdot 2} = 10$ mv

FIND: (*a*) v_oT_3 (*b*) v_oT_4 (*c*) $v_o(T_3T_4)$ (*d*) Polarity of T_3, T_4

SOLUTION:

(*a*) $v_oT_3 = A_{v\cdot 1}v_{i\cdot 1} = 200 \times 25 \times 10^{-3} = 5$ volts

(*b*) $v_oT_4 = A_{v\cdot 2}v_{i\cdot 2} = 200 \times 10 \times 10^{-3} = 2$ volts

(*c*) $v_o(T_3T_4) = A_v(v_{i\cdot 1} - v_{i\cdot 2}) = 200 \times (25 - 10)10^{-3} = 3$ volts

(*d*) As $v_{i\cdot 1} \gg v_{i\cdot 2}$, then $i_{c\cdot 1} \gg i_{c\cdot 2}$, and $i_{c\cdot 1}R_2 \gg i_{c\cdot 2}R_3$; therefore, T_4 will be positive with respect to T_3.

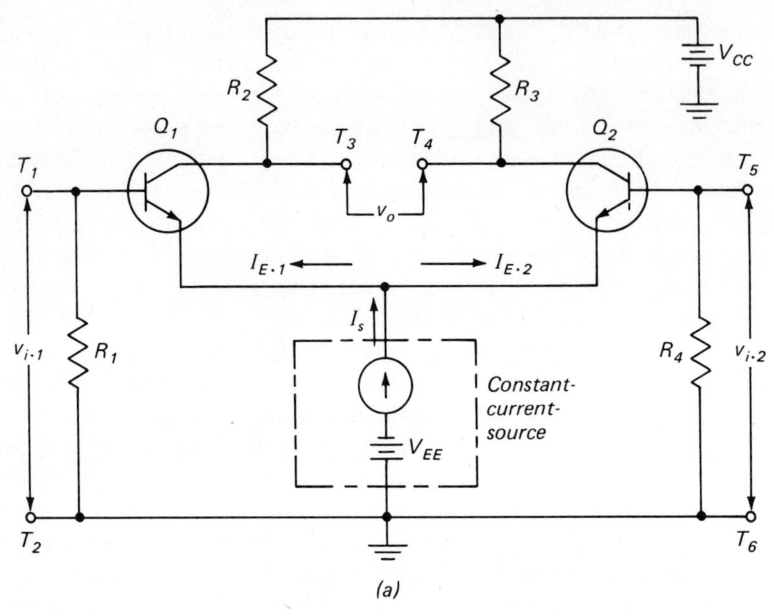

(a)

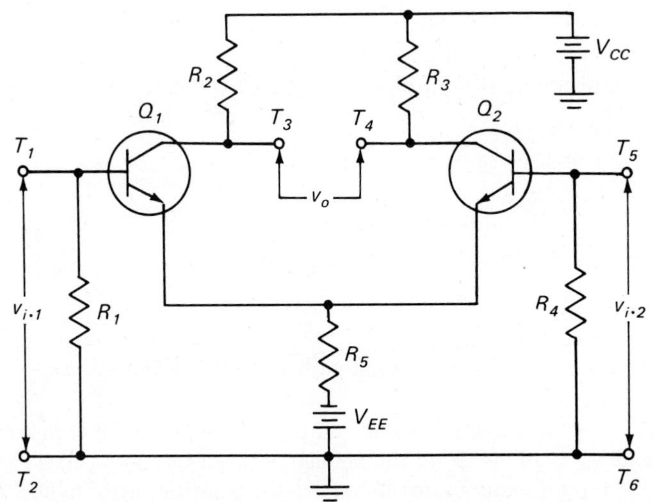

$$A_{v\cdot1} \cong \frac{R_2}{r_{e\cdot1}} \qquad A_{v\cdot2} \cong \frac{R_3}{r_{e\cdot2}}$$

$$v_o\,(T_3T_6) = A_{v\cdot1}\,v_{i\cdot1} \qquad v_o\,(T_4T_6) = A_{v\cdot2}\,v_{i\cdot2}$$

$$v_o\,(T_3T_4) = A_v\,(v_{i\cdot1} - v_{i\cdot2}) = v_o(T_3T_6) - v_o\,(T_4T_6)$$

When $R_1 = R_4$, $\beta_1 r_{e\cdot2} = \beta_2\, r_{e\cdot2}$, and $R_1 \gg \beta_1 r_{e\cdot1}$, $R_4 \gg \beta_2 r_{e\cdot2}$

$$r_i\,(T_1T_2) \cong \beta_1 r_{e\cdot1} \qquad r_i\,(T_5T_6) \cong \beta_2\, r_{e\cdot2}$$

$$r_i\,(T_1T_5) \cong 2\beta_1\, r_{e\cdot1}$$

(NOTE: R_5 has no effect on A_v and r_i when $R_1 = R_4$ and $\beta_1 r_{e\cdot1} = \beta_2 r_{e\cdot2}$.)

(b)

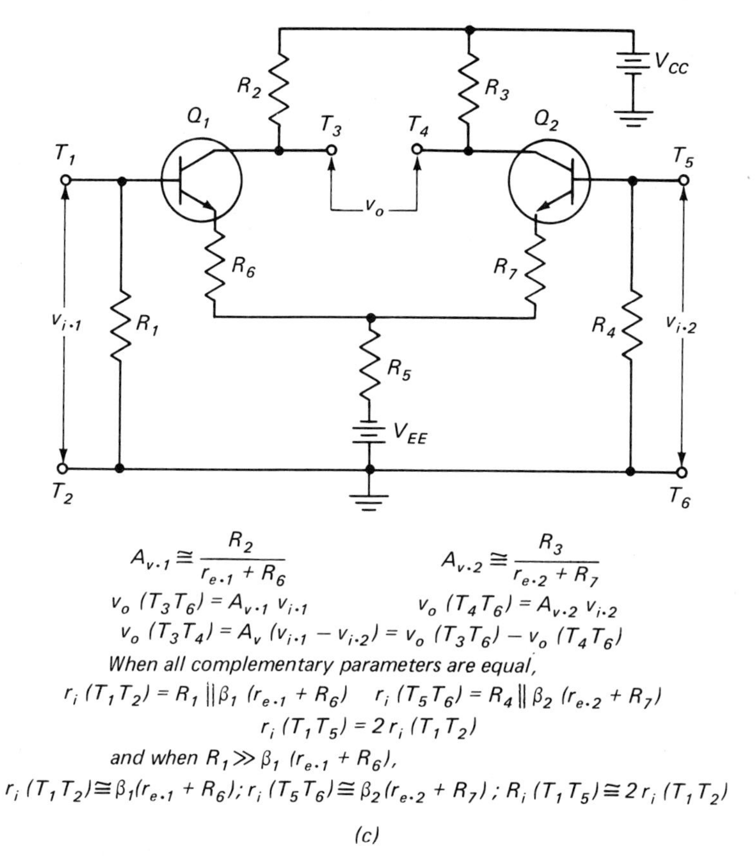

Fig. 15-24 Differential amplifier. (a) Basic circuit. (b) With constant-current source R_5V_{EE}. (c) With swamping resistors R_6R_7.

Constant-current Source. Figure 15-24 indicates that a constant-current source is required for operation of the difference amplifier. The constant-current source, also called *constant-current sink*, can be (1) a resistor and d-c power source as in Fig. 15-25*a*, or (2) a transistor circuit and d-c power source as in Fig. 15-25*b*.

Transistor-current-sink Circuit. Figure 15-25*a* shows the simplest form of constant-current source consisting only of the resistor R_5 and the power source V_{EE}. The effectiveness of this circuit is limited and a more desirable circuit is shown in Fig. 15-25*b*. In this circuit, a constant-current source is obtained from the collector output of Q_3. The diode provides more stable operation of the circuit, particularly in terms of temperature stability. The principle of this circuit is described in Art. 8-19.

Balancing the Circuit. It is possible that the parameters of the transistors and the values of the other circuit components may not match perfectly. When

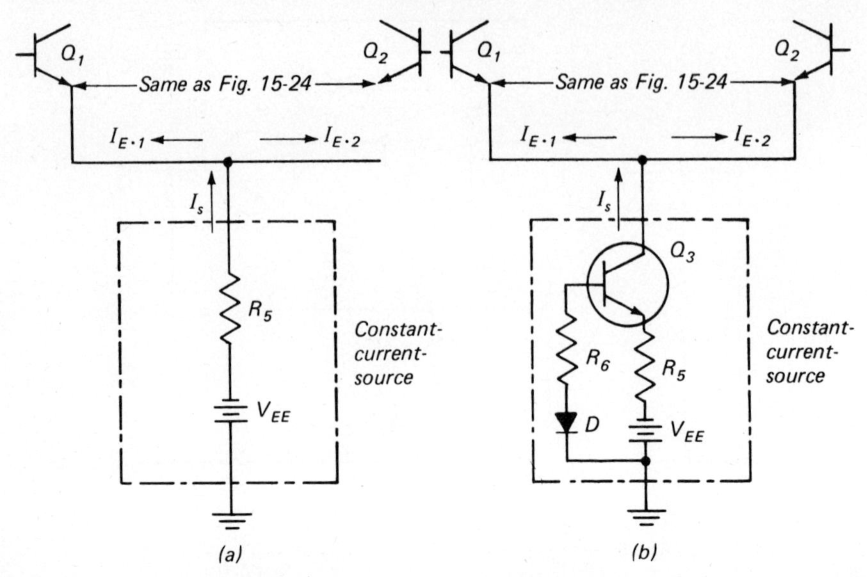

Fig. 15-25 Constant-current source for differential amplifier. (a) Resistor and d-c power source. (b) Temperature-compensated transistor circuit and d-c power source.

it is necessary to compensate for any mismatch, a low-resistance potentiometer is added to the circuit as shown in Fig. 15-26. The circuit becomes balanced when the potentiometer is set at the point that produces zero voltage output between terminals T_3 and T_4 when the inputs to T_1 and T_5 are zero.

Example 15-12 The parameters for the circuit of Figs. 15-24 and 15-25*a* are R_1, R_4 = 500,000 ohms, R_2, R_3, R_5 = 10,000 ohms, β_1, β_2 = 50, V_{CC}, V_{EE} = 20 volts, $v_{i\cdot 1}$ = 20 mv, $v_{i\cdot 2}$ = 15 mv. Find approximate values for (*a*) I_s, (*b*) $I_{E\cdot 1}$, $I_{E\cdot 2}$, (*c*) $r_{e\cdot 1}$, $r_{e\cdot 2}$, (*d*) A_v, (*e*) $v_o(T_3T_4)$. [*Note:* Assume that (1) $R_5 \gg r_{e\cdot 1}$, $r_{e\cdot 2}$, and (2) $v_{i\cdot 1}$ and $v_{i\cdot 2}$ are in phase.]

GIVEN: R_1, R_4 = 500,000 ohms R_2, R_3, R_5 = 10,000 ohms β_1, β_2 = 50
V_{CC}, V_{EE} = 20 volts $v_{i.1}$ = 20 mv $v_{i.2}$ = 15 mv

FIND: (*a*) I_s (*b*) $I_{E\cdot 1}$, $I_{E\cdot 2}$ (*c*) $r_{e\cdot 1}$, $r_{e\cdot 2}$ (*d*) A_v (*e*) $v_o(T_3T_4)$

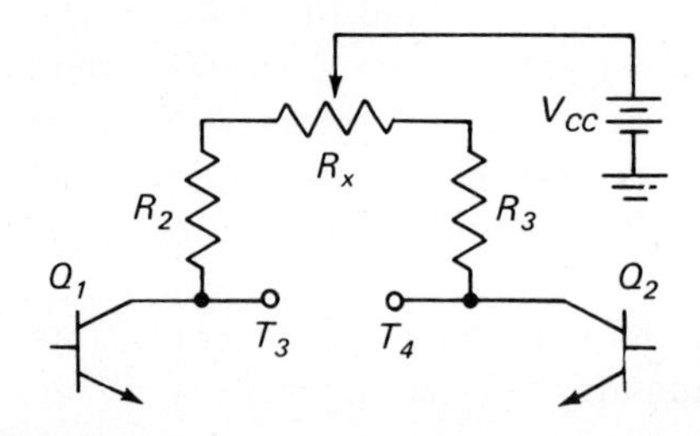

Fig. 15-26 Use of potentiometer to balance the differential amplifier.

SOLUTION:

(a) $$I_s \cong \frac{V_{EE}}{R_E} \cong \frac{20}{10{,}000} \cong 2 \text{ ma}$$

(b) $$I_{E\cdot 1} = I_{E\cdot 2} \cong \frac{I_s}{2} \cong \frac{2}{2} \cong 1 \text{ ma}$$

(c) $$r_{e\cdot 1} = r_{e\cdot 2} = \frac{25}{I_E} = \frac{25}{1} = 25 \text{ ohms}$$

(d) $$A_v = \frac{r_o}{r_e} = \frac{R_2}{r_e} = \frac{10{,}000}{25} = 400$$

(e) $$v_o(T_3T_4) = A_v(v_{i\cdot 1} - v_{i\cdot 2}) = 400(20 - 15)10^{-3} = 2 \text{ volts}$$

15-16 Operational Amplifier

An operational amplifier is the name given to certain types of direct-coupled cascade transistor amplifiers that use feedback circuits to obtain a desired response characteristic. A basic type of operational amplifier consists of two balanced-differential-amplifier stages connected in a direct-coupled cascade configuration and uses one or more feedback loops to achieve the desired response. This type of amplifier is an outcome of integrated-circuit technology and is further described in Chap. 19.

15-17 Darlington Pair

Figure 15-27*a* shows a circuit using two transistors in a configuration called the *Darlington pair*. Upon tracing the paths of the base currents of the two transistors, it can be seen that the base current of Q_1 (being part of the emitter current of Q_1) must flow through the base-emitter portion of Q_2. Consequently, the base current of Q_2 is approximately equal to the emitter current of Q_1, or

$$I_{B\cdot 2} \cong I_{E\cdot 1} \cong \beta_1 I_{B\cdot 1} \tag{15-27}$$

The emitter current of Q_2 is

$$I_{E\cdot 2} \cong \beta_2 I_{B\cdot 2} \tag{15-28}$$

Substituting Eq. (15-27) in Eq. (15-28)

$$I_{E\cdot 2} \cong \beta_2 \beta_1 I_{B\cdot 1} \tag{15-29}$$

Therefore, the beta value of the Darlington pair is

$$\beta_{Dp} = \beta_1 \beta_2 \tag{15-30}$$

and with identical transistors

$$\beta_{Dp} = \beta_1^2 \tag{15-30a}$$

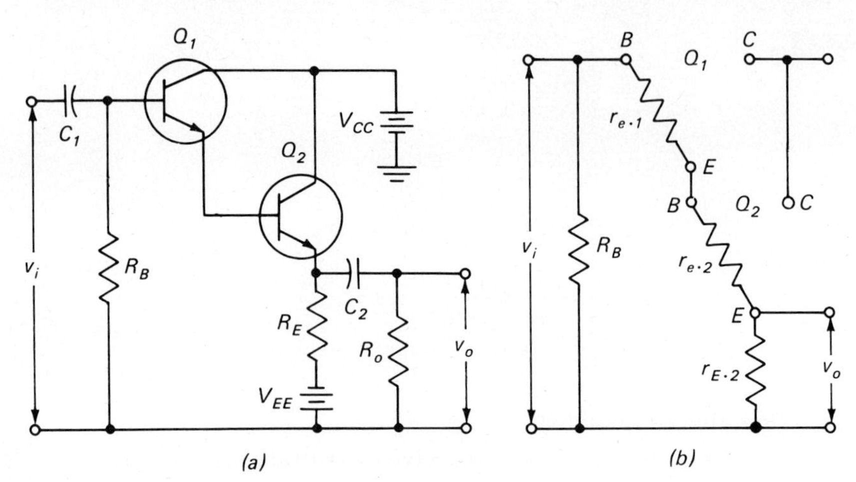

Fig. 15-27 The Darlington-pair amplifier. (a) Basic circuit. (b) A-c equivalent circuit.

From the a-c equivalent circuit of Fig. 15-27*b*, it can be shown that the a-c input resistance seen by the base of Q_1 is

$$r_iB_{Q\cdot 1} = \beta_1\beta_2(r_{e\cdot 1} + r_{e\cdot 2} + r_E) \tag{15-31}$$

where

$$r_E = R_E \| R_o \tag{15-32}$$

and, when $r_E \gg (r_{e\cdot 1} + r_{e\cdot 2})$

$$r_iB_{Q\cdot 1} \cong \beta_1\beta_2 r_E \tag{15-31a}$$

The a-c input resistance of the Darlington-pair amplifier of Fig. 15-27*a* is

$$r_iDp = R_1 \| r_iB_{Q\cdot 1} \tag{15-33}$$

From Fig. 15-27*a*, it can be seen that the output of the Darlington pair is obtained from across the resistor in the emitter circuit of Q_2 and hence this circuit is a form of the emitter-follower amplifier. Consequently, the voltage gain of the Darlington-pair amplifier will be approximately 1, as is the case with the emitter-follower amplifier.

Example 15-13 For the circuit of Fig. 15-27*a*, the circuit and transistor parameters are R_B = 1,000,000 ohms, R_E = 10,000 ohms, R_o = 500 ohms, $\beta_1, \beta_2 = 80$. Assume $R_E \gg (r_{e\cdot 1} + r_{e\cdot 2})$. Find (*a*) beta for the Darlington pair, (*b*) input resistance, (*c*) approximate voltage gain.

GIVEN: R_B = 1,000,000 ohms R_E = 10,000 ohms R_o = 500 ohms
$\beta_1, \beta_2 = 80$ $R_E \gg (r_{e\cdot 1} + r_{e\cdot 2})$

FIND: (*a*) βDp (*b*) r_i (*c*) A_v

SOLUTION:

(*a*) $\beta Dp = \beta_1^2 = 80^2 = 6{,}400$

(*b*) $r_i = R_1 \| r_i B_{Q\cdot 1} = 1{,}000{,}000 \| 3{,}040{,}000 \cong 750{,}000$ ohms

$r_i B_{Q\cdot 1} \cong \beta Dp r_E \cong 6{,}400 \times 475 \cong 3{,}040{,}000$ ohms

$r_E = R_E \| R_o = 10{,}000 \| 500 \cong 475$ ohms

(*c*) $A_v \cong 1$

Examination of Example 15-13 and Fig. 15-27*a* indicates three important characteristics of the Darlington-pair amplifier: (1) a low-impedance load (500 ohms) is transformed to a high-impedance input (750,000 ohms), (2) the circuit uses very few components, (3) a high beta value is obtainable.

Example 15-14 The circuit of Fig. 15-28*b* illustrates a two-stage amplifier consisting of (1) a common-emitter stage using Q_1, and (2) a Darlington-pair stage using Q_2 and Q_3. (The Darlington-pair stage is the same as that in Example 15-13.) Find the approximate voltage gain for the amplifier of (*a*) Fig. 15-28*a* when switch *S* is open, (*b*) Fig. 15-28*a* when switch *S* is closed, (*c*) Fig. 15-28*b*.

GIVEN: Fig. 15-28

FIND: (*a*) A_v (*b*) A_v (*c*) A_v

SOLUTION:

(*a*)

$$A_v = \frac{r_o}{r_e} = \frac{13{,}000}{26} = 500$$

$$r_o = R_3 = 13{,}000 \text{ ohms}$$

$$r_e = \frac{25}{I_E} = \frac{25}{0.96} \cong 26 \text{ ohms}$$

$$I_E = \frac{V_{EE}}{R_2 + \dfrac{R_1}{\beta_1}} = \frac{15}{15{,}000 + \dfrac{50{,}000}{80}} \cong 0.96 \text{ ma}$$

(*b*)

$$A_v = \frac{r_o}{r_e} = \frac{480}{26} \cong 18.5$$

$$r_o = R_3 \| R_4 = \frac{13{,}000 \times 500}{13{,}000 + 500} \cong 480 \text{ ohms}$$

$r_e = 26$ ohms [same as in part (*a*)]

(*c*)

$$A_v = A_{v\cdot 1} A_{v\cdot Dp} \cong 490 \times 1 \cong 490$$

$$A_{v\cdot 1} = \frac{r_o}{r_e} = \frac{12{,}800}{26} \cong 490$$

$$r_o = R_3 \| r_{i\cdot Dp} = \frac{13{,}000 \times 750{,}000}{13{,}000 + 750{,}000} \cong 12{,}800 \text{ ohms}$$

$r_{i\cdot Dp} = 750{,}000$ ohms from (*b*) of Example 15-13

$r_e = 26$ ohms same as part (*a*)

$A_{v\cdot Dp} \cong 1$ from (*c*) of Example 15-13

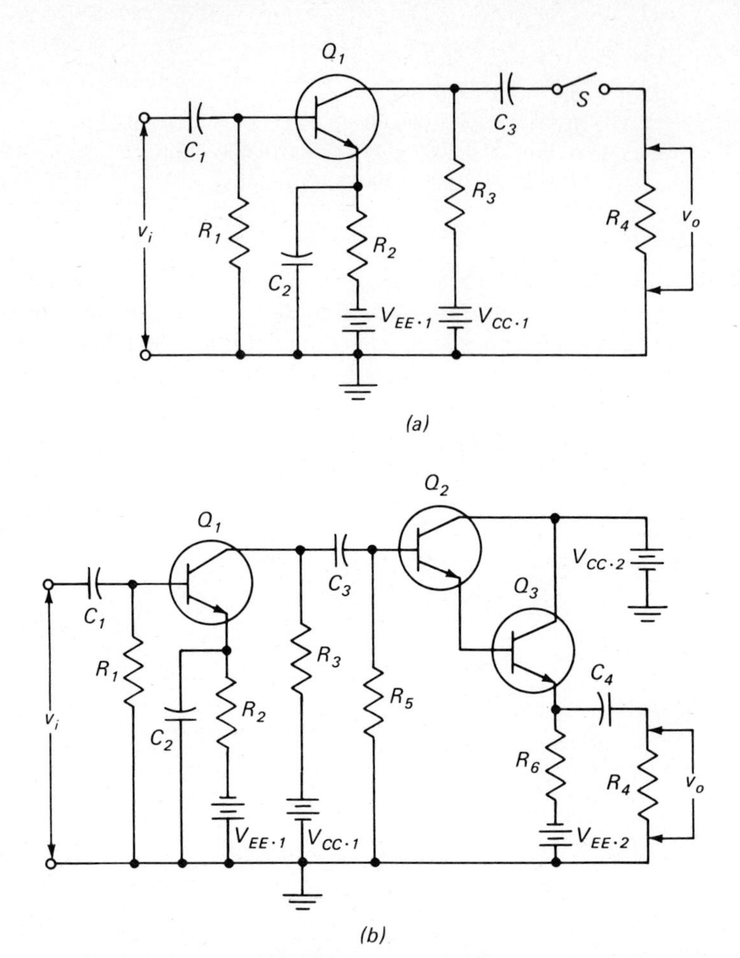

Fig. 15-28 (a) Single-stage common-emitter amplifier. (b) Two-stage amplifier consisting of: (1) common-emitter amplifier stage using Q_1, and (2) Darlington-pair stage using Q_2 and Q_3.

Example 15-14 shows that when a Darlington-pair stage is cascaded with a common-emitter amplifier stage it becomes possible to supply a low-impedance load and still retain the high gain that is characteristic of the common-emitter amplifier with a high-impedance load.

15-18 Amplifiers Using Field-effect Transistors

Characteristics of R-F Amplifiers. In the r-f stages of communication receivers, important characteristics are (1) noise figure, (2) input-signal range without overloading, (3) cross modulation, and (4) power gain, usually rated in that order. At higher frequencies, for example, the f-m broadcast band of 88 to 108 MHz, these parameters become of great importance. The selectivity of a receiver is largely dependent upon the overall noise figure of its circuit. The range of the input signal that can be accommodated without overloading depends largely upon the active device used in the circuit. Cross modulation occurs when any undesired signals of frequencies within the passband of the circuit modulate the carrier wave of the desired signals. Such spurious signals may be produced by (1) third- and higher-order odd-harmonic signals present because of nonlinearity of the active device in the amplifier, and (2) cross talk from overlapping signals on adjacent frequency bands. Achieving the maximum possible power gain may sometimes be sacrificed in order to improve the noise figure, sensitivity, and cross-modulation characteristics.

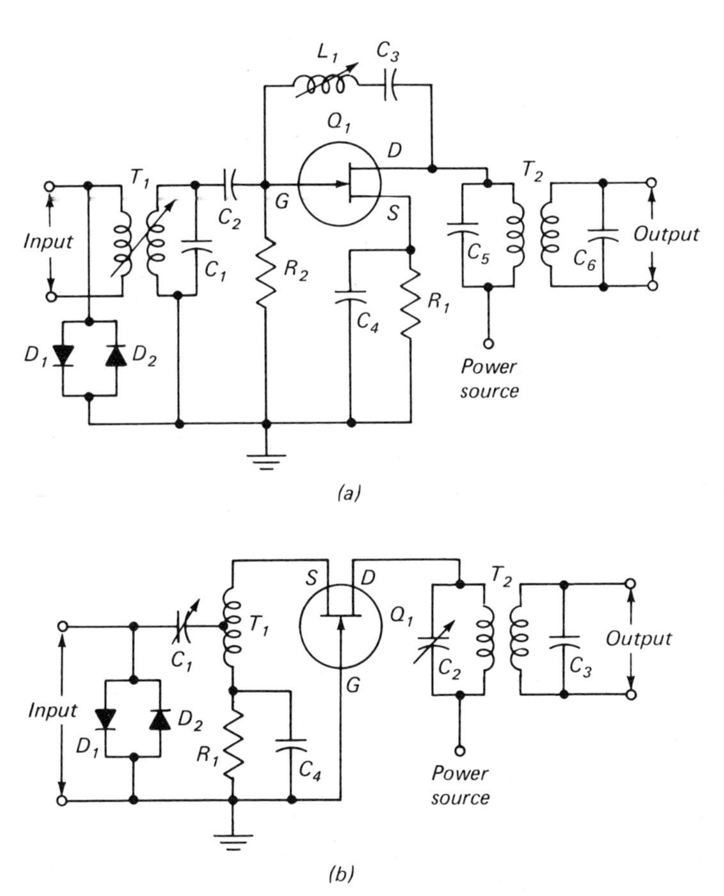

Fig. 15-29 JFET r-f amplifier. (a) Common-source circuit. (b) Common-gate circuit.

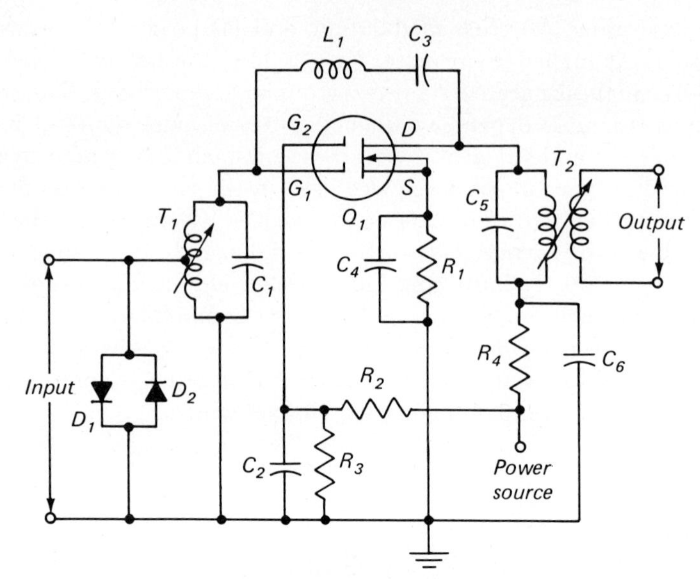

Fig. 15-30 R-f amplifier using a MOSFET.

Comparison of Active Devices. The outstanding characteristics of vacuum-tube high-frequency r-f amplifier circuits are (1) good large-input-signal performance, (2) good linearity, (3) moderate gain, (4) moderate noise figure, (5) high input impedance, (6) large physical size, (7) requires considerable amount of power, and (8) generates a considerable amount of heat.

The outstanding characteristics of the bipolar-transistor high-frequency r-f amplifier circuits are (1) relatively small input-signal-handling ability, (2) fair linearity, (3) high gain, (4) good noise figure, (5) low input impedance, (6) small size, (7) requires very little power, and (8) generates very little heat.

A high-frequency r-f amplifier circuit using a field-effect transistor combines the best characteristics of both the vacuum-tube and bipolar-transistor amplifier circuits. Thus the FET amplifier has (1) a good range of input-signal-handling ability, (2) good linearity, (3) low noise figure, (4) high gain, (5) high input impedance, (6) small size, (7) requires very little power, and (8) generates very little heat.

JFET Amplifier Circuits. An r-f amplifier using a JFET may be of (1) the common-source type as in Fig. 15-29*a*, or (2) the common-gate type as in Fig. 15-29*b*. In the circuit of Fig. 15-29*a*, L_1 and C_3 form a neutralizing circuit which must be adjusted to provide stabilized operation; the common-gate circuit does not require neutralization. The diodes D_1 and D_2, which are optional, are used to protect the transistor from damage due to abnormal amounts of voltage (and hence current) in the gate circuit which may destroy the gate-to-source junction. The components R_1C_4 establish the forward bias for the transistor.

MOSFET Amplifier Circuits. An r-f amplifier using a MOSFET may be of either the common-gate type, or the common-source type shown in Fig. 15-30. In this circuit L_1C_3, R_1C_4, and D_1D_2 perform the same functions as described in the preceding paragraph. Resistors R_2 and R_3 form a voltage divider to establish the bias voltage for G_2. The components R_4C_6 and C_2 provide filtering action to keep the r-f currents out of the power source.

QUESTIONS

1. Define (*a*) feedback, (*b*) positive feedback, (*c*) negative feedback, (*d*) feedback amplifier.
2. What are the advantages of negative feedback?
3. Where are positive-feedback circuits used?
4. Describe the principle of feedback amplifiers.
5. What are the limitations of feedback amplifiers?
6. How does the angle of lead or lag between the input voltage and feedback voltage affect the operation of a feedback amplifier?
7. Describe four types of distortion encountered in amplifiers.
8. Describe the effect of distortion in an amplifier on its output signal.
9. How does the addition of negative feedback to an amplifier circuit affect the distortion present in the amplifier?
10. How does negative feedback affect the stability of an amplifier?
11. Describe the basic voltage-controlled negative-feedback circuit and its principle of operation.
12. Describe the basic current-controlled negative-feedback circuit and its principle of operation.
13. Describe how negative feedback is obtained in the circuit of Fig. 15-6*a*.
14. Describe how negative feedback is obtained in the circuit of Fig. 15-6*b*.
15. Describe how negative feedback is obtained in the circuit of Fig. 15-7.
16. Describe how negative feedback is obtained in the circuit of Fig. 15-8*a*.
17. Describe how negative feedback is obtained in the circuit of Fig. 15-8*b*.
18. (*a*) What is a balanced-feedback amplifier? (*b*) What purpose does it serve?
19. Describe the function and principle of operation of C and R_1 in the amplifier circuit represented by Fig. 15-10.
20. Describe the function and principle of operation of L and R_2 in the amplifier circuit represented by Fig. 15-10.
21. (*a*) What is a tuned amplifier? (*b*) What purpose does it serve?
22. Describe the operation of the tuned-amplifier circuit of Fig. 15-11.
23. (*a*) What is meant by a wideband amplifier? (*b*) Name two applications that require a wide passband.
24. How is a wide passband achieved for (*a*) an *RC*-coupled amplifier? (*b*) An amplifier using tuned circuits?
25. What is meant by high-frequency compensation?
26. Describe (*a*) shunt compensation, (*b*) series compensation, (*c*) series-shunt compensation.

27. What is meant by low-frequency compensation?
28. How is low-frequency compensation achieved?
29. Describe four important requirements for a high-fidelity amplifier.
30. (*a*) Describe the emitter-follower amplifier. (*b*) What are its advantages?
31. (*a*) Describe the cathode-follower amplifier. (*b*) What are its advantages?
32. What is meant by a limiter circuit?
33. (*a*) Describe the basic series limiting circuit. (*b*) Explain how negative clipping is achieved. (*c*) Explain how positive clipping is achieved.
34. (*a*) Describe the basic shunt limiting circuit. (*b*) Explain how negative clipping is achieved. (*c*) Explain how positive clipping is achieved.
35. (*a*) What is meant by partial half-cycle limiting? (*b*) How is it achieved?
36. (*a*) What is meant by double limiting? (*b*) How is it achieved?
37. (*a*) What is a limiting amplifier? (*b*) What are its advantages?
38. For the limiting amplifier, explain how it is possible to achieve (*a*) positive clipping, (*b*) negative clipping, (*c*) double clipping.
39. Describe some uses of limiters.
40. (*a*) What is meant by a differential amplifier? (*b*) What are some of the features of this amplifier?
41. What are the advantages of the differential amplifier?
42. Describe the operation of the basic differential amplifier of Fig. 15-24.
43. Describe the operation of the constant-current source of Fig. 15-25*a*.
44. Describe the operation of the constant-current source of Fig. 15-25*b*.
45. (*a*) What is meant by a balanced differential amplifier? (*b*) Describe how a balanced condition can be achieved.
46. What is meant by an operational amplifier?
47. (*a*) What is meant by a Darlington-pair amplifier? (*b*) What are its advantages?
48. What are the advantages of FETs in high-frequency r-f amplifiers when compared with (*a*) vacuum tubes? (*b*) Bipolar transistors?
49. Describe the operation of the JFET amplifier circuit of (*a*) Fig. 15-29*a*, (*b*) Fig. 15-29*b*.
50. Describe the operation of the MOSFET amplifier circuit of Fig. 15-30.

PROBLEMS

1. The amplifier represented by Fig. 15-1*a* has a voltage gain of -120. What is the output-signal voltage if the input signal is 25 mv?
2. What gain must an amplifier represented by Fig. 15-1*a* have if it is desired to get a 4-volt output signal and the input signal available is 20 mv?
3. If a feedback loop having a 10 per cent feedback is added to the amplifier of Prob. 1, what is the magnitude of the output signal?
4. What is the voltage gain of the amplifier of Prob. 3?
5. For the amplifier of Prob. 1, what is the maximum amount of feedback that can be added if the decrease in gain for the amplifier is not to exceed 50 per cent?
6. For the amplifier of Prob. 1, what is the maximum amount of feedback that can be added if the decrease in gain for the amplifier is not to exceed 25 per cent?
7. What are the magnitude and polarity of the feedback voltage for the amplifier of Prob. 5 using (*a*) Eq. (15-5)? (*b*) Eq. (15-5*a*)?
8. What are the magnitude and polarity of the feedback voltage for the amplifier of Prob. 6 using (*a*) Eq. (15-5)? (*b*) Eq. (15-5*a*)?

9. What value of input voltage must the amplifier of Prob. 5 have in order to get the same output voltage as in Prob. 1?
10. What value of input voltage must the amplifier of Prob. 6 have in order to get the same output voltage as in Prob. 1?
11. If the amplifier of Prob. 1 has 4 per cent distortion, what is (*a*) the output-signal voltage? (*b*) The distortion voltage? (*c*) The output voltage?
12. A 200-mv signal is applied to an amplifier that has a gain of -100 and 6 per cent distortion. What is (*a*) the output-signal voltage? (*b*) The distortion voltage? (*c*) The output voltage?
13. If a feedback loop with $B = 0.1$ (as in Prob. 5) is added to the amplifier of Prob. 11, what is (*a*) the output-signal voltage? (*b*) The distortion voltage? (*c*) The output voltage?
14. If a feedback loop with $B = 0.02$ is added to the amplifier of Prob. 12, what is (*a*) the output-signal voltage? (*b*) The distortion voltage? (*c*) The output voltage?
15. If the input voltage for the amplifier of Prob. 13 is increased enough to produce the same output-signal voltage as without feedback, what is (*a*) the new input-signal voltage? (*b*) The output-signal voltage? (*c*) The distortion voltage? (*d*) The output voltage?
16. If the input voltage for the amplifier of Prob. 14 is increased enough to produce the same output-signal voltage as without feedback, what is (*a*) the new input-signal voltage? (*b*) The output-signal voltage? (*c*) The distortion voltage? (*d*) The output voltage?
17. An amplifier circuit having a voltage amplification of -100 employs a negative-feedback circuit with a feedback factor of 19. (*a*) What is the overall voltage amplification of the circuit as determined by use of Eq. (15-3*a*)? (*b*) What is the approximate overall voltage amplification of the circuit as determined by use of Eq. (15-3*c*)?
18. A certain a-f amplifier is designed to produce an overall voltage amplification of $-4{,}500$ with an input signal of 5 mv. The distortion in the amplifier is 5 per cent. (*a*) What is the output-signal voltage? (*b*) What is the distortion in volts? (*c*) If a feedback circuit with $B = 0.01$ is added, what is the value of the feedback factor? (*d*) What is the output-signal voltage with feedback added? (*e*) What is the distortion voltage with feedback added? (*f*) With this feedback, what value of input signal will be required in order to restore the output to its original value? (*g*) What is the distortion in volts with feedback and the increased input signal as found in part (*f*)?
19. The voltage gain of a certain amplifier circuit is -160 at 1,000 Hz, but drops to -20 at 30 Hz. (*a*) What is the change in gain expressed in decibels? (*b*) If a feedback circuit with $B = 0.10$ is added, what is the decibel change in gain between 1,000 and 30 Hz?
20. For the amplifier of Prob. 18, what is the change in decibels when the feedback circuit is added?
21. An amplifier represented by Fig. 15-4 has the following parameters: $e_i = 50$ mv, $A = -200$, $R_1 = 200{,}000$ ohms, and $R_f = 10{,}000$ ohms. What is (*a*) the value of B? (*b*) The gain of the amplifier? (*c*) The output voltage?
22. It is desired to change the feedback of the amplifier of Prob. 21 so that the voltage gain of the amplifier with feedback is -100. Keeping R_f at 10,000 ohms, what value is needed for R_1?
23. The parameters for an amplifier circuit similar to Fig. 15-6*a* are $R_1 = 1{,}000{,}000$

ohms, R_2 = 8,200,000 ohms, R_3 = 6,800 ohms, C_1 = 0.001 μf, β = 80, and V_{CC} = 24 volts. What is (*a*) the voltage gain without feedback? (*b*) The feedback factor with a 1,000-Hz signal? (*c*) The voltage gain with feedback at 1,000 Hz? (*d*) The feedback factor with a 20-Hz signal? (*e*) The voltage gain with feedback at 20 Hz?

24. The parameters for an amplifier circuit similar to Fig. 15-6*a* are R_1 = 390,000 ohms, R_2 = 6,800,000 ohms, R_3 = 4,700 ohms, C_1 = 500 pf, β = 100, and V_{CC} = 12 volts. What is (*a*) the voltage gain without feedback? (*b*) The feedback factor with a 1,000-Hz signal? (*c*) The voltage gain with feedback at 1,000 Hz? (*d*) The feedback factor with a 20-Hz signal? (*e*) The voltage gain with feedback at 20 Hz?

25. The parameters for an amplifier circuit similar to Fig. 15-6*b* are R_1 = 1,200,000 ohms, R_2 = 8,200 ohms, β = 80, and V_{CC} = 24 volts. What is (*a*) the voltage gain without feedback? (*b*) The feedback factor? (*c*) The voltage gain with feedback?

26. The parameters for an amplifier circuit similar to Fig. 15-6*b* are R_1 = 680,000 ohms, R_2 = 8,200 ohms, β = 80, and V_{CC} = 24 volts. What is (*a*) the voltage gain without feedback? (*b*) The feedback factor? (*c*) The voltage gain with feedback?

27. The parameters for an amplifier circuit similar to Fig. 15-7 are R_1 = 820,000 ohms, R_2 = 6,800 ohms, R_E = 100 ohms, β = 100, V_{CC} = 10 volts. What is (*a*) the voltage gain of the circuit if R_E is effectively bypassed by C_E? (*b*) The feedback factor without C_E? (*c*) The voltage gain without C_E?

28. The parameters for an amplifier circuit similar to Fig. 15-7 are R_1 = 820,000 ohms, R_2 = 6,800 ohms, R_E = 50 ohms, β = 100, V_{CC} = 10 volts. What is (*a*) the voltage gain of the circuit if R_E is effectively bypassed by C_E? (*b*) The feedback factor without C_E? (*c*) The voltage gain without C_E?

29. The resistors of the voltage-divider network of a feedback circuit similar to Fig. 15-8*a* are R_1 = 10,000 ohms and R_2 = 90,000 ohms. (*a*) What is the approximate value of B? (*b*) If the voltage amplification of the circuit is −160 without feedback, what is its gain with feedback? (*c*) What is the approximate voltage gain of the circuit with feedback using Eq. (15-3*c*)?

30. It is desired to add a feedback circuit, similar to Fig. 15-8*a*, to an output tube being worked through a 7,000-ohm load. (*a*) What minimum resistance should the feedback network have if the ratio of feedback circuit impedance to load impedance is to be approximately 20 to 1? (*b*) If a feedback factor of 3 is desired, what standard resistance values should the two resistors have? (Assume A = −20.) (*c*) What standard value of capacitor is recommended for C_3 if $(R_1 + R_2)$ should be approximately 20 times the reactance of the capacitor at a minimum frequency of 50 Hz?

31. The parameters for an amplifier circuit similar to Fig. 15-8*b* are R_1 = 150 ohms, R_2 = 2,500 ohms, r_p = 10,000 ohms, and the cathode current is 54 ma. What is (*a*) the feedback decimal portion B? (*b*) The grid-to-cathode voltage?

32. An amplifier circuit similar to Fig. 15-8*b* has a cathode current of 53 ma and a plate resistance of 11,000 ohms. What is (*a*) the resistance required at R_1 if the grid bias is to be 3.6 volts? (*b*) The feedback decimal value B?

33. The parameters for a tuned-amplifier circuit similar to Fig. 15-11 are C_3 = 125 pf, R_3 = 68,000 ohms, f_r = 455 kHz. What is (*a*) the value of inductance required for this circuit? (*b*) The value of Q? (*c*) The width of the bandpass?

34. The parameters for a tuned-amplifier circuit similar to Fig. 15-11 are L = 14.85 μh,

$C_3 = 14.85$ pf, $R_3 = 53{,}500$ ohms. What is (*a*) the resonant frequency? (*b*) The Q of the circuit? (*c*) The width of the bandpass?

35. It is desired to extend the upper frequency range f_h for the circuit of Fig. 15-13*a* to approximately 4 MHz. Parameters of the circuit are C_1, $C_3 = 60$ pf, and $C_2 = 1\ \mu$f. To satisfy this requirement, it is suggested that the value of the inductor L_1 produce resonance at a frequency of approximately 1.41 f_h. What value is recommended for L_1? (*Note:* C_2 is so large compared to C_1 and C_3 that it can be disregarded.)

36. Parameters for the circuit of Fig. 15-13*a* are C_1, $C_3 = 75$ pf, $C_2 = 1\ \mu$f, and $L_1 = 16\ \mu$h. (*a*) What is the approximate resonant frequency of C_1, C_3, L_1? (*b*) Using a ratio of $f_h/f_r = 0.707$, what is the approximate high-frequency limit of the circuit?

37. Parameters for the circuit of Fig. 15-13*b* are C_1, $C_3 = 60$ pf, $C_2 = 1\ \mu$f, and $L_2 = 53\ \mu$h. (*a*) What is the approximate resonant frequency of the tuned circuit formed by L_2, C_2, C_3? (*b*) Using a ratio of $f_h/f_r = 0.707$, what is the approximate high-frequency limit of the circuit?

38. What value is required for L_2 in Prob. 37 in order to raise the high-frequency range to 4 MHz? (*Note:* Use $f_r = 1.41\ f_h$.)

39. For the low-frequency compensating circuit of Fig. 15-14, $R_1 = 200$ ohms, $R_2 = 10{,}000$ ohms, $R_3 = 1{,}000$ ohms, and $C_1 = 5\ \mu$f. (*a*) What is the low-frequency cutoff of R_3C_1? (*b*) At what frequency will the effect of C_1 become negligible? (Assume a ratio of $X_C/R_3 = 10$.) (*c*) At what frequency will the effect of R_3 become negligible? (Assume a ratio of $R_3/X_C = 10$.)

40. For the circuit of Prob. 39, (*a*) is the impedance of R_3C_1 highest at 3.2, 32, or 320 Hz? (*b*) Is the voltage gain of the circuit highest at 3.2, 32, or 320 Hz? (*c*) What is the reason for the answer to part (*b*)?

41. The parameters of the emitter-follower circuit of Fig. 15-15 are $R_B = 50{,}000$ ohms, $R_E = 1{,}000$ ohms, $R_o = 200$ ohms, $\beta = 50$, and $V_{CC} = 12.5$ volts. Find (*a*) input resistance seen by the base of the transistor, (*b*) input resistance of the circuit, (*c*) voltage gain of the circuit, (*d*) power gain of the circuit.

42. The parameters of the emitter-follower circuit of Fig. 15-15 are $R_B = 68{,}000$ ohms, $R_E = 2{,}000$ ohms, $R_o = 680$ ohms, $\beta = 80$, and $V_{CC} = 14$ volts. Find (*a*) input resistance seen by the base of the transistor, (*b*) input resistance of the circuit, (*c*) voltage gain of the circuit, (*d*) output resistance of the amplifier seen by the load resistance R_o, (*e*) current gain of the circuit, (*f*) power gain of the circuit.

43. A certain triode is used in a cathode-follower circuit similar to Fig. 15-16*a*. The parameters of the circuit are $R_g = 220{,}000$ ohms, $R_k = 22{,}000$ ohms, $r_p = 6{,}500$ ohms, $\mu = 20$, $C_{gp} = 3.4$ pf, and $C_{gk} = 2.3$ pf. Find (*a*) voltage gain of the circuit, (*b*) approximate output resistance using Eq. (15-22*a*), (*c*) input resistance, (*d*) input capacitance.

44. If in the circuit of Prob. 43 R_k is changed to 2,200 ohms, find (*a*) voltage gain of the circuit, (*b*) output resistance using Eq. (15-22), (*c*) input resistance, (*d*) input capacitance.

45. The parameters of the cathode-follower circuit of Fig. 15-17 are $R_g = 220{,}000$ ohms, $R_2 = 700$ ohms, $R_3 = 1{,}500$ ohms, and $A_v = 0.834$. Find (*a*) input resistance, (*b*) grid bias when the plate current is 10 ma.

46. The parameters of the cathode-follower circuit of Fig. 15-17 are $R_g = 220{,}000$ ohms, $R_2 = 1{,}500$ ohms, $R_3 = 700$ ohms, and $A_v = 0.834$. Find (*a*) input resistance, (*b*) grid bias when the plate current is 6 ma.

47. For the differential amplifier of Fig. 15-24, the voltage gain of each section is 200, $v_{i\cdot 1}$ is 20 mv, and $v_{i\cdot 2}$ is 15 mv. Find the voltage between terminals (*a*) T_3 and ground, (*b*) T_4 and ground, (*c*) T_3 and T_4. (*d*) What are the polarities at terminals T_3 and T_4?

48. Repeat Prob. 47 for the condition when $v_{i\cdot 2}$ is increased to 25 mv.

49. A certain differential amplifier uses a combination of the circuits of Figs. 15-24 and 15-25*a*. The parameters of the circuit are R_1, R_4 = 50,000 ohms, R_2, R_3 = 5,000 ohms, R_5 = 10,000 ohms, β_1, β_2 = 100, V_{CC}, V_{EE} = 25 volts. Find approximate values for (*a*) $I_{R\cdot 5}$; (*b*) $I_{E\cdot 1}$, $I_{E\cdot 2}$; (*c*) $r_{r\cdot 1}$, $r_{e\cdot 2}$; (*d*) $A_{v\cdot 1}$, $A_{v\cdot 2}$; (*e*) $r_i(T_1T_5)$.

50. Repeat Prob. 49 for the condition when a 180-ohm swamping resistor is added to the emitter circuit of each transistor, and the collector circuit resistors (R_2,R_3) are changed to 10,000 ohms each.

51. The parameters for the differential amplifier with a single input (Fig. 15-31) are R_1 = 68,000 ohms, R_2, R_3 = 6,800 ohms, R_5 = 4,700 ohms, R_6, R_7 = 150 ohms, β_1, β_2 = 80, V_{CC}, V_{EE} = 14 volts, and v_i = 25 mv. Find approximate values for (*a*) $I_{R\cdot 5}$; (*b*) $I_{E\cdot 1}$, $I_{E\cdot 2}$; (*c*) $r_{e\cdot 1}$, $r_{e\cdot 2}$; (*d*) A_v; (*e*) v_o; (*f*) r_i. (*Note:* Use $30/I_E$ for r_e.)

52. The parameters for the differential amplifier with a single output (Fig. 15-32) are R_1, R_4 = 68,000 ohms, R_3 = 6,800 ohms, R_5 = 4,700 ohms, R_6, R_7 = 150 ohms, β_1, β_2 = 80, V_{CC}, V_{EE} = 14 volts, $v_{i\cdot 1}$ = 40 mv, and $v_{i\cdot 2}$ = 20 mv. Find approxi-

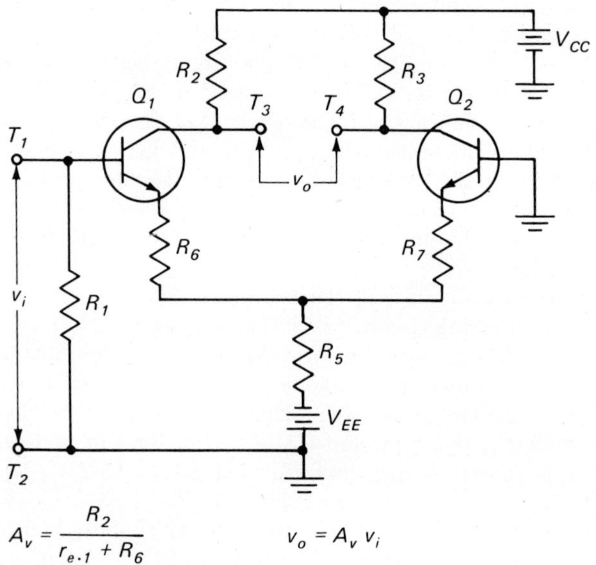

$$A_v = \frac{R_2}{r_{e\cdot 1} + R_6} \qquad v_o = A_v v_i$$

When all complementary parameters are equal,

$r_i\,(T_1T_2) = R_1 \| 2\beta_1\,(r_{e\cdot 1} + R_6)$

and when $R_1 \gg 2\beta_1\,(r_{e\cdot 1} + R_6)$,

$r_i\,(T_1T_2) \cong 2\beta_1\,(r_{e\cdot 1} + R_6)$

and when also $R_6 \gg r_{e\cdot 1}$,

$r_i\,(T_1T_2) \cong 2\beta_1\,R_6$

Fig. 15-31

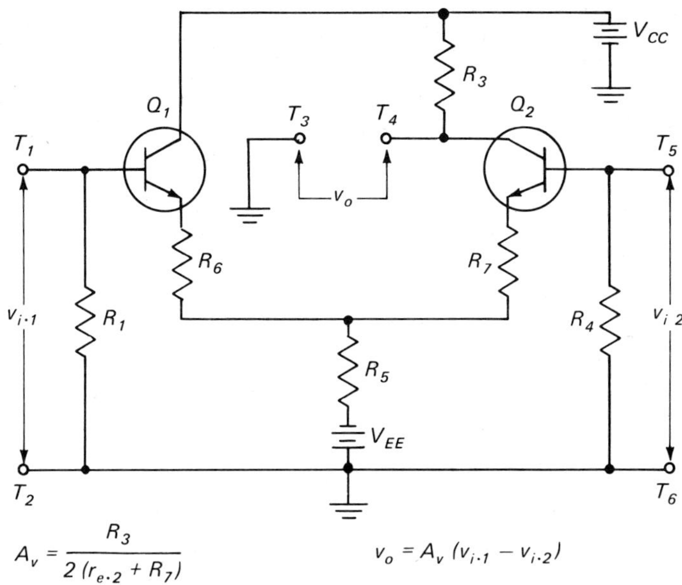

When all complementary parameters are equal,

$r_i(T_1T_2) = r_i(T_5T_6) = R_1 \| \beta_1 (r_{e\cdot1} + R_6)$; $r_i(T_1T_5) = 2r_i(T_1T_2)$

and when $R_1 \gg \beta_1 (r_{e\cdot1} + R_6)$,

$r_i(T_1T_2) = r_1(T_5T_6) \cong \beta_1 (r_{e\cdot1} + R_6)$; $r_i\,T_1T_5) \cong 2r_i(T_1T_2)$

and when also $R_6 \gg r_{e\cdot1}$,

$r_i(T_1T_2) = r_i(T_5T_6) \cong \beta_1 R_6$; $r_i(T_1T_5) \cong 2\beta_1 R_6$

Fig. 15-32

mate values for (*a*) $I_{R\cdot5}$; (*b*) $I_{E\cdot1}$, $I_{E\,2}$; (*c*) $r_{e\cdot1}$, $r_{e\cdot2}$; (*d*) v_o; (*e*) $r_i(T_1T_5)$. (*Note:* Use $30/I_E$ for $r_{re\cdot}$.)

53. The parameters of a Darlington-pair amplifier similar to Fig. 15-27 are $R_B =$ 50,000 ohms, $R_E =$ 15,000 ohms, $R_o =$ 1,000 ohms, $\beta_1 = 80$, $\beta_2 = 100$, and V_{CC}, $V_{EE} = 15$ volts. Find (*a*) beta for the Darlington pair, (*b*) a-c input resistance seen by the base of Q_1, (*c*) a-c input resistance of the amplifier, (*d*) approximate voltage gain of the amplifier.

54. The parameters of a Darlington-pair amplifier similar to Fig. 15-27 are $R_B =$ 500,000 ohms, $R_E =$ 25,000 ohms, $R_o =$ 250 ohms, β_1, $\beta_2 = 80$, and V_{CC}, $V_{EE} =$ 20 volts. Find (*a*) beta for the Darlington pair, (*b*) a-c input resistance seen by the base of Q_1, (*c*) a-c input resistance of the amplifier, (*d*) approximate voltage gain of the amplifier.

55. The parameters of a two-stage amplifier similar to the circuit of Fig. 15-28*b* are $R_1 =$ 100,000 ohms, $R_2 =$ 9,000 ohms, $R_3 =$ 10,000 ohms, $R_4 =$ 250 ohms, $R_5 =$ 500,000 ohms, $R_6 =$ 25,000 ohms, $\beta_1 = 100$, β_2, $\beta_3 = 80$, V_{CC}, $V_{EE} =$ 20 volts. (*Note 1:* Use $50/I_E$ for $r_{e\cdot1}$.) (*Note 2:* The Darlington-pair stage is the same as that in Prob. 54.) What is the approximate voltage gain of (*a*) Fig. 15-28*a* when switch *S* is open? (*b*) Fig. 15-28*a* when switch *S* is closed? (*c*) Fig. 15-28*b*?

Chapter 16

Basic Power-amplifier Circuits

Power-amplifier circuits are used to provide the power required to drive a current-operated load. This load is usually a transducer that converts electrical energy to some other form of energy. The transducer may be a loudspeaker in an audio circuit, a horizontal deflection yoke in a video circuit, a magnetic core in a computer-memory circuit, or a servomotor in an industrial control circuit. Because of the many applications of power amplifiers, the number of different circuits that are possible would be infinite. However, the principle of operation of each of the basic power-amplifier circuits is the same regardless of its application. Therefore, this chapter will describe only the basic amplifier circuits used in audio systems.

16-1 Basic Principles

Power Gain. All amplifiers provide a power gain that can be obtained with either high voltage or high power. If voltage gain is the prime factor, the amplifier is designed to produce high voltage gain, and it is called a *voltage amplifier*. If power gain is most important, the amplifier is designed to produce high power gain, and it is called a *power amplifier*. With a power amplifier the increased power must come from the d-c power source supplying the active device, or devices, used in the circuit. The purpose of the active device is to reproduce in the output circuit a large power signal that is a replica of the relatively small signal applied to its input circuit, differing only in the amplitude of power.

Active Devices. The active devices used in power-amplifier circuits may be (1) a bipolar transistor, (2) a unipolar transistor, (3) a triode vacuum tube, or (4) a pentode vacuum tube.

Modern design of power transistors has made this device highly adaptable for use in power-amplifier circuits. Because of its smaller size, greater mechanical ruggedness, higher power ratings, higher efficiency, longer-term stability, longer life, and the many circuit arrangements that can be used, the transistor has practically displaced the vacuum tube as the active device for low-frequency power amplifiers. Some practical circuit configurations and their advantages are: (1) *Direct coupling*—by eliminating the need of a coupling element it is possible to use optimum bias levels that can be set for minimum distortion, and cannot shift because of an overload. (2) *No output transformer*—

by eliminating the output transformer it is possible to obtain a wider frequency range, also eliminating any phase shift. (3) *Complementary symmetrical push-pull circuit*—this circuit eliminates the need of an input transformer or phase inverter.

Classifications. Power-amplifier circuits are used in both a-f and r-f circuits and are classified as *a-f power amplifiers* or *r-f power amplifiers.* Whereas the active device used in voltage-amplifier circuits is generally operated as Class A, the active device used in power-amplifier circuits may be operated as Class A, Class AB, Class B, or Class C. Power amplifiers can therefore be further subdivided according to the manner in which the active device is operated. Although all classes of operation may be used in both r-f and a-f power-amplifier circuits,

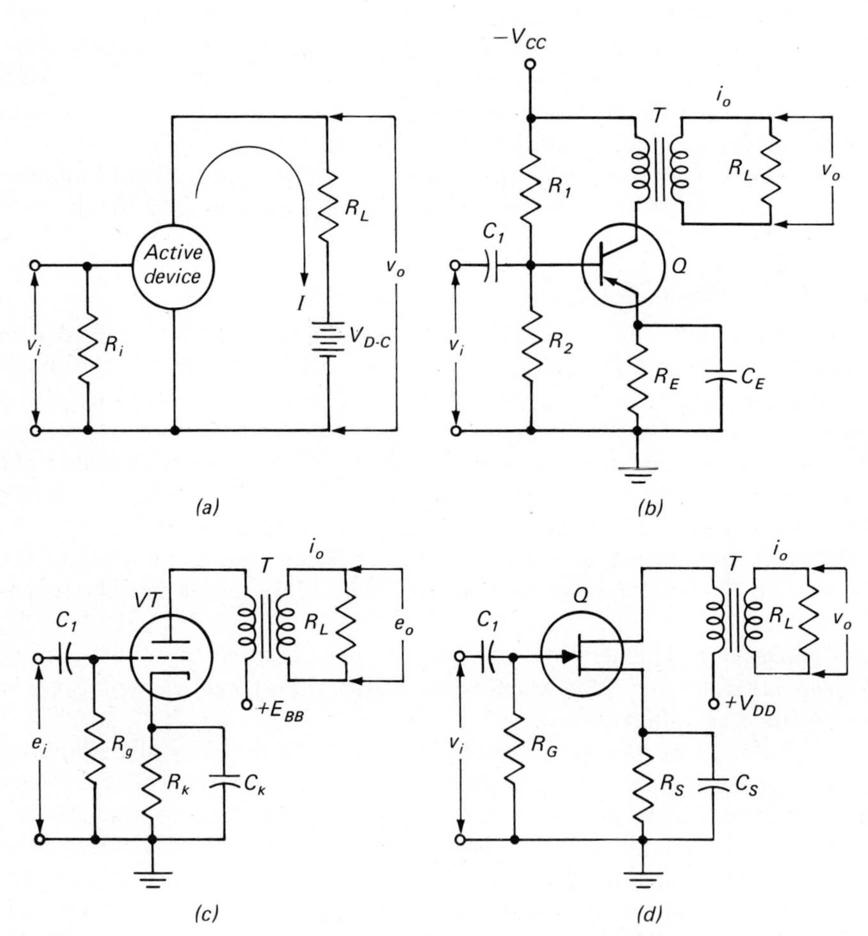

Fig. 16-1 Circuits for single-ended power amplifiers. (a) Basic amplifier circuit. (b) Bipolar-transistor power amplifier. (c) Electron-tube power amplifier. (d) Unipolar FET power amplifier.

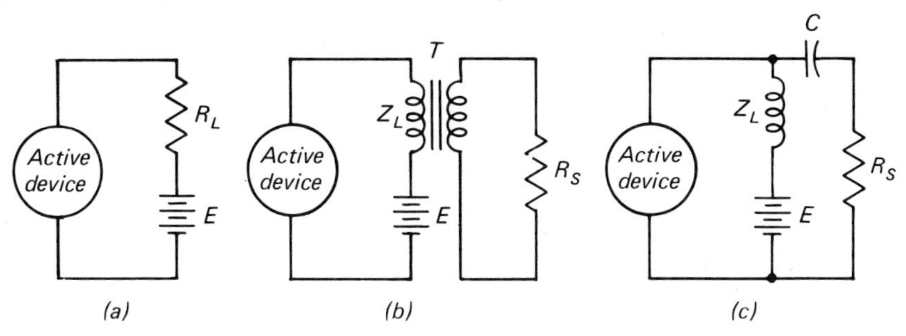

Fig. 16-2 Types of load coupling in power amplifiers. (a) Direct. (b) Transformer. (c) Shunt *LC*.

Class C is used primarily in r-f amplifiers. Depending upon the amount of power output required, the active devices used in power amplifiers may be operated (1) singly, (2) in parallel, or (3) in push-pull.

Single-ended Stage. The simplest form of power amplifier is the single-stage amplifier, commonly referred to as a *single-ended stage*. The basic single-ended stage may be represented as shown in Fig. 16-1*a*. A basic circuit for a single-ended stage using (1) a bipolar transistor is shown in Fig. 16-1*b*, (2) a triode electron tube in Fig. 16-1*c*, and (3) a unipolar transistor (FET) in Fig. 16-1*d*.

Any one of the single-stage amplifier circuits presented in Chap. 13 may be used as a power amplifier, provided (1) the active device is capable of handling the required current and voltage, and (2) the load is capable of receiving the relatively large amount of power that may be generated. Three general types of loading used are (1) direct coupling, (2) transformer coupling, and (3) shunt or *LC* coupling (see Fig. 16-2).

Power-amplifier circuits require a large driving current and/or voltage, which is usually supplied from the output of either a transistor or a vacuum-tube voltage amplifier. The operation of a single-ended-stage power amplifier is generally limited to Class A. If the active device is a transistor, the base current will be very small compared to the collector current in the load circuit. If the active device is a vacuum tube, the current taken from the driving source (current through R_g) is also very small compared to the plate current in the load circuit. The large amount of power delivered to the load is controlled by the active device and is provided by the external power source.

16-2 Power and Efficiency

Output-signal Power. When the signal output of an amplifier is an undistorted sine wave, the a-c power output can be found by use of the peak-to-peak values of the output-signal voltage v_o and the output-signal current i_o and the basic power equation $P = EI$. In the power equation the voltage and current must be the effective values; hence it is necessary to convert the peak-to-peak values of voltage and current to effective values. This is done by dividing the peak-to-peak values by the factor $2\sqrt{2}$. The a-c power output can then be expressed as

$$P_{o\cdot\text{a-c}} = \frac{(v_{o\cdot\max} - v_{o\cdot\min})}{2\sqrt{2}} \times \frac{(i_{o\cdot\max} - i_{o\cdot\min})}{2\sqrt{2}} \tag{16-1}$$

and

$$P_{o\cdot\text{a-c}} = \frac{(v_{o\cdot\max} - v_{o\cdot\min})(i_{o\cdot\max} - i_{o\cdot\min})}{8} \tag{16-1a}$$

D-C Power Input. The power delivered by the output circuit of the power amplifier is due to the power provided by the d-c power source, which is expressed by

$$P_{\text{d-c supply}} = P_{i\cdot\text{d-c}} = V_{\text{d-c supply}} I_{\text{d-c supply}} \tag{16-2}$$

However, in rating active devices used as power amplifiers the d-c power input is generally taken as the product of the quiescent values of current and voltage for the output element of the active devices.

For a common-emitter transistor amplifier

$$P_{i\cdot\text{a-d}} = V_{CE\cdot Q} I_{C\cdot Q} \tag{16-3}$$

For a common-cathode vacuum-tube amplifier

$$P_{i\cdot\text{a-d}} = E_{PC\cdot Q} I_{P\cdot Q} \tag{16-3a}$$

Efficiency. The purpose of a power amplifier is to produce at its output side an a-c signal of sufficient power to meet the demands of the load that the amplifier is to supply. This purpose is accomplished when the small amount of signal power available at the input side of the amplifier produces large variations in the a-c signal current and/or voltage at the output side of the amplifier. From Fig. 16-3, it can be observed that the amount of power released from the d-c supply is independent of the output-signal variations, since the average values of both the direct current alone and the direct current plus the a-c signal current are the same. The amount of power released from the d-c power source that is not converted to signal power generates heat. Not only is this loss

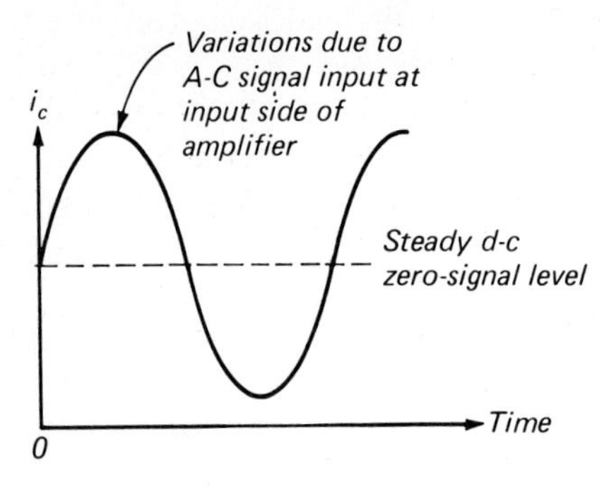

Fig. 16-3 Comparison of (1) the average value of a sine wave superimposed on a d-c level, with (2) the d-c level alone.

of power wasteful but the excessive heat generated may also be detrimental to the active device and/or circuit operation. The efficiency of a power amplifier is an important factor in the overall circuit performance. The efficiency of the active device is expressed by

$$\text{Efficiency of active device} = \frac{P_{o\cdot\text{a-c}}}{P_{i\cdot\text{a-d}}} \times 100 \tag{16-4}$$

The efficiency of the amplifier circuit may be called the *power-amplifier efficiency,* the *output-circuit efficiency,* the *collector-emitter-circuit efficiency,* or the *plate-cathode-circuit efficiency.*

$$\text{Efficiency of power amplifier} = \frac{P_{o\cdot\text{a-c}}}{P_{i\cdot\text{d-c}}} \times 100 \tag{16-4a}$$

Example 16-1 A common-emitter power-amplifier transistor is operated on its characteristic curve with quiescent values of $V_{CE\cdot Q} = 6$ volts and $I_{C\cdot Q} = 47.5$ ma. The output-signal voltage varies from 1 to 11 volts, and the output-signal current varies from 10 to 85 ma. Determine (*a*) the d-c input power to the transistor, (*b*) the a-c output power delivered to the load, and (*c*) the efficiency of the active device.

GIVEN: $V_{CE\cdot Q} = 6$ volts $\quad I_{C\cdot Q} = 47.5$ ma $\quad v_{ce\cdot\max} = 11$ volts $\quad v_{ce\cdot\min} = 1$ volt
$i_{c\cdot\max} = 85$ ma $\quad i_{c\cdot\min} = 10$ ma

FIND: (*a*) $P_{i\cdot\text{a-d}}$ (*b*) $P_{o\cdot\text{a-c}}$ (*c*) Efficiency

SOLUTION:

(*a*) $$P_{i\cdot\text{a-d}} = V_{CE\cdot Q}I_{C\cdot Q} = 6 \times 47.5 \times 10^{-3} = 0.285 \text{ watt}$$

(*b*) $$P_{o\cdot\text{a-c}} = \frac{(v_{ce\cdot\max} - v_{ce\cdot\min})(i_{c\cdot\max} - i_{c\cdot\min})}{8}$$

$$= \frac{(11-1)(85 \times 10^{-3} - 10 \times 10^{-3})}{8} = \frac{10 \times 75 \times 10^{-3}}{8} = 0.0938 \text{ watt}$$

(*c*) $$\text{Efficiency} = \frac{P_{o\cdot\text{a-c}}}{P_{i\cdot\text{a-d}}} \times 100 = \frac{0.0938}{0.285} \times 100 = 33\%$$

Example 16-2 A common-cathode power-amplifier tube is operated on its characteristic curve with quiescent values of $E_{P\cdot Q} = 250$ volts and $I_{P\cdot Q} = 65$ ma. The output-signal voltage varies from 110 to 370 volts, and the output-signal current varies from 10 to 120 ma. Determine (*a*) the d-c input power to the amplifier, (*b*) the a-c output power delivered to the load, and (*c*) the efficiency of the active device.

GIVEN: $E_{P\cdot Q} = 250$ volts $\quad I_{P\cdot Q} = 65$ ma $\quad e_{p\cdot\max} = 370$ volts $\quad e_{p\cdot\min} = 110$ volts
$i_{p\cdot\max} = 120$ ma $\quad i_{p\cdot\min} = 10$ ma

FIND: (*a*) $P_{i\cdot\text{a-d}}$ (*b*) $P_{o\cdot\text{a-c}}$ (*c*) Efficiency

SOLUTION:

(a) $$P_{i\cdot\text{a-d}} = E_{P\cdot Q} I_{P\cdot Q} = 250 \times 65 \times 10^{-3} = 16.25 \text{ watts}$$

(b) $$P_{o\cdot\text{a-c}} = \frac{(e_{p\cdot\max} - e_{p\cdot\min})(i_{p\cdot\max} - i_{p\cdot\min})}{8}$$

$$= \frac{(370 - 110)(120 \times 10^{-3} - 10 \times 10^{-3})}{8}$$

$$= \frac{260 \times 110 \times 10^{-3}}{8} \cong 3.58 \text{ watts}$$

(c) $$\text{Efficiency} = \frac{P_{o\cdot\text{a-c}}}{P_{i\cdot\text{a-d}}} \times 100 = \frac{3.58}{16.25} \times 100 \cong 22\%$$

16-3 Power Transistors

Types. Transistors used in power-amplifier circuits are usually of the bipolar junction type using either germanium or silicon for its basic elements. This type of transistor can be obtained to operate at voltages as high as 1,400 volts, for use in television deflection circuits; or to deliver 250 amp at a low voltage, for heavy-duty power service. The Darlington amplifier uses a compound-connected transistor that combines the driver and output stages in a single package.

Critical Parameters. The critical parameters of power transistors are (1) safe operating area, (2) junction temperature, and (3) maximum permissible power dissipation. In linear applications, such as amplifiers and most voltage regulators, a minimum current gain at maximum collector current and maximum gain-bandwidth-product frequency f_T are additional critical parameters to be considered. For switching applications, the additional critical parameters are saturation voltages $V_{CE\cdot\text{sat}}$ and $V_{BE\cdot\text{sat}}$, switching time, and minimum h_{FE}.

Safe Operating Area. The safe operating area of a transistor, abbreviated SOA, defines the limits of current and voltage to which a transistor can be subjected simultaneously without causing the device to fail. The maximum power that a transistor can safely dissipate at its collector is listed in transistor manuals for definite temperatures.

Since temperature is a function of energy rather than power, the SOA is dependent on the pulse width as well as voltage and current. Thus, a given transistor can handle higher currents when used in pulsed-circuit operation with fast switching times than it can when used in d-c operation.

Maximum Power-dissipation Curve. A constant-power-dissipation curve can be drawn across a family of $V_{CE}I_C$ characteristic curves to show the collector-voltage and collector-current intersections for its maximum rated power dissipation. The constant power dissipation for the transistor having the $V_{CE}I_C$ characteristics shown in Fig. 16-4 is indicated by the dashed-line curve which bends across the entire family of curves. The rated collector power dissipation for this transistor at an ambient temperature of 55°C is 30 watts. Hence, at any point of intersection of the constant-power-dissipation curve with the

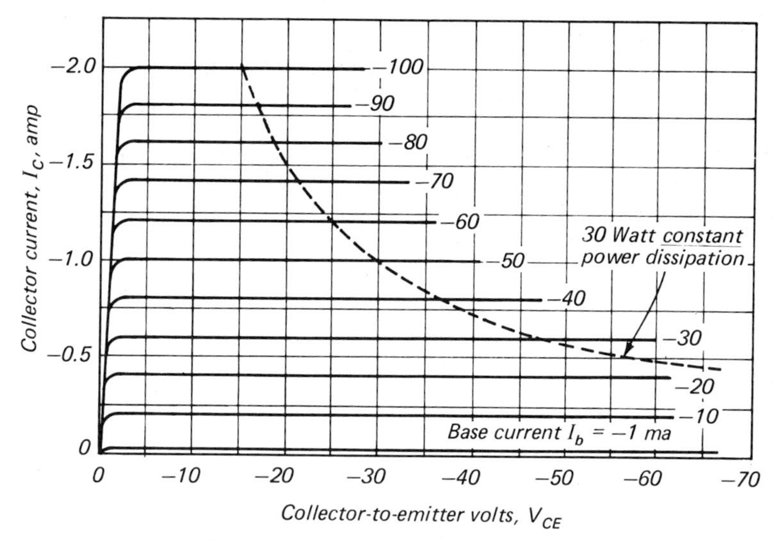

Fig. 16-4 Constant-collector power-dissipation curve for a transistor.

abscissa and ordinate, the product of V_{CE} and I_C as indicated by this intersection equals 30 watts. The area of the graph below and to the left of this curve encloses all points of operation which are within the collector power-dissipation rating of the transistor.

Junction Temperature. An excessively high temperature at the junction may cause a change in the distribution of the dopants at the junction, or it may melt the semiconductor material. The junction temperature is dependent on (1) transistor power dissipation, (2) transistor-case temperature, (3) the thermal resistance between the junction and the case, and (4) for pulsed operation, the pulse width and duty cycle.

Maximum Permissible Power Dissipation. The maximum permissible power dissipation is listed in transistor technical data as *transistor dissipation* P_T. This rating is given for the (1) ambient temperature T_A, (2) case temperature T_C, or (3) mounting-flange temperature T_{MF}, and is specified at a rated temperature stated as "up to 25°C" (or some other temperature value).

Maximum Power Dissipation versus Operating Temperature. When the operating temperature exceeds the highest value at which the maximum permissible power dissipation is rated, the permissible transistor power dissipation decreases linearly with an increase in the operating temperature up to the rated maximum operating temperature listed in the technical data under "Operating Temperature range," for example, (1) case operating temperature range $T_{C\cdot op} = -65$ to 200°C, or (2) junction operating temperature range $T_{j\cdot op} = -65$ to 175°C. At the maximum rated operating temperature, the permissible transistor power dissipation decreases to zero.

The permissible power dissipation at temperatures above that at which the maximum power dissipation can be tolerated can be found by (1) a dissipation derating factor, or (2) a dissipation derating graph (Fig. 16-5). The dissipation derating factor represents the decrease in permissible power dissipation per degree centigrade, and is abbreviated as watt/°C.

Example 16-3 A certain power transistor used in a high-fidelity amplifier has a maximum power-dissipation rating of 12.5 watts up to 81°C. The derating factor is 0.66 watt/°C, and the operating-temperature range at the junction is −65 to 100°C. Find the maximum permissible power dissipation for 90°C.

GIVEN: $P_T = 12.5$ watts up to 81°C Derating factor = 0.66 watt/°C
$T_{j\cdot op} = -65$ to 100°C

FIND: $P_{T\cdot 90}$

SOLUTION:

$$P_{T\cdot 90} = P_T - (T_D - T_R) \text{ derating factor}$$
$$= 12.5 - (90 - 81)0.66 = 12.5 - (9 \times 0.66) = 6.56 \text{ watts}$$

A derating graph (Fig. 16-5) can be constructed by connecting two points, such as *A* and *B*, with a straight line. For a given transistor, point *A* represents the highest temperature at which the maximum (100 per cent) permissible power dissipation is available, and point *B* represents the maximum permissible operating temperature listed in the technical data. When the data for points *A* and *B* are available, a derating graph can be drawn for any specific transistor.

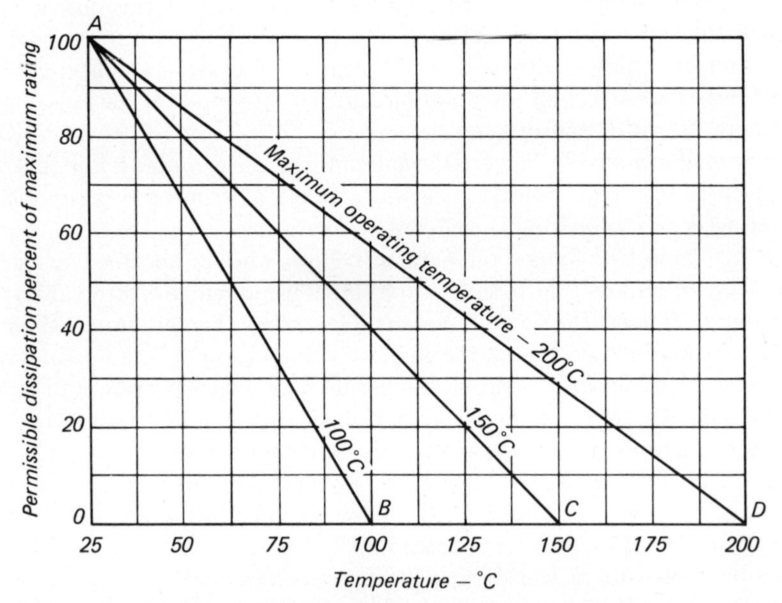

Fig. 16-5 Transistor dissipation derating graph.

Example 16-4 A certain power transistor used for direct operation from a line source in an audio-amplifier output stage has a maximum power-dissipation rating of 35 watts up to 25°C, with a temperature operating range at its junction of −65 to 200°C. Determine the maximum permissible power dissipation at a case temperature of (*a*) 90°C, (*b*) 180°C.

GIVEN: $P_T = 35$ watts up to 25°C $\quad T_{j\cdot op} = -65$ to 200°C

FIND: (*a*) $P_{T\cdot 90}$ $\quad$ (*b*) $P_{T\cdot 180}$

SOLUTION:

(*a*) A vertical line is projected from 90°C on the abscissa to the maximum junction operating temperature of 200°C (Fig. 16-5). A horizontal line drawn from this point of intersection establishes the permissible percentage of the maximum dissipation for a temperature of 90°C to be 63 per cent.

Thus,
$$P_{T\cdot 90} = \frac{\%\,P_T}{100} = \frac{63 \times 35}{100} = 22.05 \text{ watts}$$

(*b*) A vertical line is projected from the 180°C point on the abscissa to the 200°C curve. Projecting this point of intersection establishes the permissible percentage of the maximum dissipation for a temperature of 180°C to be 11 per cent.

Thus,
$$P_{T\cdot 180} = \frac{\%\,P_T}{100} = \frac{11 \times 35}{100} = 3.85 \text{ watts}$$

Heat Sink. All semiconductor devices are temperature-sensitive, and power transistors are sometimes called upon to dissipate more heat than they can normally handle safely. It is therefore necessary to provide some means of removing the excess heat from the power transistor so that the heat generated will not raise the junction temperature beyond its specified rating, or some predetermined safe value. This can be done by (1) using a fan to blow cool air across the power-handling components, and/or (2) providing additional heat-radiating surfaces at the power device. A *heat sink* is used to provide added surface area to the case of a transistor (see Fig. 16-6). The additional surface area permits the excess heat generated to be absorbed, by both convection and conduction, by the surrounding environment.

Germanium versus Silicon Power Transistors. The most important consideration in transistor operation is the junction temperature, which in effect determines the power dissipation. Many silicon transistors are made with junction operating temperatures as high as 200°C, while most germanium transistors have only 100°C as the upper limit of the junction operating-temperature range. The leakage current in silicon transistors is relatively low, often less than 1/1,000th that of germanium transistors. For high supply-voltage applications (above 50 volts) silicon is usually preferred. For low supply-voltage applications (below 12 volts) germanium is often more suitable because its saturation voltages are usually lower than those for silicon. The high saturation voltages of silicon may limit the voltage swing at low supply voltages. The low collector-to-emitter saturation voltage of germanium also

Fig. 16-6 Fin-type heat sink.

produces a lower loss and lower power dissipation at medium- and low-speed switching applications.

16-4 Power Tubes

Types. Triodes, pentodes, and beam-power tubes are used as power amplifiers, with the beam-power tube being the type most frequently used.

Critical Parameters. The critical parameters of power-amplifier tubes are (1) plate efficiency, (2) plate dissipation, and (3) power sensitivity.

Plate Efficiency. The ratio of the a-c power output to the product of the average values of plate voltage and plate current at maximum signal input is called the *plate efficiency.*

$$\text{Plate efficiency} = \frac{P_{o\cdot\text{a-c}}}{E_{P\cdot Q} I_{P\cdot Q}} \times 100 \tag{16-5}$$

Example 16-5 Determine the plate efficiency of a power-amplifier tube having an a-c power output of 3.85 watts, an operating plate voltage of 250 volts, and an operating plate current of 60 ma.

GIVEN: $E_{P\cdot Q} = 250$ volts $\quad I_{P\cdot Q} = 60$ ma $\quad P_{o\cdot\text{a-c}} = 3.85$ watts

FIND: Plate efficiency

SOLUTION:

$$\text{Plate efficiency} = \frac{P_{o\cdot\text{a-c}}}{E_{P\cdot Q} I_{P\cdot Q}} \times 100 = \frac{3.85 \times 100}{250 \times 0.06} = 25.6\%$$

Plate Dissipation. The heat given off at the plate of a tube as a result of electron bombardment is called the *plate dissipation,* and its symbol is P_p. This represents a loss in power and is equal to the difference between the power supplied to the plate of the tube and the a-c power delivered by the tube to its load.

$$P_p = P_{i\cdot\text{d-c}} - P_{o\cdot\text{a-c}} \tag{16-6}$$

Example 16-6 What is the plate dissipation of the power-amplifier tube used in Example 16-5?

GIVEN: $E_{P\cdot Q} = 250$ volts $\quad I_{P\cdot Q} = 60$ ma $\quad P_{o\cdot\text{a-c}} = 3.85$ watts

FIND: P_p

SOLUTION:

$$P_p = P_{i\text{-d-c}} - P_{o\text{-a-c}} = E_{P\cdot Q} I_{P\cdot Q} - P_{o\text{-a-c}} = (250 \times 0.06) - 3.85 = 11.15 \text{ watts}$$

Power Sensitivity. The ratio of the a-c power output to the square of the effective value of the input-signal voltage is called the *power sensitivity.* The basic unit of power sensitivity is the mho, but because of the low values of power sensitivity usually obtained, the micromho is commonly used.

$$\text{Power sensitivity} = \frac{P_{o\text{-a-c}}}{E_i^2} \times 10^6 \tag{16-7}$$

Note: E_i = effective value

Power sensitivity is ordinarily used in rating low-frequency amplifier tubes that are operated so that no current flows in the grid circuit. The power sensitivity of pentodes and beam power tubes is considerably greater than for triodes.

Example 16-7 Determine the power sensitivity of the tube used in Example 16-5 when the effective value of the input-signal voltage is 30.55 volts.

GIVEN: $P_{o\text{-a-c}} = 3.85$ watts $\quad E_i = 30.55$ volts

FIND: Power sensitivity

SOLUTION:

$$\text{Power sensitivity} = \frac{P_{o\text{-a-c}}}{E_i^2} = \frac{3.85}{(30.55)^2} \times 10^6 = 4{,}125\ \mu\text{mhos}$$

16-5 Class A Transistor Amplifier

Maximum Efficiency. Equation (16-4) represents the efficiency of the output circuit of a power amplifier. In a Class A transistor amplifier, maximum efficiency will be obtained when the maximum negative value of the varying components of the collector-to-emitter voltage V_{CE} and the collector current I_C produce the lowest possible values of collector-to emitter voltage v_{ce} and collector current i_c, theoretically zero. Under this condition, the maximum value of v_{ce} will be twice the value of the operating collector voltage $V_{CE\cdot Q}$, and the maximum value of i_c will be twice the value of the operating collector current $I_{C\cdot Q}$. The peak values of v_{ce} and i_c will thus be equal to $V_{CE\cdot Q}$ and $I_{C\cdot Q}$, respectively. As the a-c power output (for resistive loads) is equal to the product of the effective values of $v_{ce\cdot\max}$ and $i_{c\cdot\max}$, then the a-c power output for the condition of maximum efficiency may be expressed as

$$P_{o\text{-a-c}} = \frac{V_{CE\cdot Q}}{\sqrt{2}} \times \frac{I_{C\cdot Q}}{\sqrt{2}} = \frac{V_{CE\cdot Q} I_{C\cdot Q}}{2} \tag{16-8}$$

Substituting the values of $P_{o\text{-a-c}}$ of Eq. (16-8) in Eq. (16-4) shows that the efficiency of a Class A transistor cannot exceed 50 per cent. This is the theoretical maximum efficiency; however, in practical circuits the efficiency is much lower.

Load Lines. A load line can be constructed as shown in Fig. 16-7. A straight line is drawn from the collector-current ordinate when $V_{CE} = 0$ to the collector-to-emitter voltage abscissa when $I_C = 0$. The values of V_{CE} and I_C used in drawing the load line should not exceed the maximum ratings for the transistor used, and the load line should be to the left and below the constant-power-dissipation curve. It is thus possible to draw an infinite number of load lines that would conform to these conditions. However, the load line is also dependent on three circuit parameters, (1) the d-c operating point, (2) the supply voltage, and (3) the load resistance. These parameters are interrelated, and the selection of any two automatically determines the third, or

$$R_L = \frac{V_{CC} - V_{CE\cdot Q}}{I_{C\cdot Q}} \tag{16-9}$$

Maximum Power Output of a Class A Amplifier. In order to obtain the maximum power output, the load line should enclose as large an area as possible within the maximum current, voltage, and power-dissipation ratings of the transistor (see cross-hatched triangle in Fig. 16-7). Theoretically, maximum

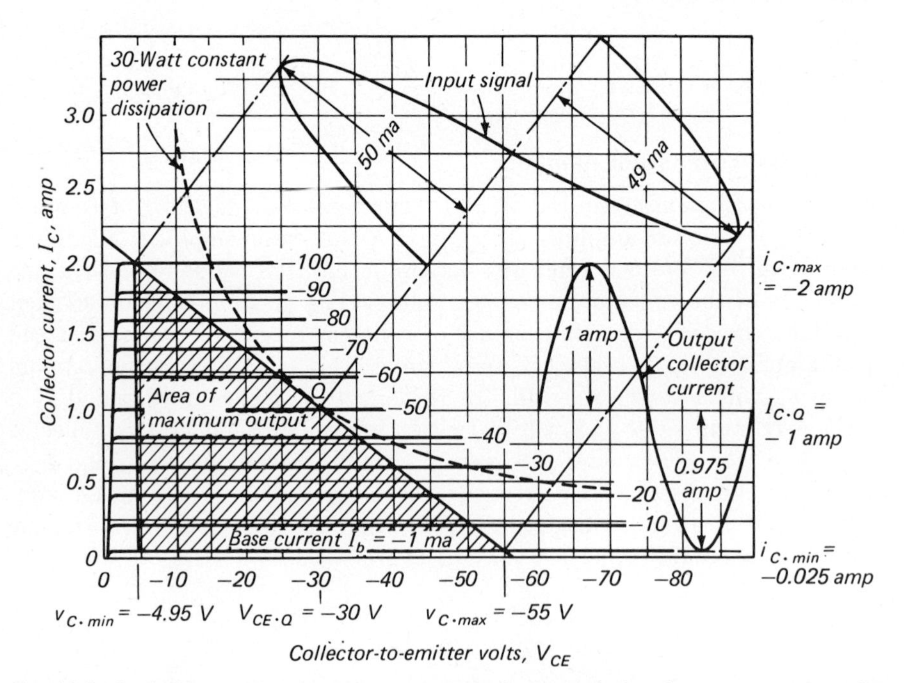

Fig. 16-7 Load line and dynamic operating characteristics for a common-emitter transistor amplifier (using curves for an ideal transistor).

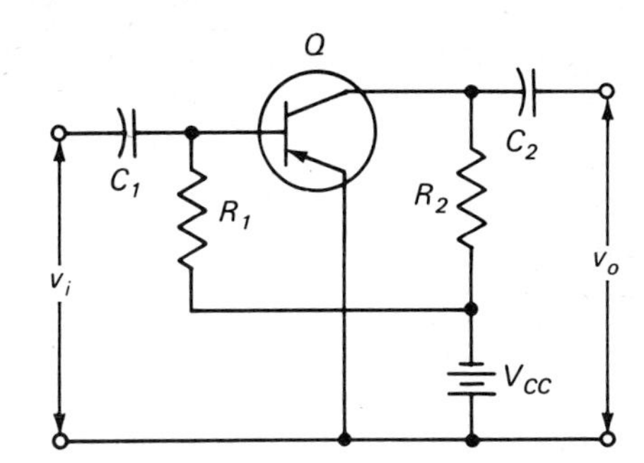

Fig. 16-8 Class A single-ended transistor power amplifier.

power output would be obtained for a load whose load line is identical to the constant-power-dissipation curve. Such a load line is curved and is therefore impractical to use, as the load line for an amplifier must be straight. A practical load line for maximum-power output may be obtained by drawing a straight line tangent to the constant-power-dissipation curve. Since this curve is nonlinear, several load lines may be drawn that will produce a practical maximum power output. The signal requirement of the circuit is the determining factor as to which load line to use.

In order to produce maximum power output with maximum signal input for a Class A amplifier it is necessary that the load line be drawn tangent to the constant-power-dissipation curve at the point where the mid-value of bias current crosses the curve (Fig. 16-7). In this case the d-c operating point is midway between the extreme limits of the base currents (−1 and −100 ma) or at the point where $V_{CE \cdot Q} = -30$ volts and $I_{C \cdot Q} = -1.0$ amp. With an input signal having a peak value of 50 ma, the output current varies uniformly from $i_{c \cdot \min} = -0.025$ amp to $i_{c \cdot \max} = -2.0$ amp. By using this graph it is possible to determine (1) the maximum input signal, (2) the value of the d-c biasing resistors, (3) the load resistance, and (4) the efficiency.

Example 16-8 Using the load line in Fig. 16-7 and the circuit of Fig. 16-8, determine (*a*) the required value of load resistance, and (*b*) the efficiency of the active device with an input signal having a peak value of 50 ma.

GIVEN: Load line (Fig. 16-7) Circuit (Fig. 16-8) $i_{b \cdot \max} = 50$ ma

FIND: (*a*) R_2 (*b*) Efficiency

SOLUTION:

(*a*) From Fig. 16-7

$$V_{CC} = -56 \text{ volts} \qquad V_{CE \cdot Q} = -30 \text{ volts} \qquad I_{C \cdot Q} = -1.0 \text{ amp}$$

$$R_2 = \frac{V_{CC} - V_{CE \cdot Q}}{I_{C \cdot Q}} = \frac{56 - 30}{1.0} = 26 \text{ ohms}$$

(*b*) From Fig. 16-7

$$V_{CE\cdot(p\text{-}p)} = -4.95 \text{ to } -55 \text{ volts} \qquad I_{C\cdot(p\text{-}p)} = -0.025 \text{ to } 2.0 \text{ amp}$$

$$P_{i\cdot a\text{-}d} = V_{CE\cdot Q} I_{C\cdot Q} = 30 \times 1.0 = 30 \text{ watts}$$

$$P_{o\cdot a\text{-}c} = \frac{V_{CE\cdot(p\text{-}p)} I_{C\cdot(p\text{-}p)}}{8} = \frac{(55 - 4.95)(2.0 - 0.025)}{8} \cong 12.4 \text{ watts}$$

$$\text{Efficiency} = \frac{P_{o\cdot a\text{-}c} \times 100}{P_{i\cdot a\text{-}d}} = \frac{12.4 \times 100}{30} \cong 41\%$$

Distortion. The family of transistor collector characteristic curves of Fig. 16-7 does not represent the characteristics of an actual transistor but is a modification of the collector-emitter characteristic curves for an actual transistor as shown in Fig. 16-9. The spacings between the collector-current—collector-emitter voltage curves of Fig. 16-7 were made the same for equal changes in base current for clarity in drawing and explanation of load lines. In the characteristic curves of Fig. 16-9 the spacing between the collector-current—collector-emitter voltage curves is not equal, as it becomes smaller for the higher values of base current. This is indicated in Fig. 16-9 by the unequal distances between AQ and QB for equal changes in base current. This crowding effect at the higher

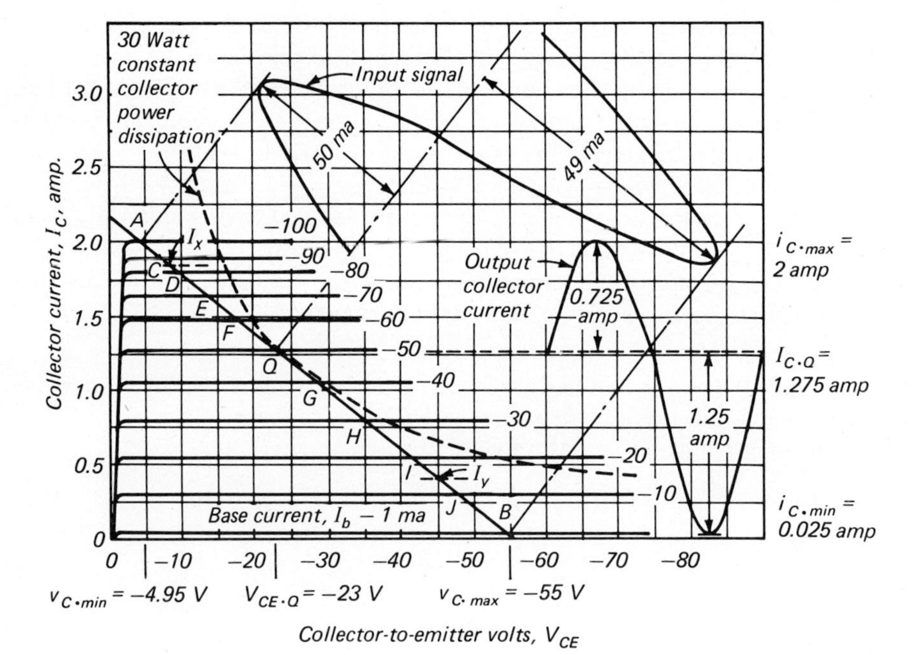

Fig. 16-9 Load line for the maximum power output plotted on the dynamic operating characteristics for a common-emitter transistor amplifier (using curves for an actual transistor).

values of base current, which is typical of most transistors, results in harmonic distortion (Fig. 16-11*a*).

Effects of Harmonic Distortion. The effects of harmonic distortion on a sinusoidal wave are shown in Fig. 16-10. A second harmonic tends to change the fundamental wave to a sawtooth wave. This type of distortion is characteristic of all even harmonics, as the addition of the fundamental and all its even harmonics will tend to produce a sawtooth wave. A third harmonic tends to change the fundamental wave to a square wave. This type of distortion is characteristic of all odd harmonics, since the addition of the fundamental and all its odd harmonics will tend to produce a square wave. The distortion produced by both the second and third harmonics results in a wave that is somewhat similar to the distorted sawtooth wave produced by the second harmonic only. However, the shape of this wave is also dependent upon the ratio of the maximum values of the second and third harmonics. For the resultant wave, shown in Fig. 16-10*c*, the ratio is 1.

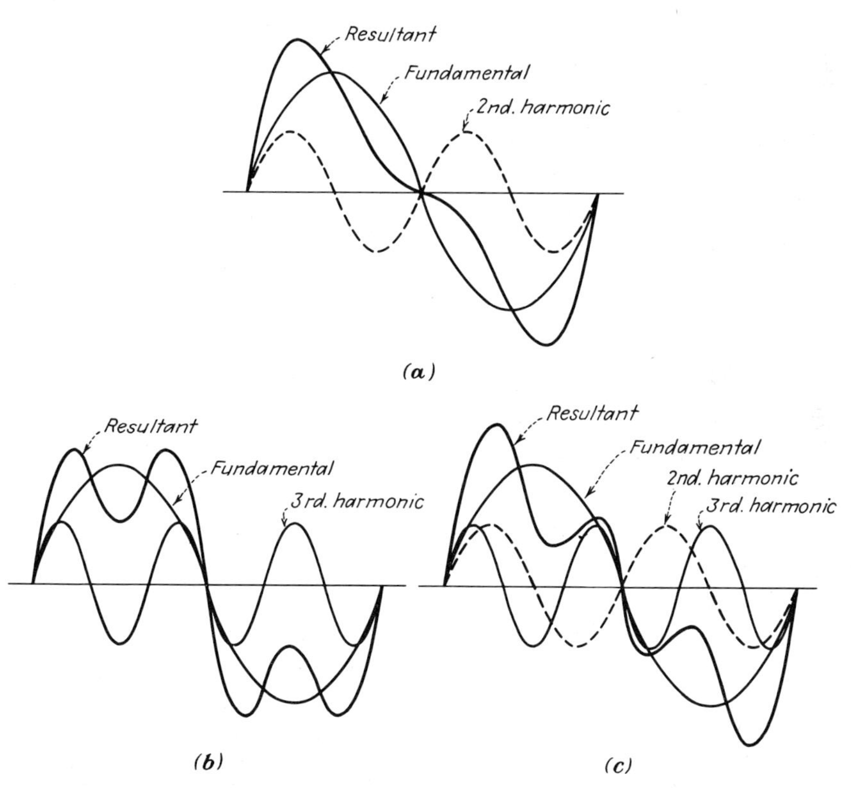

Fig. 16-10 Effects of harmonics on a sine wave. (a) Effect of second harmonic. (b) Effect of third harmonic. (c) Effect of a second and third harmonic.

Calculation of the Amount of Distortion. The per cent of harmonic distortion can be calculated from the collector-emitter characteristic curves for a given operating condition by use of Eqs. (16-10), (16-11), and (16-12). As the derivations of these equations are rather complex, they are omitted in this text.

% second-harmonic distortion

$$= \frac{i_{c\cdot\max} + i_{c\cdot\min} - 2I_{C\cdot Q}}{i_{c\cdot\max} - i_{c\cdot\min} + 1.41(I_x - I_y)} \times 100 \qquad (16\text{-}10)$$

% third-harmonic distortion

$$= \frac{i_{c\cdot\max} - i_{c\cdot\min} - 1.41(I_x - I_y)}{i_{c\cdot\max} - i_{c\cdot\min} + 1.41(I_x - I_y)} \times 100 \qquad (16\text{-}11)$$

where I_x = collector current at $1.707 I_{B\cdot Q}$
I_y = collector current at $0.293 I_{B\cdot Q}$

% total (second + third) harmonic distortion

$$= \sqrt{(\%\ \text{second-harmonic distortion})^2 + (\%\ \text{third-harmonic distortion})^2} \qquad (16\text{-}12)$$

Example 16-9 The transistor whose collector characteristics are shown in Fig. 16-9 is to be operated along a load line that will produce maximum signal output with maximum signal input with a zero-signal base current of 50 ma. Find the value of the required load resistance.

GIVEN: Collector characteristics (Fig. 16-9) $\quad I_{B\cdot Q} = 50$ ma

FIND: R_2 (Fig. 16.8)

SOLUTION: Draw a straight line from the collector-current ordinate when $V_{CE} = 0$ to the collector-emitter–voltage abscissa when $I_C = 0$ that is tangent to the constant-power-dissipation curve at the point where this curve crosses the collector characteristic curve for $I_b = 50$ ma. This load line indicates that $V_{CE} = -56$ volts, $V_{CE\cdot Q} = -23$ volts, and $I_{C\cdot Q} \cong -1.275$ amp.

$$R_2 = \frac{V_{CE} - V_{CE\cdot Q}}{I_{C\cdot Q}} = \frac{56 - 23}{1.275} \cong 26 \text{ ohms}$$

Example 16-10 The transistor of Example 16-9 is to be operated with a signal input whose peak value is 50 ma. Find the per cent of (*a*) second-harmonic distortion, (*b*) third-harmonic distortion, (*c*) total second- and third-harmonic distortion.

GIVEN: Collector characteristics (Fig. 16-9) $\quad i_{b\cdot\max} = 50$ ma

FIND: (*a*) Per cent second-harmonic distortion (*b*) Per cent third-harmonic distortion (*c*) Per cent total second- and third-harmonic distortion

SOLUTION:

From Fig. 16-9 $\quad i_{c\cdot\max} \cong 2.0$ amp $\quad i_{c\cdot\min} \cong 0.025$ amp $\quad I_{C\cdot Q} \cong 1.275$ amp
$I_x \cong 1.85$ amp $\quad I_y \cong 0.35$ amp

$$(a)\quad \text{Second-harmonic distortion} = \frac{i_{c\cdot\max} + i_{c\cdot\min} - 2I_{C\cdot Q}}{i_{c\cdot\max} - i_{c\cdot\min} + 1.41(I_x - I_y)} \times 100$$

$$= \frac{2.0 + 0.025 - 2 \times 1.275}{2.0 - 0.025 + 1.41(1.85 - 0.35)} \times 100 \cong 12.9\%$$

$$(b)\quad \text{Third-harmonic distortion} = \frac{i_{c\cdot\max} - i_{c\cdot\min} - 1.41(I_x - I_y)}{i_{c\cdot\max} - i_{c\cdot\min} + 1.41(I_x - I_y)} \times 100$$

$$= \frac{2.0 - 0.025 - 1.41(1.85 - 0.35)}{2.0 - 0.025 + 1.41(1.85 - 0.35)} \times 100 \cong 3.4\%$$

(*c*) Total (second- + third-harmonic distortion)

$$= \sqrt{(\%\ \text{second-harmonic})^2 + (\%\ \text{third-harmonic})^2}$$

$$= \sqrt{12.9^2 + 3.4^2} \cong 13.3\%$$

The equations used for calculating harmonic distortion assume that the source resistance is negligible. The resistance of the signal source in a transistor amplifier circuit is not negligible, and its value cannot be ignored. In a transistor circuit, the base-emitter voltage is not zero and the effect of this voltage together with the source resistance is to increase the overall harmonic distortion approximately 1 per cent more than the amount obtained by use of the equations. Thus, the transistor of Example 16-10 would have a total harmonic distortion of approximately 14.3 per cent.

Another type of distortion is caused by the effect of the variations in the output circuit on the input circuit of an amplifier when the input resistance is high compared to its source impedance. In this type of amplifier circuit, the input resistance decreases in the high region of the collector a-c cycle, thus causing the amplitude of the input signal to increase. In the low region of the collector a-c cycle, the input resistance increases, thus causing the amplitude of the input signal to decrease. This type of nonlinear distortion is shown in Fig. 16-11*b*. Since the effect of this type of distortion is opposite to that caused by harmonic distortion (Fig. 16-11*a*), it is possible to counteract one with the other by adjusting the source impedance.

16-6 Maximum Efficiency of Class A Vacuum-tube Amplifiers

In a similar manner to the explanation for a Class A operated transistor, it can be shown that the power output for maximum efficiency of a Class A operated vacuum tube may be expressed as

$$P_{o\cdot\text{a-c}} = \frac{E_P}{\sqrt{2}} \times \frac{I_P}{\sqrt{2}} = \frac{E_P I_P}{2} \tag{16-13}$$

and that this theoretical efficiency cannot exceed 50 per cent. As with transistors, the actual efficiency in practical vacuum-tube amplifier circuits is much lower than 50 per cent.

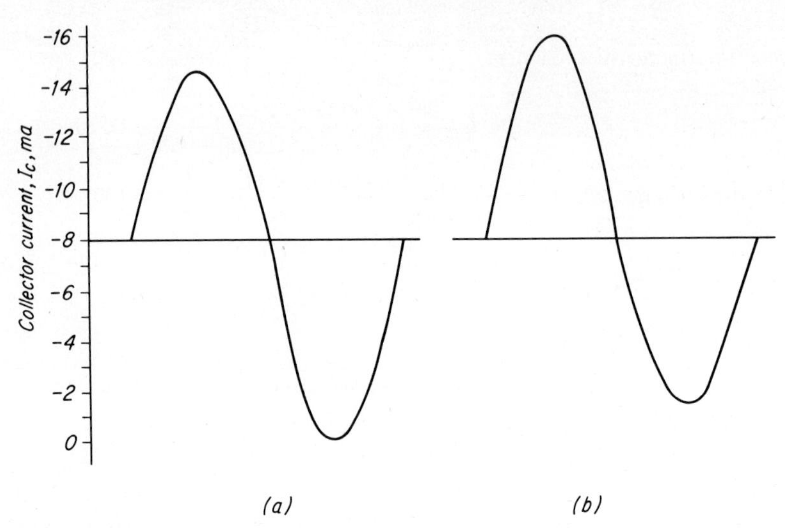

Fig. 16-11 Distortion of collector current. (a) Effect of variation of spacing between collector characteristic curves. (b) Effect of variation in the input-circuit impedance.

16-7 Class A Triode Vacuum-tube Operation

Load Lines. The location of the load line will be governed by (1) the value of plate-load resistance, (2) the voltage of the B power supply, (3) the maximum plate-dissipation rating of the tube, (4) the desired power output, (5) the allowable per cent of distortion.

An approximate value of zero-signal grid bias for a triode power-amplifier tube is

$$E_c = \frac{0.68E_P}{\mu} \tag{16-14}$$

Example 16-11 What value of grid bias is recommended for the power-amplifier tube whose characteristics are given in Fig. 16-12?

GIVEN: $E_P = 250$ volts $\quad \mu = 4.2 \quad$ Fig. 16-12

FIND: E_c

SOLUTION:

$$E_c = \frac{0.68E_P}{\mu} = \frac{0.68 \times 250}{4.2} = 40.5 \text{ volts}$$

From the curves of Fig. 16-12, when $E_c = -40.5$ volts, and $E_P = 250$ volts, then $I_P = 80$ ma. For this condition, the plate power dissipation (which is equal to the product of E_P and I_P) is equal to 250×0.08, or 20 watts. As this exceeds the maximum plate-dissipation rating of 15 watts for this tube, a new value of grid bias must be selected so that the rated value of plate dissipation is not exceeded. By simple mathematical deduction, if the plate voltage is maintained at 250 volts a plate current of 60 ma will just meet the maximum plate-dissipation rating of 15 watts. The curves show that this will occur with a grid bias of approximately 43.5 volts. Thus, the recommended value of E_c with zero-signal input is -43.5 volts.

Maximum-current Point. The maximum operating plate current is often taken as twice the value of I_P for zero-signal input. For the tube represented by Fig. 16-12 $i_{p\cdot\max}$ will be 2×60, or 120 ma.

Plotting the Load Line. The load line, which is a straight line, can now be plotted from the available data which will locate two points on the line. The quiescent or zero-signal operating point Q is located along the curve for $E_c = -43.5$ volts where this curve crosses the plate-voltage value of $E_P = 250$ volts. Point A is located where the curve for $E_c = 0$ crosses the plate-current value of 120 ma. Points A and Q may now be connected by a straight line, which is then extended through point B. Operating the tube with a grid bias of 43.5 volts limits the maximum signal-input voltage to 43.5 volts in order to avoid driving the grid positive during any part of the input cycle.

Value of the Load Resistance. The value of the plate load resistance R_o required to produce the characteristics shown in Fig. 16-12 is dependent upon the slope of the load line and is expressed by

$$R_o = \frac{e_{p\cdot\max} - e_{p\cdot\min}}{i_{p\cdot\max} - i_{p\cdot\min}} \tag{16-15}$$

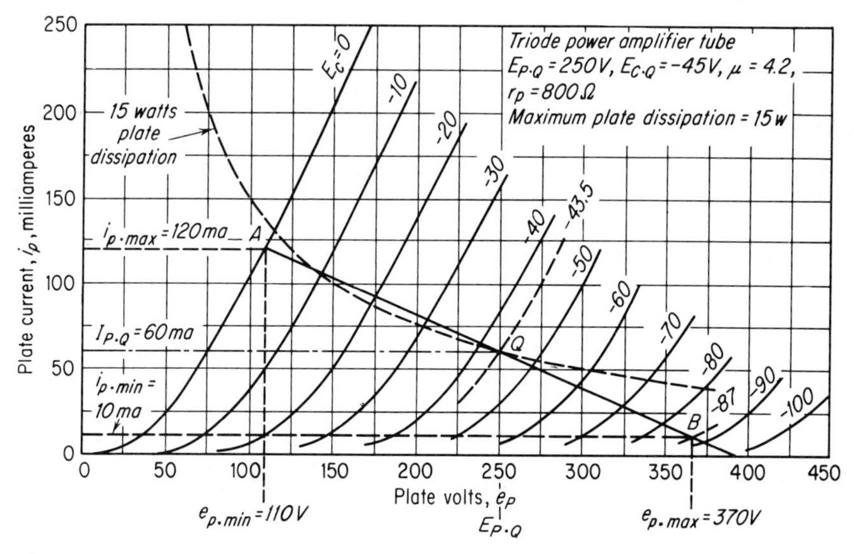

Fig. 16-12 Dynamic operating characteristics and load line for a triode vacuum tube.

Example 16-12 (*a*) What value of plate load resistance is required to produce the circuit characteristics corresponding to Fig. 16-12? (*b*) What value of B supply voltage is required with the plate-load resistance found in (*a*)?

GIVEN: Curves, Fig. 16-12

FIND: (*a*) R_o (*b*) E_{bb}

SOLUTION: From the curves of Fig. 16-12

$$\text{At point } A,\ i_{p\cdot\max} = 120 \text{ ma} \qquad e_{p\cdot\min} = 110 \text{ volts}$$
$$\text{At point } B,\ i_{p\cdot\min} = 10 \text{ ma} \qquad e_{p\cdot\max} = 370 \text{ volts}$$
$$\text{At point } Q,\ I_{P\cdot Q} = 60 \text{ ma} \qquad E_{P\cdot Q} = 250 \text{ volts}$$

(*a*)
$$R_o = \frac{e_{p\cdot\max} - e_{p\cdot\min}}{i_{p\cdot\max} - i_{p\cdot\min}} = \frac{370 - 110}{0.12 - 0.01} \cong 2{,}360 \text{ ohms}$$

(*b*)
$$E_{bb} = E_P + I_{P\cdot Q}R_o = 250 + 0.06 \times 2{,}360 \cong 392 \text{ volts}$$

Nonlinear Distortion. Examination of the static characteristic curves of Fig. 16-12 will show that all of the lines have some amount of curvature. A Class A power amplifier will therefore produce some distortion, since even the best operating portion of its characteristic curve is not actually a straight line. This distortion, caused mainly by the second harmonic, can be calculated by

$$\% \text{ second-harmonic distortion} = \frac{(i_{p\cdot\max} + i_{p\cdot\min}) - 2I_{P\cdot Q}}{2(i_{p\cdot\max} - i_{p\cdot\min})} \times 100 \qquad (16\text{-}16)$$

Example 16-13 What is the per cent of second-harmonic distortion for the amplifier circuit whose characteristics are shown in Fig. 16-12?

GIVEN: Curves, Fig. 16-12

FIND: Per cent second-harmonic distortion

SOLUTION:

$$\text{Second harmonic distortion} = \frac{(i_{p\cdot\max} + i_{p\cdot\min}) - 2I_{P\cdot Q}}{2(i_{p\cdot\max} - i_{p\cdot\min})} \times 100$$
$$= \frac{(120 + 10) - 2 \times 60}{2(120 - 10)} \times 100 = 4.54\%$$

16-8 Class A Pentode Vacuum-tube Operation

Load Lines. The load line and operating characteristics for pentodes and beam-power tubes are determined in much the same manner as for triodes. The location of the load line will be governed by the same five factors listed for triodes. Figure 16-13 shows three load lines plotted on the static characteristic curves for a typical pentode power-amplifier tube. For minimum distortion, the load line should be of such a value that $i_{p\cdot\max} - I_{P\cdot Q}$ is approximately equal to $I_{P\cdot Q} - i_{p\cdot\min}$. When operating on the 7,000-ohm load line, the difference between these two quantities is 2 ma. Increasing the load resistance to

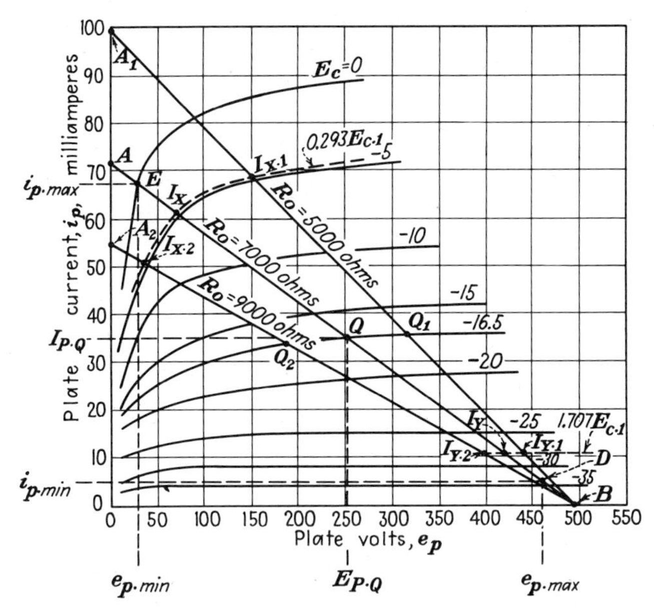

Fig. 16-13 Dynamic operating characteristics and load line for a pentode vacuum tube.

9,000 ohms increases this difference to 10 ma, and decreasing the load resistance to 5,000 ohms increases the difference to 14 ma.

The value of zero-signal grid bias for a specified operating plate voltage should be the value that will permit the greatest input-signal voltage with no more than the maximum acceptable amount of distortion. Examination of Fig. 16-13 will show that the recommended values of $E_{P\cdot Q} = 250$ volts and $E_{C\cdot Q} = -16.5$ volts will best meet these conditions. With these values $I_{P\cdot Q} = 35$ ma, and the plate dissipation is 8.75 watts, which is within the maximum rating of 11 watts. The zero-signal operating point Q is located along the curve for $E_c = -16.5$ volts where this curve crosses the plate-voltage value of $E_P = 250$ volts. Point E is located where the curve for $E_c = 0$ crosses the plate-current value of 67 ma, the maximum-current point with minimum distortion. Points E and Q may now be connected by a straight line, which is then extended to points A and B.

Value of the Load Resistance. The value of the plate load resistance R_o required to produce the load line AQB in Fig. 16-13 is dependent upon the slope of the load line and is expressed by Eq. (16-15).

Example 16-14 (*a*) What value of plate load resistance is required to produce the circuit characteristics corresponding to the load line AQB on Fig. 16-13? (*b*) What value of B supply voltage is required with the plate load resistance found in (*a*)?

GIVEN: Curves, Fig. 16-13 Load line AQB

FIND: (a) R_o (b) E_{bb}

SOLUTION: From the curves of Fig. 16-13

At point E, $i_{p\cdot max} = 67$ ma $\quad e_{p\cdot min} = 27$ volts
At point D, $i_{p\cdot min} = 5$ ma $\quad e_{p\cdot max} = 460$ volts
At point Q, $I_{P\cdot Q} = 35$ ma $\quad E_{P\cdot Q} = 250$ volts

(a)
$$R_o = \frac{e_{p\cdot max} - e_{p\cdot min}}{i_{p\cdot max} - i_{p\cdot min}} = \frac{460 - 27}{0.067 - 0.005} \cong 7{,}000 \text{ ohms}$$

(b)
$$E_{bb} = E_{P\cdot Q} + I_{P\cdot Q}R_o = 250 + 0.035 \times 7{,}000 \cong 495 \text{ volts}$$

Nonlinear Distortion. In addition to second-harmonic distortion, third-harmonic distortion is very pronounced in pentodes and beam-power tubes. The reason for this high distortion can be explained by reference to the static plate characteristic curves for these two types of tubes (see Figs. 16-13 and 16-14). It can be seen from these curves that the distances between the grid-voltage lines are not uniform. The greatest distance occurs between zero grid bias and the adjacent grid-bias line. The distance between succeeding adjacent grid-bias lines gradually decreases, reaching a minimum between the maximum grid bias and its adjacent grid-bias line. This nonuniform variation between the grid-voltage lines indicates that a change in grid voltage at the higher bias voltages will produce a smaller change in plate current than for the same grid-voltage change at the lower bias voltages. The total distortion

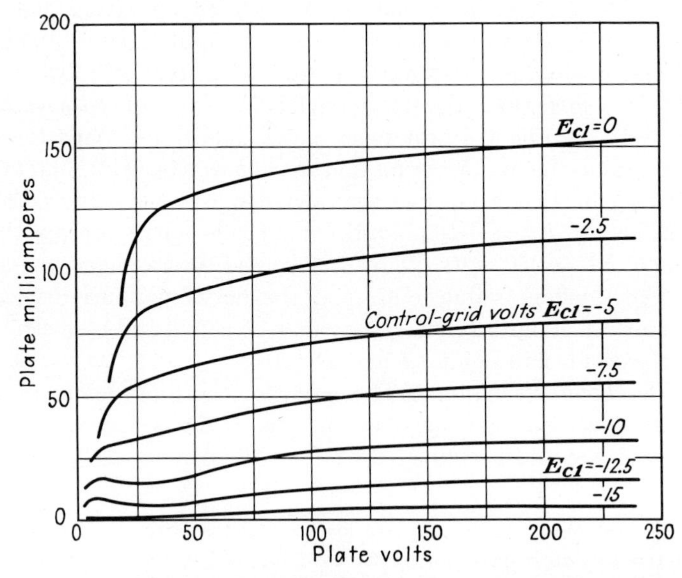

Fig. 16-14 Family of static plate characteristic curves for a beam-power tube.

produced by pentodes and beam-power tubes is therefore higher than the distortion obtained with triodes.

Calculation of the Amount of Distortion. The per cent of harmonic distortion can be calculated from the plate characteristic curves for a given operating condition by using the same methods described for transistor circuits (Art. 16-5). Substituting equivalent vacuum-tube notations, Eqs. (16-10) and (16-11) may be written as

% second-harmonic distortion

$$= \frac{i_{p\cdot\max} + i_{p\cdot\min} - 2I_{P\cdot Q}}{i_{p\cdot\max} - i_{p\cdot\min} + 1.41(I_x - I_y)} \times 100 \qquad (16\text{-}17)$$

% third-harmonic distortion

$$= \frac{i_{p\cdot\max} - i_{p\cdot\min} - 1.41(I_x - I_y)}{i_{p\cdot\max} - i_{p\cdot\min} + 1.41(I_x - I_y)} \times 100 \qquad (16\text{-}18)$$

where I_x = plate current at 0.293 E_c
I_y = plate current at 1.707 E_c

Example 16-15 A pentode power tube whose characteristics are given in Fig. 16-13 is being operated with a 7,000-ohm load resistance and a grid bias of 16.5 volts. If an input signal with a peak value of 16.5 volts is applied, what is the per cent of (*a*) second-harmonic distortion? (*b*) Third-harmonic distortion? (*c*) Total second- and third-harmonic distortion?

GIVEN: $E_{C\cdot Q} = -16.5$ volts $R_o = 7{,}000$ ohms $e_{c\cdot\max} = 16.5$ volts Load line, Fig. 16-13

FIND: (*a*) Second-harmonic distortion (*b*) Third-harmonic distortion (*c*) Total second- and third-harmonic distortion

SOLUTION: From the curves of Fig. 16-13

$i_{p\cdot\max} = 67$ ma $i_{p\cdot\min} = 5$ ma $I_{P\cdot Q} = 35$ ma $I_x = 61$ ma $I_y = 10.5$ ma

(*a*) Second-harmonic distortion

$$= \frac{i_{p\cdot\max} + i_{p\cdot\min} - 2I_{P\cdot Q}}{i_{p\cdot\max} - i_{p\cdot\min} + 1.41(I_x - I_y)} \times 100$$

$$= \frac{67 + 5 - 2 \times 35}{67 - 5 + 1.41(61 - 10.5)} \times 100 = 1.50\%$$

(*b*) Third harmonic distortion

$$= \frac{i_{p\cdot\max} - i_{p\cdot\min} - 1.41(I_x - I_y)}{i_{p\cdot\max} - i_{p\cdot\min} + 1.41(I_x - I_y)} \times 100$$

$$= \frac{67 - 5 - 1.41(61 - 10.5)}{67 - 5 + 1.41(61 - 10.5)} \times 100 = 6.91\%$$

(*c*) Total harmonic distortion

$$= \sqrt{(\%\text{ second-harmonic})^2 + (\%\text{ third harmonic})^2}$$

$$= \sqrt{1.50^2 + 6.91^2} = 7.07\%$$

The total harmonic distortion of pentodes and beam-power tubes for normal operating conditions is usually listed in standard tube manuals. Beam-power tubes have a lower percentage of distortion than pentodes and therefore are used more frequently as power-amplifier tubes.

16-9 Class A Power-amplifier Circuits

Transistor Circuits. A typical single-ended Class A power-amplifier circuit that may be used to drive one or more loudspeakers is shown in Fig. 16-15. The power output of this type of amplifier circuit is limited because (1) the maximum efficiency without using a stabilizing resistor is approximately 50 per cent, (2) adding a stabilizing resistor of a value required for maximum protection of the variations in collector current may reduce the maximum efficiency to approximately 25 per cent, and (3) in order to produce an output signal with minimum distortion the maximum value of input-signal current is limited by the crowding of the collector-current—collector-emitter voltage curves at the higher values of base current.

Example 16-16 The power-amplifier circuit of Fig. 16-15 uses a transistor whose characteristics are shown in Fig. 16-9. The transistor is to be operated Class A with an input signal having a peak value of 30 ma; other parameters are $V_{CC} = -46$ volts, collector saturation current = 2 amp, and $I_{B \cdot Q} = 30$ ma. What is the efficiency of this circuit?

GIVEN: $V_{CC} = -46$ volts $\quad I_{C \cdot \text{sat}} = 2$ amp $\quad I_{B \cdot Q} = 30$ ma $\quad i_{s \cdot \max} = 30$ ma
Fig. 16-9 Fig. 16-15

FIND: Efficiency

SOLUTION: Using Fig. 16-9, draw a load line using the two points (1) $V_{CC} = -46$ volts, $I_C = 0$; and (2) $I_C = 2$ amp, $V_{CC} = 0$. From these curves and the load line

$$V_{CE \cdot Q} = 27.5 \text{ volts} \qquad I_{C \cdot Q} = 0.8 \text{ amp} \qquad v_{c \cdot \max} = 45 \text{ volts}$$
$$v_{c \cdot \min} = 12 \text{ volts} \qquad i_{c \cdot \max} = 1.475 \text{ amp} \qquad i_{c \cdot \min} = 0.025 \text{ amp}$$

$$P_{o \cdot \text{a-c}} = \frac{(v_{c \cdot \max} - v_{c \cdot \min})(i_{c \cdot \max} - i_{c \cdot \min})}{8}$$

$$= \frac{(45 - 12)(1.475 - 0.025)}{8} = \frac{33 \times 1.45}{8} \cong 6 \text{ watts}$$

$$\text{Efficiency} = \frac{P_{o \cdot \text{a-c}}}{P_{i \cdot \text{d-c}}} \times 100 = \frac{6 \times 100}{46 \times 0.8} \cong 16.3\%$$

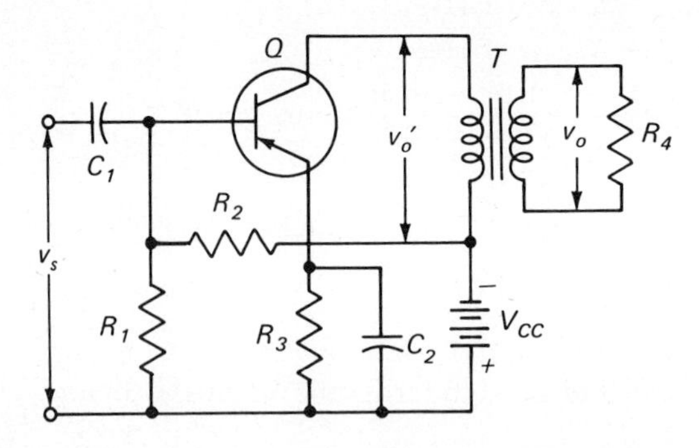

Fig. 16-15 Basic single-ended Class A transistor power-amplifier circuit.

Example 16-17 What is the total per cent of second- and third-harmonic distortion for the amplifier circuit of Example 16-16?

GIVEN: Example 16-16

FIND: Per cent total second- and third-harmonic distortion

SOLUTION:

$$I_x = i_c \text{ at } 1.707 I_{B\cdot Q} = i_c \text{ at } 1.707 \times 30 \cong 1.3 \text{ amp}$$
$$I_y = i_c \text{ at } 0.293 I_{B\cdot Q} = i_c \text{ at } 0.293 \times 30 = 0.25 \text{ amp}$$

$$\text{Second-harmonic distortion} = \frac{i_{c\cdot\max} + i_{c\cdot\min} - 2I_{C\cdot Q}}{i_{c\cdot\max} - i_{c\cdot\min} + 1.41(I_x - I_y)} \times 100$$
$$= \frac{(1.475 + 0.025) - 2 \times 0.8}{(1.475 - 0.025) + 1.41(1.3 - 0.25)} \times 100 \cong 3.4\%$$

$$\text{Third-harmonic distortion} = \frac{(i_{c\cdot\max} - i_{c\cdot\min}) - 1.41(I_x - I_y)}{(i_{c\cdot\max} - i_{c\cdot\min}) + 1.41(I_x - I_y)} \times 100$$
$$= \frac{(1.475 - 0.025) - 1.41(1.3 - 0.25)}{(1.475 - 0.025) + 1.41(1.3 - 0.25)} \times 100 \cong 1.025\%$$

Total second- and third-harmonic distortion

$$= \sqrt{(\%\text{ second-harmonic})^2 + (\%\text{ third-harmonic})^2}$$
$$= \sqrt{(3.4^2 + 1.025^2)} \cong 3.56\%$$

The power output of 12.4 watts, the efficiency of 41 per cent, and the total second- and third-harmonic distortion of 13.3 per cent as determined in Examples 16-8 and 16-10; and the power output of 6 watts, the efficiency of 16.3 per cent, and the total second- and third-harmonic distortion of 3.56 per cent as determined in Examples 16-16 and 16-17 are indicative of the characteristics of this type of amplifier circuit as described in the previous paragraph.

Output Transformer Circuit. The transformer used to couple the collector circuit of the power-output transistor to the voice coil of the loudspeaker, generally called the *output transformer*, may be considered as an impedance-matching device. The impedance reflected to the collector circuit by the voice coil can be controlled by the ratio of the primary to the secondary turns in the transformer. Expressed mathematically,

$$R_o' = R_o \left(\frac{N_P}{N_S}\right)^2 \tag{16-19}$$

and

$$\frac{N_P}{N_S} = \sqrt{\frac{R_o'}{R_o}} \tag{16-20}$$

also

$$R_o' = \frac{v_{c\cdot\max} - v_{c\cdot\min}}{i_{c\cdot\max} - i_{c\cdot\min}} \tag{16-21}$$

where R_o' = collector load resistance, ohms
R_o = resistance of voice coil, ohms
N_P = number of turns on primary winding
N_S = number of turns on secondary winding

Example 16-18 Determine the primary-to-secondary turns ratio of the output transformer used in Example 16-16 if the resistance of the voice coil of the loudspeaker is 8 ohms.

GIVEN: Example 16-16 $\qquad R_o = 8$ ohms

FIND: N_P/N_S

SOLUTION:

$$R_o' = \frac{v_{c\cdot\max} - v_{c\cdot\min}}{i_{c\cdot\max} - i_{c\cdot\min}} = \frac{45 - 12}{1.475 - 0.025} = \frac{33}{1.45} = 22.8 \text{ ohms}$$

$$\frac{N_P}{N_S} = \sqrt{\frac{R_o'}{R_o}} = \sqrt{\frac{22.8}{8}} \cong 1.69$$

Vacuum-tube Circuits. Typical single-ended Class A power-amplifier circuits used to drive a loudspeaker are shown in Fig. 16-16. The power output of a triode vacuum tube is much smaller than the power output that can be obtained from pentodes or beam-power tubes. Triodes are therefore rarely used

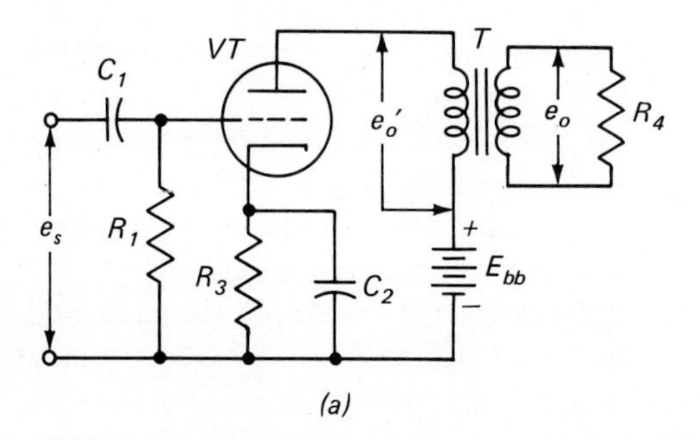

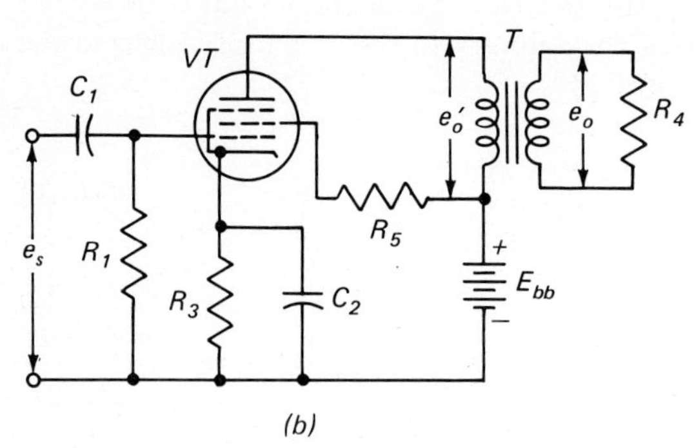

Fig. 16-16 Basic single-ended Class A vacuum-tube power-amplifier circuits (a) using a triode, (b) using a pentode.

in radio-receiver power-amplifier circuits. Because of the large amount of harmonic distortion in the output of pentodes and beam-power tubes the maximum power output that can be obtained from a single-ended Class A amplifier with minimum per cent distortion is limited. The maximum power output of pentodes and beam-power tubes is much smaller than the power output that can be obtained from transistors. When amounts of power are required that are greater than can be obtained from a single active device, two active devices are connected in parallel or push-pull.

Example 16-19 The power-amplifier circuit of Fig. 16-16*b* uses a pentode tube whose characteristics are shown in Fig. 16-13. The tube is to be operated Class A with an input signal having a peak value of 16-5 volts; other parameters are $E_{bb} = 495$ volts, $I_{P \cdot Q} = 60$ ma, and $R_o = 7{,}000$ ohms. What is the efficiency of this circuit?

GIVEN: $E_{bb} = 495$ volts $\quad I_{P \cdot Q} = 35$ ma $\quad v_{s \cdot \max} = 16.5$ volts $\quad R_o = 7{,}000$ ohms
Fig. 16-13 $\quad$ Fig. 16-16*b*

FIND: Efficiency

SOLUTION: From the curves of Fig. 16-13

$$E_{P \cdot Q} = 250 \text{ volts} \qquad e_{p \cdot \max} = 460 \text{ volts} \qquad e_{p \cdot \min} = 27 \text{ volts}$$

$$i_{p \cdot \max} = 67 \text{ ma} \qquad i_{p \cdot \min} = 5 \text{ ma}$$

$$P_{o \cdot \text{a-c}} = \frac{(e_{p \cdot \max} - e_{p \cdot \min})(i_{p \cdot \max} - i_{p \cdot \min})}{8}$$

$$= \frac{(460 - 27)(0.067 - 0.005)}{8} = \frac{433 \times 0.062}{8} = 3.36 \text{ watts}$$

$$\text{Efficiency} = \frac{P_{o \cdot \text{ac}}}{P_{i \cdot \text{d-c}}} \times 100 = \frac{3.36 \times 100}{495 \times 0.035} \cong 19.4\%$$

Example 16-20 What is the total per cent of second- and third-harmonic distortion for the amplifier circuit of Example 16-19?

GIVEN: Example 16-19

FIND: Per cent total second- and third-harmonic distortion

SOLUTION: From Fig. 16-13

$$I_x = 61 \text{ ma} \qquad I_y = 10.5 \text{ ma}$$

$$\text{Second-harmonic distortion} = \frac{(i_{p \cdot \max} + i_{p \cdot \min}) - 2I_{P \cdot Q}}{(i_{p \cdot \max} - i_{p \cdot \min}) + 1.41(I_x - I_y)} \times 100$$

$$= \frac{(67 + 5) - 2 \times 35}{(67 - 5) + 1.41(61 - 10.5)} \times 100 \cong 1.5\%$$

$$\text{Third-harmonic distortion} = \frac{(i_{p \cdot \max} - i_{p \cdot \min}) - 1.41(I_x - I_y)}{(i_{p \cdot \max} - i_{p \cdot \min}) + 1.41(I_x - I_y)} \times 100$$

$$= \frac{(67 - 5) - 1.41(61 - 10.5)}{(67 - 5) + 1.41(61 - 10.5)} \times 100 \cong 6.91\%$$

Total (second + third) harmonic distortion

$$= \sqrt{(\%\text{ second harmonic})^2 + (\%\text{ third harmonic})^2}$$
$$= \sqrt{1.5^2 + 6.91^2} = 7.07\%$$

Example 16-21 Determine the primary-to-secondary turns ratio of the output transformer used in Example 16-19 if the resistance of the voice coil of the loudspeaker is 8 ohms.

GIVEN: Example 16-19 $R_o = 8$ ohms

FIND: N_P/N_S

SOLUTION:

$$R_o' = \frac{e_{p\cdot\max} - e_{p\cdot\min}}{i_{p\cdot\max} - i_{p\cdot\min}} = \frac{460 - 27}{0.067 - 0.005} = \frac{433}{0.062} \cong 7{,}000 \text{ ohms}$$

$$\frac{N_P}{N_S} = \sqrt{\frac{R_o'}{R_o}} = \sqrt{\frac{7{,}000}{8}} \cong 30$$

16-10 Push-Pull Amplifier Circuits

Circuit Operation. A push-pull amplifier circuit employs two identical active devices operating as a single stage of amplification (Fig. 16-17). The input and output elements of the transistors or vacuum tubes are connected respectively to opposite ends of the center-tapped windings of the secondary of the input transformer T_1 and the primary of the output transformer T_2. A balanced circuit is obtained by connecting either the emitter, source, or cathode returns to center taps on the secondary and primary windings of the input and output transformers respectively. As a balanced circuit is characteristic of push-pull operation, this system is also referred to as a *balanced amplifier*. A push-pull amplifier circuit may be either resistance-capacitance-coupled or transformer-coupled.

The varying current in the primary winding of the input transformer, which is actually the output current of the previous stage, induces a voltage of corresponding waveform in the secondary. At any instant the two ends of this secondary, terminals 3 and 4 of Fig. 16-17, are of opposite polarity. Thus, the varying input voltages will be always equal and 180 degrees out of phase with each other. When the end of the secondary indicated as 3 is positive, then the other end indicated as 4 will be negative. The input of active device A will then become more positive, causing an increase in the output current flowing through section 6-7 of the primary of the output transformer. The input of active device B becomes more negative, causing the output current flowing through section 7-8 to decrease. As the two active devices are identical and the changes in their input-signal voltages are equal, the variations in output current will also be equal but 180 degrees out of phase with each other (assuming Class A operation). Consequently, one active device pushes current through one-half of the primary winding of the output transformer while the second device pulls an equal amount through the other half, hence the name *push-pull.*

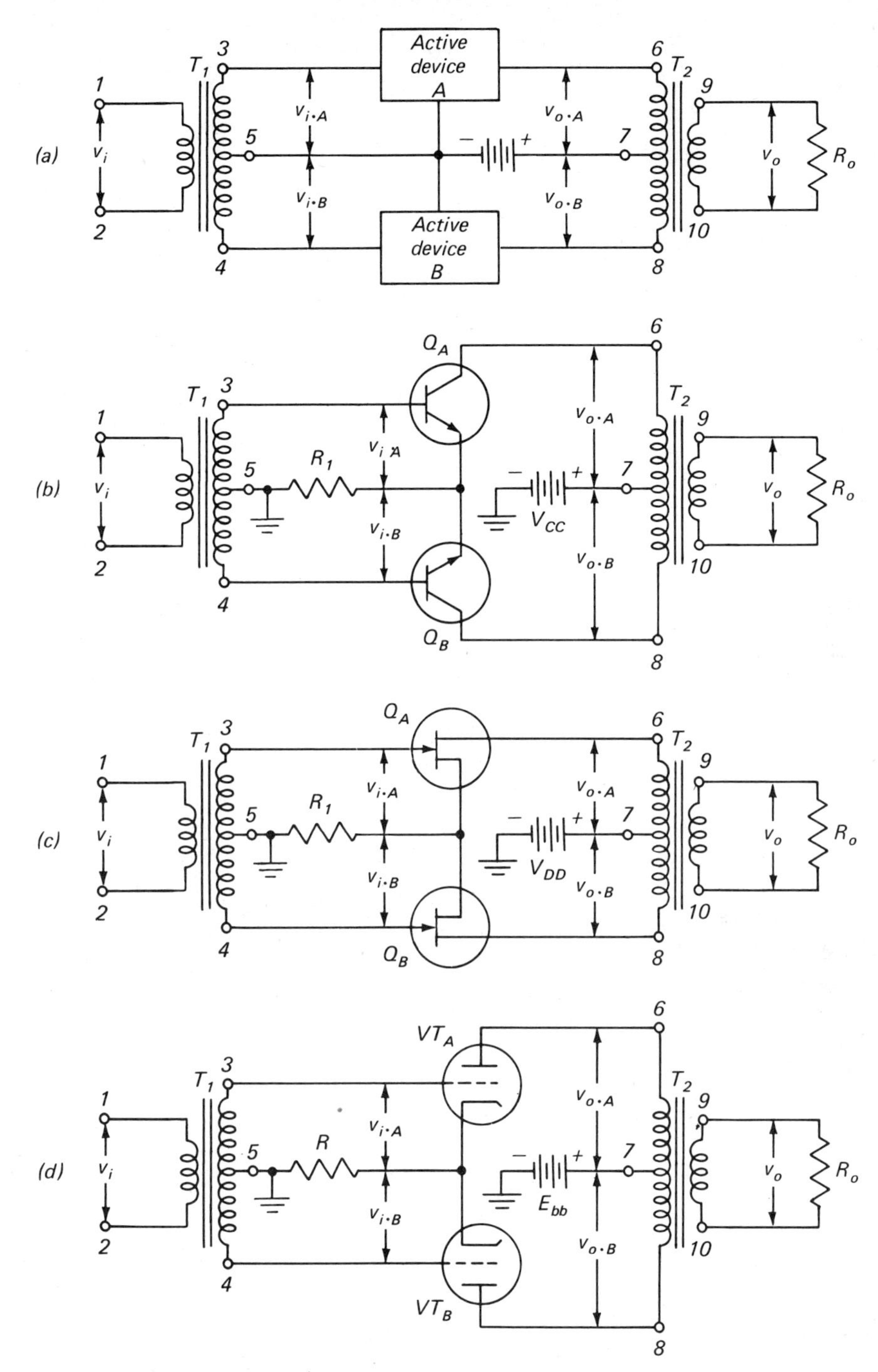

Fig. 16-17 Push-pull operation of active devices. (a) Block diagram. (b) Basic bipolar transistor circuit. (c) Basic unipolar (FET) transistor circuit. (d) Basic triode vacuum-tube circuit.

Thus, both active devices contribute alternating signal current that flows in the same direction. Both active devices carry their own a-c component of the output current. However, since these two direct currents are equal in amplitude and flow in opposite directions through the primary of T_2, the net magnetizing effect on the transformer core is the same as if the direct current flow through the primary of the output transformer was zero. Push-pull operation is not limited to any particular type of active device; thus junction or field-effect transistors, and triode, pentode, or beam-power tubes may be used in balanced amplifiers.

Classes of Operation. A power-amplifier stage may be operated as a double-ended Class A, Class AB, Class B, or Class C push-pull circuit. Regardless of the type of operation of the active device the push-pull power amplifier has distinct advantages.

One of the disadvantages of Class A single-ended or Class A push-pull amplifier circuits is that some direct current flows at all times through the active device. As a result the power dissipation in the output circuit is highest when the a-c signal current is zero. This power dissipation can be greatly reduced by using a push-pull arrangement with the two active devices connected for Class AB or Class B operation. The distortion normally produced when an active device is operated other than Class A is caused by even-order harmonics. When two active devices are connected in a push-pull circuit a 180-degree phase difference exists between the even-order harmonics produced by each active device, and they cancel each other in the output circuit (see Art. 16-11). The automatic cancellation of all even-order harmonics from the output current makes the operation of push-pull power amplifiers in Class AB or Class B desirable with communication sound equipment. The double-ended balanced power amplifier is therefore capable of accommodating more than twice the input signal of a single-ended stage using a comparable single active device. Assuming the signal-load resistance to be the same in both the single- and double-ended amplifiers, the increase in input-signal excitation of the push-pull circuit makes it possible to produce a signal power output that is more than four times the signal power output of a single-ended power amplifier.

Because of its low efficiency, Class A operation is very seldom used in audio push-pull power-amplifier circuits using transistors. Class B operation is also seldom used for this type of application because of the nonlinearity of the base-emitter input characteristics of junction transistors that produce objectionable crossover distortion in push-pull circuits. Class AB operation is usually employed in audio push-pull power-amplifier circuits using transistors. The base-emitter circuit is biased near collector cutoff so that the crossover distortion can be minimized to an acceptable value at low-signal levels. Either Class A or Class AB operation is used in push-pull audio amplifier circuits using vacuum tubes. Class C operation of either transistors or vacuum tubes is usually limited to transmitter applications where only one frequency, or a narrow band of frequencies, is to be amplified.

16-11 Transistor Class B Push-Pull Amplifier

Advantages. Class B push-pull amplifiers are used when a higher power output is required than can be obtained from the same transistors operated as a Class A push-pull amplifier. Each transistor is in an ON position and conducts signal current for one-half of each cycle of the output waveform, while during the other half-cycle it is in an OFF position or nonconducting. With zero input signal, neither transistor is in a conducting state and hence they dissipate very little power. Because of these favorable operating conditions, the d-c output efficiency is very high.

Circuit Operation. When a PNP transistor is operated as a Class B amplifier, collector current flows only during negative signal inputs. When an NPN transistor is operated as a Class B amplifier, collector current flows only during positive signal inputs. Distortion resulting from Class B operation can be reduced to a minimum by using two transistors as a push-pull amplifier. A typical push-pull audio power-amplifier circuit using two junction transistors is shown Fig. 16-18. The bias voltages for transistors Q_1 and Q_2 are obtained from the voltage-divider network R_1 and R_2, stabilization of the emitter-base resistance to temperature changes is provided by the swamping resistors R_3 and R_4, and the load resistor R_5 represents the resistance of the voice coil of the loudspeaker. The resistances of R_1 and R_2 should be of such values that the bias voltage produced causes the transistor to operate at the initial point of collector-current cutoff. The variations in collector current, collector power dissipation, and the d-c operating point caused by changes in the ambient temperature can be minimized by using (1) swamping resistors in the emitter

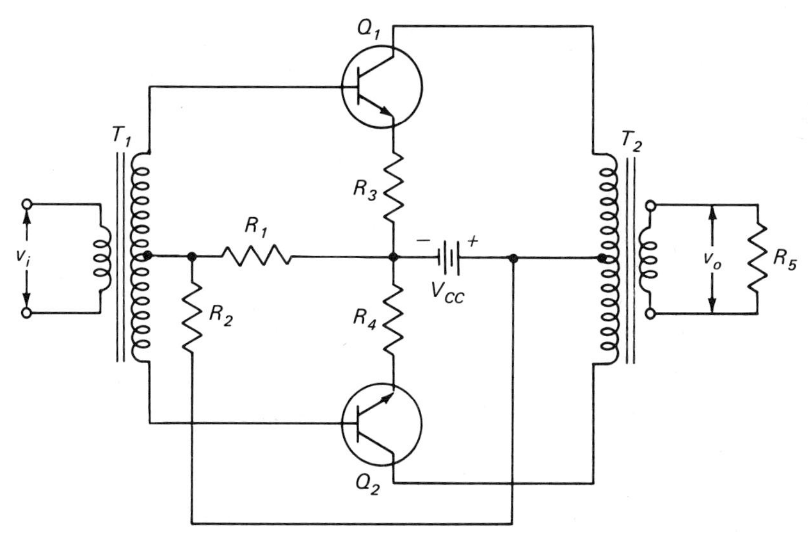

Fig. 16-18 Transistor push-pull amplifier circuit.

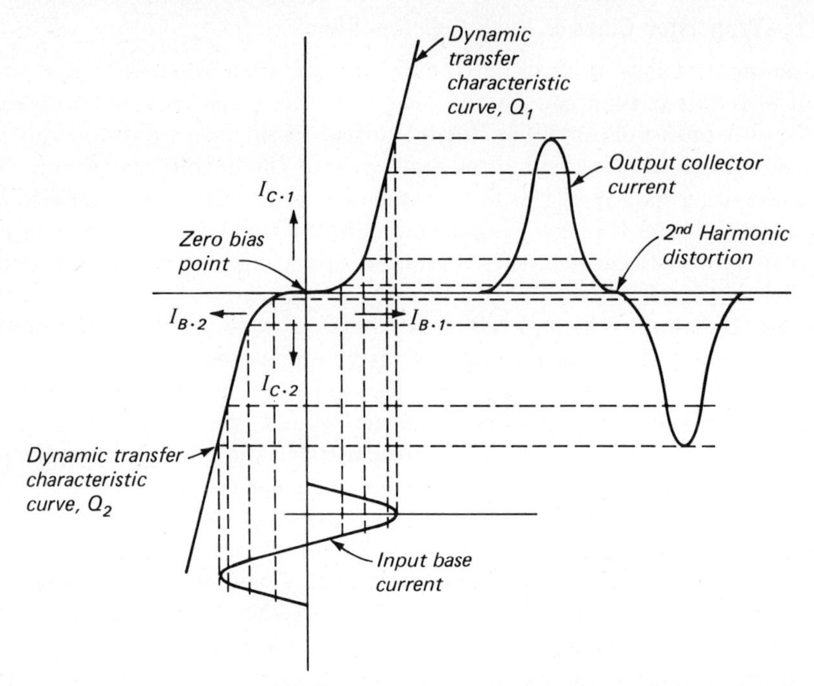

Fig. 16-19 Graphical illustration of how crossover distortion is produced.

circuits, (2) a temperature-sensitive resistor such as a thermistor, and/or (3) a bias-compensating diode in the biasing network.

Crossover Distortion. In Class B operation the transistor is operated at cutoff so that with zero input signal the collector current and collector power dissipation will be approximately zero. However, when the transistors are operated at such a d-c operating point, the output signal will be distorted at low input signals. Referring to the dynamic i_b-i_c transfer characteristic curves for the pair of power transistors of Fig. 16-19 indicates (crossover) distortion occurs during the turn-off and turn-on points of the Class B transistor amplifier. *Crossover distortion* is a result of one transistor's cutting off before the other transistor begins conduction. It should be noted that the transistor transfer characteristic curve resembles the diode curve, where a small change in voltage near cutoff does not cause any measurable change in conduction. This low-conduction-level area causes second-harmonic distortion in the waveform of the output current.

Output Transformer. In the push-pull amplifier circuit of Fig. 16-18 the output transformer T_2 is used to (1) provide the output load R_5 with the necessary current drive, and (2) transform the resistance of the voice coil to match the desired load impedance R_o' of the collector circuit. When the d-c components of the collector currents in Q_1 and Q_2 flow through the primary winding of T_2, the net magnetization effect of these currents on the magnetic core is zero. This is because the d-c component of collector current for each transistor flows in opposite directions through the primary winding of the output transformer. The partial saturation of the magnetic core material in a single-ended power amplifier, due to the d-c component of the collector current, contributes to the nonlinearity in the response of this type of amplifier. Because the d-c magnetization effect of a push-pull transformer is zero, the linearity of the response from this type of amplifier is practically uniform. Because there is no direct current in the windings to produce a large steady component of flux in the transformer, a much smaller output transformer may be used than would be required for a single-ended amplifier using the same type of transistor.

16-12 Transistor Class AB Push-Pull Amplifier

Circuit Operation. Crossover distortion produced in a Class B push-pull amplifier can be minimized by shifting the bias point for each amplifier. Each transistor is forward-biased by a small amount of base current, thus causing the circuit to operate as a Class AB push-pull amplifier (Fig. 16-20). The

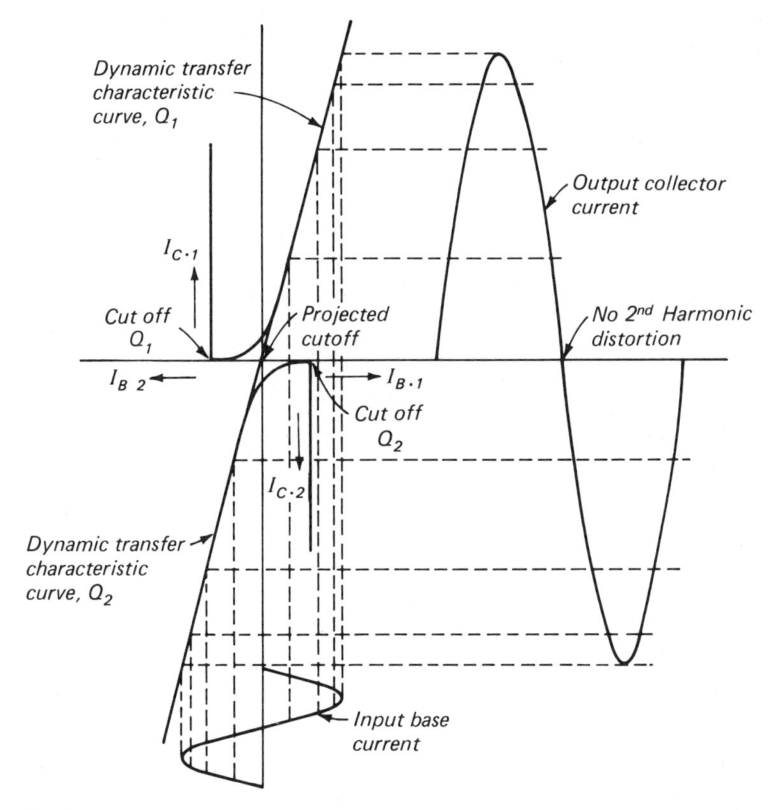

Fig. 16-20 Graphical illustration of how crossover distortion is eliminated.

Class B push-pull amplifier circuit of Fig. 16-18 can be made to operate as a Class AB push-pull amplifier by changing the values of the biasing resistors. The resistances of R_1 and R_2 should be of such values that the bias voltage produced is just sufficient to limit the amount of crossover distortion at low signal levels to a maximum permissible value.

Graphical Analysis. A graphical analysis of the operation of a Class AB push-pull transistor power amplifier can be made by arranging the family of output characteristic curves of two identical power transistors back to back (Fig. 16-21). The resultant dynamic characteristic curve *A-B* formed by these two transistors has a uniform symmetry that is not present in a single active device. The bias points selected for Class AB operation are Q_1 and Q_2, and the resultant collector family characteristics are $V_{CE \cdot Q} = 45$ volts, $I_{C \cdot Q} = 0.3$ amp, and $I_{B \cdot Q} = 10$ ma. Since the d-c resistance of the primary winding of transformer T_2 may be considered negligible, $V_{CE} \cong V_{CC}$. The d-c load lines $R_{L \cdot \text{d-c} \cdot 1}$ and $R_{L \cdot \text{d-c} \cdot 2}$ are shown as vertical lines intersecting each Q point and the V_{CE} voltage axis value. The point where the dynamic load lines $R_{L \cdot 1}'$ and $R_{L \cdot 2}'$ intersect their voltage axis at 52 volts is the only point common to both voltage axes. The maximum base-emitter signal current for each transistor is 100 ma

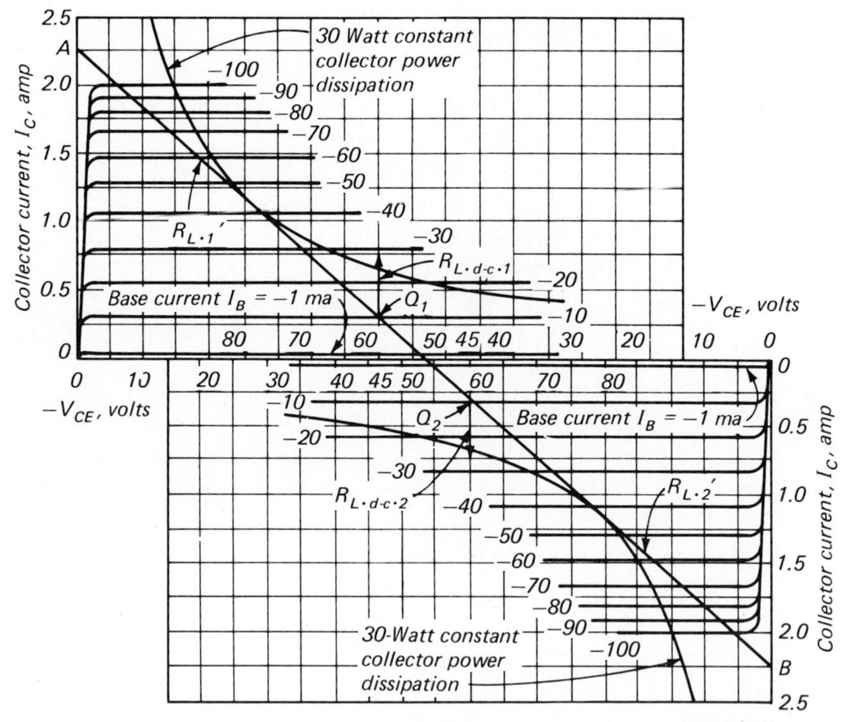

Fig. 16-21 Graphical analysis of Class AB push-pull operation.

peak. During each half-cycle of this maximum signal excitation one transistor is driven into its maximum conduction area while the other transistor is driven into cutoff. At the start of the next half-cycle the polarity of the signal current changes and both transistors are still conducting. However, after the polarity reversal of the next half-cycle occurs, the first transistor is cut off while the second transistor is driven into its maximum-conduction area.

Power Output, Efficiency, and Load Resistance. The power output, efficiency, and collector-to-collector load resistance of a Class AB push-pull transistor amplifier can be calculated for a given operating condition from its collector characteristic curves. The maximum a-c power output can be obtained by modifying Eq. (16-1*a*) to represent the output of the two active devices, thus

$$P_{o\cdot\text{a-c}\cdot\text{p-p}} = \frac{2(v_{o\cdot\max} - v_{o\cdot\min}) \times 2(i_{o\cdot\max} - i_{o\cdot\min})}{8} \tag{16-22}$$

$$= \frac{(v_{o\cdot\max} - v_{o\cdot\min})(i_{o\cdot\max} - i_{o\cdot\min})}{2} \tag{16-22a}$$

An exact equation for obtaining the d-c collector power input is rather complex as a relatively high current flows through each transistor only during the half-cycles that it is conducting, while very little or no current flows during the other half-cycles. An approximate value of the d-c input power to the active device can be obtained by

$$P_{i\cdot\text{a-d}\cdot\text{p-p}} \cong \frac{v_{c\cdot\max} i_{c\cdot\max}}{1.57} \tag{16-23}$$

The efficiency of the active devices in a push-pull amplifier can be obtained by using Eq. (16-4) with no change.

$$\text{Efficiency of active devices} = \frac{P_{o\cdot\text{a-c}}}{P_{i\cdot\text{a-d}}} \times 100 \tag{16-4}$$

The collector-to-collector load resistance for two transistors in push-pull, regardless of the class of operation, may be obtained by

$$R_L' = \frac{4v_{c\cdot\max}}{i_{c\cdot\max}} \tag{16-24}$$

Example 16-22 The push-pull power amplifier circuit of Fig. 16-18 uses two identical transistors whose characteristics are shown in Figs. 16-9 and 16-21. The transistors are operated Class AB at $V_{CE\cdot Q} = -45$ volts and $I_{B\cdot Q} = 10$ ma. Determine the (*a*) maximum a-c power output, (*b*) d-c collector-power input, (*c*) efficiency of the active devices, and (*d*) collector-to-collector load resistance.

GIVEN: $V_{CE\cdot Q} = -45$ volts $I_{B\cdot Q} = 10$ ma Figs. 16-9, 16-18, 16-21

FIND: (*a*) $P_{o\cdot\text{a-c}\cdot\text{p-p}}$ (*b*) $P_{i\cdot\text{a-d}\cdot\text{p-p}}$ (*c*) Efficiency (*d*) R_L'

SOLUTION: Using Fig. 16-21, locate the operating points Q_1 and Q_2 by determining where the -45-volt collector abscissa crosses the 10-ma base-current curve for each transistor. Draw an a-c load line through Q_1 and Q_2 that is tangent to the 30-watt constant-collector power-dissipation curve for each transistor. From these curves with an input signal having a peak value of 100 ma

$$v_{c\cdot\max} = -52 \text{ volts} \qquad v_{c\cdot\min} = -6 \text{ volts} \qquad i_{c\cdot\max} = 2 \text{ amp} \qquad i_{c\cdot\min} = 0$$

(a)
$$P_{o\cdot\text{a-c}\cdot\text{p-p}} = \frac{(v_{c\cdot\max} - v_{c\cdot\min})(i_{c\cdot\max} - i_{c\cdot\min})}{2}$$
$$= \frac{(52 - 6)(2 - 0)}{2} = \frac{46 \times 2}{2} = 46 \text{ watts}$$

(b)
$$P_{i\cdot\text{a-d}\cdot\text{p-p}} \cong \frac{v_{c\cdot\max} i_{c\cdot\max}}{1.57} \cong \frac{52 \times 2}{1.57} \cong 66.2 \text{ watts}$$

(c)
$$\text{Efficiency} = \frac{P_{o\cdot\text{a-c}\cdot\text{p-p}}}{P_{i\cdot\text{a-d}\cdot\text{p-p}}} \times 100 = \frac{46 \times 100}{66.2} = 69.5\%$$

(d)
$$R_L' = \frac{4v_{c\cdot\max}}{i_{c\cdot\max}} = \frac{4 \times 52}{2} = 104 \text{ ohms}$$

The maximum a-c power output and efficiency of a transistor, similar to the one used in Example 16-22, connected as a single-ended Class A amplifier to produce maximum power output with maximum permissible distortion is 6 watts and 16.3 per cent, respectively (Example 16-16). This single-ended Class A amplifier is also capable of delivering a maximum a-c power output of 12.4 watts with an efficiency of 41 per cent (Example 16-8) when there is no limit on the amount of distortion permitted. The power output from the Class AB push-pull amplifier in Example 16-22 is approximately eight times that of the single-ended amplifier with maximum permissible distortion, and approximately four times that of the same amplifier with no limit on the amount of distortion permitted. The efficiency of operation is also greatly improved, from 16.3 and 41 per cent, respectively, to 69.5 per cent. This large increase in a-c power output and efficiency obtained from the transistors used in Examples 16-16 and 16-22 is typical of the increased amount of a-c power output and efficiency that is obtainable using any power transistor in Class AB push-pull operation.

Distortion. When an alternating signal is applied to the imput circuits of the two transistors in Fig. 16-18 the varying collector current in each transistor is distorted. This distortion is largely due to the second harmonic, since the spacing between the collector-current—collector-emitter voltage curves over which the transistor operates is unequal. The phase relation between the second harmonics produced by each transistor are 180 degrees out of phase with each other, and thus cancel each other in the output transformer. A graphical illustration of how this is accomplished is shown in Fig. 16-22. This figure also shows that the combined output of the two transistors is a sine wave devoid of any second-harmonic distortion.

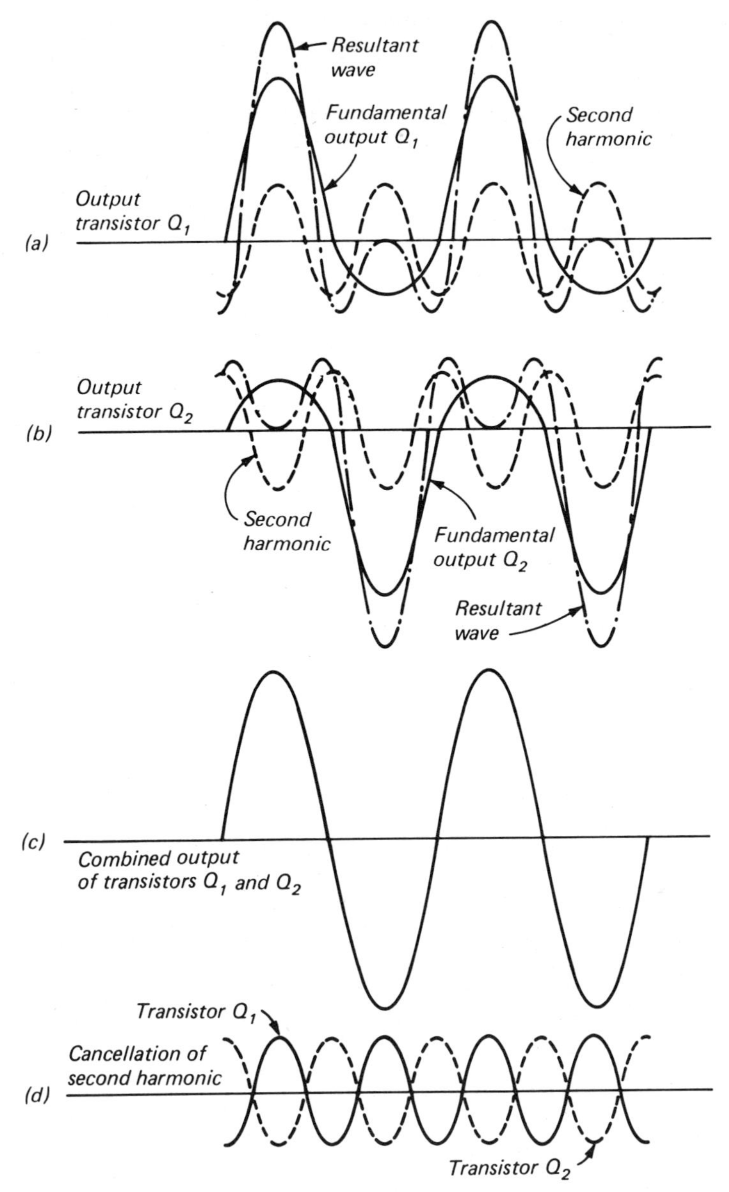

Fig. 16-22 Waveforms of a Class AB push-pull amplifier.

Since crossover distortion can be reduced to a minimum (or eliminated) by operating the transistors in Class AB, the dynamic load lines for each transistor are practically straight lines. The amount of crossover distortion produced will be determined by the amount the bias points are shifted from Class B to Class AB operation. The projected cutoff points produced by the bias-point shift should make possible an a-c load line that is practically straight from point A to point B (Fig. 16-21). A sine-wave input signal will therefore produce a sine-wave output signal across R_5 that is devoid of any third-harmonic distortion. Thus, two transistors operated as a Class AB push-pull amplifier are capable of delivering an a-c power output with practically no distortion.

16-13 Phase Inverters

Need. The analysis of push-pull operation has shown that (1) the input signals fed to the active devices A and B in Fig. 16-17 must be approximately equal in magnitude at all times, and (2) the two input-signal voltages must be 180 degrees out of phase. With transformer coupling, the 180-degree phase difference between the two input voltages is obtained by means of an input transformer having a center-tapped secondary (Figs. 16-17 and 16-18). With resistance-capacitance coupling, the 180-degree phase relation between the two input-signal voltages is obtained by employing the inverter action of certain circuits using an active device. A circuit using a transistor or vacuum tube in this manner is called a *phase inverter, phase splitter*, or *paraphase amplifier.*

Single-Stage Phase-inverter Circuit. In the basic transistor phase-inverter circuit of Fig. 16-23, two output signals of opposite phase and approximately the same amplitude are obtained from a single input signal by using only one

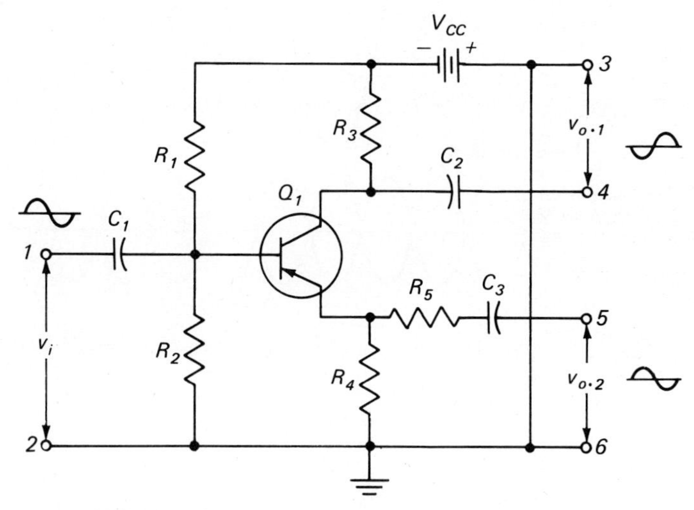

Fig. 16-23 Single-stage transistor phase-inverter circuit.

transistor stage. The base-emitter circuit is biased in a forward direction by the voltage-divider circuit R_1 and R_2. One output voltage $v_{o\cdot 1}$ is dependent upon the voltage drop across the collector load resistor R_3, while the other output voltage $v_{o\cdot 2}$ is dependent upon the voltage drop across the emitter resistor R_4. The coupling capacitors C_2 and C_3 block any direct current from the input of the following stages. When the resistances R_3 and R_4 are equal, the output voltages $v_{o\cdot 1}$ and $v_{o\cdot 2}$ are not exactly the same. The difference in voltage is small and is due to the difference between the collector and emitter currents. Distortion may occur whenever strong signal currents are present in the output circuit. This type of distortion can be eliminated by altering the values of R_3 and R_4, and also adding resistor R_5 in one line feeding the output $v_{o\cdot 2}$. The resistances of R_4 and R_5 are of such values that cause the signal-source impedances for the two output transistors in the push-pull circuit being fed by the phase inverter to be equal. The signal voltage drop across R_5 is compensated for by making the resistance of R_4 higher in value than the resistance of R_3.

When the input signal increases the effective forward bias of Q_1, the circuit actions are (1) an increase in the emitter and collector currents, (2) an increase in the voltage drop across R_3, thereby causing a decrease in the output voltage $v_{o\cdot 1}$, (3) an increase in the voltage drop across R_4, thereby causing an increase in the output voltage $v_{o\cdot 2}$. When the input signal decreases the effective forward bias of Q_1, the output voltage $v_{o\cdot 1}$ will increase and the output voltage $v_{o\cdot 2}$ will decrease. Thus, the actions of the two output voltages are opposite, and a 180-degree phase difference exists between $v_{o\cdot 1}$ and $v_{o\cdot 2}$.

A disadvantage of the single-transistor phase-inverter circuit is that the emitter-circuit resistor R_4 introduces an appreciable amount of degeneration and hence (1) requires a relatively high input signal v_i, or (2) does not provide sufficient output-signal strength to meet the requirements of a high-level output push-pull amplifier.

Two-stage Phase-inverter Circuit. The disadvantages of a single-stage phase-inverter circuit are overcome, and a more exact balance is obtained by using a two-transistor two-stage phase-inverter circuit (Fig. 16-24). In this circuit, the input signal v_i is fed to the base of Q_1 and is amplified by the transistor circuit. The output of Q_1 is inverted in phase and coupled to the output terminal 3 by the blocking capacitor C_5. A portion of the output signal of Q_1, reduced by the voltage-dropping resistor R_4, coupling capacitor C_4, R_7, R_6, is fed to the base of Q_2. The output of Q_2 is an amplified and inverted replica of the signal fed to the base of Q_2. As two-phase inversions take place between v_i and $v_{o\cdot 2}$, the output at terminal 5 is of the same phase as the input signal v_i. Resistors R_1 and R_7 provide the quiescent forward bias for Q_1 and Q_2. Resistors R_2 and R_3 are swamping resistors used to improve the operating stability. Capacitors C_2 and C_3 prevent degeneration due to R_2 and R_3. Resistors R_5 and R_6 are the collector load resistors for Q_1 and Q_2. Coupling and/or d-c blocking action are provided by capacitors C_1, C_5, and C_6.

Advantages of using a two-stage two-transistor phase-inverter circuit over

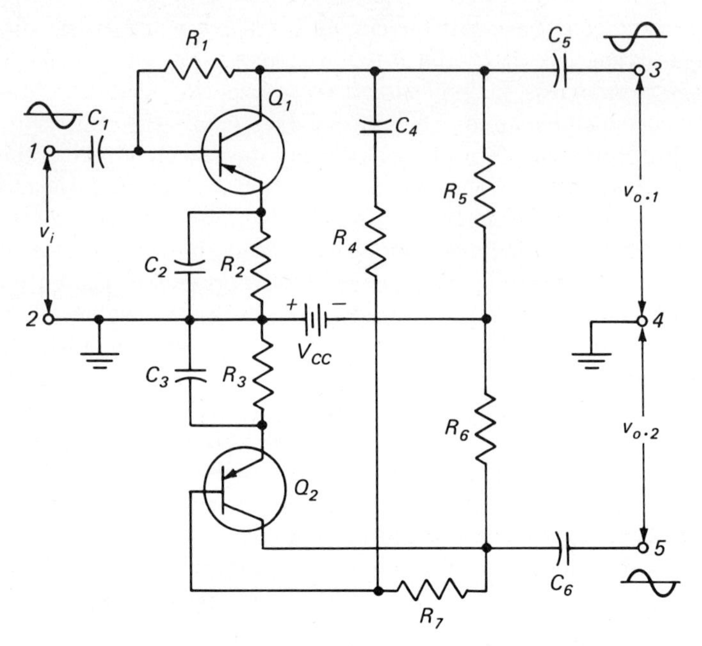

Fig. 16-24 Two-stage transistor phase-inverter circuit.

the single-transistor circuit are (1) a better balanced output is obtained, (2) a greater amount of power can be delivered to the following push-pull amplifier stage, and (3) degeneration due to the swamping resistors can be eliminated.

16-14 Complementary Symmetrical Push-Pull Amplifier

Principle of Operation. The principle of operation of a push-pull amplifier circuit using complementary symmetry is best explained with the aid of the basic circuit diagram of Fig. 16-25. Features of this circuit are (1) an NPN transistor is used for Q_1 and a PNP transistor for Q_2, (2) no d-c biasing voltages are applied and hence the transistors are operated as Class B, (3) the input circuits of the two transistors are connected in parallel to a single signal-input source, (4) the load R_2 is common to both transistors, and (5) in terms of the output circuits of the transistors they are considered as being connected in series when analyzed in terms of their d-c characteristics, and in parallel when analyzed in terms of their a-c signal characteristics.

With zero-signal input, the electron flow follows the path indicated by the solid-line arrows on Fig. 16-25, namely, $-V_{CC \cdot 1}$, $+V_{CC \cdot 2}$, $-V_{CC \cdot 2}$, collector-emitter junction of Q_2, emitter-collector junction of Q_1, $+V_{CC \cdot 1}$. It should be observed that no direct current flows through the load resistor. When an a-c signal is applied to the input circuit, the positive halves of the signal will produce a forward bias at Q_1 and a reverse bias at Q_2, thereby causing Q_1 to conduct and Q_2 to remain at cutoff. During the negative halves of the input signal, Q_2 has a forward bias and conducts while Q_1 has a reverse bias and remains at cutoff. The current-flow arrows on Fig. 16-25 show that the current in the load resistor is (1) zero when the input signal is zero, and (2) a replica of the input signal when an a-c signal is applied.

When used with sound-reproducing systems, a loudspeaker may be connected directly in place of the load resistor R_2 without producing excessive speaker-cone distortion. Because of the low resistance of the collector load, a relatively good impedance match can be obtained by (1) connecting the voice coil directly in place of R_2, or (2) connecting the voice coil in series with a resistor in place of R_2. The need for an impedance-matching output transformer is thereby eliminated. Because of the complementary action of the two symmetrical transistors, this circuit is called a *complementary symmetrical push-pull amplifier.*

Two-stage Complementary-symmetry Push-Pull Amplifier. Increased amounts of power output can be obtained by using two direct-coupled complementary-symmetry stages of amplification. The circuit of Fig. 16-26*a* uses

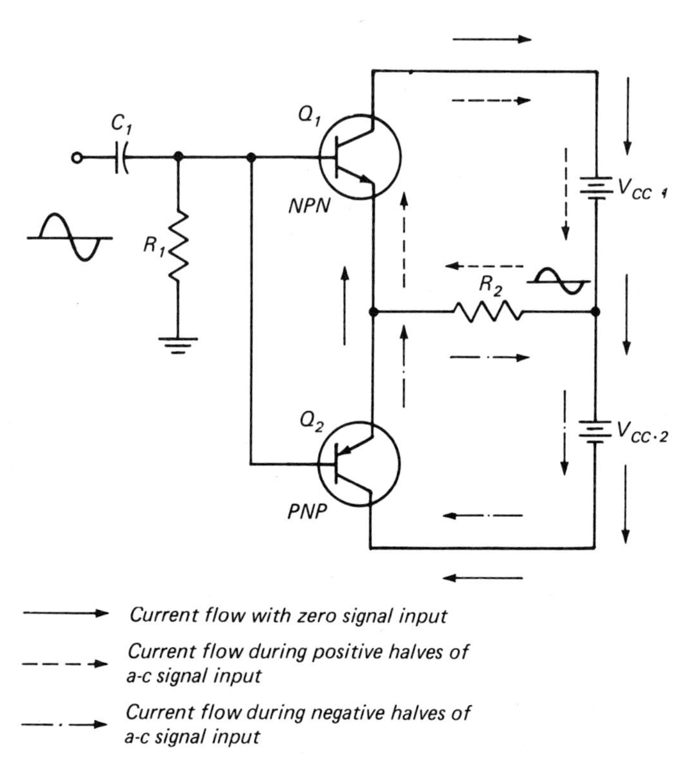

Fig. 16-25 Basic complementary symmetrical push-pull amplifier circuit.

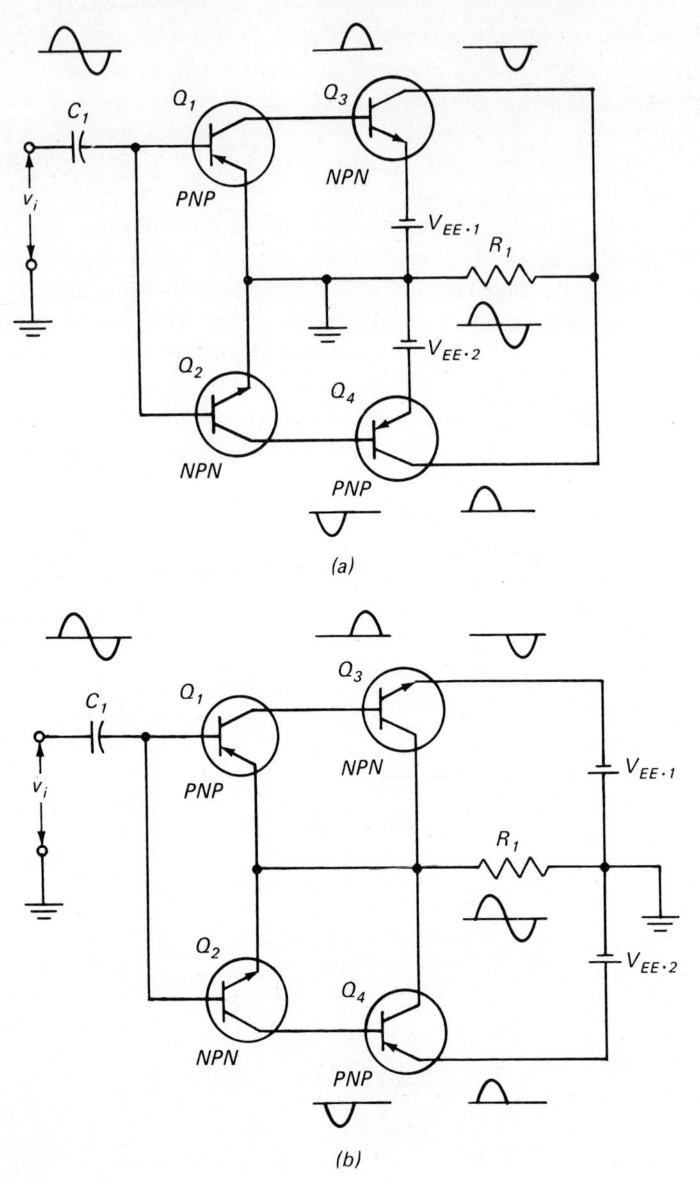

Fig. 16-26 Two-stage direct-coupled complementary symmetry amplifier (a) without feedback, (b) with feedback.

Q_1 and Q_2 in a common-emitter complementary-symmetry stage as a preamplifier for the common-emitter final amplifier stage employing Q_3 and Q_4. All transistors are assumed to be biased for Class B operation. The input signal v_i is applied to the base-emitter circuits of Q_1 and Q_2.

When the input signal is positive, the circuit actions are (1) Q_1 is reverse-biased and will not conduct, (2) the input signal to Q_3 is zero and Q_3 will remain at cutoff, (3) Q_2 is forward-biased and will conduct, (4) the output of Q_2 is reversed in phase and fed to the base of Q_4, (5) Q_4 is forward-biased and its collector current produces a positive-going signal in the load resistor R_1. When the input signal is negative, Q_2 and Q_4 will be at cutoff and the actions of Q_1 and Q_3 will produce a negative-going signal in the load resistor. The overall result is a signal current in the load resistor that is an amplified replica of the input signal v_i.

The circuit of Fig. 16-26*b* differs from that of Fig. 16-26*a* chiefly in that the grounding point for the circuit has been shifted so that the emitter currents of Q_1 and Q_2 now flow through the load resistor R_1. With this condition, negative feedback is introduced into the circuits of Q_1 and Q_2. This negative feedback increases the input resistance of Q_1 and Q_2 for those applications requiring a high input resistance. Otherwise, the actions of the circuit are the same as for the circuit of Fig. 16-26*a*.

16-15 Practical Push-Pull Amplifier Using Complementary Transistors

Using a Single Power Source. A wide variety of circuit configurations exists for amplifiers using the principle of complementary transistors to produce transformerless circuits; one example is shown in Fig. 16-27. In this circuit Q_1 and Q_2 operate as a two-stage Class A amplifier to feed Q_3 and Q_4 which are operated as Class AB transistors connected in series for complementary operation.

The two stages of the driver amplifier (Q_1Q_2) are operated in cascade. Transistor Q_1 is connected for common-emitter operation with a collector load resistor R_4 and an emitter swamping resistor R_5. The fixed forward d-c bias is established by R_1, R_2, R_3, and V_{CC}. Capacitor C_1 couples the signal from the preceding amplifier stage to Q_1 and also blocks any d-c of the preceding stage from the base of Q_1. Capacitor C_2 filters the a-c signal and any power-supply hum from the input of the amplifier. Negative feedback is provided by the emitter resistor R_5, called *d-c feedback;* and positive feedback is provided by C_3, R_6, and R_5, called *a-c feedback*. The purpose of the negative feedback is to (1) produce a high input impedance for the amplifier, and (2) add thermal stability to the amplifier. The positive feedback provides greater gain for the amplifier. Transistor Q_2 is also used in the common-emitter configuration with its collector load made up of R_5, R_6, R_7, and D_1; the bias for Q_2 is established by R_4, R_5, and Q_1.

The power-output stage consists of Q_3 and Q_4 connected in series for complementary Class AB operation. The quiescent bias for these transistors, provided by the action of R_5, R_6, R_7, D_1, and Q_2, is selected to provide the Class B operating point. Resistors R_8 and R_9 are of low values (0.5 ohm or lower) and are used to shift the operation of Q_3 and Q_4 to Class AB. The compensating diode D_1 (1) provides thermal stability, (2) eliminates variations in the bias of Q_3 and Q_4 caused by variations in resistance, power-supply voltage, etc., and (3) maintains a low-impedance path between the bases of Q_3 and Q_4. The output of the amplifier is taken from across the emitter-collector of Q_4 and is coupled to the loudspeaker voice coil by C_4, whose capacitance is in the order of 1,000 to 2,000 μf.

The operation of this circuit differs from the circuits of Figs. 16-25 and 16-26 in that (1) a single power source is used in the output stage, and (2) the output signal is taken from across one transistor, Q_4. In Fig. 16-27 the actions throughout the circuit due to changes in the input signal are as follows: (1) Q_1 and Q_2 are operated Class A, (2) Q_3 and Q_4 are operated Class B (or Class AB), (3) the NPN transistor Q_3 will conduct only with a positive-going input signal at its base, (4) the PNP transistor Q_4 will conduct only with a negative-going input signal at its base, and (5) the signal at the emitter of Q_4 (operating as an emitter follower) is of the same polarity as the input to its

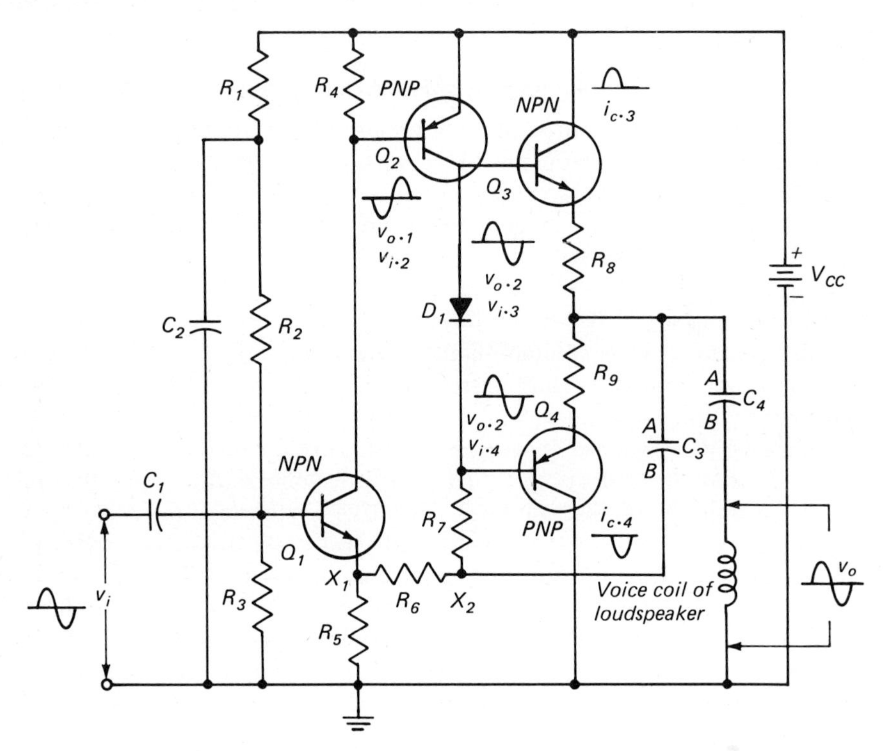

Fig. 16-27 Push-pull amplifier using complementary transistors (Q_3 and Q_4) and a single power source.

base. Therefore, a positive-going input signal at Q_1 results in a negative-going signal at the output of Q_1 and at the input of Q_2. The negative-going signal at the input of Q_2 results in a positive-going signal at the output of Q_2 and at the inputs of Q_3 and Q_4. The positive-going signals at the inputs of Q_3 and Q_4 cause Q_3 to start conducting and Q_4 to remain at cutoff. With a negative-going input signal at Q_1, Q_4 will conduct and Q_3 will remain at cutoff.

The output signal of the amplifier is produced in the following manner: When Q_3 is conducting and Q_4 is at cutoff, the path of electron flow is $-V_{CC}$, voice coil of the loudspeaker, C_4, R_8, Q_3, $+V_{CC}$; during this period C_4 (having a large capacitance) is charging. When Q_4 is conducting and Q_3 is at cutoff, the path of electron flow is $-$ terminal of C_4, voice coil of the loudspeaker, Q_4, R_9, $+$ terminal of C_4; during this period C_4 is discharging.

Positive (a-c) feedback for Q_1 is obtained in the following manner: When v_i is positive, Q_3 conducts through C_4 and the loudspeaker voice coil. Thus, point B of both C_3 and C_4 becomes more negative. While C_3 is charging through its a-c feedback circuit, the electron flow through R_5 makes point X_1 more positive, thereby increasing the effective negative feedback at Q_1. However, the electron flow through R_6 (part of which charges C_3) makes point X_1 more negative, thereby decreasing the effective negative feedback. In this circuit, R_6 is twice the value of R_5 and the net effect will be that of introducing positive feedback.

Quasi-complementary Push-Pull Amplifier. A PNP power transistor has a lower safe-area power rating than its NPN equivalent. It is also more difficult to control the base diffusion of a PNP power transistor than that of an equivalent NPN type, thus increasing the relative cost of the PNP transistor. Because of these characteristics, the complementary push-pull amplifier circuit of Fig. 16-25 is seldom used when the power-output level required exceeds 25 watts rms. One of the methods employed to compensate for the undesirable characteristics of the PNP power transistor is to use a quasi-complementary push-pull amplifier circuit (Fig. 16-28). In this circuit a low-current PNP transistor Q_4 is directly coupled to a high-current NPN transistor Q_5 to simulate a high-current PNP power transistor. This circuit requires five transistors; Q_1 in the predriver stage, Q_2 and Q_4 as complementary drivers for the pair of series-connected power-output NPN transistors Q_3 and Q_5. In order to obtain the maximum amount of power output the driver stages are connected for Class B operation. Bias voltage for Q_2 and Q_4 is obtained from R_1, R_2, R_3, Q_1, and V_{CC}. Two compensating diodes D_1 and D_2 provide thermal stability. The power-output transistors are connected in the emitter-follower configuration, and in order to minimize crossover distortion they are also connected for Class AB operation. Bias voltages for Q_3 and Q_5 are obtained from R_4 and R_5, respectively. The advantages of this circuit include (1) the power-output transistors Q_3 and Q_5 are of the more economical NPN type, (2) higher safe-area power ratings, (3) higher power output, (4) the use of Class B driver stages, and (5) lower relative cost.

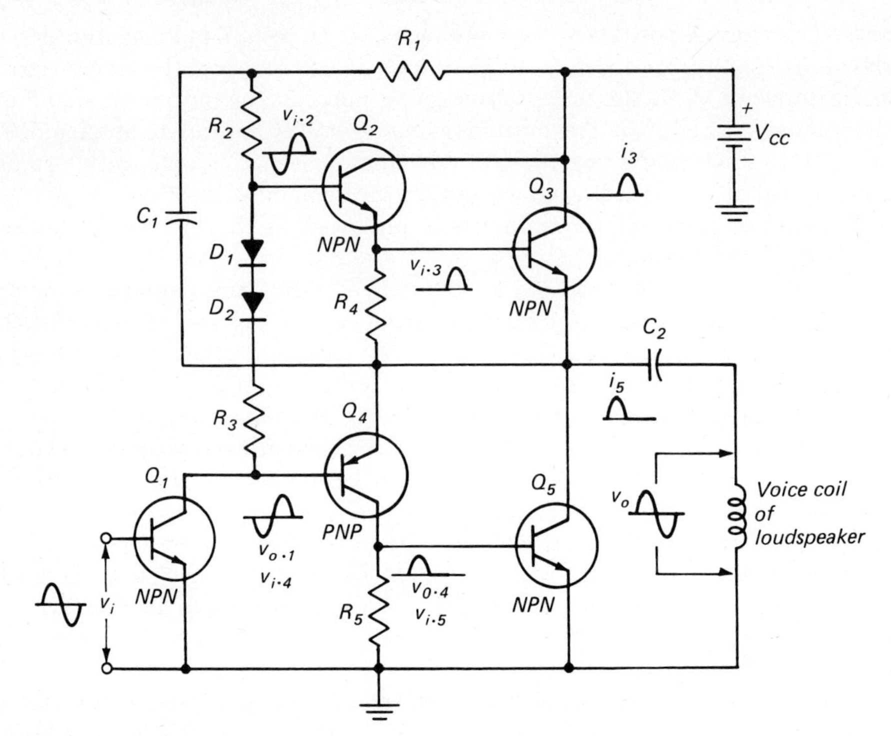

Fig. 16-28 Quasi-complementary push-pull amplifier circuit.

16-16 Vacuum-tube Push-Pull Amplifier Circuits

Basic Principles. Although triodes, pentodes, and beam-power tubes may be employed in vacuum-tube push-pull amplifier circuits, the pentode and beam-power tubes are most commonly used. A comparison of the transistor collector characteristic curves of Fig. 16-9 with the pentode plate characteristic curves of Fig. 16-13 and the beam-power tube plate characteristic curves of Fig. 16-14 will show that these three sets of family characteristic curves are similar to each other. The analysis of pentode- and beam-power-tube push-pull amplifier circuits will therefore be similar to the analysis of the transistor push-pull amplifier circuits of Arts. 16-11 to 16-13. The basic push-pull vacuum-tube amplifier circuit using both input and output transformers is similar to the circuit shown in Fig. 16-17*d*.

Phase Inverter. The input transformer in a vacuum-tube push-pull amplifier may be eliminated in a similar manner to its elimination in the transistor push-pull amplifier circuit. A resistance-capacitance-coupled push-pull power-amplifier circuit is shown in Fig. 16-29. In this circuit, (1) the input signal e_i is applied to the grid of VT_1, (2) phase inversion takes place between the input and output of VT_1, (3) the output of VT_1 is applied to the grid of VT_3, (4) a portion, determined by the voltage divider (R_1R_2), of the output signal of VT_1 is applied to the grid of VT_2, (5) phase inversion takes place between the input and output of VT_2, (6) the output of VT_2 is applied to the grid of VT_4, and (7) power output is taken from the Class B operated power-output tubes VT_3 and VT_4. When the input signal e_i is positive-going, (1) a negative-going signal is applied to VT_3 and this tube remains at cutoff, (2) the input to VT_2 is negative-going and its output is positive-going, and (3) the input to VT_4 is positive-going and the tube is conducting. When the input signal e_i is negative-going, VT_3 will be conducting and VT_4 will be cut off. The above actions occur instantaneously. In this manner, the input voltages at the two power-output tubes will always be 180 degrees out of phase. In order that the magnitude of the two voltages applied to the grids of VT_3 and VT_4 will always be equal, the voltage e_i'' applied to the input circuit of the phase inverter should always be equal to the magnitude of the input voltage e_i. This is accomplished by making the ratio of $(R_1 + R_2)$ to R_2 equal to the voltage amplification of the driving amplifier circuit of VT_1. The characteristics of VT_1 and VT_2 should be identical, and hence a twin triode is generally used for the driving amplifier and the phase inverter. As the output circuits of these two tubes should also be identi-

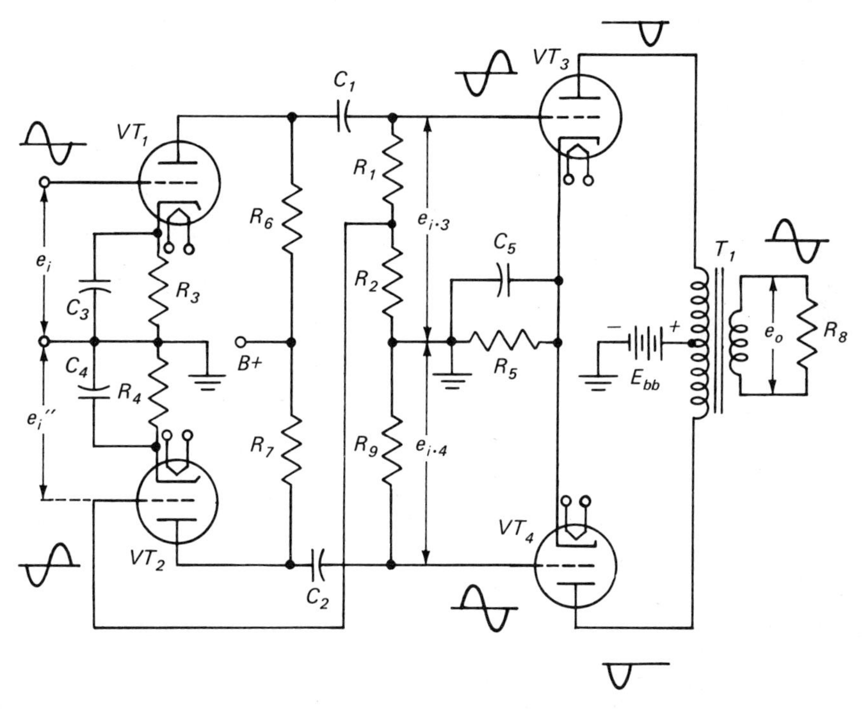

Fig. 16-29 Push-pull amplifier using vacuum tubes for phase-inverter and power-output stages.

cal, the plate-coupling resistors R_6 and R_7 are of equal values, and the grid resistor R_9 is equal to $R_1 + R_2$.

From the actions of the circuit of Fig. 16-29 it can be seen that the outputs of both VT_3 and VT_4 are negative-going signals. As the output currents of these two tubes flow in opposite directions through their respective portions of the center-tapped primary winding of T_1, the voltage e_o at the secondary winding will be an amplified replica of the input signal e_i.

16-17 Negative Feedback

Any distortion developed in a power-amplifier circuit can be reduced to a negligible amount by the use of negative feedback. The advantages, operation, limitations, and basic circuit connections for negative feedback as applied to power-amplifier circuits are similar to those for the voltage-amplifier circuits as explained in Arts. 15-1 to 15-8. The equations used in the calculations of feedback in voltage-amplifier circuits may also be used in the calculation of feedback for power-amplifier circuits. As the power output of most radio receivers exceeds the amount usually required for the average home receiver, the loss of power because of negative feedback is of little consequence.

QUESTIONS

1. (*a*) What is the purpose of a power amplifier? (*b*) Name four types of transducers that may be used as a power-amplifier load.
2. (*a*) What is the difference between a voltage amplifier and a power amplifier? (*b*) Where does the power amplifier obtain the increased power it delivers?
3. (*a*) What is the purpose of the active device in a power amplifier? (*b*) Name four types of active devices used in power-amplifier circuits.
4. (*a*) Name seven advantages that have made the transistor more desirable as an active device rather than the vacuum tube. (*b*) Describe the advantages of three circuit configurations commonly used in transistor power-amplifier circuits.
5. (*a*) What is meant by a single-ended stage? (*b*) What are its requirements? (*c*) What types of loading may be used?
6. (*a*) What is the purpose of the driver amplifier? (*b*) What are its requirements?
7. How does the loss of power released from the d-c power source that is not converted to signal power affect the (*a*) circuit efficiency? (*b*) Active device?
8. Name the critical parameters of power transistors used for (*a*) general-circuit applications, (*b*) linear-circuit applications, (*c*) switching-circuit applications.
9. (*a*) What is meant by the term safe operating area? (*b*) What are the factors that determine the SOA?
10. (*a*) What is the purpose of the maximum-power dissipation curve? (*b*) How may it be drawn?
11. (*a*) Why is it necessary to control the junction temperature of a power transistor? (*b*) What are the factors that affect the junction temperature?
12. What is meant by the following symbols and terms: (*a*) P_T? (*b*) T_A? (*c*) T_C? (*d*) T_{MF}? (*e*) Derating factor? (*f*) Derating graph?
13. (*a*) What is a heat sink? (*b*) Why is it necessary to use a heat sink?

14. Compare silicon and germanium transistors for four important operating factors.

15. (*a*) Name the critical parameters of power tubes. (*b*) Describe what is meant by each of the above parameters.

16. Explain why the theoretical maximum efficiency of a Class A transistor power amplifier cannot exceed 50 per cent.

17. (*a*) What two factors must be taken into consideration in drawing a load line? (*b*) What three circuit parameters determine the drawing of a load line?

18. What factors determine the drawing of a load line to produce maximum output with minimum signal input for a Class A amplifier?

19. (*a*) What is the determining factor about the characteristic curves of a transistor that causes harmonic distortion? What is the effect on a sine wave by the addition of (*b*) odd harmonics? (*c*) Even harmonics?

20. (*a*) How does the source resistance of a transistor affect the overall harmonic distortion? (*b*) How does having an input resistance that is much higher than its source resistance affect the overall harmonic distortion? (*c*) How may the distortion produced in (*a*) and (*b*) be eliminated?

21. (*a*) What is the maximum theoretical efficiency that can be obtained from a Class A vacuum-tube amplifier? (*b*) What is the maximum practical efficiency that can be obtained from a Class A vacuum-tube amplifier? (*c*) How do these efficiencies compare to those of a transistor Class A amplifier?

22. (*a*) What factors determine the location of the load line for a triode tube? (*b*) What two points are required to draw this load line?

23. What are the features of the characteristic curves for a triode tube that causes second-harmonic distortion?

24. What factors determine the location of the load line for a pentode tube?

25. Explain how distortion in pentodes can be minimized by varying the value of the load resistance.

26. Explain how distortion in pentodes can be minimized by (*a*) varying the zero-signal grid bias, (*b*) varying the value of the input signal.

27. Explain how nonlinear distortion is produced in pentode amplifier circuits.

28. What three factors limit the power output of a single-ended transistor Class A power amplifier?

29. (*a*) What is the purpose of the output transformer? (*b*) How is this transformer used as an impedance-matching device?

30. (*a*) Why are triode tubes rarely used in power-amplifier circuits? (*b*) What factors limit the maximum power output that can be obtained from a Class A amplifier using pentodes or beam-power tubes in sound-reproduction circuits?

31. (*a*) What is the basic principle of operation of a push-pull amplifier? (*b*) How is a balanced circuit obtained?

32. Using Fig. 16-17, explain the operation of a push-pull amplifier for a signal input at the primary of T_1 to the signal output at the secondary of T_2.

33. Why is a double-ended amplifier called (*a*) a push-pull amplifier? (*b*) A balanced amplifier?

34. Explain an important disadvantage in operating an audio power-amplifier active device in Class A.

35. (*a*) Describe two important advantages of push-pull power-amplifier circuits. (*b*) Explain how it is possible for a push-pull power amplifier to deliver approximately four times the power of a single-ended amplifier using an identical active device.

36. Explain the disadvantage of using the following type of operation of an active device for audio transistor push-pull power amplifiers: (*a*) Class A, (*b*) Class B.
37. Describe the operation of the various circuit elements of the push-pull power amplifier of Fig. 16-18.
38. In the circuit of Fig. 16-18, how are the variations in collector current, collector power dissipation, and d-c operating point minimized for changes in the ambient temperature?
39. (*a*) What is meant by crossover distortion? (*b*) Explain how this type of distortion can be eliminated.
40. (*a*) Explain why the magnetizing effects on the magnetic core of the output transformer in a push-pull circuit is zero. (*b*) Explain why it is possible to use a much smaller output transformer in a push-pull amplifier than would be required for a single-ended circuit using the same type of device.
41. Explain why the response of a power amplifier using an output transformer is (*a*) nonlinear for a single-ended circuit, and (*b*) linear for a push-pull circuit.
42. Explain graphically the operation of a Class AB push-pull transistor amplifier using the characteristic curves of Fig. 16-21.
43. Explain how the a-c power output and efficiency of a Class AB push-pull amplifier compare with a single-ended Class A amplifier using identical active devices for (*a*) maximum power output with maximum permissible distortion, (*b*) maximum power output with no limit on the amount of distortion permitted.
44. Explain (*a*) the cause for second-harmonic distortion in transistor amplifiers, (*b*) how this distortion is eliminated in a push-pull amplifier.
45. Explain how it is possible to obtain a signal output from a Class AB push-pull amplifier that is devoid of any third-harmonic distortion.
46. Why is it necessary to use a phase inverter in a resistance-capacitance-coupled push-pull power-amplifier circuit?
47. Explain the circuit operation of a single-stage transistor phase inverter.
48. Explain the circuit operation of Fig. 16-23 when the input signal (*a*) increases the effective forward bias of Q_1, (*b*) decreases the effective forward bias of Q_1.
49. What are the disadvantages of using a single-stage transistor phase inverter?
50. Explain the operation of the various circuit elements of the two-stage phase inverter of Fig. 16-24.
51. What are the advantages of using a two-stage two-transistor phase inverter?
52. Explain five important features of the basic complementary symmetrical push-pull amplifier of Fig. 16-25.
53. Explain the circuit operation of the basic complementary symmetrical push-pull amplifier of Fig. 16-25 for (*a*) zero-input signal, (*b*) positive halves of an a-c signal input, (*c*) negative halves of an a-c signal input.
54. Explain how the output transformer can be eliminated in a complementary symmetrical push-pull amplifier.
55. (*a*) What is the advantage of using a two-stage directly coupled complementary symmetrical amplifier? (*b*) What are the basic circuit elements in this type of amplifier?
56. Explain the circuit actions of the two-stage complementary push-pull amplifier of Fig. 16-26*a* for (*a*) a positive input signal, (*b*) a negative input signal.
57. (*a*) What is the purpose of using negative feedback in a two-stage complementary

push-pull amplifier? (*b*) How can the circuit of Fig. 16-26*a* be altered to provide negative feedback?

58. Explain the operation of the various circuit elements in the driver stages of the complementary push-pull amplifier of Fig. 16-27.

59. (*a*) What are the reasons for using a negative-feedback circuit in the complementary push-pull amplifier of Fig. 16-27? (*b*) How is negative feedback obtained in this circuit?

60. Explain the operation of the various circuit elements in the power-output section of the complementary push-pull amplifier of Fig. 16-27.

61. How does the operation of the complementary push-pull amplifier of Fig. 16-27 differ from the operation of the circuits of Figs. 16-25 and 16-26?

62. Explain the circuit actions of the complementary push-pull circuit of Fig. 16-27 for (*a*) a positive-going signal at the input of Q_1, (*b*) a negative-going signal at the input of Q_1.

63. Explain how the output signal is produced in the complementary push-pull circuit of Fig. 16-27.

64. (*a*) What is the advantage of using the positive feedback in the complementary push-pull amplifier of Fig. 16-27? (*b*) Explain how this positive feedback is obtained.

65. What are the disadvantages of using a PNP transistor in power-amplifier circuits?

66. Explain the functions of the various circuit elements in the quasi-complementary push-pull amplifier of Fig. 16-28.

67. What are the advantages of the quasi-complementary push-pull circuit?

68. Explain why the analysis of the operation of push-pull pentode- and beam-power-tube amplifiers is similar to the analysis of transistor push-pull amplifiers.

69. Explain the circuit operation of the phase inverter of Fig. 16-29.

70. Explain the circuit actions of the phase inverter of Fig. 16-29 when (*a*) a positive-going signal is applied to the input of VT_1, (*b*) a negative-going signal is applied to the input of VT_1.

71. How is the magnitude of the two voltages applied to the grids of VT_3 and VT_4 in Fig. 16-29 kept equal?

72. How do the advantages, operation, limitations, and basic circuit connections for negative feedback power-amplifier circuits compare to those for voltage-amplifier circuits?

PROBLEMS

1. A common-emitter power-amplifier transistor is operated on its characteristic curves with quiescent values of $V_{CE\cdot Q} = 12$ volts and $I_{C\cdot Q} = 350$ ma. The output-signal voltage varies from 2 to 22 volts, and the output-signal current varies from 150 to 550 ma. Determine (*a*) the d-c input power to the active device, (*b*) the a-c output power delivered to the load, and (*c*) the efficiency of the active device.

2. A common-emitter power-amplifier transistor is operated on its characteristic curves with quiescent values of $V_{CE\cdot Q} = 12$ volts and $I_{C\cdot Q} = 350$ ma, and the output-signal voltage varies from 2 to 22 volts. Determine (*a*) the minimum and maximum signal currents required to produce an a-c output power of 1.5 watts, (*b*) the d-c input power to the active device, and (*c*) the efficiency of the active device.

3. A common-cathode power-amplifier tube is operated on its characteristic curves with quiescent values of $E_{P \cdot Q} = 205$ volts and $I_{P \cdot Q} = 75$ ma. The output-signal voltage varies from 140 to 270 volts, and the output-signal current varies from 52 to 105 ma. Determine (*a*) the d-c input power to the active device, (*b*) the a-c output power delivered to the load, and (*c*) the efficiency of the active device.
4. A common-cathode power-amplifier tube is operated on its characteristic curves with quiescent values of $E_{P \cdot Q} = 237.5$ volts and $I_{P \cdot Q} = 65$ ma, and the output-signal current varies from 10 to 115 ma. Determine (*a*) the minimum and maximum signal voltages required to produce an a-c output power of 3.2 watts, (*b*) the d-c input power to the active device, and (*c*) the efficiency of the active device.
5. A certain power transistor has a maximum power-dissipation rating of 5 watts up to 75°C. The derating factor is 0.04 watt/°C, and the operating temperature range at the junction is −65 to 200°C. Find the maximum permissible power dissipation for 125°C.
6. Determine the operating temperature for the power transistor used in Prob. 5 for a power dissipation of 0.2 watt.
7. A certain power transistor used for audio and inverter circuits in 12-volt mobile radio equipment has a maximum power-dissipation rating of 29 watts up to 25°C, with a temperature operating range at its junction of −65 to 200°C. Using the derating chart of Fig. 16-5 determine the maximum permissible power dissipation at a junction temperature of (*a*) 100°C, (*b*) 175°C.
8. A certain power transistor used in complementary-symmetry output stages of audio-amplifier circuits has a maximum power-dissipation rating of 83 watts up to 25°C, with a temperature operating range at its junction of −65 to 150°C, and derates linearly. Determine the maximum permissible power dissipation at a junction temperature of (*a*) 75°C, (*b*) 135°C.
9. A triode power-amplifier tube has an a-c power output of 3.2 watts when operated with a plate voltage of 240 volts and a plate current of 50 ma. Determine (*a*) the plate efficiency, (*b*) the plate dissipation.
10. A pentode power-amplifier tube has an a-c power output of 3.88 watts when operated with a plate voltage of 315 volts and a plate current of 35.5 ma. Determine (*a*) the plate efficiency, (*b*) the plate dissipation.
11. Determine the power sensitivity of the tube used in Prob. 9 when the maximum value of the input signal is 43.5 volts.
12. Determine the power sensitivity of the tube used in Prob. 10 when the maximum value of the input signal is 16.5 volts.
13. Determine the a-c power output of a Class A power amplifier at maximum efficiency of the active device using the transistor and operating parameters of Prob. 1.
14. Determine the a-c power output of a Class A power amplifier at maximum efficiency of the active device using the vacuum tube and operating parameters of (*a*) Prob. 3, (*b*) Prob. 4.
15. Using the load line in Fig. 16-7, the circuit of Fig. 16-8, and $V_{CE \cdot Q} = 35$ volts, determine (*a*) the required value of load resistance, and (*b*) the efficiency of the active device with an input signal having a peak value of 40 ma.
16. Using the curves of Fig. 16-7, the circuit of Fig. 16-8, a load resistance of 21 ohms, $V_{CE \cdot Q} = -29$ volts, and an input signal having a peak value of 50 ma, determine (*a*) $I_{C \cdot Q}$, (*b*) V_{CC}, and (*c*) the efficiency of the active device.
17. The transistor whose collector characteristics are shown in Fig. 16-9 is to be oper-

ated along the load line shown on this figure with operating parameters $V_{CE \cdot Q}$ and $I_{C \cdot Q}$ at point G on this load line. Find the value of the required load resistance.

18. The transistor whose collector characteristics are shown in Fig. 16-9 is to be operated with $V_{CC} = -45.5$ volts, $V_{CE \cdot Q} = -24$ volts, and $I_{C \cdot Q} = 1.05$ amp. Find the value of the required load resistance.

19. The transistor and circuit used in Prob. 17 are to be operated with an input signal whose peak value is 40 ma. Determine *(a)* the d-c power input to the active device, *(b)* the a-c power output delivered to the load, and *(c)* the efficiency of the active device.

20. The transistor and circuit used in Prob. 18 are to be operated with an input signal whose peak value is 30 ma. Determine *(a)* the d-c power input to the active device, *(b)* the a-c power output delivered to the load, and *(c)* the efficiency of the active device.

21. For the transistor and circuit used in Probs. 17 and 19 determine the per cent of *(a)* second-harmonic distortion, *(b)* third-harmonic distortion, and *(c)* total second- and third-harmonic distortion.

22. For the transistor and circuit used in Probs. 18 and 20 determine the per cent of *(a)* second-harmonic distortion, *(b)* third-harmonic distortion, and *(c)* total second- and third-harmonic distortion.

23. What value of grid bias is recommended for the triode power-amplifier tube whose characteristics are given in Fig. 16-12 when $E_P = 200$ volts, $I_P = 60$ ma, and $\mu = 4.2$, *(a)* using Eq. 16-14? *(b)* Using the characteristic curves?

24. *(a)* What value of grid bias is recommended for a triode power-amplifier tube when $E_P = 225$ volts and $\mu = 4.2$? *(b)* Determine the operating plate current for the grid bias obtained in *(a)* using the characteristic curves of Fig. 16-12.

25. *(a)* Using the load line and characteristic curves of Fig. 16-12, determine the value of load resistance required when $E_{P \cdot Q} = 250$ volts, $I_{P \cdot Q} = 60$ ma, and the peak input signal is 33.5 volts. *(b)* What value of B supply voltage is required with the plate load resistance found in *(a)*?

26. Using the characteristic curves of Fig. 16-12 determine the load resistance required when $E_{P \cdot Q} = 200$ volts, $E_{bb} = 300$ volts, $I_{P \cdot Q} = 60$ ma, and the peak input signal is 31.5 volts, *(a)* using Eq. (16-15), *(b)* using Ohm's law.

27. For the circuit and operating characteristics for the triode tube used in Prob. 25 determine *(a)* the d-c input power to the active device, *(b)* the a-c output power delivered to the load, *(c)* the efficiency of the active device, and *(d)* the efficiency of the output circuit.

28. For the circuit and operating characteristics for the triode tube used in Prob. 26 determine *(a)* the d-c input power to the active device, *(b)* the a-c output power delivered to the load, *(c)* the efficiency of the active device, and *(d)* the efficiency of the output circuit.

29. Determine the per cent of second-harmonic distortion for the amplifier circuit of Prob. 25.

30. Determine the per cent of second-harmonic distortion for the amplifier circuit of Prob. 26.

31. The pentode tube whose characteristics are shown in Fig. 16-13 is operated at the quiescent point Q_1 along the load line A_1B with a peak input signal of 12.5 volts. Determine the plate load resistance *(a)* using Eq. (16-15), *(b)* using Ohm's law.

32. The pentode tube whose characteristics are shown in Fig. 16-13 is operated at the quiescent point Q_2 along the load line A_2B with a peak input signal of 11.5 volts. Determine the plate load resistance (*a*) using Eq. (16-15), (*b*) using Ohm's law.

33. For the circuit and operating characteristics for the pentode tube used in Prob. 31, and ignoring the power lost in the screen-grid resistor, determine (*a*) the d-c input power to the active device, (*b*) the a-c output power delivered to the load, and (*c*) the plate efficiency.

34. For the circuit and operating characteristics for the pentode tube used in Prob. 32, and ignoring the power lost in the screen-grid resistor, determine (*a*) the d-c input power to the active device, (*b*) the a-c output power delivered to the load, and (*c*) the plate efficiency.

35. For the circuit and operating characteristics for the pentode tube used in Probs. 31 and 33 with $I_{g \cdot 2} = 10.5$ ma, determine (*a*) the plate-circuit efficiency, (*b*) the output-circuit efficiency.

36. For the circuit and operating characteristics for the pentode tube used in Probs. 32 and 34 with $R_5 = 29{,}000$ ohms, determine (*a*) the plate-circuit efficiency, (*b*) the output-circuit efficiency.

37. Determine the power sensitivity of the pentode tube and circuit used in Prob. 33.

38. Determine the power sensitivity of the pentode tube and circuit used in Prob. 34.

39. For the circuit and operating characteristics for the pentode tube used in Prob. 31 determine (*a*) per cent second-harmonic distortion, (*b*) per cent third-harmonic distortion, (*c*) total per cent second- and third-harmonic distortion.

40. For the circuit and operating characteristics for the pentode tube used in Prob. 32 determine (*a*) per cent second-harmonic distortion, (*b*) per cent third-harmonic distortion, (*c*) total per cent second- and third-harmonic distortion.

41. The power-amplifier circuit of Fig. 16-15 uses a transistor whose characteristics are shown in Fig. 16-9. The transistor is to be operated Class A with an input signal having a peak value of 40 ma; other characteristics are $V_{CC} = -46$ volts, collector saturation current = 2 amp, and $I_{B \cdot Q} = 40$ ma. Determine the efficiency of the amplifier.

42. The power-amplifier circuit of Fig. 16-15 uses a transistor whose characteristics are shown in Fig. 16-9. The transistor is to be operated Class A with an input signal having a peak value of 30 ma; other characteristics are $V_{CC} = -48$ volts, collector saturation current = 2 amp, and $I_{B \cdot Q} = 30$ ma. Determine the efficiency of the amplifier.

43. What is the total per cent of second- and third-harmonic distortion for the amplifier circuit of Prob. 41?

44. What is the total per cent of second- and third-harmonic distortion for the amplifier circuit of Prob. 42?

45. Determine the primary-to-secondary turns ratio of the output transformer used in Prob. 41 if the resistance of the voice coil is 8 ohms.

46. Determine the primary-to-secondary turns ratio of the output transformer used in Prob. 42 if the resistance of the voice coil is 3.2 ohms.

47. The power-amplifier circuit of Fig. 16-16*b* uses a pentode tube whose characteristics are shown in Fig. 16-13. The tube is to be operated Class A with an input signal having a peak value of 15 volts; other parameters are $E_{bb} = 495$ volts, $E_{C \cdot Q} = -15$ volts, and $R_o = 5{,}000$ ohms. Determine the efficiency of this circuit.

48. The power-amplifier circuit of Fig. 16-16*b* uses a pentode tube whose characteristics are shown in Fig. 16-13. The tube is to be operated Class A with an input signal having a peak value of 10 volts; other parameters are $E_{bb} = 495$ volts, $E_{C \cdot Q} = -20$ volts, and $R_o = 9{,}000$ ohms. Determine the efficiency of this circuit.

49. Determine the total per cent of second- and third-harmonic distortion for the amplifier circuit of Prob. 47.

50. Determine the total per cent of second- and third-harmonic distortion for the amplifier circuit of Prob. 48.

51. Determine the primary-to-secondary turns ratio of the output transformer used in Prob. 47 if the resistance of the voice coil of the loudspeaker is 8 ohms.

52. Determine the primary-to-secondary turns ratio of the output transformer used in Prob. 48 if the resistance of the voice coil of the loudspeaker is 15 ohms.

53. The push-pull power-amplifier circuit of Fig. 16-18 uses two identical transistors whose characteristics are shown in Figs. 16-9 and 16-21. The transistors are operated Class AB at $V_{CE \cdot Q} = -45$ volts and $I_{B \cdot Q} = -10$ ma, with a signal input whose maximum positive peak voltage causes $v_{c \cdot \min}$ to be -10 volts. Determine (*a*) the a-c power output delivered to the load, (*b*) the d-c collector power input to the active devices, (*c*) the efficiency of the active devices, and (*d*) the collector-to-collector load resistance.

54. The push-pull power-amplifier circuit of Fig. 16-18 uses two identical transistors whose characteristics are shown in Figs. 16-9 and 16-21. The transistors are operated Class AB at $V_{CE \cdot Q} = -45$ volts and $I_{B \cdot Q} = -10$ ma, with a signal input whose maximum positive peak voltage causes $v_{c \cdot \min}$ to be -15 volts. Determine (*a*) the a-c power output delivered to the load, (*b*) the d-c collector power input to the active devices, (*c*) the efficiency of the active devices, and (*d*) the collector-to-collector load resistance.

Chapter 17

Oscillator Circuits

The frequency of the varying currents and voltages associated with electronic circuits may vary from a few cycles per second to millions of cycles per second. An active device, such as a transistor or vacuum tube, when used in conjunction with the proper combination of circuit elements, may be made to produce an alternating current having almost any value of frequency. The active device does not create any electrical energy; it merely changes one kind of current to another. The composite electric circuit associated with an active device when used to produce an alternating current is called an *oscillator circuit.* In addition to having a wide frequency range, the frequency and amplitude of the output of electronic oscillator circuits are comparatively easy to control. An electronic oscillator circuit can be designed so that its output is devoid of any harmonics or so that it is rich in harmonic content.

17-1 Types of Oscillator Circuits

The types and kinds of oscillator circuits are many and varied. It is beyond the scope of this text to discuss every type of oscillator circuit; however, a brief outline description of some important types used in communications and electronics will be given.

General Classifications of Oscillator Circuits. Electronic oscillator circuits may be broadly divided into two groups: (1) those circuits used to produce *nonsinusoidal waves,* and (2) those which produce *sinusoidal waves.*

Oscillators Producing Nonsinusoidal Waveform. Oscillator circuits producing a nonsinusoidal waveform are used as electronic timing and control circuits in television, radar, oscilloscope, and industrial control equipment. Nonsinusoidal voltages are generally produced by some form of *relaxation oscillator circuit.* In this type of oscillator circuit, one or more voltages or currents change abruptly one or more times during each cycle of oscillation. Among the types of relaxation oscillators are (1) multivibrator, (2) glow-tube discharge, (3) sawtooth-wave generators, (4) rectangular- or square-wave generators.

Oscillators Producing Sinusoidal Waveform. Among the types of oscillators used to produce a sinusoidal voltage are (1) negative resistance, (2) feedback, (3) heterodyne, (4) crystal, (5) magnetostriction, (6) ultrahigh frequency.

A circuit element is said to possess negative resistance when at some portion of its operating characteristic the current through the circuit decreases with an increase in the voltage and vice versa. Any circuit having a negative-resistance characteristic can be used as an oscillator. In some modes of operation, transistors, tunnel diodes, and vacuum tubes have negative-resistance characteristics that allow their use in oscillator circuits.

A tuned oscillatory circuit combined with positive feedback is the type of oscillator circuit most commonly used in communications circuits. There are numerous circuits that may be used as feedback oscillators; however, for the purpose of classification only the basic circuits are listed, as all others are merely modifications of these circuits. The basic feedback oscillator circuits are:

1. Single-tuned circuit using inductive, capacitive, or tickler feedback.
2. Hartley oscillator
3. Colpitts oscillator
4. Complex types using more than one tuned circuit
5. Resistance-capacitance tuned circuit

The heterodyne or beat-frequency oscillator consists of two oscillator circuits that operate at slightly different frequencies. The outputs of these oscillators are simultaneously applied to a common detector. By means of a filter, the higher r-f currents are removed, and the output will be of a frequency equal to the difference of the two original frequency values. The heterodyne oscillator circuit is a means of obtaining precise audio frequencies and is commonly used in test equipment operating at audio frequencies.

Crystal-controlled oscillators are used when the frequency of oscillation must be maintained at a fixed value. The crystal is not used to produce oscillations but controls the output frequency of the oscillator with which it is used.

The magnetostriction oscillator circuit is based on the principle that a change in magnetization will cause a magnetic material to expand or contract and, conversely, that a contraction or expansion of a magnetic material will cause a change in magnetization. A strong stable oscillation having a frequency of the order of 10,000 to 100,000 Hz can be obtained from this type of oscillator circuit.

The types of oscillator circuits used to generate a-f or r-f currents or voltages cannot be used to produce ultrahigh-frequency currents or voltages. Various means are used to obtain these ultrahigh frequencies, such as (1) magnetron, (2) velocity modulation, (3) resonant cavities, (4) resonant lines. The theory and analysis of ultrahigh-frequency oscillators are beyond the scope of this text.

17-2 The Amplifier as an Oscillator

Basic Principle. The block diagram of the basic oscillator (Fig. 17-1) shows that the output of the active device provides (1) a feedback signal to its input circuit, and (2) the desired a-c output signal v_o. In order to sustain an a-c output signal, the output power of the active device ($P_{o\cdot\text{a-d}}$) must equal the sum of (1) the power output demanded of the oscillator ($P_{o\cdot o}$), and (2) the power loss in the feedback network (P_f); or

$$P_{o\cdot\text{a-d}} = P_{o\cdot o} + P_f \tag{17-1}$$

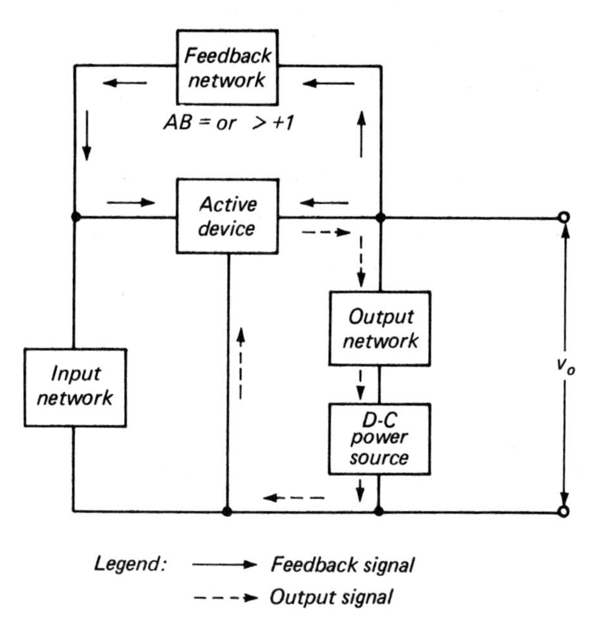

Fig. 17-1 Block diagram of a basic oscillator circuit.

Also, the resonant frequency of the frequency-controlling circuit must be within the maximum frequency limit of the active device.

Comparing Figs. 17-1 and 15-1 should lead to the conclusion that the oscillator is essentially a feedback amplifier that supplies its own input signal.

Amplifier Action of the Oscillator. In the oscillator of Fig. 17-1, a portion of the energy developed in the output circuit is returned in the proper phase relationship to the input circuit. The energy returned is amplified by the active device, and a portion of the energy developed in the output circuit is then returned to the input circuit. Each time some energy is taken from the output circuit it is regenerated in the input circuit. This cycle of operations is continually repeated. Because of the amplifying properties of the active device, the energy in the output circuit will increase with each cycle of operation until the maximum or saturation value of output current has been reached (see Fig. 17-2). The value of the output current will depend upon the characteristics of the active device and the manner in which it is operated.

As the energy consumed by the input circuit of an amplifier is considerably less than that in its output circuit, it is possible to have an amplifier supply its own input. An amplifier operated in this manner will generate oscillations at a frequency that is determined by the electrical constants of the circuit. Furthermore, since the active device operates as an amplifier, the oscillator can be made to supply power to an external circuit in addition to supplying the circuit losses required to sustain oscillations. The active device thus acts as a power converter, changing the d-c power supplied to its output circuit into a-c energy in the amplifier output circuit. It may thus be seen that the func-

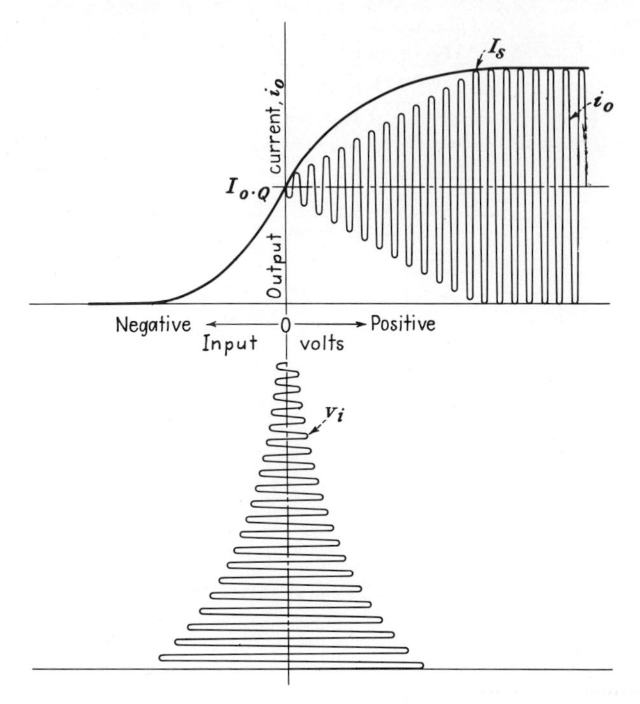

Fig. 17-2 Illustration of how the oscillations in the active device build up until the saturation value of the output current is reached.

tion of the oscillator is not to create energy but to change d-c energy to a-c energy.

Classification of Operation. The electronic oscillator can be considered as an amplifier in which part of the output voltage is returned to the input circuit in such a manner that the active device drives itself. The active device may be operated as a Class A, B, or C amplifier. Class A operation is used in high-quality a-f oscillators. Because high efficiency and low distortion are obtainable at high frequencies from Class C operation, r-f oscillators are usually operated Class C.

Oscillator Circuits. A variety of circuits may be used to produce the required oscillations, each circuit having its own advantages and disadvantages. The circuit used will depend upon the frequency and power required and its application. The oscillatory circuit may be in the input circuit, the output circuit, or both. Feedback energy can be applied to the input circuit by either capacitive, inductive, transformer, or electron coupling.

17-3 The Oscillatory Circuit

Purpose. It is often required that the waveform of the output voltage of an oscillator should be a sine wave. If the output network of the basic oscillator of Fig. 17-1 is predominantly resistive, the output voltage v_o will have abrupt changes.

A common method used to obtain a sine-wave output is to use a tuned circuit in either the output or input networks of the oscillator. In addition, the tuned circuit also controls the frequency of the oscillator output voltage. The tuned circuit consists basically of an inductor and a capacitor connected either in series or in parallel and is commonly called the *oscillatory circuit* or the *tank circuit.*

Basic Principle. In the tank circuit, oscillations of the electric current occur in accordance with the fundamental laws governing capacitor and inductor actions. The oscillatory or alternating flow of electrons in a parallel resonant tank circuit is caused by the repeated exchange of energy between the capacitor and the inductor.

Circuit Action. The operation of the oscillator circuit is explained with the aid of Fig. 17-3. It is assumed that in Fig. 17-3*a* the capacitor is fully charged and just starting to discharge (Fig. 17-3*b*). As the capacitor discharges through the inductor, the flow of electrons (indicated by the arrows) causes a magnetic field to be built up around the inductor. Electrons will continue to flow in this direction until the charges on the plates of the capacitor are equal to each other (Fig. 17-3*c*), at which time (1) the rate of electron flow is zero, and (2) the magnetic field about the inductor will start to collapse (Fig. 17-3*d*). According to Lenz's law, the collapsing magnetic field causes electrons to flow in the same direction that produced the expanding field, thus causing the capacitor to become charged with a polarity opposite to its original charge.

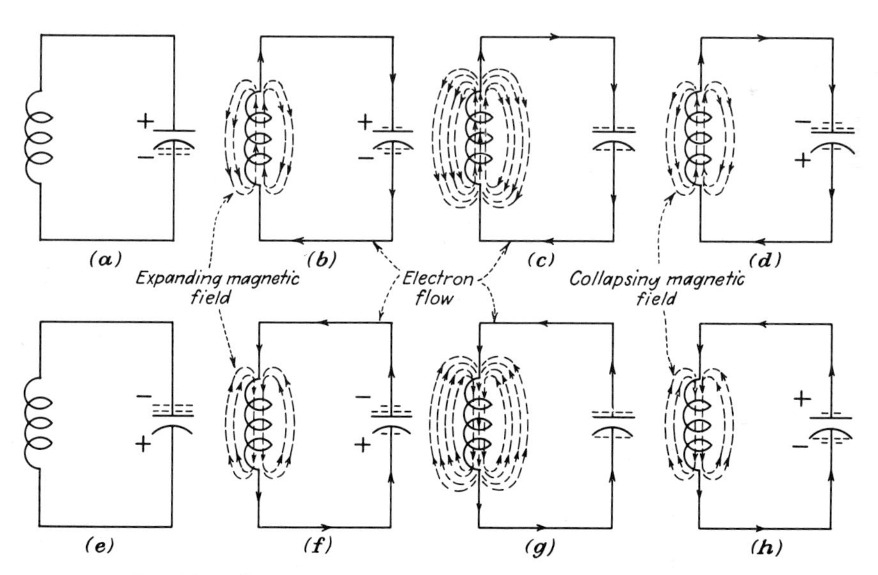

Fig. 17-3 Oscillatory action of a parallel resonant circuit.

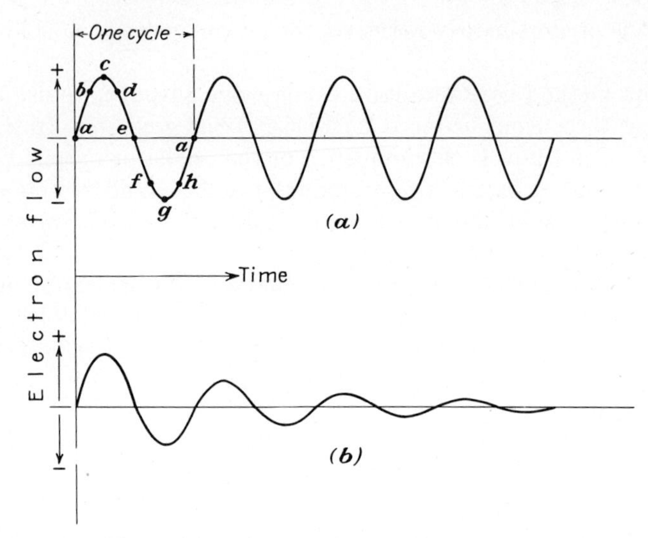

Fig. 17-4 Flow of electrons in a tank circuit. (a) Theoretical flow. (b) Actual flow.

When the field about the inductor has completely collapsed, the energy that had been stored in its magnetic field will be transferred to the electrostatic field of the capacitor (Fig. 17-3*e*).

Electron flow ordinarily ceases when the charges on the two plates of the capacitor are equal. However, because of the effects of the collapsing magnetic field, electron flow continues past this neutral point. This action of a parallel resonant circuit is sometimes referred to as the *flywheel effect.*

The capacitor will now discharge, and electrons will flow in a direction opposite to that used to charge the capacitor. This flow of electrons produces an expanding field about the inductor (Fig. 17-3*f*), until the difference in charge between the two plates is zero (Fig. 17-3*g*). As before, the magnetic field collapses, thus causing the electrons to continue to flow in the same direction (Fig. 17-3*h*). When the energy stored in the electromagnetic field has been transferred to the electrostatic field, the capacitor becomes fully charged in the opposite polarity. It is thus restored to its original state as in Fig. 17-3*a*.

This cycle of operations is repeated at a frequency approximately equal to the frequency of resonance of the parallel circuit as expressed by Eq. (11-4).

If the circuit is assumed to have zero resistance, each cycle of electron flow in the tank circuit will be similar to that shown in Fig. 17-4*a*. Thus, theoretically, a sustained alternating-current flow with constant peak magnitudes is produced. As it is impossible to construct a circuit without some amount of resistance, some energy will be lost in the form of heat during each cycle. If no energy is supplied to replace this loss, the peak magnitude of each oscillation will diminish as shown in Fig. 17-4*b*. Therefore, a simple tank circuit by itself cannot produce an alternating current of constant magnitude. The action of the active device causes the power supply to provide the energy required to sustain oscillations.

17-4 Fundamental Oscillators

Figures 17-5*a* and 17-6*a* illustrate two fundamental oscillator circuits using a transistor and vacuum tube respectively. In each circuit, the biasing networks have been omitted in order to simplify the circuit diagrams. Practical circuits include such additions as (1) biasing networks, (2) temperature-stabilizing circuit, (3) provisions for parallel power feed. The feedback networks control the amount of feedback by adjusting the position of L_2, called a *tickler coil* because it is made movable so that its mutual inductance can be varied.

17-5 Basic Transistor Oscillator

Circuit Characteristics. In the circuit of Fig. 17-5*a*, the transistor is used in the common-emitter configuration. The emitter-base diode is forward-biased by V_{BB} and the collector-base diode is reverse-biased by V_{CC}. With the tank circuit connected in the base lead of the transistor, the circuit is classed as a *tuned-base oscillator*. Both the base and the collector are series-fed. The transformer T provides positive feedback from the collector circuit to the base circuit. The correct polarity of the feedback voltage is obtained by selecting the proper terminal arrangement of the secondary winding L_2.

Characteristics Curves. Figure 17-5*b* shows the I_C versus I_B curve (base-collector transfer characteristics, Art. 8-3) for an ideal transistor. The action of the oscillator circuit can best be explained when the sections CQ and QS of the curve are symmetrical, although inverted.

Circuit Action. Upon closing switch S, current will start to flow in the collector circuit, and if the effect of the transformer T is considered to be negligible the collector current will apparently rise to a quiescent value represented by $I_{C \cdot Q}$ on the curve of Fig. 17-5*b*. However, the effect of the transformer is significant and as the current flowing through the primary winding L_1 increases in value it is accompanied by (1) an expanding magnetic field at L_1 (which also links the turns of L_2 and L_3), and (2) an induced voltage at the terminals of the secondary windings L_2 and L_3. Because oscillator operation requires positive feedback, the transformer windings L_1 and L_2 must be connected in their respective circuits in such a manner that terminal 3 is negative when the current in L_1 is increasing. Two immediate reactions of this feedback voltage are (1) an increase in the emitter-base bias voltage, and (2) a further increase in the collector current. This is followed by a succession of cycles of (1) an increase in feedback voltage, (2) an increase in the emitter-base bias, and (3) an increase in the collector current until saturation is reached (point S on Fig. 17-5*b*). During this time the tank capacitor C becomes charged. As soon as the collector current ceases to increase, the magnetic field at L_1 ceases to expand and thus no longer induces a voltage in L_2. Capacitor C, having been

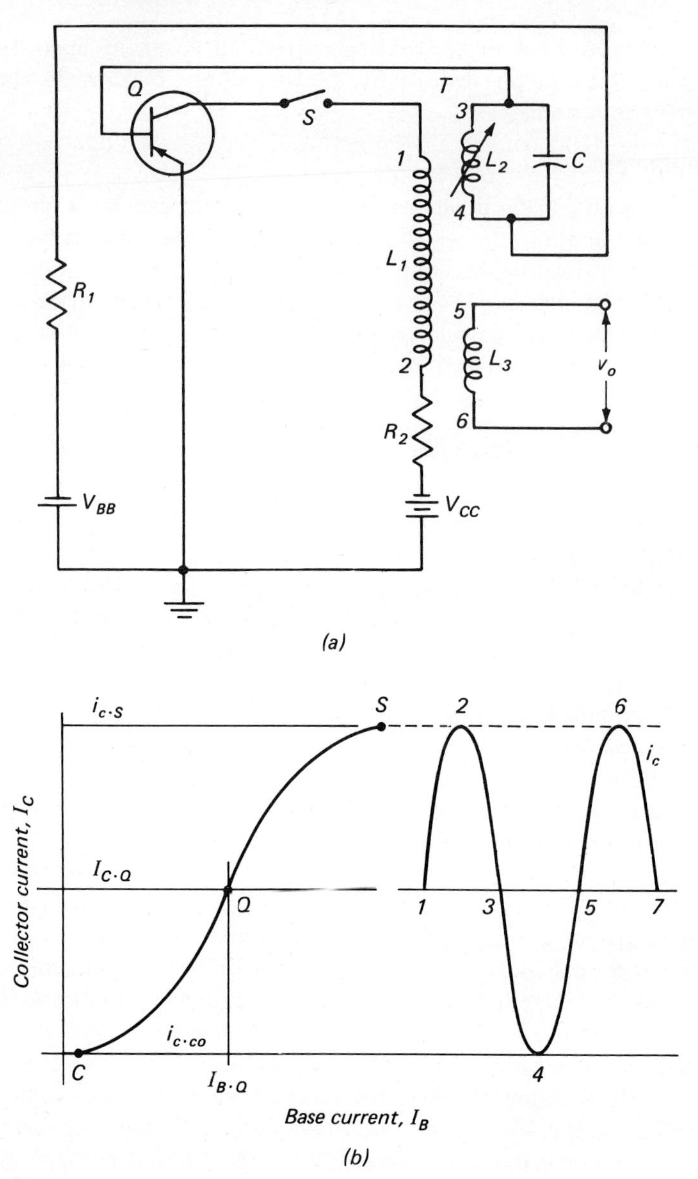

Fig. 17-5 Basic transistor oscillator. (a) Circuit diagram. (b) Collector- versus base-current characteristics.

charged to its maximum potential, now starts to discharge through L_2. The decrease in voltage across the capacitor causes the following sequence of reactions: (1) a decrease in the emitter-base bias, (2) a decrease in the collector current, (3) a collapsing magnetic field at L_1, (4) an induced voltage at L_2 with terminal 3 positive, (5) a further decrease in the emitter-base bias, (6) and so on, until collector-current cutoff is reached (point C on Fig. 17-5*b*). During this time, the tank capacitor C (1) loses its original charge, and (2) again becomes charged to its maximum potential but now with the opposite polarity. With the collector current at cutoff, the capacitor will again start to discharge through L_2. Because the polarity of the charge on the capacitor is now opposite to that when the collector current is at saturation, the sequence of reactions now will be (1) an increase in the emitter-base bias, (2) an increase in the collector current, (3) an expanding magnetic field at L_1, (4) an induced voltage at L_2 with terminal 3 negative, (5) a further increase in the emitter-base bias, (6) and so on, until collector-current saturation is again reached. This cycle of operations repeats itself as long as energy is supplied to overcome the losses of the circuit.

In the circuit of Fig. 17-5*a*, the output of the oscillator is obtained from the terminals of the secondary winding L_3 and has approximately the same waveform as the collector current i_c.

17-6 Basic Vacuum-tube Oscillator

Comparison with Basic Transistor Oscillator. Examination of the circuits of Figs. 17-5*a* and 17-6*a* will show a great similarity in the two circuits. The operation of the two circuits also is basically the same.

Circuit Characteristics. In the circuit of Fig. 17-6*a*, the vacuum tube is used in the common-cathode configuration. As with the common-cathode amplifier, the plate-cathode circuit is forward-biased and the grid-cathode circuit is reverse-biased. The transformer T provides feedback by means of the tickler coil L_2. The correct polarity of feedback voltage is obtained by selecting the proper terminal arrangement of the tickler coil L_2. The output of the oscillator is obtained from the secondary winding (L_3) of the transformer.

Characteristic Curves. Figure 17-6*b* shows the E_gI_p (grid-plate transfer characteristic) curve for ideal vacuum-tube oscillator operation. The action of the oscillator circuit can best be explained when the sections CQ and QS of the curve are symmetrical, although inverted.

Circuit Action. At the instant when switch S (Fig. 17-6*a*) is closed, plate current will start to flow and, if no feedback exists, will rise incrementally to a quiescent value $I_{P \cdot Q}$. However, feedback is present because the increasing plate current flowing through the transformer primary winding L_1 induces a voltage in the tickler coil L_2. For positive feedback, the polarity of the feedback voltage should cause a decrease in the grid bias when the plate current is increasing and an increase in the grid bias when the plate current is decreasing. Under this condition, the reactions due to an increasing plate current are (1)

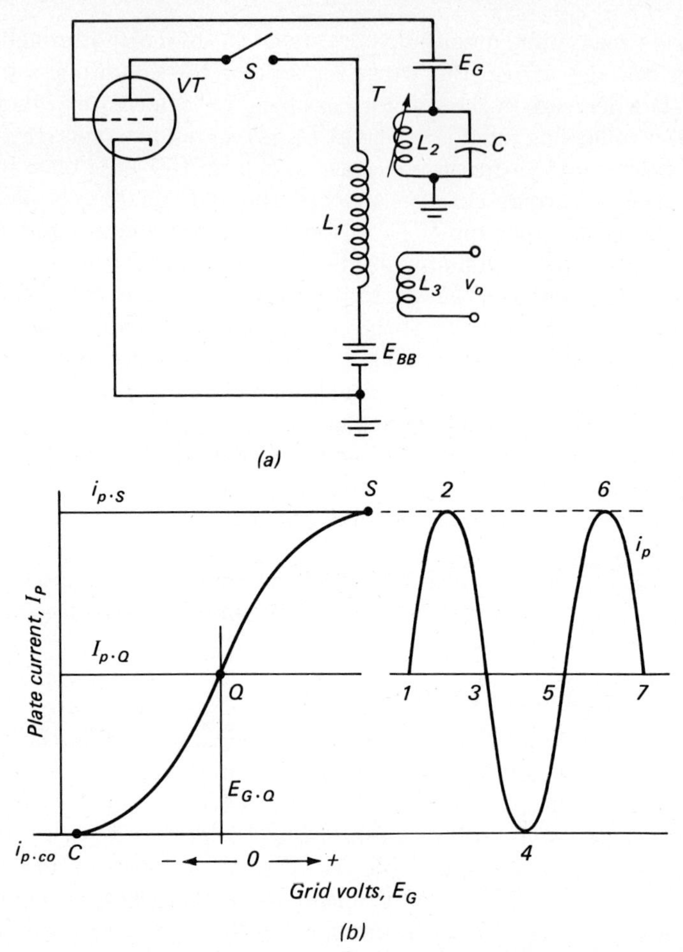

Fig. 17-6 Basic vacuum-tube oscillator. (a) Circuit diagram. (b) Plate current versus grid voltage characteristics.

expanding magnetic field in the transformer, (2) increasing feedback voltage at L_2, (3) decreasing grid bias at the tube, (4) further increase in the plate current, (5) repetition of this sequence until plate-current saturation is reached. During this time, the tank capacitor C becomes charged. At saturation, the plate current no longer increases and then (1) capacitor C starts to discharge through L_2, (2) the voltage at the capacitor decreases, (3) the grid bias at the tube increases, (4) the plate current decreases, (5) the magnetic field at L_1 decreases, (6) a negative feedback voltage is induced in L_2 which further increases the grid bias of the tube, (7) the plate current decreases, (8) this sequence repeats until the plate current decreases to the plate-current cutoff value. When plate-current cutoff is reached, (1) no induced voltage exists at

L_2, (2) capacitor C starts to discharge through L_2, (3) the grid bias at the tube decreases, (4) the plate current increases, (5) a feedback voltage is induced at L_2, (6) the grid bias at the tube decreases, (7) the plate current increases, (8) this sequence repeats until plate-current saturation is reached. This cycle of operations repeats itself as long as energy is supplied to overcome the losses in the circuit.

In the circuit of Fig. 17-6*a*, the output of the oscillator is obtained from the secondary winding L_3 of the transformer. In Fig. 17-6*b*, the sine-wave curve i_p illustrates the waveform of the oscillator output.

17-7 Circuit Considerations

Effect of the Tank Circuit on the Amplitude Stability. If the amplitude of the output of an oscillator is unstable the waveform will be distorted, thus producing undesirable harmonics. An oscillator generally operates as a Class C amplifier delivering power during less than one-half of the input cycle. Under this condition, the output-current variation will not produce a sine wave, and a tank circuit is added in order to obtain an approximate sine-wave output. In order to have an output of approximately sine-wave form, it is necessary that the tank circuit store energy during the portion of the cycle in which the active device is delivering power and to deliver this stored energy to the load during the portion of the cycle in which no power is being delivered by the active device.

Effect of the Tank Circuit Q on the Waveshape. The tank circuit generally consists of a capacitor and an inductor connected in parallel. The resistance of the inductor acts as a resistance in series with the inductance, and the Q of the tank circuit is equal to the ratio of the inductive reactance to the resistance. When the tank circuit is supplying power to a load the effect is similar to increasing the series resistance of the tank circuit and consequently reduces the value of Q. The Q of the circuit when supplying power is called the *effective* Q, designated Q_{eff}. The factor Q_{eff} is also a measure of the ratio of the energy stored during each cycle to the energy dissipated during each cycle [see Eq. (17-12*a*)]. An increase in the value of Q_{eff} indicates an increase in the amount of energy stored. Furthermore, the ability to carry each cycle past its neutral point (flywheel effect) can be increased by increasing the effective Q of the tuned circuit.

Efficiency of the Tank Circuit. The efficiency of the tank circuit may be expressed as the ratio of the power delivered to the load to the power delivered to the tank circuit. In terms of the circuit Q

$$\text{Efficiency} = \frac{Q - Q_{eff}}{Q} \times 100 \qquad (17\text{-}2)$$

where Q = tank circuit Q at no load
Q_{eff} = tank circuit Q with load

Example 17-1 The tank circuit of an oscillator has a circuit Q of 85 at no load and an effective Q of 12 when load is applied. What is the efficiency of the tank circuit?

GIVEN: $Q = 85$ $\quad Q_{eff} = 12$

FIND: Efficiency

SOLUTION:

$$\text{Efficiency} = \frac{Q - Q_{eff}}{Q} \times 100 = \frac{85 - 12}{85} \times 100 = 85.8\%$$

Analysis of the Tank Circuit. Important considerations in the study of the tank circuit are the frequency, impedance, current, power, and energy stored.

The tank circuit should be adjusted so that its resonant frequency corresponds to that of the oscillator. For the values of Q commonly used with these circuits, the results obtained by use of Eq. (11-4) are sufficiently accurate for general purposes. Accordingly, Eqs. (11-5) and (11-6) may be used for finding the values of the inductance and capacitance required in the tank circuit.

$$f_r = \frac{159}{\sqrt{LC}} \tag{11-4}$$

$$L = \frac{25{,}300}{f_r^2 C} \tag{11-5}$$

$$C = \frac{25{,}300}{f_r^2 L} \tag{11-6}$$

where f_r = frequency of resonance, kHz
L = inductance, μh
C = capacitance, μf

As the resistance of the tank circuit is small compared to the inductive reactance, a sufficiently accurate value of the impedance at resonance can be obtained by

$$Z_{t\text{-}r} = QX_L = \frac{X_L^2}{R} \tag{17-3}$$

As $X_L = X_C$ at resonance, then

$$Z_{t\text{-}r} = \frac{X^2}{R} = \frac{X_L X_C}{R} = \frac{2\pi f_r L}{2\pi f_r C R} = \frac{L}{CR} \tag{17-4}$$

When the tank circuit is in the output circuit of the active device, the voltage at the tank circuit is expressed by

$$E_t = I_{o\text{-a-c}} Z_{t\text{-}r} \tag{17-5}$$

where $I_{o\text{-a-c}}$ = alternating component of the current in the output circuit of the active device, amp

Correspondingly, the tank-circuit current may be expressed as

$$I_{t\text{-}r} \cong \frac{E_t}{X_L} \tag{17-6}$$

An approximate value of the tank-circuit current can also be found by

$$I_{t\text{-}r} \cong I_L \cong I_C \cong QI_{o\text{-a-c}} \tag{17-6a}$$

At resonance the power delivered to the tank circuit (output load plus tank-circuit losses) may be expressed as

$$P_{t\text{-}r} = I_{o\text{-a-c}}{}^2 QX_L = \frac{E_t{}^2}{QX_L} \tag{17-7}$$

Example 17-2 The tank circuit of a certain oscillator has the following circuit values: $L = 80\mu h$, $C = 365 - 35$ pf, $Q_{eff} = 20$, $E_t = 10$ volts. Find (*a*) the capacitance when the resonant frequency is 1,000 kHz, (*b*) the impedance of the tank circuit at resonance, (*c*) the alternating component of the output current required to produce the desired voltage at the tank circuit, (*d*) the current in the tank circuit, (*e*) the power supplied to the tank circuit.

GIVEN: $L = 80\mu h$ $\quad C = 365 - 35$ pf $\quad Q_{eff} = 20$ $\quad E_t = 10$ volts
$f_r = 1{,}000$ kHz

FIND: (*a*) C (*b*) $Z_{t\text{-}r}$ (*c*) $I_{o\text{-a-c}}$ (*d*) I_t (*e*) P_t

SOLUTION:

(*a*) $$C = \frac{25{,}300}{f_r{}^2 L} = \frac{25{,}300}{1{,}000^2 \times 80} = 316 \times 10^{-6}\ \mu\text{f} \doteq 316\ \text{pf}$$

(*b*) $$Z_{t\text{-}r} = Q_{eff} X_L = 20 \times 6.28 \times 10^6 \times 80 \times 10^{-6} = 10{,}048\ \text{ohms}$$

(*c*) $$I_{o\text{-a-c}} = \frac{E_t}{Z_{t\text{-}r}} = \frac{10}{10{,}048} = 0.995\ \text{ma}$$

(*d*) $$I_t = Q_{eff} I_{o\text{-a-c}} = 20 \times 0.955 = 19.9\ \text{ma}$$

(*e*) $$P_t = E_t I_{o\text{-a-c}} = 10 \times 0.995 \times 10^{-3} = 0.00995\ \text{watt}$$

Energy Stored in the Tank Circuit. An analysis of the tank circuit will show that the amount of energy stored will increase with an increase in the value of either the capacitance or the voltage. The total energy stored in a tank circuit is equal to the sum of the energy stored in the inductor and in the capacitor.

$$w_t = \frac{Li_t{}^2}{2} + \frac{Ce_t{}^2}{2} \tag{17-8}$$

where w_t = total instantaneous energy stored in the tank circuit, watt-sec
i_t = instantaneous value of current in the tank circuit, amp
e_t = instantaneous value of voltage across the tank circuit, volts
L = inductance of the tank coil, henrys
C = capacitance of the tank capacitor, farads

As the resistance of a well-designed tank circuit is very small in comparison to either of its reactances, it can generally be ignored in considering the impedance of either branch. Under this condition

$$W_t = \frac{CE_{t\text{-}m}^2}{2} \tag{17-9}$$

where W_t = energy stored in the tank circuit per cycle
Substituting $\sqrt{2}E_t$ for $E_{t\text{-}m}$ in Eq. (17-9), then

$$W_t = CE_t^2 \tag{17-9a}$$

Note: E_t = effective value of the tank-circuit voltage.

Equation (17-9*a*) shows that the total energy stored in the tank circuit can be increased by using a higher value of capacitance. In order to maintain a fixed value of frequency, an increase in the value of capacitance must be accompanied by a decrease in the value of inductance, thereby reducing the L/C ratio. An increase in the capacitance with an accompanying decrease in inductance will result in a lower value of reactance at the fixed value of resonant frequency. The effective Q of the tank circuit increases with a decrease in the reactance; this is characteristic of a loaded tank circuit and should not be confused with the no-load value of Q, which decreases with a decrease in the reactance. Thus, an increase in the capacitance (1) increases the amount of energy stored, (2) decreases the L/C ratio of the tank circuit, and (3) increases the effective Q of the tank circuit.

Increasing the effective Q of the tuned circuit increases its impedance [see Eq. (17-3)], thereby increasing the varying output voltage of the oscillator circuit. This increase in voltage will further increase the total energy stored [Eq. (17-9*a*)]. However, an increase in the voltage applied to a tuned circuit increases the circulating current, thus increasing the I^2R losses and reducing the overall efficiency of the oscillator circuit. It is therefore good practice to limit the L/C ratio to the value required to produce a satisfactory sinusoidal output.

Relation of Q_{eff} to the Ratio of Energy Stored to Energy Dissipated. The value of Q_{eff} is also a measure of the ratio of the energy stored during each cycle to the energy dissipated during each cycle. The energy stored in the tank circuit per cycle can be expressed in terms of the inductance as

$$W_t = LI_t^2 \tag{17-10}$$

The energy dissipated in the tank circuit per cycle can be expressed as

$$W_d = \frac{I_t^2 R_{eq}}{f_r} \tag{17-11}$$

where W_d = energy dissipated in the tank circuit per cycle
R_{eq} = equivalent series resistance of the tank circuit when loaded

The ratio of energy stored per cycle to energy dissipated per cycle may then be expressed as

$$\frac{\text{Energy stored per cycle}}{\text{Energy dissipated per cycle}} = \frac{W_t}{W_d} = \frac{LI_t^2}{\dfrac{I_t^2 R_{eq}}{f_r}} = \frac{f_r L}{R_{eq}} \tag{17-12}$$

Multiplying both the numerator and the denominator by 2π

$$\frac{\text{Energy stored per cycle}}{\text{Energy dissipated per cycle}} = \frac{2\pi f_r L}{2\pi R_{eq}} = \frac{Q_{eff}}{2\pi} \tag{17-12a}$$

Example 17-3 (*a*) How much energy is stored per cycle in the tank circuit of Example 17-2? (*b*) Determine the value of R_{eq} from the values of Q_{eff} and X_L in Example 17-2. (*c*) How much power is dissipated by the tank circuit? (*d*) How much energy is dissipated per cycle? (*e*) What is the ratio of the energy stored per cycle to the energy dissipated per cycle? (*f*) Check the result of part (*e*) by use of Eq. (17-12*a*).

GIVEN: $E_t = 10$ volts $\quad C = 316$ pf $\quad Q_{eff} = 20 \quad L = 80\ \mu h$
$f_r = 1{,}000$ kHz $\quad I_t = 19.9$ ma

FIND: (*a*) W_t (*b*) R_{eq} (*c*) P_{dis} (*d*) W_d (*e*) $\dfrac{W_t}{W_d}$ (*f*) $\dfrac{W_t}{W_d}$

SOLUTION:

(*a*) $W_t = CE_t^2 = 316 \times 10^{-12} \times 10^2 = 316 \times 10^{-10}$ watt per cycle

(*b*) $R_{eq} = \dfrac{X_L}{Q_{eff}} = \dfrac{6.28 \times 10^6 \times 80 \times 10^{-6}}{20} = 25.12$ ohms

(*c*) $P_{dis} = I_t^2 R_{eq} = (19.9 \times 10^{-3})^2 \times 25.12 = 9{,}947 \times 10^{-6}$ watt

(*d*) $W_d = \dfrac{P_{dis}}{f_r} = \dfrac{9{,}947 \times 10^{-6}}{10^6} = 99.47 \times 10^{-10}$ watt per cycle

(*e*) $\dfrac{W_t}{W_d} = \dfrac{316 \times 10^{-10}}{99.47 \times 10^{-10}} = 3.17$

(*f*) $\dfrac{W_t}{W_d} = \dfrac{Q_{eff}}{2\pi} = \dfrac{20}{6.28} = 3.18$

The power output of an oscillator is the useful a-c power consumed by its load. This load may be coupled by means of capacitive, inductive, transformer, or electron coupling. The frequency and amplitude stability of an oscillator will be affected by changes in the power taken by the load. In order to maintain a high degree of stability, oscillator circuits are seldom designed to deliver large amounts of power. When large amounts of power are required the oscillator circuit is used to drive a power amplifier, which in turn produces the desired amount of power output. Most oscillator circuits are therefore used as frequency-controlling devices delivering a small amount of power at a comparatively high voltage.

17-8 Frequency Stability

Factors Affecting Frequency Stability. The ability of an oscillator to maintain a constant frequency is called its *frequency stability.* Factors affecting the frequency stability are (1) operating point of the active device, (2) stability of the parameters of the active device, (3) stability of the power-supply voltages, (4) temperature variations, (5) output load, and (6) mechanical variations of the circuit components.

Operating Point. The d-c operating point for the active device is generally chosen so that the circuit operation is on the linear portion of the characteristics curve. Operation on the nonlinear portion of the curve usually results in variations of the parameters of the active device, which in turn affect the frequency stability of the oscillator.

Parameters of the Active Device. Any changes in the parameters of the active device, particularly the collector-to-emitter capacitance or the plate-to-grid capacitance, will cause a change in the oscillator output frequency, thus affecting its frequency stability. The effect of changes in the interelectrode capacitance between the elements of the active device can be counteracted by adding a swamping capacitor across the offending elements; the added capacitance can be made part of the tank circuit.

Power Source. Any variations in the operating voltages applied to the active device will cause a shift in the output frequency of the oscillator. The operating voltages can be stabilized by using a regulated power supply.

Temperature Variations. Variations in the temperature will affect (1) the parameters of the active device, and (2) the values of the resistors, inductors, and capacitors in the circuit. Any of these changes will affect the output frequency of the oscillator. Because the effects of temperature change are comparatively slow, the frequency changes are referred to as *drift.* Compensation for variations due to temperature change can be made in the same manner as described for amplifier circuits.

Output Load. A variation in the output load may affect the effective Q of the oscillator circuit and thereby cause a change in the output frequency of the oscillator.

Mechanical Variation. Mechanical vibration of the circuit elements such as transistors, tubes, capacitors, and inductors also causes their values to vary. Such changes in the values of inductance and capacitance will cause the resonant frequency to vary with the mechanical vibration. Instability due to mechanical vibration can be minimized by isolating the oscillator from the source of the mechanical vibration.

17-9 Active-device Configurations

Transistor Oscillators. The oscillator circuit of Fig. 17-5*a* shows the transistor used in the common-emitter configuration. Theoretically the common-base and common-collector configurations can also be used. The choice of configuration used depends largely on (1) requirements of the oscillator, and

(2) the characteristics of the configuration selected. The common-emitter configuration is most frequently used because of its advantages, namely, (1) moderate values of input and output impedances, (2) the current, voltage, and power gains are all greater than 1, and (3) the power gain is greater than for the common-base and common-collector configurations. The fact whether or not phase reversal occurs between the input and output must be taken into account when designing the feedback circuit.

Vacuum-tube Oscillators. The common-cathode configuration is used most frequently in the vacuum-tube oscillators. The characteristics of these oscillators are (1) high values of input and output impedances, (2) voltage and power gains can be achieved, and (3) phase reversal occurs between the input and output signals.

17-10 Other *LC* Resonant Feedback Oscillators

Basic Principle. Feedback provided by use of the basic transformer is the simplest method to understand and has been thoroughly described in Art. 17-5. There are many variations of this basic principle, and some of the more common variations are described in this chapter.

Other Types. The chief variations of the basic *LC* resonant feedback oscillator circuit are the Hartley, Colpitts, and Clapp oscillators. Other variations are (1) the tuned-base and the tuned-collector transistor oscillator, (2) the tuned-grid tuned-plate vacuum-tube oscillator, and (3) the electron-coupled vacuum-tube oscillator.

17-11 Location of the Tank Circuit

Transistor Oscillators. The tank circuit of a common-emitter transistor oscillator may be placed in (1) the base circuit as in Fig. 17-5*a*, or (2) the collector circuit as in Fig. 17-10, and the circuit is called a *tuned-base oscillator* or *tuned-collector oscillator*, respectively.

Vacuum-tube Oscillators. The tank circuit of a common-cathode vacuum-tube oscillator may be placed in (1) the grid circuit as in Fig. 17-6*a*, or (2) the plate circuit as in Fig. 17-7*a*; the tuned-grid tuned-plate oscillator uses two tank circuits, one in the grid circuit and the other in the plate circuit as in Fig. 17-7*a*.

17-12 Tuned-grid Tuned-plate Oscillator

The circuit of Fig. 17-7*a* illustrates a tuned-grid tuned-plate oscillator. Capacitive feedback, utilizing the grid-to-plate capacitance C_{gp} of the tube, is used in the tuned-grid tuned-plate oscillator. Two parallel resonant circuits are required, one in the grid circuit and the other in the plate circuit; no inductive coupling should exist between the coils in these two circuits. The frequency of the oscillator is dependent on the resonant frequency of each of the tuned circuits, therefore the name *tuned-grid tuned-plate oscillator*.

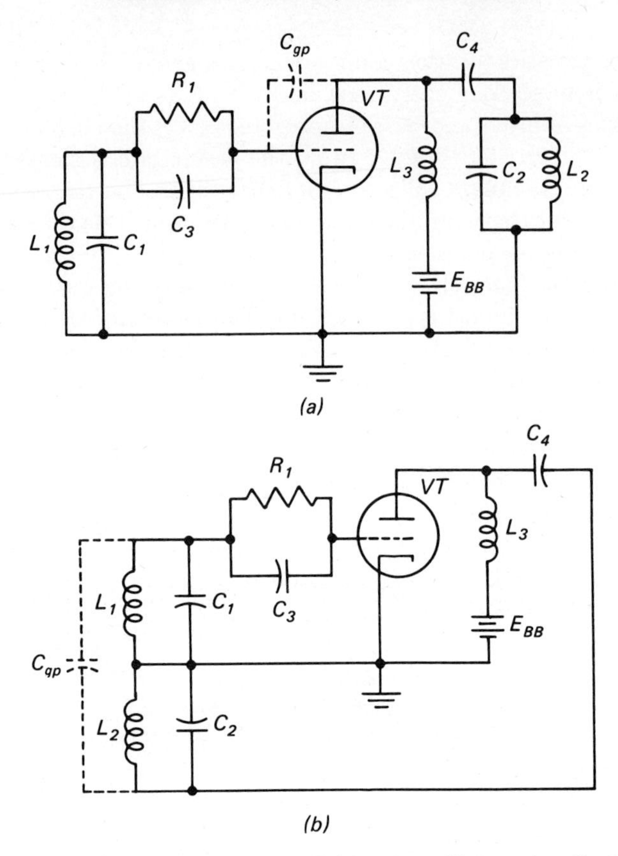

Fig. 17-7 Tuned-grid tuned-plate parallel-feed oscillator circuit. (a) Circuit diagram. (b) Equivalent circuit.

The operation of this oscillator may be more easily understood with the aid of the simplified equivalent circuit shown in Fig. 17-7*b*. This simplification is possible because C_3 and C_4 act in series with the interelectrode capacitance C_{gp} to form the feedback circuit, and since C_3 and C_4 are much larger in value than C_{gp}, they are disregarded.

In order for the system to oscillate, the plate circuit should be tuned to a slightly lower frequency than the grid circuit. The resultant inductive reactance of the two tuned circuits will then resonate with the interelectrode capacitive reactance. The frequency of the oscillator is determined by the tuned circuit having the highest Q.

The frequency stability and the voltage amplification of the tuned-grid tuned-plate oscillator are comparatively high because of the high impedance of the parallel resonant plate load. The frequency of oscillation is usually controlled by varying the resonant frequency of the plate tuned circuit, and the amount of grid excitation is controlled by varying the resonant frequency of the grid tuned circuit.

17-13 Electron-coupled Vacuum-tube Oscillator

The tank circuit of an oscillator can be isolated from its load by using electron coupling between the oscillator and the load. The impedance of the tank circuit of this type of oscillator will be practically constant, since it will not be affected by changes in plate load. As the oscillator frequency depends upon the impedance of the tank circuit, it will also be practically constant. Thus electron coupling is an effective means of making the frequency of an oscillator independent of variations in load.

Figure 17-8 illustrates an oscillator circuit employing electron coupling. A shunt-fed circuit is formed by the cathode, control grid, screen grid, and the L_1C_1 tank circuit. In this circuit the screen grid serves as a plate and the a-c path to the inductor L_1 in the tank circuit is completed through C_3. The plate of the tube serves only as an output electrode. Since C_3 blocks the direct voltage and passes the alternating voltage, the screen grid is in effect grounded for the alternating voltages. The plate is thus shielded from the oscillatory section of the tube, thereby preventing the load impedance from reacting on the oscillator.

As the plate is maintained at a higher potential than the screen grid, some of the electrons emitted by the cathode will flow through the screen-grid circuit and the remainder through the plate circuit. The frequency of the a-c component of the plate current is therefore the same as the oscillator frequency, and thus energy is delivered to the output load through an electron stream. Because the coupling medium is an electron stream, the circuit is called an *electron-coupled oscillator*.

Increasing the plate voltage of an electron-coupled oscillator will cause the

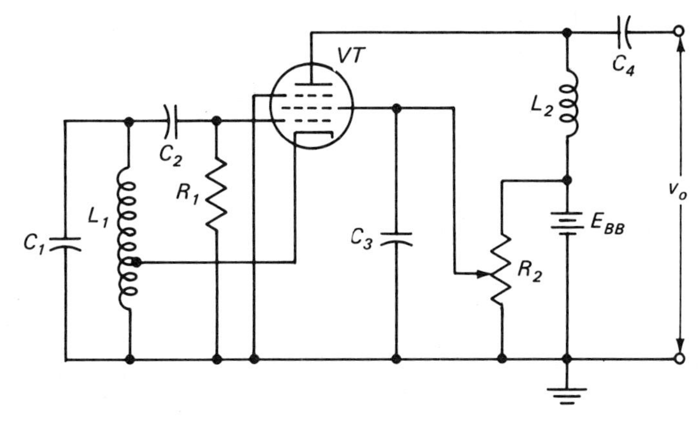

Fig. 17-8 Vacuum-tube electron-coupled oscillator circuit.

frequency of oscillation to change. Increasing the screen-grid voltage will also change the frequency of oscillation but in the opposite direction to that caused by a plate-voltage increase. If the voltage on the screen grid is made variable (R_2 in Fig. 17-8), the screen-grid voltage can be adjusted so that these two actions balance each other. The frequency of oscillation will then be practically independent of variations in the supply voltages.

17-14 Hartley Oscillator

The Hartley oscillator is one of the simplest types of oscillator circuits. As in the fundamental oscillator of Art. 17-4, the amplified energy in the output circuit is fed back to the input circuit by mutual inductive coupling. However, in the Hartley oscillator only one coil is used instead of the separate primary and secondary coils of a transformer. Figure 17-9 shows the single coil with a tap that divides the coil into two sections L_{1A} and L_{1B}. This type of coil is called a *split-tank-inductor*, or an *autotransformer*. In the block diagram of Fig. 17-9, the split inductor serves three functions: (1) it provides positive feedback to the active device, (2) together with capacitor C, it forms the resonant tank circuit of the oscillator, and (3) it determines the oscillator frequency.

17-15 Transistor Hartley Oscillator

Although any transistor configuration may be used, only the common-emitter type will be presented. With this type of circuit, the d-c power source for the collector may be either *series-fed* or *shunt-fed*.

In the shunt-fed oscillator (Fig. 17-10) only the a-c component of the collector current flows in the portion L_{1A} of the tank-circuit inductor. In the series-

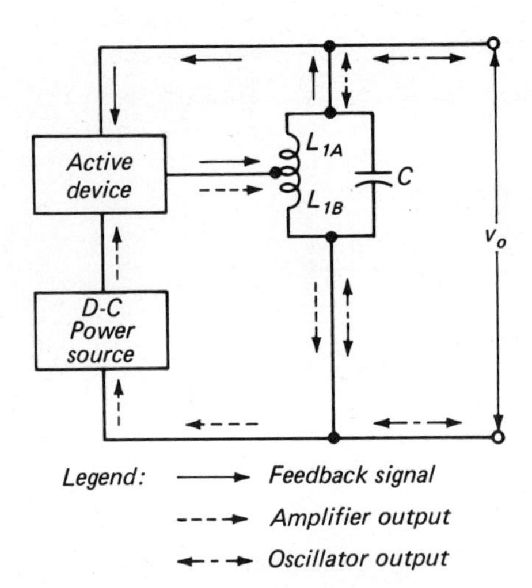

Fig. 17-9 Block diagram of basic Hartley oscillator.

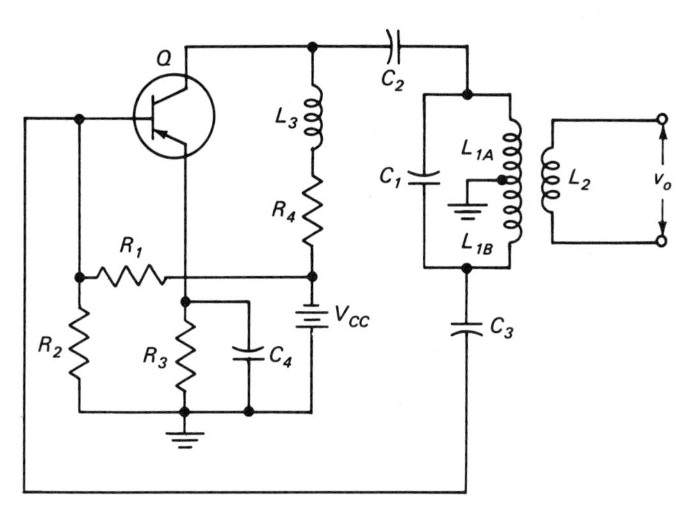

Fig. 17-10 Transistor Hartley oscillator, shunt-fed.

fed oscillator (Fig. 17-11) the d-c component of the collector current also flows through the portion L_{1A} of the tank-circuit inductor resulting in a lower value of tank-circuit Q and poorer stability than for the shunt-fed oscillator. However, the principle of operation of the shunt-fed and the series-fed oscillators is the same.

When a greater power output is required than can be obtained with a single-ended transistor oscillator, a push-pull oscillator may be used.

Shunt-fed Oscillator. The circuit of Fig. 17-10 illustrates a transistor Hartley oscillator using the common-emitter configuration and shunt feed for the collector d-c power requirement. The circuit is called a *parallel* or *shunt-fed oscillator* because the collector circuit is divided into two parallel branches, one to provide a path for the direct current and the other for the alternating current. The inductor L_3 keeps the alternating current out of the direct-current circuit, and capacitor C_2 keeps the direct current out of the alternating-current circuit. In order that most of the alternating current may flow through section L_{1A} of the tank-circuit inductor instead of through the power source, the reactance of C_2 at the resonant frequency should be small in comparison with the reactance of L_3 at the same frequency.

The frequency of the oscillator output is equal to the frequency at which the signal in the tank circuit is fed back to the base circuit. The frequency of oscillation is equal to the resonant frequency of the tuned tank circuit formed by the capacitor C_1 and the entire tank-circuit inductor L_{1A} plus L_{1B}.

Resistors R_1 and R_2 form a voltage divider to provide the base-bias voltage, and R_3 is an emitter swamping resistor to add stability to the circuit. Capacitor C_4 provides a bypass path around R_3 for the alternating current, thereby preventing any degeneration while still providing temperature stabilization.

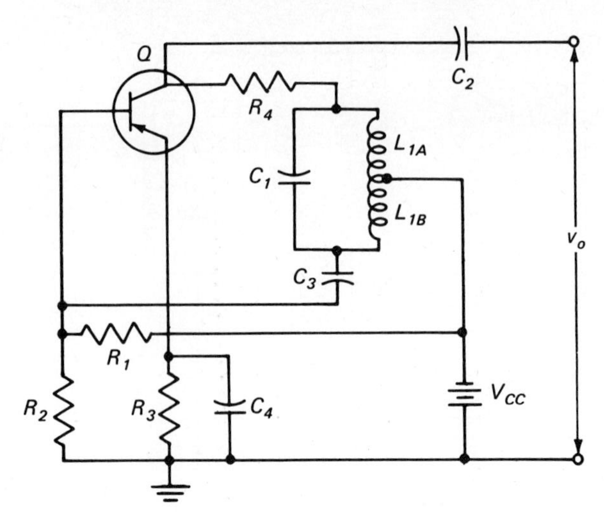

Fig. 17-11 Transistor Hartley oscillator, series-fed.

The collector load resistor R_4 is required for the amplifier action of the circuit, and inductor L_3 keeps the alternating current out of the power source V_{CC}.

The positive feedback required for oscillator operation is obtained from the tank circuit and is coupled to the base circuit by capacitor C_3. The amount of feedback is dependent on the number of turns in L_{1A} and L_{1B}. Capacitor C_3 prevents a low impedance at L_{1B} from shorting the base of the transistor to ground. The functions of capacitor C_2 are to (1) couple the a-c signal at the collector to the tank circuit, (2) block the direct current of the power source from the tank circuit, and (3) prevent the tank coil L_{1A} from shorting the collector to ground.

When V_{CC} first releases power to the circuit, an initial bias is established by the voltage divider R_1R_2 and oscillations are produced because of the feedback from the tank circuit. The amounts of bias and feedback are usually chosen to provide Class A operation, with the collector current ideally varying between saturation and cutoff. Selecting appropriate values for R_3 and R_4 will produce the proper amount of degeneration to change the base bias enough so that Class B or C operation can be achieved after several initial oscillations.

The output of the oscillator in Fig. 17-10 is obtained from the inductor L_2 coupled to the tank inductor $L_{1A}L_{1B}$. Alternate methods of obtaining an output signal are by use of (1) a capacitor connected to the collector of the transistor, or (2) a high-resistance load connected across the tank circuit.

Series-fed Oscillator. The circuit of Fig. 17-11 illustrates a transistor Hartley oscillator using the common-emitter configuration and series feed for the collector d-c power requirement. The operation of the series-fed circuit is practically the same as has been described for the shunt-fed circuit, differing only in that (1) direct current now flows through the portion L_{1A} of the tank-circuit inductor, (2) the oscillator output is coupled to the load by C_2, (3) the secondary winding L_2 is no longer needed and therefore is omitted, and (4) the inductor L_3 of the shunt-fed circuit is omitted in the series-fed circuit. All the circuit components, except C_2, serve the same function as is described for the shunt-fed oscillator.

17-16 Vacuum-tube Hartley Oscillator

The Hartley oscillator circuit using a vacuum tube operates in the same manner as the transistor oscillator circuits previously described. Shunt-fed and series-fed circuits are also applicable with the vacuum-tube Hartley oscillators.

Shunt-fed Oscillator. The circuit of Fig. 17-12 illustrates a vacuum-tube Hartley oscillator employing the common-cathode configuration and shunt feed for the plate-circuit d-c power requirement. Resistor R_1, together with capacitor C_3, establishes the grid bias for the tube. Capacitor C_3 also (1) couples a positive-feedback voltage to the grid of the tube, and (2) prevents L_{1B} from shorting out the grid resistor R_1. Additional loading for the plate circuit is provided by R_2 when the loading effect of L_3 is not sufficient. The tank circuit and capacitors C_2 and C_3 occupy the same circuit locations and function in the same manner as with the transistor oscillator circuit described in Art. 17-15.

Series-fed Oscillator. The circuit of Fig. 17-13 illustrates a vacuum-tube Hartley oscillator employing the common-cathode configuration and series feed for the plate-circuit d-c power requirement. Since both the a-c and d-c components of the plate current flow through portion L_{1A} of the tank coil, the circuit is classed as *series-fed*. The combined actions of L_3 and C_2 bypass the

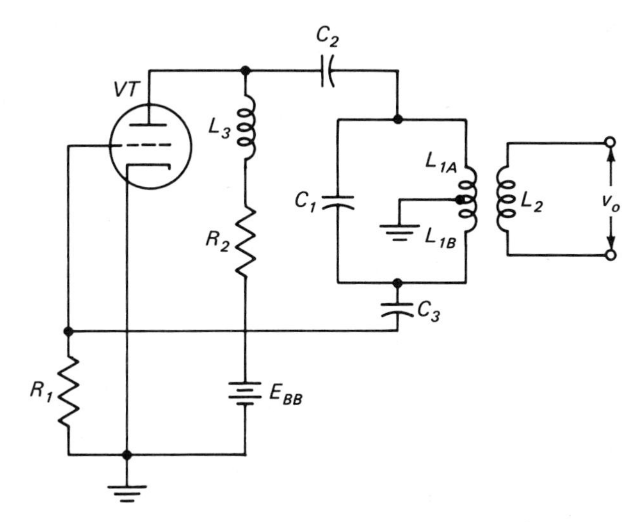

Fig. 17-12 Vacuum-tube Hartley oscillator, shunt-fed.

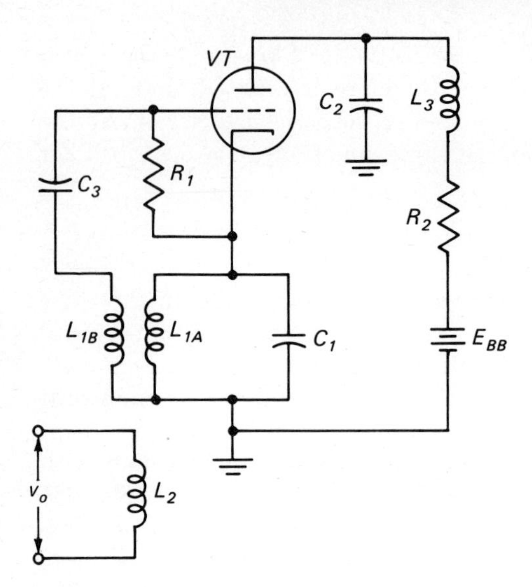

Fig. 17-13 Vacuum-tube Hartley oscillator, series-fed.

alternating current around the power source E_{BB}. Otherwise, operation of this circuit is the same as for the Hartley circuits previously described.

17-17 Colpitts Oscillator

Basic Principle. The Colpitts oscillator differs from the Hartley oscillator chiefly in that the feedback voltage is obtained from a split capacitor (actually two capacitors connected in series) in the tank circuit rather than from a split inductor.

Shunt-fed Transistor Oscillator. The circuit of Fig. 17-14*a* illustrates a transistor Colpitts oscillator using the common-emitter configuration and shunt feed for the collector d-c power requirement. Examination of Figs. 17-10 and 17-14*a* will show a great similarity between the Hartley and Colpitts oscillators. In the Colpitts oscillator circuit, capacitors C_{1A} and C_{1B} form the voltage divider used to provide the feedback signal; the voltage drop across C_{1B} constitutes the feedback voltage. All the circuit components serve the same function as with the Hartley oscillator of Fig. 17-10. In order to keep the feedback loss at a minimum, the ratio of the reactances of C_{1A} to C_{1B} should be approximately the same as the ratio of the output to input impedances of the transistor. The effect of changes in the interelectrode capacitances of the transistor on the oscillator frequency can be minimized by making the value of C_{1B} much greater than the emitter-base capacitance, and the value of C_{1A} much greater than the emitter-collector capacitance.

Shunt-fed Vacuum-tube Oscillator. The circuit of Fig. 17-14*b* illustrates a vacuum-tube Colpitts oscillator with shunt feed. Figure 17-14*a* and *b* shows the similarity of the Colpitts oscillator when using either a transistor or a vacuum tube as the active device. The operation of the circuit is the same as is explained for the preceding circuits.

17-18 Clapp Oscillator

The circuit of Fig. 17-15 illustrates the Clapp oscillator, which is a variation of the transistor Colpitts oscillator. Examination of Figs. 17-15 and 17-14*a* will show that except for the addition of capacitor C_5 to the Clapp oscillator the two circuits are the same. The advantages of adding capacitor C_5 in series

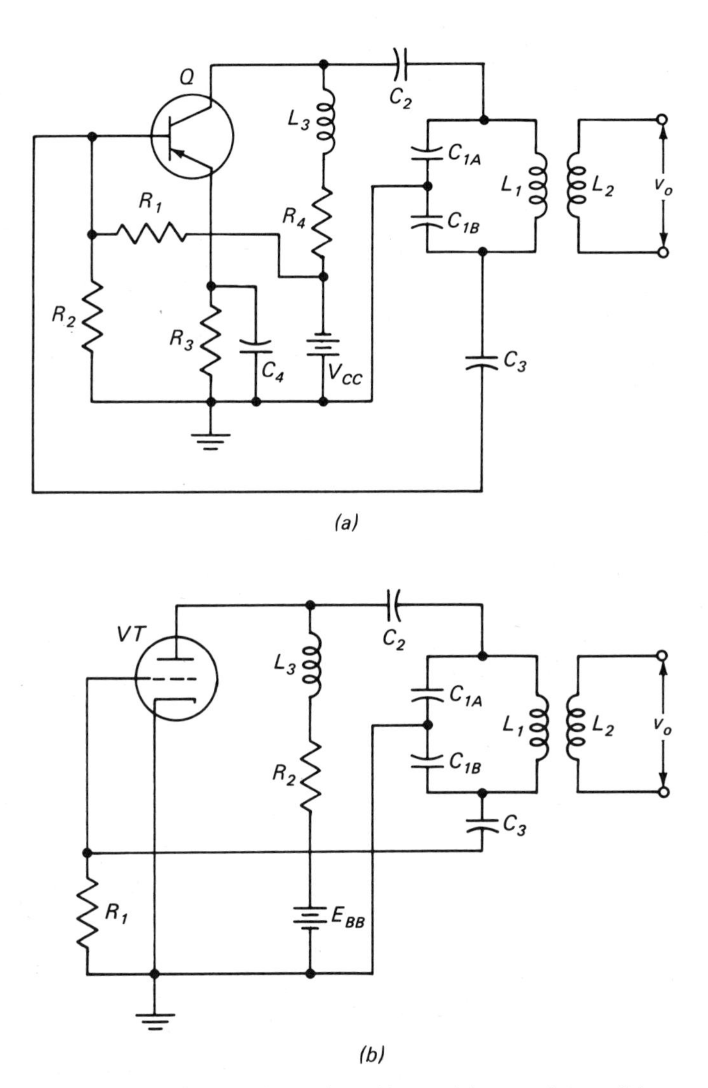

Fig. 17-14 Colpitts oscillator, shunt-fed. Using (a) transistor, (b) vacuum tube.

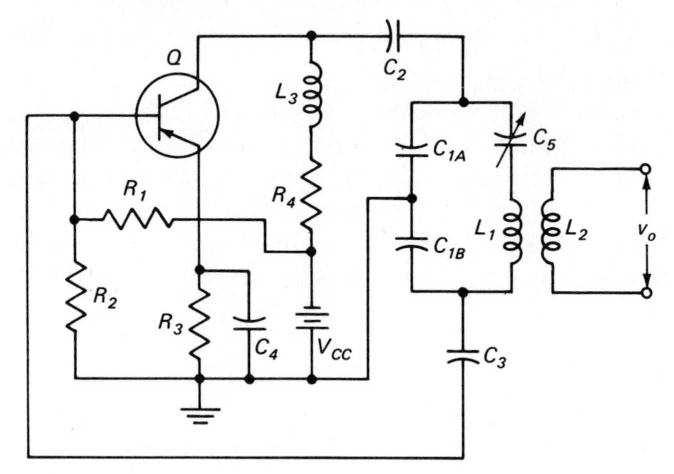

Fig. 17-15 Transistor Clapp oscillator, shunt-fed.

with the tank-circuit inductor are that the Clapp oscillator (1) provides improved stability, (2) practically eliminates the effect of the transistor's parameters on the operation of the circuit, and (3) permits capacitive tuning of the oscillator by using a variable capacitor at C_5. If the capacitance of C_5 is very low compared to that of C_{1A} and C_{1B}, then (1) the feedback ratio is not affected significantly when tuning the oscillator, and (2) the resonant frequency is determined primarily by L_1 and C_5. (Actually, the resonant frequency is determined by L_1 and the total capacitance of C_{1A}, C_{1B}, and C_5 in series.)

Because C_5 is in series with L_1, series feed cannot be used with the Clapp oscillator; also, under certain conditions the blocking capacitors C_2 and C_3 can be omitted.

17-19 Resistance-Capacitance (*RC*) Feedback Oscillators

Types. An *RC* network can be used instead of an *LC* tank circuit to produce a sinusoidal waveform output from an electronic oscillator circuit. There are two basic types of *RC* sinusoidal feedback oscillators, namely, (1) the phase-shift oscillator, and (2) the Wien-bridge oscillator. The advantage of *RC* feedback oscillators is their good stability.

Basic Principle. It has been previously established that regeneration is necessary in order to sustain oscillations in the basic amplifier circuit. With the common-emitter transistor amplifier circuit (and also with the common-cathode vacuum-tube amplifier circuit), the output signal is 180 degrees out of phase with the input signal. Accordingly, the phase of the feedback signal must be shifted 180 degrees if regeneration is to take place. A phase-shift network or a Wien-bridge circuit can be used to accomplish the 180-degree phase shift.

17-20 Phase-shift Oscillator

Basic Principle. The circuit of Fig. 17-16*a* illustrates a transistor *RC* feedback oscillator using the common-emitter configuration and a three-stage *RC* network to provide the necessary 180-degree phase shift so that the feedback signal at the base of the transistor will produce regeneration. Using additional *RC* stages improves the stability of the oscillator.

Phase-shift Principle. The manner in which the 180-degree phase shift is achieved is shown in Fig. 17-16*b* and *c*. The *RC* phase-shift portion of the oscillator circuit is redrawn in Fig. 17-16*b* to illustrate the voltages at each stage. The vector diagram (Fig. 17-16*c*) shows that with proper selection of values for *R* and *C* the phase of the voltage at the resistor will be advanced 60 degrees. Since each of the three *RC* stages advances the phase 60 degrees, the voltage $E_{f\cdot 4}$ will be advanced 180 degrees ahead of $E_{f\cdot 1}$. For a given set of *RC* values, the 180-degree phase shift occurs at only one value of frequency which becomes the output frequency (f_o) of the oscillator.

Circuit Action. The circuit is set into oscillation by any random variation of the base current such as caused by (1) noise inherent in the transistor, or (2)

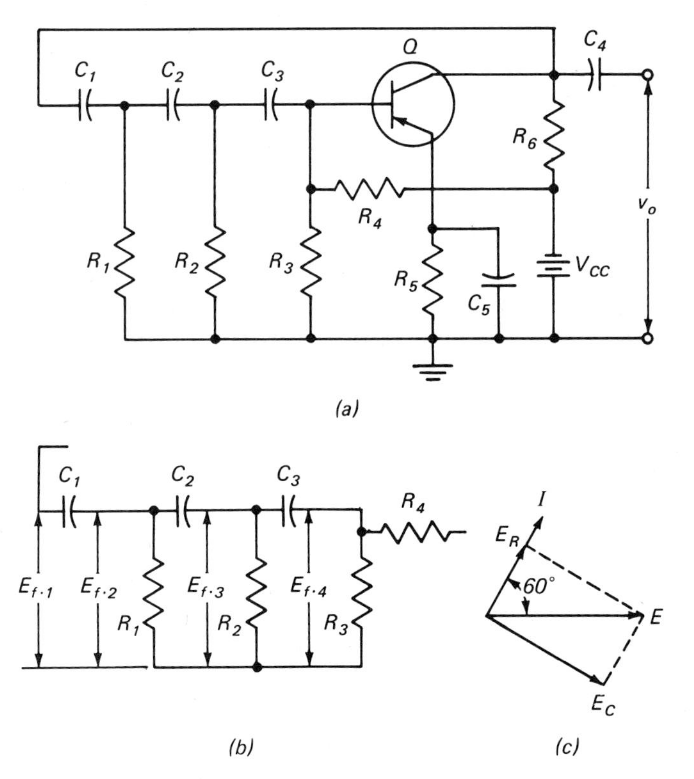

Fig. 17-16 Transistor *RC* phase-shift oscillator. (a) Circuit diagram. (b) *RC* phase-shifting network. (c) Vector analysis of phase shift produced by one *RC* stage.

minor voltage variations of the power source. Any consequent variation in the base current is (1) amplified in the collector current, (2) fed back to the RC networks C_1R_1, C_2R_2, C_3R_3, (3) reversed in phase by the RC network, (4) fed to the base of the transistor in phase with the initiating change in base current, and (5) utilized to sustain cycles of variations in collector current between saturation and cutoff values. Any deviation from a 180-degree phase shift will cause degeneration to take place, and the circuit will cease to support oscillations.

As with the preceding transistor oscillator circuits, (1) the voltage divider R_3R_4 provides a fixed emitter-base bias, (2) the collector load resistor R_6 controls the collector-emitter voltage, and (3) R_5C_5 provides temperature stabilization. The oscillator output voltage is capacitively coupled to the load by C_4. The frequency of the oscillator is determined by the time constant of the RC network instead of by the resonant frequency of an LC tank circuit.

The same principles that apply to the transistor phase-shift oscillator can be applied to an oscillator using a vacuum tube in place of the transistor as the active device.

17-21 Wien-bridge Oscillator

Basic Principle. The Wien-bridge oscillator (Fig. 17-17) uses two transistors and an RC bridge network to provide regenerative feedback. In this circuit, Q_1 serves as the amplifier oscillator and Q_2 provides phase reversal of the signal and additional amplification. The bridge circuit provides phase control of the feedback signal at the base of Q_1. Advantages of this oscillator are (1) good sine-wave output signal, (2) good frequency stability, and (3) good amplitude stability.

Phase-shift Principle. Transistor Q_1 acts as a common-emitter amplifier, and thus any base-current variations will be present in the collector current with a 180-degree shift in phase. In order to obtain a positive-feedback signal at Q_1, the Wien-bridge oscillator uses a second common-emitter amplifier Q_2 to serve as a phase inverter.

Any signal at the base-emitter diode of Q_1 is amplified by the transistor and appears at R_6 with a 180-degree phase shift. The signal at R_6 is coupled by C_4 to the input of Q_2, amplified by Q_2, and appears at R_{10} with an additional 180-degree phase shift. Thus, the signal at R_{10} is an amplified replica of the input signal at Q_1 and is of the same phase since it has twice been inverted. Because the signal at R_{10} is of the same phase as the input signal at Q_1, it can be fed back directly to the base of Q_1 to provide the regeneration needed for oscillator operation. But because Q_1 will amplify signals over a wide range of frequencies, coupling the output of Q_2 directly to the input of Q_1 would result in an oscillator having poor frequency stability. Adding the Wien-bridge circuit causes the oscillator to be sensitive only to the voltages of one frequency, thereby producing an oscillator with good frequency stability.

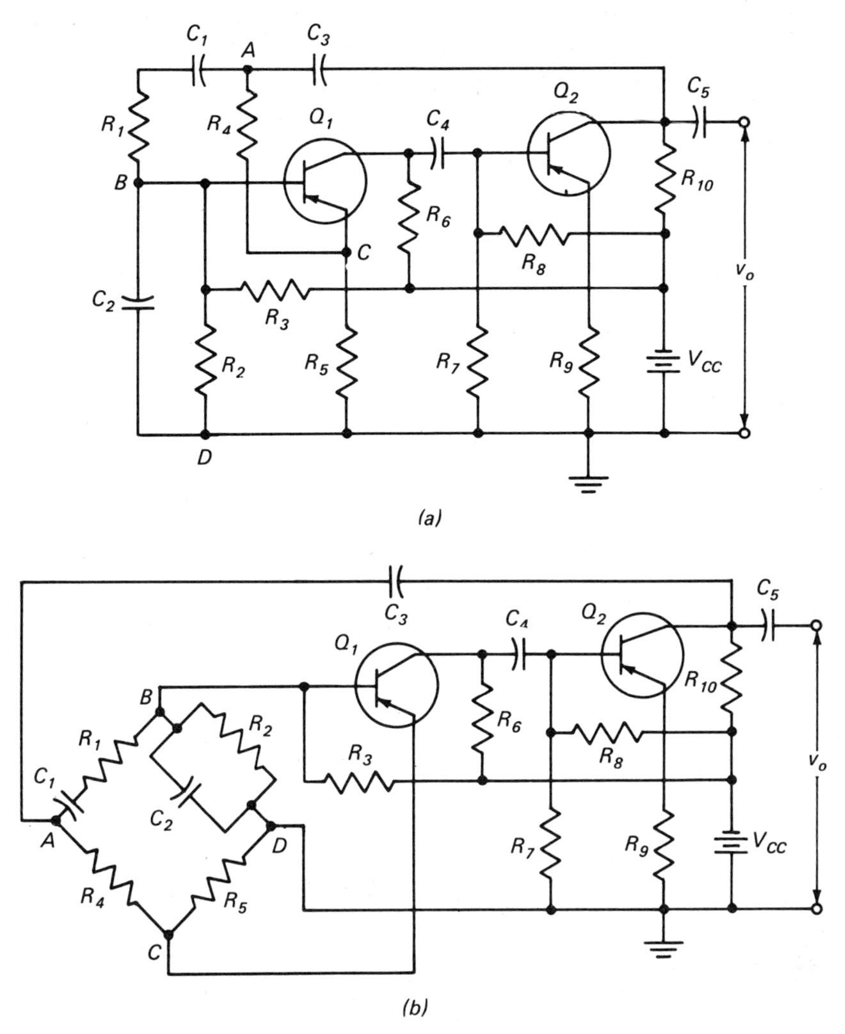

Fig. 17-17 Transistor Wien-bridge oscillator. (a) Circuit diagram. (b) Circuit redrawn to emphasize bridge portion of the circuit.

Bridge-circuit Principle. The operation of the bridge circuit is as follows: The output of Q_2 is coupled by C_3 to the bridge circuit at point A, where it divides into two paths: (1) the resistive arm ACD formed by R_4R_5, and (2) the reactive arm ABD formed by C_1R_1 and C_2R_2. The arm ACD contains an unbypassed swamping resistor (R_5) which introduces degeneration to the circuit of Q_1 because the feedback voltage introduced at the emitter is in phase with the emitter voltage and thus increases the amount of degeneration. As the arm ACD contains only resistance, any feedback it provides is not sensitive to changes in frequency. The magnitude of the feedback is determined by the voltage divider R_4R_5, and any increase or decrease in the magnitude of the feedback will result in a similar type of change in the amount of degeneration at Q_1.

The reactive arm ABD contains the base-bias resistor R_2 (shunted by C_2), which introduces an in-phase feedback signal to the base of Q_1, thereby introducing regeneration to the circuit. As this arm contains capacitors C_1 and C_2, any feedback it provides is sensitive to changes in frequency. The magnitude of the feedback signal is determined by the voltage divider formed by C_1R_1 and C_2R_2. It can be shown by vector analysis that with R_1C_1 series-connected and R_2C_2 parallel-connected the voltage E_{BD} will be in phase with the voltage E_{AD} at only one value of frequency which becomes the oscillator frequency f_o. Under this condition, the bridge is said to be balanced. For other values of frequency (1) the bridge becomes unbalanced, (2) the voltages E_{BD} and E_{AD} are no longer in phase, and (3) the amount of feedback voltage decreases.

Circuit Action. Any random change in the base current of Q_1 can initiate oscillation, as is indicated by the following sequence of changes. For example, an increase in base current at Q_1, now called the *signal*, results in (1) an amplified signal with an accompanying phase reversal at the collector of Q_1, (2) further amplification of the signal accompanied by another phase reversal at the collector of Q_2, (3) feedback of the signal at the collector of Q_2 to the input of the bridge circuit (points A–D). At the bridge circuit, a portion of the feedback signal is applied to the emitter resistor R_5, where it produces a degenerative effect. Also, a portion of the feedback signal is applied to the base-bias resistor R_2, where it produces a regenerative effect. At the rated oscillator frequency f_o, the effect of regeneration is made slightly greater than that of degeneration so that oscillation is sustained. At frequencies above and below f_o, the regenerative effect decreases; however, the degenerative effect remains constant because its source is not sensitive to frequency changes. The decrease in regeneration allows the constant value of degeneration to become the predominant factor, and the circuit will no longer sustain oscillation.

The frequency of oscillation is determined by the values of C_1, C_2, R_1, R_2 as expressed by

$$f_o = \frac{1}{2\pi \sqrt{R_1R_2C_1C_2}} \tag{17-13}$$

If, as is generally true, $R_1 = R_2$ and $C_1 = C_2$, then

$$f_o = \frac{1}{2\pi R_1C_1} \tag{17-13a}$$

From Eq. (17-13), it can be seen that the oscillator frequency can be made variable by replacing one or more of the fixed-value components R_1, R_2, C_1, C_2 by variable units.

If the swamping resistor R_5 is temperature-sensitive, such as a thermistor whose resistance increases at an optimum amount with an increase in current, the amplitude stability of the oscillator output voltage will be improved. With a thermistor at R_5, any increase in the output voltage will increase the feedback voltage and will increase the emitter current in Q_1. The increase in emitter current causes an increase in the resistance of R_5, which will in turn (1) further increase the degeneration, (2) reduce the emitter current, and (3) reduce the oscillator output voltage, thus counteracting the initial increase in the output voltage. Consequently, the amplitude of the oscillator output voltage is kept more nearly constant.

The same principles that apply to the transistor Wien-bridge oscillator can be applied to an oscillator circuit using a vacuum tube in place of the transistor as the active device.

17-22 Crystals

Uses of Crystals. The frequency of the oscillator circuits previously described is controlled by the electrical constants of the circuit. Such oscillators are called *self-controlled oscillators*. Because the values of the circuit elements will be affected by the operating conditions, the frequency of self-controlled oscillators has a tendency to drift. In order to maintain the output frequency of an oscillator at a constant value, a crystal may be used to control the frequency of oscillation. This type of oscillator is called a *crystal-controlled* oscillator.

Characteristics of Crystals. When certain crystalline materials are placed under a mechanical strain, such as compression or expansion, an electrical difference of potential will be developed across opposite faces of the crystal. This action is called the *piezoelectric effect*. Conversely, when a voltage is impressed across opposite faces of this type of crystal, it will cause the crystal to expand or contract. When alternating voltage is applied, the crystal will be set into vibration. The frequency of vibration will be equal to the resonant frequency of the crystal as determined by its structural characteristics. When the frequency of the applied voltage is equal to the resonant frequency of the crystal, the amplitude of vibration will be maximum. A high-quality crystal can maintain the frequency of oscillation so that the variation in frequency will be less than one part in a million.

Piezoelectric effects may be obtained from rochelle salts, tourmaline, and quartz; for practical reasons quartz is most commonly used. Crystals used for electronic purposes are cut into thin plates. When a plate is cut so that the flat surfaces are perpendicular to its electrical axis, it is called an *X-cut crystal*. A plate cut so that its flat surfaces are perpendicular to its mechanical axis is *called* a *Y-cut crystal*.

Temperature Characteristics. The resonant frequency of a crystal will vary with temperature changes. The number of cycles change per million cycles for a 1°C change in temperature is called the *temperature coefficient*. An *X*-cut

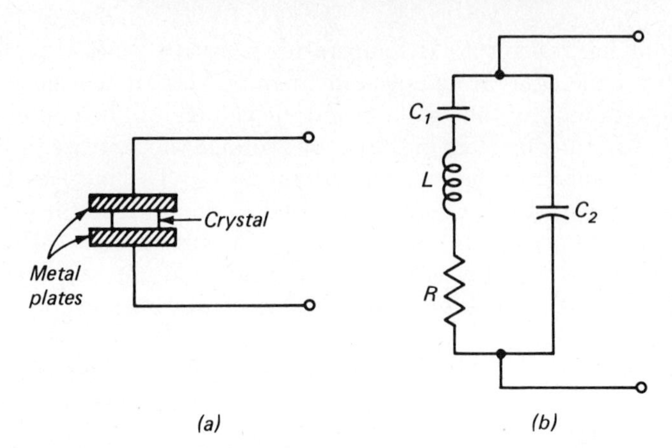

Fig. 17-18 Crystal characteristics. (a) Crystal and mounting plates. (b) Equivalent electric circuit.

crystal has a negative temperature coefficient, and thus a temperature increase will cause a decrease in the resonant frequency. A Y-cut crystal has a positive temperature coefficient, and thus a temperature increase will cause an increase in the resonant frequency.

Two methods are employed to stabilize the resonant frequency of a crystal: (1) enclosing the crystal in a thermostatically controlled container, and (2) varying the angle of cut of the crystal in order to obtain a zero temperature coefficient.

17-23 Crystal Oscillator Circuits

Equivalent Electric Circuit of a Crystal. A crystal and its equivalent electric circuit are shown in Fig. 17-18. In the equivalent electric circuit, C_1 represents the elasticity of the crystal, L represents its mass, R represents the frictional losses, and C_2 represents the parallel capacitance of the crystal (dielectric) and its two metal holding plates.

The resonant frequency of the series circuit containing L and C_1 is the same as the natural vibrating frequency of the crystal. At series resonance, maximum current will flow, thus causing the maximum amplitude of vibrations at the crystal. Conversely, maximum vibration of the crystal produces maximum voltage at the frequency of vibration.

At some frequency slightly higher than the resonant frequency, the combined effective reactance of L and C_1 is inductive and will be numerically equal to the reactance of the capacitance C_2. At this frequency the crystal circuit acts as a parallel resonant circuit, and its impedance is maximum. The circulating current in the circuit C_1LRC_2 is maximum, and therefore the vibrations of the crystal will also be maximum. Thus a parallel resonant tank circuit can be obtained by operating the crystal at the resonant frequency of the parallel tuned circuit formed by C_1LRC_2.

The equivalent inductance of a crystal is very high in comparison with either its equivalent capacitance or its equivalent resistance. Because of this high L/R ratio, the Q of a crystal circuit is many times greater than can be obtained from an electric circuit; values of Q as high as 25,000 and more can be obtained. Greater frequency stability and frequency discrimination are obtained because of the high Q and high L/R ratio of the series resonant circuit C_1LR.

Example 17-4 A certain X-cut crystal is resonant at 450 kHz. For this frequency, its equivalent inductance is 3.65 henrys, its equivalent capacitance is 0.0342 pf, and its equivalent resistance is 9,040 ohms. What is the Q of the crystal?

GIVEN: $L = 3.65$ henrys $\quad R = 9{,}040$ ohms $\quad f = 450$ kHz

FIND: Q

SOLUTION:

$$Q = \frac{2\pi fL}{R} = \frac{6.28 \times 450{,}000 \times 3.65}{9{,}040} = 1{,}141$$

17-24 Transistor Crystal Oscillator

The circuit of Fig. 17-19 illustrates the use of a crystal to stabilize the frequency output of a transistor oscillator. This circuit employs tickler-coil feedback and has a crystal in the feedback circuit. The crystal operates in its series mode and thus has minimum impedance at its resonant frequency. When the

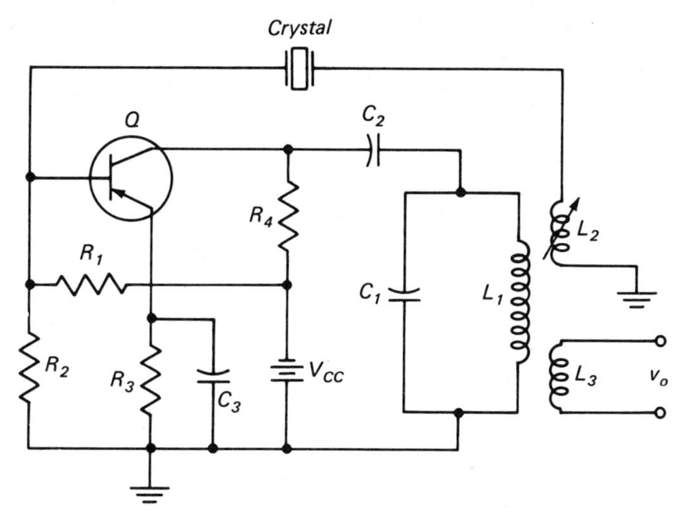

Fig. 17-19 Transistor crystal-controlled oscillator.

oscillator frequency shifts to values above or below the resonant frequency of the crystal, the impedance of the crystal increases and causes a reduction in the amount of feedback. The reduction in feedback prevents oscillation of the circuit at frequencies other than the resonant frequency of the crystal. The functions of the various resistors, inductors, and capacitors are the same as described in preceding oscillator circuits.

17-25 Vacuum-tube Crystal Oscillator

A simple crystal-oscillator circuit is shown in Fig. 17-20. This circuit is similar to the tuned-grid tuned-plate oscillator except for the substitution of the crystal for the parallel tuned-grid circuit. The voltage returned to the grid by the grid-to-plate capacitance of the tube is applied to the crystal and causes it to vibrate. The voltage set up by the crystal vibrarions is applied to the grid of the tube, thereby controlling the amount of energy released in the plate circuit.

The voltage returned to the grid circuit will be maximum when the impedance of the crystal is maximum. Maximum impedance occurs at the parallel resonant frequency of the crystal circuit. At this frequency the crystal vibrations will be maximum and thus the voltage generated by the crystal will also be maximum. The parallel resonant frequency of the crystal circuit thus determines the frequency of oscillation of a crystal-controlled oscillator. As with the tuned-grid tuned-plate oscillator, the resonant frequency of the tuned-plate circuit should be slightly greater than the parallel resonant frequency of the crystal circuit.

17-26 Frequency Multiplication

Harmonic Generation. Because the thickness of a crystal varies inversely as its frequency, at very high frequencies the crystal would have to be very thin and might easily be broken. The frequency at which a crystal becomes too thin to be practical will vary with the crystal material and the type of cut. The practical limit for quartz crystals is approximately 11 MHz; however, it is possible to grind a quartz plate to operate as high as 20 MHz. In order to

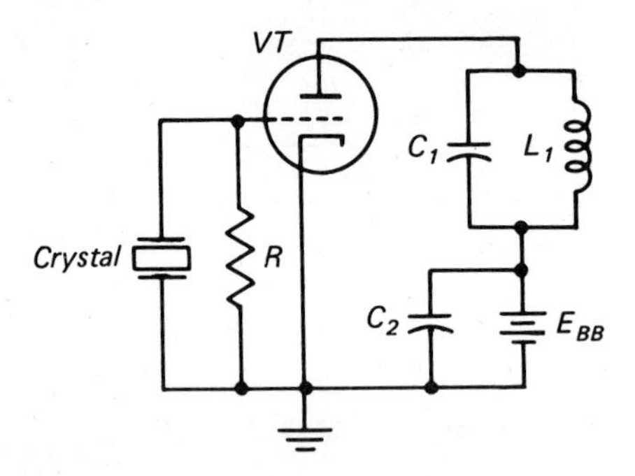

Fig. 17-20 Vacuum-tube crystal-controlled oscillator.

obtain crystal control of an oscillator at the high frequencies, frequency multiplication is employed. This is accomplished by using a crystal-controlled oscillator of a comparatively low resonant frequency and using a harmonic of the oscillator output to drive a power amplifier.

The output current of a highly biased Class C operated amplifier is not a sine wave but is made up of a fundamental frequency and a number of its harmonics. Increasing the bias on the active device will increase the intensity of the higher order of harmonics. Knowing the frequency of a crystal, the harmonic components of the output of an oscillator can be used as a frequency standard over a wide frequency range. Harmonics of a high order can be obtained from such a circuit, and it is possible that as high as a fiftieth harmonic may be used as a means of checking and calibrating high frequencies. Thus, if the fundamental frequency of an oscillator is 1,000 kHz, the twentieth harmonic is 20 MHz, and the fiftieth harmonic is 50 MHz. The crystal used to control these circuits would be required to have a resonant frequency of only 1,000 kHz.

A crystal-controlled oscillator which possesses good frequency stability is practical at only relatively low radio frequencies. In order to obtain good frequency stability at the very high and ultrahigh radio frequencies it is necessary to obtain the frequency by multiplication or harmonic generation. An oscillator circuit that produces a harmonic output is called a *frequency multiplier*.

17-27 Nonsinusoidal Oscillators

Classifications. Any waveform not of sine-wave shape is called a *nonsinusoidal waveform*. Square, rectangular, triangular, sawtooth, and pulse or spike-shaped waves are examples of nonsinusoidal waveforms. Oscillator circuits designed to produce nonsinusoidal waveforms are called *relaxation generators* or *relaxation oscillators*. A relaxation oscillator may be defined as (1) a circuit in which the voltage or current abruptly changes from one value to another and continues to oscillate or swing through these values as long as energy is applied, or (2) a circuit that depends upon the action of some circuit element to switch a low resistance automatically into or across the circuit thereby equivalently turning the current on or off and producing a desired abrupt change. Three types of relaxation oscillators are presented here: (1) sawtooth generator, (2) blocking oscillator, and (3) multivibrator.

17-28 Sawtooth Generator

Uses. Voltages having a sawtooth waveform are frequently used in electronic equipment. The most common examples are the *sawtooth* or *sweep voltages* at the picture tube of a television receiver, and the sweep voltages of the viewing screen of oscilloscopes and radar equipment.

Basic Sawtooth Generator. The simplest form of sawtooth generator requires only (1) a power source, (2) a switching device, (3) a capacitor, and (4) a resistor.

RC Circuit Action. The voltage versus time characteristic of a capacitor charging through a resistor plays an important part in the operation of the sawtooth generator. By using only the amount of charging time equal to CR, the rising portion of the sawtooth wave approximates a straight line. For periods of time greater than CR, the rising portion of the sawtooth no longer approximates a straight line because of the exponential rise in the voltage versus time characteristic (Fig. 12-43). Conversely, for periods of time less than CR, the linearity of the sawtooth wave is improved.

Frequency of Oscillation. The frequency at which the sawtooth waves occur is equal to the reciprocal of the elapsed time between the waves, for example, the time A-C of Fig. 17-21*b*. Thus,

$$f = \frac{1}{t} \tag{17-14}$$

Example 17-5 What is the frequency of the sawtooth generator of Fig. 17-21*a* if $C = 0.1\ \mu f$, $R = 10{,}000$ ohms, ionization occurs at 0.9 CR, and deionization occurs at 0.1 CR?

GIVEN: $C = 0.1\ \mu f$ $R = 10{,}000$ ohms $t_B = 0.9\ CR$ $t_A = 0.1\ CR$

FIND: f

SOLUTION:

$$f = \frac{1}{t} = \frac{1}{t_B - t_A} = \frac{1}{(0.9 - 0.1) \times 10^{-3}} = 1{,}250 \text{ Hz}$$

$$t_B = 0.9CR = 0.9 \times 0.1 \times 10^{-6} \times 10^4 = 0.9 \times 10^{-3} \text{ sec}$$

$$t_A = 0.1CR = 0.1 \times 0.1 \times 10^{-6} \times 10^4 = 0.1 \times 10^{-3} \text{ sec}$$

17-29 Neon-tube Sawtooth Generator

Neon-tube Characteristics. A neon tube can serve as a switching device because of its sensitivity to a source of variable direct voltage, namely, (1) it offers a very high resistance to current flow up to a certain voltage level, called the *ionization* or *ignition voltage level*, (2) the resistance drops abruptly to a very low value when ionization takes place, (3) a subsequent decrease in voltage produces no change in resistance until the *deionization* or *extinction voltage level* is reached, and (4) the resistance rises abruptly to a very high value when deionization takes place. A disadvantage of the neon tube as a switching device is the instability of the ionization and deionization voltage levels.

Circuit Action. For the circuit of Fig. 17-21*a* it is assumed that (1) $E = 100$ volts, (2) the ionization level of the tube is 60 volts, (3) the deionization level is 10 volts, (4) the capacitor will charge to 60 volts in CR sec, and (5) the product of $CR = 1$ sec. The sequence of actions when switch S is initially closed is: (1) the capacitor will charge at approximately a linear rate to 60 volts in 1 sec, (2) after 1 sec the tube ionizes and virtually short-circuits the capacitor, causing its voltage to drop rapidly toward zero, (3) upon reaching the deioniza-

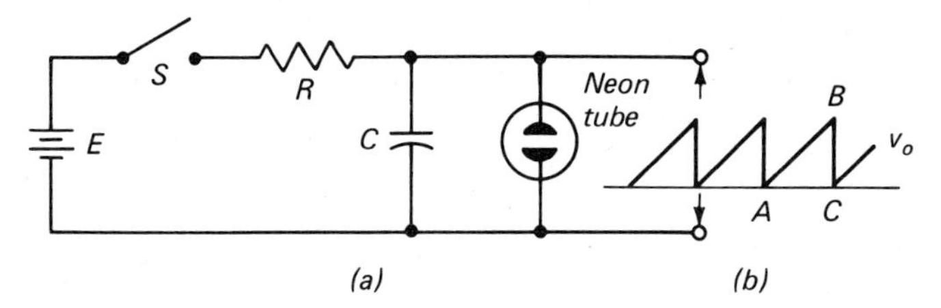

Fig. 17-21 Sawtooth generator. (a) Circuit diagram. (b) Waveform of output voltage v_o.

tion voltage level the tube again acts as an open circuit, (4) the capacitor again starts to charge and the cycle of actions repeats. After the first cycle of changes the capacitor voltage varies from 10 to 60 volts instead of 0 to 60 volts and the charging time will be reduced slightly.

The discharge portion of the cycles follows the CR time-delay pattern indicated by Fig. 12-43*b*. However, because the resistance of the tube is very low during the discharge portion of the cycle, the time constant CR is also very low, and for this presentation it is assumed to approach zero.

17-30 Transistor Sawtooth Generator

A simple sawtooth generator can be constructed using only a unijunction transistor (Art. 10-17), a capacitor, and a power source as shown in Fig. 17-22.

When switch S is initially closed, the sequence of actions is: (1) electron current flows from B_1 to B_2 and an initial reverse bias is established at the PN junction; (2) a small amount of reverse (hole) current flows as indicated by the solid-line arrows and charges the capacitor at a rate determined by the time constant produced by the capacitor and the high resistance of the reverse-biased emitter-to-base-2 junction; (3) when the charge on the capacitor exceeds

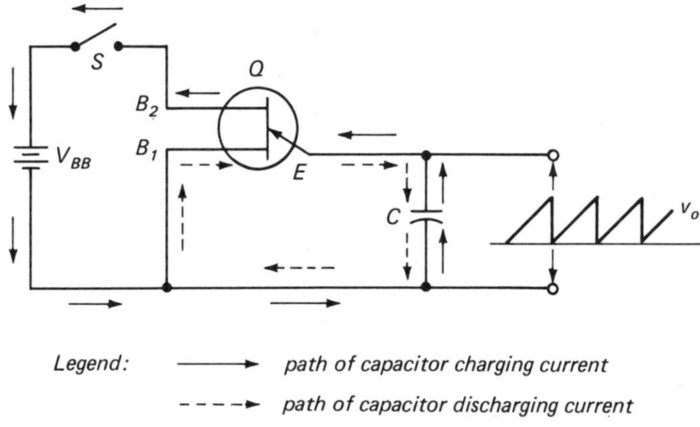

Fig. 17-22 Transistor sawtooth generator.

the voltage gradient opposite the P-type material, the PN junction changes from reverse-biased to forward-biased and its resistance decreases to a low value; (4) the capacitor discharges through the path indicated by the broken-line arrows; (5) as the capacitor voltage approaches zero, the PN junction again becomes reverse-biased and the cycle of circuit actions is repeated.

The rate at which the capacitor charges is determined by the capacitance and the relatively high resistance of the PN junction with reverse bias. The rate at which the capacitor discharges is determined by the capacitance and the relatively low resistance of the PN junction with forward bias. The resultant slow charge and fast discharge produces a sawtooth output.

17-31 Thyratron Sawtooth Generator

Basic Principle. A thyratron (Art. 5-15) can be used to perform the function of the neon tube in the sawtooth generator of Art. 17-29. Although the thyratron circuit is more complex than the neon-tube circuit, it does provide greater operating stability. Figure 17-23 illustrates a sawtooth generator using a thyratron tube.

Circuit Action. The operation of the circuit is the same as the neon-tube circuit with the additional control provided by the grid of the thyratron. Increasing the grid bias increases the ionizing potential and decreases the frequency. Changes in the grid bias have very little effect on the deionizing potential.

A disadvantage of the relaxation oscillator circuit is its poor frequency stability. If C_2 (shown in broken lines on Fig. 17-23) is added to the circuit and voltage pulses are applied to the capacitor at a constant rate, the frequency of the sawtooth-wave output will be controlled by the synchronizing voltage pulses, thereby improving the frequency stability of the oscillator.

17-32 Blocking Oscillator

Basic Principle. A blocking oscillator is a type of relaxation oscillator which cuts itself off, or blocks itself, after one or more cycles and remains blocked for a much longer period of time. A blocking oscillator that cuts off at the end of

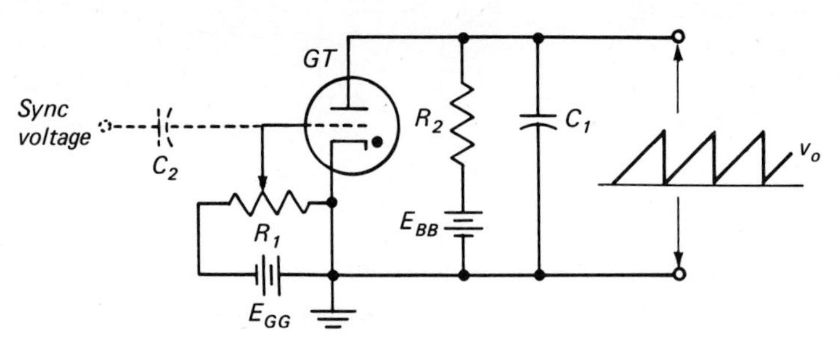

Fig. 17-23 Thyratron sawtooth oscillator.

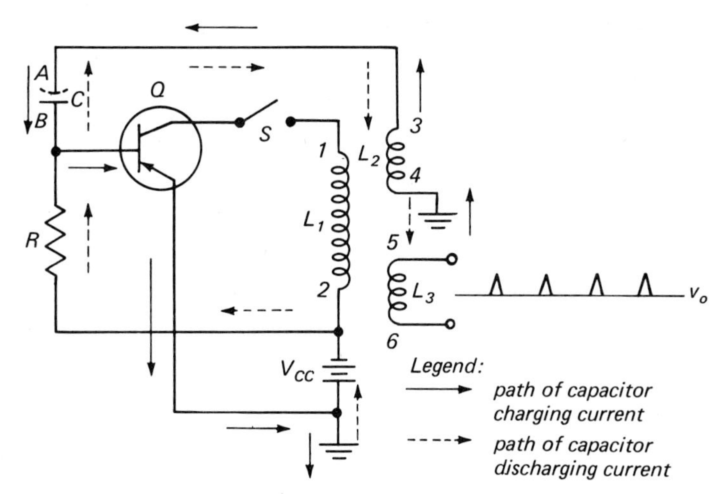

Fig. 17-24 Transistor blocking oscillator.

one cycle (or even one-half cycle) is called a *single-swing type*. A blocking oscillator that requires the completion of a number of cycles before cutoff is achieved is called a *self-pulsing type*.

17-33 Transistor Blocking Oscillator

The circuit of Fig. 17-24 illustrates a blocking oscillator using a transistor in the common-emitter configuration. The operation of this circuit is as follows: (1) The emitter-base diode is forward-biased by the power source V_{CC} and resistor R. (2) When switch S is initially closed, collector current will start flowing, and as it increases a voltage will be induced in L_2, with terminal 3 becoming more negative. (3) This induced voltage will charge capacitor C through the path indicated by the solid-line arrows. (4) This charging current increases the base current, which in turn further increases the collector current, followed by increases in induced voltage at L_2 and the capacitor charging current until collector-current saturation is reached. (5) At saturation, the induced voltage at L_2 ceases and the charging current through the base-emitter diode also ceases. (6) The capacitor, being charged with plate B positive, introduces a reverse bias at the base-emitter diode and drives the transistor to cutoff. (7) The capacitor now starts to discharge through the path indicated by the broken-line arrows. (8) When the capacitor discharges to the level where the base-emitter bias provided by V_{CC} and R again becomes predominant, the above cycle of actions is repeated.

The time required to charge the capacitor is determined by the capacitance of C and the relatively low base-emitter forward resistance and is short. The discharge time is determined by the capacitance of C and the relatively high resistance of R and therefore is much longer than the charging time. The

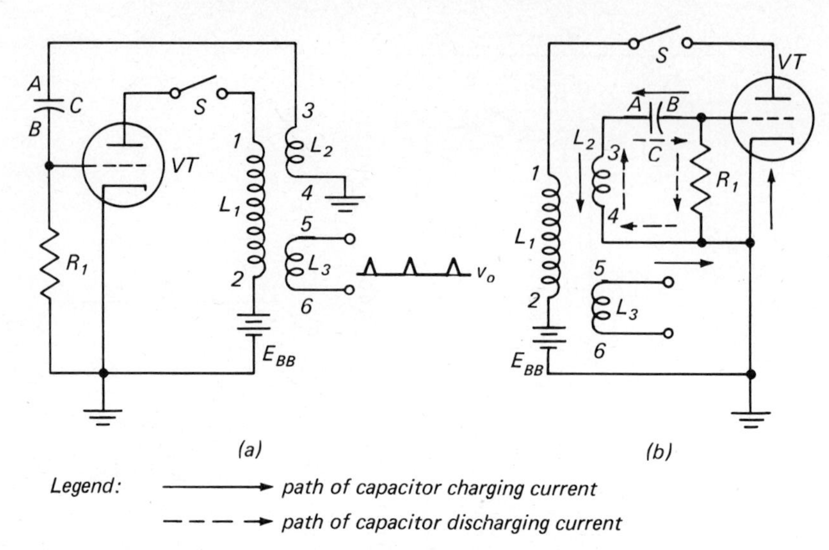

Fig. 17-25 Vacuum-tube blocking oscillator. (a) Circuit diagram. (b) Equivalent circuit showing charge and discharge paths of the capacitor.

resultant output voltage is a sharp narrow pulse during the charging period and a long blocking time during the discharge period.

17-34 Vacuum-tube Blocking Oscillator

The circuit of Fig. 17-25*a* illustrates a blocking oscillator using a vacuum tube. In Fig. 17-25*b* the circuit components have been repositioned in order to show more easily the charge and discharge paths for the capacitor. This circuit operates in much the same manner as the transistor blocking oscillator of Art. 17-33. (1) When the switch S is initially closed, plate current will start flowing, and as it increases a voltage will be induced in L_2 with terminal 3 positive. (2) This induced voltage makes the grid of the tube positive and causes (*a*) a rapid rise in plate current up to saturation, (*b*) a flow of grid current through the path indicated by the solid-line arrows on Fig. 17-25*b*, and (*c*) the capacitor C to rapidly become charged with plate B negative. (3) At saturation the plate current ceases to rise and the induced voltage at L_2 drops to zero. (4) Capacitor C is now the prime voltage source for the grid and drives the grid bias strongly negative, thereby driving the plate current very rapidly to cutoff. (5) The capacitor now slowly discharges through the path indicated by the broken-line arrows until the grid voltage reaches the point where plate current can again flow. (6) This sequence of actions is repeated.

The time required to charge the capacitor is determined by the capacitance of C and the cathode-to-grid resistance, which is much lower than R_1 when the grid is positive with respect to the cathode. The time required for the capacitor to discharge is determined by the capacitance of C and R_1 which is much lower than the grid-to-cathode resistance when the grid is negative with respect to the cathode. With the value of R_1 much greater than the cathode-grid resistance when the grid is positive and much lower than the grid-cathode resistance when the grid is negative, the resultant output voltage of the oscillator is a sharp narrow pulse during the charging period and a long blocking time during the discharge period.

17-35 Multivibrators

The multivibrator is a type of relaxation oscillator used extensively to produce various nonsinusoidal waveforms such as square, rectangular, and sawtooth. The multivibrator is basically a two-stage amplifier, or oscillator, circuit that operates in two *modes* or *states* controlled by the circuit conditions. Each amplifier stage supplies feedback to the other in such a manner that will drive the active device of one stage to saturation and the other to cutoff. This is followed by a new set of actions that causes the opposite effects, that is, the saturated stage is driven to cutoff, and the cutoff stage is driven to saturation. The successful operation of the multivibrator is based on the premise that no two active devices have exactly identical characteristics. Multivibrators are classed as (1) *astable,* (2) *bistable,* and (3) *monostable.*

The *astable* multivibrator, also called a *free-running* multivibrator, alternates automatically between the two states and continues the alternations at a rate determined by the values of the circuit components. This type of multivibrator is self-starting and sustains oscillation under normal operation; however, in some applications synchronizing, or triggering, pulses may be used in conjunction with the multivibrator.

The *bistable* multivibrator, also called an *Eccles-Jordan,* or a *flip-flop* multivibrator, has two states of operation but requires the application of an external triggering pulse to change the operation from either one state to the other. Depending upon the circuit design, the circuit remains in the existing state (1) as long as the triggering pulse is applied, or (2) until a second triggering pulse is applied.

The *monostable* multivibrator, also called a *single-shot,* a *single-swing,* or a *one-shot* multivibrator, has (1) a normal or quiescent state, and (2) a transient state induced by an external triggering pulse. Following triggering, the circuit automatically returns to its initial quiescent state after a period of time determined by the values of the circuit components.

17-36 Transistor Astable Multivibrator

The Circuit. Figure 17-26 illustrates an astable multivibrator circuit using two similar transistors. The circuit is essentially two symmetrical common-emitter amplifier stages, each providing a feedback signal to the other. In this circuit, Q_1 is forward-biased by V_{CC} and R_3, and Q_2 is forward-biased by V_{CC} and R_2. The collector load resistors R_1 and R_4, together with V_{CC}, determine the collector-emitter voltages of Q_1 and Q_2, respectively. The output of Q_1 is coupled to the input of Q_2 by C_1, and C_2 couples the output of Q_2 to the input of Q_1. The load of the multivibrator is capacitively coupled to the output of Q_2 by C_3.

Circuit Operation. At the instant that switch S is closed, the circuit actions are as follows: (1) forward bias is established at Q_1 and Q_2; (2) collector current starts to flow in Q_1 and Q_2; (3) because the characteristics of no two transistor are *exactly* alike, one transistor, say Q_1, will start conducting before the other and will have a higher amount of collector current; (4) the currents $i_{c\cdot 1}$ and $i_{c\cdot 2}$ flowing through R_1 and R_4, respectively, will cause a changing voltage, or signal, to be present at points A and B; (5) the signal at A is coupled to the base of Q_2 by C_1; and (6) the signal at B is coupled to the base of Q_1 by C_2. Continuing to examine the circuit actions, starting at Q_1 because $i_{c\cdot 1}$ is greater than $i_{c\cdot 2}$, shows that (7) an increase in $i_{c\cdot 1}$ increases the voltage drop $i_{c\cdot 1}R_1$ and makes point A less negative, which is equivalent to a positive-going signal; (8) C_1 couples this signal to the base of Q_2; (9) the positive-going signal at Q_2 decreases the forward bias of Q_2 and decreases $i_{c\cdot 2}$; (10) the decrease in $i_{c\cdot 2}$ reduces the voltage $i_{c\cdot 2}R_4$ and causes point B to become more negative, which is equivalent to a negative-going signal; (11) C_2 couples this signal to the base of Q_1; (12) the negative-going signal at Q_1 increases the forward bias of Q_1 and causes a further increase in $i_{c\cdot 1}$. This series of actions is repeated until it drives Q_1 into saturation and Q_2 into cutoff. These actions occur very rapidly and for this discussion are considered to be practically instantaneous.

When Q_1 is at saturation and Q_2 at cutoff, (1) the full voltage of V_{CC} will appear across R_1, (2) the voltage across R_4 will be zero, (3) the collector currents are stabilized, and (4) signal voltages no longer exist at A and B. The charges developed at C_1 and C_2 are sufficient to maintain the saturation and cutoff conditions at Q_1 and Q_2. This condition is represented by the time intervals 1-2 and 3-4 on Fig. 17-27*b* and *f*. However, the capacitors will not retain their charges indefinitely but will discharge through their respective circuits. The

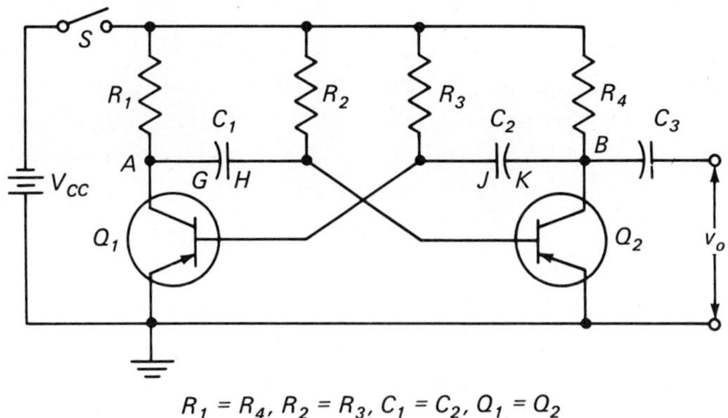

$R_1 = R_4, R_2 = R_3, C_1 = C_2, Q_1 = Q_2$

Fig. 17-26 Transistor astable multivibrator.

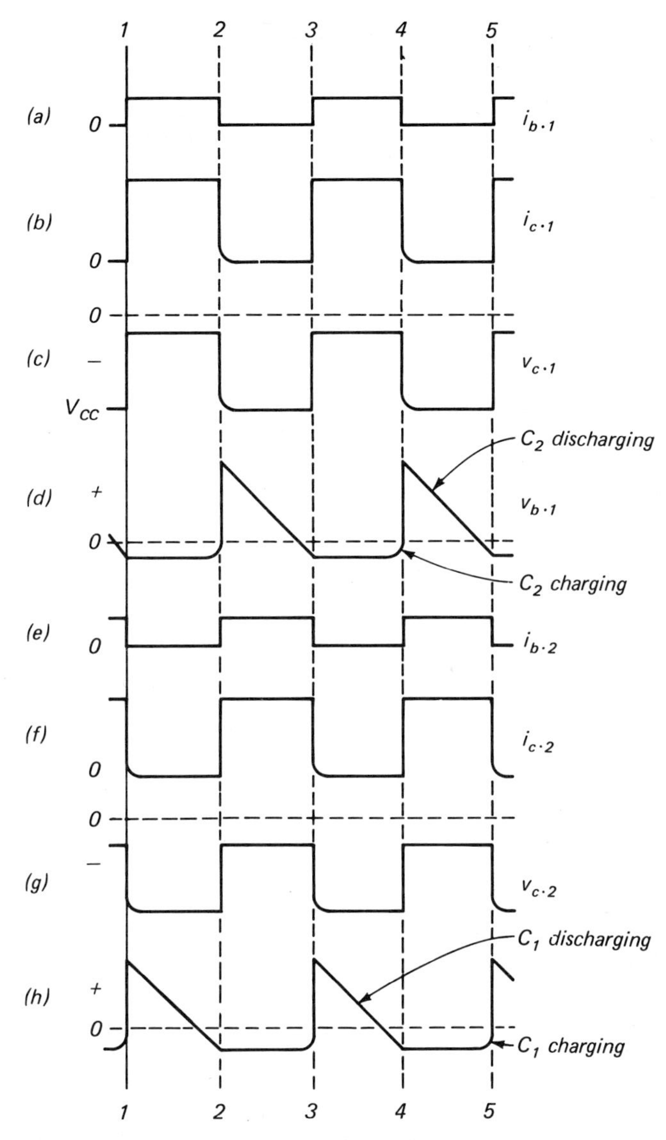

Fig. 17-27 Current and voltage waveforms with their phase relations for the multivibrator circuit of Fig. 17-26.

discharge path of C_1, with plates G negative and Q_1 conducting, is G, A, Q_1, V_{CC}, R_2, H. The discharge path for C_2, with plates K negative and Q_2 at cutoff, is K, B, R_4, R_3, J. Because the resistance of the discharge path for C_1 is lower than that for C_2, C_1 will discharge more rapidly and its effect will be the predominant one. Thus as C_1 discharges, the base bias at Q_2 becomes less positive and ultimately, at a time determined by R_2 and C_1, forward bias is reestablished at Q_2 and collector current starts to flow in Q_2. At this point C_2 has not yet discharged sufficiently to relieve Q_1 from saturation. With $i_{c \cdot 2}$ increasing, a positive-going signal is established at B and is coupled to the base of Q_1, thereby reducing the forward bias at Q_1 and causing $i_{c \cdot 1}$ to decrease. The decrease in $i_{c \cdot 1}$ establishes a negative-going signal at A which is coupled to the base of Q_2, thereby causing an increase in $i_{c \cdot 2}$. With this set of actions taking place, Q_2 is quickly driven to saturation and Q_1 to cutoff. The period of time during which Q_2 remains at saturation and Q_1 at cutoff is determined by C_2 and R_3.

Current and Voltage Characteristics. The base and collector current and voltage characteristics of Q_1 and Q_2 for the multivibrator circuit of Fig. 17-26 are illustrated in Fig. 17-27. This graph represents two cycles of operation. The time elapsed between 1 to 2 and 3 to 4 represents periods when Q_1 is conducting and Q_2 is cut off. The time elapsed between 2 to 3 and 4 to 5 represents Q_2 conducting and Q_1 cut off.

Frequency. The frequency at which a complete cycle of values repeats is represented on Fig. 17-27 by the elapsed time between intervals 1 and 3. Figure 17-27*d* and *h* shows that the time of one cycle is dependent upon the sum of the discharge times for C_1 and C_2. The frequency is also dependent on what amount of the discharging-time constant RC is required to reach the base-bias potential at which the transistors change from cutoff to conduction. Accordingly, the frequency of the multivibrator of Fig. 17-26 can be represented by

$$f = \frac{1}{(k_1 C_1 R_2) + (k_2 C_2 R_3)} \tag{17-15}$$

where k_1 = number of C_1R_2 time constants required to change Q_2 from cutoff to conduction

k_2 = number of C_2R_3 time constants required to change Q_1 from cutoff to conduction

Example 17-6 What is the frequency for the multivibrator circuit of Fig. 17-26 if C_1, $C_2 = 0.25\ \mu$f, R_2, $R_3 = 5{,}000$ ohms, k_1, $k_2 = 0.8$?

GIVEN: C_1, $C_2 = 0.25\ \mu$f $\quad$ R_2, $R_3 = 5{,}000$ ohms $\quad$ k_1, $k_2 = 0.8$

FIND: f

SOLUTION:

$$f = \frac{1}{(k_1 C_1 R_2) + (k_2 C_2 R_3)}$$

$$= \frac{1}{(0.8 \times 0.25 \times 10^{-6} \times 5 \times 10^3) + (0.8 \times 0.25 \times 10^{-6} \times 5 \times 10^3)} = 500 \text{ Hz}$$

17-37 Transistor Bistable Multivibrator

The Circuit. Figure 17-28*a* illustrates a bistable multivibrator circuit. This circuit differs from that of Fig. 17-26 in the following respects: (1) the manner

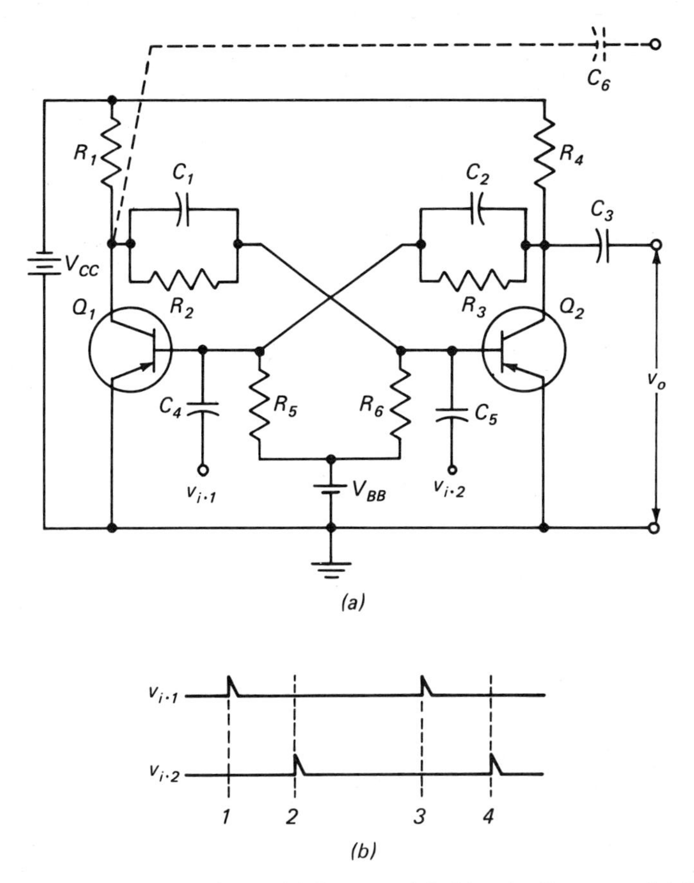

Fig. 17-28 Transistor bistable multivibrator. (a) Circuit diagram. (b) Triggering voltages.

of providing the emitter-base bias for the transistors, (2) the feedback is coupled through resistors shunted by capacitors, and (3) triggering voltages are required to change from one stable state to the other. Output signals can be taken from C_3, C_6, or both.

Circuit Operation. Base-emitter bias is established with the two power sources V_{BB} and V_{CC} and the voltage-divider circuits formed by (1) R_6, R_2, and R_1 for Q_1, and (2) R_5, R_3, and R_4 for Q_2. Although it may appear that V_{BB} will introduce a reverse bias at the transistors, the values of the resistors and the power sources are chosen to establish normal forward bias for the transistors.

Example 17-7 If for the circuit of Fig. 17-28*a* R_1, R_4 = 5,000 ohms, R_2, R_3 = 75,000 ohms, R_5, R_6 = 20,000 ohms, V_{CC} = 27 volts, and V_{BB} = 3 volts, what is the voltage at the base terminal of Q_1 when the transistor is operating at (*a*) saturation? (*b*) Cutoff?

(*Note:* At saturation the voltage drop across the collector resistor R_1 or R_4 is approximately equal to V_{CC}.)

GIVEN: R_1, R_4 = 5,000 ohms R_2, R_3 = 75,000 ohms R_5, R_6 = 20,000 ohms
V_{CC} = 27 volts V_{BB} = 3 volts

FIND: (*a*) $v_{B \cdot Q1}$ (*b*) $v_{B \cdot Q1}$

SOLUTION:
(*a*) Q_1 is at saturation and Q_2 is at cutoff
By Kirchhoff's law

$$V_{BB} - IR_5 - IR_3 - IR_4 + V_{CC} = 0$$

therefore

$$I = \frac{V_{BB} + V_{CC}}{(R_3 + R_4 + R_5)} = \frac{3 + 27}{(75 + 5 + 20) \times 10^3} = 0.3 \text{ ma}$$

and

$$v_{B \cdot Q1} = V_{BB} - IR_5 = 3 - (0.3 \times 10^{-3} \times 20 \times 10^3) = -3 \text{ volts}$$

(*b*) Q_1 is at cutoff, Q_2 is at saturation, $I_4R_4 \cong V_{CC}$ = 27 volts
By Kirchhoff's law

$$V_{BB} - IR_5 - IR_3 - I_4R_4 + V_{CC} = 0$$
$$3 - I(R_5 + R_3) - 27 + 27 = 0$$

therefore

$$I = \frac{3}{R_5 + R_3} = \frac{3}{(20 + 75) \times 10^3} = 0.0318 \text{ ma}$$

and

$$v_{B \cdot Q1} = V_{BB} - IR_5 = 3 - (0.0318 \times 10^{-3} \times 20 \times 10^3) = +2.36 \text{ volts}$$

The feedbacks between the transistors are coupled by R_2 and R_3 shunted by C_1 and C_2 to bypass any high-frequency components of the triggering voltages to ensure that rapid changes can be made from one mode to the other.

In the same manner as described for the astable multivibrator, the initial actions are established by the principle that one transistor will start conducting slightly ahead of the other. The ensuing circuit actions will drive one transistor to saturation and the other to cutoff. The voltage-divider actions for the base bias will establish a forward bias for the transistor at saturation and a reverse bias for the transistor at cutoff (Example 17-7). In the absence of any triggering pulses at C_4 and C_5, the base-emitter bias conditions remain fixed and the two transistors remain in their current modes of operation.

For further study of the circuit operation it is assumed that Q_1 is conducting and Q_2 is at cutoff; thus Q_1 is forward-biased and Q_2 is reverse-biased. If a positive triggering pulse of sufficient magnitude is applied to C_4, Q_1 will be driven to cutoff. The circuit reaction will produce (1) feedback voltages, and (2) changes in the base voltage-divider conditions that will cause Q_2 to conduct and Q_1 to be cut off. (A negative triggering pulse at C_5 will produce a similar change.) These conditions will prevail until a positive triggering pulse is applied to Q_2, or a negative pulse to Q_1.

The spacing of the triggering pulses determines the shapes of the output voltages. The time constants C_1R_2 and C_2R_3 determine the time required for the collector currents to change from conduction to cutoff for Q_1 and Q_2, respectively.

17-38 Transistor Monostable Multivibrator

The circuit of Fig. 17-29 illustrates a basic transistor monostable multivibrator. In the absence of any triggering pulse at C_2, the circuit actions are (1) the circuit values are such that V_{BB} and R_5 reverse-bias Q_1 and keep it at cutoff, (2) V_{CC} and R_2 forward-bias Q_2 into saturation, and (3) C_1 charges to approximately V_{CC}. This represents a quiescent state for the circuit. The step-by-step analysis of the actions is the same as described for the preceding multivibrator circuits.

When a negative pulse of short duration and sufficient magnitude is applied to C_2, the circuit actions are (1) Q_1 starts conducting, (2) a positive-going signal is established at the collector of Q_1, (3) this positive signal is coupled by C_1 to the base of Q_2, causing a decrease in $i_{c\cdot2}$, (4) a negative-going signal is established at the collector of Q_2, (5) this negative signal is coupled through R_3 to the base of Q_1, (6) $i_{c\cdot1}$ increases, and (7) rapid repetition of these actions drives Q_1 to saturation and Q_2 to cutoff. Capacitor C_1 discharges through the path R_2, V_{CC}, Q_1 (with Q_1 at saturation, its collector-emitter path has a low resistance). As C_1 discharges, the voltage it supplies to the base of Q_2 becomes less positive, and when it decreases to the point where the base of Q_2 is again forward-biased, Q_2 will go into conduction. This is followed by the sequence of actions: (1) a decrease in $v_{c\cdot2}$, (2) a positive-going signal is established at the collector of Q_2 which is coupled to the base of Q_1, (3) a decrease in $i_{c\cdot1}$, (4) an increase in $v_{c\cdot1}$, (5) a negative-going signal at the collector of Q_1 which is coupled to the base of Q_2, (6) a decrease in $v_{b\cdot2}$, (7) an increase in $i_{c\cdot2}$, (8) and so on until Q_1 is driven to cutoff and Q_2 to saturation. The circuit remains in this quiescent state until another triggering pulse is introduced (see Fig. 18-30).

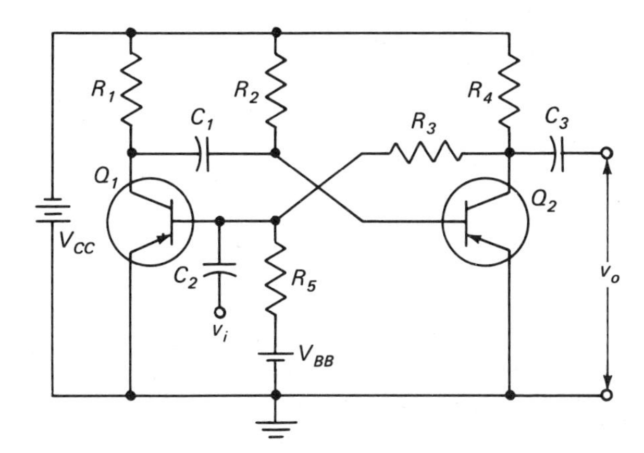

Fig. 17-29 Transistor monostable multivibrator.

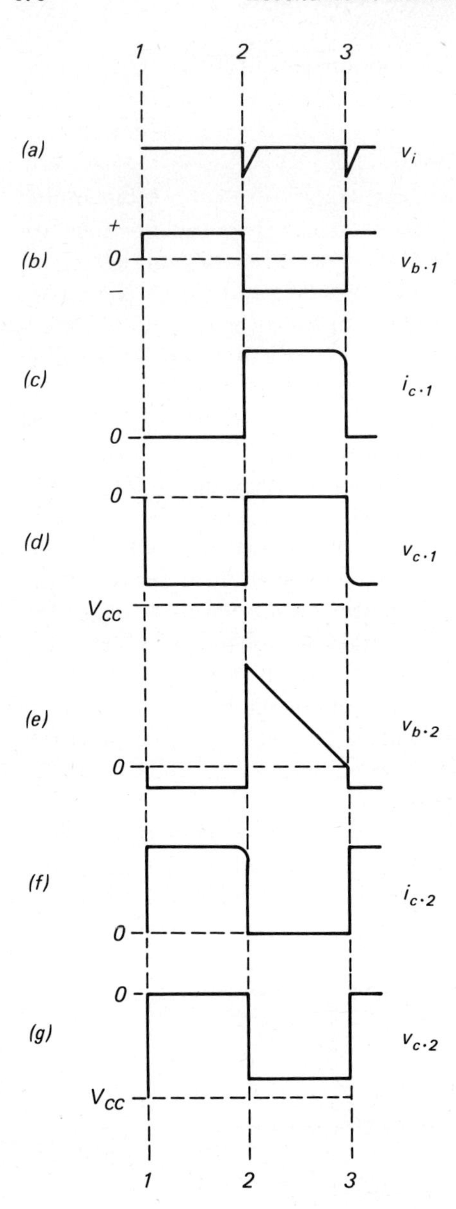

Fig. 17-30 Current and voltage waveforms with their phase relations for the multivibrator circuit of Fig. 17-29.

17-39 Vacuum-tube Multivibrators

The same principles that were presented in describing transistor multivibrators can be applied to multivibrator circuits employing vacuum tubes. Because of the similarity to their transistor counterparts only one vacuum-tube multivibrator circuit is presented.

17-40 Vacuum-tube Astable Multivibrator

The circuit of Fig. 17-31 illustrates a basic free-running multivibrator employing vacuum tubes. As with the transistor circuits, it is recognized that the characteristics of no two tubes are exactly alike. Therefore, at the instant switch S is closed, plate current will start flowing in one tube, say VT_1, before the other and will have a slightly higher amount of current. The increasing plate current at VT_1 produces a negative-going signal at the plate of the tube which is coupled by C_1 to the grid of VT_2. The negative signal at the grid of VT_2 causes a decrease in its plate current and establishes a positive-going signal at its plate. This positive signal is coupled by C_2 to the grid of VT_1 and causes a further increase in its plate current. Following a rapid succession of these changes, VT_1 is driven to saturation and VT_2 is driven to cutoff. As capacitor C_1 discharges through R_2, the grid voltage at VT_2 drops enough to change the tube operation from cutoff to conduction. A negative-going signal is established at the plate of VT_2 and is coupled to the grid of VT_1, which results in a decrease in its plate current. A positive-going signal, established at the plate of VT_1, is coupled to the grid of VT_2 and further increases its plate current. These actions continue until VT_2 is driven to saturation and VT_1 is driven to cutoff. When C_2 discharges enough to start VT_1 into conduction, the above cycle of actions is repeated.

As with the transistor astable multivibrator, the frequency of the oscillations is determined by the time constants of the RC circuits associated with the control electrode of the active device. Thus, for the circuit of Fig. 17-31, an approximate value of the frequency is expressed by

$$f_r \cong \frac{1}{R_2C_1 + R_3C_2} \tag{17-16}$$

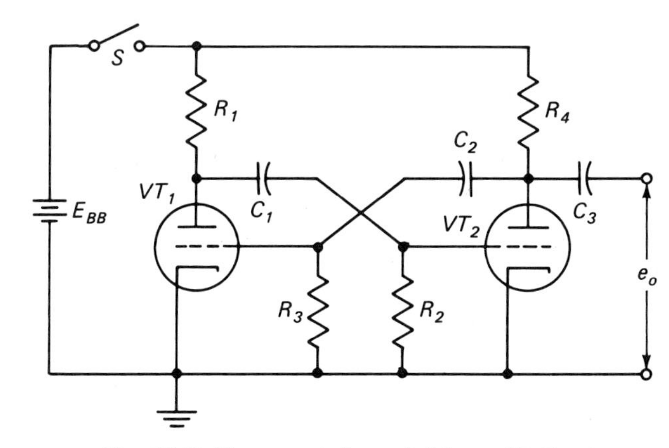

Fig. 17-31 Vacuum-tube astable multivibrator.

QUESTIONS

1. (*a*) Define the oscillator circuit. (*b*) What types of active devices are used in oscillator circuits?
2. (*a*) Name two classifications of oscillator circuits. (*b*) Define each classification.
3. Define the following types of oscillator circuits: (*a*) relaxation, (*b*) feedback, (*c*) heterodyne, (*d*) crystal.
4. Explain the relation between an amplifier and an oscillator.
5. Name four methods of obtaining feedback for oscillator circuits.
6. What is meant by (*a*) tank circuit? (*b*) Flywheel effect?
7. Explain the action of the tank circuit.
8. What is the power requirement of an oscillator if oscillation is to be sustained?
9. (*a*) What is a tickler coil? (*b*) What is its function?
10. In practical oscillator circuits, what provisions are generally added to the basic oscillator?
11. (*a*) Define the tuned-base oscillator. (*b*) Describe the action of the circuit of Fig. 17-5*a*.
12. Describe the action of the oscillator circuit of Fig. 17-6*a*.
13. What is the relation of the tank circuit of an oscillator to its (*a*) amplitude stability? (*b*) Waveshape?
14. What is the relation of the factors in a tank circuit at resonance that affect the (*a*) frequency? (*b*) Impedance? (*c*) Voltage? (*d*) Power delivered to the tank circuit?
15. Describe the effects on the tank-circuit characteristics if the value of the capacitance is increased and accompanied by a decrease in the inductance sufficient to maintain the same frequency?
16. (*a*) Name four methods of coupling a load to an oscillator. (*b*) What output characteristics of an oscillator are generally affected by changes in the power taken by the load?
17. (*a*) How are large amounts of power generally supplied to a load by an oscillator system? What function is provided by (*b*) the oscillator stage? (*c*) The amplifier stage?
18. (*a*) Define frequency stability. (*b*) What is meant by frequency drift? (*c*) What factors affect the frequency stability of an oscillator?
19. (*a*) What factors affect the choice of circuit configuration of the active device for an oscillator? What configuration is commonly used in oscillators employing (*b*) transistors? (*c*) Vacuum tubes?
20. What are the possible locations for the tank circuit in oscillators employing (*a*) transistors? (*b*) Vacuum tubes?
21. (*a*) Explain the principle of the tuned-grid tuned-plate oscillator. (*b*) What factors determine the frequency of the oscillator?
22. (*a*) Explain the principle of electron coupling. (*b*) Describe the operation of the electron-coupled oscillator of Fig. 17-8.
23. (*a*) What is meant by a split-tank inductor? (*b*) In what type of oscillator is it used? (*c*) Name three functions of the split-tank inductor.
24. What is meant by series feed and shunt feed in oscillator circuits?
25. Describe the operation of the Hartley oscillator circuit of Fig. 17-10.
26. Explain how the oscillator of Fig. 17-11 differs from that of Fig. 17-10.

27. (*a*) Explain how the oscillator of Fig. 17-12 differs from that of Fig. 17-10. (*b*) Describe the operation of the oscillator circuit of Fig. 17-12.
28. (*a*) What are the characteristics of the Colpitts oscillator? (*b*) Describe the operation of the Colpitts oscillator circuit of Fig. 17-14*a*. (*c*) Why is parallel feed necessary with the Colpitts oscillator?
29. (*a*) How does the Clapp oscillator circuit differ from the Colpitts oscillator? (*b*) What are the advantages of the Clapp oscillator? (*c*) What factors determine its resonant frequency?
30. (*a*) What is the principal difference between the *RC* and the *LC* feedback oscillators? (*b*) Name two types of *RC* feedback oscillators.
31. Explain the principle of the *RC* phase-shift circuit.
32. Describe the operation of the *RC* phase-shift oscillator circuit of Fig. 17-16*a*.
33. Explain the principle of the Wien-bridge oscillator.
34. Explain how the Wien-bridge circuit stabilizes the frequency of the oscillator.
35. (*a*) What is meant by a crystal-controlled oscillator? (*b*) What is the piezoelectric effect of a crystal? (*c*) What materials are used for crystals?
36. (*a*) What is an X-cut crystal? (*b*) What is a Y-cut crystal?
37. (*a*) Describe the effect of temperature changes on the resonant frequency of a crystal. (*b*) What methods are used to compensate for temperature changes?
38. (*a*) Describe the equivalent electric circuit of a crystal. (*b*) Why can high values of Q be achieved? (*c*) What are the upper ranges of Q for a crystal?
39. Explain the operation of a simple crystal-controlled oscillator using (*a*) a transistor, (*b*) a vacuum tube.
40. (*a*) Explain the principle of frequency multiplication. (*b*) What is meant by harmonic generation? (*c*) What are the practical frequency ranges of crystal-controlled oscillators with and without frequency multiplication?
41. (*a*) Define nonsinusoidal oscillator. (*b*) Name four types of nonsinusoidal waveforms. (*c*) Define relaxation generator.
42. (*a*) Describe the waveform of the sawtooth generator. (*b*) Explain how the action of a capacitor and resistor produces a sawtooth waveform. (*c*) What determines the frequency of the *RC* sawtooth generator?
43. What characteristics of an active device such as a neon tube, transistor, and vacuum tube make their use possible in a sawtooth generator?
44. Describe the action of the circuit of Fig. 17-21*a*.
45. For the circuit of Fig. 17-21*a*, describe the path of current when the capacitor is (*a*) charging, (*b*) discharging.
46. Describe the action of the transistor sawtooth generator of Fig. 17-22.
47. Describe the action of the circuit of Fig. 17-23.
48. (*a*) Describe the basic principle of the blocking oscillator circuit. (*b*) What is a single-swing type? (*c*) What is a self-pulsing type?
49. Describe the action of the transistor blocking oscillator circuit of Fig. 17-24.
50. Describe how a triggering pulse can be applied to the blocking oscillator of Fig. 17-24.
51. Describe the operation of the vacuum-tube blocking oscillator circuit of Fig. 17-25.
52. (*a*) Define multivibrator. (*b*) Name three classes of multivibrators.
53. Describe briefly each of the three basic classifications of multivibrators.
54. Describe the operation of the astable multivibrator circuit of Fig. 17-26.
55. (*a*) Which waveform of Fig. 17-27 also represents the output of the oscillator?

(*b*) Which circuit component values can be changed in order to make the time interval 1-2 two times that of 2-3? (*c*) What general shape of waveform would be produced?

56. Describe the operation of the bistable multivibrator circuit of Fig. 17-28*a*.

57. If the two stages of the multivibrator of Fig. 17-28*a* are symmetrical, how do the outputs at C_3 and C_6 compare?

58. Describe the operation of the multivibrator circuit of Fig. 17-29.

59. Describe the operation of the multivibrator circuit of Fig. 17-31.

60. For the multivibrator circuit of Fig. 17-31, how is the waveform of the output affected by a change in (*a*) C_1? (*b*) C_2? (*c*) R_2? (*d*) R_3?

PROBLEMS

1. The active device in a certain oscillator circuit (Fig. 17-1) has an input power level of 1 mw and amplifies the power 20 times; the power loss in the feedback network is 10 mw. Find (*a*) the output power of the amplifier, (*b*) the power required to sustain oscillations, (*c*) the power output of the oscillator.

2. The active device in a certain oscillator circuit (Fig. 17-1) has an input power level of 1 mw and amplifies the power 25 times; the loss in the feedback network is 10 dbm. Find (*a*) the output power of the amplifier, (*b*) the feedback circuit loss in watts, (*c*) the power required to sustain oscillations, (*d*) the power output of the oscillator.

3. The tank circuit of an oscillator has a circuit Q of 100 at no load and an effective Q of 25 when load is applied. What is the efficiency of the tank circuit?

4. The efficiency of a tank circuit is 90 per cent and its no-load value of Q is 150. What is the effective Q of the circuit when loaded?

5. The effective Q of a certain tank circuit is 12.5. What is the no-load value of Q if the efficiency of the circuit is 92 per cent?

6. The inductor of a certain tank circuit has an inductance of 100 μh and a resistance of 10 ohms at 1,600 kHz. What is the efficiency of the tank circuit at 1,600 kHz if the Q of the circuit when loaded is 15?

7. The tank section of a certain oscillator circuit contains an inductance of 90 μh shunted by a variable capacitor of 365-35 pf. What is the frequency range of the oscillator circuit if the distributed capacitance of the coil and wiring is disregarded?

8. What is the frequency range of the oscillator circuit of Prob. 7 if the distributed capacitance of the coil and wiring is known to be 15 pf?

9. What is the frequency range of the oscillator circuit of Prob. 7 if a 0.001-μf capacitor is connected in series with the tuning capacitor? The distributed capacitance of the coil and wiring is 15 pf.

10. If the oscillator of Prob. 9 produces a frequency 465 kHz higher than the frequency of the r-f circuit throughout its working range, what is the tuning range of the combined oscillator and tuning circuits?

11. A certain oscillator circuit has a 200-μh inductor and a 40 − 350-pf tuning capacitor for its tank circuit. The Q of the circuit when loaded is 30 and the alternating voltage across the tank circuit is 15 volts (rms). Find (*a*) the capacitance when the oscillator is adjusted to a resonant frequency of 1,500 kHz, (*b*) the impedance of the tank circuit at resonance, (*c*) the alternating component of the output current required to produce the desired voltage at the tank circuit, (*d*) the current in the tank circuit, (*e*) the power supplied to the tank circuit.

12. Repeat Prob. 11 when the oscillator is adjusted to a resonant frequency of 710 kHz.

13. An r-f Class C power amplifier is to supply 60 watts at a frequency of 2 MHz. The value of E_t is 1,000 volts, and in order to obtain a satisfactory waveform the effective Q of the tank circuit should not be less than 15. (*a*) What value of inductance should be used in the tank circuit? (*b*) What value of capacitance should be used? (*c*) What is the current in the tank circuit? (*d*) What is the output current in the active device?

14. The tank circuit of a certain Class C power amplifier is operated at a frequency of 5 MHz. The power supplied to the tank circuit is 500 watts and the value of the alternating voltage E_t is 1,500 volts. It is desired that the effective Q of the tank circuit be 12.5. Find (*a*) the value of inductance required, (*b*) the value of capacitance required, (*c*) the current in the tank circuit, (*d*) the current supplied to the tank circuit, (*e*) the amount of energy stored in the tank circuit per cycle.

15. For the circuit of Prob. 11, find (*a*) the amount of energy stored in the tank circuit during each cycle, (*b*) the amount of energy dissipated during each cycle, (*c*) the ratio of energy stored per cycle to the energy dissipated per cycle.

16. For the circuit of Prob. 12, find (*a*) the amount of energy stored in the tank circuit during each cycle, (*b*) the amount of energy dissipated during each cycle, (*c*) the ratio of energy stored per cycle to the energy dissipated per cycle.

17. A parallel-fed Hartley oscillator circuit similar to Fig. 17-10 uses an r-f choke (L_3) of 2.5 mh and a blocking capacitor (C_2) of 0.005 μf. (*a*) What is the reactance of the choke and of the capacitor at 2,200 kHz? (*b*) What is the ratio of X_L to X_C at this frequency? (*c*) What is the reactance of the choke and of the capacitor at 1,000 kHz? (*d*) What is the ratio of X_L to X_C at this frequency?

18. What is the frequency of a Hartley oscillator circuit similar to Fig. 17-11 if the total inductance of L_{1A} and L_{1B} is 80 μh and the total capacitance of the tuning circuit is 180 pf?

19. The parameters of the tank circuit of a Colpitts oscillator (Fig. 17-14) are $L_1 =$ 10 μh, $C_{1A} = 750$ pf, and $C_{1B} = 3{,}300$ pf. (*a*) What is the resonant frequency of the tank circuit? (*b*) What is the resonant frequency of the tank circuit when L_1 is shunted by an additional variable capacitance whose value is adjusted to 200 pf? What per cent of the tank-circuit voltage appears across (*c*) C_{1A}? (*d*) C_{1B}?

20. A certain Colpitts oscillator uses a tank circuit whose parameters are $L_1 = 140$ μh, $C_{1A} = 200$ pf, and $C_{1B} = 300$ pf. (*a*) What is the resonant frequency of the oscillator? (*b*) What per cent of the tank-circuit voltage is fed back to the input of the oscillator?

21. (*a*) Using Eq. (17-13*a*), what is the product of R_1C_1 for a Wien-bridge oscillator that is to operate at 20 kHz? (*b*) What value of capacitance is required if a 10,000-ohm resistor is used? (*c*) What value of resistance is required if a 0.001-μf capacitor is used?

22. (*a*) Using Eq. (17-13*a*), what is the product of R_1C_1 for a Wien-bridge oscillator that is to operate at 1,000 Hz? (*b*) What value of capacitance is required if a 10,000-ohm resistor is used? (*c*) What value of resistance is required if a 0.033-μf capacitor is used?

23. A certain X-cut quartz crystal has a resonant frequency of 3 MHz at 20°C. The crystal has a negative temperature coefficient of 25 Hz per MHz per degree C.

How much will the frequency vary if the temperature changes to (*a*) 30°C? (*b*) 15°C? (*c*) 10°C?

24. A certain Y-cut quartz crystal has a resonant frequency of 3 MHz at 20°C. The crystal has a positive temperature coefficient of 80 Hz per MHz per degree C. How much will the frequency change if the temperature changes to (*a*) 30°C? (*b*) 15°C? (*c*) 10°C?

25. A certain crystal whose resonant frequency is 450 kHz has an inductance of 3 henrys, and its effective series resistance is 2,000 ohms. What is the value of Q for this crystal?

26. If the crystal of Prob. 25 is represented by the equivalent circuit of Fig. 17-18*b*, what is the equivalent capacitance C?

27. What is the frequency of the sawtooth output of the circuit of Fig. 17-21 when the elapsed time between the waves is (*a*) 1 msec? (*b*) 0.5 msec? (*c*) 1 μsec?

28. For the circuit of Fig. 17-21, what is the elapsed time for one sawtooth wave when the repetition rate per second is (*a*) 30? (*b*) 60? (*c*) 525? (*d*) 15,750?

29. What is the natural resonant frequency of a blocking oscillator transformer (Fig. 17-24) if its inductance is 200 mh and its distributed capacitance is 23 pf?

30. What is the natural resonant frequency of a blocking oscillator transformer (Fig. 17-25) if its inductance is 150 mh and its distributed capacitance is 25 pf?

31. What is the frequency of the multivibrator circuit of Fig. 17-26 if C_1, $C_2 = 0.025$ μf, R_2, $R_3 = 5{,}000$ ohms, and k_1, $k_2 = 0.8$?

32. (*a*) What is the frequency of the multivibrator circuit of Fig. 17-26 if $C_1 = 0.25$ μf, $C_2 = 0.025$ μf, R_2, $R_3 = 5{,}000$ ohms, and k_1, $k_2 = 0.8$? (*b*) How will the waveform of the output compare with that of Fig. 17-27?

33. What is the approximate free-running frequency of the multivibrator circuit of Fig. 17-31 when C_1, $C_2 = 200$ pf, and R_2, $R_3 = 200{,}000$ ohms?

34. What is the approximate free-running frequency of the multivibrator circuit of Fig. 17-31 when $C_1 = 10$ pf, $C_2 = 400$ pf, $R_2 = 470{,}000$ ohms, and $R_3 = 88{,}000$ ohms?

Chapter 18

Modulation and Demodulation Circuits

Two necessary functions for successful transmission and reception of intelligence by use of radio waves are (1) modulation, and (2) demodulation. Modulation is performed at the transmitting end of a communications system in order to combine the low-frequency variations of intelligence signals with a relatively high-frequency carrier wave. Demodulation is performed at the receiving end of the system and represents the process of separating the intelligence-signal information from the modulated carrier-wave signal present at the input to the receiver.

18-1 Modulated Radio-frequency Waves

Need for Modulating the R-F Wave. Sound waves in air can be heard only over relatively short distances, but radio waves can be used over very long distances. Therefore, the transmission of intelligence (i.e., code, voice, music) is accomplished by combining electrical impulses corresponding to the intelligence signals with a constant high-frequency wave generated by the transmitting equipment. The process of combining the audio- and radio-frequency (a-f and r-f) waves is called *modulation.*

Modulation. The *carrier wave* of a radio transmitting system is a wave of constant value of frequency and amplitude. The carrier wave by itself will not produce any sound at the loudspeaker of a radio receiver. The transmission of intelligence occurs when the carrier wave is modulated by a signal. *Modulation* is the process by which some characteristic of a periodic wave is varied with time in accordance with a signal, and is generally accomplished by combining a signal and a carrier wave. The signal is also referred to as a *modulating wave,* and the resultant wave is called the *modulated wave.*

Although many methods of modulation can be devised, the only ones of practical value at present are amplitude modulation, frequency modulation, and phase modulation. Only amplitude modulation (abbreviated a-m) and frequency modulation (abbreviated f-m) are discussed in this chapter.

Mathematically, a carrier wave may be considered as

$$e = E_m \sin (2\pi ft + \theta) \tag{18-1}$$

From Eq. (18-1) it can be seen that the waveform can be varied by three factors, namely, E_m, f, and θ. *Amplitude modulation* occurs when modulation is obtained by varying only the voltage E_m. *Frequency modulation* occurs when modulation is obtained by varying only the frequency f. *Phase modulation*, which is very similar to frequency modulation, occurs when modulation is obtained by varying only the phase relationship represented by θ.

18-2 Amplitude Modulation

A-M Wave. An amplitude-modulated wave is one whose envelope contains a component similar to the waveform of the signal to be transmitted. In amplitude modulation the amplitude of the carrier wave is varied by the signal voltage, which is the modulating quantity. The effect of amplitude modulation is illustrated by Fig. 18-1. Figure 18-1*a* represents a high-frequency carrier

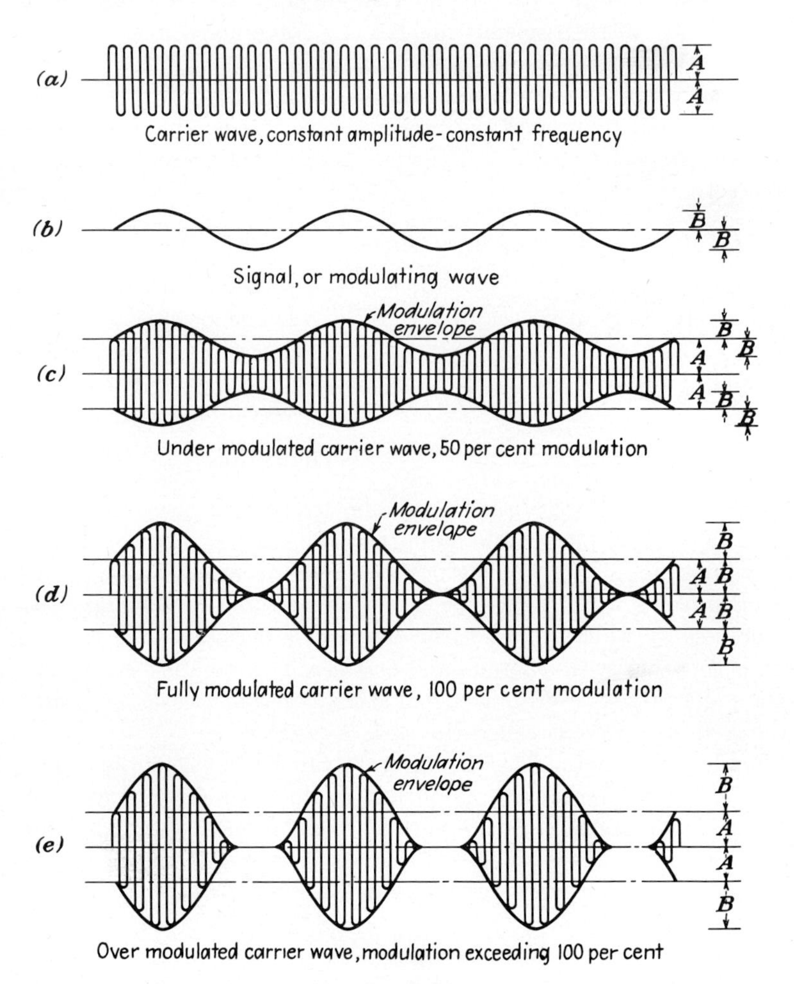

Fig. 18-1 Waveshapes of a carrier wave that is amplitude-modulated by a sine-wave signal.

wave of constant amplitude and frequency, and Fig. 18-1*b* represents an a-f signal of sine-wave form. Figure 18-1*c* shows the result obtained by modulating the carrier wave (*a*) with the modulating wave (*b*). Figure 18-1*c* shows that the outline of the modulated carrier wave is similar in form to the modulating wave; accordingly, this outline is commonly called the *modulation envelope.*

Per Cent of Modulation. In amplitude modulation it is common practice to refer to the per cent of modulation, usually designated as M. Actually this is a means of expressing the degree to which the signal modulates the carrier wave. The per cent of modulation is proportional to the ratio of the maximum values of the signal and carrier waves, or

$$M = \frac{\text{maximum value of signal}}{\text{maximum value of carrier}} \times 100 \tag{18-2}$$

This may be expressed in terms of Fig. 18-1 as

$$M = \frac{B}{A} \times 100 \tag{18-3}$$

where M = per cent of modulation
B = maximum value of modulating wave, volts
A = maximum value of carrier wave, volts

The effect of different amounts of modulation upon the carrier wave is shown by Fig. 18-1. As the maximum undistorted power output of a transmitter is obtained with 100 per cent modulation, it is generally desirable to operate with such a fully modulated carrier wave. If the modulation is less than 100 per cent, the power output is reduced, even though the power of the carrier wave has not been reduced. If the modulation exceeds 100 per cent, the output of the transmitter will be a distorted version of the original modulating wave.

Sidebands. During the process of modulation a heterodyne action takes place, and as a result two additional frequencies appear. These new frequencies are the result of the heterodyning action and are equal to the sum and difference of the carrier frequency and the modulating frequency. The sum of the two frequencies is called the *upper side frequency,* and the difference of the two frequencies is called the *lower side frequency.* In broadcasting a radio program the modulating frequency varies continually over the frequency range of the audio signal being transmitted. Accordingly, the single value of upper side frequency referred to above is replaced by a band of frequencies, called the *upper sideband,* whose width is equal to the difference between the maximum and minimum values of the modulating frequencies. Likewise, the single value of lower side frequency will be replaced by a *lower sideband* (see Fig. 18-2).

Example 18-1 If a 1,000-kHz carrier wave is modulated by audio signals varying between 100 and 5,000 Hz, what is (*a*) the frequency span of the sidebands? (*b*) The

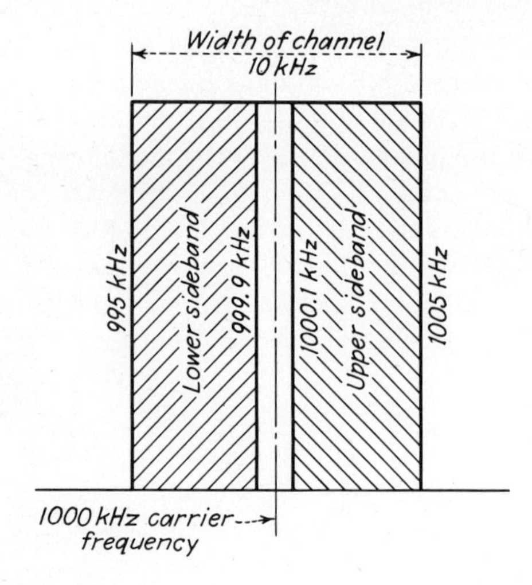

Fig. 18-2 Sidebands and channel width for a 1,000-kHz carrier wave that is being amplitude-modulated by audio frequencies ranging from 100 to 5,000 Hz.

maximum upper side frequency? (*c*) The minimum lower side frequency? (*d*) The frequency range of the channel?

GIVEN: $f_{\text{carrier}} = 1{,}000$ kHz $\quad f_{\text{audio signal}} = 100 - 5{,}000$ Hz

FIND: (*a*) f_{sideband} $\quad$ (*b*) $f_{\text{max}\cdot\text{carrier}}$ $\quad$ (*c*) $f_{\text{min}\cdot\text{carrier}}$ $\quad$ (*d*) f_{channel}

SOLUTION:

(*a*) $$f_{\text{sideband}} = (f_{\text{max}} - f_{\text{min}})_{\text{audio signal}} = 5{,}000 - 100 = 4{,}900 \text{ Hz}$$

(*b*) $$f_{\text{max}\cdot\text{upper side}} = f_{\text{carrier}} + f_{\text{max}\cdot\text{audio}} = 1{,}000 + 5 = 1{,}005 \text{ kHz}$$

(*c*) $$f_{\text{min}\cdot\text{lower side}} = f_{\text{carrier}} - f_{\text{max}\cdot\text{audio}} = 1{,}000 - 5 = 995 \text{ kHz}$$

(*d*) $$\text{Width of channel} = (f_{\text{max}} - f_{\text{min}})_{\text{carrier}} = 1{,}005 - 995 = 10 \text{ kHz}$$

18-3 Frequency Modulation

F-M Wave. In frequency modulation, the amplitude of the modulated wave is maintained at a constant strength, namely, the same as the carrier wave. The frequency of the modulated wave varies in proportion to the amplitude of the modulating signal, and at a rate determined by the frequency of the modulating signal (see Fig. 18-3). Figure 18-3 shows that the frequency of the modulated wave increases as the signal voltage increases and that it decreases as the signal voltage decreases. Comparison of Fig. 18-3*b* and *c* shows that the variation in frequency is determined only by the amplitude of the signal, and that the rate of variations in frequency is determined by the frequency of the signal. Figure 18-4 is another method of illustrating the effect of the amplitude of the modulating signal upon the frequency of the modulated wave.

Frequency Deviation. The frequency of an f-m transmitter without any signal input is called the *resting frequency* or the *center frequency* and corresponds to the assigned frequency of the transmitter. When a signal is applied, the variation in frequency either above or below the resting frequency is called the *frequency deviation,* and the total variation in frequency is called the *carrier*

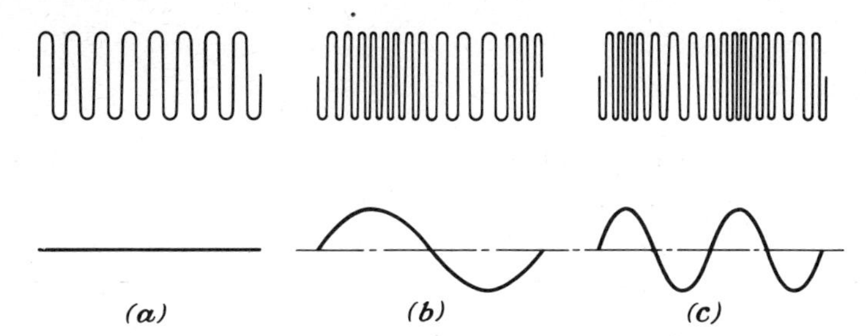

Fig. 18-3 Waveshapes of a carrier wave that is frequency-modulated by a sine-wave signal. (a) No modulating signal. (b) 500-Hz modulating signal. (c) 1,000-Hz modulating signal.

swing. For example, in Fig. 18-4 the resting frequency is 5 Hz, the deviation for the weak signal is 2 Hz, and the carrier swing is 4 Hz. With the strong signal, the deviation is 4 Hz and the carrier swing is 8 Hz. The values of frequency shown in Fig. 18-4 are not practical values, but for convenience of illustration have been made of very low values. Actually, f-m broadcast transmitters operate at frequencies of 88 to 108 MHz.

Modulation. With f-m systems, variations in the amplitude of the signal produce variations in the frequency of the modulated carrier; therefore, the degree of modulation is expressed in terms of the frequency deviation. A frequency deviation of 75 kHz is equivalent to 100 per cent modulation in the a-m system. Under this condition, the total carrier swing will be 150 kHz. To avoid interference between stations on adjacent channels, the assigned frequencies are kept at least 200 kHz apart.

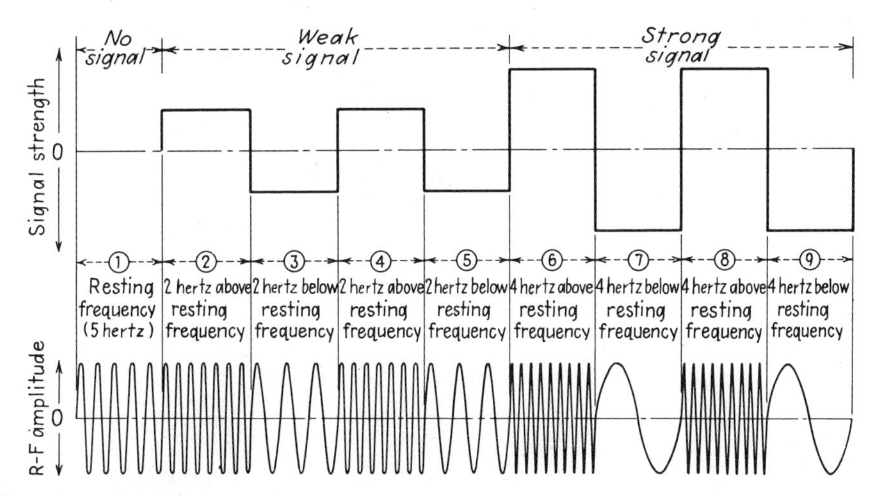

Fig. 18-4 Waveshapes of a carrier wave that is frequency-modulated by square-wave signal voltages.

The frequency deviation is sometimes expressed in terms of the ratio of maximum frequency deviation to the maximum audio frequency being transmitted; this is called the *deviation ratio.*

Sidebands. The sidebands present in f-m transmission are determined by the amplitude of the modulating signal. While the sideband frequencies are apparently unlimited, present f-m transmission is based on a 75-kHz frequency deviation. Thus, with a given signal frequency the frequency deviation, and hence the number of sidebands, is dependent upon the amplitude of the signal. If the value of the deviation ratio is approximately 5, the sidebands above the maximum frequency are so small that they may be ignored. Under this condition, the channel width for f-m transmission should be approximately double the value of the maximum frequency deviation.

Example 18-2 An f-m transmitter operates with a maximum frequency deviation of 75 kHz and reproduces audio signals up to 15 kHz. (*a*) What is the deviation ratio? (*b*) What minimum channel width is required?

GIVEN: Frequency deviation = 75 kHz Maximum audio signal = 15 kHz

FIND: (*a*) Deviation ratio (*b*) Minimum channel width

SOLUTION:

(*a*)
$$\text{Deviation ratio} = \frac{\text{max frequency deviation}}{\text{max signal frequency}} = \frac{75{,}000}{15{,}000} = 5$$

(*b*)
$$\text{Minimum channel width} = 2 \times \text{max frequency deviation} = 2 \times 75 = 150 \text{ kHz}$$

18-4 Methods of Amplitude Modulation

Basic Principle. There are two methods of performing the process of amplitude modulation: (1) amplifier modulation, and (2) oscillator modulation. Modulation may be accomplished with either a transistor or a vacuum tube. The block diagrams and waveforms shown in Figs. 18-5 and 18-7 illustrate the basic principles of the two modulating systems.

Amplitude Modulation in the Amplifier Circuit. Figure 18-5 shows (1) separate carrier-signal and modulating-signal inputs being fed to the amplifier, and (2) the resultant amplifier output waveform. The modulation process can be made to take place in any of the circuits of the active device being used for the amplifier. Thus the modulating signal may be injected into (1) the base, emitter, or collector circuit of a transistor, or (2) the grid, plate, or cathode circuit of a vacuum tube.

Figure 18-6 illustrates a modulating amplifier circuit employing a transistor used in the common-emitter configuration. The circuit differs from the common-emitter amplifier only by (1) the modulating signal input, and (2) the addition of capacitors C_1, C_2, and C_4. The carrier signal is coupled to the base of Q_1 by the transformer T_1, and the modulating signal is coupled to the base of the transistor by capacitor C_1. The voltage divider R_1R_2 establishes a quiescent

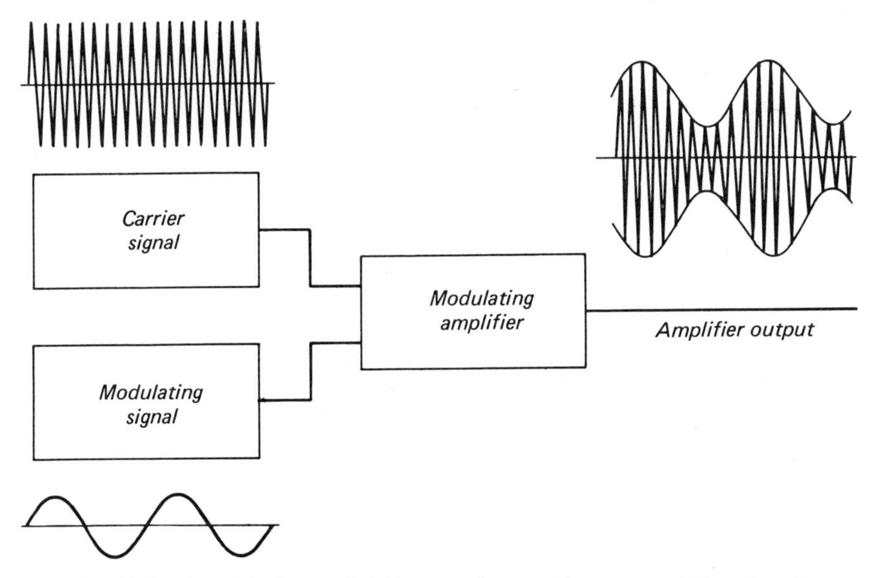

Fig. 18-5 Amplitude modulation performed in an amplifier circuit.

forward bias for the transistor. The modulating signal being applied at R_2 causes the base bias to increase and decrease with variations in the modulating signal. Capacitor C_2 bypasses the alternating currents around R_2 in order to maintain a constant quiescent operating point. Resistor R_3 is an emitter swamping resistor and C_3 bypasses the carrier signal around R_3. Capacitor C_4 tunes the primary circuit of T_2 to the carrier-signal frequency. Transformer T_2 couples the modulated carrier signal to the load on the amplifier.

Amplitude Modulation in the Oscillator Circuit. Figure 18-7 shows (1) the modulating signal being fed directly to the oscillator, and (2) the resultant oscillator output waveform. Figure 18-8 illustrates a modulating oscillator cir-

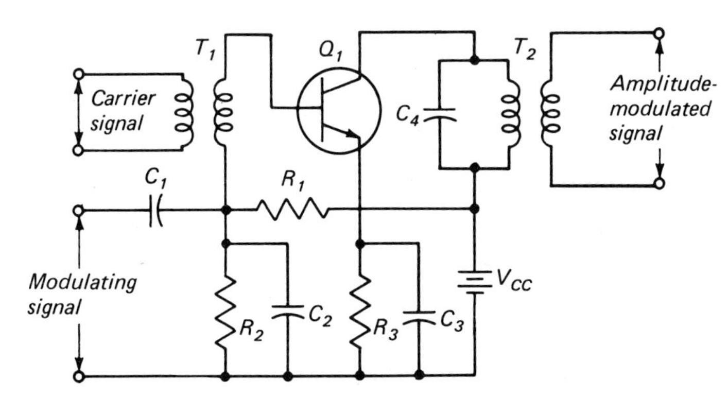

Fig. 18-6 Amplitude-modulated transistor amplifier with base injection.

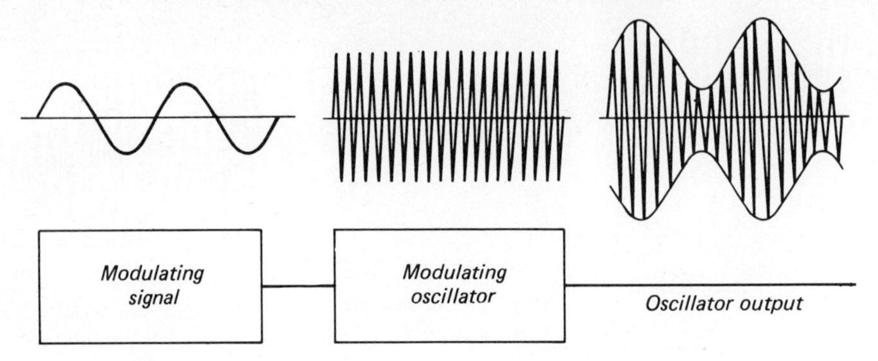

Fig. 18-7 Amplitude modulation performed in an oscillator circuit.

cuit employing a transistor used in the common-emitter configuration. The secondary winding L_3 of the transformer T_1 acts as a tickler coil to provide positive feedback for the transistor Q_1 and to establish oscillation at the carrier-wave frequency. Resistors R_1 and R_2 provide the quiescent forward bias for the transistor, and R_3C_3 provides emitter swamping action. The parallel tuned circuit L_2C_4 establishes the frequency of the carrier wave, and L_4 couples the modulated signal to the oscillator load. The series tuned circuit L_1C_2 is made resonant to the carrier frequency and thus provides a low-impedance path for the carrier-frequency signals. The impedance of L_1C_2 at the relatively low frequencies of the modulating signal is very high and thus has a negligible loading effect on the modulating signal source. The base-emitter bias of the transistor is affected by (1) a fixed bias determined by R_1, R_2, and V_{CC}, and (2) a variable bias determined by the sum of the feedback voltage at L_3 and the modulating signal voltage. An increase in the bias increases the gain of the transistor and thereby increases the amplitude of the carrier signal. A decrease in the bias decreases the gain of the transistor and thereby decreases the amplitude of the

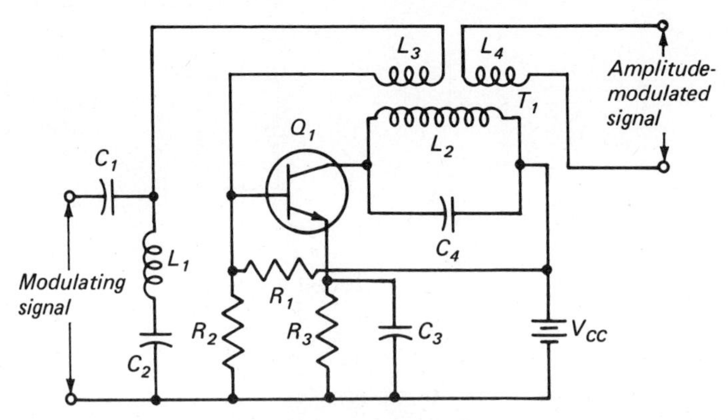

Fig. 18-8 Amplitude-modulated transistor oscillator with base injection.

carrier signal. Consequently, the carrier signal is modulated in amplitude at the frequency of the modulating signal, and therefore the oscillator output is modulated in both amplitude and frequency.

18-5 Methods of Frequency Modulation

Basic Principle. With the frequency-modulation system, the modulation of the carrier wave by the modulating signal takes place in the oscillator stage. A commonly used method of achieving frequency modulation is to use a reactance-modulator transistor (or vacuum tube). With this system, the modulating signal is fed to a separate transistor where the voltage changes of the modulating signal cause changes in the output capacitance C_{ce} of the transistor, thereby varying the reactance of its output circuit.

Frequency-modulation Oscillator Circuit. Figure 18-9 illustrates the basic circuit of a transistor frequency-modulation oscillator using Q_1 as the oscillator and Q_2 as the reactance modulator; the biasing resistors and bypass capacitors have been omitted to simplify the diagram. The oscillator transistor Q_1 receives its positive feedback from the secondary winding L_3 of transformer T_1. The frequency of the carrier wave is established by the primary winding L_1L_2 of T_1 and capacitor C_1.

The reactance modulator transistor Q_2 has the modulating signal voltage coupled to its base by the secondary winding of T_2. The output of Q_2 is coupled to a portion of the primary winding (L_2) of T_1. Any changes in the input-signal voltage at T_2 will be accompanied by changes at Q_2 in (1) the forward bias, (2) collector current, (3) collector voltage, and (4) collector-to-emitter capacitance

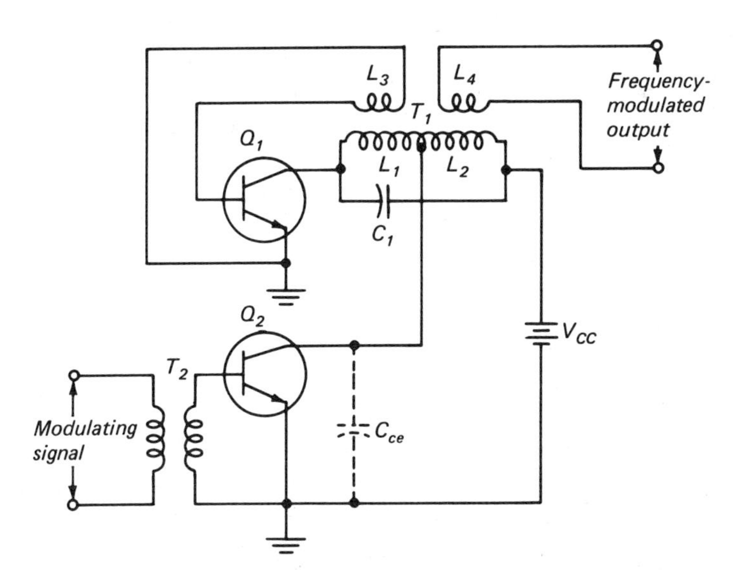

Fig. 18-9 Frequency-modulated transistor oscillator using a reactance modulator.

(C_{ce}). As the collector voltage increases, the collector-to-emitter capacitance decreases, and vice versa. When the collector-to-emitter capacitance decreases, the resonant frequency of the tuned circuit $L_1L_2C_1C_{ce}$ increases, and vice versa. Consequently, the output signal at the secondary winding L_4 of T_1 will be frequency-modulated in accordance with the voltage variations of the input signal at T_2.

18-6 Frequency Conversion

Need. High-gain amplification can best be obtained when an amplifier circuit is designed to amplify only a narrow band of frequencies. Since many communications receivers are required to select from among a wide range of r-f signals, an important function is to reduce the r-f (carrier) modulated signal to an intermediate-frequency (i-f) modulated signal that contains the same modulating signal characteristics as the original r-f carrier wave. This function can be performed in an electronic circuit by heterodyne action.

Basic Principle. A high-frequency modulated signal can be converted to a lower-frequency signal containing the original modulation characteristics by combining the high-frequency-modulated signal with a new unmodulated signal. This operation can be performed with one or two transistors (or vacuum tubes). When the process is performed in a single active device, the circuit is called a *converter*. Separate active devices can be used to (1) produce a new unmodulated signal, and (2) combine the original modulated signal with the new unmodulated signal. With this system, the circuit producing the new unmodulated signal is called the *local oscillator*, and the circuit in which the modulated and unmodulated signals are combined is called the *mixer*. The process of combining two alternating voltages of different frequencies in order to obtain a new value of frequency is called *heterodyne action*.

Principle of Heterodyne Action. Heterodyne action is the result of combining two alternating voltages of different frequencies in a common active device in order to obtain a signal of a new frequency value. Heterodyne action produces two additional signals whose frequencies are (1) the sum of the two original frequencies, and (2) the difference of the two original frequencies. Generally, only the difference value is used. When the signal waves of two frequencies are combined by heterodyne action, the envelope of the resultant wave will vary in amplitude at the new frequency rate. The range of the amplitude swing of the new voltage is determined by the sum and difference of the two voltages being combined. Figure 18-10 shows two signal waves A and B and the resultant voltage wave C. The frequency of signal A is 8 hertz, and that of signal B is 12 hertz; the frequency of the resultant wave is $12 - 8$, or 4 hertz. The maximum voltage of signal A is 2 volts, and that of B is 4 volts. The envelope of the resultant wave will therefore vary in voltage from $E_B - E_A$ to $E_B + E_A$. In this case, it will vary from $4 - 2$ to $4 + 2$, or from 2 to 6 volts.

The values of frequency and voltage used here to explain heterodyne action are not practical values but were chosen to provide an easier means of illus-

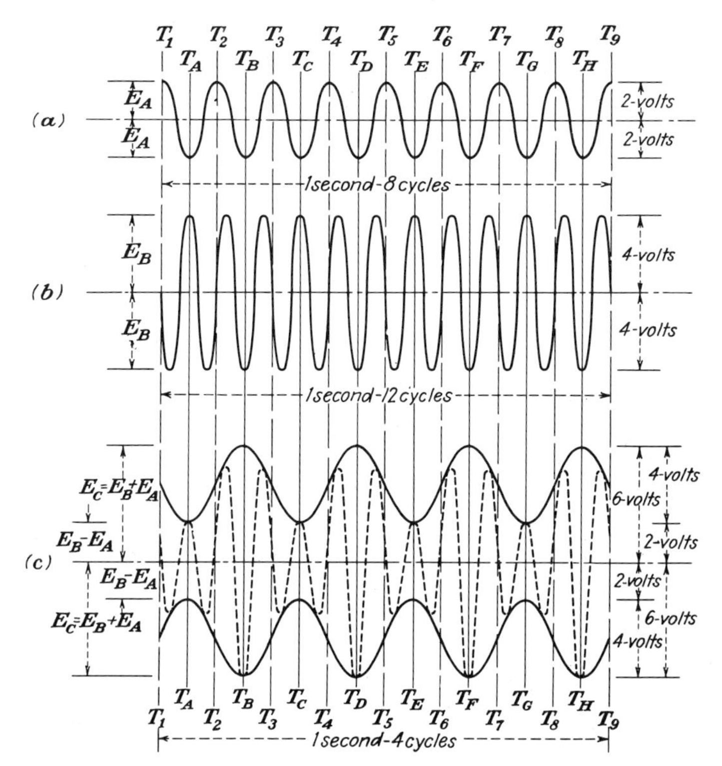

Fig. 18-10 Heterodyne action resulting when two waves of unequal amplitude and frequency are combined. (a) Waveform of signal *A*. (b) Waveform of signal *B*. (c) Waveform of signal *A* — signal *B*. (*Note:* Waveform of *A* + *B* is not shown.)

tration. This same action, however, holds for signals of higher frequencies and different voltages. The frequency at which the amplitude of the resultant wave varies is called the *beat frequency*.

Frequency Conversion Using Two Transistors. Figure 18-11 illustrates a circuit using a local oscillator and a mixer to achieve frequency conversion. The local oscillator circuit uses transistor Q_1, transformer T_1, and capacitor C_1; bias resistors and bypass capacitors have been omitted for clarity. Capacitor C_1 and the primary winding L_1 form a parallel tuned circuit to establish the oscillator frequency. The secondary winding L_2 provides the positive-feedback voltage required to sustain the oscillations. The secondary winding L_3 is used to inject the oscillator signal into the emitter circuit of the mixer transistor Q_2.

In the mixer circuit a fixed bias is provided by $R_1R_2V_{CC}$, emitter swamping is provided by R_3, and capacitors C_2 and C_3 bypass the alternating currents around R_2 and R_3, respectively. The parallel resonant circuit formed by C_4 and L_4 is tuned to the frequency of the r-f input signal, and the parallel resonant circuit formed by C_5 and L_6 is tuned to the intermediate frequency which the circuit is intended to produce. The emitter-base (bias) voltage consists of three components: (1) the fixed voltage produced across R_2, (2) the r-f input-signal voltage produced across L_5, and (3) the local oscillator voltage produced across L_3. Consequently, the signal current in the collector circuit of Q_1 contains the combined effects of both the r-f input signal and the local oscillator signal. Heterodyne action produces four signal frequencies in the output circuit, namely, (1) the r-f input-signal frequency, (2) the local oscillator signal frequency, (3) a signal frequency equal to the sum of the r-f input and local oscillator frequencies, and (4) a signal frequency equal to the difference between the r-f input and local oscillator frequencies. The parallel resonant circuit L_6C_5 selects the desired i-f signal frequency (normally the difference frequency) and couples this signal to the load connected across L_7.

Frequency Conversion Using a Single Transistor. Figure 18-12 illustrates a circuit using one transistor as a frequency converter. In this circuit, $R_1R_2V_{CC}$ provide a fixed bias for the transistor, and swamping action is produced by R_3C_3. Positive feedback for oscillator action is provided by the secondary winding L_4 of T_2, and the oscillator frequency is determined by the parallel resonant circuit L_3C_4. The desired r-f input signal is selected by the tuned circuit L_1C_1 and is coupled to the base of the transistor by the secondary winding L_2. As a result of heterodyne action, four signal frequencies are present in the output circuit of the transistor, and the tuned circuit C_5L_5 selects the one signal frequency to be coupled to the load by the secondary winding

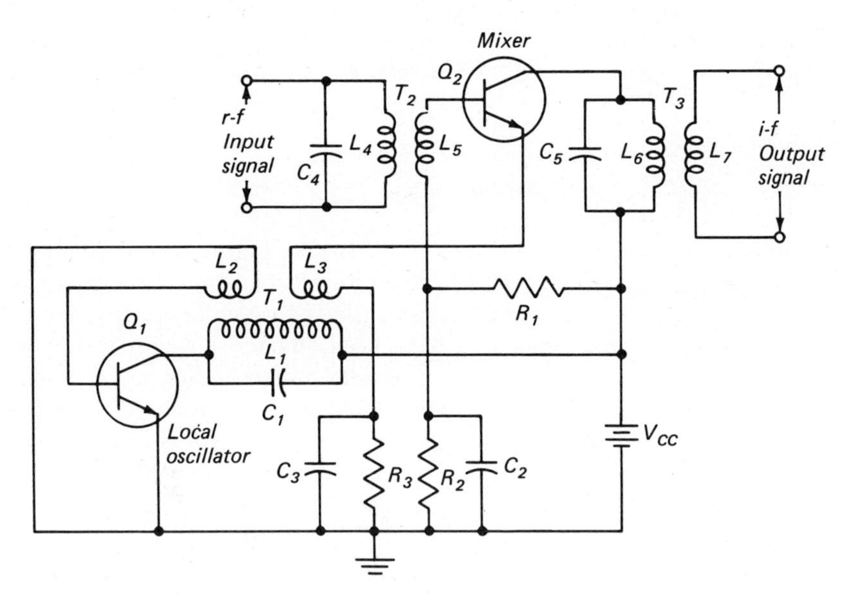

Fig. 18-11 Frequency-conversion circuit using separate transistors for local oscillator and mixer circuits.

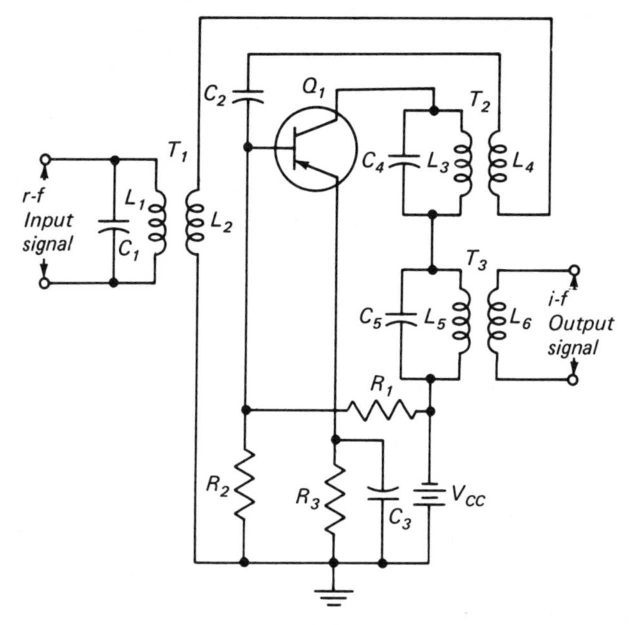

Fig. 18-12 Frequency-conversion circuit using a single transistor.

L_6 of T_2. Capacitor C_2 introduces sufficient impedance into the series path $L_2L_4C_2$ (connected in parallel with R_2) to prevent any serious loading effect on R_2.

18-7 Demodulation

Basic Principle. The process of separating the signal and carrier components of a modulated carrier wave is called *demodulation* or *detection*. Demodulation of an a-m wave generally involves two operations: (1) rectification of the modulated wave, and (2) elimination of the r-f component of the modulated wave. Demodulation of an f-m wave involves three operations: (1) converting the frequency changes corresponding to the modulating signal into corresponding amplitude changes, (2) rectification of the modulating signal and (3) elimination of the r-f component of the modulated wave.

Types of Circuits. Amplitude-modulation detector circuits may employ (1) one or two diodes, (2) a transistor, or (3) a vacuum tube. The simplest (and most frequently used) detector circuit is the half-wave rectifier using a single diode. Other detector circuits are (1) full-wave rectifier using two diodes, (2) transistor detector using either the common-emitter or common-base configuration, and (3) vacuum-tube detector with rectification achieved in either the grid or plate circuit. Other special forms of a-m detector circuits include (1) heterodyne detector, (2) regenerative detector, (3) superregeneration detector, and (4) autodyne detector.

Frequency-modulation detector circuits differ from the a-m detectors chiefly because the frequency variations corresponding to the modulating signal must first be converted to corresponding amplitude variations. Consequently, f-m detector circuits are more complex than a-m detector circuits. F-m detector circuits also may employ (1) diodes, (2) transistors, or (3) vacuum tubes. Many types of f-m detector circuits have been developed, some used for only a short time. Among the various types of f-m detectors are (1) the slope detector, (2) the Travis f-m discriminator, (3) the Foster-Seeley discriminator, (4) the ratio detector, (5) the locked-in oscillator detector, (6) the Fremodyne superregenerative detector, (7) the gated-beam detector, (8) the quadrature-grid f-m detector. Two widely used f-m detectors are (1) the phase-shift discriminator (Foster-Seeley), and (2) the ratio detector.

A-M Detector Classifications. A detector circuit in which the rectified output is proportional to the input-signal voltage is called a *linear detector*. A detector circuit in which the rectified output is proportional to the square of the input-signal voltage is called a *square-law* or *nonlinear detector*.

Detector circuits are also classified in terms of their input-signal voltage-handling capabilities. For input signals lower than one volt, the detector circuit is called a *weak-signal detector*. For input signals of one volt or more, the detector is called a *power detector*. Weak-signal detectors are generally of the square-law type, and power detectors may be of either the linear or square-law type.

Detector Action. The following explanation is based on the input to the detector being a r-f wave amplitude-modulated by an a-f wave, as in the a-m radio communication system. The r-f waves occurring during one a-f cycle are referred to as a *wave train*, and a modulated r-f wave consists of a number of consecutive wave trains. The average value of a modulated radio wave for one cycle of the a-f wave is zero, and therefore the average change of current during the same period is zero (Fig. 18-13*a*). Also, the average change of current of a modulated r-f wave will always be zero. If the modulated r-f wave is rectified, one-half of the wave is eliminated, and the average change in current for each cycle of each wave train will no longer be zero (Fig. 18-13*b*). The changes in current will be similar to the a-f signal that modulates the r-f carrier wave at the transmitter.

Because the electromechanical devices used to produce audible sound waves cannot respond to the rapid variations in current of an r-f wave, it is necessary to remove the r-f component of the demodulated a-f wave. The modulation envelope, formed by joining the peaks of each of the r-f cycles, varies in the same manner as the signal impressed upon the r-f carrier wave. The a-f component of the modulated wave is represented by the a-f variations of the modulation envelope. The current flowing through the detector circuit will be equal to the average value of the current (Fig. 18-13*b*). Its variations are identical in all respects, except intensity, to the variations in current represented by the modulation envelope.

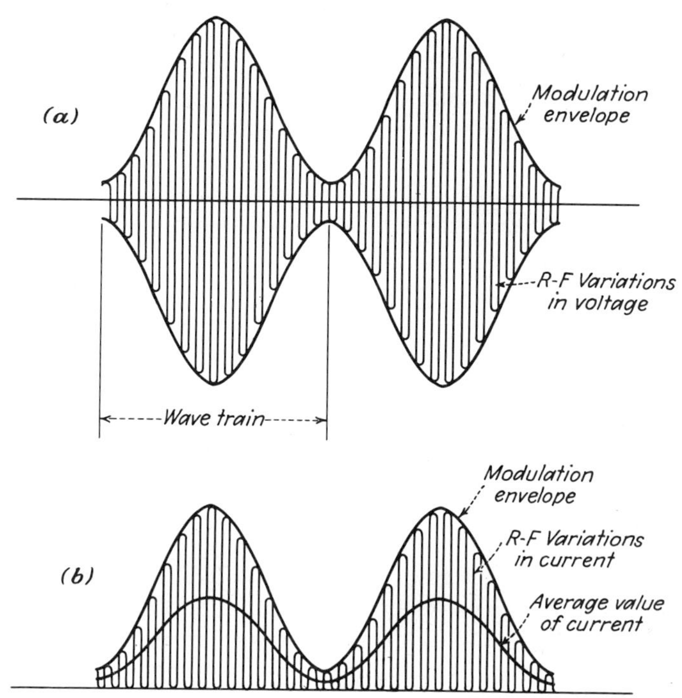

Fig. 18-13 Detector action. (a) Input-signal voltage. (b) Current through the detector.

The Active Device as a Detector. Because current can flow in an active device only when its anode is positive with respect to its cathode, diodes, transistors, or vacuum tubes can be used as rectifiers of alternating currents. As detection involves the function of rectification, diodes, transistors, or vacuum tubes may be used as detectors.

Characteristics of Detector Circuits. There is a variety of active devices and circuits that can be used to perform the function of detection. Factors to be considered when determining which active device and circuit to use are (1) sensitivity, (2) signal-handling ability, (3) fidelity of reproduction. Fidelity is a very important factor in some types of equipment (for example, a radio receiver), and the detector circuit used in this equipment should be capable of reproducing the original modulating signals with a minimum of distortion. When a detector is used in equipment that also employs high-gain amplifier circuits, sensitivity generally is not an important consideration. In portable equipment, the overall physical size is an important factor, and because one or more amplifier stages may be omitted the detector circuit may require a high degree of sensitivity. For a fixed unit, where sufficient amplifier circuits can be used, the power-handling ability of the detector circuit may become an important consideration.

18-8 A-M Diode Detectors

Half-wave Voltage Detection. Figure 18-14 illustrates the common half-wave a-m detector circuit which provides a voltage output. When the tuning circuit L_2C_2 is in resonance with a desired input signal, an r-f voltage is developed across the tuned circuit. This voltage is applied to the anode-cathode circuit of the diode D_1 through the load resistor R_1 and its bypass capacitor C_1.

When the anode of the diode is positive, it attracts electrons emitted from the cathode, and these electrons will return to the cathode through the circuit consisting of the secondary winding L_2 and the output resistor R_1, as indicated on Fig. 18-14. No current flows through the circuit during the time that the signal voltage makes the anode negative with respect to the cathode, and thus the current in R_1 will flow in only one direction.

Capacitor C_1 and resistor R_1 eliminate the r-f pulsations and increase the a-f voltage developed across R_1 in the following manner: During the initial half of the first positive half-cycle of the applied r-f voltage, shown at 0 to 1 on Fig. 18-15*a*, C_1 charges to the peak value indicated by point 1. The applied r-f voltage, continuing its cycle, then rapidly diminishes to zero. As the r-f voltage starts decreasing from its first positive peak value, C_1 starts to discharge through R_1 but at a very slow rate as indicated by points 1 to 3 on Fig. 18-15*b* and *c*. The time constant of this *RC* circuit is very long compared to the short interval required for the r-f voltage to change from the positive peak value at 1 to the next positive peak at 2. The voltage on the capacitor, therefore, will decrease only slightly during this interval. Because of this capacitor action, the voltage on the cathode will be kept more positive than the voltage applied to the anode. When the signal voltage is lower than the voltage charge on C_1, no current will flow in the diode, as is indicated during the interval 1 to 3, Fig. 18-15*d*. During the positive half of the second r-f cycle, current will again flow in the diode when the signal voltage exceeds the voltage at which the charge on the capacitor holds the cathode. The capacitor will then be charged to the peak value of the second positive half-cycle. This action will be repeated for each succeeding r-f cycle, thus causing the voltage across the capacitor to follow the peak values of the applied r-f voltage. The a-f modulation is therefore reproduced at the capacitor as indicated in Fig. 18-15*c*.

The voltage across R_1 and C_1 will be a pulsating voltage representing the

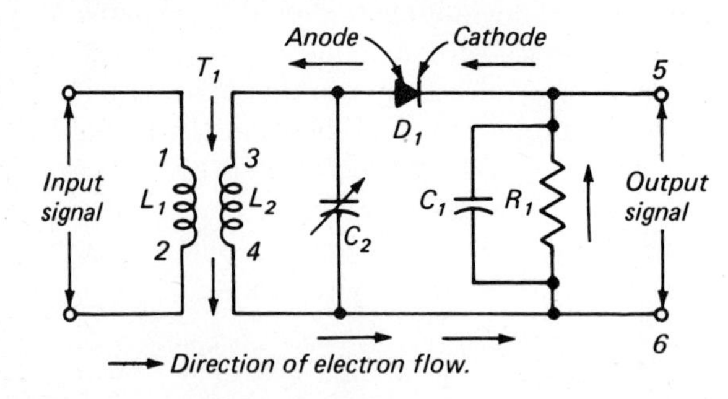

Fig. 18-14 Half-wave a-m voltage-detector circuit.

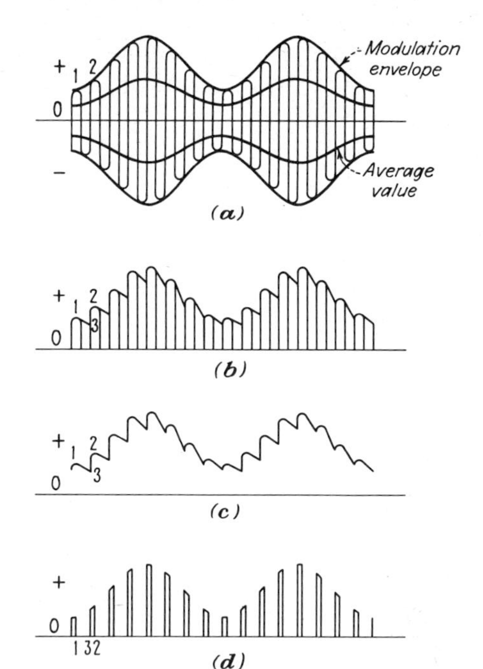

Fig. 18-15 Waveforms illustrating the operation of a diode detector circuit. (a) Modulated input voltage. (b) Charging and discharging of the capacitor. (c) Voltage across the capacitor. (d) Diode current.

positive half of the modulated r-f voltage, whose average value has been increased by C_1. The combination of the diode, C_1, and R_1 changes the r-f signal-input voltage to a pulsating voltage. Because the output of this circuit appears as a voltage at terminals 5 and 6, the circuit is sometimes called a *voltage detector*. The capacitor C_1 is called the *r-f bypass* or *r-f filter capacitor*, because it smooths out the r-f pulsations at the diode load resistor. The value of the capacitor depends on the frequency—the higher the frequency the smaller the amount of capacitance required.

Example 18-3 A detector circuit similar to Fig. 18-14 uses a 0.25-megohm resistor for R_1 and a 250-pf capacitor for C_1. (*a*) What is the time constant of R_1C_1? (*b*) If the resonant frequency of L_2C_2 is 710 kHz, what time is required for the r-f wave to complete 1 cycle? (*c*) How many times greater is the time constant of R_1C_1 than the time of 1 cycle of the r-f wave?

GIVEN: $R = 0.25$ megohm $\quad C = 250$ pf $\quad f = 710$ kHz

FIND: (*a*) t $\quad$ (*b*) t_1 $\quad$ (*c*) $\dfrac{t}{t_1}$

SOLUTION:

(*a*)
$$t = RC = 0.25 \times 10^6 \times 250 \times 10^{-12} = 62.5 \times 10^{-6} \text{ sec}$$

(*b*)
$$t_1 = \frac{1}{f} = \frac{1}{710 \times 10^3} = 1.408 \times 10^{-6} \text{ sec}$$

(*c*)
$$\frac{t}{t_1} = \frac{62.5 \times 10^{-6}}{1.408 \times 10^{-6}} = 44.3 \text{ times greater}$$

Example 18-4 The circuit of Example 18-3 also acts as a filter circuit. (*a*) What is the reactance of C_1 to a 710-kHz r-f current? (*b*) How does the reactance of C_1 at 710 kHz compare with the value of R_1? (*c*) Which path will the r-f currents take? (*d*) What is the reactance of C_1 to a 400-Hz a-f current? (*e*) How does the reactance of C_1 at 400 Hz compare with the value of R_1? (*f*) Which path will the a-f currents take?

GIVEN: $R = 0.25$ megohm $\quad C = 250$ pf $\quad$ r-f = 710 kHz $\quad$ a-f = 400 Hz

FIND: (*a*) X_C $\quad$ (*b*) $\dfrac{R}{X_C}$ $\quad$ (*c*) Path of r-f current $\quad$ (*d*) X_C $\quad$ (*e*) $\dfrac{X_C}{R}$

(*f*) Path of a-f current

SOLUTION:

(*a*) $$X_C = \frac{159{,}000}{fC} = \frac{159{,}000}{710 \times 10^3 \times 250 \times 10^{-6}} = 895 \text{ ohms}$$

(*b*) $$\frac{R}{X_C} = \frac{250{,}000}{895} = 279 \qquad (R \text{ is 279 times greater than } X_C)$$

(*c*) The r-f currents will take the capacitor path.

(*d*) $$X_C = \frac{159{,}000}{fC} = \frac{159{,}000}{400 \times 250 \times 10^{-6}} = 1{,}590{,}000 \text{ ohms}$$

(*e*) $$\frac{X_C}{R} = \frac{1{,}590{,}000}{250{,}000} = 6.36 \qquad (X_C \text{ is 6.36 times greater than } R)$$

(*f*) The a-f currents will take the resistor path.

Full-wave Voltage Detection. The detector circuit of Fig. 18-14 uses only one diode and thus utilizes only one-half of the r-f wave. Figure 18-16 illustrates an a-m detector circuit using two diodes, thus making it possible to rectify both halves of the r-f input signal. Current will flow continually in one direction through the output resistor, and full-wave detection is achieved; the direction of current flow in the remainder of the circuit during alternate half-cycles is indicated by the arrows. The full-wave detector provides more efficient filtering of the r-f wave than is obtained with the half-wave detector.

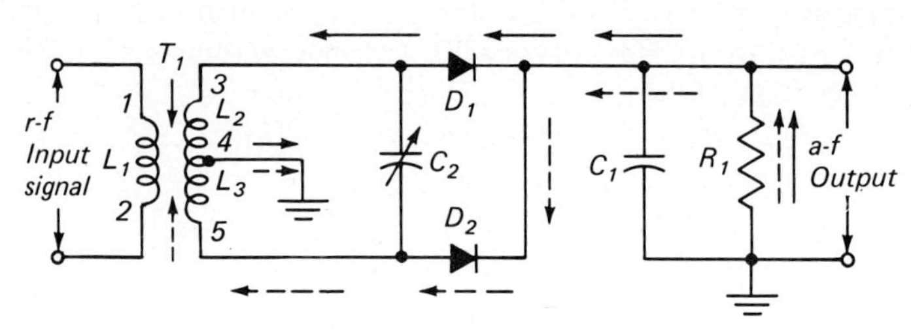

Fig. 18-16 Full-wave a-m voltage-detector circuit.

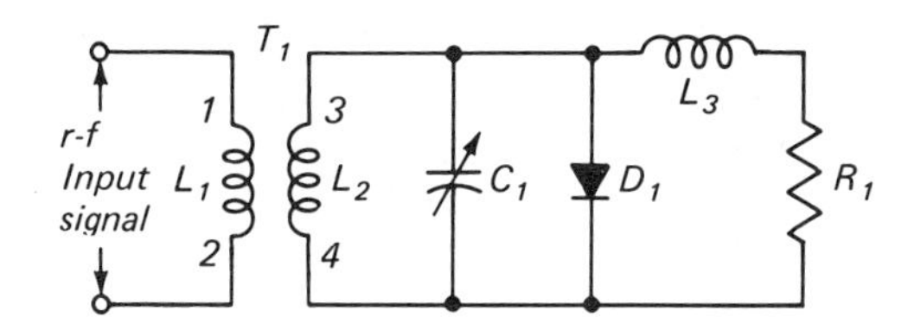

Fig. 18-17 Half-wave a-m current-detector circuit.

Half-wave Current Detection. Figure 18-17 illustrates a half-wave a-m current-detector circuit using a single diode. This type of circuit is useful with transistor circuits, since a transistor is a current-amplifying device.

When terminal 3 is positive, the impedance of D_1 is very low and virtually short-circuits the load L_3R_1; consequently, practically all of the current flows through D_1 and no appreciable amount of current flows through R_1. When terminal 3 is negative, the impedance of D_1 is very high and practically all of the current flows through L_3R_1. During the portion of the r-f cycle when the current through L_3R_1 is decreasing, the collapsing magnetic field at L_3 retards the decrease in current through R_1. Thus, inductor L_3 smooths out the variations in the output circuit in a manner similar to capacitor C_1 in the circuit of Fig. 18-14.

Characteristics of Diode Detectors. The advantages of diode detectors are (1) ability to handle input signals to well above 1 volt, (2) can be operated as a linear or a power detector, (3) low distortion and hence good linearity, and (4) adaptability to application of simple automatic-gain-control circuits. A disadvantage of diode detectors is that when the diode is conducting, power is consumed by the detector circuit, thereby reducing (1) the Q of the tuned circuit (L_2C_2), and (2) the gain and selectivity of the detector. Diode detection is used extensively in a-m radio broadcast receivers.

Practical Diode-detector Circuits. Figure 18-18 illustrates a circuit using multifiltering to prevent the r-f voltages from reaching the output. The diode load resistance is made up of R_1 and R_2; the output, however, is taken off only R_2. The filter circuits consist of R_1C_1 and R_2C_2. The additional r-f filter (R_2C_2), however, reduces the useful output of the circuit by the amount of the voltage drop at R_1. While C_1 will bypass most of the r-f current, C_2 will bypass any r-f current from R_2 that may have entered into the R_1 path. The output of this basic filtering network can be taken off R_2 at points X_1 and X_2 when the auxiliary circuits R_3C_3 and R_4C_4 are omitted. Functions of C_3 are to (1) couple the a-f output signal to the load resistor R_3, and (2) block the d-c component of the output signal from the output circuit. When an agc signal is desired, it can be provided by adding R_4 and C_4.

Example 18-5 The circuit elements of Fig. 18-18 have the following values: $R_1 =$ 50,000 ohms, $R_2 =$ 250,000 ohms, C_1, $C_2 =$ 100 pf. (*a*) What impedance does C_1 offer to a 456-kHz i-f current? (*b*) Which path will the i-f current take? (*c*) Will any of the i-f current flow into the R_1 path? (*d*) What impedance does C_2 offer to any 456-kHz i-f current? (*e*) What purpose does C_2 serve? (*f*) Neglecting the effect of C_1 and C_2, what per cent of the a-f voltage developed across R_1 and R_2 is available at the output terminals?

GIVEN: $R_1 = 50{,}000$ ohms $\quad R_2 = 250{,}000$ ohms $\quad C_1, C_2 = 100$ pf
i-f = 456 kHz.

FIND: (*a*) $X_{C\cdot 1}$ (*b*) Path of i-f current (*c*) $i_{\text{i-f}\cdot R_1}$ (*d*) $X_{C\cdot 2}$
(*e*) Purpose of C_2 (*f*) Per cent

SOLUTION:

(*a*)
$$X_{C\cdot 1} = \frac{159{,}000}{fC} = \frac{159{,}000}{456 \times 10^3 \times 100 \times 10^{-6}} = 3{,}486 \text{ ohms}$$

(*b*) The i-f current will take the capacitor path.

(*c*) Yes. A small amount.

(*d*) Same as (*a*); $X_{C\cdot 2} = 3{,}486$ ohms

(*e*) Provides a low-impedance path for any i-f current that was not bypassed by the capacitor C_1.

(*f*)
$$\% \text{ of } E_{\text{a-f}} \text{ at } R_2 = \frac{i_{\text{a-f}} R_2}{i_{\text{a-f}}(R_1 + R_2)} \times 100$$

Because the current is the same in both the numerator and the denominator, then

$$\% \text{ of } E_{\text{a-f}} \text{ at } R_2 = \frac{R_2}{R_1 + R_2} \times 100 = \frac{250{,}000}{50{,}000 + 250{,}000} \times 100 = 83.3$$

Vacuum Tubes as Diode Detectors. Although largely replaced by semiconductor diodes, the diode and duodiode vacuum tubes can be used in the circuits of Figs. 18-14 to 18-18 in place of the diodes D_1 and D_2. A multiunit

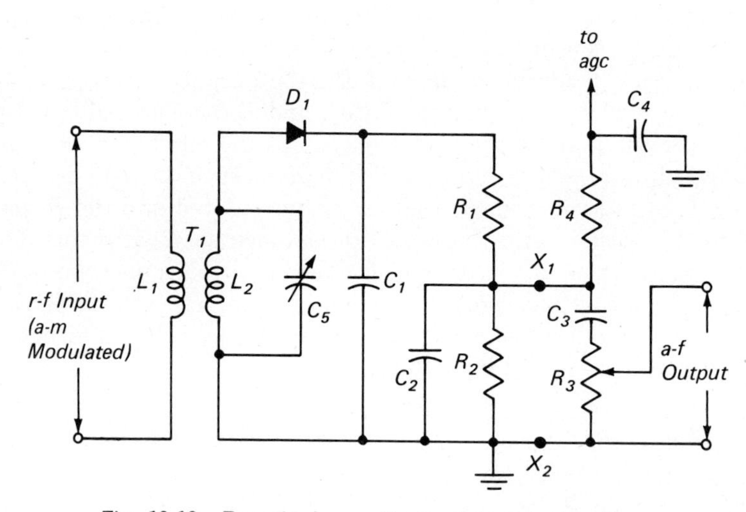

Fig. 18-18 Practical a-m diode-detector circuit.

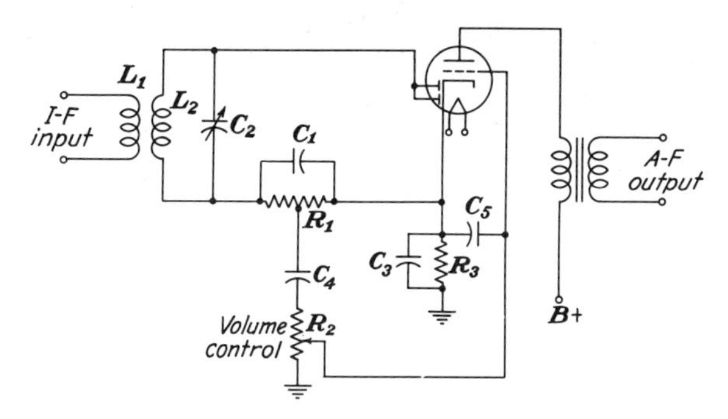

Fig. 18-19 Diode-detector circuit using a duplex-diode-triode vacuum tube.

tube, such as a duplex-diode-triode, can be used as a detector and amplifier. The two diodes may be connected together to form a simple diode half-wave detector. The triode is used to amplify the rectified signal. A circuit using a duplex-diode triode is shown in Fig. 18-19. In this circuit R_1 is the diode load resistor. The bias voltage for the triode section is obtained from the cathode-bias resistor R_3. The bypass capacitor C_3 keeps the a-f current out of the cathode-bias resistor. Capacitor C_4 blocks the d-c bias of the cathode from the grid. Capacitor C_5 bypasses any r-f current from the grid circuit to the cathode and thus prevents any r-f voltage from reaching the grid of the tube. [The principles of this circuit can be adapted to a solid-state circuit by replacing the vacuum tube with (1) discrete diode and transistor, or (2) a portion of a semiconductor integrated circuit.]

18-9 A-M Detection with Three-element Active Device

Basic Principle. Three-element active devices can be used to separate the a-f portion from the amplitude-modulated r-f signal. When a transistor or a vacuum tube is used as a detector, it performs two functions: (1) rectification, and (2) amplification.

Transistors can be used as detector-amplifiers in either the common-emitter or common-base configuration. In either case, the r-f signal is applied to the base-emitter circuit of the transistor where rectification takes place. Amplification of the recovered a-f signal takes place in (1) the emitter-collector circuit with the common-emitter configuration, and (2) the base-collector circuit with the common-base configuration.

The triode vacuum tube can be used as a detector-amplifier in either the common-grid or common-cathode configuration. In either case, the r-f signal is applied to the grid-cathode circuit of the tube where rectification takes place. Amplification of the recovered a-f signal takes place in (1) the cathode-plate circuit with the common-cathode configuration, and (2) the grid-plate circuit with the common-grid configuration.

18-10 A-M Transistor Bias-detector Circuit

Figure 18-20*a* illustrates a transistor detector circuit using the common-emitter configuration; this circuit is sometimes referred to as a *base-bias* detector, or a *bias detector*. In this circuit, the amplitude-modulated r-f input signal is selected by the parallel tuned circuit C_1L_1, and is coupled to the base-emitter circuit of the transistor by the transformer secondary winding L_2. The transistor

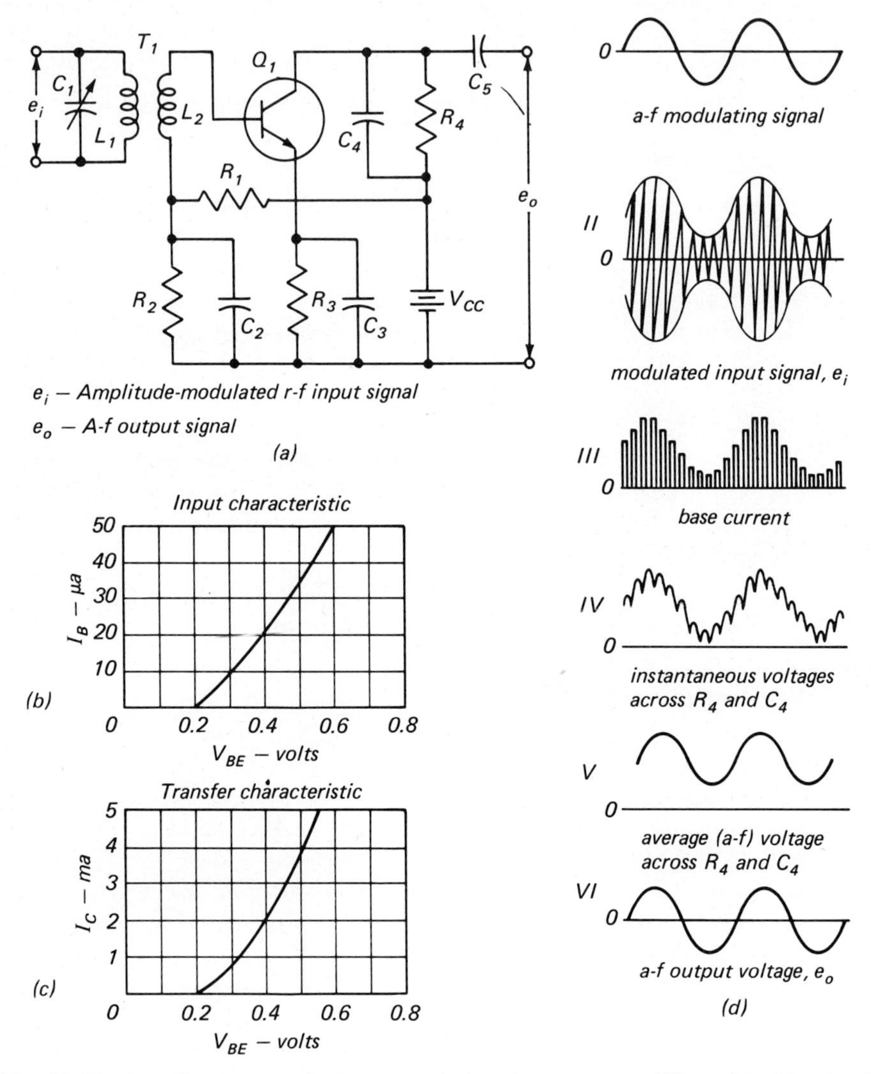

Fig. 18-20 Amplitude-modulation transistor detector-amplifier. (a) Circuit diagram. (b) Typical input characteristic curve. (c) Typical transfer characteristic curve. (d) Voltage and current waveforms.

is biased to approximately cutoff (with zero input signal) by $R_1R_2V_{CC}$; C_2 bypasses the r-f signals around R_2 so that the bias voltage established by $R_1R_2V_{CC}$ remains at a fixed value. The operation of the circuit is stabilized by the swamping resistor R_3, and C_3 bypasses the a-f and r-f currents around R_3 to prevent degeneration. Resistor R_4 serves as the load of the transistor amplifier circuit, and C_4 bypasses any r-f signals around the load resistor.

With the transistor biased at or near cutoff for zero signal input, the positive halves of the modulated r-f input signal will drive the transistor into conduction on a linear portion of the characteristic curve, and the negative halves of the input signal will be cut off. Thus, rectification of the input signal is accomplished in the base-emitter circuit. The small changes in base current produced by the variations in the positive halves of the r-f input signal cause much larger proportionate current changes in the collector-emitter current; consequently, an amplified reproduction of the a-f modulating signal appears at the load resistor R_4. Capacitor C_4 eliminates the r-f pulsations in the same manner as explained for C_1 of Fig. 18-14. The output at R_4 is a direct voltage varying at the audio frequencies of the original a-f modulating signal. Capacitor C_5 eliminates the d-c reference level, and also blocks the direct voltage from the following stage. Because the base-emitter action of the transistor is the same as that of the semiconductor diode of Fig. 18-14, the circuit characteristics shown in Fig. 18-15 will also apply to the base-emitter circuit action of the detector-amplifier circuit.

The action of the detector-amplifier circuit is further explained by analyzing the internal transistor circuit action using the characteristic curves of Fig. 18-20*b* and *c*. These curves indicate that cutoff occurs when V_{BE} is 0.2 volt, therefore, the fixed bias provided by $R_1R_2V_{CC}$ should be approximately 0.2 volt. If the r-f input signal has a peak-to-peak value of 0.4 volt, the base current will vary between zero and 20 μa during the positive halves of the input signal, and the collector current will vary between zero and 2 ma. (Since the transistor is biased near cutoff with zero input signal, no current will flow in either the base or collector circuits during the negative halves of the input signal.) Because a 20-μa change in base current produces a 2-ma change in the collector current, the amplification is equal to 2×10^{-3} divided by 20×10^{-6}, or 100.

18-11 A-M Vacuum-tube Bias-detector Circuits

Bias Detection. Figure 18-21*a* illustrates one type of detector amplifier circuit using a vacuum tube. In this circuit, the amplitude-modulated r-f input signal, selected by the parallel tuned circuit L_2C_2, is applied to the grid-cathode circuit of the tube. The cathode bias resistor R_1 provides the proper bias for the tube and capacitor C_1 bypasses any alternating currents around R_1 to prevent degeneration. The filter circuit C_3L_3 separates the a-f and r-f components of the rectified wave present in the plate circuit. The a-f signal component of the plate current is coupled to the following circuit by the transformer T_2. The grid of the tube is biased to almost cutoff so that the current in the plate

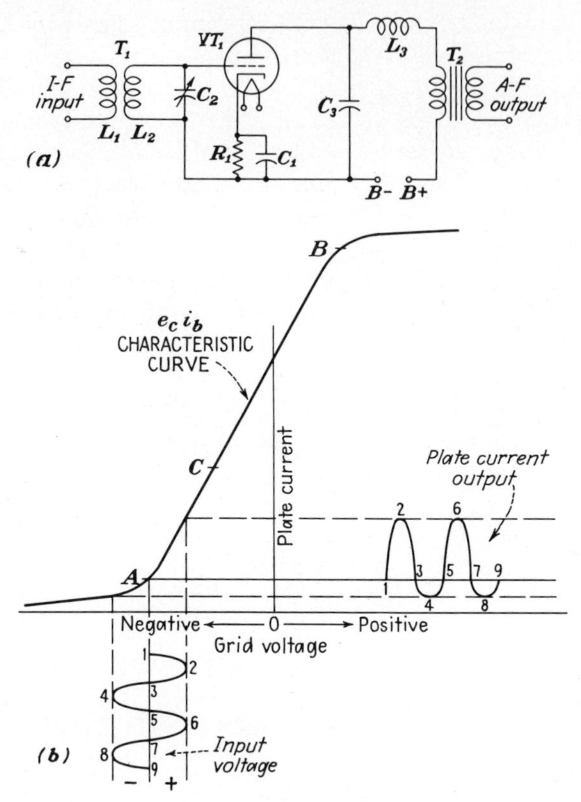

Fig. 18-21 Bias detector using a triode vacuum tube. (a) Circuit diagram. (b) Operating characteristics.

circuit is practically zero when the signal applied to the grid is zero. Positive values of input signal make the grid less negative, thus permitting plate current to flow, and negative values of input signal make the grid more negative, thereby keeping the tube at cutoff. Thus, since the circuit passes and amplifies the positive halves of the input signal and eliminates the negative halves, both detection (rectification) and amplification are accomplished. Because rectification occurs in the plate circuit and is dependent on having the proper amount of grid bias, this circuit is called (1) a *plate detector,* (2) a *grid-bias detector,* or (3) a *bias detector.*

Grid Detection. A highly sensitive vacuum-tube detector circuit is the *grid detector,* also called the *grid-leak detector.* Figure 18-22 illustrates a grid-leak-detector circuit. In this circuit, the desired amplitude-modulated r-f input signal is selected by the parallel tuned circuit L_2C_2. The grid resistor R_1 and grid capacitor C_1 form an impedance that corresponds to the load impedance C_1 and R_1 of the diode detector of Fig. 18-14. The grid circuit, consisting of the source of the input signal, the grid and cathode of the tube, and the resistor-capacitor combination R_1C_1, operates as a half-wave rectifier in the same manner as a diode detector. At the same time, the grid and cathode serve as the input circuit of the triode amplifier. The required r-f filtering is provided by the π-type low-pass filter $C_3L_3C_4$. Transformer T_2 couples the a-f signal to the following circuit.

When the input signal is zero, the voltages on the grid and cathode are zero. The voltage on the plate of the tube causes a steady stream of electrons from the cathode to the plate. As the grid is in the path of this electron flow, some electrons strike the grid and make it negative with respect to the cathode. Capacitor C_1 becomes charged, with plate A negative and plate B positive. This charge remains on the capacitor, and the resultant current flow through R_1 causes a constant bias on the grid. The amount of current flowing through R_1 is very small; therefore, the voltage drop across this resistor is also small. Consequently, when the tube is considered as a triode amplifier, it will have a small amount of grid bias.

During the positive half-cycles of the modulated r-f input signal, additional current will flow through R_1, thus increasing the grid bias. The current flowing in the plate circuit will therefore decrease. As the input-signal voltage varies with succeeding r-f cycles, the voltage across R_1 will vary accordingly. The plate current will decrease in the same manner as the voltage across R_1 increases, and vice versa.

The value of C_1 is selected so that it will pass only the high-frequency signal current, which is then rectified by the diode-detector action of the grid and cathode. The rectified signal current then flows through R_1 and develops an a-f voltage across this resistor. This voltage is applied to the input (grid-cathode) circuit of the tube, which may then be considered as operating as an amplifier.

The current in the grid circuit of the grid detector does not vary in the same proportion as the variations of the input-signal voltage; however, because of the small grid bias, the tube will operate on the curved portion of the characteristic curve. Under this condition, the variation in grid current is practically proportional to the square of the grid-voltage variations, and grid-circuit detectors are therefore often referred to as *square-law detectors*. Because

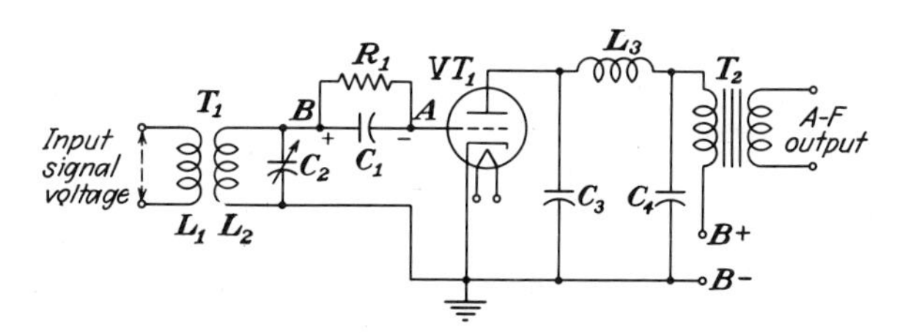

Fig. 18-22 Grid-detector circuit.

of the square-law variation, the strength of the input signal will be increased, but it will also be distorted. Grid detection is therefore characterized by its high sensitivity and its high distortion factor.

18-12 A-M Regenerative Detectors

The detector-amplifier circuits of Figs. 18-20*a* and 18-22 can be modified to operate as regenerative detectors by adding a controlled amount of positive feedback to these circuits. The circuits in Fig. 18-23 illustrate a simple method

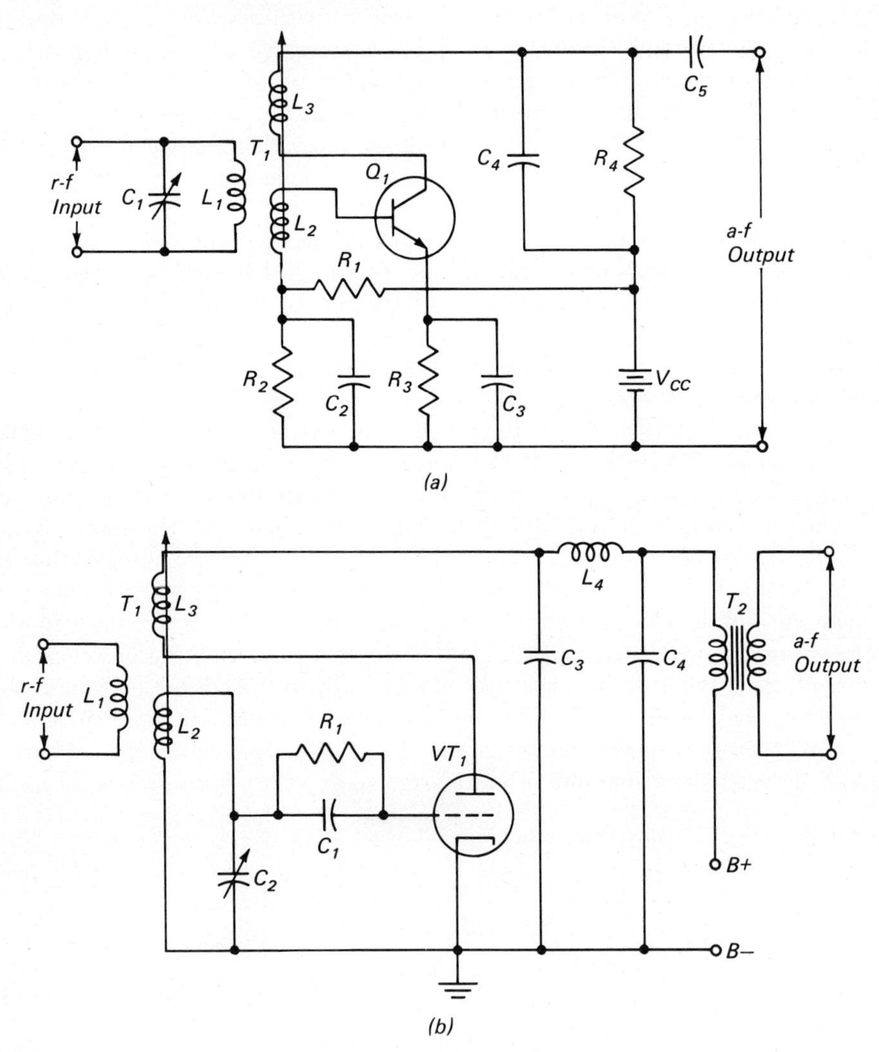

Fig. 18-23 Regenerative-detector circuits using (a) a transistor, (b) a vacuum tube.

of adding feedback by means of the magnetic coupling between L_3 and L_2, and control of the amount of feedback by varying the relative position of L_3 and L_2. Various other methods of providing controlled feedback are also possible.

When an input signal is applied to L_1, a voltage is induced in L_2 and this voltage is applied to the input circuit of the active device (Q_1 or VT_1). This signal causes a change in the output current of the active device, and it should be observed that the output current now flows through the feedback coil L_3. The varying magnetic field at L_3, caused by a change in the output current, induces a voltage in L_2 which may be further controlled by varying the relative position of L_3 and L_2. With L_3 connected to produce positive feedback, the input voltage to the active device will be the sum of the voltages induced in L_2 by L_1 and L_3, thus causing a still greater change in the amount of the output current. Because this effect is repeated, and is cumulative, it can result in the circuit's going into oscillation. The amplification of the signal voltage at the input of the active device will increase with increases in the amount of feedback, reaching a maximum at the point where the detector starts to oscillate. For proper detector operation, oscillation must be avoided, and hence the relative position of L_3 and L_2 is adjusted to the point just before causing oscillation.

The advantages of the regenerative detector are (1) its greatly increased sensitivity, and (2) an increase in the effective Q of the circuit, with an accompanying increase in selectivity. Among the disadvantages are (1) poor linearity, (2) low signal-handling ability without overloading, (3) feedback adjustment is critical, and (4) the circuit acts as a transmitter when oscillation occurs.

18-13 A-M Continuous-wave Detectors

C-W Signal Detection. Code signals are transmitted by interrupting the carrier wave at definite intervals corresponding to the dots and dashes of the Morse code. When a dot or dash is being transmitted, an r-f signal of constant amplitude is sent out. This method of transmission is called *continuous-wave* or *c-w transmission.* If a c-w signal is applied to any of the detector circuits previously described, rectification of the r-f waves will take place, and a series of interrupted currents whose average values will be of constant amplitude will appear at the output of the detector. If this output current is applied to earphones or a loudspeaker, the only sound it can produce is a click, which may occur at the start and finish of each dot or dash. In order to reproduce c-w signals, heterodyne detection may be used.

Heterodyne Detection. The principle of heterodyne action (Art. 18-6) can be used for the detection of c-w signals. Figure 18-24 illustrates the principle of the heterodyne detector. The oscillator circuit is adjusted to a frequency slightly above or below the frequency of the input signal e_i, and the oscillator signal is fed into the detector circuit where it is combined with the input signal. The heterodyne action produces two new signals, called *beat signals,* whose frequencies are (1) $f_i + f_{osc}$, and (2) $f_i - f_{osc}$. By using the difference

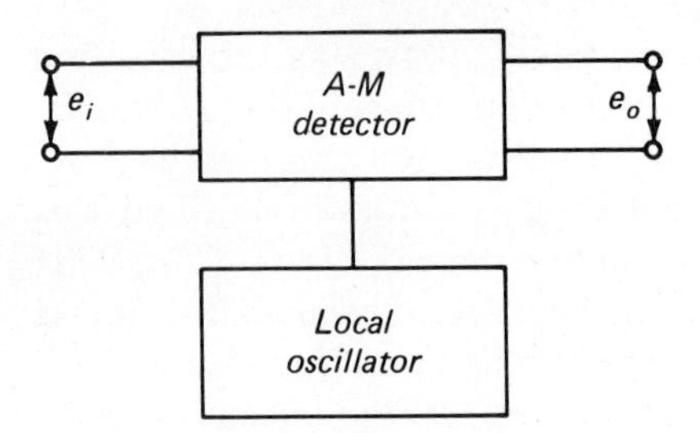

e_i = amplitude-modulated, continuous-wave, r-f input signal.

e_o = continuous wave, a-f output signal.

Fig. 18-24 Continuous-wave detection.

frequency, heterodyne detection is accomplished when the incoming signal is combined with an oscillator frequency that produces a beat signal in the a-f range, usually approximately 1,000 Hz.

Autodyne Detection. Because regeneration can be increased to a point where a circuit will break into oscillation, it is possible to obtain heterodyne detection of c-w signals with the use of a single active device. An oscillating detector used for this purpose is called an *autodyne detector*. Its circuit is the same as for the regenerative detector circuits of Fig. 18-23. By tuning the oscillating circuit to a frequency 1,000 Hz above or below the frequency of the input c-w signal, a signal with a 1,000-Hz beat frequency is obtained.

Superregeneration. In a regenerative detector, the amplification due to regeneration will build itself up to a point where the circuit breaks into oscillation. At this point, the output is the greatest because the effective resistance of the circuit is zero and the losses are at a minimum. This operating point is very critical, and therefore it is very difficult to maintain the circuit at the desired operating point. A slight change in any part of the circuit can cause the circuit to break into oscillation.

Superregeneration is a means whereby the effective resistance of the circuit is kept at zero. This is accomplished by varying the regeneration from an oscillatory to a nonoscillatory condition at a low r-f rate. During the oscillating interval, the oscillations will build up, only to be suppressed by the low frequency that is applied to the output circuit of the active device. This causes the circuit to go in and out of oscillation at a periodic rate that is equal to the frequency being applied by the low-frequency source. This frequency, called the *quench frequency,* is slightly higher than the audio frequencies, generally 20 to 25 kHz. Superregeneration may be obtained by (1) using a separate oscillator to supply the quench-frequency signal, or (2) intermittently blocking the oscillating detector circuit at the desired frequency.

The selectivity of a receiver using superregenerative detection is poor, but this is not important at the high frequencies for which this type of detection is used. The superregenerative detector circuit radiates a signal that may interfere with other nearby receivers. Superregenerative detectors are used primarily for the reception of signals whose frequencies are usually too high for other methods of detection.

18-14 Automatic Gain Control

Purpose. In some electronic equipment, for example, television and radio receivers, it is desirable to vary the gain of one or more amplifier stages automatically to compensate for wide variations in the strength of the input signals. Strong input signals require less amplification than weak signals and an automatic-gain-control (abbreviated agc) circuit provides a means of automatically decreasing the gain of an amplifier when strong signals are applied to the input circuit. These special circuits, usually associated with detectors, are classified as (1) *simple agc,* and (2) *delayed agc.* With simple agc, the gain-control action is present at all times regardless of whether the input signal is weak or strong. With delayed agc, the gain-control action occurs only with strong input signals. When the gain control ultimately affects the sound output of a radio receiver or other sound-reproducing device, the term *automatic volume control* (*avc*) is sometimes used in place of automatic gain control.

Basic Principle. The d-c bias on active devices controls to a great extent the amount of amplification produced by circuits employing these devices. Therefore, control of the d-c bias will control the gain of an amplifier. The preceding articles have shown that the output voltage of a detector circuit contains (1) an a-f signal, and (2) a d-c component. Since the d-c component is proportional to the average amplitude of the input signal, applying a portion of this direct voltage to the bias of the active device of one or more of the preceding amplifier stages will provide automatic gain control. Increases in signal strength will increase the value of this agc bias and cause a reduction in the strength of the signal applied to the detector. Conversely, a decrease in signal strength will decrease the agc bias voltage and cause an increase in the signal applied to the detector. In this manner, the agc bias tends to maintain the output of the detector circuit at a constant level automatically.

Basic AGC Voltage Source. Figure 18-25 illustrates the basic agc voltage source. The voltage at the secondary of T_1 has the waveform of the amplitude-modulated input signal. Capacitor C_3 improves the selectivity of the secondary circuit. The diode D_1 rectifies the alternating signal present at terminal 1. The unidirectional current flowing through R_1 varies at the original modulating a-f rate. Capacitor C_1 smooths out the r-f pulsations of the rectified a-f current variations in R_1. The time constant of C_1R_1 is selected to (1) eliminate the r-f variations, and (2) retain the a-f variations. The time constant of the series circuit R_2C_2 is selected to (1) eliminate the a-f variations present at terminal 2, and (2) provide a direct voltage at terminal 3 for agc use. The time constant of R_2C_2 must be long enough to eliminate the a-f variations, but still be short enough so that it will not filter out the variations in the average signal strength of the rectified a-f signal at terminal 2. With these conditions, the agc voltage

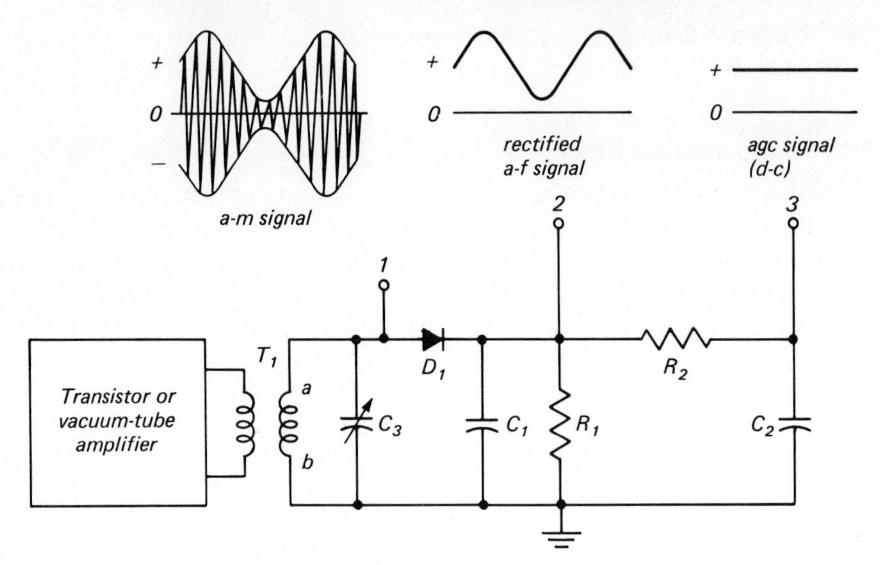

Fig. 18-25 Basic agc voltage source.

will vary only when the overall strength of the input signal at T_1 increases or decreases.

When terminal *a* of T_1 is positive, (1) diode D_1 conducts, (2) a voltage is developed across R_1, and (3) C_1 becomes charged. When terminal *a* is negative, (1) D_1 does not conduct, and (2) C_1 discharges only slightly through R_1 because of the long time constant of C_1R_1 to the r-f pulsations. Therefore, the negative halves of the r-f signal are eliminated and the a-f output at terminal 2 is positive; likewise, the agc voltage is positive. Reversing the diode D_1 will produce a negative agc voltage.

Example 18-6 In the circuit of Fig. 18-25, the agc filter (R_2C_2) consists of a 2-megohm resistor and a 0.1-μf capacitor. (*a*) What is the time constant of this circuit? (*b*) What per cent of the maximum voltage charge possible will be attained in the time that the current from 1 cycle of a 50-Hz a-f signal will flow in the detector circuit (assume that the signal voltage is constant)? (*c*) How many cycles of the 50-Hz signal will be completed in one time constant? (*d*) If a sine-wave voltage, whose maximum value is equal to the constant voltage assumed in part (*b*), is applied to the detector circuit, will it require the same amount of time, more time, or less time for the capacitor to reach 63.2 per cent of its final charge? Why?

GIVEN: $R = 2$ megohms $\quad C = 0.1\ \mu$f $\quad f = 50$ Hz

FIND: (*a*) t $\quad$ (*b*) % E $\quad$ (*c*) Cycles $\quad$ (*d*) t

SOLUTION:

(*a*) $$t = RC = 2 \times 10^6 \times 0.1 \times 10^{-6} = 0.2 \text{ sec}$$

(*b*) As current can flow only during the positive half of the signal, the maximum time of current flow in one cycle of the 50-Hz a-f signal is 0.01 sec. This corresponds to one-twentieth of a time constant, and from Table 12-1 the charge produced will be approximately 5 per cent of the maximum voltage charge possible when $k = 1/20$ or 0.05.

(*c*) Time for 1 cycle of the 50-Hz a-f signal,

$$t_1 = \frac{1}{f} = \frac{1}{50} = 0.02 \text{ sec}$$

Cycles completed in one time constant,

$$\text{No. of cycles} = \frac{t}{t_1} = \frac{0.2}{0.02} = 10$$

(*d*) A much greater amount of time will be required because the value of the sine-wave voltage is equal to the voltage of part (*b*) at its maximum point only and then only for a very small period of time.

Basic Delayed AGC. The purpose of delayed agc is to defer or *delay* the application of an agc voltage to the controlled circuits until a specific level of agc voltage is reached. Figure 18-26 illustrates the principle of delayed agc circuits. In this circuit the agc voltage cannot cause current to flow through R_4 until e_{agc} is greater than e_k (plus the starting voltage—or potential barrier—of the diode D_2). For example, if the voltage at the cathode-bias resistor R_4 is 2 volts and the potential barrier of D_2 is 0.2 volt, then current cannot flow from the agc source (terminal 3) until the agc voltage is greater than -2.2 volts. With agc voltages of more than 2.2 volts (1) the current in R_4 increases, (2) the grid bias of VT_1 increases, and (3) the gain of VT_1 decreases.

18-15 Transistor AGC Applications

Types. Automatic gain control is often used with transistor tuned-amplifier circuits in the r-f and i-f ranges. There are two commonly used methods of achieving automatic-gain control, namely, by (1) controlling the d-c emitter

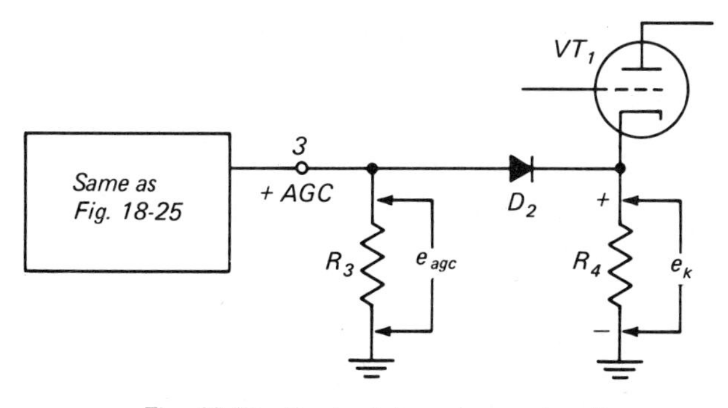

Fig. 18-26 Basic delayed agc circuit.

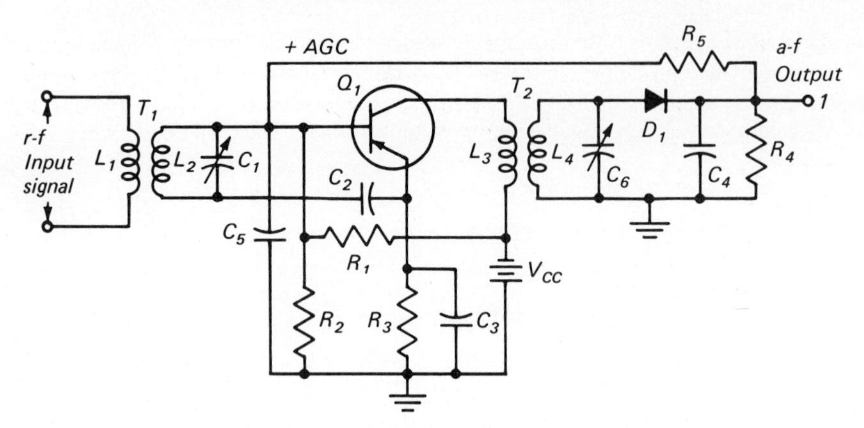

Fig. 18-27 Transistor amplifier with simple reverse agc.

current, referred to as *reverse agc,* and (2) controlling the d-c collector voltage, referred to as *forward agc.*

Reverse AGC. The circuit of Fig. 18-27 illustrates the application of simple reverse agc to a common-emitter-type transistor amplifier. In this circuit an amplitude-modulated r-f signal is applied to L_1. Selectivity is achieved with the tuned circuit L_2C_1. The r-f signal is coupled to the input circuit of Q_1 by L_2 and C_2. Forward bias for Q_1 (with zero modulation of the r-f input signal) is provided by $R_1R_2V_{CC}$. Amplifier stability is improved by the swamping action of R_3 and C_3. The amplified r-f signal at L_3 is coupled to the detector-agc circuit by L_4, which is tuned to resonance by C_6. The components D_1, R_4, and C_4 form a diode-detector circuit with an a-f signal available at terminal 1. The series circuit of R_5C_5 forms a filter network to provide an agc voltage which is fed to the input circuit of Q_1. The time constant of R_5C_5 should be of such a value that R_5C_5 will (1) prevent the a-f variations from reaching the base of Q_1, and (2) pass any slow changes in the average d-c strength of the a-f signal on to the base of the transistor.

The fixed forward bias provided by $R_1R_2V_{CC}$ establishes a negative voltage at the base of Q_1, and the agc voltage applied to the base is positive. When the agc voltage increases, as occurs with an increase in the signal strength at L_1, the effective forward-bias voltage at the base of the transistor decreases and causes a decrease in the emitter current. The decrease in emitter current is accompanied by a decrease in the gain of the amplifier. Conversely, a decrease in the signal strength at L_1 results in (1) a decrease in the agc voltage, (2) an increase in the effective forward bias of Q_1, (3) an increase in the emitter current, and (4) an increase in the gain of the amplifier.

Forward AGC. Figure 18-28 illustrates the application of simple forward agc to a common-emitter transistor amplifier. This circuit differs from the reverse agc circuit of Fig. 18-27 only in that (1) diode D_1 is reversed, and (2) collector resistor R_6 and its bypass capacitor C_7 are added. Reversing D_1 changes the polarity of the agc voltage from positive to negative. Adding R_6 causes changes in the agc voltage to produce changes in the magnitude of the collector voltage and thereby alter the gain of the amplifier. Capacitor C_7 bypasses the alternating current around R_6 and on to ground.

When the agc voltage increases, because of an increase in the input-signal strength, the effective forward-bias voltage at the base of Q_1 increases and causes an increase in the emitter and collector currents. While this effect alone will increase the gain of the amplifier, it should be observed that the increase in collector current flowing through R_6 will increase the voltage drop across R_6 and thus decrease the collector voltage. The decrease in collector voltage causes a decrease in the gain of the amplifier that is greater than the increase in gain due to the increase in the emitter and collector currents. When the input-signal strength at L_1 decreases, the circuit actions are (1) a decrease in the agc voltage, (2) a decrease in the effective forward bias of the transistor, (3) a decrease in the emitter and collector currents, (4) a decrease in the voltage drop across R_6, (5) an increase in the collector voltage, and (6) an increase in the gain of the amplifier.

Characteristics. The characteristics of reverse agc are: (1) simpler to use than forward agc, (2) can be applied to most transistors, (3) increases the input and output resistance and hence has greater selectivity with strong signals, (4) only slight changes in the input and output capacitances, hence little effect on the tuned circuits, (5) only slight changes in the loading effect on the tuned circuits.

The characteristics of forward agc are: (1) greater reduction in gain than with reverse agc, (2) can be used only with transistors having special remote current-cutoff characteristics, (3) decreases the input and output resistances, hence reduces the selectivity with stronger signals, (4) appreciable changes in the input and output capacitances, hence some detuning with strong signals.

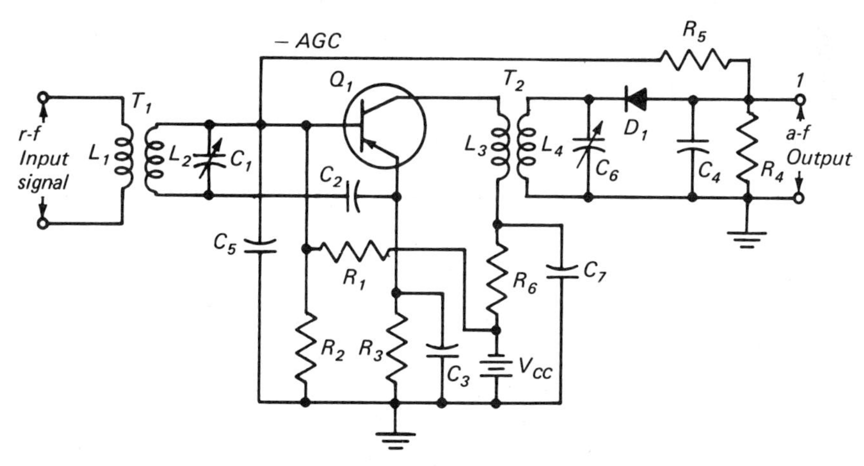

Fig. 18-28 Transistor amplifier with simple forward agc.

18-16 Vacuum-tube AGC Applications

Although single-diode or twin-diode vacuum tubes can be substituted for the semiconductor diodes of Figs. 18-25 and 18-26, the small size and low cost of semiconductor diodes result in extensive use of these components. If the stage following the detector employs a triode or pentode tube, it may be advantageous to use a single multiunit tube, such as a twin-diode-triode or a diode-pentode, to perform detection, amplification, and agc functions.

In the circuit of Fig. 18-29, the detector and agc functions are performed by part of a twin-diode-triode-type tube. Plate D_1 of the twin-diode section acts as the plate for the diode detector. Plate D_2, which is fed by capacitor C_5, supplies the agc bias. The cathode-biasing resistor R_4 provides the bias for the triode section of the tube, and also makes the agc diode plate D_2 negative with respect to the cathode, thereby providing the means of obtaining the delayed agc. The diode load resistor R_1 is connected directly to the cathode, and hence the detector plate D_1 is at cathode potential. Because of this, rectification will take place in the detector circuit during part of the positive half of each cycle of the input signal, whether from a weak station or from a strong one. The rectified current flowing through R_1 produces a unidirectional voltage at R_1, which varies in the same manner as the input signal. This output voltage is applied to the triode or amplifier portion of the tube through the coupling capacitor C_6 and the manual volume control R_5.

The agc plate D_2 is connected to ground through its load resistor R_3. With zero input-signal voltage, D_2 will be negative with respect to the cathode by the amount of cathode bias produced by R_4, usually 2 or 3 volts. Under this condition, no current will flow in the D_2 plate circuit and hence no agc voltage is developed. When a signal is being received, a voltage will be applied to the agc plate D_2 by means of the coupling capacitor C_5. When this voltage exceeds the value of the cathode bias, D_2 will be positive with respect to the cathode, and a rectified current will flow in the agc circuit consisting of D_2, R_3, and R_4. A voltage will then be developed across the agc load resistor R_3, point A being

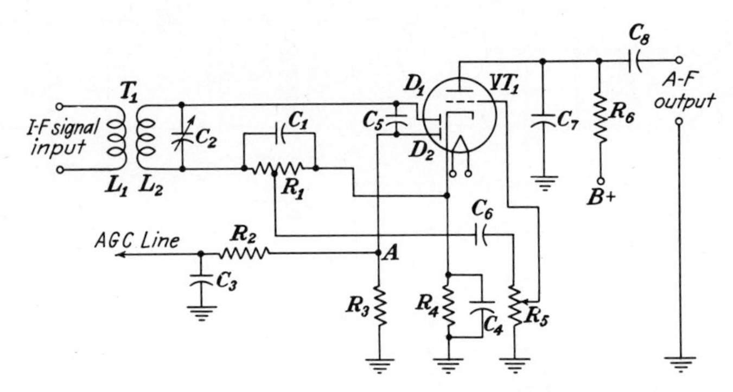

Fig. 18-29 Diode detector using a multiunit vacuum tube. Delayed agc obtained from cathode bias.

negative with respect to ground. The agc filter network R_2C_3 eliminates the a-f voltage variations and passes on to the agc line any variations due to fading and changes in signal strength when selecting a new station signal. Thus, the delayed agc circuit results in having the maximum signal received from weak stations and automatic gain control applied only to the strong stations.

18-17 F-M Detection

Basic Principle. Detector circuits for both a-m and f-m signals depend on the basic principle that amplitude or frequency changes of the r-f signal caused by modulation must produce a proportional change in the rectified current flowing through diodes. The manner in which the diode detects the a-f signal in the a-m system has already been presented. To utilize this same principle to recover the modulation from an f-m signal, the signal must first be converted so that it will appear as a modulated r-f voltage across the anode to the cathode of the diode.

A simple method of converting frequency variations to voltage variations makes use of the principle that reactance varies with the frequency. An alternating current, such as an r-f or i-f signal, flowing through an inductor will remain at a constant value if neither the voltage nor the frequency is varied. However, since reactance varies with frequency, the current flowing through an inductor will vary in amplitude when the frequency of the applied signal varies, even though the amplitude of the voltage remains constant. The amount of change in the amplitude of the current is dependent upon the amount of shift in frequency. Since an f-m signal varies in frequency above and below the center frequency by an amount depending upon the amplitude of the modulating signals, applying a frequency-modulated signal to an inductor will convert the frequency deviations to amplitude changes in current. These amplitude changes in current when made to flow through a resistor will produce corresponding voltage changes across the terminals of the resistor. The same principle can be illustrated when a capacitor is substituted for the inductor.

The ideal response curve of an f-m detector (Fig. 18-30) shows that each frequency variation produces a definite value of voltage and also that a linear relationship exists between the frequency and the voltage. This linear relationship is essential in order to produce distortionless conversion.

18-18 F-M Phase-shift Discriminator

Figure 18-32 illustrates a commonly used f-m detector circuit called the *phase-shift discriminator*, also called the *balanced phase-shift discriminator* or the *Foster-Seeley discriminator*. Because it is also sensitive to amplitude modulations, it must be preceded by a limiter stage. This type of circuit requires a transformer with a center-tapped secondary, generally called the *discriminator transformer*, two diodes, and a number of resistors and capacitors.

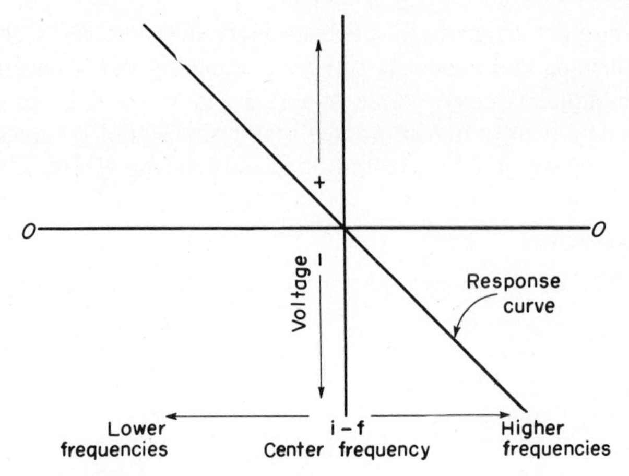

Fig. 18-30 Ideal response curve for an f-m detector.

Action of the Transformer. Understanding the operation of the phase-shift discriminator depends upon understanding the relations among the currents and voltages at the discriminator transformer. Important characteristics of the transformer, T_1 in the circuit of Fig. 18-32, are: (1) it is an air-core transformer and therefore has a low coefficient of coupling, (2) the primary (L_1) and secondary windings (L_2L_3) are parts of tuned circuits that are resonant at the center frequency of the input voltage e_i, and (3) the secondary winding is center-tapped to provide two equal voltages ($e_{s\cdot 1}$ and $e_{s\cdot 2}$) that are of opposite phase and hence are always equal in magnitude and always differ in phase by 180°.

For input signals at the center-frequency value, the phase relations among the voltages and currents are shown in Fig. 18-31*a*. An important characteristic of the transformer with a tuned secondary (or a tuned primary and tuned secondary) is that at resonance the voltage developed at the secondary lags the primary voltage by 90°, which is explained in the following manner: The series resonant secondary circuit reflects very little impedance into the primary, and therefore the primary circuit acts as a highly inductive circuit which results in the primary current i_p and the primary flux ϕ_p lagging the input voltage e_i by approximately 90°. The flux ϕ_p induces a voltage e_s in the total secondary winding L_2L_3 that lags the flux by 90°, and hence lags e_i by 180°. Because the secondary tuned circuit is at resonance, the secondary current i_s is in phase with e_s. The magnetic flux ϕ_s due to the current i_s develops a voltage in the secondary windings L_2L_3 that lags the secondary flux by 90°. Because the secondary winding is center-tapped, the voltage developed across this winding appears as two voltages $e_{s\cdot 1}$ and $e_{s\cdot 2}$ (Figs. 18-31 and 18-32). It is important to observe that because of the center-tapped secondary (1) $e_{s\cdot 1}$ leads e_i by 90°, and (2) $e_{s\cdot 2}$ lags e_i by 90°.

For input signals above the center-frequency value, the phase relations among the voltages and currents are shown in Fig. 18-31*b*. At frequencies higher than the resonant frequency, the total reactance of L_2 and L_3 increases and the reactance of C_3 decreases, thereby causing a shift in the phase angle between i_s and e_s, with i_s now lagging e_s. It should be observed that (1) the phase relation among $e_{s\cdot1}$, $e_{s\cdot2}$, and i_s remains at 90°, and (2) the phase relation among $e_{s\cdot1}$, $e_{s\cdot2}$, and e_i is shifted by the same amount as the angle between i_s and e_s.

For input signals below the center-frequency value, the phase relations among the voltages and currents are shown in Fig. 18-31*c*. At frequencies lower than the resonant frequency, the reactance of L_2L_3 decreases and the reactance of C_3 increases, thereby causing a shift in the phase angle between i_s and e_s, with i_s now leading e_s. It should be observed that (1) the phase relation among $e_{s\cdot1}$, $e_{s\cdot2}$, and i_s remains at 90°, and (2) the phase relation among $e_{s\cdot1}$, $e_{s\cdot2}$, and e_i is shifted by the same amount as the angle between i_s and e_s.

Input Reference Voltage. In order to utilize the effect of the phase shift of the voltages $e_{s\cdot1}$ and $e_{s\cdot2}$ with respect to e_i, the voltage e_i must be available at the center tap of the secondary winding. This requirement is met by the series circuit consisting of $C_2L_4C_5$ (Fig. 18-32) which (1) is connected across the source of the input voltage e_i, and (2) has the high side of L_4 connected to the center tap (terminal 4) of the secondary winding L_2L_3. The values of C_2, L_4, and C_5 are such that for the frequency range of the input signal (1) the reactance of L_4 is very high compared to the total reactance of C_2 and C_5 connected in series, (2) the voltage drops at C_2 and C_5 are negligible compared with the voltage drop at L_4, and (3) the voltage drop across L_4 is approximately equal in magnitude and phase to the input voltage e_i.

Circuit Analysis. For the circuit of Fig. 18-32, the input signal e_i is applied to the primary of the discriminator transformer T_1 and produces an alternating voltage across terminals 3 to 5 of the secondary. During one-half of each cycle

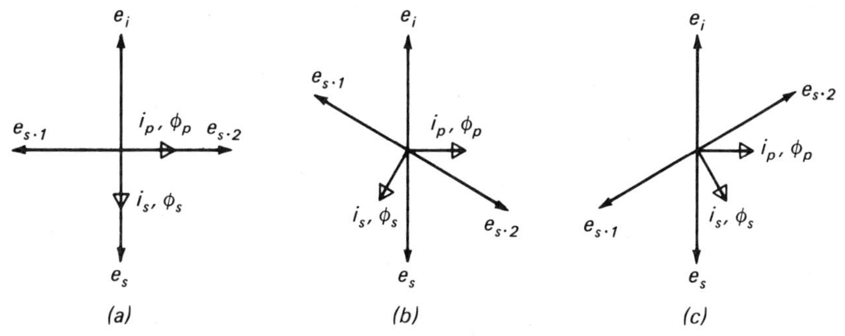

Fig. 18-31 Phase relation among voltages, currents, and magnetic flux in the discriminator transformer of Figs. 18-32 and 18-34. (a) For center frequency value. (b) For frequencies above the center frequency. (c) For frequencies below the center frequency.

terminal 3 is positive and terminal 5 is negative; also, during this time terminal 4 is negative with respect to terminal 3 and positive with respect to terminal 5. For this half-cycle the anode of D_1 is positive with respect to its cathode and D_1 conducts; also, the anode of D_2 is negative with respect to its cathode and D_2 is cut off. The path of electron flow is L_2, L_4, C_4, D_1, L_2. During the other half of each cycle terminal 3 is negative, terminal 5 is positive, and terminal 4 is positive with respect to terminal 3 and negative with respect to terminal 5. For these half-cycles D_2 conducts and D_1 is cut off; the path of electron flow is L_3, L_4, C_5, D_2, L_3.

When D_1 is conducting, the magnitude of the current is proportional to e_1, which is equal to the vector sum of $e_i + e_{s \cdot 1}$. With an appropriate time constant for R_1C_4, the charge on C_4 will follow the a-f waveform of the modulated input signal e_i in a manner similar to that shown in Fig. 18-15. A similar action takes place in the circuit of D_2 during the alternate half-cycles with e_2 being the vector sum of $e_1 + e_{s \cdot 2}$. The voltages E_1 and E_2 are pulsating direct voltages of opposite polarities. The output voltage e_o, which is the algebraic sum of E_1 and E_2, is an alternating voltage that follows the changes in magnitude and frequency of the a-f modulating signals present in the modulated input signal e_i.

Behavior of the Circuit for Center-frequency Signals. When the center (unmodulated) frequency of the input signal coincides with the resonant frequencies of the transformer primary and secondary circuits, $e_{s \cdot 1}$ and $e_{s \cdot 2}$ will be equal and 180° apart, and also 90° out of phase with e_i (Fig. 18-31*a*). The vector additions of $(e_i + e_{s \cdot 1})$ and $(e_i + e_{s \cdot 2})$ are illustrated in Fig. 18-33*a*, and show that e_1 and e_2 are equal in magnitude. Consequently, E_1 and E_2 also are equal, and with the polarities of E_1 and E_2 shown in Fig. 18-32, the output voltage e_o will be zero. Although the voltages E_1 and E_2 may vary in value, as long as they remain equal (and of opposite polarities) the output voltage of the discriminator will be zero.

Behavior of the Circuit with Signals above the Center Frequency. When the modulations produce frequencies higher than the center value, the phase angle

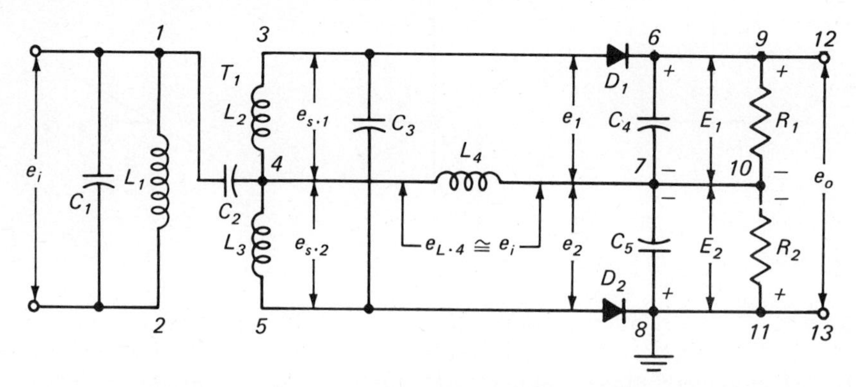

Fig. 18-32 Basic phase-shift discriminator circuit.

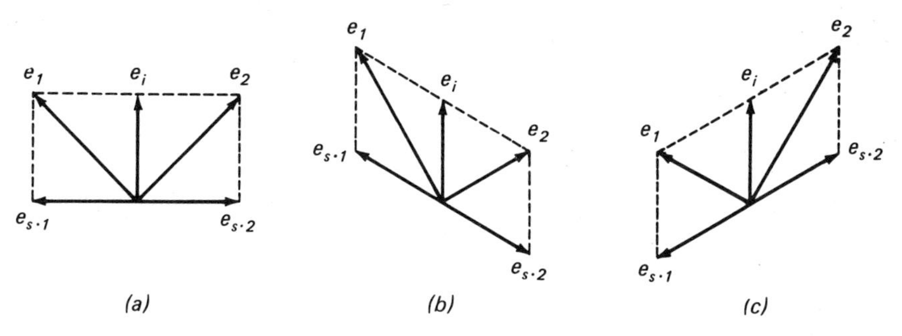

Fig. 18-33 Phase relation among voltages in the phase-shift discriminator circuit and the ratio detector. (a) For center-frequency value. (b) For frequencies above the center frequency. (c) For frequencies below the center frequency.

between $e_{s\cdot 1}$ and e_i becomes less than 90° and the phase angle between $e_{s\cdot 2}$ and e_i becomes more than 90° (Fig. 18-31*b*). Figure 18-33*b* shows that for this condition e_1 and e_2 are no longer equal even though the values of $e_{s\cdot 1}$, $e_{s\cdot 2}$, and e_i have not changed. The values of E_1 and E_2 will vary in the same manner as e_1 and e_2, and as E_1 becomes greater than E_2, e_o will have a significant value, with point 12 being positive with respect to ground. As the frequency increases, the phase angle between $e_{s\cdot 1}$ and e_i decreases and the phase angle between $e_{s\cdot 2}$ and e_i increases. Thus, the value of e_1 increases and the value of e_2 decreases, causing the value of e_o to increase.

Behavior of the Circuit with Signals below the Center Frequency. When the modulations produce frequencies lower than the center value, the phase angle between $e_{s\cdot 1}$ and e_i becomes more than 90° and the phase angle between $e_{s\cdot 2}$ and e_i becomes less than 90° (Fig. 18-31*c*). Figure 18-33*c* shows that for this condition (1) e_2 becomes greater than e_1, (2) E_2 becomes greater than E_1, and (3) e_o will have a significant value, with point 12 being negative with respect to ground. As the frequency decreases, the phase angle between $e_{s\cdot 1}$ and e_i increases and the phase angle between $e_{s\cdot 2}$ and e_i decreases. Thus, the value of e_1 decreases and the value of e_2 increases, causing the value of e_o to increase.

Behavior of the Circuit with an F-M Signal. When the i-f signal applied to the discriminator is frequency-modulated, deviations in frequency above and below the center i-f value will take place continually. The output signal e_o will be an alternating voltage whose value is dependent upon the frequency modulations. The rate of voltage changes will be the same as the a-f signals that produced the frequency modulations at the transmitting station.

18-19 The Ratio Detector

Comparison of the Phase-shift Discriminator and the Ratio Detector. The ratio detector shown in Fig. 18-34, which is a modification of the basic phase-shift discriminator circuit, is inherently insensitive to amplitude modulations. Because noiseless operation can be obtained without limiter stages, the ratio

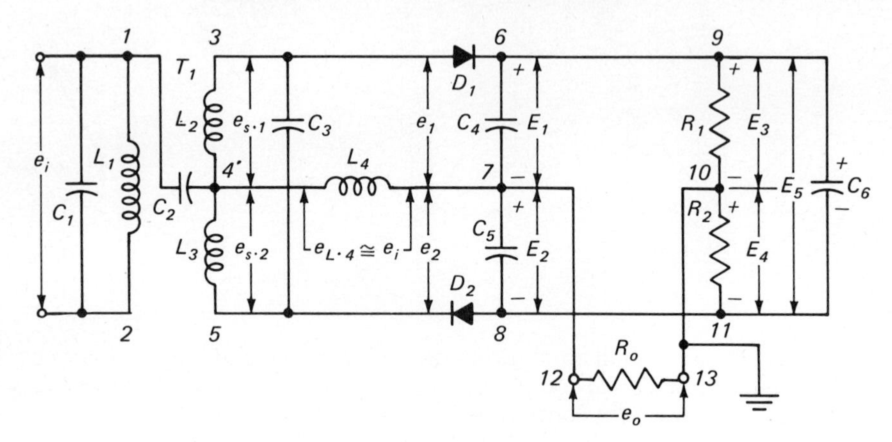

Fig. 18-34 Basic ratio-detector circuit.

detector is frequently used. The circuit of the ratio detector differs from the phase-shift discriminator circuit in that (1) diode D_2 is reversed, (2) the output (terminals 12 to 13) is taken at a different location, and (3) the capacitor C_6 (2 to 10 μf) is added to the circuit.

Basic Principles. The ratio detector derives its name from the fact that the ratio of the voltages across the diodes varies only with the f-m signals. The magnitudes of the voltages e_1 and e_2 (developed in the same manner as described for the phase-shift discriminator circuit) depend upon the magnitudes and the relative phase of the voltages e_i, $e_{s\cdot1}$, and $e_{s\cdot2}$. The ratio of e_1 to e_2 (1) remains constant as long as the phase relation among e_i, $e_{s\cdot1}$, and $e_{s\cdot2}$ remains constant, and (2) changes when the phase angle changes.

Circuit Operation. When terminal 3 of T_1 is positive, diodes D_1 and D_2 will conduct and current will flow through the parallel paths of (1) C_5C_4, (2) R_2R_1, and (3) C_6. During this time C_4, C_5, and C_6 will be charging. With C_4 and C_5 in the order of 300 pf, these capacitors will charge rapidly to the peak values of e_1 and e_2. With C_6 in the order of 2 μf, it will charge to the sum of the peak values that e_1 and e_2 would assume for the unmodulated carrier of the input signal e_i, but will require many cycles of operation before attaining full charge.

When terminal 3 of T_1 is negative: (1) diodes D_1 and D_2 will be cut off, (2) C_4 and C_5 will discharge through R_1, R_2, and R_o, and (3) C_6 will lose only a negligible portion of its charge because of (*a*) its high capacitance, (*b*) the long time constant of $R_1R_2C_6$, and (*c*) the short period of time that terminal 3 is negative.

Because the voltage E_5 acts as a bias on the diodes, only very small amounts of rectified current will flow to C_6 after it has reached the operating value for a given carrier-signal strength. Also, the long time constant of $R_1R_2C_6$ keeps amplitude variations at a-f rates from altering the voltage E_5. Under this condition the ratio of E_3 to E_4 will remain fixed, and with equal values for R_1 and R_2 the ratio will be unity. For any rapid decrease of short duration in the carrier-signal strength, the amount of discharge of C_6 through R_1R_2 is so small that E_5 remains practically constant. Likewise, a rapid increase of short duration will fail to produce any significant change in E_5. Therefore, no change is produced in E_5 by rapid fluctuations in the amplitude of the input signal such as may occur because of noise and other disturbances.

Behavior of the Circuit for Center-frequency Signals. When an unmodulated f-m signal is being received and it produces an i-f signal of the same value as the resonant frequency of the discriminator transformer, e_1 and e_2 are equal (Fig. 18-33*a*), and hence the ratio of e_1 to e_2 is unity. For this condition, E_1, E_2, E_3, and E_4 are all of the same magnitude. The output voltage e_o, which is equal to $E_4 - E_2$, is zero.

Behavior of the Circuit with Signals above the Center Frequency. When the modulations produce frequencies higher than the center value, the phase angle will vary in the same manner as with the phase-shift discriminator circuit and will cause e_1 to increase and e_2 to decrease (Fig. 18-33*b*). With e_1 now greater than E_1: (1) D_1 conducts, (2) the resultant current flowing through the path $L_2L_4C_4D_1$ increases the charge on C_4, and (3) the voltage E_1 increases. With e_2 now lower than E_2: (1) D_2 remains at cutoff, (2) C_5 discharges through the path $C_5R_2R_o$, and (3) the voltage E_2 decreases. As E_3, E_4, and E_5 remain constant, the output e_o (namely, $E_4 - E_2$) will have a significant value with terminal 12 negative with respect to ground.

Behavior of the Circuit with Signals below the Center Frequency. When the modulations produce frequencies lower than the center value, the phase angle will vary in the same manner as with the phase-shift discriminator circuit and will cause e_1 to decrease and e_2 to increase (Fig. 18-33*c*). With e_2 now higher than E_2: (1) D_2 conducts, (2) the resultant current flowing through the path $L_3D_2C_5L_4$ increases the charge on C_5, and (3) the voltage E_2 increases. With e_1 now lower than E_1: (1) D_1 remains at cutoff, (2) C_4 discharges through the path $C_4R_oR_1$, and (3) the voltage E_1 decreases. As E_3, E_4, and E_5 remain constant, the output voltage e_o will have a significant value with terminal 12 positive with respect to ground.

Behavior of the Circuit with an F-M Signal. When the input i-f signal e_i applied to the detector is frequency-modulated, deviations in frequency above and below the center-frequency value will take place continually. From the preceding explanations, it is evident that the output signal e_o will be an alternating voltage whose value is dependent upon the frequency modulations. The rate of the voltage changes will be the same as the a-f signals that produced the frequency modulations at the transmitting station. The a-f output signal e_o will not be affected by changes in amplitude of the f-m signal applied to the detector; hence the ratio detector can be used without a limiter.

18-20 Automatic Frequency Control

Need. A common fault of basic oscillator circuits is unwanted variations (called *drift*) in the frequency of its output. In some electronic-circuit applica-

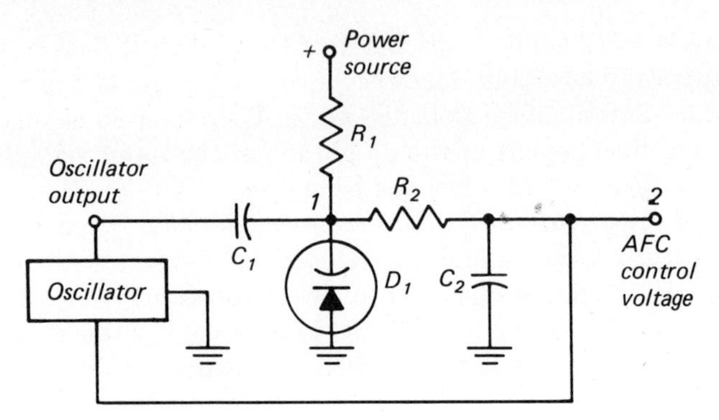

Fig. 18-35 Basic circuit for afc voltage.

tions an appreciable amount of frequency drift cannot be tolerated and some means of automatic frequency control (afc) becomes necessary. When only one signal source is involved in the functioning of an electronic device, the crystal-controlled oscillator (Arts. 17-23 to 17-25) serves adequately. When frequency drift occurs in a signal that is not within local control, such as an f-m signal at a radio receiver, automatic frequency control of the local oscillator of the receiver is highly desirable.

Basic Principle. The basic principle of automatic frequency control is to provide a frequency-sensitive control voltage to the active device in an oscillator circuit. Variations in this control voltage produce changes in the interelectrode capacitances of the active device and thus automatically alter the frequency of the oscillator.

Figure 18-35 illustrates a basic circuit for providing an afc voltage. In this circuit D_1 is a variable-capacitance diode (see *varactor*, Art. 10-3). Capacitor C_1 couples the oscillator output to the afc circuit formed by D_1, R_1, R_2, and C_2. Resistor R_1 provides a small amount of reverse bias to D_1 to obtain linear operation of the diode. Any increase in the frequency of the oscillator output from its rated value decreases (1) the impedance of the diode, and (2) the voltage at point 1. Conversely, a decrease in the oscillator frequency results in an increase in the voltage at point 1. Thus, changes in frequency are converted into voltage changes. The simple filter circuit R_2C_2 is designed to eliminate from the afc output at point 2 any rapidly repeating frequency changes that occur in the modulation, but pass on to the oscillator those voltage changes due to slower variations in frequency such as undesired frequency drift.

Other AFC Sources. The outputs of the phase-shift discriminator and the ratio detector can provide a control voltage for afc use. If a drift occurs in the center-frequency value of the input signal for the circuits of Figs. 18-32 and 18-34, the voltages E_1 and E_2 will no longer be equal with zero modulation. Under this condition, the output voltage e_o will have a significant value of direct voltage whose polarity will depend on whether the frequency drift causes an increase or a decrease in the center-frequency value. With the use of an appropriate simple filter circuit, this voltage can be used to provide automatic frequency control.

QUESTIONS

1. Define (*a*) modulation, (*b*) demodulation, (*c*) detection.
2. (*a*) Why are the processes of modulation and demodulation important to communications systems? Which part of a communications system includes the process of (*b*) modulation? (*c*) Demodulation?
3. Define (*a*) signal wave, (*b*) carrier wave, (*c*) modulating wave, (*d*) modulated wave.
4. Describe (*a*) amplitude modulation, (*b*) frequency modulation, (*c*) phase modulation.
5. Define (*a*) amplitude-modulated wave, (*b*) modulation envelope, (*c*) per cent modulation, (*d*) sidebands, (*e*) upper sideband, (*f*) lower sideband.
6. Define (*a*) frequency-modulated wave, (*b*) center frequency, (*c*) frequency deviation, (*d*) carrier swing, (*e*) deviation ratio.
7. (*a*) Describe the sidebands present in f-m transmission. (*b*) Compare the sideband characteristics of the f-m system with those of the a-m system.
8. (*a*) Name two methods of performing amplitude modulation. (*b*) Describe briefly each method in terms of Figs. 18-5 and 18-7.
9. Describe the operation of the circuit of Fig. 18-6.
10. Describe the operation of the circuit of Fig. 18-8.
11. Describe briefly a method of frequency modulation.
12. Describe the operation of the circuit of Fig. 18-9.
13. (*a*) Explain the need for frequency-conversion circuits. Define (*b*) converter, (*c*) local oscillator, (*d*) mixer, (*e*) heterodyne action.
14. (*a*) Explain the principle of heterodyne action. (*b*) What is meant by beat frequency? (*c*) What frequencies are applied to the circuit? (*d*) What frequencies appear in the output of the circuit?
15. (*a*) Describe the principle of operation of the circuit of Fig. 18-11. (*b*) Describe the circuit operation in detail.
16. (*a*) Describe the principle of operation of the circuit of Fig. 18-12. (*b*) Describe the circuit operation in detail.
17. Describe briefly the basic principle of demodulation of (*a*) an amplitude-modulated signal, (*b*) a frequency-modulated signal. (*c*) How does a-m detection differ from f-m detection?
18. Define the following detector classifications: (a) linear, (*b*) square-law, (*c*) weak signal, (*d*) power.
19. Define (*a*) wave train, (*b*) modulation envelope. (*c*) Name three important characteristics of a detector circuit.
20. Describe the operation of the detector circuit of Fig. 18-14.
21. Describe how filtering action is accomplished at R_1C_1 in the circuit of Fig. 18-14.
22. Describe the operation of the full-wave detector of Fig. 18-16.
23. (*a*) Describe the principle of half-wave current detection. (*b*) Where is current detection used? (*c*) How does the circuit of Fig. 18-17 differ from that of Fig. 18-14?
24. (*a*) Name four advantages of diode detection. (*b*) Name three disadvantages of diode detection.

25. Describe the principle and action of the multifiltering network used in the circuit of Fig. 18-18.
26. Describe the operation of the vacuum-tube detector-amplifier circuit of Fig. 18-19.
27. Describe the principle of detection using a three-element active device.
28. Describe the operation of an a-m bias-detector circuit using a transistor.
29. Describe the function of each component in the circuit of Fig. 18-20.
30. Describe the principle of an a-m bias-detector circuit using a vacuum tube.
31. Describe the operation of the bias-detector circuit of Fig. 18-22.
32. (*a*) What is the principle of regenerative detection? (*b*) How is regeneration achieved in the detector circuit of Fig. 18-23? (*c*) What are its advantages? (*d*) What are its disadvantages?
33. (*a*) What is meant by c-w transmission? (*b*) What is its application? (*c*) Name three methods of c-w signal detection.
34. Describe (*a*) heterodyne action, (*b*) autodyne detection, (*c*) superregenerative detection.
35. (*a*) What is the purpose of automatic gain control? Define (*b*) simple agc, (*c*) delayed agc.
36. (*a*) Describe the basic principle of automatic gain control. (*b*) How is the agc voltage derived in the circuit of Fig. 18-25?
37. (*a*) What is the meaning of *delayed* in the delayed agc circuit? (*b*) Explain the operation of the basic agc circuit of Fig. 18-26.
38. Describe the operation of the transistor reverse agc circuit of Fig. 18-27.
39. Describe the operation of the transistor forward agc circuit of Fig. 18-28.
40. What are the operating characteristics of (*a*) reverse agc? (*b*) Forward agc?
41. Explain how the agc voltage is derived in the circuit of Fig. 18-29.
42. (*a*) Describe the basic principle of f-m detection. (*b*) How does it differ from a-m detection? (*c*) Describe the ideal response curve for an f-m detector.
43. Describe the principle of operation of the phase-shift discriminator detector circuit.
44. Describe the action of the phase-shift discriminator for the following signal conditions: (*a*) center-frequency signal, (*b*) signals above the center frequency, (*c*) signals below the center frequency, (*d*) an f-m signal.
45. (*a*) Describe the principle of operation of the ratio detector. (*b*) Compare the circuit configuration of the ratio detector with that of the phase-shift discriminator detector.
46. (*a*) How does the ratio detector react to variations in amplitude? (*b*) How does this affect the need for a limiter circuit?
47. Describe the action of the ratio detector for the following signal conditions: (*a*) center-frequency signal, (*b*) signals above the center frequency, (*c*) signals below the center frequency, (*d*) an f-m signal.
48. (*a*) Explain the use for an automatic-frequency-control circuit. (*b*) What is the basic principle of automatic frequency control? (*c*) Describe the operation of the afc circuit of Fig. 18-35.

PROBLEMS

1. What is the ratio of the carrier-wave frequency to the audio-signal frequency when a 1,000-Hz signal is broadcast from a 710-kHz transmitter?

2. What is the ratio of the carrier-wave frequency to the audio-signal frequency when a 1,000-Hz signal is broadcast from a 98.7-MHz transmitter?
3. A certain radio station operating on an assigned frequency of 1,050 kHz is transmitting a violin solo. (*a*) If the notes of the violin range from 200 to 8,000 vibrations per second, what is the range of the a-f current impulses? (*b*) How many cycles does the carrier-wave current make for each cycle of the lowest frequency note? (*c*) How many cycles does the carrier-wave current make for each cycle of the highest-frequency note?
4. A certain radio station operating on an assigned frequency of 105.1 MHz is transmitting a violin solo. (*a*) If the notes of the violin range from 200 to 8,000 vibrations per second, what is the range of the a-f current impulses? (*b*) How many cycles does the carrier-wave current make for each cycle of the lowest-frequency note? (*c*) How many cycles does the carrier-wave current make for each cycle of the highest-frequency note?
5. What is the per cent of modulation of an a-m wave if the maximum values of the signal and carrier waves are 4 and 5 volts, respectively?
6. What is the per cent of modulation of an a-m wave if the maximum values of the signal and carrier waves are 25 and 30 volts, respectively?
7. What maximum value of signal-wave voltage is permissible if the maximum carrier-wave voltage is 7.5 volts and the modulation is not to exceed 75 per cent?
8. What maximum value of carrier-wave voltage is permissible if the maximum signal-wave voltage is 7.5 volts and the modulation is not to exceed 75 per cent?
9. If a 710-kHz carrier wave is amplitude-modulated by audio signals varying between 50 and 7,000 Hz, what is (*a*) the frequency span of the sidebands? (*b*) The maximum upper side frequency? (*c*) The minimum lower side frequency? (*d*) The frequency range of the channel?
10. If a 1,650-kHz carrier wave is amplitude-modulated by audio signals varying between 70 and 6,000 Hz, what is (*a*) the frequency span of the sidebands? (*b*) The maximum upper side frequency? (*c*) The minimum lower side frequency? (*d*) The frequency range of the channel?
11. What is the carrier-frequency swing of an f-m signal when the frequency deviation is (*a*) 40 kHz? (*b*) 60 kHz?
12. What is the carrier-frequency swing of an f-m signal when the frequency deviation is (*a*) 35 kHz? (*b*) 70 kHz?
13. If the deviation ratio of an f-m signal is 5, what are the corresponding maximum a-f signals in Prob. 11?
14. If the deviation ratio of an f-m signal is 5, what are the corresponding maximum a-f signals in Prob. 12?
15. An f-m transmitter operates with a maximum frequency deviation of 72 kHz and reproduces audio signals up to 12,000 Hz. (*a*) What is the deviation ratio? (*b*) What minimum channel width is required?
16. An f-m transmitter operates with a maximum carrier swing of 140 kHz and reproduces audio signals up to 14 kHz. (*a*) What is the deviation ratio? (*b*) What minimum channel width is required?
17. A 6-Hz sine-wave voltage whose maximum value is 4 volts is combined by heterodyne action with a 9-Hz sine-wave voltage whose maximum value is 8 volts. (*a*) What are the two beat frequencies? (*b*) What is the voltage swing of the beat frequencies?

18. A 710-kHz, 50-mv (peak) sine-wave voltage is combined by hererodyne action with a 1,165-kHz, 100-mv (peak) sine-wave voltage. (*a*) What are the two beat frequencies? (*b*) What is the voltage swing of the beat frequencies?

19. The frequency of the input signal to the circuit of Fig. 18-11 is 1,140 kHz, and it is desired that the output-signal frequency shall be 465 kHz. (*a*) What are the possible frequency values for the oscillator? (*b*) What frequencies are present in the output of Q_2?

20. The frequency of the input signal to the circuit of Fig. 18-11 is 105.1 MHz, and it is desired that the output-signal frequency shall be 10.7 MHz. (*a*) What are the possible frequency values for the oscillator? (*b*) What frequencies are present in the output of Q_2?

21. A certain diode-detector circuit, similar to Fig. 18-14, uses a 0.5-megohm resistor for R_1 and a 100-pf capacitor for C_1. (*a*) What is the time constant of this RC circuit? (*b*) If the resonant frequency of the tuned circuit L_2C_2 is 1,500 kHz, how much time is required for the r-f wave to complete 1 cycle? (*c*) How many times greater is the time constant of R_1C_1 than the time of 1 cycle of the r-f wave?

22. The circuit of Prob. 21 also acts as a filter circuit. (*a*) What is the reactance of C_1 to a 1,500-kHz r-f current? (*b*) How does this reactance compare with the resistance of R_1? (*c*) What path will the r-f currents take? (*d*) What is the reactance of C_1 to a 500-Hz a-f current? (*e*) How does the reactance of C_1 to the 500-Hz current compare with the resistance of R_1? (*f*) What path will the a-f currents take?

23. A certain diode-detector circuit, similar to Fig. 18-14, uses a 300,000-ohm resistor for R_1 and a 250-pf capacitor for C_1. (*a*) What is the time constant of this RC circuit? (*b*) If the resonant frequency of the tuned circuit L_2C_2 is 500 kHz, how much time is required for the r-f wave to complete 1 cycle? (*c*) How many times greater is the time constant of R_1C_1 than the time of 1 cycle of the r-f wave?

24. The circuit of Prob 23 also acts as a filter circuit. (*a*) What is the reactance of C_1 to a 500-kHz r-f current? (*b*) How does this reactance compare with the resistance of R_1? (*c*) What path will the r-f currents take? (*d*) What is the reactance of C_1 to a 500-Hz a-f current? (*e*) How does the reactance of C_1 to the 500-Hz current compare with the resistance of R_1? (*f*) What path will the a-f currents take?

25. The current-detector circuit of Fig. 18-17 is tuned to a 455-kHz input signal that is amplitude-modulated by an a-f signal having a maximum frequency of 7,000 Hz; the circuit has a 10,000-ohm resistor (R_1) and a 175-mh inductor (L_3). What is (*a*) the time constant of R_1L_3? (*b*) The time required for 1 cycle of the input signal? (*c*) The time required for 1 cycle of the highest-frequency a-f signal? (*d*) The ratio of the R_1L_3 time constant to the time of 1 cycle of the input signal? (*e*) The ratio of the time of 1 cycle of the a-f signal to the R_1L_3 time constant? (*f*) The effect of the variations of the individual cycles of the input signal on the current in the R_1L_3 path? (*g*) The effect of the individual cycles of the a-f signals on the current in the R_1L_3 path?

26. The current-detector circuit of Fig. 18-17 is tuned to a 465-kHz input signal that is amplitude-modulated by an a-f signal having a maximum frequency of 5,000 Hz. What is the amount of time required for 1 cycle of (*a*) the input signal? (*b*) The highest a-f signal? (*c*) When $R_1 = 20{,}000$ ohms, what value of inductance (L_3) is required if the minimum ratio of 10 to 1 is desired for (1) the time constant of R_1L_3 to the time for 1 cycle of the input signal, and (2) the time for 1 cycle of the highest a-f signal to the time constant of R_1L_3?

27. The circuit elements in Fig. 18-18 have the following values: R_1 = 100,000 ohms, R_2 = 400,000 ohms, C_1, C_2 = 100 pf. (*a*) What impedance does C_1 offer to a 465-kHz current? (*b*) Which path will the 465-kHz current take? (*c*) Will any 465-kHz current flow into the R_1 path? (*d*) What impedance does C_2 offer to any 465-kHz current? (*e*) What purpose does C_2 serve? (*f*) Neglecting the effects of C_1 and C_2, what per cent of the a-f voltage developed across R_1 and R_2 is available at the output terminals?

28. The circuit elements in Fig. 18-18 have the following values: R_1 = 50,000 ohms, R_2 = 200,000 ohms, C_1, C_2 = 250 pf. (*a*) What impedance does C_1 offer to a 455-kHz current? (*b*) Which path will the 455-kHz current take? (*c*) Will any 455-kHz current flow into the R_1 path? (*d*) What impedance does C_2 offer to any 455-kHz current? (*e*) What purpose does C_2 serve? (*f*) Neglecting the effects of C_1 and C_2, what per cent of the a-f voltage developed across R_1 and R_2 is available at the output terminals?

29. The r-f filter circuit of Fig. 18-21*a* has a 100-pf capacitor at C_3 and a 30-mh choke at L_3. (*a*) What impedance does the capacitor offer to a 1,500-kHz r-f current? (*b*) What impedance does the choke offer to a 1,500-kHz r-f current? (*c*) What impedance does the capacitor offer to a 500-Hz a-f current? (*d*) What impedance does the choke offer to a 500-Hz a-f current? (*e*) Which path will the r-f currents take? (*f*) Which path will the a-f currents take?

30. The r-f filter circuit of Fig. 18-21*a* has a 500-pf capacitor at C_3 and a 125-mh choke at L_3. (*a*) What impedance does the capacitor offer to a 455-kHz i-f current? (*b*) What impedance does the choke offer to a 455-kHz i-f current? (*c*) What impedance does the capacitor offer to a 500-Hz a-f current? (*d*) What impedance does the choke offer to a 500-Hz a-f current? (*e*) Which path will the i-f currents take? (*f*) Which path will the a-f currents take?

31. A certain tube is operated with a grid bias of 3 volts. What is the grid-voltage swing when the a-c signal applied to the input circuit is (*a*) 1 volt? (*b*) 3 volts? (*c*) 4 volts?

32. A certain tube is operated with a grid bias of 6 volts. What is the grid-voltage swing when the a-c signal applied to the input circuit is (*a*) 2 volts? (*b*) 4 volts? (*c*) 6 volts?

33. A certain triode tube used as a bias detector is operated with 250 volts on its plate, and the recommended grid bias is approximately 17 volts. What is the highest signal voltage that can be applied to the input circuit without causing distortion?

34. A certain pentode tube used as a bias detector is operated with 250 volts on its plate, 100 volts on the screen grid, and a grid bias of 4.3 volts. What is the highest signal voltage that can be applied to the input circuit without causing distortion?

35. A certain triode used as a bias detector is operated with a grid bias of 12 volts. What value of cathode resistance is required if with zero input signal the plate current is (*a*) 0.1 ma? (*b*) 0.25 ma? (*c*) 0.4 ma?

36. A certain triode used as a bias detector is operated with a grid bias of 14 volts. What value of cathode resistance is required if with zero input signal the plate current is (*a*) 0.1 ma? (*b*) 0.2 ma? (*c*) 0.35 ma?

37. It is recommended that the plate current of the detector of Prob. 33 be adjusted to 0.2 ma when the input signal is zero. (*a*) What value of cathode resistor is necessary to provide the recommended grid-bias voltage? (*b*) How much power is consumed by this resistor? (*c*) What power rating should the resistor have?

38. It is recommended that the cathode current of the detector of Prob. 34 be adjusted to 0.43 ma when the input signal is zero. (*a*) What value of cathode resistor is necessary to provide the recommended grid-bias voltage? (*b*) How much power is consumed by this resistor? (*c*) What power rating should the resistor have?

39. The detector of Prob. 37 uses a cathode resistor of 85,000 ohms. (*a*) What value of bypass capacitor should be used with this resistor if it is desired that the resistor offer at least 100 times more impedance to a 500-Hz a-f current than the capacitor? (*b*) What standard rating and type of capacitor is recommended?

40. The detector of Prob. 38 uses a cathode resistor of 10,000 ohms. (*a*) What value of bypass capacitor should be used with this resistor if it is desired that the resistor offer at least 100 times more impedance to a 500-Hz a-f current than the capacitor? (*b*) What standard rating and type of capacitor is recommended?

41. The operating characteristics of a certain detector are provided in Probs. 33, 37, and 39. If a 250,000-ohm load resistor is connected in its plate circuit, how much voltage must the B power supply provide in order to maintain 250 volts between the cathode and plate when the plate current is 0.2 ma?

42. The operating characteristics of a certain detector are provided in Probs. 34, 38, and 40. If a 220,000-ohm load resistor is connected in its plate circuit, how much voltage must the B power supply provide in order to maintain 250 volts between the cathode and plate when the cathode current is 0.43 ma and the plate current is 0.35 ma?

43. A c-w signal transmitted on a frequency of 2,000 kHz is being received by an autodyne detector. At what frequency must the oscillator circuit of the receiver be set in order that the frequency in the output circuit will be (*a*) 500? (*b*) 1,000? (*c*) 1,500 Hz?

44. A c-w signal transmitted on a frequency of 2,500 kHz is being received by an autodyne detector. At what frequency must the oscillator circuit of the receiver be set in order that the frequency in the output circuit will be (*a*) 400? (*b*) 800? (*c*) 1,600 Hz?

45. The agc filter circuit shown in Fig. 18-25 has the following constants: $R_2 = 1$ megohm, $C_2 = 0.1$ μf. (*a*) What is the time constant of this circuit? (*b*) What per cent of the maximum voltage charge possible will be attained in the time that the current from 1 cycle of a 100-Hz a-f signal will flow in the detector circuit (assuming that a constant voltage was being applied)? (*c*) How many cycles of the 100-Hz signal would be completed in one time constant?

46. The agc filter circuit shown in Fig. 18-25 has the following constants: $R_2 = 2$ megohms, $C_2 = 0.05$ μf. (*a*) What is the time constant of this circuit? (*b*) What per cent of the maximum voltage charge possible will be attained in the time that the current from 1 cycle of a 50-Hz a-f signal will flow in the detector circuit (assuming that a constant voltage was being applied)? (*c*) How many cycles of the 50-Hz signal would be completed in one time constant?

47. In the agc filter of Fig. 18-27, $R_5 = 12{,}000$ ohms, and $C_5 = 10$ μf. (*a*) What is the time constant of this filter? (*b*) What is the time required for 1 cycle of a 50-Hz a-f signal? (*c*) What is the ratio of the filter-circuit time constant to the time of 1 cycle of the 50-Hz signal? (*d*) Will the a-f signal affect the voltage at the base of Q_1 and why? (*e*) Using a ratio of 10 to 1 for RC and the time between changes in the average strength of the input signals, what is the minimum fading interval of the input signal for which the agc circuit will provide control?

48. In the agc filter of Fig. 18-27, $R_5 = 12{,}000$ ohms, and $C_5 = 0.1$ μf. (*a*) What is the time constant of this filter? (*b*) What is the time required for 1 cycle of a 50-Hz a-f signal? (*c*) What is the ratio of the filter-circuit time constant to the time of 1 cycle of the 50-Hz signal? (*d*) Will the a-f signal affect the voltage at the base of Q_1 and why? (*e*) Using a ratio of 10 to 1 for RC and the time between changes in the average strength of the input signals, what is the minimum fading interval of the input signal for which the agc circuit will provide control?

49. The delayed agc circuit of Fig. 18-29 has the following constants: $R_1 = 0.5$ megohm, $R_2 = 1$ megohm, $R_3 = 2$ megohms, $R_4 = 2{,}500$ ohms, $C_1 = 250$ pf, $C_3 = 0.1\ \mu$f, $C_4 = 10\ \mu$f. When the signal input is zero, the cathode current is 1.2 ma. (*a*) What are the magnitude and polarity of the voltage between D_2 and the cathode with zero input signal? (*b*) How much current flows in R_3 under the condition in part (*a*)? (*c*) What is the voltage of the agc line under the condition of part (*a*)? (*d*) To what voltage must the charge on C_5 raise the plate D_2 in order to produce a current flow in R_3?

50. For the circuit of Prob. 49: (*a*) What are the magnitude and polarity of the voltage developed at point A when 1 μa flows through R_3? (*b*) What voltage will the agc line have under the condition in part (*a*)? (*c*) What is the time constant of the agc filter? (*d*) What is the time constant of the diode load resistor R_1 and the diode bypass capacitor C_1? (*e*) What impedance does C_4 offer to the lowest a-f current of 50 Hz? (*f*) What purpose does C_4 serve?

51. In the f-m detector circuit of Fig. 18-32, capacitors C_2 and C_3 are each 33 pf. What is the impedance of these capacitors to signals of (*a*) 10.7 MHz? (*b*) 7,000 Hz?

52. In the f-m detector circuit of Fig. 18-32, capacitors C_2 and C_3 are each 100 pf. What is the impedance of these capacitors to signals of (*a*) 41.25 MHz? (*b*) 5,000 Hz?

53. The voltages of an f-m detector represented by Fig. 18-33 are $e_i = e_{s\cdot 1} = e_{s\cdot 2} =$ 10 volts. If the input-signal modulations cause $e_{s\cdot 2}$ to lag e_i by 120°: (*a*) what is the phase angle between $e_{s\cdot 1}$ and e_i? What is the magnitude of (*b*) e_1? (*c*) e_2? (*d*) e_o of Fig. 18-32?

54. The voltages of an f-m detector represented by Fig. 18-33 are $e_i = e_{s\cdot 1} = e_{s\cdot 2} =$ 10 volts. If the input-signal modulations cause $e_{s\cdot 1}$ to lead e_i by 120°: (*a*) What is the phase angle between $e_{s\cdot 2}$ and e_i? What is the magnitude of (*b*) e_1? (*c*) e_2? (*d*) e_o of Fig. 18-32?

Chapter 19

Integrated Circuits

The ever-increasing demands of the communications, computer, and aerospace industries for electronic equipment of smaller size, lighter weight, low power requirement, high degree of reliability, and the ability to operate at extreme values of temperature have led to the development of various types of microminiaturized circuits. One type of circuit, called an *integrated circuit* and generally abbreviated IC, is actually a combination of components such as transistors, diodes, resistors, capacitors, and their interconnections, formed into one or more complete circuits designed to perform one or more specific functions. The components of an integrated circuit are produced on a small common base which is generally mounted in one of three standard packages; one type of IC package resembles the familiar small-signal transistor in both size and shape. Since this method of fabrication combines both active and passive components in a monolithic structure, the complete unit is called an *integrated circuit.*

The fabrication of integrated circuits involves (1) the knowledge of photographic thin-film technology, (2) slicing of very thin wafers of silicon crystals, and (3) use of materials such as tantalum, nichrome, and various oxides. The detailed explanation of the fabrication of integrated circuits is therefore beyond the scope of this text. However, regardless of the method used to fabricate an active or passive circuit element, the basic characteristics and circuit operation of an IC are the same as for any of their counterparts in a similar circuit using separate circuit elements as described throughout this text.

Integrated circuits may be fabricated to produce (1) a basic circuit operation such as rectification, amplification, oscillation, etc., or (2) a complete circuit to perform a desired function. By means of sophisticated manufacturing processes, ICs are mass-produced as components for (1) computers, (2) radio and television receivers, (3) communications equipment, (4) hearing aids, (5) aerospace equipment, (6) etc.

19-1 IC Classifications

Construction. Four basic types of construction are employed in the manufacture of integrated circuits, namely, (1) monolithic, (2) thin-film, (3) thick-film, and (4) hybrid.

In the *monolithic IC,* all of the circuit components and their interconnections are formed on a single thin wafer called the *substrate*. This type of construction is the most common and is the only one considered in this text.

In the *thin-film IC*, appropriate materials for forming resistors and/or capacitors are deposited in thin layers of approximately one mil (0.001 inch) on the substrate material. This type of construction usually does not include transistors among its components.

In the *thick-film IC*, fabrication is similar to the thin-film method. However, thick-film fabrication is not used extensively.

In the *hybrid IC*, features of both the monolithic and thin-film construction are used.

Mode of Operation. Integrated circuits are classified according to their mode of operation as (1) linear, or (2) digital or nonlinear.

With linear ICs, the output of the circuit is proportional to the input, and hence the output varies linearly with the input. Among the applications of linear ICs are (1) a-f, i-f, and r-f amplifiers, (2) video amplifiers, (3) d-c amplifier, (4) differential amplifier, (5) Darlington-pair amplifier, (6) operational amplifier, (7) diode and transistor arrays, (8) modulator, and (9) voltage regulator.

With digital ICs, the output is nonlinear and hence nonsinusoidal. This type of IC frequently involves some form of bistable action. Digital ICs are common in computer circuits and are used in such applications as (1) data distributor, (2) data selector, (3) decoder, (4) counter, (5) logic elements, (6) memory cells, (7) adder, (8) subtractor, (9) gate, (10) frequency divider, (11) translator, (12) inverter.

Since the principles of ICs can be shown with either linear or digital ICs, only linear ICs are discussed in this text.

19-2 Construction of Monolithic ICs

Basic Principles. The base or foundation, called the *substrate*, of the monolithic IC is generally a wafer (also called a *chip*) of P-type silicon. By the process of diffusion, appropriate materials are added to the substrate to produce diodes, transistors, resistors, and capacitors. In IC construction, *diffusion* is the technique of deliberately adding controlled minuscule amounts of impurities at specific locations on the substrate by thermal processes. Resistors are formed by adding nichrome or tin oxide, capacitors are formed by adding tantalum, and diodes and transistors are formed by adding N-type and P-type semiconductor materials.

Diodes. One or more diodes are formed by diffusing one or more small N-type deposits at appropriate locations on the substrate. A thin-film insulating oxide layer is deposited over the entire surface to provide (1) protection to the IC, and (2) a means of attaching terminals. The terminals are processed by (1) etching the oxide layer at the desired locations, and (2) depositing the metal terminal material on top of the oxide layer at the specific locations. Figure 19-1*a* illustrates a two-diode IC with a common cathode.

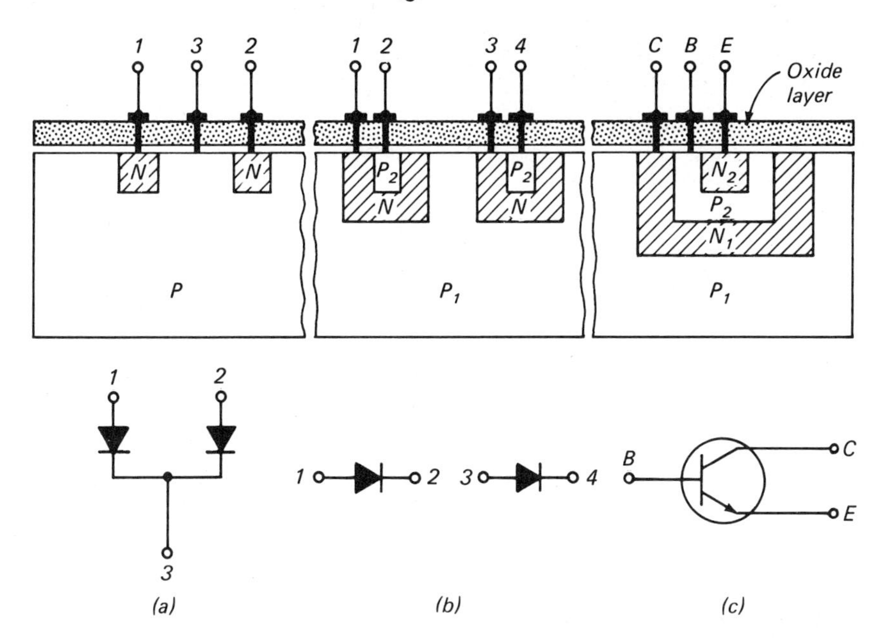

Fig. 19-1 Fabrication of active-type components on a monolithic integrated circuit. (a) Two diodes with a common cathode. (b) Two diodes with individual cathodes. (c) Bipolar transistor.

Figure 19-1*b* shows a two-diode IC with individual cathodes; this construction provides isolation of the diodes from other components of the IC.

Transistors. Transistors are formed by using the same principle as for diodes. Figure 19-1*c* shows how a transistor is formed on a portion of the substrate of a monolithic IC. The procedures of construction are: (1) the large N-type area N_1 is diffused into the P-type substrate P_1 and becomes the collector; (2) a P-type area P_2 is diffused into the collector area N_1 and becomes the base; (3) a small N-type area N_2 is diffused into the base P_2 and becomes the emitter; (4) the oxide layer is applied; (5) the oxide layer is etched for locating the terminals; and (6) the terminals are added. The entire transistor occupies only a very small fraction of an inch as must be apparent when it is possible to accommodate 20 or more transistors and a number of resistors and capacitors on a chip approximately 40 mils square. One commercial integrated circuit contains 524 different components on a chip 50 by 70 mils.

Although the transistor represented in Fig. 19-1*c* is the common bipolar type, additional construction techniques have been developed for the fabrication also of FET- and MOSFET-type transistors.

Resistors. The processes in the construction of a resistor in the monolithic IC (Fig. 19-2*a*) are: (1) an N-type area is diffused into the P-type substrate; (2) a P-type material P_2 (nichrome or tin oxide) is diffused into the N-type area; (3) an oxide layer is applied; (4) the oxide layer is etched; and (5) the terminals are added. Diffusing the resistance material P_2 into the N-type area isolates the resistor from the substrate and any other components of the IC. The value of the resistor is determined by (1) the material, and (2) its length and cross-section area. The high-resistance resistors are long and narrow, and low-resistance resistors are short and of greater cross-section area.

Capacitors. The processes in the construction of a capacitor in the monolithic IC (Fig. 19-2*b*) are: (1) an N-type area is diffused into the substrate and forms one plate of the capacitor; (2) an oxide layer is applied and becomes the dielectric of the capacitor; (3) the oxide layer is etched and terminal 1 is added; (4) a large (compared to the electrode at terminal 1) metallic electrode is deposited on the oxide layer opposite the capacitor plate formed by the N-type material and becomes the second plate of the capacitor; and (5) terminal 2 is added. The value of the capacitor is determined by (1) the dielectric constant of the oxide layer, (2) the thickness of the dielectric, and (3) the area of the smaller of the two plates.

Basic Composite Circuit. Figure 19-3 illustrates a hypothetical monolithic integrated circuit consisting of a capacitor, resistor, diode, and transistor connected in series. Actual ICs contain a larger number of components as indicated by Fig. 19-14.

Examination of Figs. 19-1 to 19-3 will show that the IC chip of Fig. 19-3*a* contains one of each type of element shown separately in Figs. 19-1 and 19-2. The etching process must provide minute openings in the oxide layer at points 1, 3, 4, 5, 6, 7, 8, and 9. No etching is done at point 2 in order that the oxide layer under the area surrounding 2 may act as the dielectric separating the two plates of the capacitor. The interconnection of the circuit elements is accomplished by extending the metallic deposits from terminal to terminal of

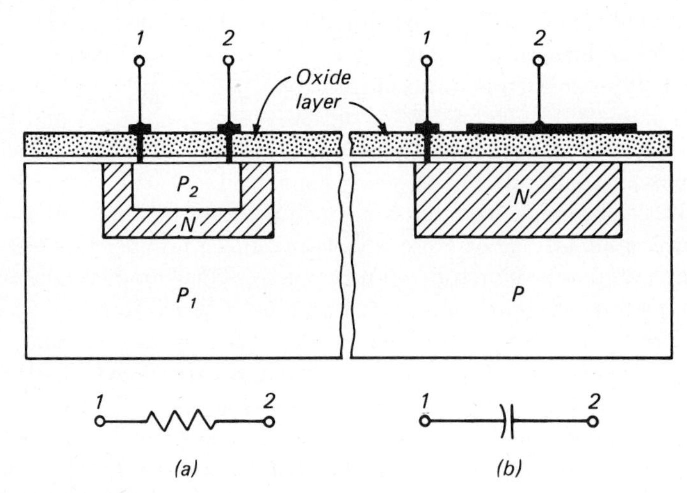

Fig. 19-2 Fabrication of passive-type components on a monolithic integrated circuit. (a) Resistor. (b) Capacitor.

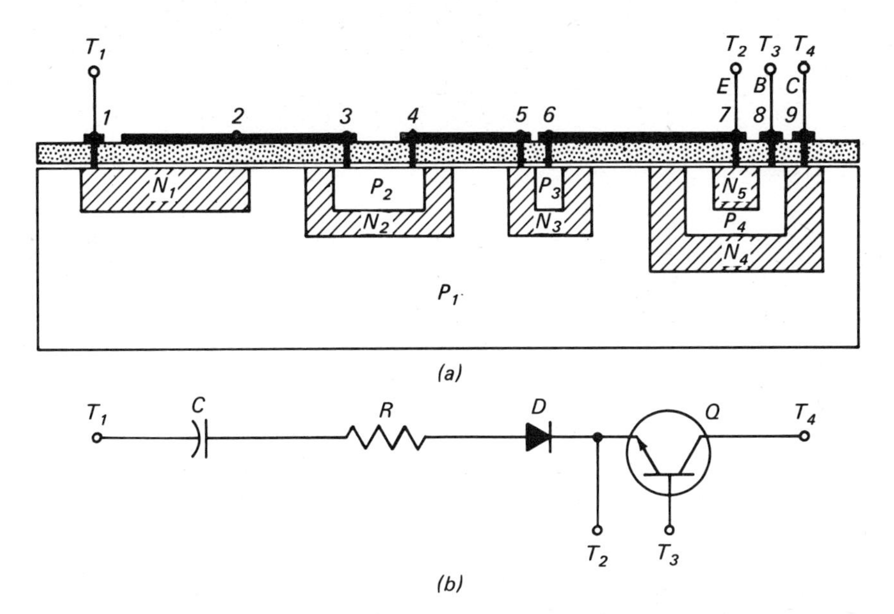

Fig. 19-3 Simple monolithic integrated circuit. (a) Construction. (b) Electric circuit.

adjacent components, as is indicated on Fig. 19-3*a* from 2 to 3, 4 to 5, and 6 to 7.

Large-scale Integration. Combining a number of separate integrated-circuit units, discrete components, and their associated wiring into a single wafer is referred to as *large-scale integration*, commonly abbreviated as LSI.

19-3 IC Packaging

Three methods commonly used for mounting integrated circuits in containers are (1) TO-5 glass-metal can-type unit, (2) ceramic flat pack, and (3) dual-in-line ceramic or plastic flat pack (see Fig. 19-4).

The TO-5 container resembles the common small-signal transistor in both appearance and size but differs in that it has either 8, 10, or 12 pigtail-type leads. The lead numbers are arranged with the highest numbered lead located at the centering notch (Fig. 19-5*a*). The flat-pack containers with 14 leads, seven on each side, with an index or reference notch (Fig. 19-5*b* and *c*).

19-4 Symbols

Linear ICs. Figure 19-6 illustrates the simplicity of the symbols used with integrated circuits. Figure 19-6*a* represents an integrated-circuit r-f amplifier containing 3 transistors, 3 resistors, and 8 terminals. Figure 19-6*b* represents an integrated-circuit audio amplifier which contains 6 transistors, 2 diodes, and 17 resistors, and has 12 terminals.

No set pattern is used for locating the terminals on the IC symbol; instead,

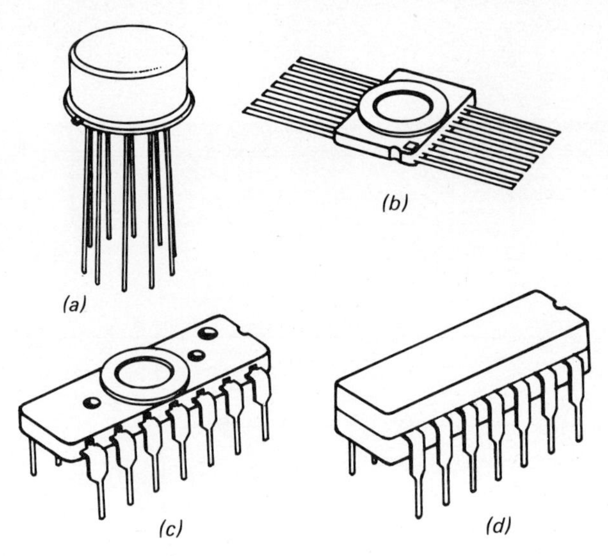

Fig. 19-4 Integrated-circuit packages. (a) TO-5 with 10 terminals. (b) Ceramic flat pack. (c) Dual-in-line ceramic flat pack. (d) Dual-in-line plastic flat pack.

the terminal numbers are placed to provide the most convenient arrangement for each circuit diagram when the external components are included. Any need to study the internal connections of an IC is generally fulfilled by reference to the technical manual of the IC manufacturer.

Digital ICs. The symbols for digital ICs are shown in Fig. 19-7 and indicate the function as well as the terminal numbers.

19-5 Technical Data

Design fundamentals, applications, and technical data are generally provided in the technical data manuals issued by the manufacturers. Among the

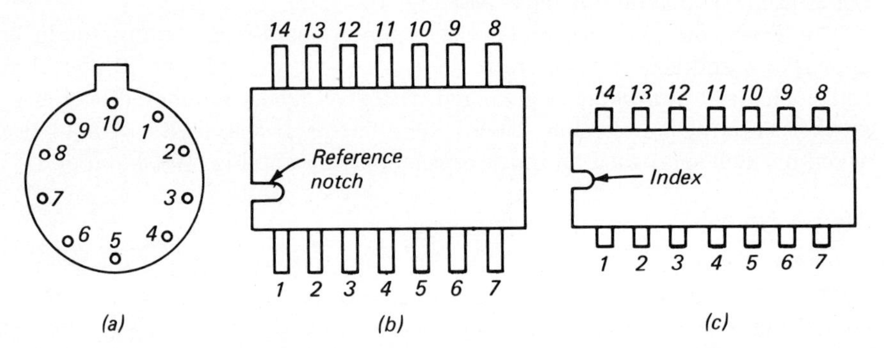

Fig. 19-5 Terminal markings for linear integrated-circuit units. (a) Bottom view of 10-terminal TO-5 package. (b) Top view of flat pack. (c) Top view of dual-in-line package.

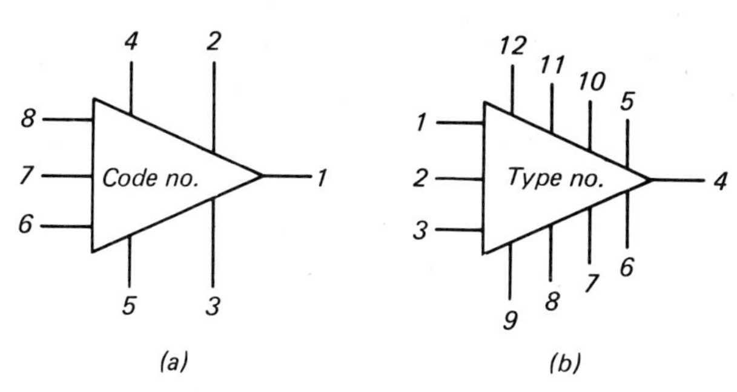

Fig. 19-6 Symbols for linear integrated circuits with (a) 8 terminals, (b) 12 terminals.

technical data provided are (1) circuit diagrams, (2) applications, (3) packaging, (4) maximum current, voltage, power, and temperature ratings, (5) typical operating characteristics, (6) frequency ratings, (7) input and output impedances, (8) voltage and/or power gain.

19-6 Commercial ICs

Description. One of the simplest integrated-circuit units is the RCA type CA-3028A r-f amplifier; it is housed in a TO-5 package and has 8 leads. This unit consists of 3 transistors and 3 resistors connected to form a differential amplifier and includes a constant-current sink (Art. 15-15). The locations at which terminals are provided permit connecting the unit to be operated as a (1) differential amplifier, (2) cascode amplifier, or (3) single-stage amplifier. By use of proper terminals and appropriate external circuit components other functions can be performed such as (1) wide- or narrowband amplification, (2) mixing, (3) limiting, (4) frequency generation, (5) pulse and digital waveform generation.

As a balanced differential amplifier, the CA-3028A can produce a voltage

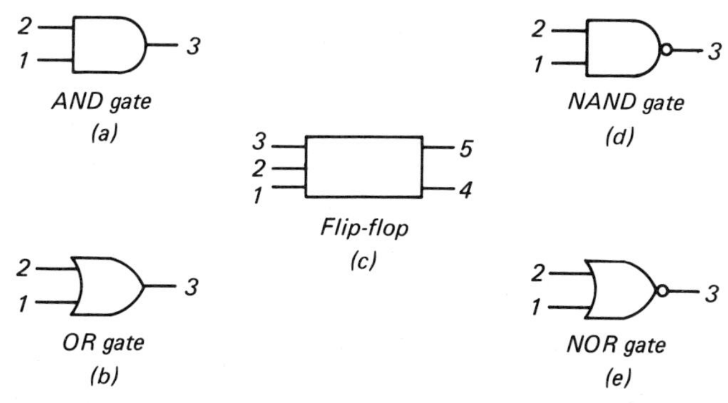

Fig. 19-7 Symbols for digital integrated circuits.

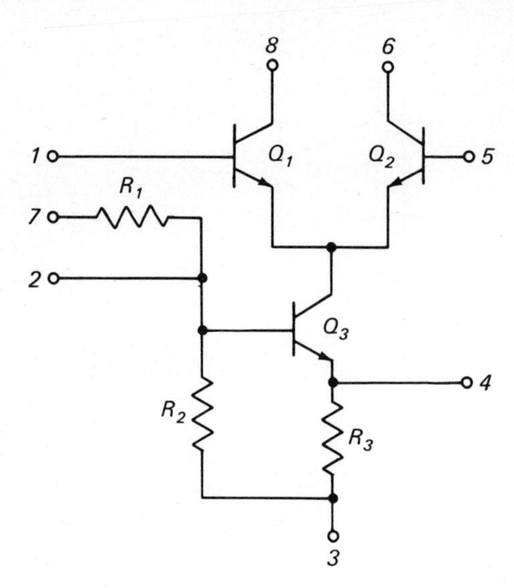

Fig. 19-8 Schematic diagram of RCA type CA-3028A integrated-circuit r-f amplifier.

gain of 32 db at 10.7 MHz or a power gain of 32 db. As a cascode amplifier, the maximum voltage and power gains are 98 and 39 db, respectively.

The Circuit. Diagrams of the internal circuit of ICs are not generally shown in practical applications because use of the simplified IC symbol provides sufficient information for construction and servicing the circuits using the ICs. The complete circuit diagram can be obtained from manufacturers' manuals. Figure 19-8 shows the schematic diagram of the CA-3028A integrated circuit.

Differential Amplifier. Figure 19-9*a* illustrates the application of the CA-3028A IC to form a complete r-f amplifier stage. This figure shows the addition of (1) tuned circuits T_1 and T_2 at the input and output of the stage, (2) a voltage divider R_4R_5, and (3) a bypass capacitor C_3 at terminal 5. Figure 19-9*b* is a detailed diagram showing (1) the components and connections of the integrated circuit, and (2) the external components indicated in Fig. 19-9*a*.

Circuit Operation. The desired input signal v_i is selected by the tuned circuit consisting of C_1 and the primary winding of T_1 and is coupled to the input terminals 1 and 5 of the integrated circuit by the input transformer T_1. Transistors Q_1 and Q_2 are connected for operation as a differential amplifier and amplify the input signal. The output signal is coupled to the load by the transformer T_2 whose primary circuit is tuned by C_2. Transistor Q_3 provides a constant-current source for the emitters of Q_1 and Q_2. Resistors R_1 and R_2 form a voltage divider to provide the forward bias for Q_3. Resistors R_4 and R_5 form a voltage divider to provide fixed bias voltage for Q_2. The unbypassed resistor R_3 in the emitter circuit of Q_3 introduces an amount of degeneration sufficient to improve the linearity of the circuit and increase its signal-handling capability.

Cascode Amplifier. Figure 19-10*a* illustrates a conventional circuit diagram for using a CA-3028A integrated circuit as a cascode amplifier. A detailed circuit diagram is shown in Fig. 19-10*b*.

The input signal is coupled by T_1 and C_3 to the base of Q_3. The secondary of T_1 is tuned to the desired frequency by C_1. Resistors R_1 and R_2 establish the forward bias for Q_3. Resistor R_3 stabilizes the emitter current of Q_3, and capacitor C_4 prevents degeneration due to the emitter resistor R_3. The output of Q_3 is directly coupled to Q_2, thus operating Q_3 and Q_2 as a cascode amplifier. The output of Q_2 is coupled to the load by T_2 whose primary circuit is tuned by C_2.

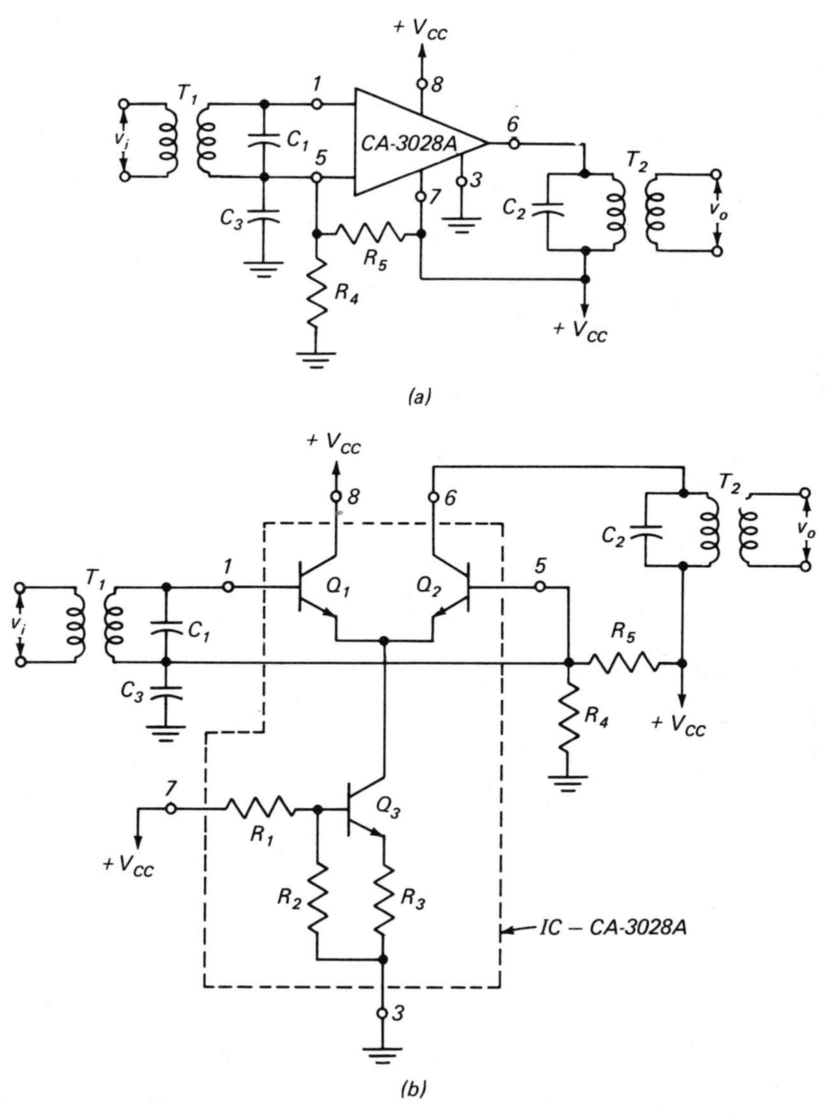

Fig. 19-9 Use of a type CA-3028A integrated circuit in an r-f amplifier stage. (a) Conventional diagram. (b) Detailed diagram.

Resistors R_4 and R_5 form a voltage divider to provide the forward bias for Q_2. Transistor Q_1 is not used in this application and is disabled by connecting together terminals 1 and 8.

Oscillator. Figure 19-11 illustrates the conventional circuit diagram for an oscillator using the CA-3028A integrated circuit. After having presented detailed circuit diagrams in Figs. 19-9*b* and 19-10*b*, it is assumed that the reader

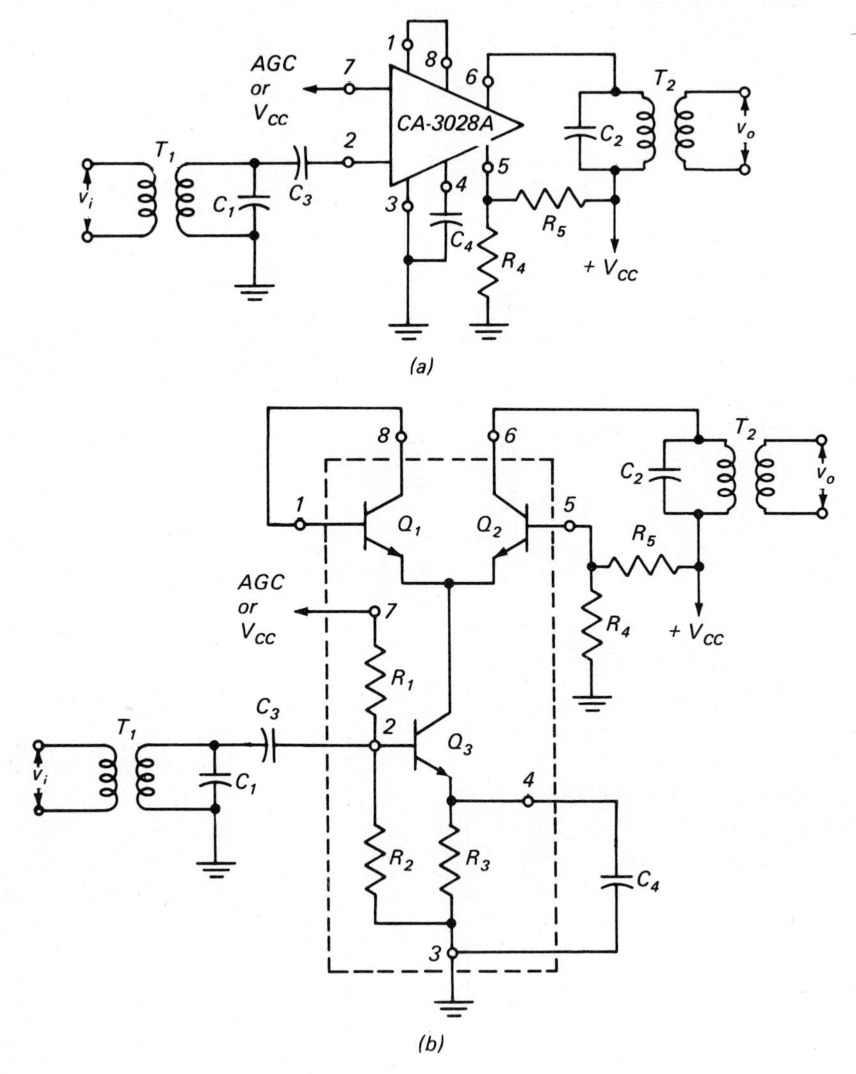

Fig. 19-10 Use of type CA-3028A integrated circuit as a cascode amplifier. (a) Conventional diagram. (b) Detailed diagram.

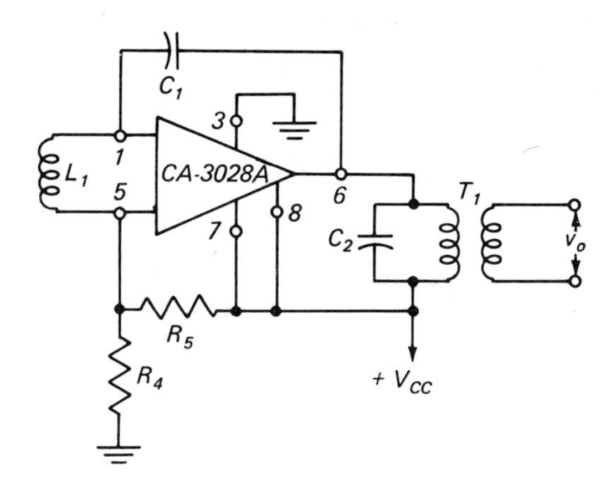

Fig. 19-11 Use of type CA-3028A integrated circuit as an oscillator.

should be able to visualize the detailed circuit diagram for Fig. 19-11 by coordinating the circuits of Figs. 19-8 and 19-11.

With terminal 7 connected to V_{CC}, the voltage divider R_1R_2 establishes the forward bias for Q_3. As the initial collector current in Q_3 increases, it is amplified by Q_1Q_2 and an output voltage becomes available at T_1. Capacitor C_1 couples some of the output signal to the input of Q_1, causing regeneration, and produces oscillations. Resistors R_4 and R_5 establish forward bias for Q_2. The unbypassed resistor R_3 in the emitter circuit of Q_3 introduces sufficient degeneration to add stability to the oscillator.

This circuit can readily be adapted to other oscillator types such as (1) transformer feedback, (2) *RC* phase shift, and (3) Wien bridge.

19-7 Complex ICs

The type CA-3028A integrated circuit was used in Art. 19-6 because its simple circuit provided an excellent means of illustrating the applications of ICs. Most integrated circuits have more complex circuitry.

The RCA type CA-3000 (Fig. 19-12) illustrates a more complex IC circuitry. This unit contains 5 transistors, 2 diodes, and 11 resistors, and is supplied in a TO-5 package with 10 terminals. The circuit is basically a single-stage differential amplifier using transistors Q_2 and Q_4, and a constant-current sink Q_3. The emitter resistors R_4 and R_5 provide sufficient degeneration to improve the linearity and stability of the circuit. (Note the similarity of this circuit to that described in Art. 15-15.) The input to the differential amplifier is through two emitter-follower amplifiers Q_1 and Q_5.

Terminals 3, 4, and 5 provide the means of changing the mode of operation by changing the base and emitter bias conditions of Q_3. For example, (1) connecting terminals 3 and 5 together cuts D_1, D_2, and R_{11} out of the circuit; (2) connecting terminals 3 and 4 together cuts R_{10} out of the circuit and decreases the emitter bias resistance; and (3) connecting terminals 3, 4, and 5 together cuts D_1, D_2, R_{10}, and R_{11} out of the circuit. Each mode of operation produces different circuit characteristics as described in the manufacturer's technical manual.

Applications. Among the applications of the type CA-3000 integrated circuit are: (1) amplifier for d-c to 1 MHz, (2) higher frequencies by use of external resistors or coils, (3) a-f amplifier, (4) narrowband i-f or r-f amplifier, (5) oscillator, (6) mixer, (7) modulated oscillator, (8) RC-coupled feedback amplifier.

Narrowband Amplifier. Figure 19-13*a* shows a conventional circuit diagram for operating the type CA-3000 integrated circuit as a 10-MHz narrowband amplifier. The mode of operation used leaves terminals 4 and 5 open. The external circuit provides a tuned input by means of L_1C_1 and a tuned output by means of L_2C_2. With the parameters indicated on Fig. 19-13*a*, the circuit has an input Q of 26, an output Q of 25, and a total effective Q of 37; the voltage gain at 10 MHz is 29.6 db. The response curve of the amplifier is shown in Fig. 19-13*b*.

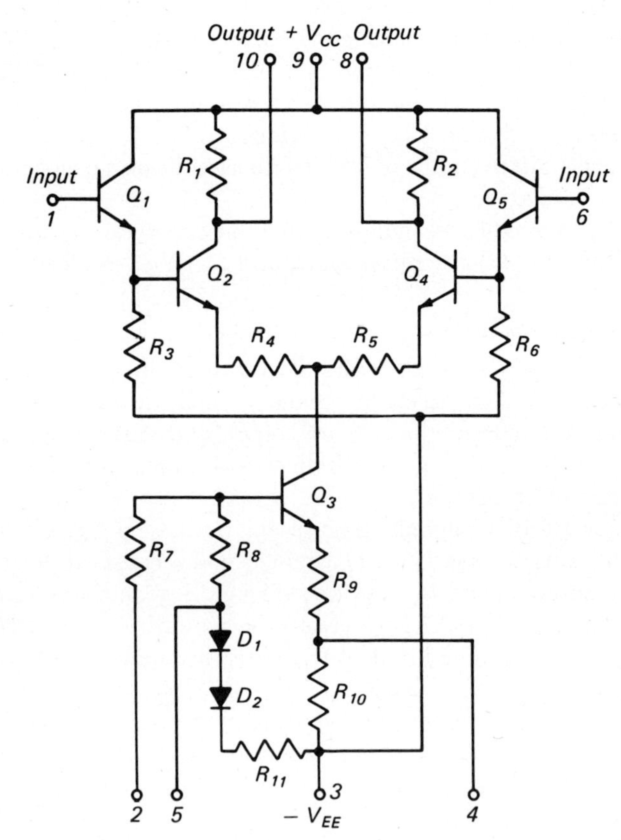

Fig. 19-12 Schematic diagram of RCA type CA-3000 integrated-circuit d-c amplifier.

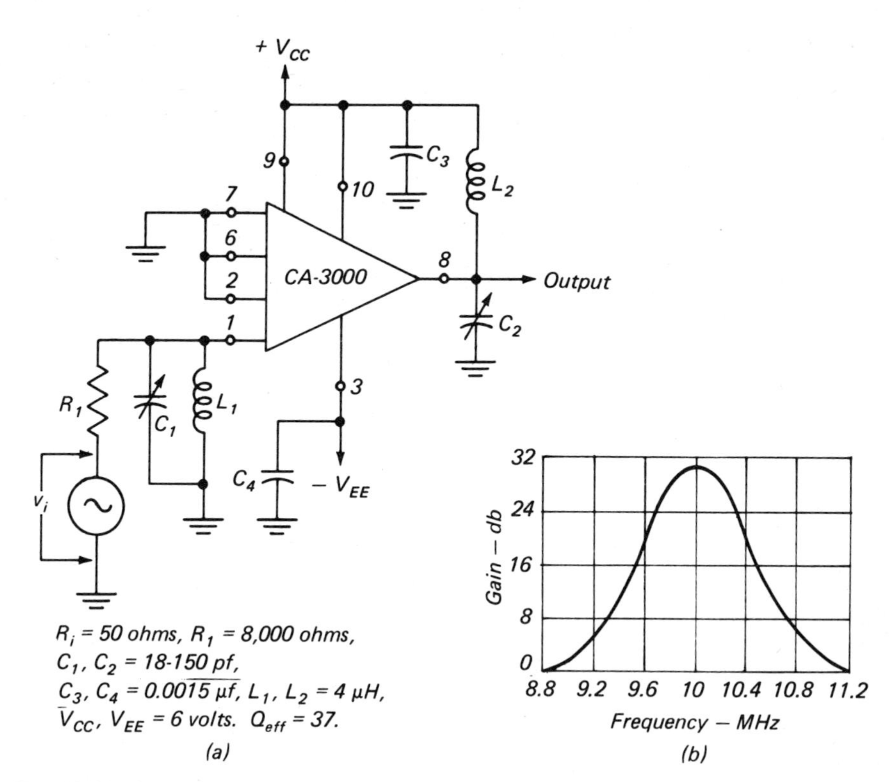

Fig. 19-13 Use of type CA-3000 integrated circuit as a 10-MHz narrowband amplifier. (a) Schematic diagram. (b) Response curve.

19-8 Operational Amplifiers

Basic Principles. The balanced differential amplifier (Fig. 19-8) is the basis of a great majority of the general-purpose linear integrated circuits. The operational amplifier is a widely used extension of the basic differential amplifier. Although there is a variety of operational amplifiers (sometimes abbreviated *opamp*), they generally consist of two balanced differential-amplifier stages connected in cascade, with the first stage supplying a push-pull input to the second stage which is coupled to an appropriate output stage (see Fig. 19-14).

Originally, the operational amplifier was intended to perform a variety of mathematical functions such as summation, subtraction, integration, and differentiation. However, a vast variety of other applications have been developed such as signal amplification, wave forming, impedance transformation, servo controls, to mention but a few. These ICs are useful in communications equipment, telemetry, instrumentation, and data processing.

Characteristics. The characteristics of an ideal operational amplifier include (1) infinite open-loop (i.e., without feedback) gain, (2) zero noise, (3) zero drift, and (4) zero offset. Practical operational amplifiers do not attain these ideals;

however, they do have good operating characteristics, as is indicated by (1) d-c voltage gains in the order of 10^3 to 10^6, (2) broad bandwidths starting at d-c with rolloff to unity gain in the range of 1 to 100 MHz, (3) high input impedance, (4) very low drift with temperature change, and (5) very low input d-c offset.

Inverting and Noninverting Inputs. Figure 19-14 shows two inputs: (1) terminal 3, labeled *inverting input,* and (2) terminal 4, labeled *noninverting input.* When an input signal is applied to terminal 3, an amplified inverted signal will appear at the output (terminal 12, Fig. 19-15*a*). If the same signal is applied to terminal 4, an amplified signal with the same phase as the input signal will appear at the output (terminal 12, Fig. 19-15*b*). This feature permits a wider variety of applications than with the single-stage differential

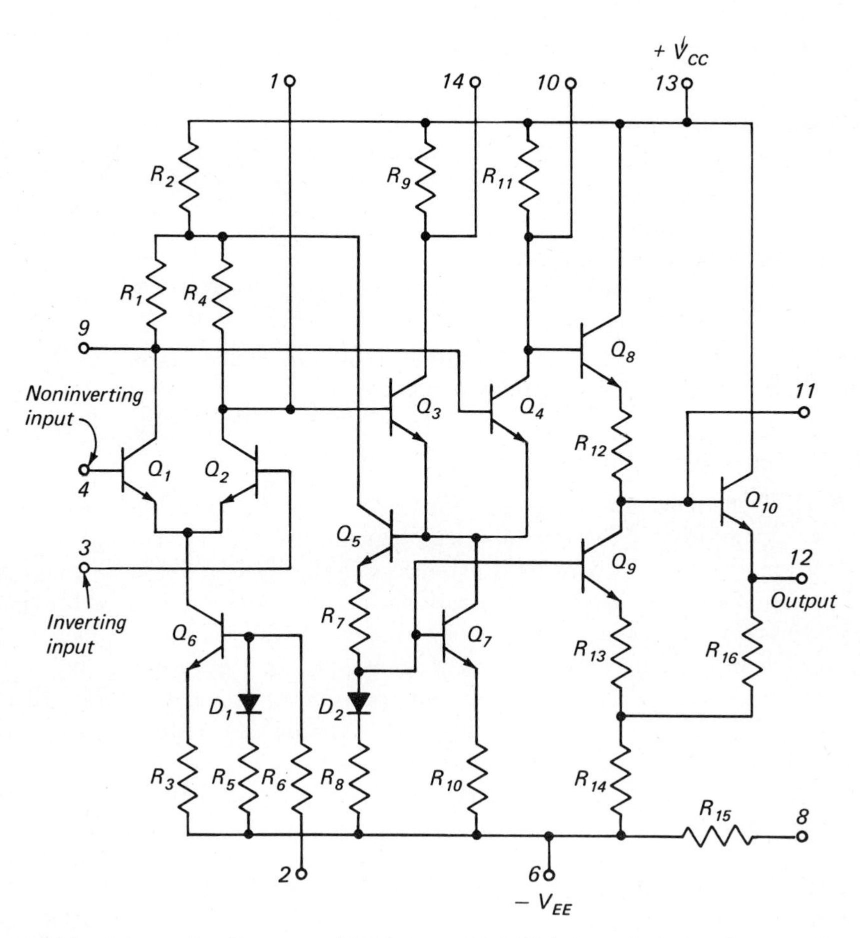

Fig. 19-14 Schematic diagram of RCA type CA-3008 integrated-circuit operational amplifier.

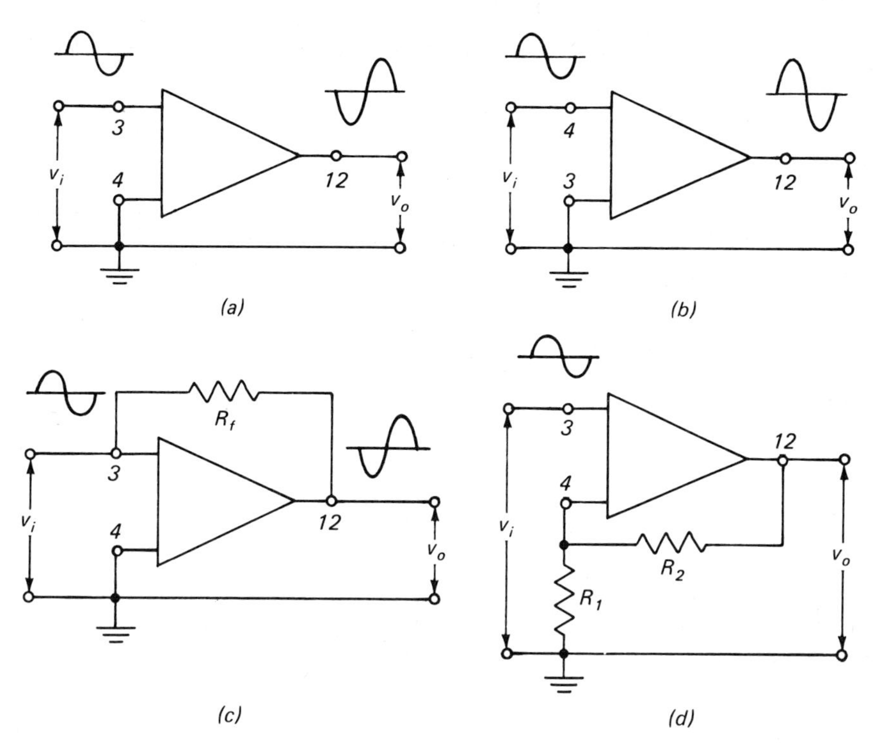

Fig. 19-15 Methods of connecting operational amplifiers. (a) Open-loop inverting amplifier. (b) Open-loop noninverting amplifier. (c) Closed-loop amplifier with negative feedback. (d) Closed-loop amplifier with positive feedback.

amplifier of Fig. 19-12, especially when feedback is used. Figure 19-15*c* and *d* illustrates two methods of connecting operational-amplifier ICs for negative and positive feedback, respectively.

Phase Compensation. In any feedback amplifier, the phase of the feedback signal must be taken into consideration. A phase shift from the desired phase of the feedback signal will alter the characteristics of the amplifier. An extreme example occurs when in a negative-feedback amplifier the phase shift is great enough to introduce positive feedback and cause the amplifier to go into oscillation. Thus, some IC applications require the inclusion of external circuit components to compensate for any undesired phase shift.

19-9 Typical Integrated-circuit Operational Amplifier

Characteristics. Figure 19-14 is the schematic diagram of the RCA type CA-3008 integrated circuit. Functions that can be performed by this IC include (1) narrowband amplifier, (2) wideband amplifier, (3) oscillator, (4) multivibrator, (5) integrator, (6) differentiator, (7) comparator, and (8) scaling adder. The unit is packaged in a 14-terminal ceramic flat pack. Typical operating characteristics are (1) open-loop voltage gain of 60 db at 1 kHz, (2) useful frequency range of from d-c to 15 MHz, (3) open-loop −3-db bandwidth of 300 kHz, (4) input impedance of 14 kilohms at 1 kHz, and (5) output impedance of 200 ohms at 1 kHz.

Circuit. This amplifier consists basically of (1) two differential amplifiers using Q_1Q_2 and Q_3Q_4, respectively, (2) a single-ended output stage Q_8Q_{10}, (3) constant-current sinks Q_6, Q_7, and Q_9, and (4) a feedback-signal source Q_5. The two differential-amplifier stages and the output stage are connected in cascade, with most of the voltage gain provided by the differential amplifiers.

The input signals are applied to the bases of Q_1 and Q_2 which form an emitter-coupled differential amplifier; Q_6 provides the constant-current source for this amplifier stage. The output of this stage is directly coupled to the input (bases) of the second differential-amplifier stage Q_3Q_4; Q_7 provides the constant-current source for this amplifier stage. The output of this stage (collector of Q_4) is directly coupled to the base of Q_8. Transistors Q_8, Q_9, and Q_{10} form an emitter-follower single-ended output circuit. The input to the base of Q_{10} is supplied from the emitter circuit of Q_8. Transistor Q_9 provides (1) a constant-current sink for Q_8, and (2) a small amount of voltage gain by means of the emitter-to-emitter connection between Q_9 and Q_{10}. Transistor Q_5 develops a negative-feedback signal used to (1) reduce any error signals devel-

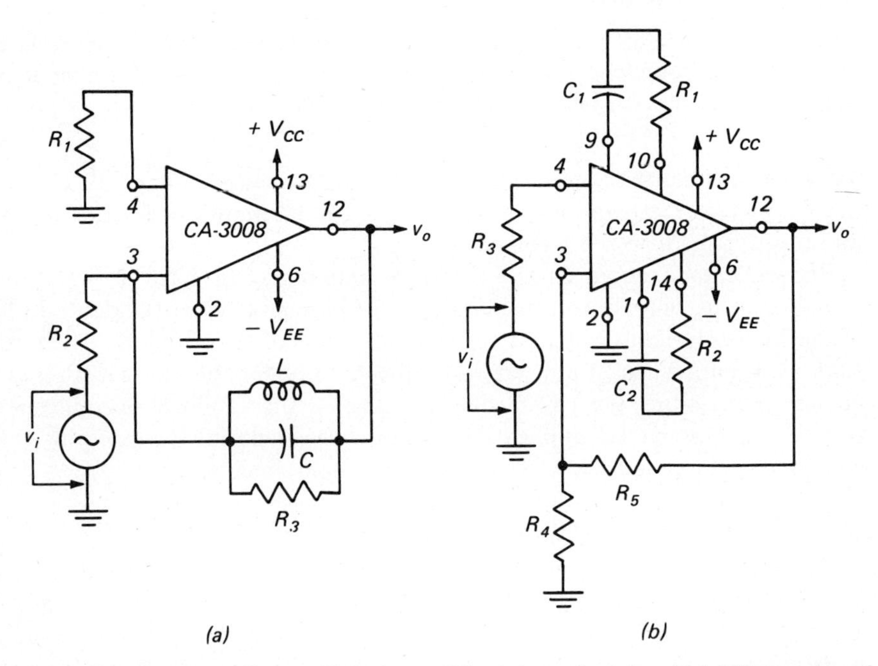

Fig. 19-16 Applications of operational-amplifier integrated circuit. (a) Narrowband tuned amplifier. (b) Video amplifier.

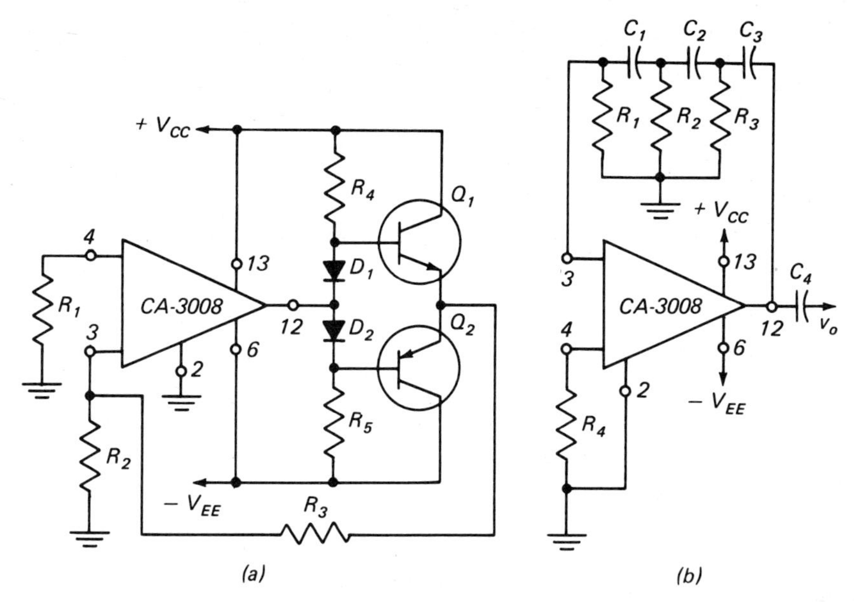

Fig. 19-17 Applications of operational-amplifier integrated circuit. (a) Use with external push-pull stage to increase the output capability. (b) An *RC* phase-shift oscillator.

oped when the same input signal is applied to terminals 3 and 4, and (2) reduce effects produced by variations in the supply voltages.

19-10 Operational-amplifier Applications

Narrowband Amplifier. Figure 19-16*a* shows the operational-amplifier IC used as a narrowband tuned amplifier. Negative feedback is provided by R_2 and the parallel tuned circuit L,C,R_3. With circuit components selected to provide a resonant frequency of 100 kHz, a 3-kHz bandwidth can be attained at −3 db.

Wideband Amplifier. By applying negative feedback and suitable phase compensation, a flat gain versus frequency response from d-c up to as high as 10 MHz can be obtained from an operational amplifier. Figure 19-16*b* shows the operational-amplifier IC used as a closed-loop, noninverting wideband amplifier. Resistors R_4 and R_5 supply the feedback signal, and phase compensation is provided by C_1R_1 and C_2R_2.

Power-amplifier Driver. Figure 19-17*a* shows an operational amplifier IC used to drive a Class B push-pull amplifier stage. It may be observed that except for R_1, R_2, and R_3 (and the power supplies), the IC is used as a complete self-contained amplifier. Resistors R_1 and R_2 control the base bias of the input transistors in the IC, and R_3 provides a negative-feedback signal to the IC.

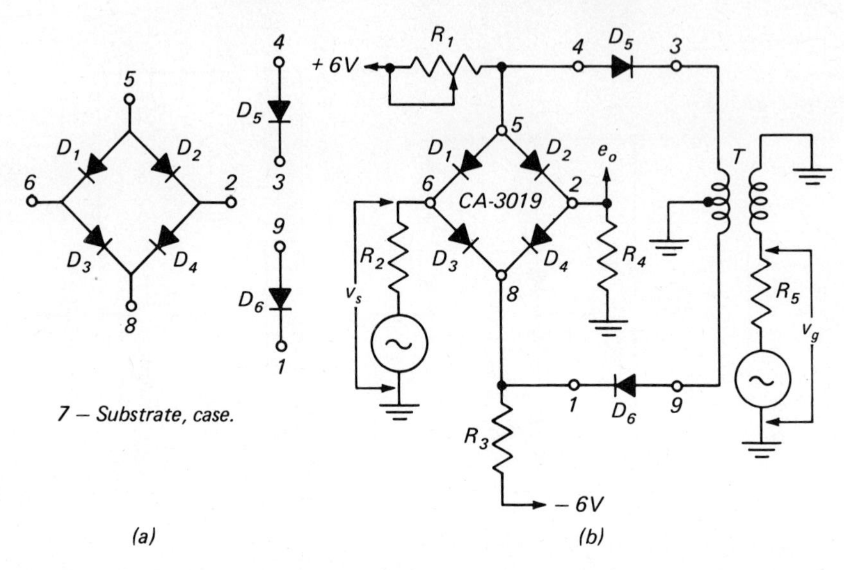

Fig. 19-18 RCA type CA-3019 integrated-circuit diode array. (a) Schematic diagram. (b) Application in a gating circuit.

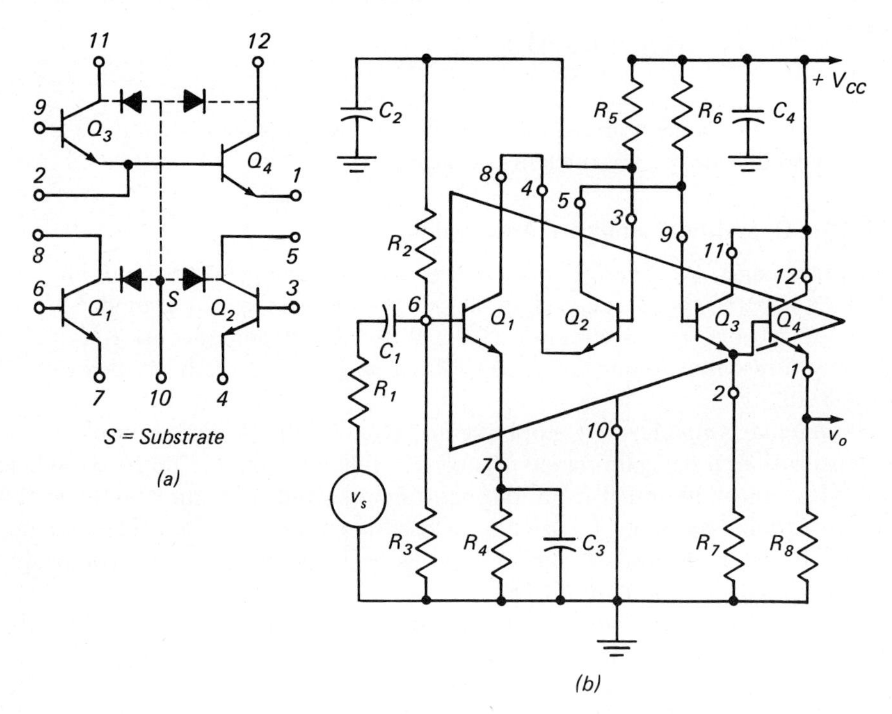

Fig. 19-19 RCA type CA-3018 integrated-circuit transistor array. (a) Schematic diagram. (b) Application in a wideband amplifier.

Oscillator. Because of the high gain and wideband capabilities, the operational-amplifier ICs can be used in oscillator circuits such as the *LC*, *RC*, Wien bridge, and crystal-controlled oscillators. Furthermore, the dual inverting and noninverting inputs make it possible to use positive feedback for producing oscillation and negative feedback for improving the waveform. Figure 19-17*b* shows the circuit diagram for a phase-shift *RC* oscillator using an operational-amplifier integrated circuit.

19-11 Integrated-circuit Arrays

Integrated-circuit arrays provide groups of (1) unconnected transistors, (2) diode quads, (3) transistors connected in Darlington pairs, and (4) individual circuit stages. Advantages of arrays are (1) closely matched device and/or circuit characteristics, and (2) usefulness in conjunction with circuit components available only in discrete form.

Diode Array. Figure 19-18*a* shows the schematic diagram of the type CA-3019 integrated-circuit diode array. Diodes D_1 to D_4 are connected to form a diode quad with four terminals brought out. Diodes D_5 and D_6 are left independent with their individual terminals brought out. The circuit of Fig. 19-18*b* illustrates the use of a diode array in a gating circuit.

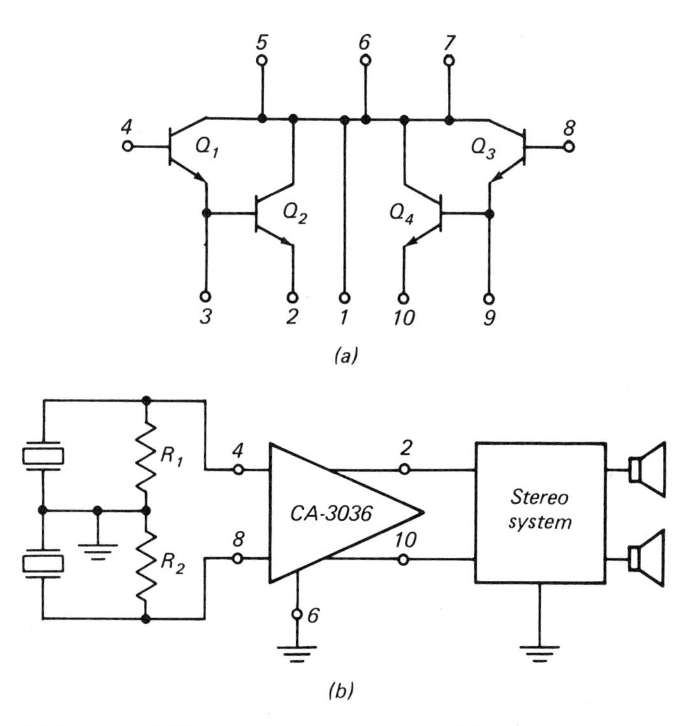

Fig. 19-20 RCA type CA-3036 integrated-circuit dual Darlington array. (a) Schematic diagram. (b) Application as a preamplifier for a stereo-phono system.

Transistor Array. Figure 19-19*a* shows the schematic diagram of the type CA-3018 integrated-circuit transistor array. Transistors Q_1 and Q_2 have their leads brought out separately and thus can be used independently. The emitter of Q_3 and the base of Q_4 are permanently connected. The substrate lead acts as a shield between the base and collector of Q_3.

The circuit of Fig. 19-19*b* illustrates the use of the CA-3018 transistor array in a wideband amplifier. The input signal is fed to the base of Q_1 and is amplified by the cascade-connected pair Q_1Q_2. The signal voltage at the collector of Q_2 is fed to the base of the emitter-follower stage Q_3 and amplified. The signal is then directly coupled from the emitter of Q_3 to the base of Q_4 and further amplified by the emitter-follower stage Q_4. The last two stages (Q_3Q_4) are cascaded. Typical frequency versus gain characteristics obtainable with this integrated-circuit application are (1) frequency range of 6 kHz to 11 MHz with variation in gain not exceeding −3 db, and (2) 37-db voltage gain at mid-frequency.

Darlington Pairs. Figure 19-20*a* shows the schematic diagram of the type CA-3036 integrated-circuit dual Darlington array. The four transistors are connected to form two independent Darlington pairs. Figure 19-20*b* illustrates the use of the Darlington-pair array as a preamplifier for a stereo-phono system.

QUESTIONS

1. (*a*) Name five advantages of integrated circuits. (*b*) What types of components can be included in an integrated circuit?
2. Describe four basic types of construction used with integrated circuits.
3. (*a*) Name two modes of operating integrated circuits. (*b*) Describe the characteristics of each mode. (*c*) Give some applications of each mode of operation.
4. Define (*a*) substrate, (*b*) chip, (*c*) diffusion.
5. What types of materials are used to form (*a*) resistors? (*b*) Capacitors? (*c*) Diodes? (*d*) Transistors?
6. Describe the basic procedure used in forming a diode in a monolithic IC.
7. Describe the basic procedure used in forming a transistor in a monolithic IC.
8. Describe the basic procedure used in forming a resistor in a monolithic IC.
9. Describe the basic procedure used in forming a capacitor in a monolithic IC.
10. Draw a diagram showing the fabrication of a monolithic IC consisting of a resistor, transistor, capacitor, and diode connected in series in that order.
11. What is meant by large-scale integration?
12. (*a*) Name three types of IC packaging. (*b*) Describe each type.
13. (*a*) Describe the symbol used to represent a linear IC. (*b*) Describe the method of numbering the leads on the symbol.
14. Describe the symbols used to represent the digital ICs.
15. (*a*) How much circuit information is provided with the IC symbol in a schematic diagram? (*b*) How is detailed circuit information obtained?
16. In the circuit of Fig. 19-9, how can the degeneration due to R_3 be eliminated?
17. Redraw the schematic diagram of Fig. 19-10*a* showing how Q_1 and Q_3 can be used to perform the same duty.

18. Draw a detailed circuit diagram for the oscillator of Fig. 19-11.
19. Prepare a schematic circuit diagram for a Wien-bridge oscillator using a CA-3028A integrated circuit.
20. The schematic diagram of Fig. 19-13*a* shows terminal 7 grounded, yet the circuit diagram of Fig. 19-12 does not include a terminal 7. What logical conclusion can be drawn as to the location and purpose of terminal 7?
21. For the amplifier represented by Fig. 19-13*b*, what is the approximate bandwidth when the gain drops to 3 db below the maximum gain?
22. What is the basic circuitry of an operational amplifier?
23. Name four characteristics of an ideal operational amplifier.
24. What are the operating characteristics of practical operational amplifiers?
25. For the operational-amplifier integrated circuit of Fig. 19-14, what is the operating feature of (*a*) the inverting input? (*b*) The noninverting input?
26. Why is phase compensation important with some operational-amplifier integrated-circuit applications?
27. Name some applications of integrated-circuit operational amplifiers.
28. (*a*) Name three types of integrated-circuit arrays. (*b*) What are the advantages of integrated-circuit arrays?

Chapter 20

Types and Sources of Waveforms

In the study of communication electronic circuits, reference is frequently made to (1) sawtooth waves, (2) square waves, (3) rectangular waves, (4) synchronizing pulses, (5) triggering pulses, (6) blanking pulses, (7) sweep circuits, (8) separator circuits, (9) etc. To understand the many functions that waveforms may perform and how circuits use these functions, it is necessary to study the characteristics of various waveforms of voltage and current and how the desired waveforms may be produced.

20-1 Waveforms

The Sine Wave. The basic and most frequently used waveform is the sine wave, which is one that varies in amplitude in proportion to the sine of an angle and varies continuously through 360° (Fig. 20-1). Sine waves may be regarded as basic building blocks, since any desired wave shape may be obtained by adding sine waves of different frequencies, amplitudes, or phase. Conversely, any nonsinusoidal wave may be broken down into simple sine-wave components of different frequencies, amplitude, and phase.

The Square Wave. A square wave is one in which the ON time and OFF time are of equal duration. This waveform may consist of (1) only positive half-cycles, (2) only negative half-cycles, or (3) both positive and negative half-cycles (Fig. 20-2).

The Rectangular Wave. When the duration of and the interval between pulses are not equal, the waveform is called *rectangular*. These waveforms may

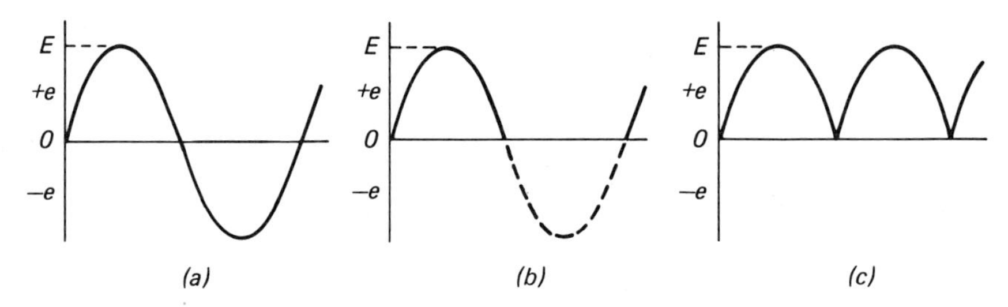

Fig. 20-1 Sine-wave forms. (a) Full sine wave. (b) Half sine wave due to half-wave rectification of full sine waves. (c) Half sine waves due to full-wave rectification of full sine waves.

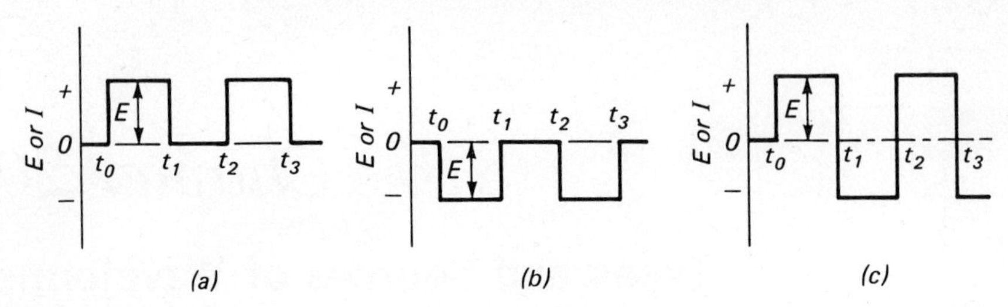

Fig. 20-2 Square waves. (a) With only positive half-cycles. (b) With only negative half-cycles. (c) With positive and negative half-cycles.

have only positive or negative values, or both positive and negative values (Fig. 20-3).

The Sawtooth or Triangular Wave. Sawtooth waves are used extensively in sweep circuits. A sawtooth wave is one whose amplitude increases at a constant linear rate; that is, at each successive instant of time it is increasing at the same rate as in the previous instant until it reaches its maximum value, after which it very abruptly decreases to the starting value and the action is repeated (Fig. 20-4).

Other Composite Waves. By combining sine, square, rectangular, and/or sawtooth waves, other waveforms can be derived. An example of a more complex waveform is the blanking and sync pulse of a television receiver, as illu-

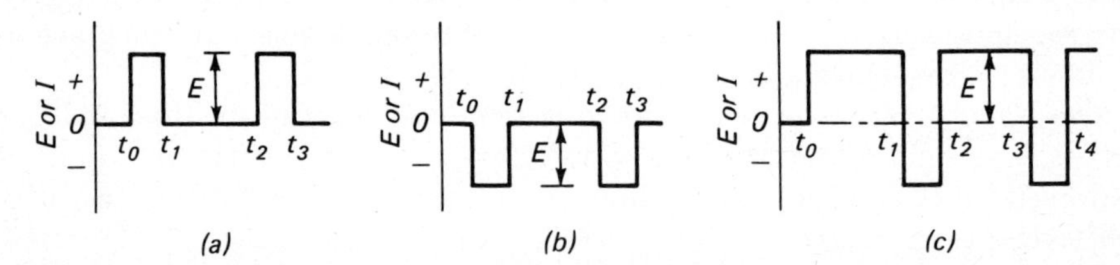

Fig. 20-3 Rectangular waves. (a) With only positive values. (b) With only negative values. (c) With positive and negative values.

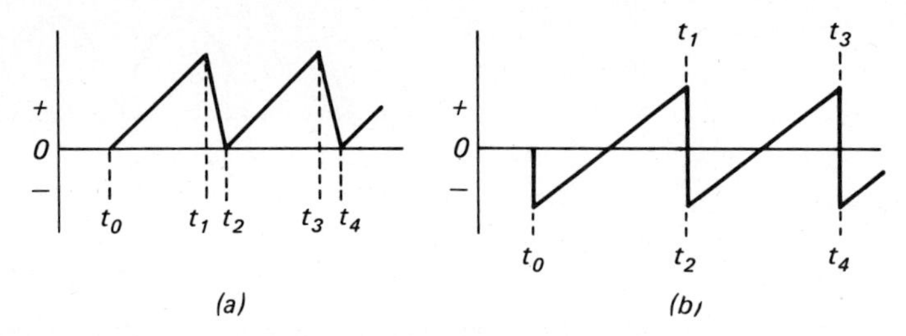

Fig. 20-4 Sawtooth or triangular waves. (a) Sawtooth wave with only positive values. (b) Sawtooth wave with positive and negative values.

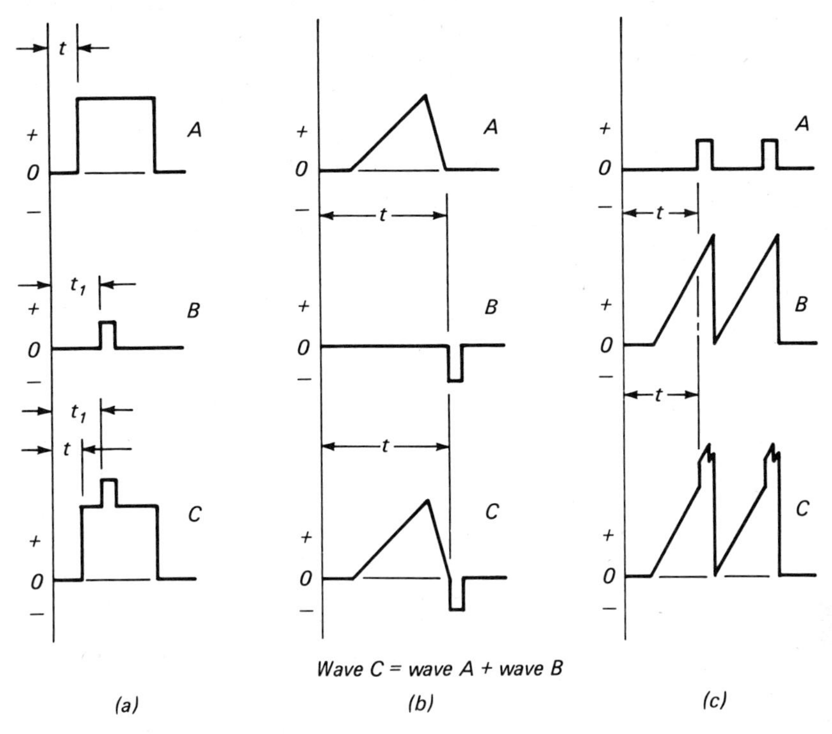

Fig. 20-5 Complex waveforms. (a) Combining a square and rectangular wave having different rise and duration times. (b) and (c) Combining sawtooth and rectangular waves.

strated in Fig. 20-5*a*; this wave is a combination of a square wave and a rectangular wave. Two other complex waveforms are shown in Fig. 20-5*b* and *c*.

20-2 Waveform Terminology

Duration. The length of time that a voltage or current exists is called its *duration.* A wave may have either or both positive and negative duration (see Fig. 20-2). In electronics, duration is frequently expressed in microseconds.

Interval. The time that elapses before the occurrence of another voltage or current wave (either positive or negative pulse) is called the *interval.* Sometimes the interval is taken to mean the time that elapses before the occurrence of another pulse of like polarity. In Fig. 20-2*a* and *b*, the duration is represented by $t_0 - t_1$ and $t_2 - t_3$, while the interval is represented by $t_1 - t_2$. In the wave of Fig. 20-2*c* there is no interval between the positive and negative pulses, but between pulses of like polarity there is an interval equal to the duration of the pulse of the opposite polarity.

Polarity. Whether a wave is positive or negative depends upon the reference point. With a common reference point, such as a chassis, the polarity of the waveform is considered positive when the negative side of the power source is connected to the chassis and negative when the positive side of the power source is connected to the chassis.

Rise Time. The time required for the voltage or current to reach a particular value is called the *rise time.* For a sine wave, it is the time required for the wave to go from zero to either its positive or negative peak value. For an ideal square waveform, the rise time is zero. For a nonideal square waveform, it is the time required for the wave to go from zero to a maximum positive or negative value. The rise time for a sawtooth wave, such as Fig. 20-4, is considered as $t_0 - t_1$.

Decay Time. The time required for a voltage or current to fall from its maximum value to its initial value is called the *decay time.* The decay time of the sawtooth wave of Fig. 20-4 is $t_1 - t_2$ and $t_3 - t_4$.

20-3 Harmonic Analysis of Waveforms

Sine Wave as the Fundamental. Waveforms and their applications in electronics can best be understood by analyzing the composition of these waveforms. A replica of any waveform may be constructed by adding a fundamental periodic wave and one or more of its harmonics. The *fundamental* periodic wave is that varying quantity of the lowest frequency which produces a sine waveform. This waveform (Fig. 20-1) is called the *fundamental* or the *first harmonic.*

Harmonics. A harmonic is a sine-wave voltage or current whose frequency is an integral multiple of the fundamental wave. Thus, a 60-Hz fundamental wave may have harmonics of 120, 180, 240, 300, 360, etc., Hz called respectively the second, third, fourth, fifth, sixth, etc., harmonics of the fundamental.

Harmonic Content of Waveforms. The harmonic content of various waveforms generally conforms to the following pattern: A square wave consists of its fundamental frequency and its odd harmonics only; a waveform with seven odd harmonics is considered as being reasonably close to a square wave (see Fig. 16-10*b*).

A sawtooth or triangular wave consists of its fundamental frequency and a number of even and odd harmonics whose amplitudes decrease inversely as the order of the harmonic; a proper phase relation among the harmonics is also required. A fair approximation of a sawtooth wave is produced with 15 to 20 harmonics of the fundamental (see Fig. 16-10*a*).

20-4 Sources of Waveforms

Sine Wave. Three ways of producing a sine wave (or an acceptable approximation thereof) are (1) the alternator, (2) an electronic oscillator circuit, and (3) tuned circuits.

Square Waves. Three ways of producing a square wave (or an acceptable approximation thereof) are (1) a d-c power source and a switching arrangement, (2) clipping a sine wave with a limiting amplifier, and (3) electronic multivibrator circuits.

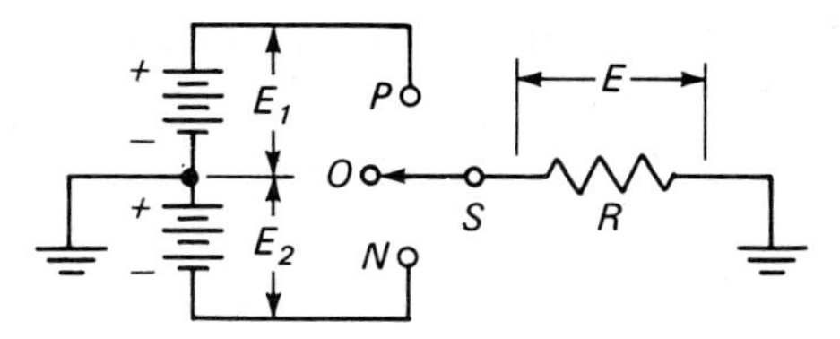

Fig. 20-6 Simple method of producing square or rectangular waves.

A simple method of producing square (or rectangular) waves is shown in Fig. 20-6. The square wave of Fig. 20-2*a* can be produced by connecting the switch S to points P and O for equal periods of time. Alternating the switch between points N and O will produce the square wave of Fig. 20-2*b*. With $E_1 = E_2$, connecting the switch to points P and N for equal periods of time will produce a square wave having symmetrical positive and negative portions (Fig. 20-2*c*). By using different voltages for E_1 and E_2 and using different time intervals for connecting the switch to various combinations of P, O, and N, a wide variety of waveforms can be obtained.

The limiting amplifier, described in Art. 15-14, can also be used to produce reasonably approxmiate square waves. Likewise, the multivibrator circuits described in Arts. 17-35 to 17-37 can be designed to produce square waves.

Sawtooth Waves. Sawtooth or triangular waves can be produced by (1) a simple RC or RL circuit and a switching mechanism, (2) a relaxation oscillator circuit, and (3) using square or rectangular waves to control the action of special circuits.

The principle of obtaining a sawtooth wave with a resistor and a capacitor is based on the manner in which a voltage charge builds up on a capacitor (Art. 12-12). The rate of voltage change for a charging capacitor (Fig. 12-41*b*) roughly approximates the rise portion of a sawtooth wave. However, if the time of the rise portion is limited to one-tenth of a time constant (0.1 RC), the rise portion of the sawtooth wave will approximate a straight line. In the circuit of Fig. 20-7, when the switch S is in position 1, e_C is zero, and upon changing the switch to position 2, e_C will rise in accordance with the curve of Fig. 12-41*b*. However, if after 0.1 RC second the switch is returned to position 1, the voltage e_C will instantly return to zero (assuming the resistance R_C of the capacitor is zero). By alternately placing the switch in position 2 for 0.1 RC second and in position 1 for only an instant, the sawtooth wave of Fig. 20-7*b* will be produced. When the effect of R_C is included, the voltage decay time becomes significant, and the waveform is then as shown in Fig. 20-7*c*. The equivalent action of alternating the switch between positions 1 and 2 may be performed (1) mechanically, or (2) electronically.

Producing a sawtooth wave by use of a relaxation oscillator has been described in Arts. 17-27 to 17-34. Relaxation circuits can easily be synchronized and can produce a wide range of frequencies by varying either or both the resistance or the reactance of the circuit.

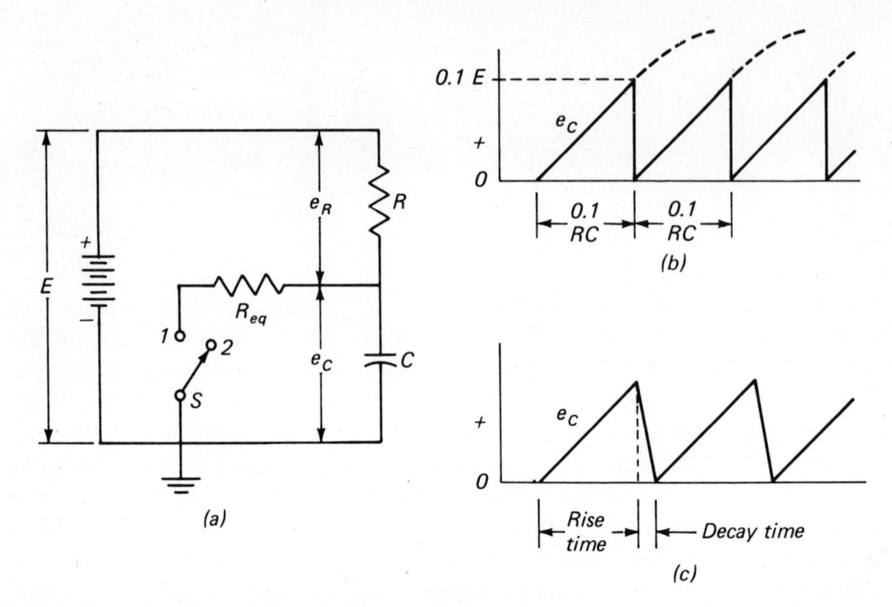

Fig. 20-7 Principle of producing a sawtooth wave. (a) Circuit. (b) When R_{eq} is zero. (c) When R_{eq} has a significant value.

20-5 Integrator and Differentiator Circuits

RC and RL Circuits. Integrator and differentiator circuits are simple *RC* or *RL* voltage-divider networks designed so that the output waveform is a greatly distorted version of the input waveform. The basic integrator circuit consists of either (1) a series resistor and a parallel capacitor as in Fig. 20-8*a*, or (2) a series inductor and a parallel resistor. The basic differentiator circuit consists of either (1) a series capacitor and a parallel resistor as in Fig. 20-10*a*, or (2) a series resistor and a parallel inductor. The basic integrator circuit is essentially a low-pass filter whose time constant is one-tenth or less than the period of the input signal. The basic differentiator circuit is essentially a high-pass filter whose time constant is 10 or more times the period of the input signal.

Because the basic principles of *RC* and *RL* integrator and differentiator circuits are quite similar, and because the *RC* circuits are more frequently used, only the *RC* circuits are described.

Integrator Circuits. In some *RC* circuits, the capacitor stores some of the charge on each cycle; that is, it integrates or adds the charges on each pulse until it becomes charged to some definite value. *RC* circuits used to integrate pulses are called *RC integrator circuits*. *RC* integrators are frequently used in electronic systems in agc, detector, filter, and sync circuits. Because of its slow response to voltage changes, the integrator may also be used as a pulse delay network. Integrators are also used as phase changers and low-pass filters.

Differentiator Circuits. In some *RC* circuits, the waveform of the voltage across the capacitor is substantially the same as that of the applied voltage, whereas the voltage across the resistor has a new waveform which is dependent upon how fast the applied voltage changes. Because of this ability to differentiate between rates of change of the applied voltage, these circuits are called *differentiator circuits* and the waveform appearing across the resistor is called the *differentiated wave.*

20-6 *RC* Integrator Circuits

Basic Circuit. In an *RC* integrator circuit, the output voltage is taken from across the capacitor and the waveform of the output is dependent upon the time constant and the frequency and shape of the input waveform (see Fig. 20-8). Merely using the voltage across the capacitor does not make the circuit an integrator; it is also necessary to proportion the time constant properly. The *RC* product should be many times greater than the duration and the interval of the pulses to be integrated; a ratio of 10 to 1 is usually sufficient to provide

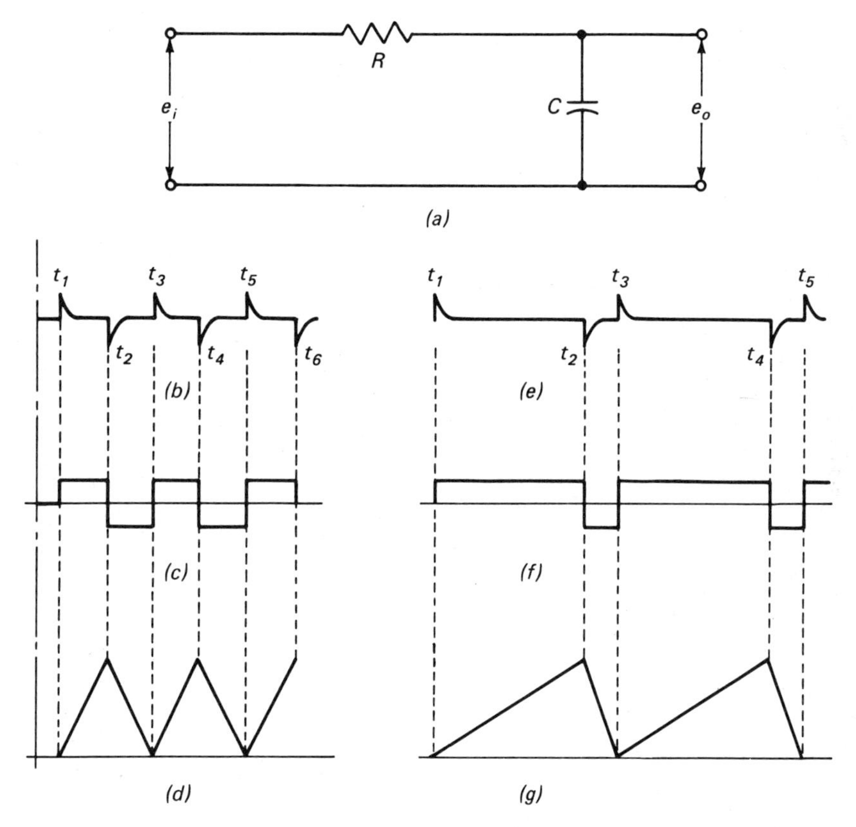

Fig. 20-8 *RC* integrator. (a) The circuit. (b) to (g) Input and output waveforms. Input (b) produces output (c); input (c) produces output (d); input (e) produces output (f); input (f) produces output (g).

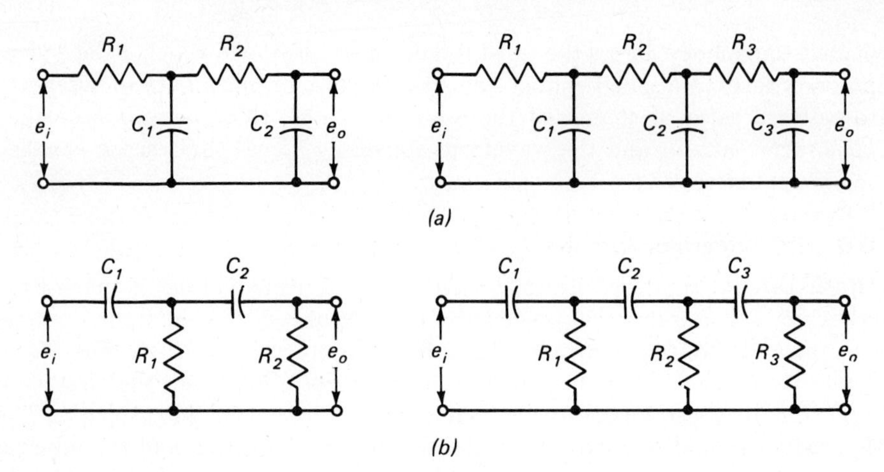

Fig. 20-9 (a) Multisection integrator circuits. (b) Multisection differentiator circuits.

good integration. When an RC circuit has a long time constant, each succeeding pulse adds to the charge on the capacitor until it reaches an average value of the applied voltage. For circuits with relatively short time constants, the capacitor is likely to charge and discharge so rapidly that the output waveform is substantially the same as the input waveform with perhaps the exception of the edges being rounded off.

Figure 20-8 shows that for an integrator circuit with a long time constant $[RC \gtrsim 10(t_1 - t_3)]$: (1) an input signal with equally spaced pulses (Fig. 20-8*b*) will produce a square-wave output (Fig. 20-8*c*); (2) an input signal with unequally spaced pulses (Fig. 20-8*e*) will produce a rectangular-wave output (Fig. 20-8*f*); (3) a square-wave input (Fig. 20-8*c*) will produce a triangular-wave output (Fig. 20-8*d*); (4) a rectangular-wave input (Fig. 20-8*f*) will produce a sawtooth-wave output (Fig. 20-8*g*). No transformation of waveform occurs when a sine-wave signal is applied to the input as the voltage drops across the capacitor and the resistor will also be sine waves; however, the phase relation among the voltages and current will be altered.

Figure 20-11 shows the waveforms obtained with a square-wave input when (1) $RC = 0.1t$, (2) $RC = t$, and (3) $RC = 10t$; where t is the period $(1/f)$ of the input signal.

Multisection Integrator Circuits. Two or more basic integrator circuits may be cascaded (Fig. 20-9*a*) to provide more pronounced effects and to produce additional variation in the output waveforms.

20-7 *RC* Differentiator Circuits

In an RC differentiator circuit, the output voltage is taken from across the resistor, and the waveform of the output is dependent upon the time constant and the frequency and shape of the input waveform (see Fig. 20-10). The RC product should be much smaller than the duration of the interval of the pulses to be differentiated; a ratio of 1 to 10 is usually sufficient to provide good differentiation.

The amplitude of the differentiated wave at any instant depends on the rate of change of the applied voltage and (1) increases when the rate of change increases, (2) decreases when the rate of change decreases, and (3) is zero when the rate of change is zero. Also, the polarity of the differentiated wave reverses whenever the applied voltage decreases or changes to the opposite direction.

The following rules can be used to determine the shape of any differentiated wave: (1) When there is no change in the amplitude of the applied wave, the differentiated wave has zero amplitude. (2) When the amplitude of the applied

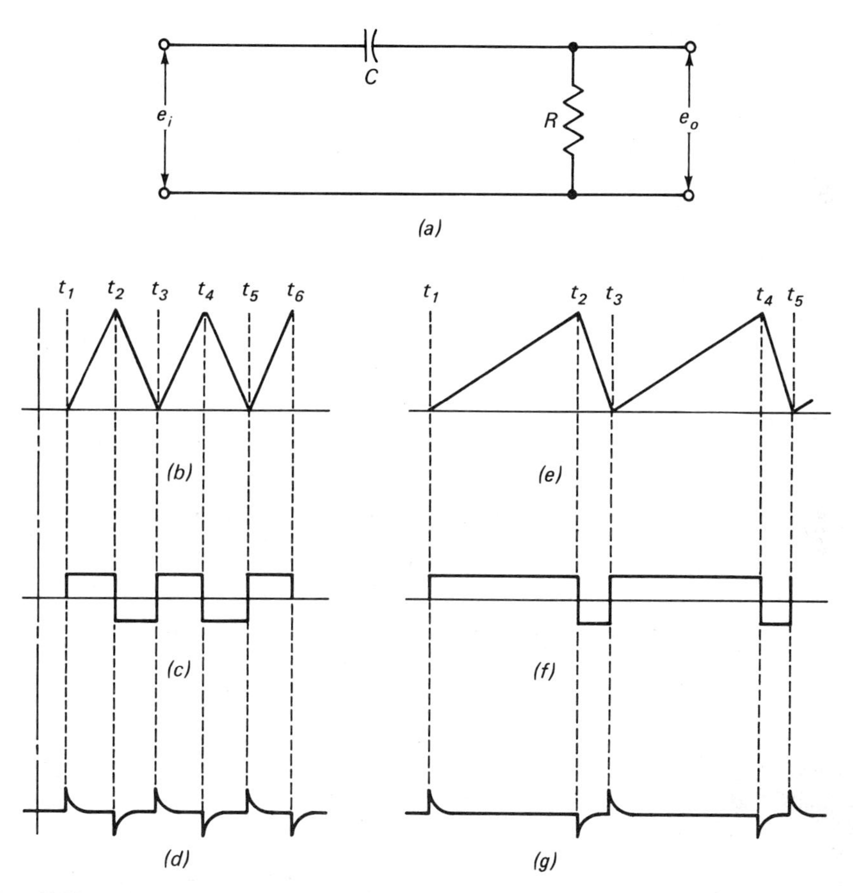

Fig. 20-10 RC differentiator. (a) The circuit. (b) to (g) Input and output waveforms. Input (b) produces output (c); input (c) produces output (d); input (e) produces output (f); input (f) produces output (g).

wave changes abruptly, the differentiated wave is a sharp narrow pulse for each abrupt change. (3) When the change is less abrupt, the differentiated wave is a pulse approaching a very narrow rectangular form. (4) When the amplitude of the applied wave changes at a constant rate, the differentiated wave has a constant value for each constant rate of change, and for a sawtooth wave the differentiated wave is a succession of rectangular waves of equal or unequal duration depending upon the shape of the applied wave.

Figure 20-10 shows that for a differentiator circuit with a short time constant $[RC \lessapprox 0.1(t_1 - t_3)]$: (1) an input signal with a triangular waveform (Fig. 20-10*b*) will produce a square-wave output (Fig. 20-10*c*); (2) an input signal with a sawtooth waveform (Fig. 20-10*e*) will produce a rectangular-wave output (Fig. 20-10*f*); (3) a square-wave input signal (Fig. 20-10*c*) will produce a series of equally spaced spikes (Fig. 20-10*d*); (4) a rectangular-wave input signal (Fig. 20-10*f*) will produce a series of spikes with alternately long and short spacing (Fig. 20-10*g*). A decrease in the time interval ($t_2 - t_3$, Fig. 20-10*f*) will increase the magnitude of the voltage pulses (called *spikes* or *pips*) in the output waveform shown in Fig. 20-10*g*. No transformation of waveform occurs when a sine-wave signal is applied to the input, as the voltage drops across the resistor and the capacitor will also be sine waves; however, the phase relations among the voltages and current will be altered.

Figure 20-11 shows the waveforms obtained with a square-wave input when (1) $RC = t$, (2) $RC = 10t$, and (3) $RC = 0.1t$; where t is the period ($1/f$) of the input signal.

Multisection Differentiator Circuits. Two or more basic differentiator circuits may be cascaded (Fig. 20-9*b*) to provide sharper differentiation.

20-8 Illustration of the Effect of the Time Constant

The preceding discussion established that the ratio of the RC time constant to the period of the input waveform should be in the order of 10 to 1 for an integrator circuit and 1 to 10 for a differentiator circuit. The simple circuit of Fig. 20-11*a* can represent both the integrator and differentiator circuits of Figs. 20-8 and 20-10, with the integrator characteristics indicated by the voltage e_C and the differentiator characteristics indicated by the voltage e_R.

Figures 20-11*b*, *c*, and *d* illustrates the voltage characteristics when the ratio of RC to t equals 1, 10, and 0.1, respectively; it should be observed that the input signal E_i is the same for all three cases. When the ratio is 1, then (1) the integrator output e_C does not provide uniform voltage variation, (2) the differentiator output e_R does not produce a very significant voltage spike, and (3) the circuit does not function well as either an integrator or a differentiator.

When the ratio of RC to t is 10 to 1, then (1) the voltage variation of e_C is uniform and produces a triangular waveform, (2) the voltage waveform of e_R closely resembles the input-signal waveform, and (3) the circuit functions well as an integrator and poorly as a differentiator.

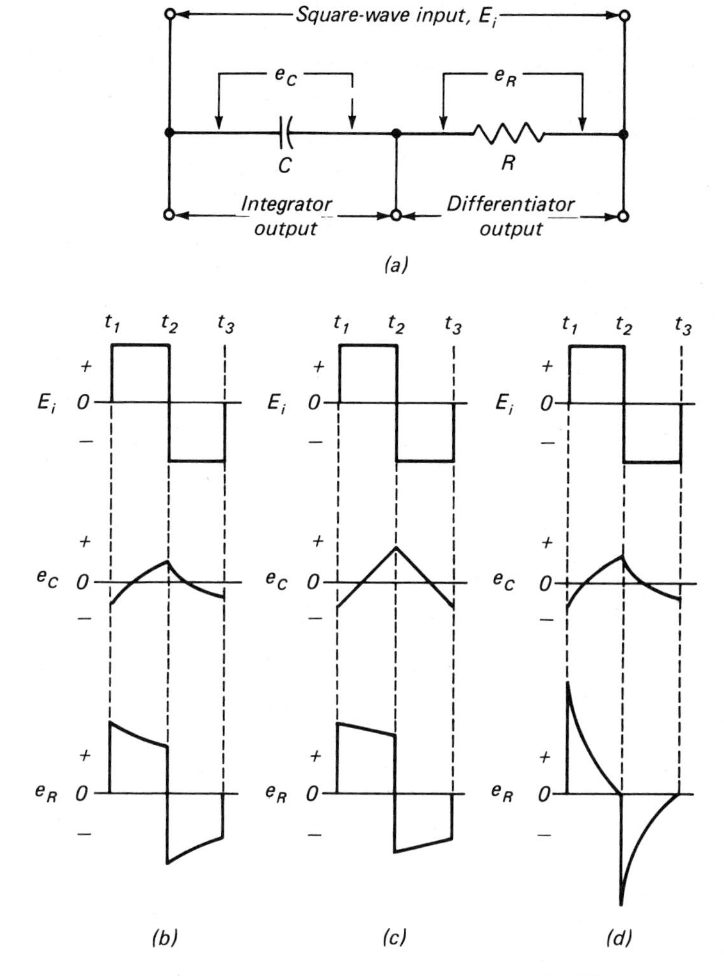

Fig. 20-11 Effect of time constant on integrator and differentiator circuits. (a) Circuit with integrator and differentiator outputs. Waveforms when ratio of RC to $(t_1 - t_3)$ is: (b) 1, (c) 10, (d) 0.1.

When the ratio of RC to t is 1 to 10, then (1) the voltage variation of e_C is not uniform, (2) the voltage waveform of e_R has sharp peaks whose magnitudes are in the order of $2E_i$, and (3) the circuit functions well as a differentiator and poorly as an integrator.

QUESTIONS

1. What is the relation between a sine wave and a nonsinusoidal wave?
2. Define the following terms: (*a*) a rectified sine wave, (*b*) a square wave, (*c*) a rectangular wave, (*d*) a sawtooth wave, (*e*) a complex wave.

3. Explain the following waveform terms: (*a*) duration, (*b*) interval, (*c*) polarity, (*d*) rise time, (*e*) decay time.
4. Describe the harmonic content of (*a*) a square wave, (*b*) a sawtooth wave.
5. Name three methods for producing a sine wave.
6. Explain a simple method for producing a square wave having (*a*) only negative half-cycles, (*b*) only positive half-cycles, (*c*) both positive and negative half-cycles.
7. Name two other methods used to produce a square waveform.
8. Name three methods used to produce a sawtooth waveform.
9. Explain the principle of producing a square waveform using an *RC* circuit.
10. Describe how the frequency of a relaxation oscillator can be varied.
11. Describe (*a*) the basic components and their arrangement in an integrator circuit, (*b*) the filter action of this circuit.
12. Describe (*a*) the basic components and their arrangement in a differentiator circuit, (*b*) the filter action of this circuit.
13. What is meant by (*a*) an *RC* integrator circuit? (*b*) An integrated wave?
14. What is meant by (*a*) an *RC* differentiator circuit? (*b*) A differentiated wave?
15. (*a*) In an *RC* integrator circuit what factors affect the shape of the output waveform? (*b*) What ratio is required between the time constant and the duration and interval periods?
16. Describe the actions of an *RC* integrator circuit having a long time constant and driven by an input signal (*a*) having equally spaced pulses, (*b*) having unequally spaced pulses, (*c*) with a square waveform, (*d*) with a rectangular waveform.
17. Describe the effects produced by cascading two or more basic integrator circuits.
18. (*a*) In an *RC* differentiator circuit what factors affect the shape of the output waveform? (*b*) What ratio is required between the time constant and the duration and interval periods?
19. How does the amplitude of a differentiated wave vary with the following rates of change in the applied voltage to its input: (*a*) an increase? (*b*) A decrease? (*c*) Zero?
20. Describe four rules used to determine the shape of a differentiated wave.
21. Describe the actions of an *RC* differentiator circuit having a short time constant and driven by an input signal with (*a*) a triangular waveform, (*b*) a sawtooth waveform, (*c*) a square waveform, (*d*) a rectangular waveform.
22. What are the effects produced by cascading two or more basic differentiator circuits?
23. Describe the shape of the waveform across the resistor and capacitor of Fig. 20-11 and its circuit function when the ratio of the time constant to the time of one cycle of the square-wave input is 1.
24. Describe the shape of the waveform across the resistor and capacitor of Fig. 20-11 and its circuit function when the ratio of the time constant to the time of one cycle of the square-wave input is 10 to 1.
25. Describe the shape of the waveform across the resistor and capacitor of Fig. 20-11 and its circuit function when the ratio of the time constant to the time of one cycle of the square-wave input is 1 to 10.

Appendix 1

Trigonometry

The solution of a-c problems frequently involves adding or subtracting quantities such as voltages, currents, and ohmages by means of vectors. The mathematical solution of these problems requires the use of trigonometry. The method of solution presented in the text makes it possible to solve all such problems by the use of right triangles. The following statements apply to any right triangle and are illustrated in Fig. A-1.

1. A right triangle is one in which one of the angles is a right angle (90 degrees).
2. The hypotenuse is the side opposite the right angle.
3. The legs of a right triangle are the two sides that form the right angle.
4. The sine of any angle θ is equal to the side opposite that angle divided by the hypotenuse.
5. The cosine of any angle θ is equal to the side adjacent to that angle divided by the hypotenuse.
6. The square of the hypotenuse is equal to the sum of the squares of the two legs of the triangle. (This is also commonly known as the *theorem of Pythagoras.*)

$$\sin A = \frac{a}{c} \qquad a = c \sin A \qquad c = \frac{a}{\sin A}$$

$$\cos A = \frac{b}{c} \qquad b = c \cos A \qquad c = \frac{b}{\cos A}$$

$$\sin B = \frac{b}{c} \qquad b = c \sin B \qquad c = \frac{b}{\sin B}$$

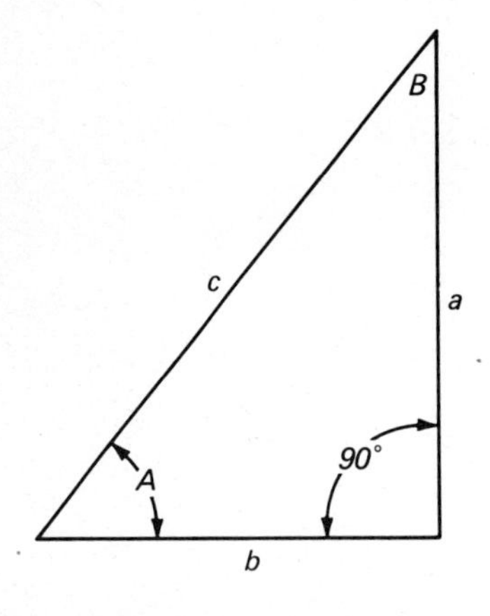

Fig. A-1

$$\cos B = \frac{a}{c} \qquad a = c \cos B \qquad c = \frac{a}{\cos B}$$

$$c^2 = a^2 + b^2 \qquad a^2 = c^2 - b^2 \qquad b^2 = c^2 - a^2$$

The tables of Appendix 2 list the values of sine and cosine for angles between 0 and 90 degrees. In some instances, it is desired to obtain the sine of angles greater than 90 degrees, and they may be obtained in the following manner:

When θ is between 90 and 180 degrees

$$\sin \theta = \cos (\theta - 90)$$

Example What is the sine of 137 degrees?

$$\sin 137° = \cos (137 - 90) = \cos 47° = 0.682$$

When θ is between 180 and 270 degrees

$$\sin \theta = - \sin (\theta - 180)$$

Example What is the sine of 218 degrees?

$$\sin 218° = - \sin (218 - 180) = - \sin 38° = -0.616$$

When θ is between 270 and 360 degrees

$$\sin \theta = - \cos (\theta - 270)$$

Example What is the sine of 336 degrees?

$$\sin 336° = - \cos (336 - 270) = - \cos 66° = -0.407$$

Appendix 2

Sine and Cosine Tables

DEGREES	SIN	COS	DEGREES	SIN	COS
0.0	0.000	1.000	19.5	0.334	0.942
0.5	0.009	1.000	20.0	0.342	0.940
1.0	0.017	0.999	20.5	0.350	0.937
1.5	0.026	0.999	21.0	0.358	0.933
2.0	0.035	0.999	21.5	0.366	0.930
2.5	0.043	0.999	22.0	0.374	0.927
3.0	0.052	0.998	22.5	0.383	0.924
3.5	0.061	0.998	23.0	0.391	0.920
4.0	0.070	0.997	23.5	0.399	0.917
4.5	0.078	0.997	24.0	0.407	0.913
5.0	0.087	0.996	24.5	0.415	0.910
5.5	0.096	0.995	25.0	0.422	0.906
6.0	0.104	0.994	25.5	0.430	0.902
6.5	0.113	0.993	26.0	0.438	0.899
7.0	0.122	0.992	26.5	0.446	0.895
7.5	0.130	0.991	27.0	0.454	0.891
8.0	0.139	0.990	27.5	0.462	0.887
8.5	0.148	0.989	28.0	0.469	0.883
9.0	0.156	0.988	28.5	0.477	0.879
9.5	0.165	0.986	29.0	0.485	0.875
10.0	0.173	0.985	29.5	0.492	0.870
10.5	0.182	0.983	30.0	0.500	0.866
11.0	0.191	0.981	30.5	0.507	0.862
11.5	0.199	0.980	31.0	0.515	0.857
12.0	0.208	0.978	31.5	0.522	0.853
12.5	0.216	0.976	32.0	0.530	0.848
13.0	0.225	0.974	32.5	0.537	0.843
13.5	0.233	0.972	33.0	0.544	0.839
14.0	0.242	0.970	33.5	0.552	0.834
14.5	0.250	0.968	34.0	0.559	0.829
15.0	0.259	0.966	34.5	0.566	0.824
15.5	0.267	0.963	35.0	0.574	0.819
16.0	0.275	0.961	35.5	0.581	0.814
16.5	0.284	0.959	36.0	0.588	0.809
17.0	0.292	0.956	36.5	0.595	0.804
17.5	0.301	0.954	37.0	0.602	0.798
18.0	0.309	0.951	37.5	0.609	0.793
18.5	0.317	0.948	38.0	0.616	0.788
19.0	0.325	0.945	38.5	0.622	0.783

DEGREES	SIN	COS	DEGREES	SIN	COS
39.0	0.629	0.777	65.0	0.906	0.423
39.5	0.636	0.772	65.5	0.910	0.415
40.0	0.643	0.766	66.0	0.913	0.407
40.5	0.649	0.760	66.5	0.917	0.399
41.0	0.656	0.755	67.0	0.920	0.391
41.5	0.663	0.749	67.5	0.924	0.383
42.0	0.669	0.743	68.0	0.927	0.375
42.5	0.675	0.737	68.5	0.930	0.366
43.0	0.682	0.731	69.0	0.934	0.358
43.5	0.688	0.725	69.5	0.937	0.350
44.0	0.695	0.719	70.0	0.940	0.342
44.5	0.701	0.713	70.5	0.943	0.334
45.0	0.707	0.707	71.0	0.945	0.326
45.5	0.713	0.701	71.5	0.948	0.317
46.0	0.719	0.695	72.0	0.951	0.309
46.5	0.725	0.688	72.5	0.954	0.301
47.0	0.731	0.682	73.0	0.956	0.292
47.5	0.737	0.675	73.5	0.959	0.284
48.0	0.743	0.669	74.0	0.961	0.276
48.5	0.749	0.663	74.5	0.964	0.267
49.0	0.755	0.656	75.0	0.966	0.259
49.5	0.760	0.649	75.5	0.968	0.250
50.0	0.766	0.643	76.0	0.970	0.242
50.5	0.772	0.636	76.5	0.972	0.233
51.0	0.777	0.629	77.0	0.974	0.225
51.5	0.783	0.622	77.5	0.976	0.216
52.0	0.788	0.616	78.0	0.978	0.208
52.5	0.793	0.609	78.5	0.980	0.199
53.0	0.798	0.602	79.0	0.982	0.191
53.5	0.804	0.595	79.5	0.983	0.182
54.0	0.809	0.588	80.0	0.985	0.174
54.5	0.814	0.581	80.5	0.986	0.165
55.0	0.819	0.574	81.0	0.988	0.156
55.5	0.824	0.566	81.5	0.989	0.148
56.0	0.829	0.559	82.0	0.990	0.139
56.5	0.834	0.552	82.5	0.991	0.130
57.0	0.839	0.544	83.0	0.992	0.122
57.5	0.843	0.537	83.5	0.994	0.113
58.0	0.848	0.530	84.0	0.994	0.104
58.5	0.853	0.522	84.5	0.995	0.096
59.0	0.857	0.515	85.0	0.996	0.087
59.5	0.862	0.507	85.5	0.997	0.078
60.0	0.866	0.500	86.0	0.997	0.070
60.5	0.870	0.492	86.5	0.998	0.061
61.0	0.875	0.485	87.0	0.998	0.052
61.5	0.879	0.477	87.5	0.999	0.043
62.0	0.883	0.469	88.0	0.999	0.035
62.5	0.887	0.462	88.5	0.999	0.026
63.0	0.891	0.454	89.0	0.999	0.017
63.5	0.895	0.446	89.5	1.000	0.009
64.0	0.899	0.438	90.0	1.000	0.000
64.5	0.903	0.430			

Appendix 3

Common Logarithms of Numbers

N	0	1	2	3	4	5	6	7	8	9
10	0000	0043	0086	0128	0170	0212	0253	0294	0334	0374
11	0414	0453	0492	0531	0569	0607	0645	0682	0719	0755
12	0792	0828	0864	0899	0934	0969	1004	1038	1072	1106
13	1139	1173	1206	1239	1271	1303	1335	1367	1399	1430
14	1461	1492	1523	1553	1584	1614	1644	1673	1703	1732
15	1761	1790	1818	1847	1875	1903	1931	1959	1987	2014
16	2041	2068	2095	2122	2148	2175	2201	2227	2253	2279
17	2304	2330	2355	2380	2405	2430	2455	2480	2504	2529
18	2553	2577	2601	2625	2648	2672	2695	2718	2742	2765
19	2788	2810	2833	2856	2878	2900	2923	2945	2967	2989
20	3010	3032	3054	3075	3096	3118	3139	3160	3181	3201
21	3222	3243	3263	3284	3304	3324	3345	3365	3385	3404
22	3424	3444	3464	3483	3502	3522	3541	3560	3579	3598
23	3617	3636	3655	3674	3692	3711	3729	3747	3766	3784
24	3802	3820	3838	3856	3874	3892	3909	3927	3945	3962
25	3979	3997	4014	4031	4048	4065	4082	4099	4116	4133
26	4150	4166	4183	4200	4216	4232	4249	4265	4281	4298
27	4314	4330	4346	4362	4378	4393	4409	4425	4440	4456
28	4472	4487	4502	4518	4533	4548	4564	4579	4594	4609
29	4624	4639	4654	4669	4683	4698	4713	4728	4742	4757
30	4771	4786	4800	4814	4829	4843	4857	4871	4886	4900
31	4914	4928	4942	4955	4969	4983	4997	5011	5024	5038
32	5051	5065	5079	5092	5105	5119	5132	5145	5159	5172
33	5185	5198	5211	5224	5237	5250	5263	5276	5289	5302
34	5315	5328	5340	5353	5366	5378	5391	5403	5416	5428
35	5441	5453	5465	5478	5490	5502	5514	5527	5539	5551
36	5563	5575	5587	5599	5611	5623	5635	5647	5658	5670
37	5682	5694	5705	5717	5729	5740	5752	5763	5775	5786
38	5798	5809	5821	5832	5843	5855	5866	5877	5888	5899
39	5911	5922	5933	5944	5955	5966	5977	5988	5999	6010
40	6021	6031	6042	6053	6064	6075	6085	6096	6107	6117
41	6128	6138	6149	6160	6170	6180	6191	6201	6212	6222
42	6232	6243	6253	6263	6274	6284	6294	6304	6314	6325
43	6335	6345	6355	6365	6375	6385	6395	6405	6415	6425
44	6435	6444	6454	6464	6474	6484	6493	6503	6513	6522
45	6532	6542	6551	6561	6571	6580	6590	6599	6609	6618
46	6628	6637	6646	6656	6665	6675	6684	6693	6702	6712
47	6721	6730	6739	6749	6758	6767	6776	6785	6794	6803

N	0	1	2	3	4	5	6	7	8	9
48	6812	6821	6830	6839	6848	6857	6866	6875	6884	6893
49	6902	6911	6920	6928	6937	6946	6955	6964	6972	6981
50	6990	6998	7007	7016	7024	7033	7042	7050	7059	7067
51	7076	7084	7093	7101	7110	7118	7126	7135	7143	7152
52	7160	7168	7177	7185	7193	7202	7210	7218	7226	7235
53	7243	7251	7259	7267	7275	7284	7292	7300	7308	7316
54	7324	7332	7340	7348	7356	7364	7372	7380	7388	7396
55	7404	7412	7419	7427	7435	7443	7451	7459	7466	7474
56	7482	7490	7497	7505	7513	7520	7528	7536	7543	7551
57	7559	7566	7574	7582	7589	7597	7604	7612	7619	7627
58	7634	7642	7649	7657	7664	7672	7679	7686	7694	7701
59	7709	7716	7723	7731	7738	7745	7752	7760	7767	7774
60	7782	7789	7796	7803	7810	7818	7825	7832	7839	7846
61	7853	7860	7868	7875	7882	7889	7896	7903	7910	7917
62	7924	7931	7938	7945	7952	7959	7966	7973	7980	7987
63	7993	8000	8007	8014	8021	8028	8035	8041	8048	8055
64	8062	8069	8075	8082	8089	8096	8102	8109	8116	8122
65	8129	8136	8142	8149	8156	8162	8169	8176	8182	8189
66	8195	8202	8209	8215	8222	8228	8235	8241	8248	8254
67	8261	8267	8274	8280	8287	8293	8299	8306	8312	8319
68	8325	8331	8338	8344	8351	8357	8363	8370	8376	8382
69	8388	8395	8401	8407	8414	8420	8426	8432	8439	8445
70	8451	8457	8463	8470	8476	8482	8488	8494	8500	8506
71	8513	8519	8525	8531	8537	8543	8549	8555	8561	8567
72	8573	8579	8585	8591	8597	8603	8609	8615	8621	8627
73	8633	8639	8645	8651	8657	8663	8669	8675	8681	8686
74	8692	8698	8704	8710	8716	8722	8727	8733	8739	8745
75	8751	8756	8762	8768	8774	8779	8785	8791	8797	8802
76	8808	8814	8820	8825	8831	8837	8842	8848	8854	8859
77	8865	8871	8876	8882	8887	8893	8899	8904	8910	8915
78	8921	8927	8932	8938	8943	8949	8954	8960	8965	8971
79	8976	8982	8987	8993	8998	9004	9009	9015	9020	9025
80	9031	9036	9042	9047	9053	9058	9063	9069	9074	9079
81	9085	9090	9096	9101	9106	9112	9117	9122	9128	9133
82	9138	9143	9149	9154	9159	9165	9170	9175	9180	9186
83	9191	9196	9201	9206	9212	9217	9222	9227	9232	9238
84	9243	9248	9253	9258	9263	9269	9274	9279	9284	9289
85	9294	9299	9304	9309	9315	9320	9325	9330	9335	9340
86	9345	9350	9355	9360	9365	9370	9375	9380	9385	9390
87	9395	9400	9405	9410	9415	9420	9425	9430	9435	9440
88	9445	9450	9455	9460	9465	9469	9474	9479	9484	9489
89	9494	9499	9504	9509	9513	9518	9523	9528	9533	9538
90	9542	9547	9552	9557	9562	9566	9571	9576	9581	9586
91	9590	9595	9600	9605	9609	9614	9619	9624	9628	9633
92	9638	9643	9647	9652	9657	9661	9666	9671	9675	9680
93	9685	9689	9694	9699	9703	9708	9713	9717	9722	9727
94	9731	9736	9741	9745	9750	9754	9759	9763	9768	9773
95	9777	9782	9786	9791	9795	9800	9805	9809	9814	9818
96	9823	9827	9832	9836	9841	9845	9850	9854	9859	9863
97	9868	9872	9877	9881	9886	9890	9894	9899	9903	9908
98	9912	9917	9921	9926	9930	9934	9939	9943	9948	9952
99	9956	9961	9965	9969	9974	9978	9983	9987	9991	9996

Appendix 4

Answers to Odd-numbered Problems

Note 1: Answers are provided for approximately 50 per cent of the problems. Instructors using this text may obtain an answer book containing answers to the remaining problems.

Note 2: As far as is practicable, all answers are accurate to three significant figures.

Note 3: Answers to problems involving values obtained from curves are generally difficult to check accurately because of variations in reading the curves. In preparing the answer book, enlarged drawings of the curve were used to aid in obtaining greater accuracy. In many cases the values obtained from the curves for use in solving the problems have been included with the answers.

CHAPTER 1

1. 250 m
3. 30 MHz
5. (*a*) 0.469 m
   (*b*) 1.54 ft
   (*c*) 18.5 in.
7. 561.25 MHz
9. (*a*) 3.14 m
   (*b*) 10.3 ft
   (*c*) 123.6 in.
11. 0.197 in.
13. (*a*) 0.0268 sec
    (*b*) 0.0537 sec
15. 0.000538 sec
17. (*a*) 1.345 in.
    (*b*) 4.414 ft
19. 4,687.5 Hz
21. 0.0753 ft
23. (*a*) 4,687.5 Hz
    (*b*) 80 Hz
25. (*a*) 0.0885 sec
    (*b*) 0.00107 sec
    (*c*) The home listener
27. 1,646 miles
29. (*a*) 500 Hz
    (*b*) 1,000 Hz
31. (*a*) 100 Hz
    (*b*) 50 Hz
    (*c*) 50 Hz
    (*d*) 100 Hz
33. (*a*) 2.2553
    (*b*) 3.4393
    (*c*) 0.942
    (*d*) 1.0969
    (*e*) 0.699
    (*f*) 1.9934
    (*g*) 4.5441
    (*h*) 7.4871 − 10
    (*i*) 1.2625
    (*j*) 6.9938 − 10
35. (*a*) 300
    (*b*) 755,000
    (*c*) 6
    (*d*) 0.0107
    (*e*) 0.000218
37. 2.9 db
39. 5.52 watts
41. (*a*) 26 db
    (*b*) 35.56 db
43. −6.02 db
45. −1.86 db
47. (*a*) 50 ft-c
    (*b*) 4 ft-c
    (*c*) 0.5 ft-c
49. (*a*) 0.0000666 cm
    (*b*) 0.0000262 in.
    (*c*) 26.2 μin.

CHAPTER 2

1. (*a*) 5 ohms
   (*b*) 400 megohms
3. (*a*) 10 ohms
   (*b*) 100,000 ohms
5. (*a*) 0.2775 ohm
   (*b*) 0.555 ohm
   (*c*) 5.55 ohms
7. (*a*) 0.375 ohm
   (*b*) 0.5 ohm
   (*c*) 0.625 ohm
9. (*a*) 0.5 ohm
   (*b*) 1 ohm
   (*c*) 10 ohms
11. (*a*) 3.61 ohms
    (*b*) 3.89 ohms
    (*c*) 8.88 ohms
13. (*a*) 5.375 ohms
    (*b*) 5.5 ohms
    (*c*) 5.625 ohms
15. (*a*) 10.5 ohms

(*b*) 11.0 ohms
(*c*) 20.0 ohms
17. (*a*) 3.61 to 8.88 ohms
(*b*) 3.73 to 11.33 ohms
19. (*a*) 17 ma
(*b*) 75 ma
21. (*a*) 28 ohms
(*b*) 12 ohms
23. 937.5 ohms
25. (*a*) 52.65 volts
(*b*) 105.3 volts
(*c*) 105.3 volts

CHAPTER 4

1. (*a*) 7.85 ma
(*b*) 5.9 ma
3. (*a*) −4 ma
(*b*) −1 ma
5. (*a*) −4.2 ma
(*b*) −2.8 ma
7. (*a*) $I_C = 4.9$ ma
$I_B = 0.1$ ma
(*b*) $I_C = 4.91$ ma
$I_B = 90\ \mu a$
9. 0.975
11. 15
13. 19
15. 0.952
17. (*a*) 11.37 ma
(*b*) −1.37 ma
19. 14
21. $I_C = 1.355$ ma
$I_B = 45\ \mu a$
23. 1,900
25. 1,175
27. 0.875
29. 32.5 db
31. 44.4 db
33. 14.8 db
35. (*a*) Curve
(*b*) Load line
37. (*a*) 115
(*b*) 192
(*c*) 43.4 db
39. 125 mmhos
41. 83.3 mmhos
43. 79.5 MHz
45. 8.5 MHz
47. 4 MHz
49. (*a*) 20
(*b*) 13 db
51. 4
53. 650 nsec
55. (*a*) 13 nsec
(*b*) 16 nsec

CHAPTER 5

5. (*a*) 135 volts
(*b*) 400 volts
(*c*) 240 volts
7. (*a*) −7 volts
(*b*) −4.2 volts
(*c*) −16 volts
9. (*a*) 400 volts
(*b*) 80 volts
(*c*) 300 volts
11. 80
13. 20
15. 50,000 ohms
17. 7,500 ohms
19. 1,400 μmhos
21. (*a*) 2,650 μmhos
(*b*) 2,660 μmhos
23. 1,400 μmhos
25. 74
27. 32,000 ohms
29. (*a*) 8 ma
(*b*) 4 ma
(*c*) 4.5 ma
(*d*) 6 ma
31. (*a*) 2 volts
(*b*) 10 volts
33. (*a*) 150 ma
(*b*) 105 ma
(*c*) 75 ma
(*d*) 45 ma
35. (*a*) −8.7 volts
(*b*) −7.5 volts
(*c*) −4.5 volts
(*d*) −2.8 volts

CHAPTER 6

1. (*a*) 7.5 volts
(*b*) 12 volts
(*c*) 22.5 volts
(*d*) 30 volts
(*e*) 60 volts
3. (*a*) 6.6 volts
(*b*) 13.2 volts
(*c*) 33 volts
(*d*) 88 volts
(*e*) 132 volts
5. (*a*) 6
(*b*) 30
(*c*) 45
(*d*) 80
(*e*) 100
7. (*a*) 400 ma
(*b*) 600 ma
(*c*) 1 amp
(*d*) 1.6 amp
(*e*) 2.4 amp
9. (*a*) 3
(*b*) 5
(*c*) 8
(*d*) 10
11. (*a*) 8 cells
(*b*) 5 cells
(*c*) 40 cells
(*d*) cct. diag.
13. (*a*) 5 cells
(*b*) 3 cells
(*c*) 15 cells
(*d*) cct. diag.
15. (*a*) 5.55%
(*b*) 0.0555
17. 0.18 volt
19. 3.74%
21. (*a*) 1,500 ohms
(*b*) 998 μf
(*c*) 2.66 ohms
(*d*) 62 amp
(*e*) Tube will burn out
(*f*) No
23. (*a*) 7.48%
(*b*) 0.134%
25. (*a*) 12.4%
(*b*) 0.452%
(*c*) 0.022%
27. (*a*) 7.8%
(*b*) 0.517%
30. 50%

32. 96 volts
33. 2%
35. (*a*) 1%
    (*b*) 0.0162%
37. 11.5%
39. (*a*) 1.44%
    (*b*) 0.0337%
40. 4.8 henrys
42. $R_1 = 100$ ohms
    $R_2 = 100$ ohms
    $R_3 = 150$ ohms
    $R_4 = 30$ ohms
43. (*a*) 0.95 watt
    (*b*) $P_1 = 0.09$ watt
    $P_2 = 0.0225$ watt
    $P_3 = 0.015$ watt
    $P_4 = 0.075$ watt
    (*c*) 0.2025 watt
46. $R_1 = 3{,}367$ ohms
    $R_2 = 1{,}900$ ohms
    $R_3 = 5{,}618$ ohms
    $R_4 = 24.7$ ohms
    $R_5 = 6.2$ ohms
    $R_6 = 191$ ohms
47. (*a*) 145.6 watts
    (*b*) $P_1 = 6$ watts
    $P_2 = 2.6$ watts
    $P_3 = 3.6$ watts
    $P_4 = 0.324$ watt
    $P_5 = 0.081$ watt
    $P_6 = 2.5$ watts
50. (*a*) 323 volts
    (*b*) 343 volts
52. (*a*) 99.8 volts
    (*b*) 116.8 volts

CHAPTER 7

1. (*a*) 340 volts
   (*b*) 120 Hz
   (*c*) 170 volts
   (*d*) 340 volts
3. (*a*) 510 volts
   (*b*) 120 Hz
   (*c*) $C_1 = 170$ volts
   $C_2 = 340$ volts
   $C_3 = 170$ volts
   (*d*) 340 volts
5. (*a*) 680 volts
   (*b*) 60 Hz
   (*c*) $C_1 = 170$ volts
   $C_2 = C_3 = C_4$
   $= 340$ volts
   (*d*) 340 volts
7. (*a*) 6
   (*b*) Diagram
   (*c*) $C_1 = 170$ volts
   $C_2 = C_3 = C_4$
   $= C_5 = C_6$
   $= 340$ volts
   (*d*) 340 volts
9. 22.8%
11. 20 ohms
13. $R_1 = 133$ ohms
    $R_2 = 72.8$ ohms
15. 15 ohms

CHAPTER 8

3. (*a*) 2.65 ma
   (*b*) 3.8 ma
   (*c*) 1.5 ma
   (*d*) 1.15 ma
   (*e*) 1.15 ma
   (*f*) No
   (*g*) Class A
5. (*a*) 1.5 ma
   (*b*) 3.8 ma
   (*c*) 0
   (*d*) 2.3 ma
   (*e*) 1.5 ma
   (*f*) Yes
   (*g*) Class AB
7. (*a*) 0
   (*b*) 3.25 ma
   (*c*) 0
   (*d*) 3.25 ma
   (*e*) 0
   (*f*) Yes
   (*g*) Class B
9. From Fig. 8-3
   $V_{CE} = 3.3$ volts
   53,500 ohms
11. (*a*) 12 μa
    (*b*) 1.2 ma
    (*c*) 1.212 ma
    (*d*) 12 volts
13. 400,000 ohms
15. (*a*) 2 ma
    (*b*) 1.92 ma
    (*c*) 0.08 ma
    (*d*) 10.8 volts
17. 7,800 ohms
19. (*a*) 0.705 ma
    (*b*) 0.705 ma
    (*c*) 8.8 μa
    (*d*) 9.9 volts
21. (*a*) 24 μa
    (*b*) 2.4 ma
    (*c*) 0
23. 2,000,000 ohms
25. (*a*) 10 volts
    (*b*) 1 ma
    (*c*) 1 ma
    (*d*) 5 volts
27. 3,000 ohms
29. (*a*) 1.25 ma
    (*b*) 1.25 ma
    (*c*) 15.6 μa
    (*d*) 7.5 volts
31. 10.9 volts
33. 1
35. 14.45 ma
37. 321.2 ma
39. 400%
41. 300%
43. 1.25
45. 0.5
48. 0
49. 50
51. (*a*) 30 μa
    (*b*) 1.8 ma
    (*c*) 4.5 volts
53. $R_B = 225{,}000$ ohms
    $R_L = 2{,}000$ ohms
55. (*a*) 20 ma
    (*b*) 0.5 ma
    (*c*) 8.6 volts
    (*d*) 1.28
    (*e*) 0.031
57. (*a*) 21.3 ma
    (*b*) 0.533 ma
    (*c*) 7.2 volts
    (*d*) 1.28
    (*e*) 0.031

59. (*a*) 0.01
 (*b*) 0.00052
61. (*a*) 5.8 ma
 (*b*) 116 μa
 (*c*) 12.4 volts
 (*d*) 0.5
 (*e*) 50
63. (*a*) 1.6 ma
 (*b*) 8 volts
 (*c*) 0.01
 (*d*) 0.667
65. (*a*) 2.9 ma
 (*b*) 3.4 ma
 (*c*) 5.2 ma
67. (*a*) 5 volts
 (*b*) 4 volts
 (*c*) 6 volts
69. (*a*) 2 ma
 (*b*) 4 volts
 (*c*) 0.333
 (*d*) 50
71. 40 volts
73. (*a*) 2 volts
 (*b*) 3 volts
 (*c*) 4 volts
 (*d*) 4 volts
75. (*a*) 0.5 volt
 (*b*) 33.3%
77. (*a*) 2 volts
 (*b*) 2.5 mv
79. (*a*) 4 μf
 (*b*) 4 μf, 25 volts, electrolytic

CHAPTER 9

1. 2×10^{-4} mho
3. 150
5. 1,850 ohms
7. 175×10^{-5}
9. (*a*) 1,135 ohms
 (*b*) 1,135 ohms
11. 17,200 ohms
13. 68
15. 360
17. (*a*) 24,500
 (*b*) 43.9 db
19. 1,420 ohms
21. 1,120,000 ohms
23. 17,200 ohms
25. 100
27. 80
29. 1,790
31. $A_P = 56{,}000$
 $G_P = 47.5$ db
33. 90.9%
35. 66.6%
37. 50%
39. 33.3%
41. $\dfrac{\beta r_e}{R_g + \beta r_e}$
43. 9.1%
45. d-c load line
 a-c load line
47. (*a*) 7.55 volts
 (*b*) 25 mv (peak)
49. (*a*) 1.12 megohms
 (*b*) 8.55 volts
 (*c*) 25 mv (peak)
51. (*a*) 312.5 ohms
 (*b*) 3,330 ohms
 (*c*) 532
 (*d*) 13.3 volts
 (*e*) 31.4
 (*f*) 42.2 db
53. (*a*) 1 ma
 (*b*) 1,205 ohms
 (*c*) 1,430 ohms
 (*d*) 57
 (*e*) 1 volt
55. (*a*) 4.5 ma
 (*b*) 2,940 ohms
 (*c*) 124 ohms
 (*d*) 530
 (*e*) 7.1
 (*f*) 35.8 db
57. (*a*) 0.5 ma
 (*b*) 49.8 ohms
 (*c*) 3,500 ohms
 (*d*) 0.98
 (*e*) 70
 (*f*) 1.75 volts (peak)
 (*g*) 18.4 db
59. $\dfrac{r_e}{R_g + r_e}$
61. $h_{ob} = 1.32 \times 10^{-6}$ mho
 $h_{fb} = -0.993$
 $h_{ib} = 12.2$ ohms
 $h_{rb} = 70 \times 10^{-5}$
63. (*a*) 14.6 ohms
 (*b*) −0.99
 (*c*) 237
 (*d*) 23.7 db
65. (*a*) 1 ma
 (*b*) 915 ohms
 (*c*) 38,800 ohms
 (*d*) 80
 (*e*) 0.973
 (*f*) 18.9 db
67. $h_{oc} = 200$ μmhos
 $h_{fc} = -151$
 $h_{ic} = 1{,}850$ ohms
 $h_{rc} = 1$
69. (*a*) 915 ohms
 (*b*) 118,000 ohms
 (*c*) 128
 (*d*) 0.99
 (*e*) 21 db
71. (*a*) 47
 (*b*) 62,500 ohms
73. (*a*) 0.1 ma
 (*b*) 9.4 volts
75. 31
77. (*a*) 18.6 volts
 (*b*) 1.5 ma
79. 54
81. 22.5
83. 7,000 ohms
85. 444
87. (*a*) 1 ma
 (*b*) 200 volts
89. 667
91. (*a*) 800
 (*b*) 2,000
 (*c*) 444 vs. 800
 667 vs. 2,000
 (*d*) Eq. (9-49) useful only when $r_p \gg Z_L$
93. (*a*) 333 ohms
 (*b*) 6
95. 0.934
97. 0.962

CHAPTER 10

1. (*a*) 200 ohms
   (*b*) 300 ohms
3. (*a*) 1,670 ohms
   (*b*) 172 ohms
5. (*a*) $dV_G = -1$ to -2
   $dI_D = 0.38$ to 1.75
   $g_{fs} = 1{,}370$ μmhos
   (*b*) $dV_G = -1$ to 0
   $dI_D = 1.75$ to 4.1
   $g_{fs} = 2{,}350$ μmhos
7. (*a*) 27
   (*b*) 36
9. (*a*) 33.2
   (*b*) 44.8
11. 15.11
13. 21.37 db
15. 1.5
17. 0.51 db
19. 0.83
21. 21.4 db
23. $V_{DS} = 10$ $I_D = 4.2$
    $V_{DS} = 15$ $I_D = 4.6$
    $r_{os} = 12{,}500$ ohms
25. $V_{GS} = -1.5$ $I_D = 5.0$
    $V_{GS} = -2.5$ $I_D = 3.0$
    $g_{fs} = 2.0$ mmhos
27. $V_{DS} = 12.5$ $I_D = 3.0$
    $V_{DS} = 7.5$ $I_D = 2.8$
    $r_{os} = 25{,}000$ ohms
29. $V_{GS} = 4.5$ $I_D = 2.9$
    $V_{GS} = 3.5$ $I_D = 1.4$
    $g_{fs} = 1.5$ mmhos

CHAPTER 11

1. (*a*) 320 pf
   (*b*) 35.5 pf
3. (*a*) 289 μh
   (*b*) 2,416 kHz
5. (*a*) 478 pf
   (*b*) 46
   (*c*) 10 kHz
7. (*a*) 5,000 Hz
   (*b*) 10,000 Hz
9. (*a*) $E_R = 10$ mv
   $E_L = 1.987$ volts
   $E_C = 1.987$ volts
   (*b*) 198.7
11. (*a*) Curve
    (*b*) Curve
13. (*a*) 86 ohms (cap.)
    (*b*) 46 ohms (cap.)
    (*c*) 7 ohms (cap.)
    (*d*) 33 ohms (ind.)
    (*e*) 73 ohms (ind.)
15. (*a*) 1,001.8 kHz
    (*b*) 1,000 kHz
17. Curve
19. (*a*) 5,000 Hz
    (*b*) 5,039 Hz
21. (*a*) 996 μh
    (*b*) 292
23. 840,457 ohms
25. (*a*) 80,000 ohms
    (*b*) 135,000 ohms
    (*c*) 83,000 ohms
    (*d*) 137,000 ohms
27. Curve
29. (*a*) $Z_{pr} = 790{,}000$ ohms
    $I_{pr} = 126$ μa
    (*b*) $Z_{pr} = 395{,}000$ ohms
    $I_{pr} = 252$ μa
31. (*a*) 50 ma
    (*b*) 50 ma
33. 489 to 1,707 kHz
35. (*a*) 194 MHz
    (*b*) 204 MHz
37. 209.6 pf
39. (*a*) $L = 199$ μh
    (*b*) 1,083
    (*c*) 1.2 kHz
41. (*a*) 70 pf
    (*b*) 3.014 ohms
43. (*a*) 30 to 335 pf
    (*b*) 490 to 1,632 kHz
45. 480 to 1,420 kHz
47. (*a*) 5.4 pf
    (*b*) Negligible, 500 to 494 kHz
49. (*a*) 8.57 to 18.8 pf
    (*b*) 2.06 to 3.06 MHz
51. (*a*) 710 pf
    (*b*) 2,092 kHz
53. (*a*) 1,145 pf
    (*b*) 1,890 kHz
55. (*a*) 25 μh
    (*b*) 8.2 MHz
57. (*a*) 24 μh
    (*b*) 5.92 MHz
59. (*a*) 15.86 pf
    (*b*) 1.96 MHz

CHAPTER 12

1. (*a*) 3,290 ohms
   (*b*) 1,645 ohms
   (*c*) 31.7 ohms
   (*d*) 3.56 ohms
   (*e*) 31.6 ohms
   (*f*) cct. diag.
   (*g*) 3,259 ohms
   (*h*) 3.1 ma
   (*i*) 0.71 volt
   (*j*) 0.43 ma
3. 1.19 henrys
5. (*a*) 4,000 Hz audio freq.
   (*b*) 375 times
   (*c*) 4,000 Hz audio freq.
7. (*a*) 63,600 ohms
   (*b*) 628 ohms
   (*c*) 424 ohms
   (*d*) 94,200 ohms
9. (*a*) 6,625 ohms
   (*b*) 1,507.2 ohms
   (*c*) 331.25 ohms
   (*d*) 30,144 ohms
11. 0.6
13. 0.163 pf
15. 9.3 kHz
17. 187.5
19. (*a*). 00219
    (*b*) 451 to 461 kHz
21. (*a*) $R_1 = 1{,}500$ ohms
    $R_2 = 360$ ohms
    $K = 6$
    (*b*) 1,800 ohms
    (*c*) $e_o = 10$ volts
    $K = 6$

23. (*a*) $e_o = 30$ volts
    $R_1 = 80{,}000$ ohms
    $R_2 = 50{,}000$ ohms
    $R_3 = 80{,}000$ ohms
  (*b*) 120,000 ohms
25. (*a*) $K = 1.5$
    $R_1 = 60{,}000$ ohms
    $R_2 = 5{,}000$ ohms
    $R_3 = 60{,}000$ ohms
  (*b*) 12,000 ohms
27. (*a*) 0.025 sec
  (*b*) 0.04 sec
29. (*a*) 2 megohms
  (*b*) 1.33 megohms
  (*c*) 0.2 μf
31. (*a*) $t_{50} = 0.005$ sec
    $t_{100} = 0.0125$ sec
    $t_{200} = 0.04$ sec
  (*b*) 500 μa
  (*c*) 100 μa
  (*d*) 500 μa
  (*e*) 92.6 volts
  (*f*) 0.0175 sec

CHAPTER 13

1. Cct. diagram
3. (*a*) 180
  (*b*) 1,250 ohms
  (*c*) 2.7 volts
  (*d*) 8.75 volts
5. 61.5
7. (*a*) 200
  (*b*) 50
  (*c*) 80 db
  (*d*) 11 volts
  (*e*) 12.5 volts
9. (*a*) $A_{v\cdot 2}$ decreases 25%
  (*b*) $A_{v\cdot 1}$ decreases 6%
  (*c*) A_v decreases 29.5%; down 3 db
  (*d*) $V_{CE\cdot 1}$ decreases 25%
  (*e*) $V_{CE\cdot 2}$ decreases 25%
11. (*a*) Decreases to less than 1
  (*b*) No change
  (*c*) Decreases almost 100%
  (*d*) No change
  (*e*) No change
13. (*a*) 65.5
  (*b*) 6
  (*c*) 393
  (*d*) 52 db
15. (*a*) 140
  (*b*) 16
  (*c*) 2,240
  (*d*) 67 db
17. (*a*) 420
  (*b*) 59.5
  (*c*) 25,000
  (*d*) 88 db
19. (*a*) 35
  (*b*) 19.7
  (*c*) 690
  (*d*) 56.8 db
21. 25 to 1
23. (*a*) 3,200 ohms
  (*b*) 120 mv
25. (*a*) 160
  (*b*) 6.4 volts (p-p)
  (*c*) 640 mv (p-p)
  (*d*) 16
  (*e*) 8 volts
  (*f*) 10 volts
  (*g*) d-c a-c load lines
27. 200,000 ohms
29. (*a*) 20
  (*b*) 10 volts (p-p)
  (*c*) 50 mv (p-p)
  (*d*) d-c a-c load lines
31. (*a*) 64.6
  (*b*) 0.234
  (*c*) 15.1
  (*d*) 755 mv
33. (*a*) 1,250
  (*b*) 1
  (*c*) 120
  (*d*) 120
  (*e*) 51.8 db
35. (*a*) 9.4
  (*b*) 3
  (*c*) 28.2
  (*d*) 13,500 ohms
37. (*a*) 1 ma
  (*b*) 25 ma
  (*c*) 1,250
  (*d*) 240
  (*e*) 55 db
39. (*a*) 80 μf
  (*b*) 240
  (*c*) 2
  (*d*) 480
41. (*a*) 16 volts
  (*b*) 6 volts
  (*c*) 6,000 ohms
43. (*a*) $I_{E\cdot 1} = 7.5$ ma
    $r_e = 3.33$ ohms
  (*b*) $I_{E\cdot 2} = 375$ ma
    $r_e = 0.0667$ ohm
  (*c*) 1
  (*d*) 240
  (*e*) 240
45. (*a*) 1 ma
  (*b*) 1 ma
  (*c*) 1
  (*d*) 160
  (*e*) 160
  (*f*) 25 ohms
47. (*a*) 0.5 ma
  (*b*) 25 ma
  (*c*) 0.95
  (*d*) 19.2
  (*e*) 18.2
  (*f*) 50 ohms
49. (*a*) $I_{E\cdot 1} = I_{E\cdot 2} = I_{E\cdot 3} = 2$ ma
  (*b*) $r_{e\cdot 1} = r_{e\cdot 2} = r_{e\cdot 3} = 12.5$ ohms
  (*c*) 200
  (*d*) 80
  (*e*) 80
  (*f*) 122 db
51. (*a*) 2 ma
  (*b*) 12.5 ohms
  (*c*) 49
  (*d*) 8.9
  (*e*) 8.9
  (*f*) 71.8 db
53. (*a*) 270,000 ohms
  (*b*) 265,000 ohms
55. (*a*) 2,000 ohms
  (*b*) 53 μf
  (*c*) 5 volts
57. (*a*) 2,500 ohms

(b) 20 μf
(c) 70 volts
59. (a) 0.158
(b) 6.7 μf

CHAPTER 14

1. 0.333
3. 38 Hz
5. 0.707
7. 0.236
9. 0.312
11. −0.56 db
13. 1.5 MHz
15. 0.235
17. 0.316
19. Down 0.44 db
21. $f_{l\cdot co}$ = 59 Hz
$f_{h\cdot co}$ = 960 kHz
23. Change C_4 to 100 μf
25. (a) 23.6 Hz
(b) 50 Hz
27. 28.5 ohms
29. (a) 1.2 MHz
(b) 370 kHz
(c) 4.15 MHz
31. Increases to 2.37 MHz
33. 478 kHz
35. 568 kHz
37. 13.2
39. (a) 15.7 Hz
(b) 55.5 Hz
(c) 44 Hz
41. (a) 6.9 MHz
(b) 2.42 MHz
43. (a) 0.384
(b) 5.1
(c) −8.3 db
45. (a) 0.99
(b) 13.1
(c) −0.1 db
47. (a) 360
(b) 53 Hz
(c) 83 kHz
49. (a) 12.5 henrys
(b) 0.5 μf
(c) 0.707
51. (a) 80
(b) 422
(c) 33,760; 90.6 db
53. (a) 0.812
(b) 0.77
55. (a) 46 Hz
(b) 70 kHz
57. 159 Hz
59. (a) 32 Hz
(b) 53.5 kHz
(c) 35.5 kHz
61. (a) 2.5
(b) 1.77
(c) 1.77
(d) 2.77
63. (a) 1.53
(b) 5.2
65. $R_{eq}^2 = 2X_{CT}^2 - X_{eq}^2$
67. (a) 54 kHz
(b) 54 vs. 53.5 kHz
69. (a) 6.4 Hz
(b) 45 kHz
(c) 29 kHz
(d) 6.85

CHAPTER 15

1. −3 volts
3. −0.23 volt
5. 0.0083
7. (a) −12.5 mv
(b) −12.5 mv
9. 50 mv
11. (a) −3 volts
(b) −0.12 volt
(c) −3.12 volts
13. (a) −0.23 volt
(b) −0.0092 volt
(c) −0.2392 volt
15. (a) 325 mv
(b) −3 volts
(c) 0.00924 volt
(d) −3.009 volts
17. (a) −5
(b) 5.26
19. (a) −18 db
(b) −3 db
21. (a) 0.0475
(b) −19
(c) −0.95 volt
23. (a) −523
(b) 57
(c) −9
(d) −43
(e) −11.9
25. (a) −340
(b) −0.545
(c) −220
27. (a) −327
(b) −4.75
(c) −57
29. (a) 0.10
(b) −9.4
(c) −10
31. (a) 0.0148
(b) −8.1 volts
33. (a) 975 μh
(b) 24.5
(c) 18.5 kHz
35. 6.6 μh
37. (a) 2.82 MHz
(b) 2 MHz
39. (a) 32 Hz
(b) 3.2 Hz
(c) 320 Hz
41. (a) 8,550 ohms
(b) 7,300 ohms
(c) 0.976
(d) 48.8
43. (a) 0.94
(b) 320 ohms
(c) 220,000 ohms
(d) 3.54 pf
45. (a) 510,000 ohms
(b) 7 volts
47. (a) 4 volts
(b) 3 volts
(c) 1 volt
(d) T_3 is negative
T_4 is positive
49. (a) 2.5 ma
(b) 1.25 ma
(c) 20 ohms
(d) 250
(e) 4,000 ohms

51. (a) 3 ma
    (b) 1.5 ma
    (c) 20 ohms
    (d) 40
    (e) 1 volt
    (f) 19,400 ohms
53. (a) 8,000
    (b) 7.5 megohms
    (c) 50,000 ohms
    (d) 1
55. (a) 400
    (b) 10
    (c) 390

CHAPTER 16

1. (a) 4.2 watts
   (b) 1 watt
   (c) 23.8%
3. (a) 15.375 watts
   (b) 0.861 watt
   (c) 5.6%
5. 3 watts
7. (a) 16.5 watts
   (b) 4 watts
9. (a) 18.3%
   (b) 9.8 watts
11. 2,340 μmhos
13. 2.1 watts
15. From Fig. 16-7
    $V_{CC} = -56$ volts
    $I_{C\cdot Q} = -0.8$ amp
    $V_{CE(\text{p-p})} = -15$ to -55 volts
    $I_{C(\text{p-p})} = -0.025$ to -1.6 amp
    (a) 26.2 ohms
    (b) 28.2%
17. From Fig. 16-9
    $V_{CE\cdot Q} = -28.5$ volts
    $V_{CC} = -56$ volts
    $I_{C\cdot Q} = -1.05$ amp
    26 ohms
19. From Fig. 16-9
    $V_{CE(\text{p-p})} = 9.5$ to 55 volts
    $I_{C(\text{p-p})} = 0.025$ to 1.8 amp
    (a) 30 watts
    (b) 10 watts
    (c) 33.3%
21. From Fig. 16-9
    $i_{c\cdot\max} = 1.8$ amp
    $i_{c\cdot\min} = 0.025$ amp
    $I_{C\cdot Q} = 1.05$ amp
    $I_x = 1.65$ amp
    $I_y = 0.35$ amp
    (a) 7.64%
    (b) 1.53%
    (c) 7.8%
23. (a) -32.4 volts
    (b) -31.5 volts
25. From Fig. 16-12
    $e_{p\cdot\max} = 340$ volts
    $e_{p\cdot\min} = 140$ volts
    $i_{p\cdot\max} = 105$ ma
    $i_{p\cdot\min} = 20$ ma
    (a) 2,360 ohms
    (b) 392 volts
27. From Fig. 16-12
    $e_{p\cdot\max}$, $e_{p\cdot\min}$
    $i_{p\cdot\max}$, $i_{p\cdot\min}$
    Same as Prob. 25
    $E_{P\cdot Q} = 250$ volts
    $I_{P\cdot Q} = 60$ ma
    $E_{bb} = 392$ volts
    $R_o = 2,360$ ohms
    (a) 15 watts
    (b) 2.125 watts
    (c) 14.2%
    (d) 9%
29. 2.94%
31. From Fig. 16-13
    $e_{c\cdot\max} = -29$ volts
    $e_{c\cdot\min} = -4$ volts
    $i_{p\cdot\max} = 70$ ma
    $i_{p\cdot\min} = 8$ ma
    $e_{p\cdot\max} = 450$ volts
    $e_{p\cdot\min} = 144$ volts
    $E_{P\cdot Q} = 315$ volts
    $I_{P\cdot Q} = 36$ ma
    $E_{bb} = -495$ volts
    (a) 4,940 ohms
    (b) 5,000 ohms
33. (a) 11.3 watts
    (b) 2.37 watts
    (c) 21%
35. (a) 13.3%
    (b) 10.3%
37. 30,000 μmhos
39. From Fig. 16-13
    $E_C = -16.5$ volts
    $I_x = 70$ ma
    $I_y = 10$ ma
    (a) 4.1%
    (b) 15.4%
    (c) 15.8%
41. From Fig. 16-9
    $v_{c\cdot\max} = 45$ volts
    $v_{c\cdot\min} = 4.5$ volts
    $i_{c\cdot\max} = 1.8$ amp
    $i_{c\cdot\min} = 0.025$ amp
    $V_{CE\cdot Q} = 21.5$ volts
    $I_{C\cdot Q} = 1.05$ amp
    18.6%
43. From Fig. 16-9
    $I_{B\cdot Q} = 40$ ma
    $I_x = 1.65$ amp
    $I_y = 0.35$ amp
    7.75%
45. 1.7
47. From Fig. 16-13
    $e_{p\cdot\max} = 455$ volts
    $e_{p\cdot\min} = 90$ volts
    $i_{p\cdot\max} = 81$ ma
    $i_{p\cdot\min} = 8$ ma
    $E_{P\cdot Q} = 288$ volts
    $I_{P\cdot Q} = 41$ ma
    $E_{bb} = 495$ volts
    16.4%
49. From Fig. 16-13
    $E_C = -15$ volts
    $I_x = 70$ ma
    $I_y = 15$ ma
    5.5%
51. 25
53. From Fig. 16-21
    $v_{c\cdot\max} = -52$ volts
    $v_{c\cdot\min} = -10$ volts
    $i_{c\cdot\max} = -1.85$ amp
    $i_{c\cdot\min} = 0$
    (a) 38.8 watts
    (b) 61 watts
    (c) 63.5%
    (d) 112 ohms

CHAPTER 17

1. (a) 20 mw
   (b) 11 mw
   (c) 9 mw
3. 75%
5. 156
7. 880 to 2,830 kHz
9. 1,000 to 2,400 kHz
11. (a) 56 pf
    (b) 56,500 ohms
    (c) 0.265 ma
    (d) 7.95 ma
    (e) 3.97 mw
13. (a) 88.4 μh
    (b) 71.5 pf
    (c) 0.9 amp
    (d) 60 ma
15. (a) 126×10^{-10} watts per cycle
    (b) 26.4×10^{-10} watts per cycle
    (c) 4.77 to 1
17. (a) X_L = 34,540 ohms
    X_C = 14.5 ohms
    (b) 2,380 to 1
    (c) X_L = 15,700 ohms
    X_C = 31.8 ohms
    (d) 494 to 1
19. (a) 2,040 kHz
    (b) 1,770 kHz
    (c) 81.5%
    (d) 18.5%
21. (a) 7.96×10^{-6}
    (b) 796 pf
    (c) 7,960 ohms
23. (a) 750 Hz (decrease)
    (b) 375 Hz (increase)
    (c) 750 Hz (increase)
25. 4,240
27. (a) 1,000 Hz
    (b) 2,000 Hz
    (c) 1 MHz
29. 74 kHz
31. 5,000 Hz
33. 12.5 kHz

CHAPTER 18

1. 710 to 1
3. (a) 200 to 8,000 Hz
   (b) 5,250 cycles
   (c) 131.25 cycles
5. 80%
7. 5.625 volts
9. (a) 6,950 Hz
   (b) 717 kHz
   (c) 703 kHz
   (d) 14 kHz
11. (a) 80 kHz
    (b) 120 kHz
13. (a) 8 kHz
    (b) 12 kHz
15. (a) 6
    (b) 144 kHz
17. (a) $f_{b\cdot 1}$ = 3 Hz
    $f_{b\cdot 2}$ = 15 Hz
    (b) 4 to 12 volts
19. (a) 675 kHz, 1,605 kHz
    (b) 465 kHz, 675 kHz 1,140 kHz, 1,605 kHz
21. (a) 50 μsec
    (b) 0.667 μsec
    (c) 75
23. (a) 75 μsec
    (b) 2 μsec
    (c) 37.5
25. (a) 17.5 μsec
    (b) 2.2 μsec
    (c) 143 μsec
    (d) 8 to 1
    (e) 8 to 1
    (f) Negligible
27. (a) 3,420 ohms
    (b) Path through C_1
    (c) Yes, a small amount
    (d) 3,420 ohms
    (e) Additional i-f filtering
    (f) 80%
29. (a) 1,060 ohms
    (b) 282,600 ohms
    (c) 3,180,000 ohms
    (d) 94.2 ohms
    (e) The capacitor path
    (f) The inductor path
31. (a) −4 to −2 volts
    (b) −6 to 0 volts
    (c) −7 to 1 volt
33. 17 volts
35. (a) 120,000 ohms
    (b) 48,000 ohms
    (c) 30,000 ohms
37. (a) 85,000 ohms
    (b) 0.0034 watt
    (c) 0.25 watt
39. (a) 0.37 μf
    (b) 0.5 μf, 200-volt paper capacitor
41. 317 volts
43. (a) 2,000.5 or 1,999.5 kHz
    (b) 2,001 or 1,999 kHz
    (c) 2,001.5 or 1,998.5 kHz
45. (a) 0.1 sec
    (b) 9.5%
    (c) 10 cycles
47. (a) 120×10^{-3} sec
    (b) 20×10^{-3} sec
    (c) 6 to 1
    (d) No, because the charge on C_5 charges much more slowly than the slowest a-f value
    (e) 1.2 sec
49. (a) 3 volts; D_2 is negative
    (b) Zero
    (c) Zero, or ground potential
    (d) Above 3 volts
51. (a) 450 ohms
    (b) 690,000 ohms
53. (a) $e_{s\cdot 1}$ leads e_i by 60°
    (b) 17.3 volts
    (c) 10 volts
    (d) 7.3 volts

Index